Fluorine in Medicinal Chemistry and Chemical Biology

Fluorine in Medicinal Chemistry and Chemical Biology

Edited by
Iwao Ojima
Institute of Chemical Biology & Drug Discovery,
State University of New York at Stony Brook, New York, USA

A John Wiley & Sons, Ltd., Publication

This edition first published 2009

Registered office
John Wiley & Sons Ltd, The Atrium, Southern Gate, Chichester, West Sussex, PO19 8SQ, United Kingdom

For details of our global editorial offices, for customer services and for information about how to apply for permission to reuse the copyright material in this book please see our website at www.wiley.com.

Library of Congress Cataloging-in-Publication Data

Fluorine in medicinal chemistry and chemical biology / Iwao Ojima (editor)
p. cm.
Includes bibliographical references and index.
ISBN 978-1-4051-6720-8 (cloth : alk. paper)
1. Organofluorine compounds. 2. Fluorination – Physiological effect. 3. Bioorganic chemistry. 4. Pharmaceutical chemistry. I. Ojima, Iwao. II. Taguchi, Takeo.
QP535.F1F58 2009
615′.19–dc22

2008049835

A catalogue record for this book is available from the British Library.
Set in 10 on 12 pt Times by SNP Best-set Typesetter Ltd., Hong Kong
Printed and bound in Great Britain by CPI Antony Rowe, Chippenham, Wiltshire

Contents

Preface

This book presents the recent developments and future prospects of fluorine in medicinal chemistry and chemical biology. The extraordinary potential of fluorine-containing biologically relevant molecules in peptide/protein chemistry, medicinal chemistry, chemical biology, pharmacology, and drug discovery as well as diagnostic and therapeutic applications, was recognized by researchers who are not in the traditional fluorine chemistry field, and thus the new wave of fluorine chemistry has been rapidly expanding its biomedical frontiers. In 1996, I edited a book, *Biomedical Frontiers of Fluorine Chemistry* (American Chemical Society's Symposium Series), together with Dr James R. McCarthy and Dr John T. Welch, covering the emerging new aspects of fluorine chemistry at the biomedical interface. Since then, remarkable progress has been made. For example, in 2006, the best-selling and second-best-selling drugs in the world were Lipitor® (atorvastatin calcium) (by Pfizer/Astellas; $14.4 billion/year) and Advair®(USA)/seretide®(EU) (a mixture of fluticasone propionate and salmeterol) (by GlaxoSmithKline; $6.1 billion/year), which contain one and three fluorine atoms, respectively. Also, risperidone with a fluorine atom, for schizophrenia (by Janssen; $4.2 billion/year) and lansoprazole, a proton pump inhibitor, with a CF_3 moiety (by Takeda/Abbott; $3.4 billion/year) were ranked 10th and 17th, respectively. As such, it is not an exaggeration to say that, at present, every new drug discovery and development, without exception, explores fluorine-containing drug candidates. Also, applications of ^{18}F PET (positron emission tomography), a powerful *in vivo* imaging technology in oncology, neurology, psychiatry, cardiology, and other medical specialties have already become an important part of medical care. In addition, ^{18}F PET has emerged as an important tool in drug development, especially for accurate measurements of pharmacokinetics and pharmacodynamics.

Accordingly, I believe that it is the right time for us to review the recent advances and envision the new and exciting developments in the future. This book has a focus on the unique and significant roles that fluorine plays in medicinal chemistry and chemical biology, but also covers new and efficient synthetic methods for medicinal chemistry, ^{18}F PET, and expanding applications of ^{19}F NMR spectroscopy to biomedical research.

The book starts with an introductory chapter, summarizing and discussing the unique properties of fluorine and their relevance to medicinal chemistry and chemical biology. Then, several hot topics in medicinal chemistry as well as discovery and development of fluorine-containing drugs and drug candidates are described, including fluorinated prostanoids (for glaucoma), fluorinated conformationally restricted glutamate analogues (for CNS disorder), fluorinated MMP inhibitors (e.g. for cancer metastasis intervention), fluoro-taxoids (for cancer), trifluoro-artemisinin (for malaria), and fluorinated nucleosides (for viral infections).

Development of efficient synthetic methods is crucial for medicinal chemistry and optimization of fluorine-containing drug candidates. Thus, the subsequent section presents the recent advances in synthetic methodology for *gem*-difluoromethylene sugar nucleosides, trifluoroalanine oligopeptides, fluoroalkene dipeptide isosteres, fluorinated heterocyclic systems, and oligo-*gem*-difluorocyclopropanes. In addition, an emerging "fluorous technology" for the efficient synthesis and purification of biomolecules is discussed.

As mentioned above, there has been a remarkable advance in the biomedical applications of ^{18}F PET imaging. Thus, the next chapter deals with ^{18}F radiopharmaceuticals. On the other hand, fluorinated amino acids and peptides have been playing a key role in structural biology, chemical biology, and biological chemistry. Therefore, the subsequent section presents the recent advances in this growing field of research, including the structural and chemical biology of fluorinated amino acids and peptides, protein design using fluorinated amino acids, and fluorinated methionines as probes in biological chemistry.

^{19}F NMR spectroscopy, both in solution and in the solid state, has been expanding its utility in chemical biology as well as diagnostic tools. Accordingly, the last two chapters discuss the structure analysis of membrane-active peptides using fluorinated amino acids by solid-state ^{19}F NMR and the applications of *in vivo* ^{19}F-magnetic resonance spectroscopy to metabolism, biodistribution and neuropharmacology studies.

Although several recent reviews and books claim that a large number of fluorine-containing drugs have been approved by the Federal Drug and Food Administration (FDA) of the United States, there has not been an actual list of those drugs with structures. Accordingly, I organized a team to go through the FDA records to collect all FDA-approved fluorine-containing drugs, to date, for humans and for animals, and they have produced a very useful list of those drugs, which is compiled in the Appendix.

I believe that this book is extremely informative for researchers who want to take advantage of the use of fluorine in biomedical research such as rational drug design, theory and synthesis, the use of fluorine labels for chemical and structural biology, metabolism and biodistribution studies, protein engineering, clinical diagnosis, etc. This book will serve as an excellent reference book for graduate students as well as scientists at all levels in both academic and industrial laboratories.

I am grateful to Professor Takeo Taguchi, Tokyo University of Pharmacy and Life Sciences for his valuable advice for the selection of world-leading contributors with cutting-edge research projects, especially from Asia. I also acknowledge the very helpful editorial assistance of Dr Qing Huang, Institute of Chemical Biology & Drug Discovery, State University of New York at Stony Brook. I also would like to thank the Wiley–Blackwell editorial team, Richard Davies, Rebecca Stubbs, Sarah Hall, and Sarahjayne Sierra, for their productive cooperation.

Iwao Ojima
Editor

Contributors

David Armstrong, Department of Chemistry, Durham University, Durham, UK

Jean-Pierre Bégué, BIOCIS-CNRS, Centre d'Etudes Pharmaceutiques, Châtenay-Malabry, France

Danièle Bonnet-Delpon, BIOCIS-CNRS, Centre d'Etudes Pharmaceutiques, Châtenay-Malabry, France

Gabriele Candiani, C.N.R. – Istituto di Chimica del Riconoscimento Molecolare, Milan, Italy

Matthew W. Cartwright, Department of Chemistry, Durham University, Durham, UK

John A. Christopher, GlaxoSmithKline R&D, Stevenage, Hertfordshire, UK

Ginevra A. Clark, Department of Chemistry, Tufts University, Medford, Massachusetts, USA

Manisha Das, Department of Chemistry, State University of New York at Stony Brook, Stony Brook, New York, USA

John F. Honek, Department of Chemistry, University of Waterloo, Waterloo, Ontario, Canada

Florent Huguenot, C.N.R. – Istituto di Chimica del Riconoscimento Molecolare, Milan, Italy

Masahiro Ikejiri, Faculty of Pharmaceutical Sciences at Kagawa Campus Tokushima Bunri University, Sanuki City, Kagawa, Japan

Toshiyuki Itoh, Department of Chemistry and Biotechnology, Tottori University, Tottori, Japan

Kunisuke Izawa, AminoScience Laboratories, Ajinomoto Co., Inc., Kawasaki-ku, Kawasaki, Japan

Monika Jagodzinska, C.N.R. – Istituto di Chimica del Riconoscimento Molecolare, Milan, Italy

Michael R. Kilbourn, Department of Radiology, University of Michigan Medical School, Ann Arbor, Michigan, USA

Beate Koksch, Institute of Chemistry and Biochemistry – Organic Chemistry, Freie Universität Berlin, Berlin, Germany

Krishna Kumar, Department of Chemistry, Tufts University, Medford, Massachusetts, USA

Larisa Kuznetsova, Brookhaven National Laboratory, Biology Department, Upton, New York, USA

Roger Lin, Case Western Reserve University, School of Medicine, Cleveland, Ohio, USA

Raffaella Maffezzoni, C.N.R. – Istituto di Chimica del Riconoscimento Molecolare, Milan, Italy

Tokumi Maruyama, Faculty of Pharmaceutical Sciences at Kagawa Campus, Tokushima Bunri University, Sanuki City, Kagawa, Japan

Yasushi Matsumura, Research Center, Asahi Glass Company, Ltd., Kanagawa-ku, Yokohama, Japan

He Meng, Department of Chemistry, Tufts University, Medford, Massachusetts, USA

Wei-Dong Meng, College of Chemistry, Chemical Engineering and Biotechnology, Donghua University, Shanghai, China

David D. Miller, GlaxoSmithKline R&D, Stevenage, Hertfordshire, UK

Atsuro Nakazato, Medicinal Research Laboratories, Taisho Pharmaceutical Co., Ltd., Kita-ku, Saitama-shi, Japan

Iwao Ojima, Department of Chemistry and Institute of Chemical Biology & Drug Discovery (ICB&DD), State University of New York at Stony Brook, Stony Brook, New York, USA

Tomoyuki Onishi, AminoScience Laboratories, Ajinomoto Co. Inc., Kawasaki-ku, Kawasaki, Japan

Emma L. Parks, Department of Chemistry, Durham University, Durham, UK

Graham Pattison, Department of Chemistry, Durham University, Durham, UK

Antonella Pepe, Institute for Therapeutics Discovery and Development, Department of Medicinal Chemistry, College of Pharmacy, University of Minnesota, Minneapolis, Minnesota, USA

Elizabeth Pollina Cormier, Office of New Animal Drug Evaluation, Center for Veterinary Medicine, United States Food & Drug Aministration, Rockville, USA

Feng-Ling Qing, Key Laboratory of Organofluorine Chemistry, Shanghai Institute of Organic Chemistry, Chinese Academy of Sciences, Shanghai, China

Mario Salwiczek, Institute of Chemistry and Biochemistry – Organic Chemistry, Freie Universität Berlin, Berlin, Germany

Graham Sandford, Department of Chemistry, Durham University, Durham, UK

Monica Sani, C.N.R. – Istituto di Chimica del Riconoscimento Molecolare, Milan, Italy

Erika Schneider, Imaging Institute, The Cleveland Clinic, Department of Diagnostic Radiology, Cleveland, Ohio, USA

Xia Shao, Department of Radiology, University of Michigan Medical School, Ann Arbor, Michigan, USA

Roberta Sinisi, C.N.R. – Istituto di Chimica del Riconoscimento Molecolare, Milan, Italy

Rachel Slater, Department of Chemistry, Durham University, Durham, UK

Paul W. Smith, GlaxoSmithKline R&D, Harlow, Essex, UK

Erik Strandberg, Karlsruhe Institute of Technology, Institute for Biological Interfaces, Forschungszentrum Karlsruhe, Karlsruhe, Germany

Liang Sun, Department of Chemistry, State University of New York at Stony Brook, Stony Brook, New York, USA

Takeo Taguchi, Tokyo University of Pharmacy and Life Sciences, Hachioji, Tokyo, Japan

Kenji Uneyama, Department of Applied Chemistry, Okayama University, Tsushimanaka, Okayama, Japan

Toni Vagt, Institute of Chemistry and Biochemistry – Organic Chemistry, Freie Universität Berlin, Berlin, Germany

Alessandro Volonterio, C.N.R. – Istituto di Chimica del Riconoscimento Molecolare, Milan, Italy

Antonio Vong, GlaxoSmithKline R&D, Harlow, Essex, UK

Parvesh Wadhwani, Karlsruhe Institute of Technology, Institute for Biological Interfaces, Forschungszentrum Karlsruhe, Karlsruhe, Germany

Ian Wilson, Department of Chemistry, Durham University, Durham, UK

Takashi Yamazaki, Department of Applied Chemistry, Graduate School of Engineering, Tokyo University of Agriculture and Technology, Koganei, Japan

Hikaru Yanai, Tokyo University of Pharmacy and Life Sciences, Hachioji, Tokyo, Japan

Matteo Zanda, C.N.R. – Istituto di Chimica del Riconoscimento Molecolare, Milan, Italy

Wei Zhang, Department of Chemistry, University of Massachusetts Boston, Boston, Massachusetts, USA

Abbreviations

acac	acetylacetonate
AIBN	azobisbutyronitrile
Bn	benzyl
Boc t-Boc tBoc	*tert*-butoxycarbonyl
BSA	*O,N*-bistrimethylsilyl acetamide
Bz	benzoyl
Cbz	carbobenzyloxy
Cp	cyclopentadienyl
CYP	cytochrome C P-450
DAST	diethylaminosulfur trifluoride
DBU	1,5-diazabicyclo[4.3.0]non-5-ene
DCC	*N,N′*-dicyclohexylcarbodiimide
DCM	dichloromethane
DDQ	2,3-dichloro-5,6-dicyanobenzoquinone
DEAD	diethyl azodicarboxylate
DFM	difluoromethionine
DHQ	decahydroquinoline
DIBAL, DIBAL-H	diisobutylaluminum hydride
DIC	*N,N′*-diisopropylcarbodiimide
DIPE	diisopropylethylamine
DMAP	4-(*N,N*-dimethylamino)pyridine
DMF	*N,N*-dimethylformamide
DMPC	1,2-dimyristoyl-*sn*-glycero-3-phosphocholine
DMSO	dimethylsulfoxide
dppe	1,2-bis(diphenylphosphino)ethane
DPPP, dppp	1,3-bis(diphenylphosphino)propane
EDC	1-(3-dimethylaminopropyl)-3-ethylcarbodiimide hydrochloride
Fmoc	9-fluorenylmethoxycarbonyl
HATU	*O*-(7-azabenzotriazol-1-yl)-*N,N,N′,N′*-tetramethyluronium hexafluorophosphate
HMPA, HMPT	hexamethylphosphoramide
HOAt	1-hydroxy-7-azabenzotriazole
HOBt	1-hydroxybenzotriazole
HYTRA	(*R*)-(+)-2-hydroxy-1,2,2-triphenylethyl acetate
IBX	*o*-iodoxybenzoic acid
Im	Imidazolyl

KHMDS	potassium hexamethyldisilazide
LDA	lithium diisopropylamide
LHMDS	lithium hexamethyldisilazide
LTMP	lithium 2,2,6,6-tetramethylpiperidine
MAP	methionine aminopeptidase
mCPBA, *m*-CPBA	*meta*-chloroperbenzoic acid
MEM	2-methoxyethoxymethyl
MFM	monofluoromethionine
Ms	mesyl, methanesulfonyl
NBS	*N*-bromosuccinimide
NIS	*N*-iodosuccinimide
NMM	*N*-methylmorpholine
NMNO	*N*-methylmorpholine-*N*-oxide
NMP	*N*-methylpyrrolidone
PDC	pyridinium dichromate
PG	protecting group
Phth	phtharyl
PMB	*para*-methoxybenzyl, 4-methoxybenzyl
PMP	*para*-methoxyphenyl, 4-methoxyphenyl
pPTS, PPTS	pyridinium p-toluenesulfonate
Pv	pivaloyl
Py	pyridine
PyBroP	bromotripyrrolidinophosphonium hexafluorophosphate
SEM	2-(trimethylsilyl)ethoxymethyl
TASF	tris(dimethylamino)sulfonium difluorotrimethylsilicate
TBAF	tributyl ammonium fluoride
TBAI	tributyl ammonium iodide
TBDMS, TBS	*tert*-butyldimethylsilyl
TBDPS	*tert*-butyldiphenylsilyl
TDS	thexyldiemethylsilyl
TEA	triethylamine
TEMPO	2,2,6,6-tetramethylpiperidine-1-oxyl
TES	triethylsilyl
Tf	triflate, trifluoromethanesulfonyl
TFAA	trifluoroacetic anhydride
TFA	trifluoroacetic acid
TFM	trifluoromethionine
THF	tetrahydrofuran
THP	tetrahydropyran
TIPS	triisopropylsilyl
TMAI	tetramethylammonium iodide
TMS	trimethylsilyl

Introduction: Basic Aspects of Fluorine in Chemistry and Biology

1

Unique Properties of Fluorine and Their Relevance to Medicinal Chemistry and Chemical Biology

Takashi Yamazaki, Takeo Taguchi, and Iwao Ojima

1.1 Fluorine-Substituent Effects on the Chemical, Physical and Pharmacological Properties of Biologically Active Compounds

The natural abundance of fluorine as fluorite, fluoroapatite, and cryolite is considered to be at the same level as that of nitrogen on the basis of the Clarke number of 0.03. However, only 12 organic compounds possessing this special atom have been found in nature to date (see Figure 1.1) [1]. Moreover, this number goes down to just five different types of compounds when taking into account that eight ω-fluorinated fatty acids are from the same plant [1]. [*Note:* Although it was claimed that naturally occurring fluoroacetone was trapped as its 2,4-dinitrohydrazone, it is very likely that this compound was fluoroacetaldehyde derived from fluoroacetic acid [1]. Thus, fluoroacetone is not included here.]

In spite of such scarcity, enormous numbers of synthetic fluorine-containing compounds have been widely used in a variety of fields because the incorporation of fluorine atom(s) or fluorinated group(s) often furnishes molecules with quite unique properties that cannot be attained using any other element. Two of the most notable examples in the field of medicinal chemistry are 9α-fluorohydrocortisone (an anti-inflammatory drug) [2] and 5-fluorouracil (an anticancer drug) [3], discovered and developed in 1950s, in which the introduction of just a single fluorine atom to the corresponding natural products brought about remarkable pharmacological properties. Since then, more than half a century has

Fluorine in Medicinal Chemistry and Chemical Biology Edited by Iwao Ojima

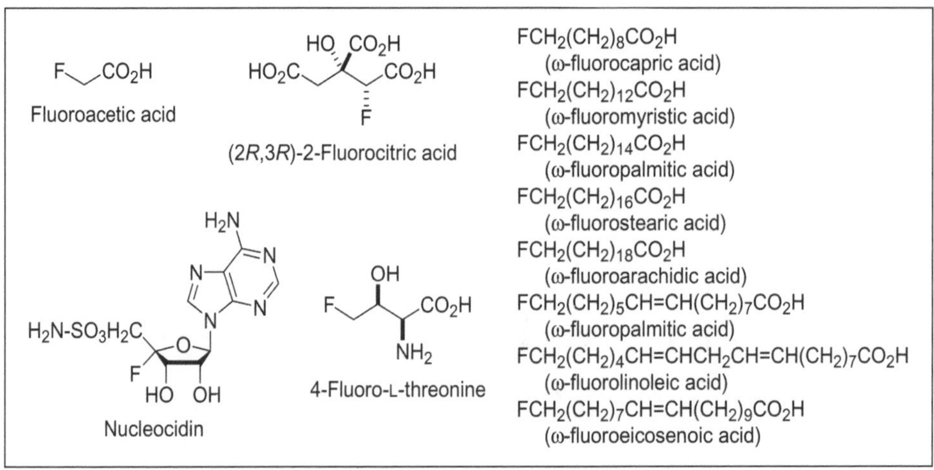

Figure 1.1 *Naturally occurring fluorine-containing compounds.*

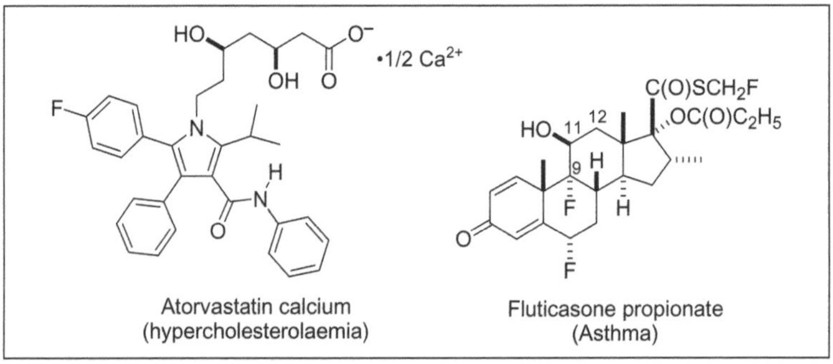

Atorvastatin calcium (hypercholesterolaemia)

Fluticasone propionate (Asthma)

Figure 1.2 *Two fluorine-containing drugs.*

passed, and the incorporation of fluorine into pharmaceutical and veterinary drugs to enhance their pharmacological properties has become almost standard practice. In 2006, the best- and the second-best-selling drugs in the world were Lipitor® (atorvastatin calcium) by Pfizer/Astellas ($14.4 billion/year) and Advair®(U.S.A.)/Seretide®(E.U.) (a mixture of fluticasone propionate and salmeterol; only the former contains fluorine) by GlaxoSmithKline ($6.1 billion/year) which contain one and three fluorine atoms, respectively (see Figure 1.2) [4]. These fluorine-containing drugs are followed by risperidone (rank 10th, with one fluorine and $4.2 billion/year) for schizophrenia by Janssen, and lansoprazole (rank 17th, with a CF_3 moiety and $3.4 billion/year), a proton pump inhibitor, by Takeda/Abbott [4].

These huge successes of fluorine-containing drugs continue to stimulate research on fluorine in medicinal chemistry for drug discovery. It would not be an exaggeration to say

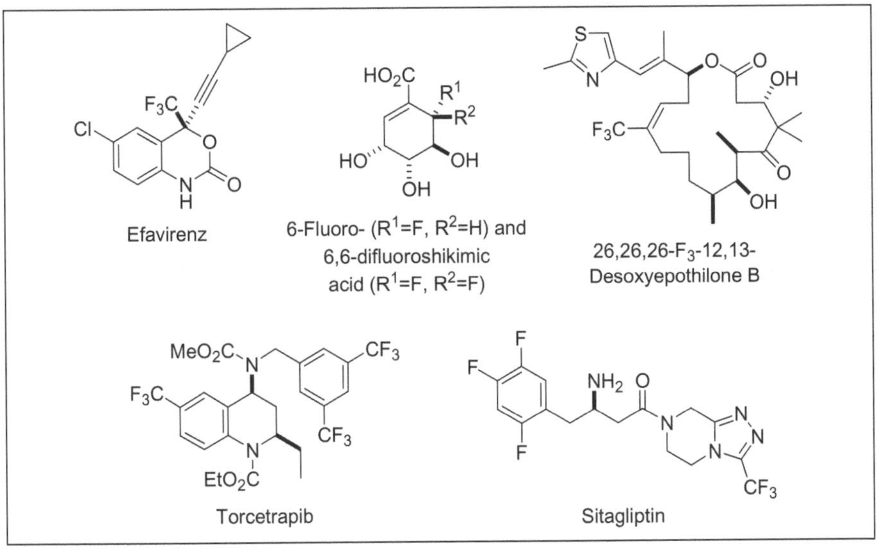

Figure 1.3 *Examples of fluorine-containing drugs and drug candidates.*

that currently every new drug discovery and development program without exception explores fluorine-containing drug candidates. Accordingly, a wide variety of fluorine-containing compounds based either on known natural products or on new skeletons have been synthesized and subjected to biological evaluation. In most cases, 1–3 fluorines are incorporated in place of hydroxyl groups or hydrogen atoms. Representative examples include efavirenz (CF_3) (HIV antiviral) [5], fluorinated shikimic acids (CHF or CF_2) (antibacterial) [6], and epothilone B analogue (CF_3) (anticancer) [7] (see Figure 1.3). In addition to these fluorine-containing drugs or drug candidates, a smaller number of fluorine-containing drugs include 6–9 fluorines in a molecule. For example, torcetrapib (a potent inhibitor of cholesterol ester transfer protein) possesses three CF_3 groups (Pfizer) [8], and sitagliptin (an antidiabetic for type 2 diabetes) has three fluorine atoms and one CF_3 group (Merck) [9] (see Figure 1.3).

In order to synthesize a variety of fluorine-containing biologically active compounds, development of efficient synthetic methods applicable to fluorine-containing organic compounds is necessary [10–19]. There is a strong demand for expansion of the availability of versatile fluorine-containing synthetic building blocks and intermediates to promote target-oriented synthesis as well as diversity-oriented synthesis. The limited availability of fluorochemicals for bioorganic and medicinal chemistry as well as pharmaceutical and agrochemical applications is mainly due to the exceptional properties and hazardous nature of fluorine and fluorochemical sources. Also, in many cases, synthetic methods developed for ordinary organic molecules do not work well for fluorochemicals because of their unique reactivity [10–19].

In this chapter, characteristic properties associated with the incorporation of fluorines or fluorine-containing groups to organic molecules are described in detail based on the most updated literature sources [20–32].

1.1.1 Mimic Effect and Block Effect

Table 1.1 summarizes representative physical properties of fluorine in comparison with other selected elements. As Table 1.1 clearly shows, the van der Waals (vdW) radius of fluorine is 1.47 Å [33] which is only 20% larger than that of hydrogen and much smaller than those of other halogens (Cl and Br are 46% and 54% larger than hydrogen, respectively). The C–F bond length in CH_3F is 1.382 Å, which is 0.295 Å longer than the methane's C–H bond, but 0.403 and 0.551 Å shorter than the C–Cl and C–Br bonds, respectively. Because of this similarity in size to hydrogen, it has been shown that microorganisms or enzymes often do not recognize the difference between a natural substrate and its analogue wherein a C–H bond of the substrate is replaced with a C–F bond. This observation is the basis of what is regarded as the "mimic effect" of fluorine for hydrogen.

One of the best-known examples is the behavior of fluoroacetic acid (CH_2FCO_2H) in the TCA (citrate acid or Krebs) cycle (see Scheme 1.1) [36]. As Scheme 1.1 illustrates, fluoroacetic acid is converted to fluoroacetyl-CoA (**1b**), following the same enzymatic transformation to the original substrate, acetic acid. Then, fluoroacetic acid is converted to (2*R*,3*R*)-fluorocitric acid (**2b**) by citrate synthase since this enzyme does not distinguish **1b** from acetyl-CoA (**1a**). In the next step, **2b** is dehydrated by aconitase to give (*R*)-fluoro-*cis*-aconitic acid (**3b**) in the same manner as that for the natural substrate **2a** to afford *cis*-aconitic acid (**3a**). In the normal TCA cycle, **3a** is hydroxylated to give isocitric acid (**4**), while **3b** undergoes hydroxylation-defluorination in an S_N2' manner to give (*R*)-hydroxy-*trans*-aconitic acid (**5**). It has been shown that the high affinity of **5b** for aconitase shuts down the TCA cycle, which makes **5b** as well as its precursors, fluoroacetic acid and **2b,** significantly toxic [36].

A number of protein structures determined by X-ray crystallography have been reported to date for the complexes of various enzymes with fluorine-containing substrate mimics or inhibitors bearing multiple fluorines. This strongly suggests that not only single fluorine displacement but also various fluorine-substitution patterns in substrate analogues

Table 1.1 *Representative physical data of selected elements [33–35]*

	Element (X)					
	H	C	O	**F**	Cl	Br
Electronegativity[a]	2.20	2.55	3.44	**3.98**	3.16	2.96
van der Waals radius[b] (Å)	1.20	1.70	1.52	**1.47**	1.75	1.85
H_3C–X bond length[a] (Å)	1.087	1.535[d]	1.425[e]	**1.382**	1.785	1.933
H_3C–X dissociation energy[c] (kcal/mol)	103.1	88.0[d]	90.2[e]	**108.1**	81.1	67.9
Ionization potential[a] (kcal/mol)	313.9	259.9	314.3	**402.2**	299.3	272.7
Electron affinity[a] (kcal/mol)	17.42	29.16	3.73	**78.52**	83.40	77.63

[a] Ref. [34].
[b] Ref. [33].
[c] Ref. [35].
[d] X = CH_3.
[e] X = OH.

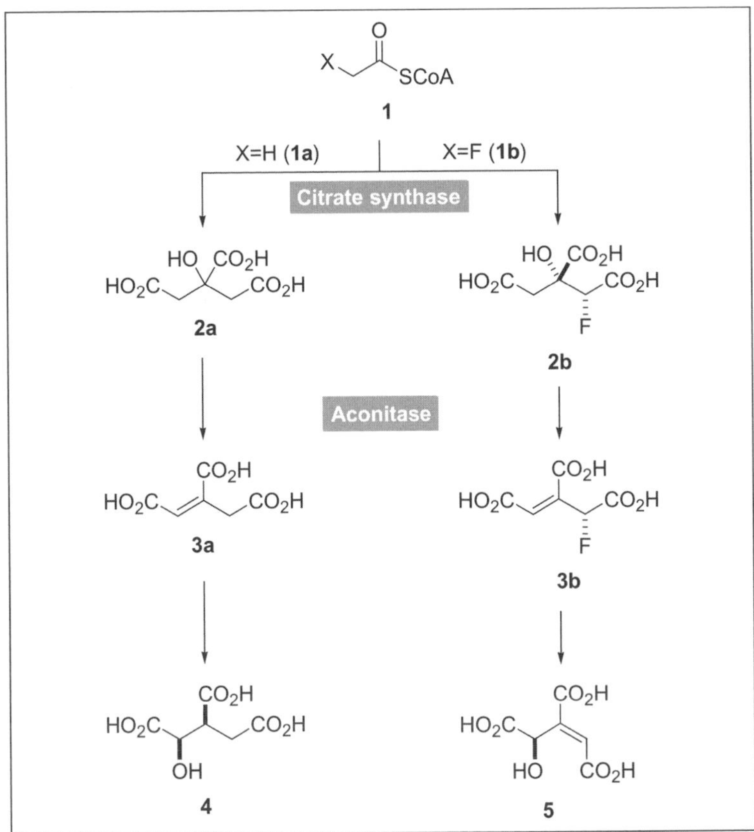

Scheme 1.1 *Conversion of acetyl-CoA and fluoroacetyl-CoA in the TCA cycle.*

or inhibitors could be adapted to biological systems in a molecular-recognition mode similar to that of the natural substrates [22, 37, 38].

Since the H_3C–*F* bond is stronger than that of H_3C–*H* by 5.0 kcal/mol (see Table 1.1), the replacement of a specific C–H bond with a C–F bond can effectively block metabolic processes via hydroxylation of C–H bonds, predominantly by the cytochrome P-450 family of enzymes. This function is referred to as the "block effect," and the strategic incorporation of fluorine(s) into metabolism site(s) has been widely used to prevent deactivation of biologically active substances *in vivo*. For example, it was found that the hydroxylation of the methylene moiety at the 24-position in the side-chain of vitamin D_3 (**6a**) was a critical deactivation step prior to excretion [39]. To block this undesirable metabolism, a strategic fluorine substitution was used; that is, a difluoromethylene group was introduced to the 24-position of **6b** [40], which indeed effectively blocked the hydroxylation at this site. The resultant **6c** was then hydroxylated enzymatically at the 1 position to give **7b**. 24,24-Difluoro-25-hydroxyvitamin D_3 (**6c**) was found to be slightly more potent than **6b**, and 24,24-difluoro-1,25-dihydroxyvitamin D_3 (**7b**) exhibited 5–10 times higher potency than **7a** [41] (see Scheme 1.2).

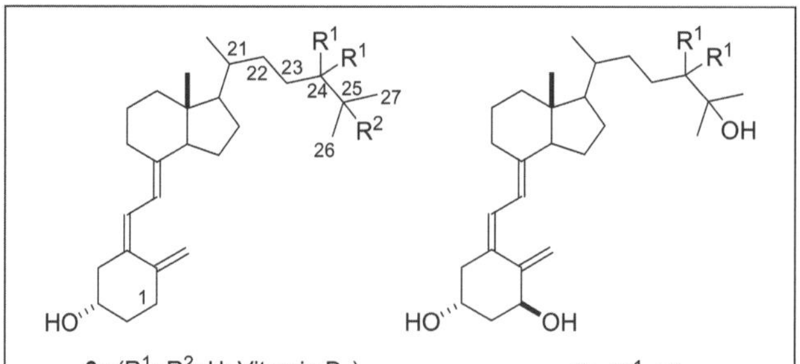

6a (R^1=R^2=H; Vitamin D_3)
6b (R^1=H, R^2=OH)
6c (R^1=F, R^2=OH)

7a (R^1=H)
7b (R^1=F)

Scheme 1.2 *24,24-Difluoro-25-OH-vitamin D_3 (**6c**) and 24,24-difluoro-1,25-dihydroxyvitamin D_3 (**7b**).*

In addition to the "mimic effect" and "block effect," introduction of just one fluorine substituent can induce electronic effects on its neighbors by affecting the electron density of functional groups such as hydroxyl and amino groups. This electronic effect decreases the pK_a value and Lewis basicity of these functional groups and retards their oxidation. For example, fluticasone propionate (see Figure 1.2) and related anti-inflammatory steroids contain fluorine at the 9α position. The role of the 9α-fluorine is to increase the acidity of the hydroxyl group at the 11 position, which promotes better binding to the enzyme active site and inhibits undesirable oxidation [42, 43].

1.1.2 Steric Effect of Fluorine and Fluorine-containing Groups

As Table 1.1 shows, fluorine is the second smallest element, with size approximately 20% larger than the smallest element, hydrogen. Table 1.2 summarizes four steric parameters for various elements and groups: (i) Taft steric parameters E_s [44], (ii) revised Taft steric parameters E_s' [45], (iii) Charton steric parameters υ [46], and (iv) A values [47]. The steric parameters, E_s, E_s', and υ are determined on the basis of relative acid-catalyzed esterification rates, while the A values are derived from the Gibbs free energy difference calculated from the ratios of axial and equatorial conformers of monosubstituted cyclohexanes by NMR.

As Table 1.2 shows, a stepwise substitution of a methyl group with fluorine gradually increases its bulkiness. For the bulkiness of CH_2F compared to CH_3, E_s, E_s', and υ values all show about 20% increase in size. For the bulkiness of CHF_2 and CF_3, however, the E_s values indicate 50% increase for CHF_2 and 90% increase for CF_3 in size as compared to CH_3, while the E_s' and υ values indicate only 30% and 70% increase in size, respectively.

In the case of A values, it is interesting to note that the first introduction of a fluorine atom into a methyl group leads to higher axial preference than that of a methyl group (i.e., CH_2F is regarded as smaller than CH_3), and the second fluorine substitution (i.e.,

Table 1.2 *Selected steric parameters of various elements and groups [44, 46–49]*

	$-E_s$ [a]	$-E'_s$ [a]	υ [b]	A [c] (kcal/mol)
H	0.00	0.00	0.00	0.00
F	0.46	0.55	0.27	0.15
CH_3	1.24	1.12	0.52	1.70
Cl	0.97	1.14	0.55	0.43
CH_3CH_2	1.31	1.20	0.56	1.75
CH_2F	1.48 (1.19)[d]	1.32 (1.18)[d]	0.62 (1.19)[d]	1.59[e] (0.94)[d]
Br	1.16	1.34	0.65	0.38
CHF_2	1.91 (1.54)[d]	1.47 (1.31)[d]	0.68 (1.31)[d]	1.85[e] (1.09)[d]
$(CH_3)_2CH$	1.71	1.60	0.76	2.15
CF_3	2.40 (1.94)[d]	1.90 (1.70)[d]	0.91 (1.75)[d]	2.37[f] (1.39)[d]
$(CH_3)_2CHCH_2$	2.17	2.05	0.98	–
$(CH_3)_3C$	2.78	2.55	1.24	4.89[e]

[a] Ref. [44].
[b] Ref. [46].
[c] Ref. [47].
[d] The ratio on the basis of the value of a CH_3 groups is shown in parentheses.
[e] Ref. [49].
[f] Ref. [48].

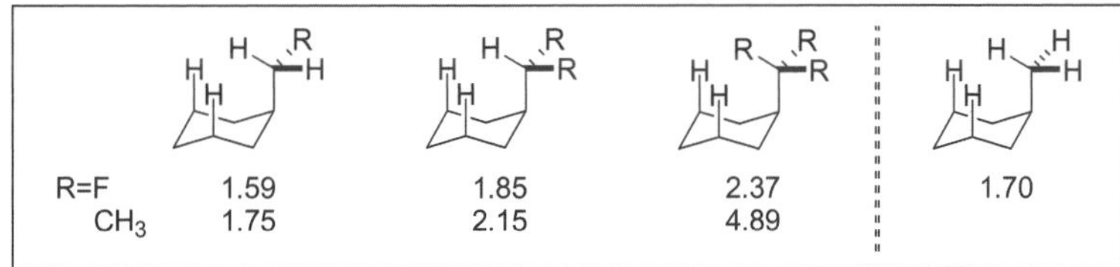

Figure 1.4 *Correlation of* A *values (kcal/mol) with 1,3-diaxial strains.*

CHF_2) causes a rather small increase in the *A* value. The *A* values of CH_2F and CHF_2 can be explained by taking into account a specific conformation of the axial conformers of monosubstituted cyclohexanes bearing these two substituents. As Figure 1.4 illustrates, the C–H of CH_2F and CHF_2 substituents occupies the *endo* position and bulkier fluorine(s) take(s) *exo* position(s) to minimize 1,3-diaxial strain. However, in the case of a spherical CF_3 substituent, a C–F inevitably takes the *endo* position, which increases the 1,3-diaxial strain. This would be the reason why the difference in *A* values between CF_3 and CHF_2 is considerably larger than that between CHF_2 and CH_2F. The same trend is observed for CH_3, CH_2CH_3, $CH(CH_3)_2$, and $C(CH_3)_3$ (see Figure 1.4) [50]. *A* values for other substituents have recently been reported: C_2F_5 (2.67), CF_3S (1.18), CF_3O (0.79), and CH_3O (0.49) [49].

The bulkiness of a CF_3 group has been estimated on the basis of comparison of rotational barriers along the biphenyl axis of 1,1′-disubstituted biphenyls **8a** to **8c** [51] as well as **9a** and **9b** [52]. These data clearly indicate that the bulkiness of a CF_3 group is similar to that of a $(CH_3)_2CH$ group (see Figure 1.5) [51–53].

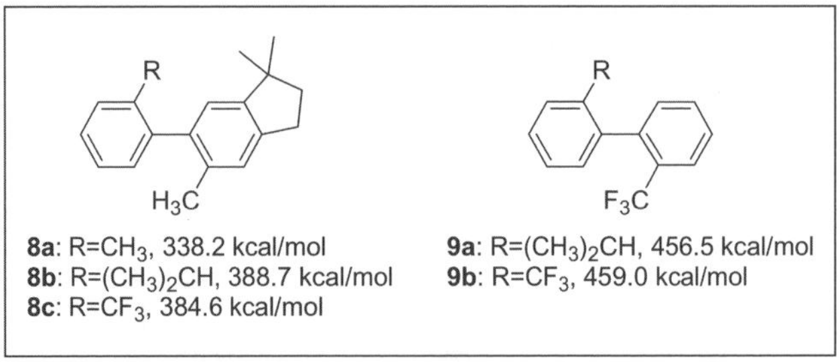

Figure 1.5 *Rotational barriers for **8** and **9**.*

Table 1.3 *Hansch hydrophobicity parameters for monosubstituted benzenes [54, 55]*

X in C_6H_5-X	π_X[a]	X in C_6H_5-X	π_X	X in C_6H_5-X	π_X
F	0.14	OCH_3	−0.02	$CH_3C(O)NH$-	−1.63
Cl	0.71	OCF_3	1.04	$CF_3C(O)NH$-	0.55
OH	−0.67	$CH_3C(O)$-	−1.27	CH_3SO_2-	−1.63
CH_3	0.56	$CF_3C(O)$-	0.08	CF_3SO_2-	0.55
CF_3	0.88				

[a] π_X: $\log P_X - \log P_H$ (octanol–water).

1.1.3 Lipophilicity of Fluorine and Fluorine-containing Groups

The absorption and distribution of a drug molecule *in vivo* are controlled by its balance of lipophilicity and hydrophilicity as well as ionization. Enhanced lipophilicity together with change in amine pK_a often leads to increase in blood–brain barrier (BBB) permeability or binding free energy through favorable partition between the polar aqueous solution and the less-polar binding site. It is generally conceived that incorporation of fluorine or fluorinated groups increases the lipophilicity of organic compounds, especially aromatic compounds. Table 1.3 shows selected Hansch π_X parameters [54, 55] for monosubstituted benzenes.

As Table 1.3 shows, fluorobenzene is slightly more lipophilic than benzene, but chlorobenzene is much more lipophilic. In a similar manner, CF_3-Ph is 57% more lipophilic than CH_3-Ph. The comparison of CF_3-Y-Ph and CH_3-Y-Ph (Y = O, CO, CONH, SO_2) reveals that CF_3-Y-Ph is substantially more lipophilic than CH_3-Y-Ph. In these CF_3-containing compounds, a strongly electron-withdrawing CF_3 group significantly lowers the electron density of the adjacent polar functional groups Y, which compromises the hydrogen-bonding capability of these functional groups with water molecules, and hence decreases hydrophilicity. Along the same lines, CF_3-benzenes with Lewis basic functionalities such as amine, alcohol, ether, carbonyl, and amide at the *ortho* or *para* position, decreases the hydrogen-bond accepting capability of these functionalities in an aqueous phase, which leads to increase in hydrophobicity and thus lipophilicity.

In contrast, the introduction of fluorine into aliphatic compounds results in decrease in lipophilicity. For example, pentane ($\log P = 3.11$) is more lipophilic than

Table 1.4 *log*P *values of straight-chain alkanols [57]*

Alcohols	X = H	X = F	$\Delta \log P_F - \log P_H$
	$\log P_H$	$\log P_F$	
CX_3CH_2OH	−0.32	0.36	0.68
$CX_3(CH_2)_2OH$	0.34	0.39	0.05
$CX_3(CH_2)_3OH$	0.88	0.90	0.02
$CX_3(CH_2)_4OH$	1.40	1.15	−0.25
$CX_3(CH_2)_5OH$	2.03	1.14	−0.89

1-fluoropentane ($\log P = 2.33$). Likewise, (3-fluoropropyl)benzene ($\Delta\log P = -0.7$) and (3,3,3-trifluoropropyl)benzene ($\Delta\log P = -0.4$) are considerably less lipophilic than propylbenzene [34]. In fact, the Hansch parameters π for CF_3 and CH_3 are 0.54 and 0.06, respectively, in aliphatic systems [56].

Table 1.4 shows selected $\log P$ values of straight-chain alkanols bearing a terminal CF_3 groups and comparisons with those of the corresponding nonfluorinated counterparts [57]. Trifluoroethanol is more lipophilic than ethanol ($\Delta\log P = -0.68$), which can be ascribed to the significant decrease in the basicity of the hydroxyl group by the strong electron-withdrawing effect of the CF_3 moiety. This strong through-bond σ-inductive effect of the CF_3 moiety extends up to three methylene inserts but diminishes beyond four. When the inductive effect of the CF_3 moiety does not affect the basicity of the hydroxyl group, reversal of relative lipophilicity is observed for 4-CF_3-butanol ($\Delta\log P = -0.25$) and 5-CF_3-pentanol ($\Delta\log P = -0.89$) as compared with butanol and pentanol, respectively. In the case of amines, a large enhancement of lipophilicity is observed, in general, when fluorine is introduced near to an amino group. This is attributed to the decrease in the amine basicity through the σ-inductive effect of fluorine, resulting in increase in the neutral amine component as opposed to the ammonium ion in equilibrium [58].

1.1.4 Inductive Effect of Fluorine and Fluorine-containing Groups

Since fluorine is the most electronegative element, it is natural that groups containing fluorine have unique inductive effects on the physicochemical properties of the molecules bearing them. For example, substantial changes in pK_a values of carboxylic acids, alcohols, or protonated amines are observed upon incorporation of fluorine into these molecules. Thus, when fluorine(s) and/or fluorine-containing group(s) are incorporated into bioactive compounds, these substituents will exert strong effects on the binding affinity for the receptors or target enzymes, biological activities, and pharmacokinetics.

Halogen substitution at the 2-position of acetic acid decreases the pK_a values in the order Br > Cl > F, which is qualitatively parallel to electronegativity (see Table 1.5). Thus, fluorine exerts the strongest effect ($\Delta pK_a = -2.17$ compared to 4.76 for acetic acid). Further substitutions with two fluorines ($\Delta pK_a = -3.43$) and three fluorines ($\Delta pK_a = -4.26$) at this position increase the acidity.

Since a CF_3 group withdraws electrons only in an inductive manner, insertion of a methylene group between CF_3 and CO_2H moieties naturally diminishes its inductive effect.

Table 1.5 *Selected pK_a values of various fluorinated compounds [35, 59]*

Compound	pK_a	Compound	pK_a	Compound	pK_a
CH_3CO_2H	4.76	$CH_3CH_2CO_2H$	4.87	$(CH_3)_2CHOH$	17.1[a]
CH_2FCO_2H	2.59	$CF_3CH_2CO_2H$	3.06	$(CF_3)_2CHOH$	9.3[a]
CH_2ClCO_2H	2.87	$C_6H_5CO_2H$	4.21[a]	$(CH_3)_3COH$	19.0[a]
CH_2BrCO_2H	2.90	$C_6F_5CO_2H$	1.7[a]	$(CF_3)_3COH$	5.4[a]
CHF_2CO_2H	1.33	CH_3CH_2OH	15.93[a]	C_6H_5OH	9.99
CF_3CO_2H	0.50	CF_3CH_2OH	12.39[a]	C_6F_5OH	5.5[a]

[a] Ref. [59].

Nevertheless, the pK_a of 3,3,3-trifluoropropanoic acids is 3.06, which is still substantially more acidic than propanoic acid (pK_a = 4.87) (ΔpK_a = −1.81). Introduction of CF_3 groups to methanol dramatically increases the acidity of the resulting alcohols. Thus, pK_a values of CF_3CH_2OH, $(CF_3)_2CHOH$ and $(CF_3)_3COH$ are 12.39, 9.3 and 5.4, respectively. The pK_a value of $(CF_3)_3COH$ is only 0.7 larger than that of acetic acid.

As discussed above, the introduction of fluorine(s) to alkyl amines decreases their amine basicity, which results in higher bioavailability, in some cases, due to the increase in lipophilicity [60]. For example, the pK_a values of ethylamines decrease linearly upon successive fluorine introductions: $CH_3CH_2NH_2$ (10.7), $FCH_2CH_2NH_2$ (9.0), $F_2CHCH_2NH_2$ (7.3), and $F_3CCH_2NH_2$ (5.8) [58]. Based on experimental data, a practical method has been developed for the prediction of pK_a values of alkyl amines through the "σ-transmission effect" (i.e., inductive effect) of fluorine [58]. The inductive effects of fluorine and fluorine-containing groups confer favorable properties on some enzyme inhibitors. For example, a significant difference in potency was observed between $CF_3SO_2NH_2$ (pK_a = 5.8; $K_i = 2 \times 10^{-9}$ M) and $CH_3SO_2NH_2$ (pK_a = 10.5; K_i in 10^{-4} M range) for the inhibition of carbonic anhydrase II (a zinc metalloenzyme) [61, 62]. This can be attributed to the substantial increase in the acidity of the sulfonamide functionality by the introduction of a trifluoromethyl group, which facilitates deprotonation and better binding to the Zn(II) ion in the catalytic domain of the enzyme [61, 62].

Perfluorination of benzoic acid and phenol increases their acidity by 2.5 and 4.5 pK_a units, respectively (see Table 1.5). In these compounds, the π-electrons are localized at the center of the perfluorobenzene ring because of strong electronic repulsion between the lone pairs of five fluorines on the ring. When fluorine is bonded to an sp^2-hybridized carbon, a significant cationic charge distribution is observed at the carbon with fluorine (C^1), while a substantial negative charge develops at C^2 (see **10b** and **10c**), as shown in Figure 1.6 [*Note:* DFT computation was performed using Gaussian 03 (Revision B.03) by one of the authors (T.Y.)]. This remarkable polarization of a carbon–carbon double bond is attributed to the strong electronic repulsion between the π-electrons of the carbon–carbon double bond and the lone pairs of fluorine.

This phenomenon is unique to fluoroethenes (**10b** and **10c**) and the corresponding dichloroethene (**10d**) shows much weaker "p–π repulsion." This can be ascribed to the facts that (i) chlorine is a third-row element and its 3p lone pairs do not effectively interact with the $2p_z$ olefinic π-electrons, and (ii) the C–Cl bond is considerably longer than the C–F bond (1.744 Å for **10d** vs. 1.326 Å for **10c**, see also Table 1.1). Another unique structural feature of **10c** is its unusually small F–C^1–F bond angle (109.5°), which is 10.5°

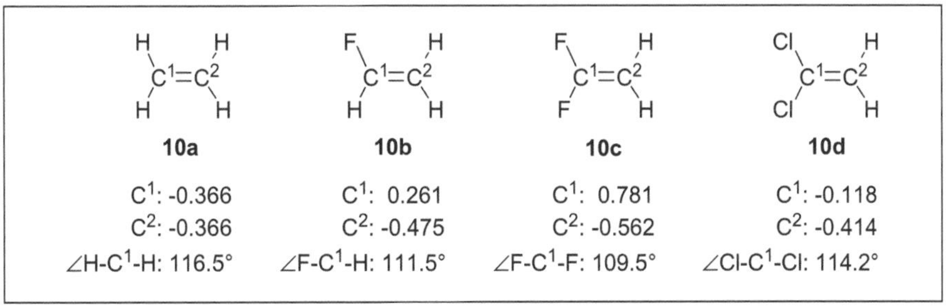

10a	**10b**	**10c**	**10d**
C^1: -0.366	C^1: 0.261	C^1: 0.781	C^1: -0.118
C^2: -0.366	C^2: -0.475	C^2: -0.562	C^2: -0.414
∠H-C^1-H: 116.5°	∠F-C^1-H: 111.5°	∠F-C^1-F: 109.5°	∠Cl-C^1-Cl: 114.2°

Figure 1.6 *Estimated charges on carbons in ethylenes,* ***10a-d****, by* ab initio *calculations.*

smaller than the angle expected for the ideal sp^2 hybridization [63, 64]. This might suggest a substantial contribution of an $F^{(+)}$=C(F)–$^{(-)}CH_2$ resonance structure and an attractive electronic interaction between $F^{(+)}$ and F: [64].

1.1.5 Gauche Effect

The strong electronegativity of fluorine renders its related molecular orbitals relatively lower-lying. For example, a C–F bond can readily accept electrons to its vacant σ^*_{C-F} orbital from a vicinal electron-donating orbital, while the electron-occupied σ_{C-F} orbital is reluctant to donate electrons. Such characteristics often influence the three-dimensional shape of a molecule in a distinct manner. Figure 1.7 illustrates basic examples.

Two conformations, *gauche* and *anti*, are possible for 1,2-difluoroethane [14]. Based on the fact that fluorine is about 20% larger and much more electronegative than hydrogen, it is quite reasonable to assume that *anti*-**14** should be more stable than *gauche*-**14** from both steric and electrostatic points of view. However, this is not the case, and analyses by infrared spectroscopy [65], Raman spectroscopy [65], NMR [66] and electron diffraction [67, 68] have led to the unanimous conclusion that the latter conformation is preferred by approximately 1.0 kcal/mol, which was also confirmed by *ab initio* calculations [69, 70]. This phenomenon, termed the "*gauche effect*," is rationalized by taking into account stabilization through the critical donation of electrons from the neighboring σ_{C-H} orbital to the lower-lying vacant σ^*_{C-F} orbital as shown in Figure 1.7, which is not possible in the corresponding *anti* isomer. In the case of vicinal 1,2-dihaloethanes, this preference is observed specifically for 1,2-difluoroethane (**14**), because the increasing steric hindrance caused by two other halogens in the *gauche* geometry exceeds the energy gain by the orbital interaction using the energetically lowered $\sigma^*_{C-Halogen}$ [69, 70]. For example, an exchange of one fluorine in 1,2-difluoroethane (**14**) with chlorine (i.e., 1-fluoro-2-chloroethane) prefers the *anti* conformation on the basis of computational analysis as well as experimental results [71].

2-Fluoroethanol (**15**) also takes the *gauche* conformation predominantly [72–74]. Initially this preference was attributed to its possible formation of intramolecular *F*···*H*–O hydrogen bonding in addition to the *gauche* effect [72–74]. However, it was later found that 1-fluoro-2-methoxyethane, which could not form a hydrogen bond, also took the *gauche* conformation as its predominant structure [75, 76]. This finding clearly eliminated

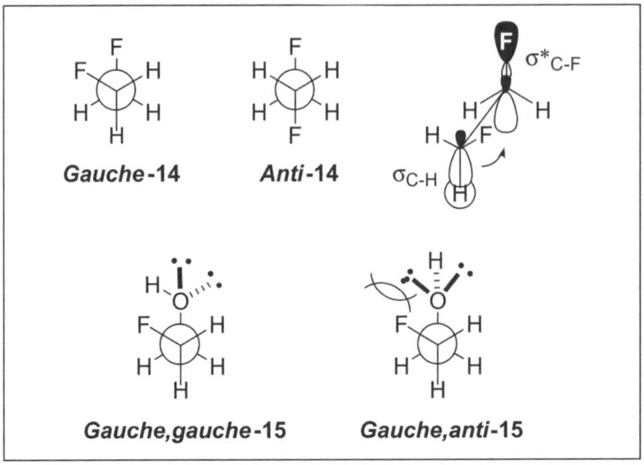

Figure 1.7 *Orbital interaction of 1,2-difluoroethane (**14**) and conformation of 2-fluoroethanol (**15**).*

the contribution of the *F···H–O* hydrogen bonding as the major reason for the dominant *gauche* conformation of **15** wherein the O–H is in parallel to the C–F bond. Thus, it is most likely that **15** takes this particular conformation to minimize unfavorable electronic repulsion between the lone pairs of the oxygen and fluorine [75]. The absence of *F···H–O* and *F···H–N* hydrogen bonding was also confirmed for α-fluorocarboxamides [77] and 2-fluoroethyl trichloroacetate [78].

It has been shown that the *gauche* effect plays a key role in controlling the conformation of various acyclic compounds [79]. For example, the x-ray crystallographic analysis of difluorinated succinamides, *syn*-**16** and *anti*-**16** shows unambiguously that these compounds take *gauche* conformation with *F–C–C–F* dihedral angles of 49.0° and 75.2°, respectively (see Figure 1.8). *Syn*-**16** and *anti*-**16** possess respectively two and one $\sigma_{C\text{-}H}\cdots\sigma^*_{C\text{-}F}$ interactions which are reflected in the lengths of their (F)*C–C*(F) bonds: 1.495 Å and 1.538 Å, respectively, the former being 0.043 Å shorter than the latter (see Figure 1.7). Although the conformation of *syn*-**16** might look reasonable by taking into account the *antiperiplanar* placement of two bulky amide moieties, it is not the case for *anti*-**16**, indicating the importance of the *gauche effect*. In the latter case, however, a 7-membered ring hydrogen bond between two amide groups might also make some contribution. Similar *gauche* effects have been reported in other systems [80–84].

The *gauche* effect is also clearly observed in the conformational preference of 4-fluoroprolines (**17b** and **17c**) determined by ^{1}H NMR analysis [85]. As Table 1.6 shows, (4*R*)-fluoroproline (**17b**: $R^1 = F$, $R^2 = H$) strongly prefers the C^{γ}-*exo* conformation in which an electron-releasing C^{γ}–H bond should occupy the antiperiplanar position to the electron-accepting C–F group to maximize the *gauche* effect. On the other hand, its epimer (4*S*)-fluoroproline (**17c**: $R^1 = H$, $R^2 = F$) takes the C^{γ}-*endo* conformation almost exclusively to optimize the $\sigma_{C\text{-}H}\cdots\sigma^*_{C\text{-}F}$ interaction for the *gauche* effect. It should be noted that the observed preferences are independent of the *trans* or *cis* amide linkage by calculation [85].

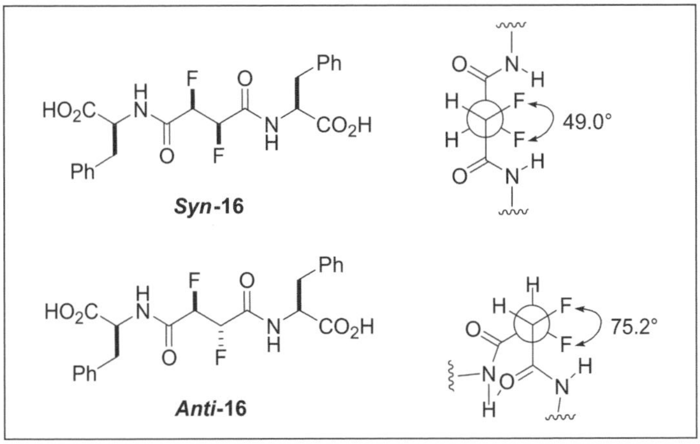

Figure 1.8 *Conformation of diastereomeric **16** by X-ray crystallographic analysis.*

Table 1.6 *Conformational preference of Ac-Xaa-OMe (Xaa = proline and its derivative) [85]*

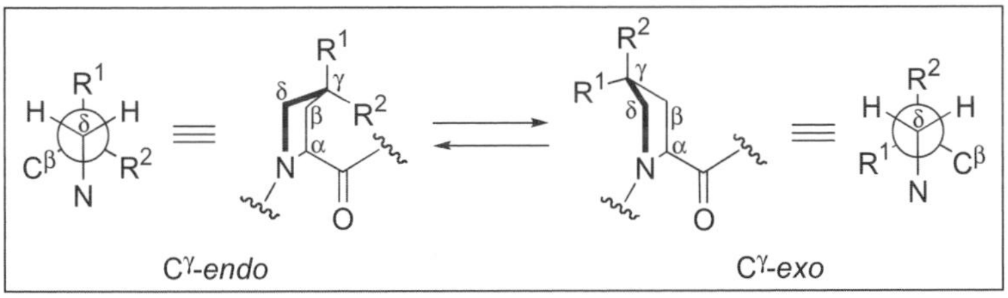

Xaa in Ac-Xaa-OMe	R^1	R^2	Conformation			$\Delta E_{endo} - E_{exo}$ (kcal/mol)	
			endo	:	*exo*	*trans*	*cis*
Proline (**17a**)	H	H	66	:	34	−0.41	−0.60
(4*R*)-4-Fluoroproline (**17b**)	F	H	14	:	86	0.85	1.18
(4*S*)-4-Fluoroproline (**17c**)	H	F	95	:	5	−0.61	−1.99

1.1.5.1 Unique Electronic Effects of Fluorine Related to the Origin of the Gauche Effect

The interaction of the $\sigma^*_{C\text{–}F}$ orbital with the lone pairs of fluorine in fluorinated methanes is substantial [83]. In these compounds, the C–H bond length is almost constant regardless of the number of fluorines in a molecule. In sharp contrast, the C–F bond length decreases as the number of fluorines increases due to substantial n_F–$\sigma^*_{C\text{–}F}$ interaction mentioned above [83]. Alternatively, a significantly strong positive charge developed on carbon may play a key role in strengthening C–F bonds in an electrostatic manner [82].

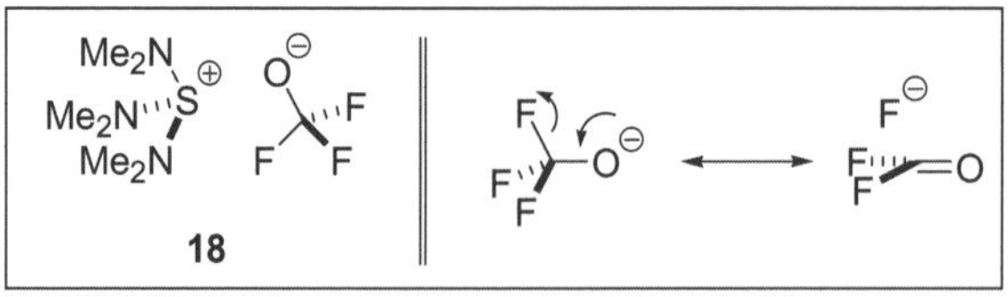

*Figure 1.9 Compound **18** and negative hyperconjugation in CF_3–O^- anion.*

"Negative hyperconjugation" of fluorine [86–89], essentially the same electron donation pattern as that of the *gauche* effect, that is, interaction of an electron-rich bond with the lower-lying vacant orbital of a polarized neighboring C–F bond (σ^*_{C-F}), has been clearly observed in the X-ray structure of $[(CH_3)_2N]_3S^+CF_3O^-$ (**18**) (see Figure 1.9) [90]. The counter-anion, CF_3O^-, of **18** possesses a significantly short C–O bond (1.227 Å) and long C–F bonds (1.390 and 1.397 Å). For comparison, the gas-phase structures of electronically neutral counterparts, CF_3OR (R = F, Cl, CF_3), show 1.365–1.395 Å and 1.319–1.327 Å for the C–O and C–F distances, respectively [90]. It is worthy of note that the C–O single bond length in CF_3O^- (1.227 Å) is close to that of a C=O bond length (e.g., 1.171 Å for F_2C=O) and the F–C–F bond angle is extraordinarily small (101.7° and 102.2°) compared with the ideal sp^3 bond angle (109.5°). This phenomenon strongly indicates the effective orbital interaction of the electron rich n_O orbital with lower-lying σ^*_{C-F} orbital, i.e., negative hyperconjugation.

1.1.6 Hydrogen Bonding

Fluorine can share three sets of lone-pair electrons with electron-deficient atoms intramolecularly or intermolecularly, in particular with a relatively acidic hydrogen bound to a heteroatom. In addition, as described in section 1.4, strongly electron-withdrawing perfluoroalkyl groups increase the acidity of proximate functional groups such as alcohol, amine, amide, and carboxylic acid.

It is readily anticipated that the acidity of CF_3-containing benzylic alcohol **19** is as high as or higher than that of phenol (see Table 1.5 for hexafluoro-2-propanol). Moreover, a fluorine atom of the CF_3 groups at the 3- and 5-positions should increase its anionic character by negative hyperconjugation (see above). Thus, it is reasonable to assume that the benzylic hydroxyl group would form a hydrogen bond with a proximate CF_3 group [91, 92]. In fact, the X-ray crystallographic analysis of **19** shows that **19** forms a unique dimer structure in the solid state through two strong intermolecular hydrogen bonds (H···F distance is 2.01 Å), as illustrated in Figure 1.10 [91]. The strength of this hydrogen bond is obvious by comparing the sum of the van der Waals radii of H and F (2.67 Å) with the observed H···F bond length (2.01 Å). On the other hand, **19** appears to form an intramolecular hydrogen bonding between the same benzylic hydroxyl CF_3 groups in a hexane solution on the basis of low-temperature ^{13}C NMR analysis, as illustrated in Figure 1.10. Although the *C*F_3 carbon appeared as a normal quartet (J_{C-F} = 274 Hz) at 24 °C, the coupling pattern was changed to the doublet of triplet (J_{C-F} = 261, 279 Hz, respectively) at −96 °C. The result appears to indicate the nonequivalence of three fluorine atoms in the CF_3 groups, and only one of the three fluorine atoms participates in the hydrogen bonding.

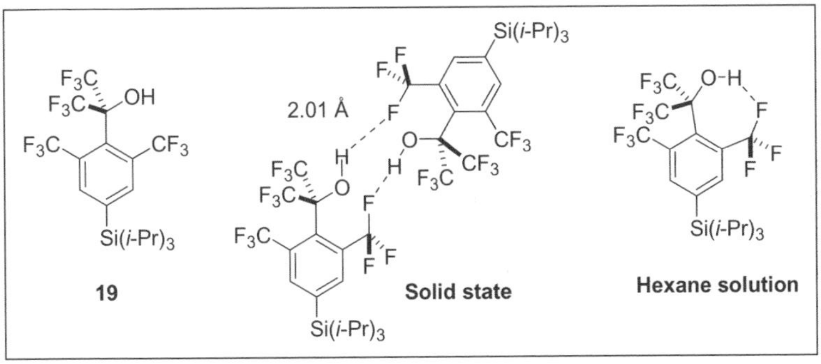

Figure 1.10 *Intermolecular and intramolecular H···F hydrogen bonding patterns of* ***19****.*

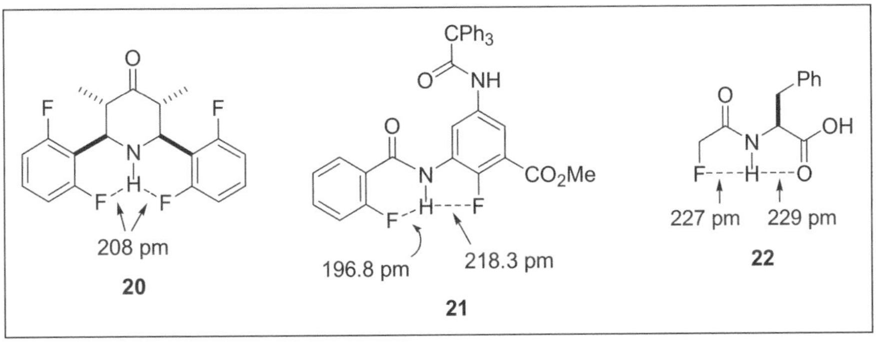

Figure 1.11 *Representative intramolecular NH···F hydrogen-bonding interactions.*

However, very recently, a counterargument on this hydrogen bonding suggested that the nonequivalence of three fluorine atoms of the CF_3 groups should be attributed to steric crowding and not H–F hydrogen bonding [93]. Accordingly, the interpretation of the NMR data is still unsettled.

Similar hydrogen bonding has been observed between fluorine and amine or amide hydrogen. The NH···F hydrogen bond is especially favorable with an amide hydrogen because of its acidity compared with that of an amine. Figure 1.11 illustrates three representative compounds, **20**, **21**, and **22** whose structures were determined by X-ray crystallography. The H···F distance of **20** is 2.08 Å, which is 0.59 Å shorter than the sum of the van der Waals radii [94]. Compound **21** includes two hydrogen bonds with six- and five-membered ring systems, bearing NH···F distances 1.97 Å and 2.18 Å, respectively [95]. The NH moiety of *N*-fluoroacetylphenylalanine (**22**) also formed a bifurcated intramolecular hydrogen bonds with F and O with practically the same bond lengths: 2.27 Å and 2.29 Å, respectively [96].

Fluorinated benzenes, in general, have been shown to form intermolecular hydrogen bonds in the solid state. Among those fluorobenzenes, the X-ray crystal structure of 1,2,4,5-tetrafluorobenzene (**24**) exhibited the shortest intermolecular H···F distance of

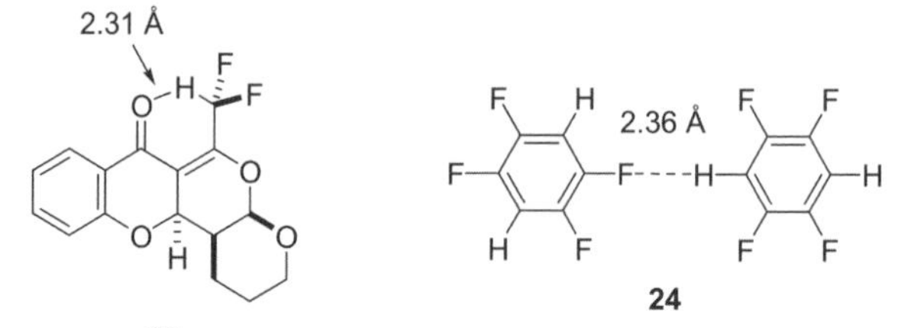

Figure 1.12 *Other hydrogen-bonding patterns.*

2.36 Å [97] (see Figure 1.12). Although it is not H···F bonding, a related hydrogen bonding pattern has been reported. As Figure 1.12 illustrates, the X-ray crystallographic analysis of chromone **23** shows the *O*···*H*–CF_2 distance of 2.31 Å (the sum of the van der Waals radii of H and O is 2.72 Å) [98, 99]. Other C=*O*···*H*–CF_2 bonding examples have been reported [100].

1.1.7 Orthogonal multipolar C–F Bond–Protein Interactions

It has recently been shown that polar C–F bond–protein interactions play a critical role in the stabilization of fluorine-containing drugs and their target proteins [101]. These polar interactions are found in the X-ray crystal structures of drug-protein complexes compiled in the Cambridge Structural Database (CSD) and the Protein Data Bank (PDB) [20–32, 101]. A large number of examples for the polar C–F bond–protein interactions, found in the CSD and PDB through database mining, include those between a C–F bond and polar functional groups such as carbonyl and guanidinium ion moieties in the protein side chains, that is, C–*F*···*C*=O and *C*–*F*···$C(NH_2)(=NH)$. The majority of examples from the protein crystallographic database indicate that a C–F bond unit serves as a poor hydrogen-bond acceptor. Instead of hydrogen bonding, however, a C–F bond forms polar interactions with a large number of polarizable bonds in the manner as C–F···X(H) (X = O, N, S) and C–F···C_α–H (C_α = α-carbon of the α-amino acid), in which the F···H–X separation is well beyond hydrogen-bonded contact distance [101].

For example, a thrombin inhibitor **25** (K_i = 0.25 μM) is more potent than the nonfluorinated counterpart (K_i = 1.6 μM) and the X-ray crystal structure of the inhibitor–enzyme complexes showed remarkable conformational differences between the two inhibitors. This conformational change is caused by the dipolar C–F···N(H)(Gly216) interaction with a F–N distance of 3.5 Å, as illustrated in Figure 1.13a [25].

A large number of examples for the orthogonal dipolar C–F···C=O interactions have been found in the CSD and PDB [101]. For example, Figure 1.13b illustrates a rather unique double interaction, that is, C–F···$(C{=}O)_2$, in the inhibitor–enzyme complex of **26** with p38 MAP kinase, wherein the fluorine atom of the 4-fluorophenyl moiety of **26** interacts with the amide carbonyl carbons of Leu104 and Val105 with equal distance of 3.1 Å [102].

Table 1.7 summarizes systematic SAR studies of tricyclic thrombin inhibitors, *rac*-**27** and *rac*-**28**, through "fluorine scan" to map the effects of fluorine introduction on

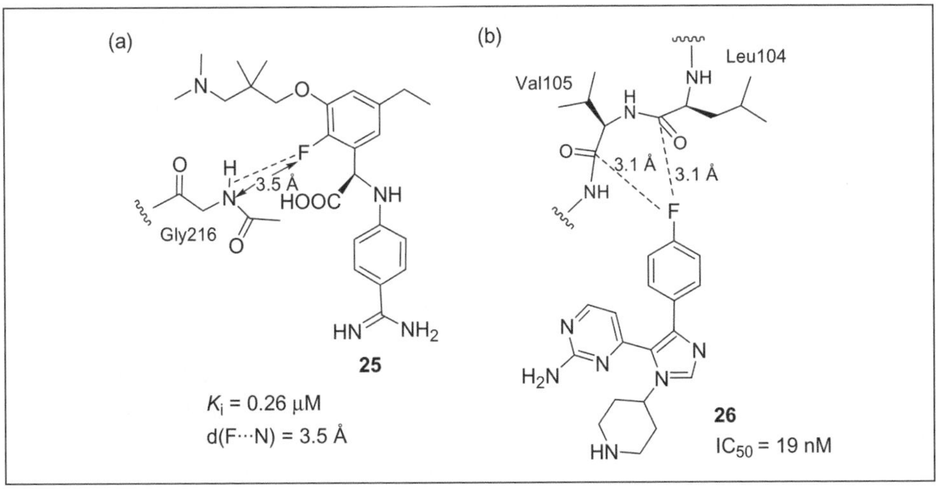

Figure 1.13 *(a) Interaction of C–F···N(H)(Gly216) in the thrombin-inhibitor **25** complex. (b) C–F···C=O multipolar interactions of p38 MAP kinase inhibitor **26** with the kinase.*

Table 1.7 *Enzyme inhibitory activity of tricyclic thrombin inhibitors [105]*

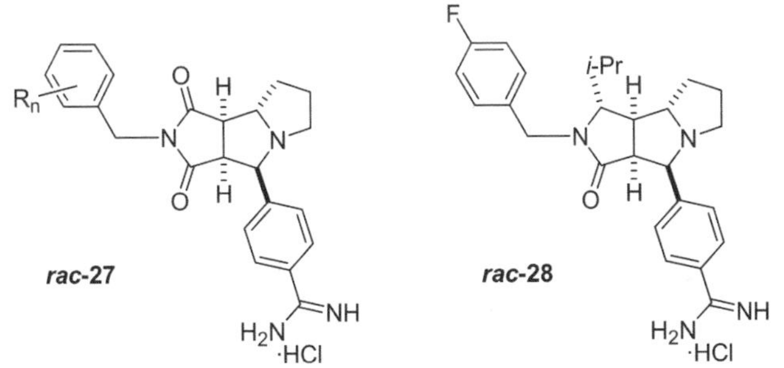

Inhibitor	Substituent[a]	K_i^b (mM)	Selectivity[c]	log *D*
27a	–	0.31	15	−1.24
27b	2-F	0.50	9.8	<−1.00
27c	3-F	0.36	26	−1.24
27d	4-F	0.057	67	−1.08
27e	2,3-F_2	0.49	18	–[d]
27f	2,6-F_2	0.61	9.0	–[d]
27g	3,4-F_2	0.26	29	–[d]
27h	3,5-F_2	0.59	25	−1.25
27i	2,3,4,5,6-F_5	0.27	44	−1.14
27j	4-Cl	0.19	30	–[d]
28	4-F	0.005	413	–[d]

[a] Substituents on the benzene ring of the benzylimide moiety.
[b] With ±20% uncertainty.
[c] K_i (trypsin)/K_i (thrombin).
[d] Not determined.

inhibitory activity, change of amine basicity, and favorable interactions of C–F bonds with the protein [103–106].

Inhibitory activity, selectivity between thrombin and trypsin and lipophilicity (log *D*) of *rac*-**27** are shown in Table 1.7, which indicates that 4-monofluorinated analog *rac*-**27d** is the most potent inhibitor in this group (K_i = 0.057 mM; K_i(trypsin)/K_i(thrombin) selectivity = 67) [105]. The inhibitory activity K_i is further optimized to 5 nM by changing the imide ring of *rac*-**27** to a lactam bearing an isopropyl group, *rac*-**28**, wherein the isopropyl group fits better in the P-pocket (see Figure 1.14). The trypsin/thrombin selectivity of *rac*-**28** is also dramatically improved to 413 [105].

The X-ray crystallographic analysis of the enantiopure **27d**–thrombin complex indicates that the H–C_α–C=O fragment of the Asn98 residue possesses significant "fluorophilicity" [105]. As Figure 1.14 illustrates, the C–F residue of enantiopure **27d** has strong multipolar C–F···C_α–H [*d*(F–C_α) = 3.1 Å] and C–F···C=O [*d*(F–C=O) = 3.5 Å] interactions with the Asn98 residue in the distal hydrophobic pocket ("D pocket") of thrombin. It is worthy of note that the C–F bond and the electrophilic carbonyl group are positioned in a nearly orthogonal manner along the pseudotrigonal axis of the carbonyl group [105]. This preferred geometry for the C–F···C=O interactions is further corroborated by the X-ray crystal structure analyses of fluorine-containing small molecules [103–106] and the database mining of PDB and CSD [107–110]. The latter furnished numerous cases

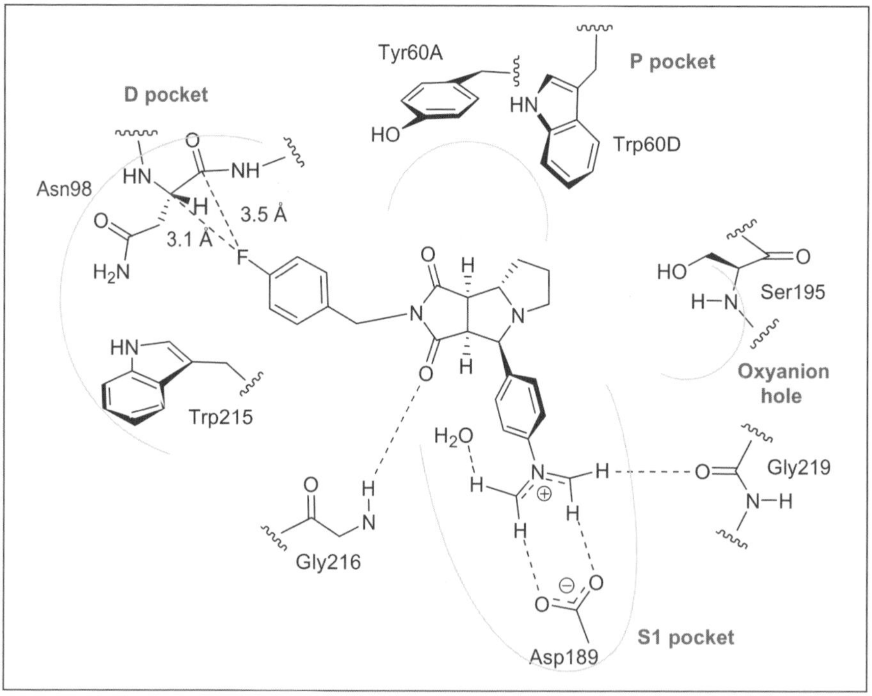

Figure 1.14 *Binding mode of tricyclic thrombin inhibitor* ***27d*** *on the basis of X-ray crystallographic analysis of its protein complex.*

of C–F···C=O contacts with the F···C distance below the sum of the van der Waals radii and the C–F···C=O torsion angle toward 90°. A similar but intramolecular orthogonal polar interaction between a CF_3 groups and a spatially close amide C=O group, which plays a role in the folding of an extended indole molecule, has recently been reported [111].

Figure 1.15 illustrates an example of the polar interaction of a C–F bond with the guanidinium carbon of the Arg residue in an enzyme, revealed by the X-ray crystal structure of atorvastatin-HMG-CoA reductase (HMG-CoA = 3-hydroxy-3-methylglutaryl-coenzyme A; the rate-determining enzyme in the biosynthesis of cholesterol). As mentioned above (see section 1.1), atorvastatin is the best-selling cholesterol-lowering drug, which binds to HMG-CoA reductase tightly (IC_{50} = 8 nM). In the early stage of the drug discovery, a series of phenylpyrrole-hydroxylactones were evaluated for their inhibitory activity against HMG-CoA reductase, as shown in Figure 1.15a. Then, the SAR study on the 4-substituted phenyl analogues identified that the 4-fluorophenyl analogue (R = F) was the most active; that is, fluorine was better than hydrogen, hydroxyl or methoxy group at this position [112]. This finding eventually led to the discovery of atorvastatin. The X-ray crystal structure of atorvastatin–HMG-CoA reductase [113] shows a strong C–F···C(NH_2)(=NH) polar interaction between the 4-fluorophenyl moiety and the Arg590 residue of the enzyme with a very short F···C distance (2.9 Å), as illustrated in Figure 1.15b [101].

Several other examples of similar polar interactions between a C–F bond in a ligand and the guanidinium ion moiety of an Arg residue in protein have been found in PDB. However, no linear C–F···H–N interactions are identified, reflecting the poor hydrogen-bond accepting ability of a C–F bond. Instead of hydrogen bonding, a C–F bond is found to orient either parallel or orthogonal to the plane of the guanidinium ion moiety, bearing a delocalized positive charge. These examples confirm the fluorophilic character of the guanidinyl side-chain of an Arg residue.

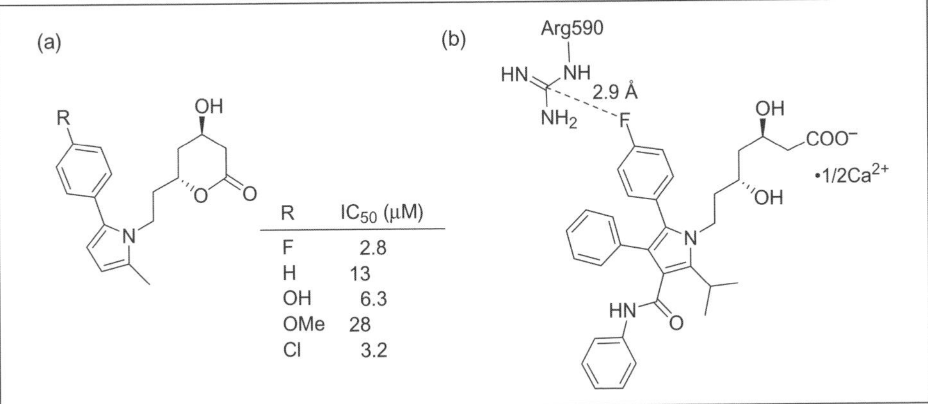

R	IC_{50} (μM)
F	2.8
H	13
OH	6.3
OMe	28
Cl	3.2

Figure 1.15 *(a)* In vitro *activity of HMG-CoA reductase inhibitors. (b) Orthogonal polar interaction of the fluorine substituent of atorvastatin with Arg590 in HMG-CoA reductase.*

1.1.8 Isostere

A variety of natural and synthetic peptides exhibit biological activities, which can be developed as pharmaceutical drugs or biochemical agents. However, those peptides consisting of amide linkages are, in most cases, easily deactivated through cleavage of key amide bonds by hydrolytic enzymes. Accordingly, it is very useful if a key amide bond is replaced with noncleavable bond, keeping the characteristics of the amide functionality.

An amide linkage in a peptide such as **29a** has an imidate-like zwitterioninc resonance structure **29b**. Since the contribution of this resonance structure is significant, free rotation around the C–N bond is partially restricted because of the substantial double-bond character of the C–N bond (see Figure 1.16). Accordingly, it might be possible to replace an amide linkage with its isostere (i.e., a molecule having the same number of atoms as well as valence electrons [114]), regarded as a "peptide isostere."

The first and simplest attempt at the "peptide isostere" was made by replacing an amide bond of enkephalin with a *trans*-olefin unit [115], but it did not give a desirable effect. However, computational analysis of a model amide, *N*-methylacetamide (**30b**) and its isosteres, by semiempirical molecular orbital calculation revealed that fluoroolefin **30c** resembled **30a** much more closely than **30b** [116]. To confirm this finding based on a semiempirical method, the *ab initio* analysis of **30a**, (*E*)-2-butene (**30b**), (*Z*)-2-fluoro-2-butene (**30c**) and (*Z*)-1,1,1-trifluoro- 2-methyl-2-butene (**30d**) was carried out using the Gaussian 03 program (B3LYP/6–311++G**) [117] by one of the authors (T. Y.), which gave electrostatic potentials of these compounds. The results are illustrated in Figure 1.17.

As readily anticipated, **30a** has a very negative oxygen, a highly positive carbonyl carbon, a highly negative nitrogen, and a very positive NH hydrogen. In sharp contrast, **30b** has a nonpolarized negative C=C bond and only weakly positive CH hydrogen. Thus, it is apparent that **30b** does not mimic **30a** electronically. In contrast, **30c** has an appropriately polarized C=C double bond, a negative fluorine atom in place of the oxygen atom of amide **30a**, and a positive CH in place of the NH moiety of amide **30a**. Thus, **30b** indeed mimics **30a** electronically. Finally, **30d** has a nonpolarized C=C bond, a modestly positive CH hydrogen, and three negative fluorine atoms. Thus, **30d** mimics **30a** electronically to some extent, although sterically the CF_3 group is much bulkier than an oxygen atom.

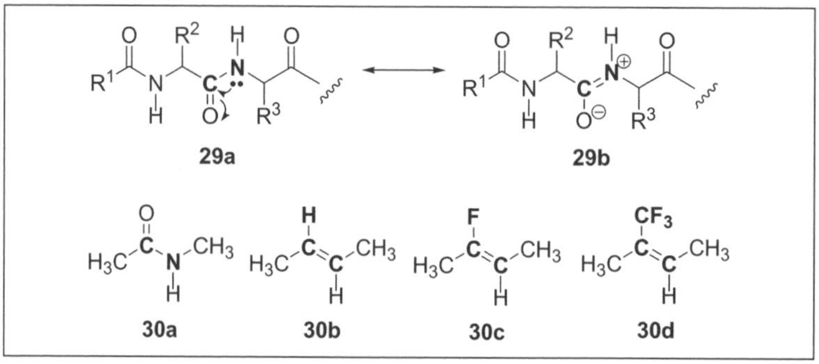

Figure 1.16 *Resonance structures of amide **29**, amide **30a**, and its isosteres **30b–d**.*

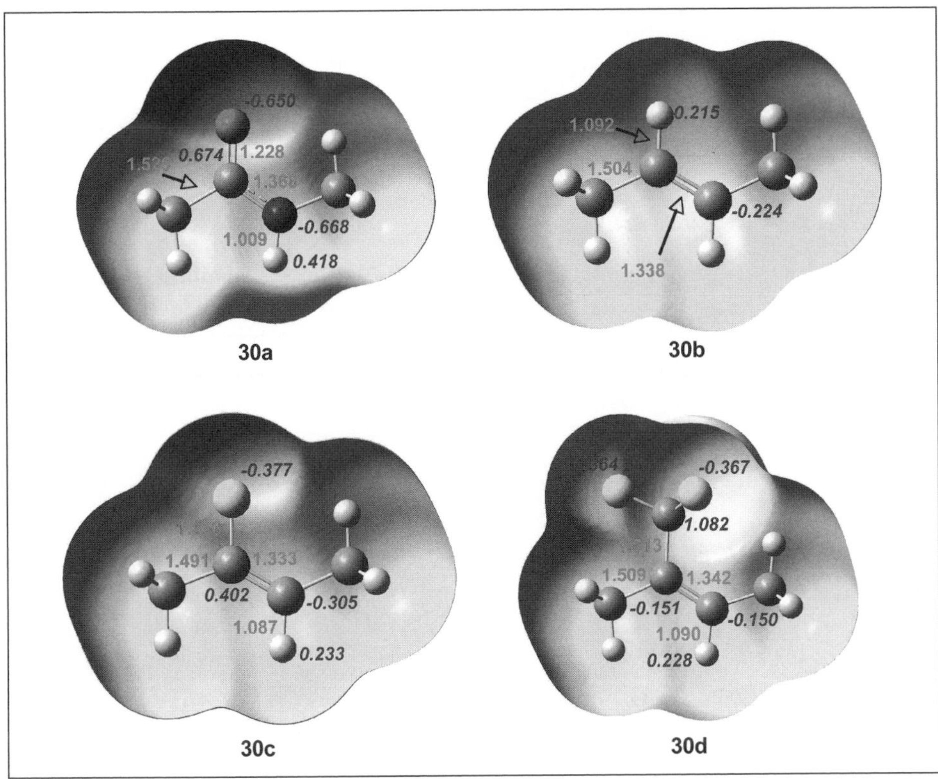

Figure 1.17 *Electrostatic potential of the model compounds* ***30a*** *to* ***30d*** *(color alteration from red to blue describes the shift of the electronically rich to deficient circumstance). See color plate 1.17.*

A fluoroolefin isostere was successfully introduced to a physiologically important neuropeptide, Substance P (**31**) [118]. As Figure 1.18 shows, the (*E*)-fluoroethene isostere replaced the peptide linkage between the Phe^8-Gly^9 residues of **31** to give Substance P analogue **32a** and its epimer **32b** [119]. The neurokinin-1 receptor binding assay of **32a** disclosed that **32a** was almost as potent as the original peptide **31**. On the other hand, its epimer **32b** was 10 times less potent than **32a**, as anticipated.

Another fluoroolefin isostere, (*E*)-trifluoromethylethene, was also introduced to antibiotic Gramicidin S (**33a**), replacing the peptide linkage between Leu and Phe (two sites) to give an analogue **33b**, as illustrated in Figure 1.19 [120]. Variable-temperature NMR measurements of **33a** and **33b** indicated that the NH(Leu) and NH(Val) protons were forming intramolecular hydrogen bonds. In addition, NOESY experiments of **33a** and **33b** showed interstrand NOEs between NH(Leu) and NH(Val) as well as NH(Leu) and H_α(Orn) in both compounds. Moreover, X-ray crystallographic analysis of **33c** confirmed the presence of a pair of intramolecular hydrogen bonds between NH(Leu) and C=O(Val) (1.96 Å and 2.00 Å). However, the replacement of the Leu-Phe amide linkage with the bulky (*E*)-trifluoromethylethene isostere caused a 70° twist in the plane of the C=C double bond

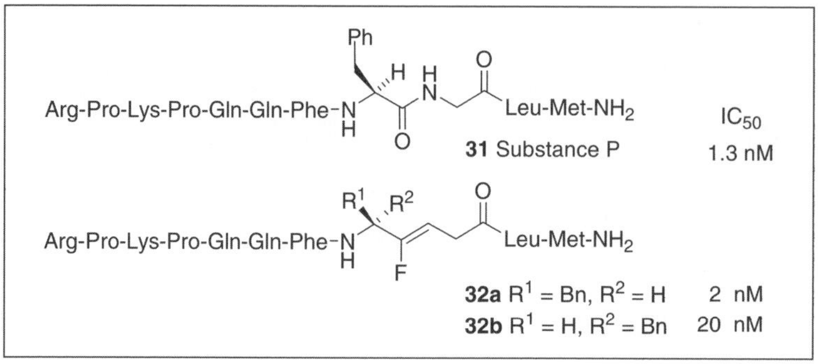
Ph
Arg-Pro-Lys-Pro-Gln-Gln-Phe-N
Leu-Met-NH2
31 Substance P
IC50
1.3 nM
R1 R2
Arg-Pro-Lys-Pro-Gln-Gln-Phe-N
F
Leu-Met-NH2
32a R1 = Bn, R2 = H 2 nM
32b R1 = H, R2 = Bn 20 nM

Figure 1.18 *Substance P (**31**) and its fluoroolefin dipeptide isosteres (**32a, b**).*

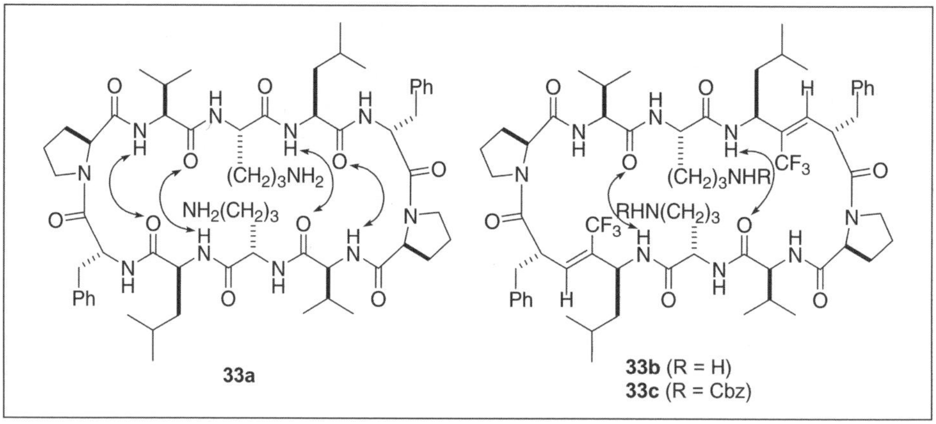
(CH2)3NH2
NH2(CH2)3
Ph
33a
(CH2)3NHR
RHN(CH2)3
CF3
33b (R = H)
33c (R = Cbz)

Figure 1.19 *Gramicidin S (**33a**) and its CF_3-containing isostere (**33b**).*

compared with the amide moiety in the original β-turn structure of **33a**. This is very likely attributable to the bulkiness of the CF_3 groups, which are difficult to be accommodated inside the β-turn hairpin structure. In spite of some conformational difference, **33b** was found to possess high antibiotic activity against *Bacillus subtilis* equivalent to that of **33a** [120]. The use of the (*E*)-trifluoromethylethene peptide isostere as β-turn promoter and peptide mimetics in general has also been studied [121]. A fairly large dipole moment of this isostere ($\mu = 2.3$ D) appears to play a key role in the improved mimicry of the electrostatic potential surface of the amide linkage (dipole moment of the amide unit: $\mu = 3.6$ D) [120].

Although fluoroethene and trifluoroethene peptide isosteres were successfully employed as nonhydrolyzable amide substitutes, more recently trifluoroethylamines have emerged as highly promising peptide isosteres. The trifluoroethylamine isostere has been studied for partially modified retro- and retro-inverso Ψ[NHCH(CF_3)]Gly peptides [122, 123]. Furthermore, it has been demonstrated that the CF_3 group of the trifluoroethylamine

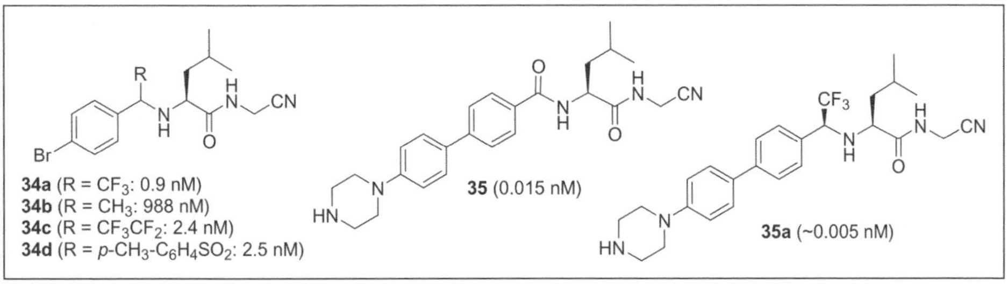

34a (R = CF_3: 0.9 nM)
34b (R = CH_3: 988 nM)
34c (R = CF_3CF_2: 2.4 nM)
34d (R = *p*-CH_3-$C_6H_4SO_2$: 2.5 nM)
35 (0.015 nM)
35a (~0.005 nM)

Figure 1.20 *Cathepsin K inhibitors **34** and **35**, bearing trifluoroethylamine isosteres. Numbers in parentheses are IC_{50} values.*

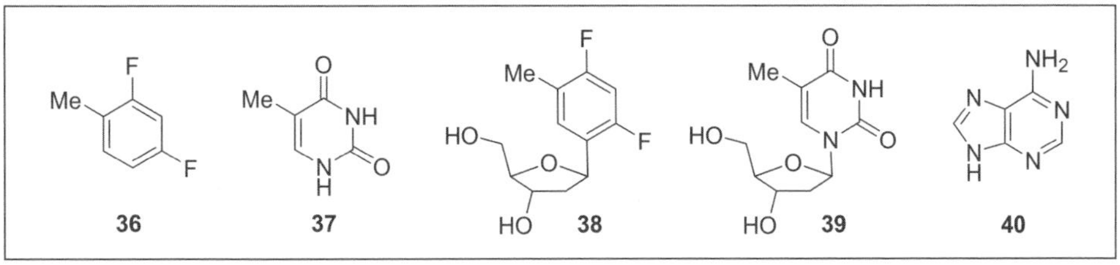

Figure 1.21 *Difluorotoluene (**36**), difluorotoluene deoxyriboside (**38**), thymine (**37**), thymidine (**39**) and adenine (**40**).*

isostere can function as the carbonyl group of an amide to provide a metabolically stable, nonbasic amine that has excellent hydrogen-bonding capability due to the strong inductive effect of the CF_3 group [124, 125]. These unique characteristics of the trifluoroethylamine isostere have been exploited and quite successfully applied to the optimization of cathepsin K inhibitors, which are promising antiresorptive agents for treatment of osteoporosis [124]. As Figure 1.20 shows, trifluoroethylamine analogue **34a** exhibits 1000 time better potency than the corresponding ethylamine analogue **34b**. Pentafluoroethylamine analogue **34c** possesses a somewhat reduced activity, but still in a comparable level to **34a**. Tosylmethylamine analogue **34d** shows virtually the same activity as that of **34c**. The docking study using the cathepsin K crystal structure has revealed that **34a** forms a critical hydrogen bond with the carbonyl oxygen of Gly66, as anticipated. The fact that the CF_3 group of this isostere is attached to an sp^3 carbon provides flexibility in the orientation of the NH bond to form hydrogen bonding in the most favorable manner [124]. Optimization of the aromatic group of **34a** led to the discovery of the exceptionally potent inhibitor **35a**, exhibiting an IC_{50} of 5 pM or less. The corresponding amide **35** is also extremely potent (IC_{50} = 0.015 nM) but is metabolically labile. Thus, **35a** has been identified as the most potent and promising drug candidate in this study [124].

It has been shown that 2,3-difluorotoluene (**36**) serves as a nonpolar shape mimic of thymine (**37**) and difluorotoluene deoxyriboside (**38**) is, surprisingly, an excellent nonpolar nucleoside isostere of thymidine (**39**) (see Fig. 1.21) [126, 127]. Thus, the nucleotide of **36** was incorporated into DNA by several high-fidelity DNA polymerases [126, 128], forming a positional pair with adenine (**40**) in place of **37** without perturbing the double helix structure, which was confirmed by X-ray crystallography (see Fig. 1.21) [127]. It has also been shown that difluorotoluene does not form appreciable hydrogen bonds with adenine, and hence it is nonselective in base pairing and destabilizing DNA *in the absence of* DNA polymerases [128]. The probability of thymidine triphosphate choosing a wrong partner is less than 0.1% and that of the triphosphate of **38** is less than 0.3% [126]. Thus, the results clearly indicate that **36** is virtually a perfect shape mimic of **37** for DNA polymerases.

Since these surprising findings, first made in 1997, challenge the Watson–Click base-pairing principle in DNA (although the difluorotoluene isostere is only a minor part of DNA), there has been a debate on the interpretation of the results, including the polarity of **36** and possible F···HN as well as CH···N hydrogen bonding interactions [129]. However, the X-ray crystallographic data [127] as well as extensive computational analysis [130] unambiguously confirmed that **36** is indeed a nonpolar thymine isostere. Thus, these findings have opened a new avenue of research on a variety of nonpolar nucleoside isosteres [128].

1.1.9 Difluoromethylene as Isopolar Mimic of the Oxygen Component of P–O–C Linkage

Difluoromethylene has been recognized as an isopolar mimic of the oxygen component of P–O–C or P–O–P linkage in phosphates, which can be used to generate nonhydrolyzable phosphate analogues of nucleotides, enzyme substrates, and enzyme inhibitors [131–137]. For example, a difluoromethylene linkage was successfully introduced to a protein-tyrosine phosphatase (PTP) inhibitor of insulin receptor dephosphorylation [132] (see Figure 1.22). A comparison of a hexamer peptide inhibitor bearing a phosphonomethylphenylalanine (Pmp) residue (**41a**) with that having a phosphonodifluoromethylphenylalanine (F_2Pmp) residue (**41b**) revealed that **41b** was 1000 times more potent than **41a**, retaining high affinity for the SH2 domain of PTP [132]. The marked difference in potency observed for **41a** and **41b** can be ascribed to the fact that the difluoromethylene group increases the acidity of the phosphonic acid moiety and maintains appropriate polarity.

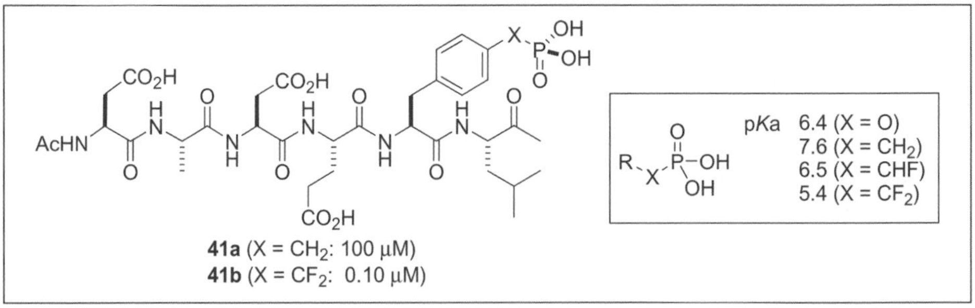

41a (X = CH_2: 100 μM)
41b (X = CF_2: 0.10 μM)

pKa 6.4 (X = O)
7.6 (X = CH_2)
6.5 (X = CHF)
5.4 (X = CF_2)

Figure 1.22 *Nonhydrolyzable phosphate mimics **41** and their estimated pKa values. Numbers within the parentheses are IC_{50} values.*

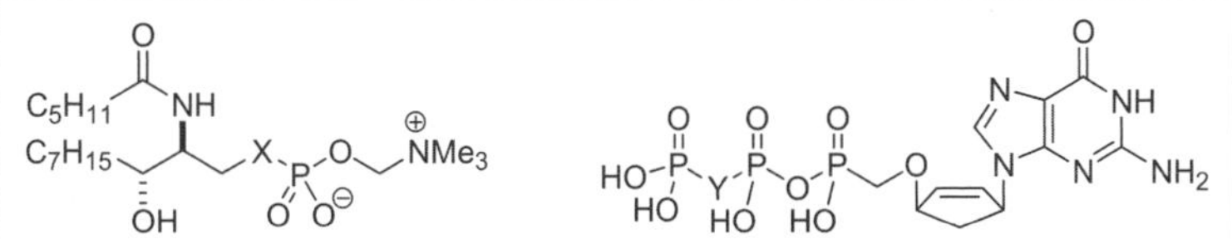

42a (X = CH_2: 120 μM)
42b (X = CF_2: 57 μM)

43a (Y = CH_2: >100 μM)
43b (Y = CHF: 34.8 μM)
43c (Y = CF_2: 5.8 μM)
43d (Y = O: 0.5 μM)

Figure 1.23 *SMase inhibitors (**42**) and FKBP 12 rotamase inhibitors (**43**). Numbers within the parentheses are IC_{50} values.*

A similar incorporation of a P–CF_2 unit as a P–O bond surrogate to sphingomyelin was reported [134] for the development of nonhydrolyzable inhibitors of sphingomyelinase (SMase), which cleaves sphingomyelin to release ceramide. As Figure 1.23 shows, difluoromethylene analogue **42b** is twice as potent as methylene analogue **42a** in the inhibition of SMase from *B. cereus* [134].

Norcarbovir triphosphate (**43d**) is a potent inhibitor of HIV reverse transcriptase, comparable to AZT and carvovir, but is amenable to enzymatic hydrolysis *in vivo* [135]. Thus, methylenediphosphonate, fluoromethylenediphosphonate, and difluoromethylenediphosphonate analogues of **43d** were synthesized to block the enzymatic hydrolysis and evaluated for their potency and stability in human fetal blood serum [135]. As Figure 1.23 shows, the methylenediphosphonate analogue (**43a**) is inactive, and the fluoromethylenediphosphonate analogue (**43b**) shows only a moderate potency. Difluoromethylenediphosphonate analogue **43c** is the most active among these nonhydrolyzable analogues with IC_{50} of 5.8 μM, which is 10 times weaker than that of **43d**. Nevertheless, **43c** possesses >40 times longer half-life (45 h) than the natural enantiomer of **43d** (65 min) and >500 times more stable than AZT triphosphate (5 min) [135].

The fluoromethylene and difluoromethylene linkages have also been incorporated into an aspartyl phosphate, providing the first synthetic inhibitors of aspartate semialdehyde dehydrogenase [136] as well as lysophosphatidic acid analogues, which increased the half-lives of analogues in cell culture [137].

The use of a difluoromethylene unit as a surrogate of a carbonyl group is another logical extension. In fact, difluoromethylene analogues of the inhibitors of the rotamase activity of FK506 binding protein 12 (FKBP12), which catalyzes *cis–trans* isomerization of a peptidyl-prolyl bond, have been investigated [138]. As Table 1.8 shows, the K_i values of difluoromethylene analogues, **44a** and **44d**, for the FKBP12 inhibition (rotamase activity) are comparable to or better than those of the carbonyl counterparts, **44b** and **44e**. It should be noted that the corresponding simple methylene analogues, **44c** and **44f** do not show any appreciable activity, which suggests that the *gem*-difluoromethylene group participates in some specific interactions with the protein. After optimization of the proline and its ester moieties, the most potent inhibitor **45** (K_i 19 nM) was developed. The X-ray crystal structure of the **45**–FKBP12 complex strongly suggests that the two fluorine atoms participate in the moderate-to-weak hydrogen bonding interactions with the Phe36 phenyl ring as well as Tyr26 hydroxyl group.

Table 1.8 *FKBP12 rotamase inhibition [138]*

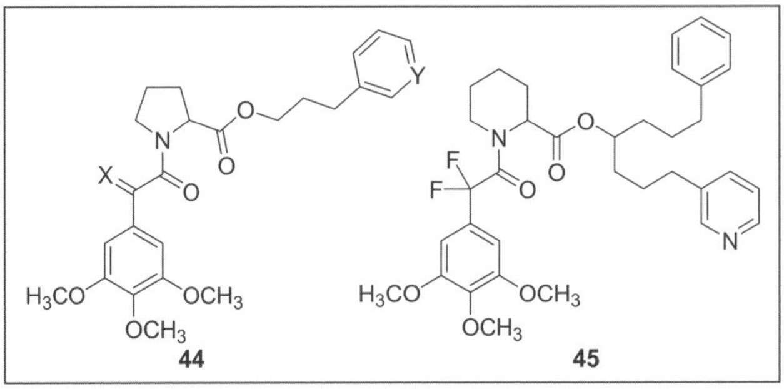

Compound	X	Y	K_i (μM)
44a	F_2	N	0.872
44b	O	N	4.00
44c	H_2	N	–[a]
44d	F_2	CH	1.30
44e	O	CH	2.20
44f	H_2	CH	–[a]
45			0.019

[a] No appreciable inhibition at 10 μM.

1.1.10 High Electrophilicity of Fluoroalkyl Carbonyl Moieties

Successive substitution of hydrogen atoms in acetone with fluorine atoms clearly lowers the frontier orbital energy levels by 0.2–0.6 eV on the basis of *ab initio* computation (see Table 1.9). It is worthy of note that the positive charge at the carbonyl carbon and the negative charge at the oxygen atoms in these ketones, *decrease* as the number of fluorine atoms increases. The validity of the computational analysis is confirmed experimentally by the ^{13}C NMR chemical shifts of these carbons, which decrease from 206.58 ppm for acetone to 172.83 ppm for hexafluoroacetone, in good agreement with the carbonyl carbon charges, as shown in Table 1.9.

The substantial decrease in the HOMO (highest occupied molecular orbital) energy level of trifluoroacetophenone (**46b**), compared with acetophenone (**46a**) was observed [139] by ^{13}C NMR when a 1:1 mixture of **46a** and **46b** was treated with $BF_3 \cdot OEt_2$. Thus, a 16.7 ppm downfield shift took place only for the carbonyl carbon atom of **46a**, showing the coordination of the carbonyl group to the Lewis acid, while no change in the ^{13}C chemical shift was observed for **46b** (see Scheme 1.3). This marked difference in reactivity is attributed to the substantially decreased basicity of the carbonyl oxygen in **46b** caused by the strong inductive effect of the trifluoromethyl moiety. This selectivity was demonstrated in the highly chemoselective reduction of **46a** and **46b** under two different reaction conditions, as illustrated in Scheme 1.3. Thus, the reduction of a 1:1 mixture of **46a** and **46b** with tributyltin hydride gave 1-phenyl-2,2,2-trifluoroethanol (**48b**) as the sole product in 40% yield, while the same reaction in the presence of one equivalent of $BF_3 \cdot OEt_2$ at −78 °C afforded 1-phenylethanol (**48a**) exclusively in 82% yield.

Table 1.9 *Calculated HOMO and LUMO levels of fluorinated acetones*[a]

Compound	Energy Level (eV)		Charges		^{13}C NMR chemical shift of C=O
	HOMO	LUMO	C=O	C=O	
$CH_3C(O)CH_3$	−7.054	−0.784	0.573	−0.553	206.58[b]
$CH_2FC(O)CH_3$	−7.507	−1.247	0.544	−0.547	205.62[c]
$CHF_2C(O)CH_3$	−7.911	−1.873	0.518	−0.513	197.38[c]
$CF_3C(O)CH_3$	−8.278	−2.098	0.505	−0.486	189.33[c]
$CF_3C(O)CF_3$	−9.384	−3.476	0.421	−0.419	172.83[b]

[a] Computation was carried out by one of the authors (T.Y.) using the Gaussian 03W at the B3LYP/6–311++G** level.
[b] Ref. [132].
[c] Ref. [140].

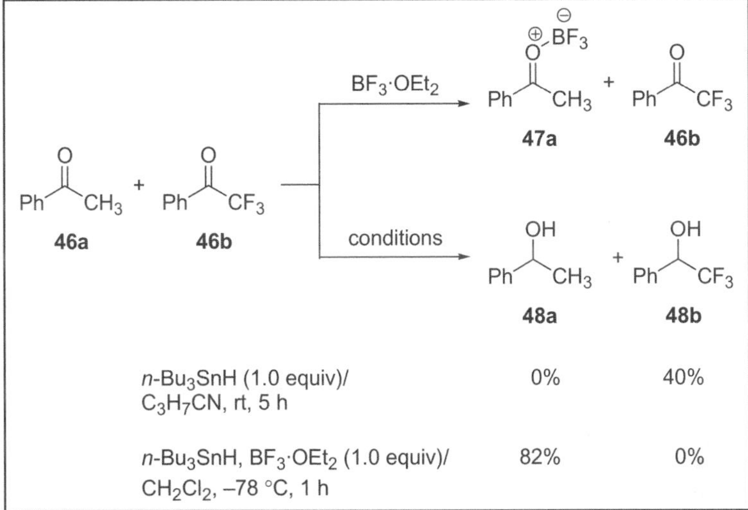

Scheme 1.3 *Reactions of* ***46a*** *and* ***46b*** *with* $BF_3 \cdot OEt_2$ *and n-*Bu_3SnH.

1.1.11 Use of Difluoromethyl and Trifluoromethyl Ketones as Transition State Analogues in Enzyme Inhibition

Tetrahedral transition state analogues of ester and amide substrates are known to function as efficient enzyme inhibitors of hydrolytic enzymes such as serine and aspartyl proteases as well as metalloproteinases (see Fig. 1.24) [141–144]. Although ketals of alkyl or aryl ketones are usually not stable, those of difluroalkyl or trifluoromethyl ketones have considerable stability, as exemplified by their facile formations of the corresponding stable hydrates [145, 146]. Therefore, substrate analogues, containing difluoroalkyl or trifluoromethyl ketone moiety in appropriate positions, have been studied as effective transition state inhibitors of hydrolytic enzymes [141, 147–150].

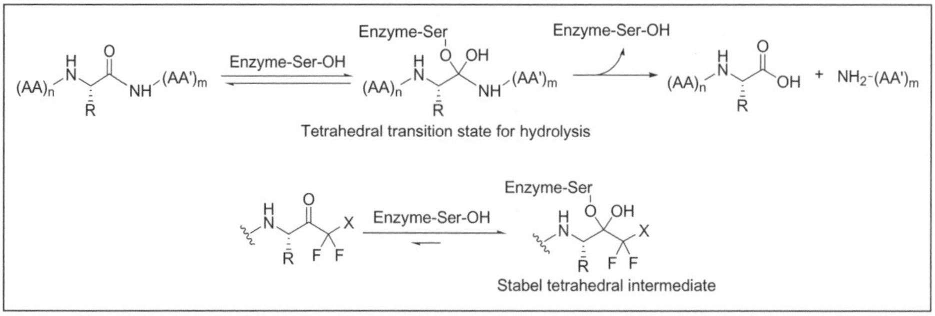

Figure 1.24 *Mechanism of peptide hydrolysis by a serine protease and enzyme inhibition by forming stable tetrahedral intermediate.*

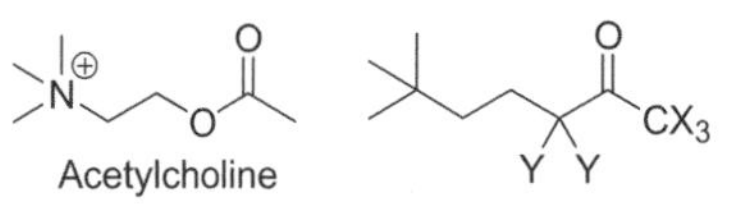

	K_i (nM)
49a (X = H, Y = H)	310,000
49b (X = F, Y = H)	16
49c (X = H, Y = F)	1.6

Figure 1.25 *Transition state inhibitors of AchE based on difluoroalkyl and trifluoromethyl ketone substrate analogues.*

For example, the rather simple difluoroalkyl ketone **49c** and trifluoromethyl ketone **49b** were designed as inhibitors of acetylcholinesterase (AchE), a serine esterase [151]. As Figure 1.25 shows, these fluoroketones exhibit excellent AchE inhibitory activities with nanomolar level K_i values, and **49c** possesses 200,000 times higher potency (K_i = 1.6 nM) than the corresponding alkyl ketone **49a** [151]. This remarkable result can be readily rationalized by taking into account the strong electrophilicity of the carbonyl group of difluoroalkyl and trifluoromethyl ketones, described in the preceding section.

In a similar manner, a variety of transition state analogues bearing a trifluoroacetyl moiety as the key functional group have been designed and synthesized as the inhibitors of human neutrophil elastase [152, 153], human cytomegalovirus protease [154, 155], human leukocyte elastase [156, 157], and other enzymes. Some of these inhibitors exhibit extremely high potency. For example, **51** has a K_i of 3.7 pM against AchE [158] and **50** has an IC_{50} of 0.88 nM against insect juvenile hormone esterase [159] (see Figure 1.26).

Difluoroalkyl ketones have also been successfully employed as the key structure of renin inhibitors. Renin is an aspartyl protease and transforms angiotensinogen (452 amino acid residues for human) to angiotensin I, which is further cleaved by angiotensin-converting enzyme (ACE) to angiotensin II, which exerts a vasoconstricting effect [160]. Accordingly, renin inhibitors have been extensively studied for their use in controlling hypertension. Although human angiotensinogen has 452 amino acid residues, the first dodecapeptide sequence is the most important for its activity (see Figure 1.27) [160].

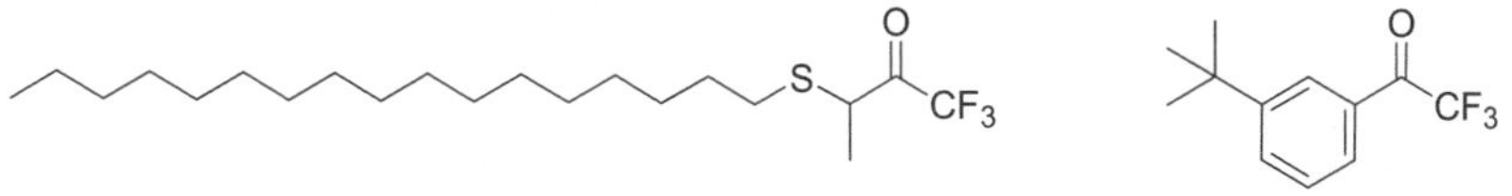

50 (IC_{50} = 0.88 nM; insect JH esterase) **51** (K_i = 3.7pM; AchE)

Figure 1.26 *Highly active transition state inhibitors bearing a trifluoroacetyl moiety.*

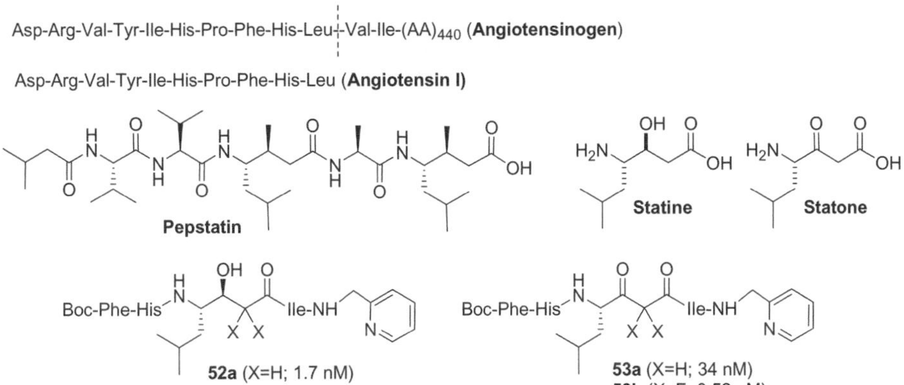

52a (X=H; 1.7 nM)
52b (X=F; 12 nM)
53a (X=H; 34 nM)
53b (X=F; 0.52 nM)

Figure 1.27 *Renin inhibitors containing difluorostatine and difluorostatone residues.*

Renin cleaves the Leu^{10}-Val^{11} linkage of angiotensinogen to generate angiotensin I. Thus, the tetrahedral transition state analogues of this peptide sequence should act as excellent renin inhibitors. In addition, the naturally occurring aspartyl peptidase inhibitor pepstatin [161], active against renin, contains a unique non-proteinogenic amino acid residue, "statine," which mimics the tetradedral transition state (see Figure 1.27) [162]. Accordingly, peptidomimetics, containing difluorostatine residue and its oxidized form, difluorostatone residue, have been synthesized and their activities evaluated. For example, difluorostatine-containing **52b** and difluorostatone-containing **53b** represent this class of renin inhibitors (see Figure 1.27) [163, 164]. In comparison with the corresponding hydrocarbon counterparts **52a** and **53a, 52b** shows several times reduced potency, while **53b** exhibits 65 times enhanced potency [163]. Apparently, the difluoromethylene group of **52b** reduces the basicity of the hydroxyl group, which would be counterproductive if this hydroxyl group acts as a hydrogen-bond acceptor, which explains the reduced potency of **52b** as compared to **52a**. In contrast, the difluorostatone moiety of **53b** should form a stable ketal with a water molecule, which mimics the tetrahedral transition state very well, which is the most likely reason for the two orders of magnitude enhancement in potency as compared to **53a**.

1.2 The Use of ^{19}F and ^{18}F Probes in Biophysical and Analytical Methods Relevant to Bioorganic and Medicinal Chemistry

1.2.1 ^{19}F NMR Spectroscopy

Fluorine-19 is an attractive probe nucleus for NMR because of its relatively small size, a nuclear spin of 1/2, natural abundance, high NMR sensitivity, and pronounced long-range effects. Solid-state ^{19}F NMR is an excellent tool for studying membrane-active peptides in their lipid-bound environment. The prerequisite of ^{19}F NMR experiments is the inclusion of a specific ^{19}F-reporter group into peptides and proteins. In structural studies, the most preferable labeling is with CF_3-groups that are rigidly attached to the peptide backbone. Problems of ^{19}F-labeled amino acid incorporation, including HF-elimination, racemization, and slow coupling, can easily be overcome. ^{19}F NMR data analysis will provide information about conformational changes, the structure and orientation of peptides, or the kinetics of ligand binding, properties that are intimately related to the function of the peptide. For example, solid-state ^{19}F NMR spectroscopy revealed that Trp41 participates in the gating mechanism of the M2 proton channel of influenza A virus [165] (see Figure 1.28). The integral membrane protein M2 of influenza A virus assembles as a tetrameric bundle to form a proton-conducting channel that is activated by low pH. A synthetic 25-residue peptide containing the M2 transmembrane (TM) domain was labeled with 6F-Trp41 and studied in lipid membranes by solid-state ^{19}F NMR. Through this solid-state ^{19}F NMR analysis in combination with computational energy calculations, it was demonstrated that the side-chain conformation and position of Trp41 differs significantly in the two pH-dependent states of the proton channel of the TM domain of the M2 protein, suggesting that the four tryptophan residues actively participate in activating the channel mechanically.

In another application, ^{19}F NMR studies have provided critical information on the bioactive conformation of taxoids. Fluorine-containing taxoids have been used as probes for NMR analysis of the conformational dynamics of paclitaxel in conjunction with molecular modeling [166]. The dependence of the ^{19}F chemical shifts and the $J_{H2'–H3'}$ values of these fluorinated analogues is examined through ^{19}F and ^{1}H variable-temperature (VT) NMR measurements. The experiments clearly indicate highly dynamic behavior of these molecules and the existence of equilibrium between conformers. The analysis of the VT NMR data in combination with molecular modeling, including restrained molecular dynamics (RMD), has identified three key conformers, which were further confirmed by the ^{19}F–^{1}H heteronuclear NOE measurements.

Fluorine-probe protocol has been applied to solid-state magic-angle spinning (SSMAS) ^{19}F NMR analysis with the radiofrequency-driven dipolar recoupling (RFDR) method to measure the F–F distance in the microtubule-bound conformation of F_2-10-Ac-docetaxel (see Figure 1.29a) [167]. Moreover, five intramolecular distances of the key atoms in the microtubule-bound ^{19}F/^{2}H/^{13}C-labeled paclitaxel were determined by the rotational echo double resonance (REDOR) method (see Figure 1.29b and c) [168, 169].

1.2.2 *In vivo* ^{19}F Magnetic Resonance Spectroscopy

Magnetic resonance (MR) spectroscopy is used to measure the levels of different metabolites in body tissues. With the increasing popularity of fluorinated compounds in

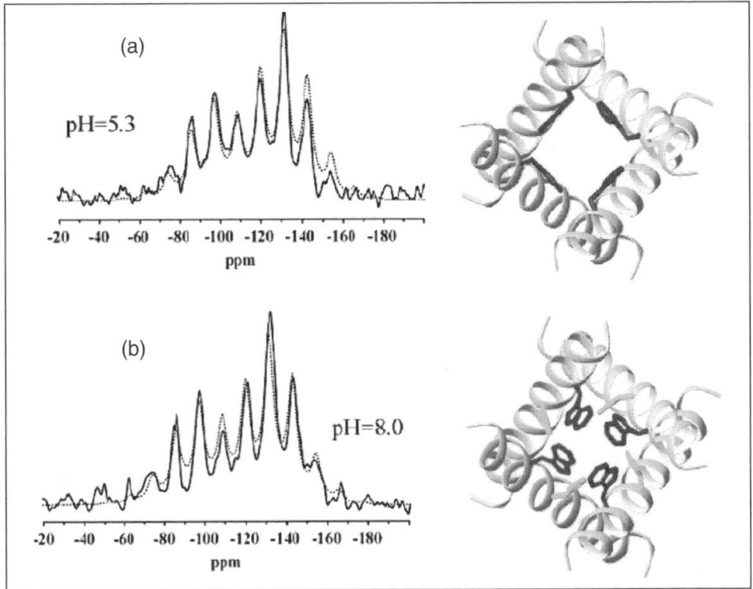

Figure 1.28 *Experimental (solid lines) and simulated (dashed lines) spin-echo spectra of 6F-Trp41-M2TMD at 6.5 kHz MAS at pH 5.3 (a) and pH 8.0 (b). Side-chain conformations (bottom view) of Trp41 (blue) and His37 (green) in the TM channel structure of the homo-tetrameric M2 protein are shown to the right side of the spectra. At pH 8.0, the structural parameters implicate an inactivated state, while at pH 5.3 the tryptophan conformation represents the activated state. See color plate 1.28.*
*(*Source: *Reprinted with permission from Witter, R., Nozirov, F., Sternberg, U., Cross, T. A., Ulrich, A. S., and Fu, R. Solid-state* ^{19}*F NMR spectroscopy reveals that Trp41 participates in the gating mechanism of the M2 proton channel of influenza A virus,* J. Am. Chem. Soc. *(2008)* **130***, 918–924. Copyright (2008) American Chemical Society.)*

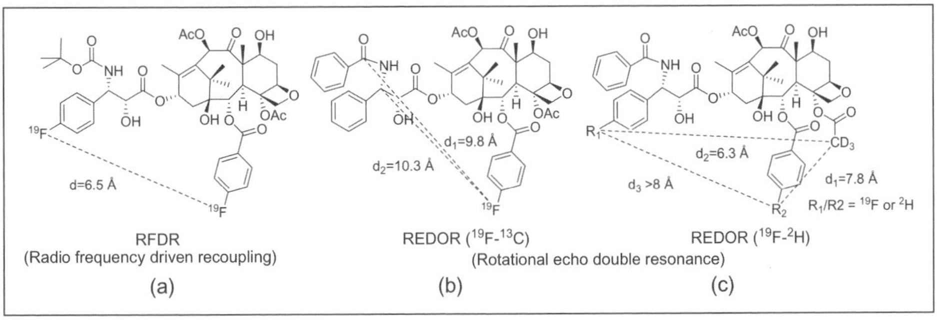

d=6.5 Å; d_2=10.3 Å; d_1=9.8 Å; d_2=6.3 Å; d_3 >8 Å; d_1=7.8 Å; R_1/R2 = ^{19}F or ^{2}H

RFDR (Radio frequency driven recoupling) (a)

REDOR (^{19}F-^{13}C) (b)

REDOR (^{19}F-^{2}H) (c)

(Rotational echo double resonance)

Figure 1.29 *Solid-state NMR studies on the microtubule-bound conformation of taxoids using fluoro-taxoid probes.*

pharmaceutical research, *in vivo* ^{19}F MR spectroscopy has become a valuable tool for identifying and monitoring fluorinated compounds and metabolites. Although almost all ^{19}F MR examinations are limited by the relatively low signal–noise ratio (SNR) of the spectra, ^{19}F MR spectroscopy has a unique role in clinical research applications and is complementary to other structural and functional imaging tools. It is able to identify and measure fluorine-containing drugs and their metabolites in biofluids due in part to the lack of naturally occurring MR-visible fluorine metabolites. In addition, it is noninvasive and therefore provides options for investigating the long-term disposition of pharmaceuticals. A wide range of fluorine-containing pharmaceuticals and metabolites have been evaluated in patients using *in vivo* ^{19}F MR spectroscopy. The most frequently examined organs have been the brain and the liver, although heart and extremity muscle, as well as bone marrow, have also been evaluated. Assessment of fluorine-containing drugs using human *in vivo* fluorine (^{19}F) MR spectroscopy has proved to be an important technique in drug design and preclinical studies. One good example is seen in the application of ^{19}F NMR to metabolic studies of capecitabine, the recently developed oral prodrug of 5-fluorouracil [170].

The degradation pathway of capecitabine, incorporating the new fluorinated metabolites found in urine using ^{19}F NMR, is depicted in Figure 1.30, and a typical ^{19}F NMR spectrum of urine is presented in Figure 1.31. The mean percentage of dose excreted in urine as parent drug and its fluorinated metabolites up to 48 h post dosing, measured by ^{19}F NMR, was 84.2%, close to 92.3% of the radioactivity recovered at that time. This difference of less than 10% demonstrates that ^{19}F NMR spectroscopy is a suitable technique for quantitative studies. ^{19}F NMR was also able to measure the relevant level of each fluorinated metabolite.

1.2.3 Positron Emission Tomography

Positron emission tomography (PET) is a powerful *in vivo* imaging technique that produces three-dimensional images or maps of functional processes in the body. It is used heavily in clinical research and diagnosis, as well as in drug development. Figure 1.32 shows an example of assessing dopamine D_2 receptor availability in cocaine abusers by the use of PET [171]. When compared with normal controls, cocaine abusers showed significant decreases in dopamine D_2 receptor availability that persisted 3–4 months after detoxification, as indicated by the marked reduction in the binding of [^{18}F]*N*-methylspiroperidol in PET scans.

Since PET works by detecting pairs of γ-rays emitted directly by a positron-emitting radioisotope introduced into the body on a metabolically active molecule, the use of PET is driven by the characteristics and availability of appropriately labeled radiopharmaceuticals that are specifically designed for measurement of targeted biochemical processes. Fluorine-18 has particularly appealing properties as positron-emitting radionuclide for PET imaging: (i) addition of a fluorine atom to a large molecule can often (but not always) be accomplished without significant changes to the physiochemical or biological properties of the compound; (ii) fluorine-18 has a relatively long half-life, which permits longer synthesis times, thus opening up possibilities for multistep radiolabeling procedures, as well as the option for commercial manufacture at an offsite location.

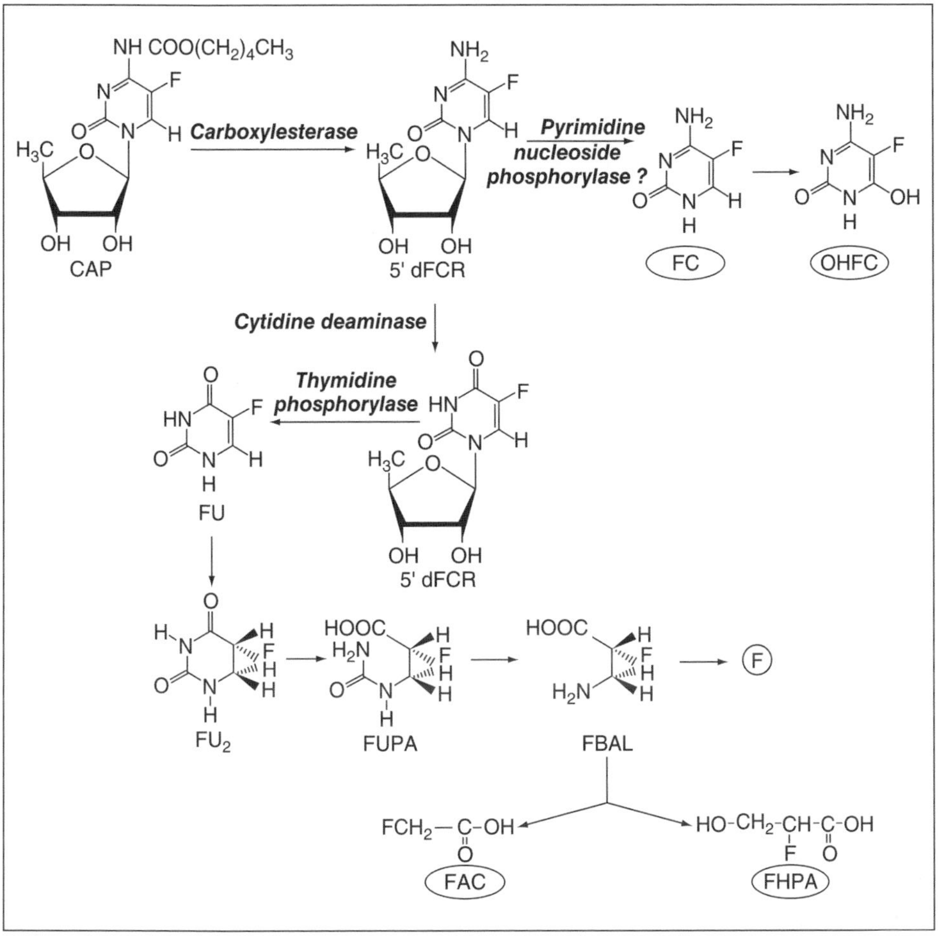

Figure 1.30 *Catabolic pathway of capecitabine from* ^{19}F *NMR analysis of patients' urine. All the compounds are represented in neutral form. CAP, capecitabine; 5′dFCR, 5′-deoxy-5-fluorocytidine; FC, 5-fluorocytosine; OHFC, 6-hydroxy-5-fluorocytosine; 5′dFUR, 5′-deoxy-5-fluorouridine; FU, 5-fluorouracil;* FUH_2*, 5,6-dihydro-5-fluorouracil; FUPA, α-fluoro-β-ureidopropionic acid; FBAL, α-fluoro-β-alanine;* F^-*, fluoride ion; FHPA, 2-fluoro-3-hydroxypropanoic acid; FAC, fluoroacetic acid. Metabolites identified for the first time in urine of patients with* ^{19}F *NMR are represented in ellipses.*

*(*Source: *Reprinted with permission from Malet-Martino, M., Gilard, V., Desmoulin, F., and Martino, R. Fluorine nuclear magnetic resonance spectroscopy of human biofluids in the field of metabolic studies of anticancer and antifungal fluoropyrimidine drugs,* Clin. Chim. Acta *(2006)* **366***, 61–73. Copyright (2006) Elsevier.)*

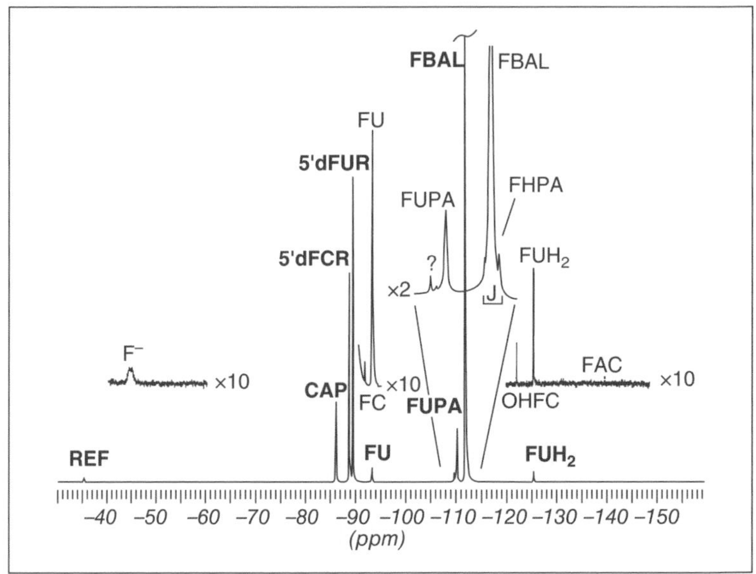

Figure 1.31 *^{19}F NMR spectrum at 282 MHz with proton decoupling of a urine sample from a patient receiving oral capecitabine at a dose of 3800 mg/day, administered twice daily at 12-hour interval, as a second treatment 3 months after the first one. Urine fraction 0–12 h collected after the first dose of 1900 mg and 10-fold concentrated, pH of the sample: 5.45. The chemical shifts are expressed relative to the resonance peak of TFA (5% w/v aqueous solution) used as external reference.*
(Source: *Reprinted with permission from Malet-Martino, M., Gilard, V., Desmoulin, F., and Martino, R. Fluorine nuclear magnetic resonance spectroscopy of human biofluids in the field of metabolic studies of anticancer and antifungal fluoropyrimidine drugs,* Clin Chim. Acta *(2006)* ***366****, 61–73. Copyright (2006) Elsevier.)*

Development of methods for fluorine-18 labeling has been actively pursued in the last two decades (see Figure 1.33) [172, 173]. Nucleophilic and electrophilic fluorinations have been successfully applied to generate an increasingly diverse assortment of chemical structures. Recent application of "click" chemistry and the use of protic solvents in nucleophilic fluorinations will significantly impact both the types of compounds labeled and the yields obtained.

1.3 Summary

This introductory chapter concisely summarizes the unique properties of fluorine and fluorine-containing substituents and functional groups from a physical organic chemistry point of view and then the applications of those characteristics to organic, bioorganic, and medicinal chemistry as well as chemical biology and biomedical research. The following chapters will further elaborate and discuss a wide range of topics in the rapidly expanding scope of fluorine in medicinal chemistry and chemical biology.

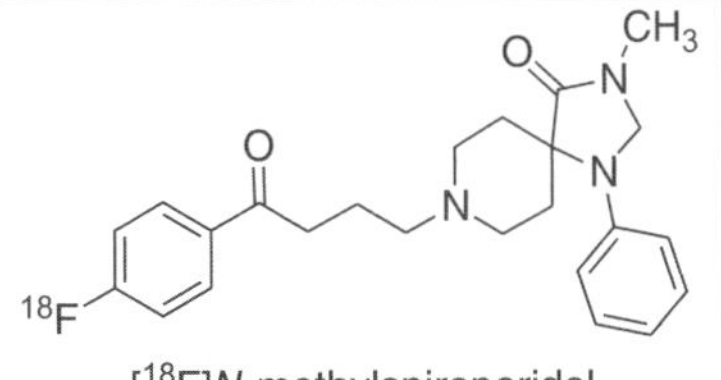

[^{18}F]*N*-methylspiroperidol

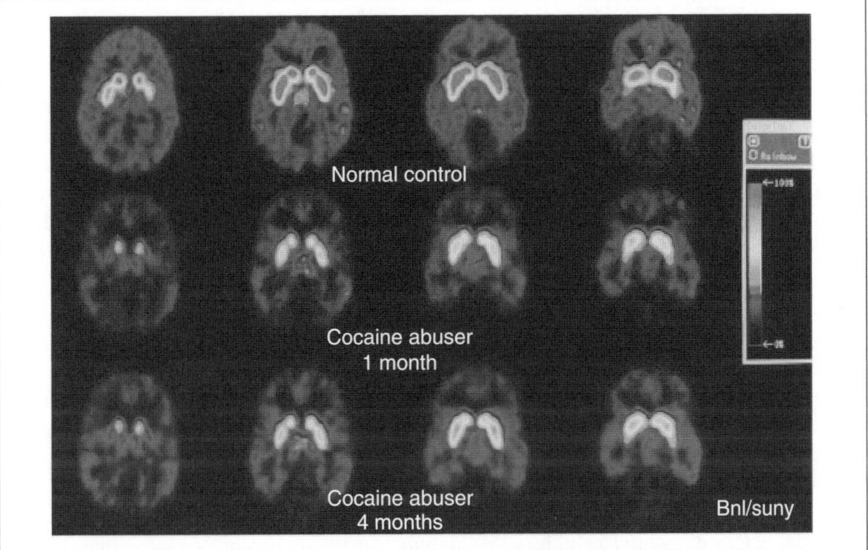

Figure 1.32 *[^{18}F]*N-methylspiroperidol *images in a normal control and in a cocaine abuser tested 1 month and 4 months after last cocaine use. The images correspond to the four sequential planes where the basal ganglia are located. The color scale has been normalized to the injected dose. See color plate 1.32.*
*(*Source: *Volkow, N. D., Fowler, J. S., Wang, G. J., Hitzemann, R., Logan, J., Schlyer, D. J., Dewey, S. L., and Wolf, A. P. Decreased dopamine D2 receptor availability is associated with reduced frontal metabolism in cocaine abusers,* Synapse *(1993)* ***14****, 169-177. Reprinted with permission of Wiley-Liss Inc., a subsidiary of John Wiley & Sons, Inc. Copyright (1993) Wiley Interscience.)*

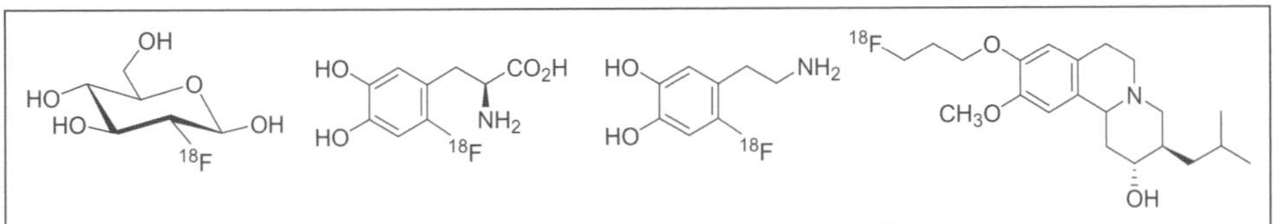

Deoxy-[^{18}F]fluoro-D-glucose [^{18}F]Fluoro-DOPA [^{18}F]Fluorodopamine [^{18}F]Fluoropropyldihydrotetrabenazine

Figure 1.33 *Examples of ^{18}F-radiopharmaceuticals.*

References

1. O'Hagan, D. and Harper, D. B. (1999) Fluorine-containing natural products. *J. Fluorine Chem.*, **100**, 127–133.
2. Fried, J. and Sabo, E. F. (1953) Synthesis of 17α-hydroxycorticosterone and its 9α-halo derivatives from 11-epi-17α-hydroxycorticosterone. *J. Am. Chem. Soc.*, **75**, 2273–2274.
3. Heidelberger, C., Chaudhuri, N. K., Danneberg, P., *et al.* (1957) Fluorinated pyrimidines, a new class of tumor-inhibitory compounds. *Nature*, **179**, 663–666.
4. *MedAdNews 200 – World's Best-Selling Medicines, MedAdNews, July 2007* (http://en.wikipedia.org/wiki/List_of_top_selling_drugs).
5. Tan, L.-S., Chen, C.-Y., Tillyer, R. D., *et al.* (1999) A novel, highly enantioselective ketone alkynylation reaction mediated by chiral zinc aminoalkoxides. *Angew. Chem. Int. Ed.*, **38**, 711–713.
6. Humphreys, J. L., Lowes, D. J., Wesson, K. A. and Whitehead, R. C. (2006) Arene *cis*-dihydrodiols – useful precursors for the preparation of antimetabolites of the shikimic acid pathway: application to the synthesis of 6,6-difluoroshikimic acid and (6*S*)-6-fluoroshikimic acid. *Tetrahedron*, **62**, 5099–5108.
7. Chou, T.-C., Dong, H.-J., Rivkin, A., *et al.* (2003) Design and total synthesis of a superior family of epothilone analogues, which eliminate xenograft tumors to a nonrelapsable state. *Angew. Chem. Int. Ed.*, **42**, 4762–4767.
8. Damon, D. B., Dugger, R. W., Magnus-Aryitey, G., *et al.* (2006) Synthesis of the CETP inhibitor torcetrapib: The resolution route and origin of stereoselectivity in the iminium ion cyclization. *Org. Process Res. Dev.*, **10**, 464–471.
9. Kim, D., Wang, L., Beconi, M., *et al.* (2005) (2*R*)-4-Oxo-4-[3-(trifluoromethyl)-5,6-dihydro [1,2,4]triazolo[4,3-a]pyrazin-7(8*H*)-yl]-1-(2,4,5-trifluorophenyl)butan-2-amine: A potent, orally active dipeptidyl peptidase IV inhibitor for the treatment of type 2 diabetes. *J. Med. Chem.*, **48**, 141–151.
10. Uneyama, K. (2006) *Organofluorine Chemistry*, Blackwell Publishing, Oxford.
11. Soloshonok, V. A., Mikami, K., Yamazaki, T., *et al.* (2006) *Current Fluoroorganic Chemistry: New Synthetic Directions, Technologies, Materials, and Biological Applications*, ACS Symposium Series 949, American Chemical Society, Washington, D. C.
12. Percy, J. M. (2006) *Science of Synthesis*, Vol. **34**. Georg Thieme Verlag, Stuttgart.
13. Soloshonok, V. A. (2005) *Fluorine-Containing Synthons*. ACS Symposium Series 911, American Chemical Society, Washington, D. C.
14. Kirsch, P. (2004) *Modern Fluoroorganic Chemistry: Synthesis, Reactivity, Applications*, Wiley-VCH Verlag, GmbH, Stuttgart.
15. Hiyama, T. (2000) *Organofluorine Compounds: Chemistry and Applications*, Springer Verlag, Stuttgart.
16. Soloshonok, V. A. (1999) *Enantiocontrolled Synthesis of Fluoroorganic Compounds – Stereochemical Challenges and Biomedicinal Targets*, John Wiley & Sons, Inc., New York.
17. Kitazume, T. and Yamazaki, T. (1998) *Experimental Methods in Organic Fluorine Chemistry*, Kodansha, Gordon and Breach Science Publisher, Tokyo.
18. Hudlicky, M. and Pavlath, A. E. (1995) *Chemistry of Organic Fluorine Compounds II – A Critical Review*, American Chemical Society, Washington, D. C.
19. Kukhar', V. P. and Soloshonok, V. A. (1995) *Fluorine-containing Amino Acids – Synthesis and Properties*, John Wiley & Sons, Inc., New York.
20. Müller, K., Faeh, C. and Diederich, F. (2007) Fluorine in Pharmaceuticals: Looking beyond intuition. *Science*, **317**, 1881–1886.

21. Isanbor, C. and O'Hagan, D. (2006) Fluorine in medicinal chemistry: A review of anti-cancer agents. *J. Fluorine Chem.*, **127**, 303–319.
22. Bégué, J.-P. and Bonnet-Delpon, D. (2006) Recent advances (1995–2005) in fluorinated pharmaceuticals based on natural products. *J. Fluorine Chem.*, **127**, 992–1012.
23. Prakesch, P., Grée, D., Chandrasekhar, S. and Grée, R. (2005) Synthesis of fluoro analogues of unsaturated fatty acids and corresponding acyclic metabolites. *Eur. J. Org. Chem.*, 1221–1232.
24. Natarajana, R., Azerada, R., Badetb, B. and Copin, E. (2005) Microbial cleavage of C–F bond. *J. Fluorine Chem.*, **126**, 425–436.
25. Böhm, H.-J., Banner, D., Bendels, S., *et al.* (2004) Fluorine in medicinal chemistry. *ChemBioChem*, **5**, 637–643.
26. Jeschke, P. (2004) The unique role of fluorine in the design of active ingredients for modern crop protection. *ChemBioChem*, **5**, 570–589.
27. Ojima, I. (2004) Use of fluorine in the medicinal chemistry and chemical biology of bioactive compounds – A case study on fluorinated taxane anticancer agents. *ChemBioChem*, **5**, 628–635.
28. Yoder, N. C. and Kumar, K. (2002) Fluorinated amino acids in protein design and engineering. *Chem. Soc. Rev.*, **31**, 335–341.
29. Smart, B. E. (2001) Fluorine substituent effects (on bioactivity). *J. Fluorine Chem.*, **109**, 3–11.
30. Dax, C., Albert, M., Ortner, J. and Paul, B. J. (2000) Synthesis of deoxyfluoro sugars from carbohydrate precursors. *Carbohydr. Res.*, **327**, 47–86.
31. Schlosser, M. (1998) Parametrization of substituents: Effects of fluorine and other heteroatoms on OH, NH, and CH acidities. *Angew. Chem. Int. Ed.*, **37**, 1496–1513.
32. O'Hagan, D. and Rzepa, H. S. (1997) Some influences of fluorine in bioorganic chemistry. *Chem. Commun.*, 645–652.
33. Bondi, A. (1964) van der Waals volumes and radii. *J. Phys. Chem.*, **68**, 441–451.
34. Lide, D. R. (2005) *Handbook of Chemistry and Physics*, 86th edn., CRC Press, New York.
35. Dean, J. A. (1999) *Lange's Handbook of Chemistry*, 15th edn., McGraw-Hill, New York.
36. (a) Peters, R. and Wakelin, R. W. (1953) Biochemistry of fluoroacetate poisoning; the isolation and some properties of the fluorotricarboxylic acid inhibitor of citrate metabolism. *Proc R. Soc. Ser. B*, **140**, 497–507. (b) Lauble, H., Kennedy, M. C., Emptage, M. H., Beinert, H., Stout, C. D., Sout, C. D. (1996) The reaction of fluorocitrate with aconitase and the crystal structure of the enzyme-inhibitor complex. *Proc. Natl. Acad. Sci. USA*, **93**, 13699–13703.
37. Black, W. C., Bayly, C. I., Davis, D. E., *et al.* (2005) Trifluoroethylamines as amide isosteres in inhibitors of cathepsin K. *Bioorg. Med. Chem. Lett.*, **15**, 4741–4744.
38. Barker, M., Clackers, M., Demaine, D. A., *et al.* (2005) Design and synthesis of new nonsteroidal glucocorticoid modulators through application of an "agreement docking" method. *J. Med. Chem.*, **48**, 4507–4510.
39. DeLuca, H. F. and Schnoes, H. K. (1984) Vitamin D: metabolism and mechanism of action. *Annu. Rep. Med. Chem.*, **19**, 179–190.
40. Kobayashi, Y. and Taguchi, T. (1985) Fluorine-modified vitamin D_3 analogs. *J. Synth. Org. Chem. Jpn.*, **43**, 1073–1082.
41. Okamoto, S., Tanaka, Y., DeLuca, H. F., *et al.* (1983) Biological activity of 24,24-difluoro-1,25-dihydroxyvitamin D_3. *Am. J. Physiol.*, **244**, E159–163.
42. Bush, I. E. and Mahesh, V. B. (1964) Metabolism of 11-oxygenated steroids. III. Some 1-dehydro and 9α-fluoro steroids. *Biochem. J.*, **93**, 236–255.
43. Wettstein, A. (1972) Chemistry of fluorosteroids and their hormonal properties, in *A Ciba Foundation Symposium: Carbon–Fluorine Compounds*, Elsevier Excerpta Medica, Amsterdam, pp. 281–301.
44. Taft, J. R. W. (1956) *Steric Effects in Organic Chemistry*, John Wiley & Sons, Inc., New York.

45. MacPhee, J. A., Panaye, A. and Dubois, J.-E. (1978) Steric effects – I: A critical examination of the Taft steric parameters–Es. Definition of a revised, broader and homogeneous scale. Extension to highly congested alkyl groups. *Tetrahedron*, **34**, 3553–3562.
46. Charton, M. (1975) Steric effects. I. Esterification and acid-catalyzed hydrolysis of esters. *J. Am. Chem. Soc.*, **97**, 1552–1556.
47. Hirsch, J. A. (1967) Table of conformational energies–1967. *Top. Stereochem.*, **1**, 199–222.
48. Carcenac, Y., Diter, P., Wakselman, C. and Tordeux, M. (2006) Experimental determination of the conformational free energies (A values) of fluorinated substituents in cyclohexane by dynamic ^{19}F NMR spectroscopy. Part 1. Description of the method for the trifluoromethyl group. *New J. Chem.*, **30**, 442–446.
49. Carcenac, Y., Tordeux, M., Wakselman, C. and Diter, P. (2006) Experimental determination of the conformational free energies (A values) of fluorinated substituents in cyclohexane by dynamic ^{19}F NMR spectroscopy. Part 2. Extension to fluoromethyl, difluoromethyl,pentafluoroethyl, trifluoromethylthio and trifluoromethoxy groups. *New J. Chem.*, **30**, 447–457.
50. Juaristi, E. (1991) *Introduction to Stereochemistry and Conformational Analysis*, John Wiley & Sons, Inc., New York.
51. Bott, G., Field, L. D. and Sternhell, S. (1980) Steric effect. A study of a rationally designed system. *J. Am. Chem. Soc.*, **102**, 5618–5626.
52. Wolf, C., König, W. A. and Roussel, C. (1995) Influence of substituents on the rotational energy barrier of atropisomeric biphenyls – Studies by polarimetry and dynamic gas chlomatography. *Liebigs Ann.*, 781–786.
53. Leroux, F. (2004) Atropisomerism, biphenyls, and fluorine: A comparison of rotational barriers and twist angles. *ChemBioChem*, **5**, 644–649.
54. Hansch, C., Leo, A. and Hoekman, D. H. (1995) *Exploring QSAR: Fundamentals and Applications in Chemistry and Biology*, American Chemical Society, Washington, D. C.
55. Hansch, C., Leo, A. and Hoekman, D. H. (1995) *Exploring QSAR: Hydrophobic, electronic, and steric constants*, American Chemical Society, Washington, D. C.
56. Hansch, C. and Leo, A. (1979) *Substituent Constants for Correlation Analysis in Chemistry and Biology*, John Wiley & Sons, Inc., Hoboken.
57. Ganguly, T., Mal, S. and Mukherjee, S. (1983) Hydrogen bonding ability of fluoroalcohols. *Spectrochim. Acta, Part A* **39A**, 657–660.
58. Morgenthaler, M., Schweizer, E., Hoffmann-Roder, A., *et al.* (2007) Predicting and tuning physicochemical properties in lead optimization: amine basicities. *ChemMedChem*, **2**, 1100–1115.
59. Abraham, M. H., Grellier, P. L., Prior, D. V., *et al.* (1989) Hydrogen bonding. Part 7. A scale of solute hydrogen-bond acidity based on log K values for complexation in tetrachloromethane. *J. Chem. Soc. Perkin Trans.* **2**, 699–711.
60. Van Niel, M. B., Collins, I., Beer, M. S., *et al.* (1999) Fluorination of 3-(3-(piperidin-1-yl)propyl)indoles and 3-(3-(piperazin-1-yl)propyl)indoles gives selective human 5-HT1D receptor ligands with improved pharmacokinetic profiles. *J. Med. Chem.*, **42**, 2087–2104.
61. Maren, T. and Conroy, C. W. (1993) A new class of carbonic anhydrase inhibitor. *J. Biol. Chem.*, **268**, 26233–26239.
62. Kim, C.-Y., Chang, J.-S., Doyon, J. B., *et al.* (2000) Contribution of fluorine to protein-ligand affinity in the binding of fluoroaromatic inhibitors to carbonic anhydrase II. *J. Am. Chem. Soc.*, **122**, 12125–12134.
63. Smart, B. E. (1986) Fluorinated organic molecules, in *Molecular Structure and Energetics* (eds. F. Liebman and A. Greenberg), Wiley–VCH Verlag GbmbH, Weinheim.
64. Bock, C. W., George, P., Mains, G. J. and Trachtman, M. (1979) An ab initio study of the stability of the symmetrical and unsymmetrical difluoroethylenes relative to ethylene and monofluoroethylene. *J. Chem. Soc., Perkin Trans.* **2**, 814–821.

65. Klaboe, P. and Nielsen, J. R. (1960) Infrared and Raman spectra of fluorinated ethanes. XIII. 1,2-Difluoroethane. *J. Chem. Phys.*, **33**, 1764–1774.
66. Hirano, T., Nonoyama, S., Miyajima, T., *et al.* (1986) Gas-phase fluorine-19 and proton high-resolution NMR spectroscopy: application to the study of unperturbed conformational energies of 1,2-difluoroethane. *J. Chem. Soc., Chem. Commun.*, 606–607.
67. Fernholt, L. and Kveseth, K. (1980) Conformational analysis. The temperature effect on the structure and composition of the rotational conformers of 1,2-difluoroethane as studied by gas electron diffraction. *Acta Chem. Scand. A.*, **34**, 163–170.
68. Friesen, D. and Hedberg, K. (1980) Conformational analysis. 7. 1,2-Difluoroethane. An electron-diffraction investigation of the molecular structure, composition, trans-gauche energy and entropy differences, and potential hindering internal rotation. *J. Am. Chem. Soc.*, **102**, 3987–3994.
69. Dixon, D. A., Matsuzawa, N. and Walker, S. C. (1992) Conformational analysis of 1,2-dihaloethanes: a comparison of theoretical methods. *J. Phys. Chem.*, **96**, 10740–10746.
70. Wiberg, K. B., Keith, T. A., Frisch, M. J. and Murcko, M. (1995) Solvent effects on 1,2-dihaloethane gauche/trans ratios. *J. Phys. Chem.*, **99**, 9072–9079.
71. Rablen, P. R., Hoffmann, R. W., Hrovat, D. A. and Borden, W. T. (1999) Is hyperconjugation responsible for the "gauch effect" in 1-fluoropropane and other 2-substituted-1-fluoroethanes? *J. Chem. Soc., Perkin Trans. 2*, 1719–1726.
72. Hagen, K. and Hedberg, K. (1973) Conformational analysis. III. Molecular structure and composition of 2-fluoroethanol as determined by electron diffraction. *J. Am. Chem. Soc.*, **95**, 8263–8266.
73. Huang, J.-F. and Hedberg, K. (1989) Conformational analysis. 13. 2-Fluoroethanol. An investigation of the molecular structure and conformational composition at 20, 156, and 240°. Estimate of the anti-gauche energy difference. *J. Am. Chem. Soc.*, **111**, 6909–6913.
74. Chitale, S. M. and Jose, C. I. (1986) Infrared studies and thermodynamics of hydrogen bonding in 2-halogenoethanols and 3-halogenopropanols. *J. Chem. Soc., Faraday Trans. 1*, **82**, 663–679.
75. Bakke, J. M., Bjerkeseth, L. H., Rønnow, T. E. C. L. and Steinsvoll, K. (1994) The conformation of 2-fluoroethanol – is intramolecular hydrogen bonding important? *J. Mol. Struct.*, **321**, 205–214.
76. Hoppilliard, Y. and Solgadi, D. (1980) Conformational analysis of 2-haloethanols and 2-methoxyethyl halides in a photoelectron-spectrometer. The interpretation of spectra by "ab initio" calculations. *Tetrahedron*, **36**, 377–380.
77. Michel, D., Witschard, M. and Schlosser, M. (1997) No evidence for intramolecular hydrogen bonds in α-fluorocarboxamides. *Liebigs Ann.*, 517–519.
78. Abraham, R. J. and Monasterios, J. R. (1973) Rotational isomerism. XVI. AA′BB′X NMR spectrum and rotational isomerism of 2-fluoroethyl trichloroacetate. *Org. Magn. Reson.*, **5**, 305–310.
79. Schüler, M., O'Hagan, D. and Slawin, A. M. Z. (2005) The vicinal F–C–C–F moiety as a tool for influencing peptide conformation. *Chem. Commun.*, 4324–4326.
80. Tavasli, M., O'Hagan, D., Pearson, C. and Petty, M. (2002) The fluorine gauche effect. Langmuir isotherms report the relative conformational stability of (±)-erythro- and (±)-threo-9,10-difluorostearic acids. *Chem. Commun.*, 1226–1227.
81. Briggs, C. R. S., O'Hagan, D., Howard, J. A. K. and Yufit, D. S. (2003) The C–F bond as a tool in the conformational control of amides. *J. Fluorine Chem.*, **119**, 9–13.
82. Nicoletti, M., O'Hagan, D. and Slawin, A. M. Z. (2005) α,β,γ-Trifluoroalkanes: A stereoselective synthesis placing three vicinal fluorines along a hydrocarbon chain. *J. Am. Chem. Soc.*, **127**, 482–483.
83. Hunter, L., O'Hagan, D. and Slawin, A. M. Z. (2006) Enantioselective synthesis of an all-syn vicinal fluorine motif. *J. Am. Chem. Soc.*, **128**, 16422–16423.

84. Hunter, L., Slawin, A. M. Z., Kirsch, P. and O'Hagan, D. (2007) Synthesis and conformation of multi-vicinal fluoroalkane diastereomers, *Angew. Chem. Int. Ed.*, **46**, 7887–7890.
85. DeRider, M. L., Wilkens, S. J., Waddell, M. J., *et al.* (2002) Collagen stability: Insights from NMR spectroscopic and hybrid density functional computational investigations of the effect of electronegative substituents on prolyl ring conformations. *J. Am. Chem. Soc.*, **124**, 2497–2505.
86. Apeloig, Y. (1981) Negative fluorine hyperconjugation. A theoretical re-examination. *J. Chem. Soc., Chem. Commun.*, 396–398.
87. Rahman, M. M. and Lemal, D. M. (1988) Negative hyperconjugation. The rotation-Inversion barrier in α-fluoroamines. *J. Am. Chem. Soc.*, **110**, 1964–1966.
88. Schneider, W. F., Nance, B. I. and Wallington, T. J. (1995) Bond strength trends in halogenated methanols: Evidence for negative hyperconjugation? *J. Am. Chem. Soc.*, **117**, 478–485.
89. Raabe, G., Gais, H.-J. and Fleischhauer, J. (1996) Ab initio study of fluorination upon the structure and configurational stability of α-sulfonyl carbanions: The role of negative hyperconjugation. *J. Am. Chem. Soc.*, **118**, 4622–4630.
90. Farnham, W. B., Smart, B. E., Middleton, W. J., *et al.* (1985) Crystal and molecular structure of $[(CH_3)_2N]_3S^+CF_3O^-$. Evidence for negative fluorine hyperconjugation. *J. Am. Chem. Soc.*, **107**, 4565–4567.
91. Barbarich, T. J., Rithner, C. D., Miller, S. M., *et al.* (1999) Significant inter- and intramolecular O–H···FC hydrogen bonding. *J. Am. Chem. Soc.*, **121**, 4280–4281.
92. Katagiri, T. and Uneyama, K. (2001) Chiral recognition by multicenter single proton hydrogen bonding of trifluorolactates. *Chem. Lett.*, 1330–1331.
93. Bartolomé, C., Espinet, P. and Martín-Alvarez, J. M. (2007) Is there any bona fide example of O–H···F–C bond in solution? The cases of $HOC(CF_3)_2(4\text{-}X\text{-}2,6\text{-}C_6H_2(CF_3)_2)$ (X = Si(i-Pr)$_3$, CF_3). *Chem. Commun.*, 4384–4386.
94. Pham, M., Gdaniec, M. and Polonski, T. (1998) Three-center CF···HN intramolecular hydrogen bonding in the 2,6-bis(2,6-difluorophenyl)piperidine systems. *J. Org. Chem.*, **63**, 3731–3734.
95. Li, C., Ren, S.-F., Hou, J.-L., *et al.* (2005) F···H–N Hydrogen bonding driven foldamers: Efficient receptors for dialkylammonium ions. *Angew. Chem. Int. Ed. Engl.*, **44**, 5725–5729.
96. Banks, J. W., Batsanov, A. S., Howard, J. A. K., *et al.* (1999) The preferred conformation of α-fluoroamides. *J. Chem. Soc., Perkin Trans. 2*, 2409–2411.
97. Thalladi, V. R., Weiss, H.-C., Bläser, D., *et al.* (1998) C–H···F Interactions in the crystal structures of some fluorobenzenes. *J. Am. Chem. Soc.*, **120**, 8702–8710.
98. Sosnovskikh, V. Y., Irgashev, R. A., Khalymbadzha, I. A. and Slepukhin, P. A. (2007) Stereoselective hetero-Diels–Alder reaction of 3-(trifluoroacetyl)chromones with cyclic enol ethers: synthesis of 3-aroyl-2-(trifluoromethyl)pyridines with ω-hydroxyalkyl groups. *Tetrahedron Lett.*, **48**, 6297–6300.
99. Wagner, T., Afshar, C. E., Carrell, H. L., *et al.* (1999) Difluoromethylcobalamin: Structural aspects of an old tree with a new branch. *Inorg. Chem.*, **38**, 1785–1794.
100. Erickson, J. A. and McLoughlin, J. I. (1995) Hydrogen bond donor properties of the difluoromethyl group. *J. Org. Chem.*, **60**, 1626–1631.
101. Paulini, R., Müller, K. and Diederich, F. (2005) Orthogonal multipolar interactions in structure chemistry and biology. *Angew. Chem. Int. Ed.*, **44**, 1788–1805.
102. Wang, Z., Canagarajah, B. J., Boehm, J. C., *et al.* (1998) Structural basis of inhibitor selectivity in MAP kinases. *Structure*, **6**, 1117–1128.
103. Schweizer, E., Hoffmann-Roder, A., Scharer, K., *et al.* (2006) A fluorine scan at the catalytic center of thrombin: C–F, C–OH, and C–OMe bioisosterism and fluorine effects on pK_a and log *D* values. *Chem. Med. Chem.*, **1**, 611–621.

104. Schweizer, E., Hoffmann-Roeder, A., Olsen, J. A., *et al.* (2006) Multipolar interactions in the D pocket of thrombin: large differences between tricyclic imide and lactam inhibitors. *Org. Biomol. Chem.*, **4**, 2364–2375.
105. Olsen, J. A., Banner, D. W., Seiler, P., *et al.* (2004) Fluorine interactions at the thrombin active site: Protein backbone fragments H–Cα–C=O comprise a favorable C–F environment and interactions of C–F with electrophiles. *ChemBioChem*, **5**, 666–675.
106. Olsen, J. A., Banner, D. W., Seiler, P., *et al.* (2003) A fluorine scan of thrombin inhibitors to map the fluorophilicity/fluorophobicity of an enzyme active site: Evidence for C–F···C=O interactions. *Angew. Chem. Int. Ed.*, **42**, 2507–2511.
107. Berman, H. M., Westbrook, J., Feng, Z., *et al.* (2000) The protein data bank. *Nucleic Acids Res.*, **28**, 235–242.
108. Hughes, D. L., Sieker, L. C., Bieth, J. and Dimicoli, J. L. (1982) Crystallographic study of the binding of a trifluoroacetyl dipeptide anilide inhibitor with elastase. *J. Mol. Biol.*, **162**, 645–658.
109. Adler, M., Davey, D. D., Phillips, G. B., *et al.* (2000) Preparation, characterization, and the crystal structure of the inhibitor ZK-807834 (CI-1031) complexed with factor Xa. *Biochemistry*, **39**, 12534–12542.
110. Mattos, C., Giammona, D. A., Petsko, G. A. and Ringe, D. (1995) Structural analysis of the active site of porcine pancreatic elastase based on the X-ray crystal structures of complexes with trifluoroacetyl-dipeptide-anilide inhibitors. *Biochemistry*, **34**, 3193–3203.
111. Fischer, F. R., Schweizer, W. B. and Diederich, F. (2007) Molecular torsion balance: Evidence for favorable orthogonal dipolar interactions between organic fluorine and amide groups. *Angew. Chem. Int. Ed.*, **46**, 8270–8273.
112. Roth, B. D., Ortwine, D. F., Hoefle, M. L., *et al.* (1990) Inhibitors of cholesterol biosynthesis., 1. trans-6-(2-Pyrrol-1-ylethyl)-4-hydroxypyran-2-ones, a novel series of HMG-CoA reductase inhibitors. 1. Effects of structural modifications at the 2- and 5-positions of the pyrrole nucleus. *J. Med. Chem.*, **33**, 21–31.
113. Istvan, E. S. and Deisenhofer, J. (2001) Structural mechanism for statin inhibition of HMG-CoA reductase. *Science*, **292**, 1160–1164.
114. Langmuir, I. (1919) Isomorphism, isosterism and covalence. *J. Am. Chem. Soc.*, **41**, 1543–1559.
115. Hann, M. M., Sammes, P. G., Kennewell, P. D. and Taylor, J. B. (1980) On double bond isosteres of the peptide bond: an enkephalin analog. *J. Chem. Soc., Chem. Commun.*, 234–235.
116. Allmendinger, T., Furet, P. and Hungerbuhler, E. (1990) Fluoroolefin dipeptide isosteres. – I. The synthesis of GlyΨ(CF = CH)Gly and racemic PheΨ(CF = CH)Gly. *Tetrahedron Lett.*, **31**, 7297–7300.
117. Frisch, M. J., Trucks, G. W., Schlegel, H. B., *et al.* (2004) Gaussian 03 Release Notes, http://www.gaussian.com/g_tech/g03_rel.htm, Gaussian, Inc., Wallingford CT.
118. Datar, P., Srivastava, S., Coutinho, E. and Govil, G. (2004) Substance P: structure, function, and therapeutics. *Curr. Top. Med. Chem.*, **4**, 75–103.
119. Allmendinger, T., Felder, E. and Hungerbuhler, E. (1990) Fluoroolefin dipeptide isosteres. – II. Enantioselective synthesis of both antipodes of the Phe-Gly dipeptide mimic. *Tetrahedron Lett.*, **31**, 7301–7304.
120. Xiao, J.-B., Weisblum, B. and Wipf, P. (2005) Electrostatic versus steric effects in peptidomimicry: Synthesis and secondary structure analysis of gramicidin S analogues with (*E*)-alkene peptide isosteres. *J. Am. Chem. Soc.*, **127**, 5742–5743.
121. Wipf, P., Henninger, T. C. and Geib, S. J. (1998) Methyl- and (trifluoromethyl)alkene peptide isosteres: synthesis and evaluation of their potential as β-turn promoters and peptide mimetics. *J. Org. Chem.*, **63**, 6088–6089.

122. Volonterio, A., Bravo, P. and Zanda, M. (2000) Synthesis of Partially Modified Retro and Retroinverso [NHCH(CF3)]-Peptides. *Org. Lett.*, **2**, 1827–1830.
123. Volonterio, A., Bellosta, S., Bravin, F., *et al.* (2003) Synthesis, structure and conformation of partially-modified retro- and retro-inverso [NHCH(CF_3)]Gly peptides. *Chem. Eur. J.*, **9**, 4510–4522.
124. Black, W. C., Baylya, C. I., Davisc, D. E., *et al.* (2005) Trifluoroethylamines as amide isosteres in inhibitors of cathepsin K Bioorg. *Med. Chem. Lett.*, **15**, 4741–4744.
125. Sani, M., Volonterio, A. and Zanda, M. (2007) The trifluoroethylamine function as peptide bond replacement. *ChemMedChem*, **2**, 1693–1700.
126. Moran, S., Ren, R. X.-F., Rumney IV, S. and Kool, E. T. (1997) Difluorotoluene, a nonpolar isostere for thymine, codes specifically and efficiently for adenine in DNA replication. *J. Am. Chem. Soc.*, **119**, 2056–2057.
127. Guckian, K. M. and Kool, E. T. (1997) Highly precise shape mimicry by a difluorotoluene deoxynucleoside, a replication-component substitute for thymidine. *Angew. Chem. Int. Ed. Engl.*, **36**, 2825–2828.
128. Kool, E. T. and Sintim, H. O. (2006) The difluorotoluene debate – a decade later. *Chem. Commun.*, 3665–3675.
129. Evans, T. A. and Seddon, K. R. (1997) Hydrogen bonding in DNA – a return to the status quo. *Chem. Commun.*, 2023–2024.
130. Wang, X. and Houk, K. N. (1998) Difluorotoluene, a thymine isostere, does not hydrogen bond after all. *Chem. Commun.*, 2631–2632.
131. Blackburn, G. M., Kent, D. E. and Kolkmann, F. (1984) The synthesis and metal binding characteristics of novel, isopolar phosphonate analogs of nucleotides. *J. Chem. Soc., Perkin Trans. 1*, 1119–1125.
132. Burke, J., T. R., Kole, H. K. and Roller, P. P. (1994) Potent inhibition of insulin receptor dephosphorylation by a hexamer peptide containing the phosphotyrosyl mimetic F2pmp. *Biochem. Biophys. Res. Commun.*, **204**, 129–134.
133. Nieschalk, J. and O'Hagan, D. (1995) Monofluorophosphonates as phosphate mimics in bio-organic chemistry: a comparative study of CH_2-, CHF-, and CF_2-phosphonate analogues of sn-glycerol-3-phosphate as substrates for sn-glycerol-3-phosphate dehydrogenase. *J. Chem. Soc., Chem. Commun.*, 719–720.
134. Hakogi, T., Yamamoto, T., Fujii, S., *et al.* (2006) Synthesis of sphingomyelin difluoromethylene analogue. *Tetrahedron Lett.*, **47**, 2627–2630.
135. Hamilton, C. J., Roberts, S. M. and Shipitsin, A. (1998) Synthesis of a potent inhibitor of HIV reverse transcriptase. *Chem. Commun.*, 1087–1088.
136. Cox, R. J., Hadfield, A. T. and Mayo-Martín, M. B. (2001) Difluoromethylene analogues of aspartyl phosphate: the first synthetic inhibitors of aspartate semi-aldehyde dehydrogenase. *Chem. Commun.*, 1710–1711.
137. Xu, Y., Aoki, J., Shimizu, K., *et al.* (2005) Structure–activity relationships of fluorinated lysophosphatidic acid analogues. *J. Med. Chem.*, **48**, 3319–3327.
138. Dubowchik, G. M., Vrudhula, V. M., Dasgupta, B., *et al.* (2001) 2-Aryl-2,2-difluoroacetamide FKBP12 ligands: Synthesis and X-ray structural studies. *Org. Lett.*, **3**, 3987–3990.
139. Asao, N., Asano, T. and Yamamoto, Y. (2001) Do more electrophilic aldehydes/ketones exhibit higher reactivity toward nucleophiles in the presence of Lewis acids? *Angew. Chem. Int. Ed.*, **40**, 3206–3208.
140. Abraham, R. J., Jones, A. D., Warne, M. A., *et al.* (1996) Conformational analysis. Part 27. NMR, solvation and theoretical investigation of conformational isomerism in fluoro- and 1,1-difluoroacetone. *J. Chem. Soc., Perkin Trans. 2*, 533–539.
141. Maryanoff, B. E. and Costanzo, M. J. (2008) Inhibitors of proteases and amide hydrolases that employ an α-ketoheterocycle as a key enabling functionality. *Bioorg. Med. Chem.*, **16**, 1562–1595.

142. Westerik, J. O. and Wolfenden, R. (1971) Aldehydes as inhibitors of papain. *J. Biol. Chem.*, **247**, 8195–8197.
143. Thompson, R. C. (1973) Use of peptide aldehydes to generate transition-state analogs of elastase. *Biochemistry*, **12**, 47–51.
144. Shah, D. O., Lai, K. and Gorenstein, D. G. (1984) Carbon-13 NMR spectroscopy of "transition-state analog" complexes of *N*-acetyl-l-phenylalaninal and α-chymotrypsin. *J. Am. Chem. Soc.*, **106**, 4272–4273.
145. Hauptschein, M. and Braun, R. A. (1955) The reaction of ethyl perfluorobutyrate with sodium. An improved synthesis of perfluoroheptan-4-one. *J. Am. Chem. Soc.*, **77**, 4930–4931.
146. Husted, D. R. and Ahlbrecht, A. H. (1952) The chemistry of the perfluoro acids and their derivatives. III. The perfluoro aldehydes. *J. Am. Chem. Soc.*, **74**, 5422–5426.
147. Takahashi, L. H., Radhakrishnan, R., Rosenfield Jr., R. E., *et al.* (1989) Crystal-structure of the covalent complex formed by a peptidyl α,α-difluoro-β-ketop amide with porcine pancreatic elastase at 1.78-Å resolution. *J. Am. Chem. Soc.*, **111**, 3368–3374.
148. Brady, K., Wei, A.-Z., Ringe, D. and Abeles, R. H. (1990) Structure of chymotrypsin-trifluoromethyl ketone inhibitor complexes: Comparison of slowly and rapidly equilibrating inhibitors. *Biochemistry*, **29**, 7600–7607.
149. Takahashi, L. H., R., R., Rosenfield Jr., R. E., Meyer, Jr. E. F., *et al.* (1988) X-ray diffraction analysis of the inhibition of porcine pancreatic elastase by a peptidyl trifluoromethylketone. *J. Mol. Biol.*, **201**, 423–428.
150. Veale, C. A., Bernstein, P. R., Bryant, C., *et al.* (1995) Nonpeptidic inhibitors of human leukocyte elastase. 5. Design, synthesis, and X-ray crystallography of a series of orally active 5-aminopyrimidin-6-one-containing trifluoromethyl ketones. *J. Med. Chem.*, **38**, 98–108.
151. Gelb, M. H., Svaren, J. P. and Abeles, R. H. (1985) Fluoro ketone inhibitors of hydrolytic enzymes. *Biochemistry*, **24**, 1813–1817.
152. Peet, N. P., Burkhart, J. P., Angelastro, M. R., *et al.* (1990) Synthesis of peptidyl fluoromethyl ketones and peptidyl α-keto esters as inhibitors of porcine pancreatic elastase, human neurophil elastase, and rat and human neurophil cathepsin. *J. Med. Chem.*, **33**, 394–407.
153. Angelastro, M. R., Baugh, L. E., Bey, P., *et al.* (1994) Inhibition of human neutrophil elastase with peptidyl electrophilic ketones. 2. Orally active PG-Val-Pro-Val pentafluoroethyl ketones. *J. Med. Chem.*, **37**, 4538–4553.
154. Ogilvie, W., Bailey, M., Poupart, M.-A., *et al.* (1997) Peptidomimetic inhibitors of the human cytomegalovirus protease. *J. Med. Chem.*, **40**, 4113–4135.
155. LaPlante, S. R., Bonneau, P. R., Aubry, N., *et al.* (1999) Characterization of the human cytomegalovirus protease as an induced-fit serine protease and the implications to the design of mechanism-based inhibitors. *J. Am. Chem. Soc.*, **121**, 2974–2986.
156. Warner, P., Green, R. C., Gomes, B. and Strimpler, A. M. (1994) Non-peptidic inhibitors of human leukocyte elastase. 1. The design and synthesis of pyridone-containing inhibitors. *J. Med. Chem.*, **37**, 3090–3099.
157. Veale, C. A., Bernstein, P. R., Bohnert, C. M., *et al.* (1997) Orally active trifluoromethyl ketone inhibitors of human leukocyte elastase. *J. Med. Chem.*, **40**, 3173–3181.
158. Nair, H. K. and Quinn, D. M. (1993) m-Alkyl α,α,α-trifluoroacetophenones: A new class of potent transition state analog inhibitors of acetylcholinesterase. *Bioorg. Med. Chem. Lett.*, **3**, 2619–2622.
159. Linderman, R. J., Graves, D. M., Garg, S., *et al.* (1993) Unique inhibition of a serine esterase. *Tetrahedron Lett.*, **34**, 3227–3230.
160. Basso, N. and Terragno, N. A. (2001) History about the discovery of the renin-angiotensin system. *Hypertension*, **38**, 1246–1249.
161. Umezawa, H., Aoyagi, T., Morishima, H., *et al.* (1970) Pepstatin, a new pepsin inhibitor produced by Actinomycetes. *J. Antibiot.*, **23**, 259–262.

162. Marciniszyn, J., Hartsuck, J. A. and Tang, J. (1976) Mode of inhibition of acid proteases by pepstatin. *J. Biol. Chem.*, **251**, 7088–7094.
163. Thaisrivongs, S., Pals, D. T., Kati, W. M., *et al.* (1986) Design and synthesis of potent and specific renin inhibitors containing difluorostatine, difluorostatone, and related analogues. *J. Med. Chem.*, **29**, 2080–2087.
164. Fearon, K., Spaltenstein, A., Hopkins, P. B. and Gelb, M. H. (1987) Fluoro ketone containing peptides as inhibitors of human renin. *J. Med. Chem.*, **30**, 1617–1622.
165. Witter, R., Nozirov, F., Sternberg, U., *et al.* (2008) Solid-state ^{19}F NMR spectroscopy reveals that Trp_{41} participates in the gating mechanism of the M2 proton channel of influenza A virus. *J. Am. Chem. Soc.*, **130**, 918–924.
166. Ojima, I., Kuduk, S. D., Chakravarty, S., *et al.* (1997) A novel approach to the study of solution structures and dynamic behavior of paclitaxel and docetaxel using fluorine-containing analogs as probes. *J. Am. Chem. Soc.*, **119**, 5519–5527.
167. Ojima, I., Kuduk, S. D. and Chakravarty, S. (1998) Recent advances in the medicinal chemistry of taxoid anticancer agents. *Adv. Med. Chem.*, **4**, 69–124.
168. Li, Y., Poliks, B., Cegelski, L., *et al.* (2000) Conformation of Microtubule-Bound Paclitaxel Determined by Fluorescence Spectroscopy and REDOR NMR. *Biochemistry*, **39**, 281–291.
169. Paik, Y., Yang, C., Metaferia, B., *et al.* (2007) Rotational-echo double-resonance NMR distance measurements for the tubulin-bound paclitaxel conformation, *J. Am. Chem. Soc.*, **129**, 361–370.
170. Malet-Martino, M., Gilard, V., Desmoulin, F. and Martino, R. (2006) Fluorine nuclear magnetic resonance spectroscopy of human biofluids in the field of metabolic studies of anticancer and antifungal fluoropyrimidine drugs. *Clin. Chim. Acta*, **366**, 61–73.
171. Volkow, N. D., Fowler, J. S., Wang, G. J., *et al.* (1993) Decreased dopamine D2 receptor availability is associated with reduced frontal metabolism in cocaine abusers. *Synapse*, **14**, 169–177.
172. Ding, Y.-S. and Fowler, J. S. (1996) Fluorine-18 labeled tracers for PET studies in the neurosciences, in *Biomedical Frontiers of Fluorine Chemistry*, ACS Symposium Series 639, American Chemical Society, Washington, D. C., pp 328–343.
173. Snyder, S. E. and Kilbourn, M. R. (2003) Chemistry of fluorine-18 radiopharmaceuticals, in *Handbook of Radiopharmaceuticals* (eds. M. J. Welch and C. S. Redvanly), pp. 195–227, John Wiley & Sons, Ltd, Chichester.

Medicinal Chemistry

2

Fluorinated Prostanoids: Development of Tafluprost, a New Anti-glaucoma Agent

Yasushi Matsumura

2.1 Introduction

2.1.1 Background

Prostanoids, consisting of prostaglandins (PGs) and thromboxanes (TXs), are members of the lipid mediators derived enzymatically from fatty acids. Arachidonic acid, a C_{20} essential fatty acid for most mammalians, is freed from the phospholipid molecule by phospholipase A_2, which cleaves off the fatty acid precursor. Prostanoids are produced in a wide variety of cells throughout the body from the sequential oxidation of arachidonic acid by cyclooxygenase, PG hydroperoxidase, and a series of prostaglandin synthases (Figure 2.1).

Prostanoids have extensive pharmacological actions in various tissues in order to maintain homeostasis [1]. They play an important role as local hormones that mediate inflammation, pain, fever, vasoconstriction or vasodilation, coagulation, calcium regulation, cell growth, and so on. The prostanoids have been thought to exert their multiple physiological actions via specific protein receptors on the surface of target cells. The prostanoid receptors were pharmacologically identified from the functional data and classified into DP, EP, FP, IP and TP receptors, which are specific for PGD_2, PGE_2, $PGF_{2\alpha}$, PGI_2 (prostacyclin), and TXA_2, respectively [2]. The recent molecular cloning of cDNAs that encode the prostanoid receptors has identified the receptor structures and confirmed

Fluorine in Medicinal Chemistry and Chemical Biology Edited by Iwao Ojima

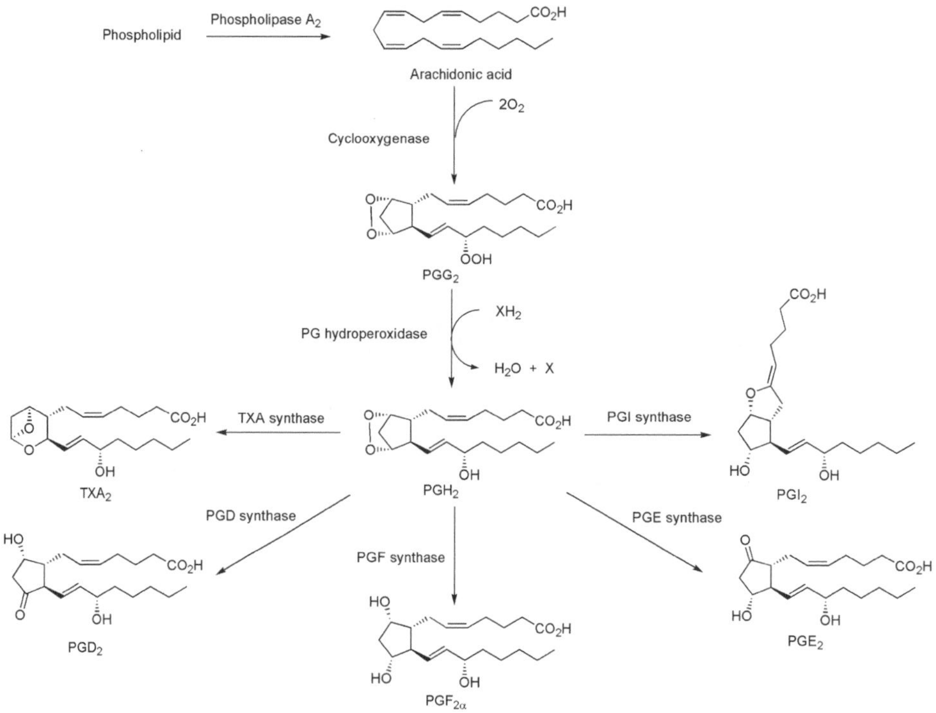

Figure 2.1 *Biosynthesis of prostanoids.*

the receptor classification including further subdivision for the EP receptor subtypes, EP1–EP4 [3, 4].

The prostanoid receptors belong to the G-protein-coupled rhodopsin-type receptors with seven transmembrane domains. They include an extracellular amino group terminus, an intracellular carboxyl group terminus, three extracellular loops, and three intracellular loops (Figure 2.2) [3]. The overall homology among the receptors is not high, though these receptors conserve important amino acid sequences in several regions, especially in the seventh transmembrane domain. It is proposed that the conserved arginine residue in the domain serves as the binding site for the terminal carboxyl group of the prostanoid molecules. Information obtained from studies on the receptor structures and properties has been utilized recently for computer-aided molecular modeling and structure-based drug design. The use of expressed receptors aids in screening of compounds for the specific receptor agonists or antagonists.

The properties of the prostanoid receptors, such as second messengers in signal transduction pathways, and localization in the eye are summarized in Table 2.1. Prostanoid receptors are widely distributed in the monkey and human eyes [5]. The expression and localization of the FP and EP receptor subtypes in the tissues was studied intensely by in-situ hybridization and immunohistochemistry to gain a better understanding of the ocular effects of the prostanoids and their analogues. This work suggests a wide distribution but differential expression of FP and EP receptor subtypes in human ocular tissues. The highest expression of FP receptor mRNA and protein was found in the corneal epithelium, ciliary

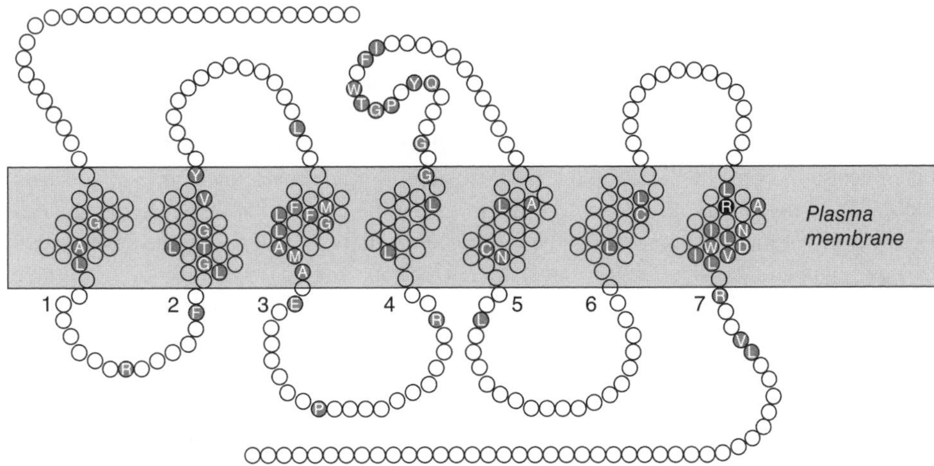

● : Conserved amino acid residues

● : Arginine residue (binding site to CO_2H group)

Figure 2.2 *Prostanoid receptor.*

Table 2.1 *Properties of prostanoid receptors*

PGs	Receptor	Signal transduction	Localization in the human eye
PGD_2	DP	cAMP ↑	Retina
PGE_2	EP_1	Ca^{2+}↑	Ciliary body, trabecular meshwork, retina, iris, lens, cornea, conjunctiva
	EP_2	cAMP ↑	Trabecular meshwork, cornea, choroid
	EP_3	cAMP ↓	Ciliary body, trabecular meshwork, retina, iris, cornea, conjunctiva
	EP_4	cAMP ↑	Ciliary body, trabecular meshwork, retina, iris, cornea, conjunctiva
$PGF_{2\alpha}$	FP	PI response	Ciliary body, trabecular meshwork, iris, cornea
PGI_2	IP	cAMP ↑, PI response	Trabecular meshwork
TXA_2	TP	PI response	Trabecular meshwork, corneal epithelium, ciliary processes, retina

PI response: phosphatidylinositol response.

epithelium, ciliary muscle, and iris muscles by immunohistochemistry [6]. In the trabecular meshwork, the gene expression of the EP2 receptor is more abundant than that of other receptors [7].

2.1.2 Fluorinated Prostanoids Research

It is well known that introduction of fluorine atoms into biologically active substances may lead to improvements in pharmacological properties and an increase in therapeutic efficacy [8]. These advantageous pharmacological effects of fluorinated molecules are mainly derived from the following physicochemical characters of fluorine: (1) relatively small atomic size, (2) high carbon–fluorine bond energy, (3) high electronegativity, and (4) enhancement of lipophilicity.

The main problems with the use of natural prostanoids utilized as drugs have been perceived to be both chemical and metabolic instability, and separation of side-effects from the multiple physiological actions. In order to overcome these difficulties, chemical modifications of natural prostanoids have been studied extensively along with the development of new synthetic methodologies since the 1970s [9]. Taking advantage of the unique characteristics of fluorine, a large number of fluorinated prostanoids have also been reported (Figure 2.3) [10]. For example, fluprostenol, with a *m*-trifluoromethylphenoxy group in the ω-chain, emerged in 1974 was one of the first successfully marketed analogues with application as a potent luteolytic agent in veterinary medicine [11]. The compound is known as a selective FP receptor agonist and is widely used as a pharmacological tool. The strong inductive effect and enhancement in lipophilicity caused by the CF_3 group should contribute to improvement of the biological profile. The 16,16-difluoro-PGE_2 was reported in 1975 to be metabolically stabilized by 15-dehydrogenase inhibition of the degradation pathways *in vitro* [12]. The inhibition of enzymatic oxidation is explained by the destabilization effects of electron-withdrawing fluorine atoms causing a shift to the reduced form between the allyl alcohol and the enone in equilibrium. Fried *et al.* reported that 10,10-difluoro-13,14-dehydro-PGI_2 [13] and 10,10-difluoro-TXA_2 [14] showed an increase in the stability against hydrolysis in comparison with PGI_2 and TXA_2, respectively. Our group studied a 7,7-difluoro-PGI_2 derivative (AFP-07) for modification of the physical and physiological properties of natural PGI_2 by the inductive effects of fluorine atoms [15]. AFP-07 showed not only higher stability in aqueous media of at least 10 000

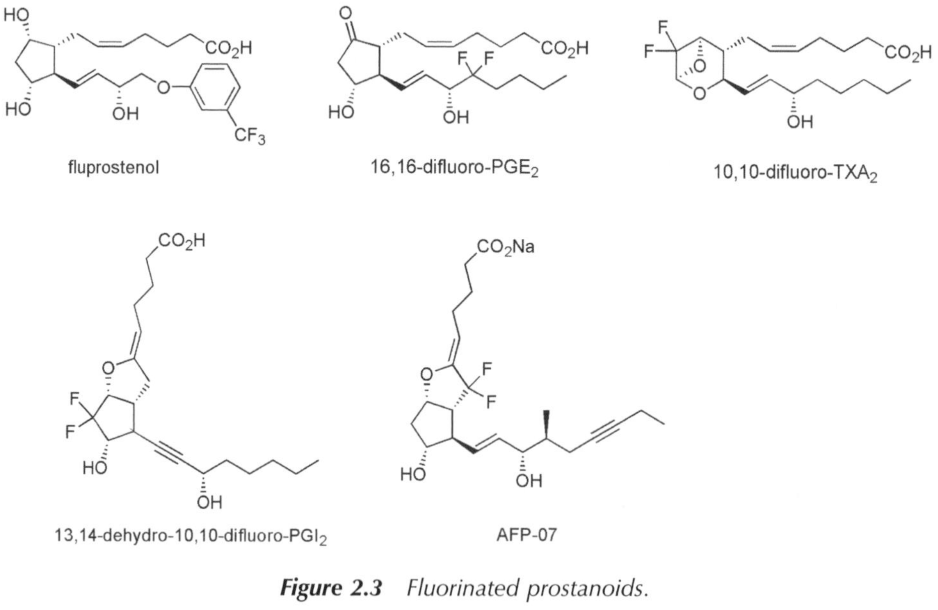

fluprostenol

16,16-difluoro-PGE_2

10,10-difluoro-TXA_2

13,14-dehydro-10,10-difluoro-PGI_2

AFP-07

Figure 2.3 *Fluorinated prostanoids.*

times that of the natural compound, but also potent and selective affinity for the IP receptor [16]. These instances demonstrate the high potential of chemical modification of prostanoids with fluorine for lead discovery and optimization in drug development, if fluorine atoms can be introduced into the right positions of the prostanoid structure on the basis of rational drug design.

2.2 Therapy of Glaucoma

2.2.1 Glaucoma

Glaucoma is one of the most common but serious eye diseases and that can damage the optic nerve and result in loss of vision and blindness. There may be no symptoms in the early stages of the disease. A recent epidemiological survey of glaucoma conducted in Japan (the Tajimi study) showed that the prevalence of glaucoma in residents aged 40 years and older is about 5.0%, which is higher than that of previous surveys and demonstrates that the number of patients with glaucoma has been increasing in Japan [17]. Worldwide, it is the second leading cause of blindness, according to the World Health Organization.

It is thought that high pressure within the eye (intraocular OP) is the main cause of the optic nerve damage. Although elevated IOP is clearly a risk factor, other factors must also be involved because even people with normal levels of pressure can experience vision loss from glaucoma. The Tajimi study in Japan revealed that a substantial number of glaucoma patients are diagnosed as suffering normal tension glaucoma (NTG). The evidence suggests that in patients with NTG a 30% reduction in IOP can slow the rate of progressive visual field loss [18]. IOP can be lowered with medication, usually eye drops. There are several classes of drugs for treating glaucoma, with several medications in each class. The first-line therapy of glaucoma treatment is currently prostanoids, which will allow better flow of fluid within the eye. IOP-lowering eye drops, by acting on their respective receptors to decrease the secretion of aqueous humor such as β-blockers or α_2 agonists, also might be considered, although these may not be used in people with heart conditions, because they can affect cardiovascular and pulmonary functions.

2.2.2 Prostanoids in the Therapy of Glaucoma

Since the discovery in 1980s that $PGF_{2\alpha}$ reduces IOP in an animal model [19], extensive efforts have been devoted to developing FP receptor agonists as promising new antiglaucoma agents [20]. Most reported analogues are esters used as prodrugs that are rapidly hydrolyzed by corneal enzymes to the free acids to account for the ocular hypotensive effects by activation of the FP prostanoid receptor (Figure 2.4).

Unoprostone [21], a docosanoid, a structural analogue of an inactive biosynthetic metabolite of $PGF_{2\alpha}$,was developed by Ueno *et al.* and first marketed in Japan for the treatment of glaucoma in 1994. Clinical studies showed that in patients with mean baseline IOP of 23 mmHg, it lowered IOP by approximately 3–4 mmHg throughout the day. The recommended dosage is one drop in the affected eyes twice daily.

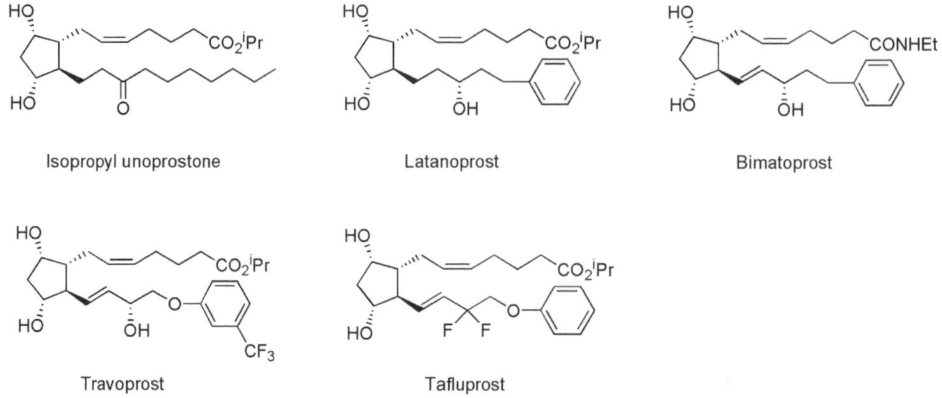

Isopropyl unoprostone   Latanoprost   Bimatoprost

Travoprost   Tafluprost

Figure 2.4 *Anti-glaucoma prostanoids.*

Stjernschantz *et al.* later developed latanoprost [22], which has potent IOP-reducing effects with topical administration once daily in the evening. Compared with a representative β-blocker, timolol, it demonstrated superior efficacy in clinical studies in reducing IOP by approximately 27–34% from baseline. Latanoprost is the FP receptor agonist most widely used worldwide as an anti-glaucoma drug.

Bimatoprost and travoprost are recently approved prostanoids with high efficacy in reducing IOP. The chemical structure of bimatoprost differs from that of latanoprost only in a double bond of the ω-chain and an ethylamide in the C-1 position. Although the classification of bimatoprost is still subject of controversy, it has been demonstrated as a "prostamide," a class of drugs distinct from PGs [23]. Travoprost is an isopropyl ester of a single enantiomer of fluprostenol, which was already known as a potent and selective FP receptor agonist [24]. A new synthetic route of travoprost from a tricyclic ketone with ring cleavage by the attack of a vinyl cuprate has recently been reported [25].

The aqueous humor flows out the eye mainly via the conventional route of the trabecular meshwork and Schlemm's canal. However, about 10–20% of outflow is via the uveoscleral (nonconventional) route whereby the aqueous humor passes between the ciliary muscle bundles and into the episcleral tissues, where it is reabsorbed into orbital blood vessels and drained via the conjunctival vessels. The IOP-lowering effect of these prostanoid drugs occurs predominantly through enhancement of uveoscleral outflow [26], although unoprostone and bimatoprost also increase flow via the trabecular meshwork to a lesser extent.

The prostanoids have been widely used for the treatment of ocular hypertension in many countries because they have good IOP-reducing effects without causing serious systemic side-effects. However, the drugs cause local adverse effects, such as hyperemia and iris/skin pigmentation [27]. Moreover, the existing ocular hypotensive drugs, even latanoprost, do not produce satisfactory IOP control in all patients. It is therefore hoped that a new-generation prostanoid having powerful and prolonged IOP-reducing efficacy together with improvement in ocular circulation, and causing fewer side-effects, will become available for patients with glaucoma.

2.3 Development of Tafluprost

2.3.1 Screening and Discovery

Prostanoids are generally flexible molecules that change their conformation in response to changes in their environment through the intramolecular hydrogen bonding between the terminal carboxylic acid and the hydroxyl group at C-9, C-11, or C-15 [28]. In the drug–receptor complex, the prostanoids can adopt a preferred conformation through the forces involved ionic interactions and dipole–dipole interactions, including hydrogen bonding, between these functional groups and the corresponding amino acid residues of receptors [29]. If a specific position of the prostanoids is substituted by fluorine, it should affect not only the molecular conformation but also the drug–receptor complex through possible participation of the fluorine in the interactions.

The substitution of the hydroxyl group at C-15 of $PGF_{2\alpha}$ with fluorine atoms and the biological effects of the molecule has not been well-studied [30] because the 15-hydroxyl group is believed to be essential for pharmacological activity of PGs [31]. The carbon–fluorine bond (van der Waals radius = 1.47 Å) is nearly isosteric with the carbon–oxygen bond (van der Waals radius = 1.52 Å). Compared to the hydroxylated carbon, the fluorinated atom should be much more electronegative because of the strong electron-withdrawing effect of fluorine. In contrast to the hydroxyl group, the fluorine cannot be a donor in hydrogen bonding; it can be a weak acceptor for hydrogen bonding, although this is still a matter of controversy [32]. In addition, the enhancement of lipophilicity on introducing fluorine atoms in a position close to a rigid pharmacophore such as an aromatic functionality in the ω-chain may be an effective way to increase the specific affinity for a hydrophobic pocket of the receptors.

Our research group at Asahi Glass Co., Ltd. has collaborated with Santen Pharmaceutical Co., Ltd. to find a new FP receptor agonist having more potent IOP-reducing activity and weaker side-effects. We have recently discovered a 15-deoxy-15,15-difluoro-17,18,19,20-tetranor-16-phenoxy-$PGF_{2\alpha}$ isopropyl ester, tafluprost (AFP-168), which shows highly potent and selective affinity for the FP receptor [33]. We have synthesized newly designed $PGF_{2\alpha}$ derivatives and investigated their prostanoid FP receptor-mediated functional activities both *in vitro* and *in vivo*. A functional prostanoid FP-receptor-affinity assay was performed using iris sphincter muscle isolated from cat eyes, which predominantly expresses the prostanoid FP receptor. The results on constrictions induced by PG-derivatives are shown in Table 2.2. A carboxylic acid of latanoprost induced constriction with an EC_{50} value of 13.6 nM. In the functional FP receptor affinity assay, we found that 15-deoxy-15-fluoro-16-aryloxy-tetranor-$PGF_{2\alpha}$ derivatives (AFPs-159 and 120) caused strong constriction of the isolated cat iris sphincter [34]. This suggested that exchanging the 15-hydroxy group for fluorine preserved agonistic activities on FP receptor. In contrast, the diastereomers of these derivatives with the fluorine atom attached at C-15 showed much weaker binding affinities (data not shown). Interestingly, 15,15-difluorinated analogues – AFPs-164, 157, 162, and 172 – demonstrated much more potent agonistic activities than the monofluorinated derivatives. The introduction of a chlorine atom into the *meta*-position of the benzene ring of these difluorinated derivatives reduced the prostanoid FP-receptor functional activities. The 13,14-dihydro analogues AFP-164 and AFP-162 had a weaker affinities than the unsaturated ones, AFPs-157 and 172.

Table 2.2 *Functional assay of PG derivatives on FP receptor*[a]

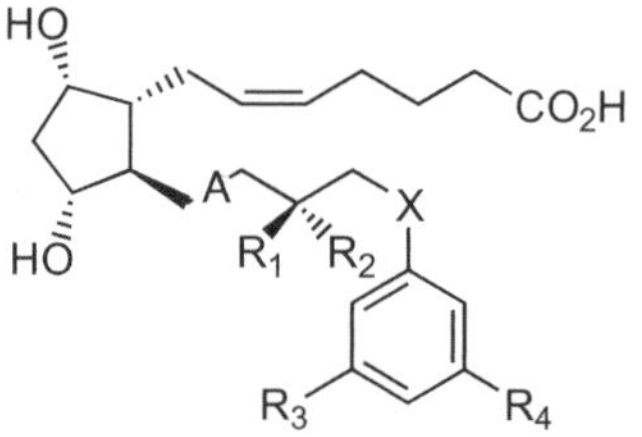

Compounds	A	X	R^1	R^2	R^3	R^4	EC_{50} (nM)
Latanoprost acid form	Single bond	CH_2	H	OH	H	H	13.6
AFP-159	Single bond	O	H	F	H	H	6.6
AFP-120	Double bond	O	H	F	Cl	Cl	37.9
AFP-164	Single bond	O	F	F	H	Cl	9.4
AFP-157	Double bond	O	F	F	H	Cl	1.9
AFP-162	Single bond	O	F	F	H	H	2.4
AFP-172	Double bond	O	F	F	H	H	0.6

[a] Constriction effects of PG derivatives on cat iris sphincters.

Overall, AFP-172, the active carboxylic acid form of tafluprost, displayed the most potent activity.

2.3.2 Synthesis of Tafluprost

A synthetic route for tafluprost is shown in Scheme 2.1 [33]. The synthesis was started from the Corey aldehyde **1**, which was converted to enone **2** by Horner–Emmons reaction. Since a general method to prepare allyl difluorides from enones had not been reported, we studied the fluorination reaction. It was found that the reaction of the enone **2** with morpholinosulfur trifluoride **3** and successive deprotection gave the desired geminal difluoride **4** in good yield. Reduction of the lactone **4** with diisobutylaluminum hydride in THF–toluene at −78 °C afforded the lactol **5**. The Wittig reactions of the lactol **5** with the ylide prepared from 4-carboxybutyltriphenylphosphonium bromide with various bases yielded the 15-deoxy-15,15-difluoro-$PGF_{2\alpha}$ derivative as a mixture of 5*Z* and 5*E* isomers. The Wittig reaction using sodium bis(trimethylsilyl)amide as base gave the best result for stereoselectivity (5*Z*/5*E* = 99/1). The esterification of the crude acid treated with isopropyl iodide and 1,8-diazabicyclo[5.4.0]undec-7-ene (DBU) afforded the desired 15-deoxy-15,15-difluoro-$PGF_{2\alpha}$ derivative (tafluprost).

2.3.3 Pharmacology of Tafluprost

2.3.3.1 Prostanoid Receptor Affinities

The affinity of the corresponding carboxylic acid of tafluprost for the recombinant human FP receptor expressed in clonal cells was 0.4 nM, which was 12 times and 1700 times

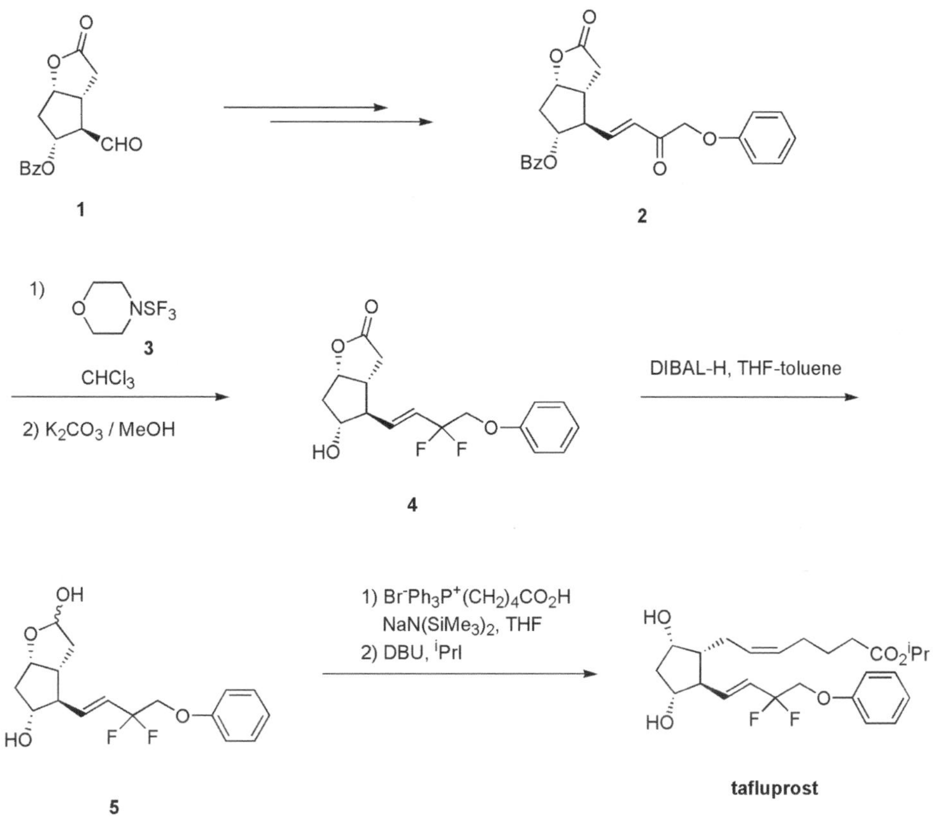

Scheme 2.1 *Synthesis of tafluprost.*

Table 2.3 *Affinities of prostanoids for the human prostanoid FP receptor*

Compound (acid form)	K_i (nM)	Ratio (tafluprost = 1)
Tafluprost	0.40	1
Latanoprost	4.7	12
Unoprostone	680	1700

higher than those of latanoprost and isopropyl unoprostone, respectively (Table 2.3) [35]. It should be noted that substitution of difluoro-moiety for the hydroxyl group at C-15 of $PGF_{2\alpha}$ derivatives increases binding to the FP receptor to such a large extent because the hydroxyl group was thought to be indispensable to exhibit biological activity [31].

Since the acid form of tafluprost did not show significant affinities for other prostanoid receptors, the drug proved to be a highly selective compound [35]. Compared with the other prostanoids [26b, 36], tafluprost is regarded as one of the most selective prostanoid FP receptor agonists.

2.3.3.2 IOP-Reducing Effects

Tafluprost has a potent IOP-reducing effect in animal models. For example, the maximal IOP reduction achieved with tafluprost at 0.0025% was greater than with latanoprost at 0.005% in both laser-induced glaucomatous and ocular normotensive monkeys [35]. The effects of tafluprost and latanoprost on IOP reduction in conscious ocular normotensive monkeys are indicated in Figure 2.5. The peak time for the IOP reduction induced by tafluprost was 6–8 h after its application, similar to that of latanoprost. The duration of the IOP reduction seen with tafluprost was greater than that seen with latanoprost. Once-daily applications of tafluprost led to progressive increases in the daily maximal IOP reduction and in the IOP reduction at the trough time-point (just before the next application), while these effects at the trough time-point were not observed with latanoprost in the monkey study. These results indicate that the IOP-lowering effect of tafluprost is stronger and more continuous than that of latanoprost.

The mechanism underlying the IOP-lowering effect of tafluprost was investigated in ocular normotensive monkeys (Table 2.4) [35]. The methods used in this study were validated by their ability to reveal the effects of positive controls, such as timolol, $PGF_{2\alpha}$-isopropyl ester, and pilocarpine. Tafluprost decreased the flow to blood (FTB, conventional outflow) and increased the uveoscleral outflow. The effect of tafluprost on aqueous humor formation (AHF) was similar by the different methods, increases of 10% by fluorophotometry (not significant) and 14% by isotope perfusion ($p < 0.05$). Compared with the increase in uveoscleral outflow, this increase in AHF is relatively small. Thus, tafluprost may affect AHF slightly, as do the other $PGF_{2\alpha}$ analogues. Tafluprost also decreased FTB and the mechanism may due to rerouting of flow to the uveoscleral pathway. Thus, the primary mechanism underlying the IOP-reducing effect of tafluprost is via an increase in uveoscleral outflow, as with other PG derivatives [26, 37].

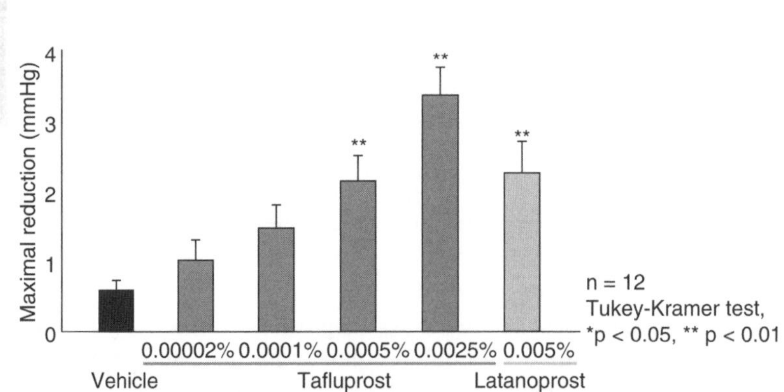

Figure 2.5 *Effects of tafluprost and latanoprost on maximal reduction of intraocular pressure (IOP) in conscious ocular normotensive monkeys.*
(Source: *Reprinted from Takagi, Y., Nakajima, T., Shimazaki, A.,* et al. *Pharmacological characteristics of AFP-168 (tafluprost), a new prostanoid FP receptor agonist, as an ocular hypotensive drug. Exp. Eye Res., (2004)* **78**, *767–776, with permission from Elsevier)*

Table 2.4 *Effects of tafluprost on aqueous humor dynamics in anesthetized ocular normotensive monkeys*

Experiments/treatments (*n*)	Control (contralateral eye)	Treayed eye	Ratio of treated/control
Fluorophotometry for aqueous humor formation (AHF, µl/min)			
Baseline (8)	1.49 ± 0.14	1.43 ± 0.12	0.97 ± 0.03
Tafluprost (8)	1.73 ± 0.13	1.88 ± 0.12	1.10 ± 0.04
Baseline (8)	1.55 ± 0.14	1.64 ± 0.17	1.06 ± 0.05
Timolol-gel (8)	1.39 ± 0.15	1.06 ± 0.10	0.77 ± 0.02**
Isotope perfusion for AHF (µl/min), flow to blood (FTB, µl/min), and uveoscleral outflow (Fu, µl/min)			
AHF tafluprost (12)	1.54 ± 0.12	1.73 ± 0.15	1.14 ± 0.06*
FTB tafluprost (12)	0.78 ± 0.16	0.61 ± 0.14	0.78 ± 0.06**
Fu tafluprost (10)	0.92 ± 0.17	1.22 ± 0.14	1.65 ± 0.24*
AHF $PGF_{2\alpha}$-ie (8)	1.45 ± 0.17	1.54 ± 0.19	1.11 ± 0.14
FTB $PGF_{2\alpha}$-ie (8)	0.43 ± 0.12	0.14 ± 0.03	0.41 ± 0.08**
Fu $PGF_{2\alpha}$-ie (8)	1.01 ± 0.22	1.40 ± 0.20	2.31 ± 0.99
Two-level constant-pressure perfusion for total outflow facility (µl/min/mmHg)			
Tafluprost (12)	0.45 ± 0.08	0.57 ± 0.11	1.33 ± 0.13*
$PGF_{2\alpha}$-ie (8)	0.60 ± 0.10	0.58 ± 0.09	1.15 ± 0.23
Pilocarpine (8)	0.83 ± 0.16	2.23 ± 0.40	2.84 ± 0.33**

$PGF_{2\alpha}$-ie: prostaglandin $F_{2\alpha}$-isopropyl ester.
Data represent the mean ± SEM. For ratio values, *$p < 0.05$,
**$p < 0.01$ for difference from 1.0 (two-tailed paired *t*-test).
Source: Reprinted from Ref. 35, Copyright (2004) with permission from Elsevier Limited.

2.3.3.3 IOP-Lowering Effects in Prostanoid Receptor-Deficient Mice

Ota *et al.* reported the IOP-lowering effects of tafluprost in wild-type mice [38] and prostanoid receptor-deficient mice [39], topically administered by a microneedle method. The IOP-lowering effect of tafluprost was compared with that of latanoprost in ddY mice over a 24 h period. By area-under-the-curve analysis, tafluprost was more effective in reducing mouse IOP, and its ocular hypotensive effect lasted longer than that of latanoprost [38]. In B6 mice, both tafluprost and latanoprost lowered IOP in a dose-dependent manner from 1 to 6 h after administration, but the magnitude of IOP reduction induced by tafluprost was significantly greater than that induced by latanoprost. The more effective IOP reduction of tafluprost may be the result of its higher affinity for FP receptor. In EP1KO and EP2KO mice, there was no significant difference in IOP reduction induced by tafluprost and latanoprost as compared with B6 mice. Although tafluprost and latanoprost significantly lowered IOP in EP3KO mice, the magnitude of IOP reduction was significantly less than the effect in B6 mice. The EP3 receptor may play a role in IOP reduction induced by tafluprost and latanoprost. In FPKO mice, tafluprost and latanoprost had no obvious IOP reduction. These results suggest that tafluprost lowers IOP and produces endogenous PG via mainly prostanoid FP receptor, and the endogenous PG may lower IOP via prostanoid EP3 receptor, similarly to the findings with travoprost, bimatoprost, and unoprostone in a previous study [40].

2.3.3.4 Increase of Ocular Blood Flow

Tafluprost significantly increases retinal blood flow and blood velocity in animal models. The improvement of ocular blood flow is thought to be relevant in glaucoma therapy, especially for normal-tension glaucoma patients since it is assumed that optic nerve damage is involved not only in mechanical compression caused by IOP but also in impairment of ocular blood flow.

The effects of tafluprost on IOP and retinal blood flow (RBF) were studied in adult cats [41]. A single drop of tafluprost was placed in one eye and IOP, vessel diameter, blood velocity, and RBF were measured simultaneously by laser Doppler velocimetry. Measurements carried out at 30 and 60 min after dosing showed 16.1% and 21.0% IOP reduction, respectively, as well as 1% and 2.4% reduction in mean vessel diameter, respectively. The mean blood velocity increases were 17.4% and 13.7%, respectively, and the mean RBF increases were 20.7% and 18.8%, respectively, 30 and 60 min after dosing.

Another study aimed to evaluate and compare the effect of tafluprost, latanoprost, and travoprost on optic nerve head (ONH) blood flow in rabbits [42]. A quantitative index of blood flow, squared blur rate (SBR), was determined with the laser speckle method, when 50 μl of 0.0015% tafluprost, 0.005% latanoprost, or 0.004% travoprost were topically administrated once a daily for 28 days. After 28 days' administration of tafluprost, latanoprost, and travoprost, the trough SBR values became $111.9 \pm 3.9\%$, $107.2 \pm 4.3\%$ and $106.7 \pm 3.5\%$, respectively, compared with the value before administration. Sixty minutes after final administration on day 28, the SBR value with tafluprost, latanoprost, and travoprost become $116.1 \pm 3.5\%$, $106.1 \pm 3.0\%$, and $104.2 \pm 3.7\%$, respectively, compared with the value before administration. These results indicate that topical administrations of these compounds stably increase the ONH blood flow in rabbits. The magnitude of increase in ONH blood flow produced by tafluprost was greater than that of latanoprost or travoprost.

2.3.3.5 Protective Effect of Tafluprost on Glutamate-Induced Cytotoxicity

The protective effect of tafluprost on the cytotoxicity and intracellular Ca^{2+} increase induced by l-glutamate (Glu) using primary cultures obtained from the fetal rat retina has been reported [43]. Tafluprost acid form significantly prevented Glu-induced cytotoxicity in a concentration-dependent manner of more than 10 nM. However, latanoprost acid form did not show any effect on Glu-induced cytotoxicity. Tafluprost acid form showed the cell protective effect on Glu-induced cytotoxicity through the inhibition of intracellular Ca^{2+} increase in retinal cells.

Glaucoma is a progressive neuropathy characterized by loss of the visual field resulting from neuronal cell death [44]. These results suggest that tafluprost is an effective therapy for glaucoma to prevent the retinal cell damage in addition to its effects of lowering IOP and increasing the activity of ocular blood flow.

2.3.3.6 Melanogenesis

In long-term clinical use, prostanoids are known to cause iris pigmentation as a characteristic side-effect; this has been observed in 5–15% of patients treated [27]. In cultured melanoma cells, a carboxylic acid of latanoprost has been reported to increase melanogenesis [45]. However, a carboxylic acid of tafluprost did not have the stimulatory effects on melanin content in cultured B16-F10 melanoma cells [34, 35]. The melanogenesis-promoting effects of latanoprost acid and tafluprost acid *in vitro* are compared in Figure 2.6. This finding implies that the application of tafluprost may cause less iris pigmentation than that of latanoprost.

2.3.4 Pharmacokinetics and Metabolism

To evaluate the distribution and metabolism of [^{3}H]tafluprost in ocular tissues and to study the IOP-lowering effects of the major metabolites of tafluprost, single ocular doses of [^{3}H]tafluprost were administered to male/female cynomolgus monkeys (1 µg/eye for tissue distribution studies and 10 µg/eye for metabolic studies) [46]. Tafluprost was rapidly absorbed into ocular tissues and subsequently entered the systemic circulation. The highest concentrations of radioactivity were observed in the bulbar conjunctiva and the palpebral conjunctiva (323 and 180 ng-eq/g, respectively) at 0.083 h after administration, and in the cornea (784 ng-eq/g) at 0.25 h after administration. Nonvolatile radioactivity in plasma peaked (0.907 ng-eq/g) at 0.083 h after administration and then declined steadily. Three major metabolites shown in Figure 2.7, a carboxylic acid of tafluprost (AFP-172), 1,2-

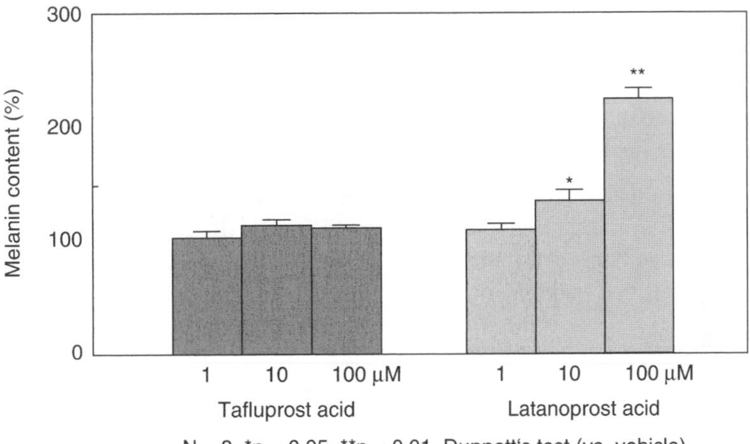

Figure 2.6 *Effects of tafluprost acid and latanoprost acid on melanin contents of cultured B16-F10 melanoma cells.*

HO CO$_2$H HO F F O — Tafluprost acid (AFP-172)

HO CO$_2$H HO F F O — 1,2-dinor-AFP-172

HO CO$_2$H HO F F O — 1,2,3,4-tetranor-AFP-172

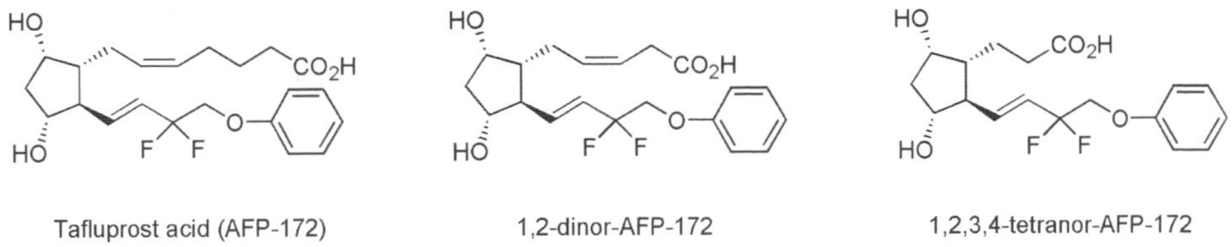

Figure 2.7 *Metabolites of tafluprost.*

dinor-AFP-172, and 1,2,3,4-tetranor-AFP-172, accounted for most of the radioactivity in the aqueous humor and other ocular tissues. AFP-172 was demonstrated to be the most abundant and the only pharmacologically active metabolite in ocular tissues. A small amount of tafluprost was detected in the ciliary body, cornea, and iris.

2.4 Conclusion

A novel 15,15-difluorinated prostanoid, tafluprost was discovered as a highly potent and selective prostanoid FP receptor agonist. Tafluprost demonstrates powerful and prolonged IOP-lowering effects in animal models. The maximal IOP reduction achieved with tafluprost was greater than that with latanoprost in both normotensive and glaucomatous monkeys. Tafluprost showed significantly increasing efficacy of ocular blood flow and protective effect on glutamate-induced cytotoxicity. In its pharmacological characteristics, tafluprost may be superior to latanoprost: potent IOP-reducing efficacy, effective increase in ocular blood flow, and weak melanogenetic side-effect. Tafluprost has completed clinical trials, and new drug applications for tafluprost have been filed in Japan and the EU. Tafluprost is expected to become a new-generation prostanoid FP agonist that strongly reduces IOP and effectively improves ocular circulation in patients with glaucoma.

Acknowledgments

The author thanks the collaborators at Research & Development Center, Santen Pharmaceutical Co., Ltd. for pharmacological studies and fruitful discussions in the development of tafluprost. The author is also grateful to coworkers at Asahi Glass Co., Ltd. whose names are shown in published papers for their essential contributions.

References

1. Vane, J. R. and O'Grady, J. (1993) *Therapeutic Applications of Prostaglandins*, Edward Arnold, London.
2. Coleman, R. A., Kennedy, I., Humphery, P. P. A., *et al.* (1990) *Prostanoids and their receptors*. In: *Comprehensive Medicinal Chemistry*, vol. **3** (ed. J. C. Emmett)–, Pergamon, Oxford, pp. 643–714.
3. (a) Coleman, R. A., Smith, W. L. and Narumiya, S. (1994) Classification of prostanoid receptors: Properties, distribution, and structure of the receptors and their subtypes. *Pharmacological Reviews*, **46**, 205–229; (b) Narumiya, S., Sugimoto, Y. and Ushikubi, F. (1999) Prostanoid receptors: Structures, properties, and functions. *Physiological Reviews*, **79**, 1193–1226; (c) Hata, A. N. and Breyer, R. M. (2004) Pharmacology and signaling of prostaglandin receptors: Multiple roles in inflammation and immune modulation. *Pharmacology & Therapeutics*, **103**, 147–166; (d) Murota, S. and Yamamoto, S. (2001) *New Development of Prostaglandin Research*, vol.**38**, Tokyo Kagaku Dojin, Tokyo.

4. [A second PGD_2 receptor having different functions, chemoattractant receptor-homologous molecule expressed on T helper (Th) 2 cells (CRTH2), was also identified recently.] (a) Hirai, H., Tanaka, K., Yoshie, *et al.* (2001) Prostaglandin D_2 selectively induces chemotaxis in T helper type 2 cells, eosinophils, and basophils via seven-transmembrane receptor CRTH2. *Journal of Experimental Medicine*, **193**, 255–261; (b) Nagata, K., Tanaka, K., Ogawa, K., *et al.* (1999) Selective expression of a novel surface molecule by human Th2 cells *in vivo. Journal of Immunology*, **162**, 1278–1286.
5. (a) Woodward, D. F., Regan, J. W., Lake, S. and Ocklind, A. (1997) Molecular biology and ocular distribution of prostanoid receptors. *Survey of Ophthalmology.*, **41** Suppl 2, S15–21; (b) Ocklind, A., Lake, S., Wentzel, P., *et al.* (1996) Localization of the prostaglandin F2 alpha receptor messenger RNA and protein in the cynomolgus monkey eye. *Investigative Ophthalmology & Visual Science*, **37**, 716–726.
6. (a) Schlötzer-Schrehardt, U., Zenkel, M. and Nüsing, R. M. (2002) Expression and localization of FP and EP prostanoid receptor subtypes in human ocular tissues. *Investigative Ophthalmology & Visual Science*, **43**, 1475–1487; (b) Anthony, T. L., Pierce, K. L., Stamer, W. D. and Regan, J. W. (1998) Prostaglandin $F_{2\alpha}$ receptors in the human trabecular meshwork. *Investigative Ophthalmology & Visual Science*, **39**: 315–321; (c) Mukhopadhyay, P., Bian, L., Yin, H., *et al.* (2001) Localization of EP_1 and FP receptors in human ocular tissues by in situ hybridization. *Investigative Ophthalmology & Visual Science*, **42**: 424–428.
7. Kamphuis, W., Schneemann, A. van Beek, L. M., *et al.* (2001) Prostanoid receptor gene expression profile in human trabecular meshwork: A quantitative real-time PCR approach. *Investigative Ophthalmology & Visual Science*, **42**, 3209–3215.
8. (a) Filler, R. and Kobayashi, Y. (1982) *Biomedicinal Aspects of Fluorine Chemistry*, Kodansya and Elsevier Biomedical, Tokyo and Amsterdam; (b) Welch, J. T. and Eswarakrishman, S. (1991) *Fluorine in Bioorganic Chemistry*, John Wiley & Sons, New York; (c) Filler, R., Kobayashi, Y. and Yagupolskii, L. M. (1993) *Organofluorine Compounds in Medicinal Chemistry and Biomedical Applications*, Elsevier, Amsterdam; (d) Kukhar, V. P. and Soloshonok, V. A. (1995) *Fluorine-containing Amino Acids: Synthesis and Properties*, John Wiley & Sons Ltd, Chichester; (e) Ojima, I., McCarthy, J. R. and Welch, J. T. (1996) *Biomedical Frontiers of Fluorine Chemistry*, American Chemical Society, Washington, D.C.; (f) Hiyama, T. (2000) *Organofluorine Compounds, Chemistry and Applications*, Springer, Berlin; (g) Kirsch, P. (2004) *Modern Fluoroorganic Chemistry*, Wiley-VCH Verlag GmbH, Weinheim; (h) Uneyama, K. (2006) *Organofluorine Chemistry*, Blackwell, Oxford; (i) Begue, J-P. and Bonnet-Delpon, D. (2006) Recent advances (1995–2005) in fluorinated pharmaceuticals based on natural products. *Journal of Fluorine Chemistry*, **127**, 992–1012; (j) Kirk, K. L. (2006) Fluorine in medicinal chemistry: Recent therapeutic applications of fluorinated small molecules. *Journal of Fluorine Chemistry*, **127**, 1013–1029.
9. (a) Roberts, S. M. and Scheimann, F. (1982) *New Synthetic Routes to Prostaglandins and Thromboxanes*, Academic Press, London; (b) Collins, P. W. and Djuric, S. W. (1993) Synthesis of therapeutically useful prostaglandin and prostacyclin analogs. *Chemical Reviews*, **93**, 1533–1564; (c) Das, S., Chandrasekhar, S., Yadav, J. S. and Grée, R. (2007) Recent developments in the synthesis of prostaglandins and analogues. *Chemical Reviews*, **107**, 3286–3337.
10. (a) Barnette, W. E. (1984) The synthesis and biology of fluorinated prostacyclins. *CRC Critical Rev. Biochem.*, **15**, 201–235; (b) Yasuda, A. The fluoroarachidonic acid cascade, in Ref [8c], pp. 275–307; (c) Matsumura, Y. (2005) Recent developments in fluorinated prostanoids *Journal of Synthetic Organic Chemistry, Japan*, **63**, 40–50; (d) Matsumura, Y. (2008) Synthesis and pharmacological properties of fluorinated prostanoids, in *Fluorine and Health: Molecular Imaging, Biomedical Materials and Pharmaceuticals* (eds A. Tressaud and G. Haufe), Elsevier, Amsterdam, pp. 623–659.

11. Dukes, M., Russel, W. and Walpole, A. L. (1974) Potent luteolytic agents related to prostaglandin $F_{2\alpha}$ *Nature*, **250**, 330–331.
12. Magerlein, B. J. and Miller, W. L. (1975) 16-Fluoroprostaglandins. *Prostaglandins*, **9**, 527–530.
13. Fried, J., Mitra, M., Nagarajan, M. and Mehrotra, M. M. (1980) 10,10-Difluoro-13-dehydroprostacyclin: a chemically and metabolically stabilized potent prostacyclin. *Journal of Medicinal Chemistry*, **23**, 234–237.
14. (a) Morinelli, T. A., Okwu, A. K., Mais, *et al.* (1989) Difluorothromboxane A2 and stereoisomers: stable derivatives of thromboxane A2 with differential effects on platelets and blood vessels. *Proceedings of the National Academy of Sciences of the USA*, **86**, 5600–5604; (b) Fried, J., John, V., Szwedo, M. J., *et al.* (1989) Synthesis of 10,10-difluorothromboxane A2, a potent and chemically stable thromboxane agonist. *Journal of the American Chemical Society*, **111**, 4510–4511.
15. (a) Nakano, T., Makino, M., Morizawa, Y. and Matsumura, Y. (1996) Synthesis of novel difluoroprostacyclin derivatives: unprecedented stabilizing effect of fluorine substituents. *Angewandte Chemie, International Edition English*, **35**, 1019–1021; (b) Matsumura, Y., Nakano, T., Mori, N. and Morizawa, Y. (2004) Synthesis and biological properties of novel fluoroprostaglandin derivatives: highly selective and potent agonists for prostaglandin receptors. *Chimia*, **58**, 148–152; (c) Matsumura, Y., Nakano, T., Asai, T. and Morizawa, Y. Synthesis and properties of novel fluoroprostacyclins. Potent and stable prostacyclin agonists, in Ref. [8e], pp. 83–94.
16. Chan, C-S., Negishi, M., Nakano, T., *et al.* (1997) 7,7-Difluoroprostacyclin derivative, AFP-07, a highly selective and potent agonist for the prostacyclin receptor. *Prostaglandins*, **53**, 83–90.
17. (a) Iwase, A., Suzuki, Y., Araie, M., *et al.* (2004) The prevalence of primary open-angle glaucoma in Japanese. 1. The Tajimi Study. *Ophthalmology*, **111**, 1641–1648; (b) Yamamoto, T., Iwase, A., Araie, M., *et al.* (2005) The Tajimi Study report 2: prevalence of primary angle closure and secondary glaucoma in a Japanese population. *Ophthalmology*, **112**, 1661–1669.
18. Collaborative Normal-Tension Glaucoma Study Group, (1998) Comparison of glaucomatous progression between untreated patients with normal-tension glaucoma and patients with therapeutically reduced intraocular pressures. *American Journal of Ophthalmology*. **126**, 487–497.
19. (a) Camras, C. B. and Bito, L. Z. (1981) Reduction of intraocular pressure in normal and glaucomatous primate (*Aotus trivirgatus*) eyes by topically applied prostaglandin $F_{2\alpha}$. *Current Eye Research*, **1**, 205–209; (b) Bito, L. Z. (1984) Comparison of the ocular hypotensive efficacy of eicosanoids and related compounds. *Experimental Eye Research*, **40**, 181–194.
20. (a) Stjernschantz, J. and Resul, B. (1992) Phenyl substituted prostaglandin analogs for glaucoma treatment. *Drugs of the Future*, **17**, 691–704; (b) Ishida, N., Odani-Kawabata, N., Shiazaki, A. and Hara, H. (2006) Prostanoids in the therapy of glaucoma. *Cardiovascular Drug Reviews*, **24**, 1–10.
21. Sakurai, M., Araie, M., Oshika, T., *et al.* (1991) Effects of topical application of UF-021, a novel prostaglandin derivative, on aqueous humor dynamics in normal human eyes. *Japanese Journal of Ophthalmology.*, **35**, 156–165.
22. Resul, B., Stjernschantz, J., No, K., *et al.* (1993) Phenyl-substituted prostaglandins: Potent and selective antiglaucoma agents. *Journal of Medicinal Chemistry*, **36**, 243–248.
23. (a) Wand, M., Ritch, R., Isbey, E. K. J. and Zimmerman, T. J. (2001) Latanoprost and periocular skin color changes. *Archives of Ophthalmology*, **119**, 614–615; (b) Woodward, D. F., Krauss, A. H., Chen, J., *et al.* (2003) Pharmacological characterization of a novel antiglaucoma agent, bimatoprost (AGN 192024). *The Journal of Pharmacology and Experimental Therapeutics* **305**, 772–785.
24. (a) Netland, P. A., Landry, T., Sullilvan, E. K., *et al.* (2001) Travoprost compared with latanoprost and timolol in patients with open-angle glaucoma or ocular hypertension. *American Journal of Ophthalmology*, **132**, 472–484; (b) Hellberg, M. R., Sallee, V. L., Mclaughlin, M.

A., *et al.* (2001) Preclinical efficacy of travoprost, a potent and selective FP prostaglandin receptor agonist. *Journal of Ocular Pharmacology & therapeutics* **17**, 421–432.

25. Boulton, L. T., Brick, D., Fox, M. E., *et al.* (2002) Synthesis of the potent antiglaucoma agent, travoprost. *Organic Process Research & Development*, **6**, 138–145.
26. (a) Gabelt, B. T. and Kaufman, P. L. (1989) Prostaglandin $F_{2\alpha}$ increases uveoscleral outflow in the cynomolgus monkey. *Experimental Eye Res* **49**, 389–402; (b) Stjernschantz, J., Selén, G., Sjöquist, B. and Resul, B. (1995) Preclinical pharmacology of latanoprost, a phenyl-substituted $PGF_{2\alpha}$ analogue. *Advances in Prostaglandin, Thromboxane, and Leukotriene Research*, **23**, 513–518.
27. (a) Wand, M., Ritch, R., Isbey, E. K. J. and Zimmerman, T. J. (2001) Latanoprost and periocular skin color changes. *Archlves of Ophthalmology*, **119**, 614–615; (b) Alexander, C. L., Miller, S. J. and Abel, S. R. (2002) Prostaglandin analog treatment of glaucoma and ocular hypertension. *Annals of Pharmacotherapy*, **36**, 504–511.
28. Takasuka, M., Kishi, M. and Yamakawa, M. (1994) FTIR spectral study of intramolecular hydrogen bonding in thromboxane A2 receptor agonist (U-46619), prostaglandin $(PG)E_2$, PGD_2, $PGF_{2\alpha}$, prostacyclin receptor agonist (carbacyclin) and their related compounds in dilute carbon tetrachloride solution: structure-activity relationships. *Journal of Medicinal Chemistry*, **37**, 47–56.
29. [A report of PGF derivatives on ligand docking experiments to a human FP receptor model] Wang, Y., Wos, J. A., Dirr, M. J., *et al.* (2000) Design and synthesis of 13,14-dihydro prostaglandin $F_{1\alpha}$ analogues as potent and selective ligands for the human FP receptor. *Journal of Medicinal Chemistry*, **43**, 945–952.
30. (a) [Synthesis of 15-deoxy-15-fluoro-$PGF_{2\alpha}$] Bezuglov, V. V. and Bergelson, L. D. (1980) Synthesis of fluoroprostacyclins. *Doklady Akademii Nauk SSSR*, **250**, 468–469; (b) [Recently, the ocular hypotensive effects of 15-deoxy-15-fluoro-$PGF_{2\alpha}$ derivatives have been independently reported] Klimko, P., Hellberg, M., McLaughlin, M., *et al.* (2004) 15-Fluoro prostaglandin FP agonists: a new class of topical ocular hypotensives. *Bioorg. Med. Chem.*, **12**, 3451–3469.
31. Resul, B. and Stjernschantz, J. (1993) Structure-activity relationships of prostaglandin analogues as ocular hypotensive agents. *Expert Opinion on Therapeutic Patents*, **3**, 781–795.
32. (a) Howard, J. A. K., Hoy, V. J., O'Hagen, D. and Smith, G. T. (1996) How good is fluorine as a hydrogen bond acceptor? *Tetrahedron*, **52**, 12613–12622; (b) Dunitz, J. D. and Taylor, R. (1997) Organic fluorine hardly ever accepts hydrogen bonds. *Chemistry A European Journal*, **3**, 89–98; (c) Ref. [8h], Chapter 4.
33. Matsumura, Y., Mori, N., Nakano, T., *et al.* (2004) Synthesis of the highly potent prostanoid FP receptor agonist, AFP-168: a novel 15-deoxy-15,15-difluoroprostaglandin $F_{2\alpha}$ derivative. *Tetrahedron Letters*, **45**, 1527–1529.
34. Nakajima, T., Matsugi, T., Goto, W., *et al.* (2003) New fluoroprostaglandin $F_{2\alpha}$ derivatives with prostanoid FP-receptor agonistic activity as potent ocular-hypotensive agents. *Biological & Pharmaceutical Bulletin*, **26**, 1691–1695.
35. Takagi, Y., Nakajima, T., Shimazaki, A., *et al.* (2004) Pharmacological characteristics of AFP-168 (tafluprost), a new prostanoid FP receptor agonist, as an ocular hypotensive drug. *Experimental; Eye Research*, **78**, 767–776.
36. Sharif, N. A., Kelly, C. R., Crider, J. Y., *et al.* (2003) Ocular hypotensive FP prostaglandin (PG) analogs: PG receptor subtype binding affinities and selectivities, and agonist potencies at FP and other PG receptors in cultured cells. *Journal of Ocular Pharmacology & Therapeutics* **19**, 501–515.
37. (a) Brubaker, R. F., Schoff, E. O., Nau, C. B., *et al.* (2001) Effects of AGN 192024, a new ocular hypotensive agent, on aqueous dynamics. *American Journal of Ophthalmology* **131**, 19–24; (b) Brubaker, R. F., Schoff, E. O., Nau, C. B., *et al.* (2005) Effects of travoprost on aqueous humor dynamics in monkeys. *Journal of Glaucoma*, **14**, 70–73.

38. Ota, T., Murata, H., Sugimoto, E., *et al.* (2005) Prostaglandin analogues and mouse intraocular pressure: effects of tafluprost, latanoprost, travoprost, and unoprostone, considering 24-hour variation. *Investigative Ophthalmology & Visual Science.* **46**, 2006–2011.
39. Ota, T., Aihara, M., Saeki, T., *et al.* (2007) The IOP-lowering effects and mechanism of action of tafluprost in prostanoid receptor-deficient mice. *British Journal of Ophthalmolology* **91**, 673–676.
40. (a) Ota, T., Aihara, M., Narumiya, S. and Araie, M. (2005) The effects of prostaglandin analogues on IOP in prostanoid FP-receptor-deficient mice. *Investigative Ophthalmology & Visual Science.* **46**, 4159–4163; (b) Ota, T., Aihara, M., Saeki, T., *et al.* (2006) The effects of prostaglandin analogues on prostanoid EP1, EP2, and EP3 receptor-deficient mice. *Investigative Ophthalmology & Visual Science.* **47**, 3395–3399.
41. Izumi, N., Nagaoka, T., Sato, E., *et al.* (2004) DE–085 increases retinal blood flow. *Investigative Ophthalmology & Visual Science.* **45**, Abst. 2340.
42. Akaishi, T., Kurashima, H., Odani, N., *et al.* (2007) Comparison of the effects of tafluprost, latanoprost and travoprost on the optic nerve head blood flow in rabbits. *Investigative Ophthalmology & Visual Science.* **48**, Abst. 2278.
43. Odani, N., Seike, H., Kurashima, H., *et al.* (2007) Protective effect of tafluprost on glutamate-induced cytotoxicity. *World Glaucoma Congress*, Singapore, 2007, Geneva Medical publishers, Geneva, Abst. P327, p.184.
44. Ferrer, E. and Bozzo, J. (2006) Understanding optic nerve degradation: key to neuroprotection inglaucoma. *Drugs of the Future*, **31**, 355–363.
45. Bergh, K., Wentzel, P. and Stjernschantz, J. (2002) Production of prostaglandin E_2 by iridial melanocytes exposed to latanoprost acid, a prostaglandin $F_{2\alpha}$ analogue. *Journal of Ocular Pharmacology & therapeutics*, **18**, 391–400.
46. Kawazu, K. and Fukano, Y. (2006) Distribution and metabolism of [^{3}H]tafluprost in ocular tissues following administration of a single ocular dose to *Cynomolgus* monkey. *Investigative Ophthalmology & Visual Science.* **47**, Abst. 5102.

3

Fluorinated Conformationally Restricted Glutamate Analogues for CNS Drug Discovery and Development

Atsuro Nakazato

3.1 Introduction

(1*S*,2*R*,5*R*,6*S*)-2-Aminobicyclo[3.1.0]hexane-2,6-dicarboxylic acid (**4**; see Figure 3.2), a conformationally restricted glutamate analogue, is a selective group II metabotropic glutamate receptor (mGluR2/3) agonist. Orally active mGluR2/3 agonists and antagonists have been successfully discovered by the introduction of fluorine atoms onto the bicyclo[3.1.0]hexane ring of **4**. (1*S*,2*S*,3*S*,5*R*,6*S*)-2-Amino-3-fluorobicyclo[3.1.0]hexane-2,6-dicarboxylic acid ((+)-**6a**, MGS0008), (1*R*,2*S*,5*R*,6*R*)-2-amino-6-fluorobicyclo[3.1.0] hexane-2,6-dicarboxylic acid ((−)-**7a**, MGS0022) and (1*R*,2*S*,5*S*,6*S*)-2-amino-6-fluoro-4-oxobicyclo[3.1.0]hexane-2,6-dicarboxylic acid monohydrate ((+)-**7b**, MGS0028) are typical fluorinated mGluR2/3 agonists. Among these agonists, the orally administered compounds (+)-**6a** (MGS0008) and (+)-**7b** (MGS0028), in particular, show potent antipsychotic-like effects in laboratory animals. (1*R*,2*R*,3*R*,5*R*,6*R*)-2-Amino-3-(3,4-dichlorobenzyloxy)-6-fluorobicyclo[3.1.0]hexane-2,6-dicarboxylic acid (**14a**, MGS0039) is a typical fluorinated mGluR2/3 antagonist, and its *n*-heptyl 6-carboxylate **15a**, a compound that was designed by utilizing the chemical characteristics of α-fluoro carboxylic acid, is the most effective prodrug for **14a** (MGS0039).

In this chapter, the design strategy, synthesis, pharmacology, and pharmacokinetics of fluorinated conformationally restricted glutamate analogues are described in comparison with those of the corresponding hydrocarbon-based compounds.

Fluorine in Medicinal Chemistry and Chemical Biology Edited by Iwao Ojima

3.2 Metabotropic Glutamate Receptors

L-Glutamate (**2**) acts as a neurotransmitter at the vast majority of excitatory synapses in the brain. The normal functioning of glutamatergic synapses is required for all major brain functions [1, 2]. Glutamate receptors are broadly classified into two types: ionotropic glutamate receptors (iGluRs), in which the receptors have an ion channel structure, and metabotropic glutamate receptors (mGluRs), which are coupled to G-proteins. mGluRs are classified into eight subtypes, identified as subtypes 1 through 8, which are further classified into three groups (I–III) on the basis of sequence homology, signal transduction mechanisms, and pharmacology [3–9]. Group I mGluRs (mGluR1/5) are positively coupled to phospholipase C, and their activation produces phosphoinositide turnover and diacylglycerol within target neurons. In contrast, both group II mGluRs (mGluR2/3) and group III mGluRs (mGluR4/6/7/8) are located in glutamatergic terminals and are negatively coupled to the activity of adenyl cyclase (see Figure 3.1) [9–11].

mGluRs have been implicated in the pathology of major psychiatric disorders such as depression, anxiety, and schizophrenia [12] because of their critical role as modulators of synaptic transmission, ion channel activity, and synaptic plasticity [7]. Indeed, the efficacy of mGluR2/3 agonists in animal models and in clinical trials suggests that, by inhibiting the glutamatergic system, agonists of mGluR2/3 may be useful for the treatment of many diseases and conditions including schizophrenia [3–17], anxiety [13, 14, 18–21], and panic disorder [22].

In contrast, clarification of the efficacy of mGluR2/3 antagonists in animal models has taken longer than that of mGluR2/3 agonists. This delay might be due to the lack of potent and selective mGluR2/3 antagonists capable of penetrating the blood–brain barrier (BBB). The antidepressant-like and anxiolytic effects of mGluR2/3 antagonists, **14a** (MGS0039) [23–31] and (2*S*)-amino-2-[(1*S*,2*S*)-2-carboxycycloprop-1-yl]-3-(9-xanthyl)propionic acid (**11**, LY341495) [32, 33], have been reported. Compound **14a** (MGS0039) exerted dose-dependent antidepressant-like effects in the rat forced-swim test

Ionotropic glutamate receptors (iGluRs)

- NMDA
- AMPA
- Kainate

Metabotropic glutamate receptors (mGluRs)

- Group I { mGluR1, mGluR5 } - Positively coupled to phospholipase C
- Group II { mGluR2, mGluR3 }
- Group III { mGluR4, mGluR6, mGluR7, mGluR8 }

(Groups II and III:) - Located in glutamatergic terminals - Negatively coupled to the activity of adenyl cyclase

Figure 3.1 Glutamate receptors (GluRs).

and the mouse tail suspension test, but had no apparent effects on classical models of anxiety, such as the rat social interaction test and the rat elevated plus-maze (0.3–3 mg/kg, i.p.). However, compound **14a** (MGS0039) showed dose-dependent anxiolytic-like activities when evaluated using the conditioned fear stress test in rats, the marble-burying behavior test in mice, and the stress-induced hyperthermia test in single housed mice. These findings suggest that mGluR2/3 antagonists may be useful for the treatment of depression and anxiety.

3.3 mGluR2/3 Agonists and Antagonists

Typical mGluR2/3 antagonists that have been derived from the corresponding agonists are shown in Figure 3.2. The 2-methyl analogue (**8**, MAP-4) of agonist **1** is a weak but selective antagonist for group III mGluRs [34]. A potent and relatively selective antagonist for mGluR2/3, compound **9**, was obtained by introducing a 4,4-diphenybutyl group at the C-4 position of glutamic acid (**2**). Furthermore, the introduction of a phenyl group to the C-2′ position of **3** or a 9-xanthylmethyl group at the C-2 position of **3** yielded moderately potent and very potent mGluR2/3 antagonists **10** [35] and **11** (LY341495) [32, 33], respectively. Compound **11** (LY341495) binds to rat brain mGluRs with a very high affinity (IC_{50} = 2.9 nM) and is a very potent functional antagonist (IC_{50} = 23 nM and 10 nM for human mGluR2 and mGluR3, respectively). Additionally, the intraperitoneal administration of **11** (LY341495) resulted in good plasma and acceptable brain concentrations, but the oral bioavailability of **11** (LY341495) was low (<5%) [33]. A selective mGluR2/3 agonist **4** (LY354740) exhibits oral activity [21] and has been chemically modified to discover better drug candidates. Among these compounds, compound **5** (LY404039) is a typical agonist for mGlu2/3 receptors with a good pharmacokinetic profile in laboratory animals [14]. In contrast, the introduction of a hydroxyl group or a methyl group at the C-3 position of a typical group II mGluR agonist **4** (LY354740) yielded mGluR2/3 antagonists **12a** (Ro 65–3479) [36] and **12b** [37], respectively.

On the basis of the chemical properties of fluorine, the introduction of a fluorine atom into compound **4** (LY354740) was an interesting endeavor, resulting in compounds **6** and **7**. Fluorine is not a sterically demanding substituent, since its small van der Waals radius resembles that of hydrogen. In molecules where conformational recognition is important, minimal steric disturbance by a substituent is especially critical. The electronegativity of fluorine can have pronounced effects on the electron density within molecules, affecting the basicity or acidity of neighboring groups, dipole moments within molecules, and the overall reactivity and stability of neighboring functional groups. Once introduced, the high carbon–fluorine bond energy renders the substituent relatively resistant to metabolic transformation [38]. Given the nature of fluorine, the C-3 or C-6 position of the conformationally restricted glutamate analogue **4** (LY354740) – an orally active, potent, and selective mGluR2/3 agonist – was selected for the introduction of fluorine, since its introduction was expected to influence the functional groups (carboxylic acid and the amino group at the C-2 position or the carboxylic acid at the C-6 position) in a direct manner. Fluoro-compounds **6a** (MGS0008) and **7a** (MGS0022) exhibited the best inhibitory effects on cAMP formation, with the same EC_{50} values as that of compound **4** (LY354740). In

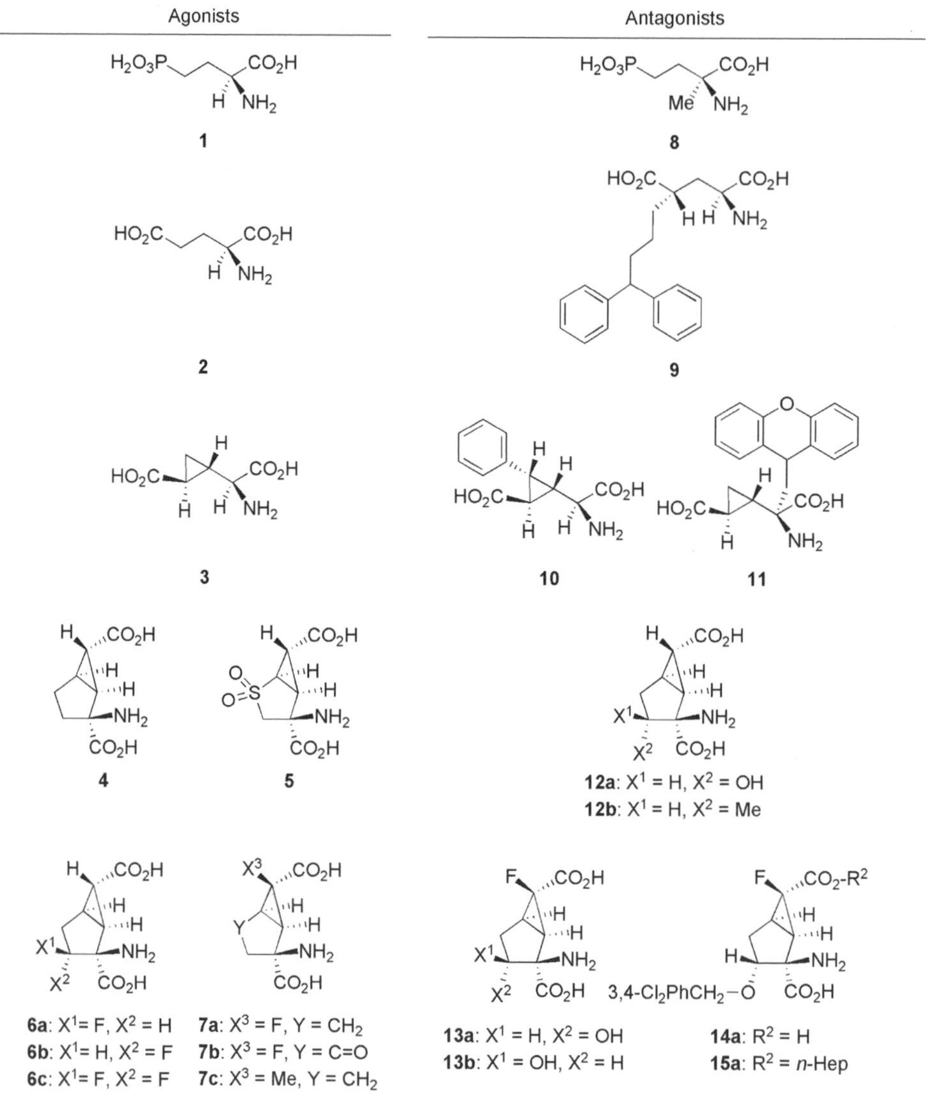

Figure 3.2 *Group II mGluR agonists and antagonists.*

addition, **6a** (MGS0008) had a higher oral activity in laboratory animals than compound **4** (LY354740) [15]. Moreover, chemical modification based on the replacement of the CH_2 group at the C-4 position of compound (−)-**7a** (MGS0022) with a carbonyl group produced (+)-**7b** (MGS0028), which displayed the best mGluR2/3 agonist properties in both *in vitro* and *in vivo* pharmacological profiles [15]. In contrast, the introduction of an alkoxy group at the C-3 position of (−)-**7a** (MGS0022) yielded a potent and selective mGluR2/3 antagonist, **14a** (MGS0039) [25]. The antagonist **14a** (MGS0039) showed acceptable BBB penetration, and its low oral bioavailability in laboratory animals has been improved by the discovery of prodrug **15a** (MGS0210), which was designed by applying the characteristics of α-fluoro carboxylic acid [39, 40].

3.4 mGluR2/3 Agonists (+)-6a (MGS0008), (–)-7a (MGS0022), and (+)-7b (MGS0028)

In this section, the original syntheses and pharmacological profiles of racemic and optically active compounds **6a**, **6b**, **6c**, **7a,** and **7b** are presented [15]. The process chemistry for (+)-**6a** (MGS0008) [41] and (+)-**7b** (MGS0028) [42–45] is also described. Structures in the following schemes illustrate the absolute configurations corresponding to (+)-**6a** (MGS0008) and (+)-**7b** (MGS0028) for simplicity, Thus, only one structure is shown for racemic compounds and the structures of enantiomers corresponding to (–)-**6a** and (–)-**7b** are not depicted.

3.4.1 Synthesis of (+)-6a (MGS0008)

Racemic compounds (±)-**6a**, (±)-**6b**, and (±)-**6c**, and optically pure compounds (+)-**6a** (MGS0008) and (–)-**6a**, were originally synthesized via racemic fluorinated intermediates, (±)-**17a-1**, (±)-**17b**, and (±)-**17c**, respectively, which were synthesized by the fluorination of the racemic compound (±)-**16** (see Scheme 3.1) [15].

Racemic compound (±)-**16** was reacted with trimethylsilyl chloride (TMSCl) and lithium hexamethyldisilazide (LHMDS) to give a silyl enolate, which was electrophilically fluorinated with *N*-fluorobenzenesulfonamide (NFSI) [46, 47] to yield a mixture of monofluoro compounds (59% yield), (±)-**17a-1** and (±)-**17b**, as well as the difluoro compound (±)-**17c** (25% yield)after column chromatography on silica gel. The mixture of (±)-**17a-1** and (±)-**17b** was subjected to the Bucherer–Bergs conditions (ammonium carbamate and potassium cyanide) to yield β-fluorohydantoin (±)-**18a-1** (8% yield) and α-fluorohydantoin (±)-**18b** (6% yield), respectively, after column chromatography on silica gel and recrystallization. In a similar manner, difluorohydantoin (±)-**18c** was prepared from compound (±)-**17c** under the Bucherer–Bergs conditions (47% yield). Hydantoins (±)-**18a-1**, (±)-**18b**, and (±)-**18c** were hydrolyzed under acidic or basic conditions to yield the corresponding amino acids, (±)-**6a** (25% yield), (±)-**6b** (21% yield), and (±)-**6c** (56% yield), respectively (see Scheme 3.1).

Optically pure (+)-**6a** (MGS0008) (64% yield) and its enantiomer (–)-**6b** (73% yield) were produced by acidic hydrolysis of carboxylic acids (+)-**19** and (–)-**19**, respectively. Compounds (+)-**19** and (–)-**19** were obtained through selective saponification of the ester moiety of (±)-**18** under basic conditions (93% yield), followed by optical resolution of the resulting (±)-**19** with (*R*)- and (*S*)-1-phenylethylamine [(+)-**19**: 39% yield, (–)-**19**: 36% yield], respectively (see Scheme 3.1).

Next, the process chemistry for a large-scale synthesis of (+)-**6a** (MGS0008) is presented (see Scheme 3.2) [41]. The reaction of racemic (±)-**16** with TMSCl and LHMDS, followed by dehydrosilylation with palladium acetate afforded enone (±)-**20** (90% yield), which was stereoselectively epoxidized using *tert*-butyl hydroperoxide (TBHP) in the

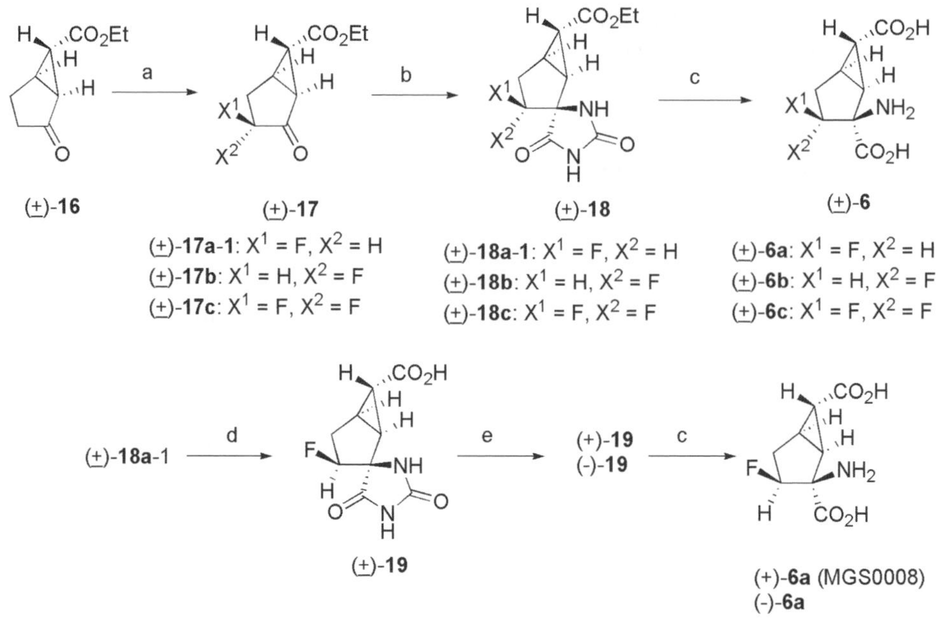

(±)-**16** →(a) (±)-**17** →(b) (±)-**18** →(c) (±)-**6**

(±)-**17a-1**: X^1 = F, X^2 = H
(±)-**17b**: X^1 = H, X^2 = F
(±)-**17c**: X^1 = F, X^2 = F

(±)-**18a-1**: X^1 = F, X^2 = H
(±)-**18b**: X^1 = H, X^2 = F
(±)-**18c**: X^1 = F, X^2 = F

(±)-**6a**: X^1 = F, X^2 = H
(±)-**6b**: X^1 = H, X^2 = F
(±)-**6c**: X^1 = F, X^2 = F

(±)-**18a**-1 →(d) (±)-**19** →(e) (+)-**19**, (-)-**19** →(c) (+)-**6a** (MGS0008), (-)-**6a**

Scheme 3.1 *Synthesis of (+)-**6a** (MGS0008) and its congeners (1).* Reagents and conditions: *(a) (i) LHMDS, TMSCl, THF, (ii) $(PhSO_2)_2NF$, CH_2Cl_2; (b) $(NH_4)_2CO_3$, KCN, EtOH, H_2O; (c) 60% aq. H_2SO_4, 2.5 or 3.0 M aq. NaOH; (d) 2 M aq. NaOH; (e) (R)-PhCH(NH_2)Me, acetone, H_2O and then 1 M aq. HCl.*

presence of Triton B to yield epoxide (±)-**21-1**. Ethyl ester (±)-**21-1** was transformed to 2,4-dimethyl-3-pentyl ester (±)-**21-2** in two steps to avoid hydrolysis and transesterification of the ethyl ester under the fluorination conditions. Thus, (±)-**21-1** was hydrolyzed under basic conditions (92% yield), followed by esterification with 2,4-dimethyl-3-pentanol in the presence of dicyclohexylcarbodiimide (DCC) to give (±)-**21-2**, which was used in the next step without purification. The nucleophilic fluorination of epoxide (±)-**21-2** with potassium hydrogen difluoride [48] in ethylene glycol yielded fluoroenone (±)-**22**. The hydrogenation of (±)-**22** over palladium on carbon proceeded in a highly stereoselective manner to give (±)-**17a-2** (48% yield for three steps). Fluoroketone (±)-**17a-2** was converted to optically pure (+)-**6a** in a four-step process: (i) formation of hydantoin (±)-**18a-2** under the Bucherer–Bergs conditions (75% yield); (ii) selective hydrolysis of the ester moiety with 48% aqueous HBr to give (±)-**19a-2** (82% yield); (iii) optical resolution of (±)-**19a-2** with (*R*)-1-phenylethylamine afforded optically pure hydantoin (+)-**19a-2** (48% yield, >99% ee); and (iv) hydrolysis of (+)-**19a-2** under acidic conditions to give (+)-**6a** (MGS0008) (74% yield).

3.4.2 Synthesis of (–)-7a (MGS0022)

Racemic (±)-**7a** and optically pure (+)-**7a** and (–)-**7a** (MGS0022) were originally synthesized from fluorobicyclo[3.1.0]cyclohexanecarboxylates (±)-**25**, (+)-**25**, and (–)-**25**,

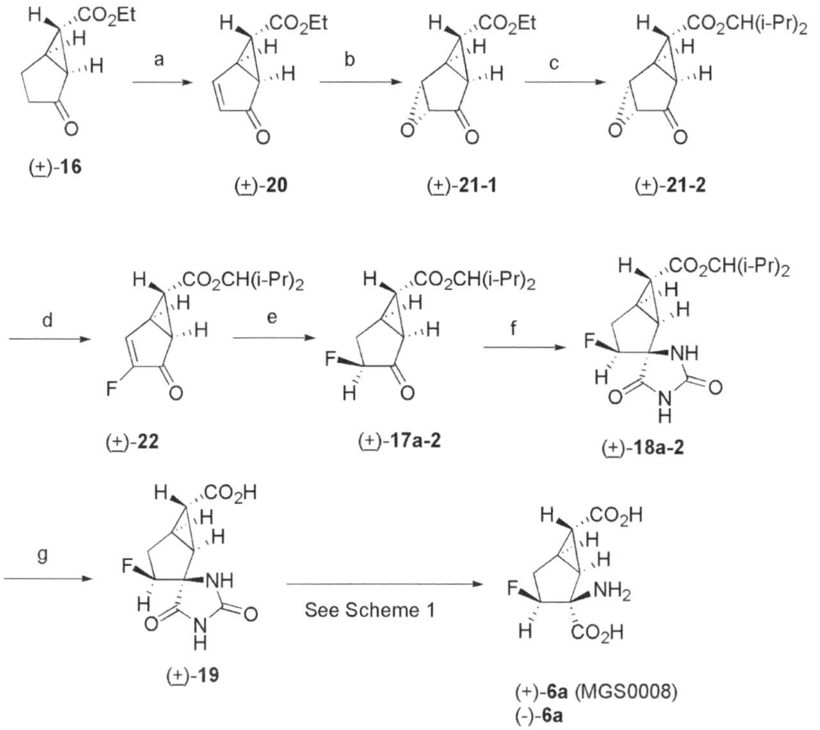

Scheme 3.2 *Synthesis of (+)-***6a** *(MGS0008) and its congeners (2).* Reagents and conditions: *(a) (i) LHMDS, TMSCl, TFA, (ii) Pd(OAc)$_2$, MeCN; (b) TBHP, Triton B, PhMe; (c) (i) 2 M aq. NaOH, (ii) (i-Pr)$_2$CHOH, DCC, DMAP, CHCl$_3$; (d) KF-HF, ethylene glycol, (e) H$_2$, 5% Pd/C, EtOH; (f) (NH$_4$)$_2$CO$_3$, KCN, EtOH, H$_2$O; (g) 48% HBr.*

respectively, through (i) hydrolysis of the ethyl ester moieties of (±)-**25**, (+)-**25**, and (−)-**25**, (ii) formation of hydantoins under Bucherer–Bergs conditions (99%, 88%, and 90% yields (two steps), respectively), and (iii) hydrolysis of hydantoins (±)-**25** (34% yield), (+)-**25** (72% yield), and (−)-**25** (73% yield), under acidic conditions (see Scheme 3.2) [15]. The key intermediate (±)-**25** was synthesized by Cu(TBS)-catalyzed (TBS = *N-tert*-butylsalicylaldimine) intramolecular cyclopropanation of diazoketo-fluoroalkenoate **24**, which was prepared from **23** through reaction with oxalyl chloride and then diazomethane (27% total yield from **23** to **25**) [49, 50]. Effective optical resolution of (±)-**25** was achieved using chiral HPLC. Enantioselective intramolecular cyclopropanation of **24** was also examined using chiral copper catalysts. Among the catalysts examined, the Cu(II) complex with (*S*)-(−)-2,2′-isopropylidenebis(4-phenyl-4,5-dihydro-1,3-oxazole) was found to be the best, yielding (−)-**26** with up to 65% ee [42].

3.4.3 Synthesis of (+)-7b (MGS0028)

Racemic (±)-**7b** as well as optically pure (+)-**7b** (MGS0028) and (−)-**7b** were originally synthesized via hydroxy-keto compounds (±)-**29**, (+)-**29**, and (−)-**29**, respectively (see

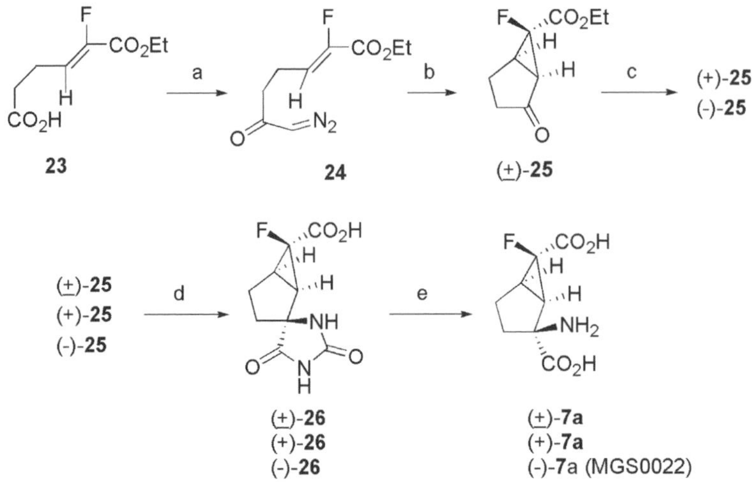

Scheme 3.3 *Synthesis of (–)-7 (MGS0022) and its congeners.* Reagents and conditions: *(a) (i) $(COCl)_2$, hexane, (ii) CH_2N_2, Et_2O; (b) $Cu(TBS)_2$, PhH; (c) chiral HPLC; (d) (i) 1 M aq. NaOH, (ii) KCN, $(NH_4)_2CO_3$, EtOH, H_2O; (e) 60% H_2SO_4.*

Scheme 3.4) [15]. The key intermediates (±)-**29** and (+)-**29** were prepared from the corresponding (±)-**25** and (+)-**25** in three steps: (i) reaction with TMSCl and LHMDS, followed by dehydrosilylation catalyzed by palladium acetate (89% and 89% yields), (ii) stereoselective epoxidation by TBHP in the presence of Triton B (98% and 88% yields), and (iii) regioselective reduction of α,β-epoxy ketone (±)-**28** with benzeneselenol, which was generated *in situ* from diphenyl diselenide $(PhSe)_2$ and sodium borohydride in the presence of acetic acid (76% and 71% yields). Hydroxyketones (±)-**29** and (+)-**29** were converted to their *tert*-butyldimethylsilyl (TBS) ethers, followed by thioketalization (the TBS–O bond was cleaved during the work-up) to give (±)-**30** (85% yield) and (+)-**30** (94% yield), respectively. Hydroxythioketals (±)-**30** and (+)-**30** were oxidized with dimethyl sufoxide (DMSO) and DCC in the presence of pyridine and trifluoroacetic acid to afford the corresponding ketones (85% and 76% yields, respectively). The resulting ketone (±)-**31**, was converted to (±)-**7b** (12% yield from (±)-**32**), in three steps via (±)-**32** (79% yield) using a protocol similar to that described above for (±)-**6** (see Scheme 3.2). Also, optically active (+)-**7b** (MGS0028) (45% yield, from highly polar isomer **33**) and (–)-**7b** (28% yield, from slightly polar isomer **33**) were obtained by separating the diastereomers of hydantoinamides **33** (highly polar isomer **33**: 46% yield and slightly polar isomer **33**: 46% yield, from (±)-**32**), which were prepared by coupling (±)-**32** with (*R*)-(+)-1-phenylethylamine (see Scheme 3.4).

Next, process chemistry for the practical synthesis of **7b** (MGS0028) is discussed (Schemes 3.5–3.7) [42–45]. First, the synthesis of key intermediate (+)-**29** from racemic acetoxycyclopentene (**34**) is shown in Scheme 3.5 [43]. The key reaction in this approach was Trost's asymmetric allylic alkylation reaction of ethyl 2-fluoroacetoacetate with **34**, which afforded **35** in high yield and high enantioselectivity, especially when a bulky tetra-*n*-hexyl ammonium bromide was used as a phase-transfer reagent (89% yield, 94–96%

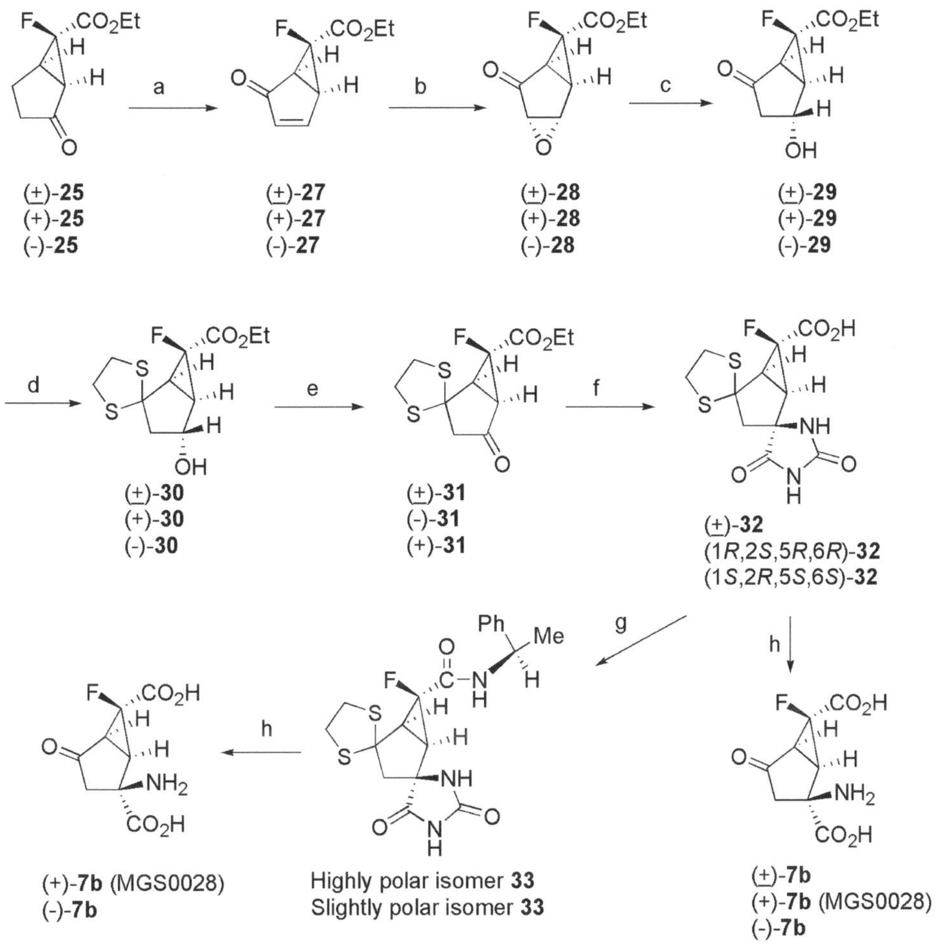

(±)-**25**
(+)-**25**
(-)-**25**

(±)-**27**
(+)-**27**
(-)-**27**

(±)-**28**
(+)-**28**
(-)-**28**

(±)-**29**
(+)-**29**
(-)-**29**

(±)-**30**
(+)-**30**
(-)-**30**

(±)-**31**
(-)-**31**
(+)-**31**

(±)-**32**
(1*R*,2*S*,5*R*,6*R*)-**32**
(1*S*,2*R*,5*S*,6*S*)-**32**

(+)-**7b** (MGS0028)
(-)-**7b**

Highly polar isomer **33**
Slightly polar isomer **33**

(±)-**7b**
(+)-**7b** (MGS0028)
(-)-**7b**

Scheme 3.4 *Synthesis of (+)-**7b** (MGS0028) and its congeners.* Reagents and conditions: *(a) (i) LHMDS, TMSCl, THF, (ii) $Pd(OAc)_2$, MeCN; (b) TBHP, Triton B, PhMe; (c) $(PhSe)_2$, $NaBH_4$, AcOH, EtOH; (d) (i) TBSCl, imidazole, DMF, (ii) $HS(CH_2)_2SH$, BF_3-Et_2O, $CHCl_3$; (e) DMSO, DCC, Py, TFA; (f) (i) 1 M aq. NaOH, (ii) KCN, $(NH_4)_2CO_3$, EtOH, H_2O; (g) (i) (R)-(+)-$PhCH(NH_2)CH_3$, EDC-HCl, HOBt, DMF, (ii) chromatography on silica gel. (h) 60% H_2SO_4.*

ee). Cleavage of the acetyl group of **35**, diastereoselective epoxidation via bromohydrin (*cis* : *trans* = 8 : 1) and intramolecular epoxide opening, followed by oxidation with 1-hydroxy-1,2-benziodoxal-3(1*H*)-one-1-oxide (IBX) gave intermediate (+)-**25** [43]. Ketone (+)-**25** was converted to enone (+)-**27** in five steps: (i) bromination of (+)-**25** (100% yield), (ii) azeotropic ketalization (92% yield), (iii) dehydrobromination, (iv) deketalization (92% yield for two steps), and (v) treatment with diazomethane (90% yield). Finally, enone (+)-**27** was transformed to β-hydroxy ketone **29** through epoxidation, followed by reduction with benzeneselenol generated *in situ*.

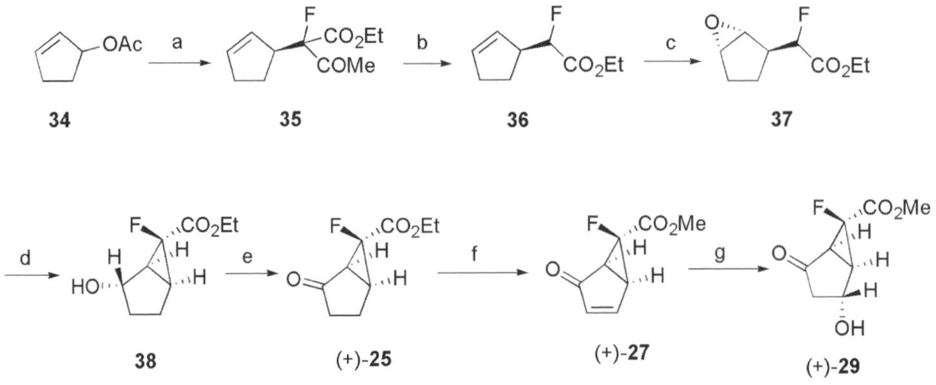

Scheme 3.5 Synthesis of the key intermediate (+)-**29** for (+)-**7b** (MGS0028) (1). Reagents and conditions: (a) (R,R)-Trost ligand, allylpalladium chloride dimer, $(n\text{-hex})_4NBr$, NaH, $AcCHFCO_2Et$, CH_2Cl_2; (b) 0.1 M, EtONa, EtOH; (c) (i) NBS, acetone, H_2O, (ii) DBU, CH_2Cl_2; (d) LHMDS, Et_3Al, THF; (e) IBX, DMSO, PhMe; (f) (i) Br_2, CH_2Cl_2, (ii) ethylene glycol, p-TsOH-H_2O, PhMe, (iii) t-BuOK, H_2O (1 eq.) THF, (iv) 1 M HCl; (v) CH_2N_2, (g) (i) TBHP, Triton B, PhMe, (ii) $(PhSe)_2$, $NaBH_4$, AcOH, EtOH.

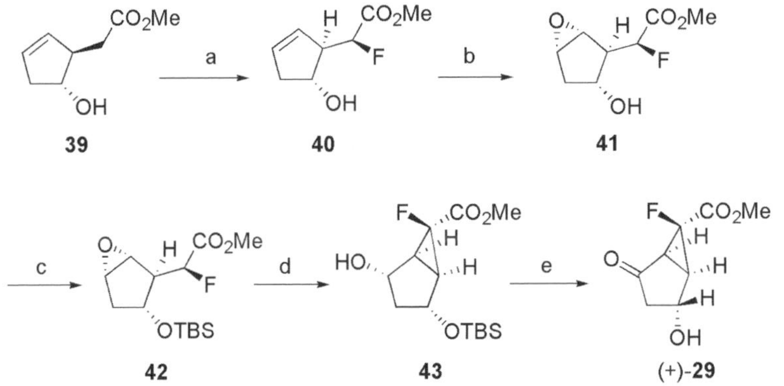

Scheme 3.6 Synthesis of key intermediate (+)-**29** for (+)-**7b** (MGS0028) (2). Reagents and conditions: (a) LDA, NFSI, THF; (b) TBHP, $VO(acac)_2$, PhMe; (c) TBSCl, imidazole, DMF; (d) Et_3Al, LHMDS; (e) (i) NaClO, $RuCl_3$ (1 mol%), MeCN, (ii) 1 M HCl.

Next, the most efficient route to (+)-**7b** (MGS0028), to date, via monoprotected fluorinated diol **43** and ketal **45** as key intermediates is discussed (see Schemes 3.6 and 3.7) [45]. Key intermediate **43** was synthesized from hydroxycyclopentenylacetate **39** in four steps: (i) fluorination of the dianion of **39** (85% yield) to form **40**, (ii) vanadium-mediated epoxidation of **40** to give epoxide **41** (93% yield), (iii) protection of the hydroxyl moiety of **41** to form TBS-ether **42** (95% yield), and (iv) cyclopropanation of **42** through ring-opening of the epoxide moiety to afford key intermediate **43** (96% yield). It should be noted that the epoxidation of **40** proceeded with excellent stereoselectivity to give the

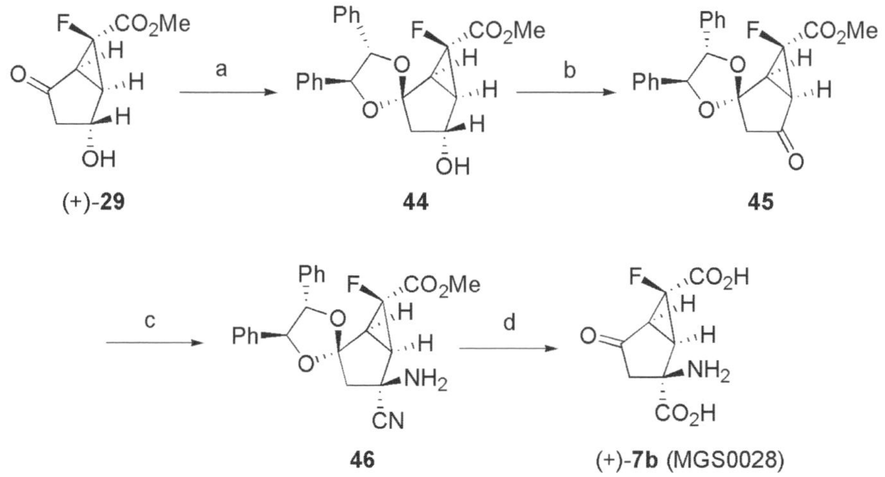

Scheme 3.7 *Synthesis of (+)-**7b** (MGS0028) from key intermediate (+)-**29**.* Reagents and conditions: *(a) (S,S)-PhCH(OTMS)-CH(OTMS)Ph, TfOH, CH_2Cl_2; (b) NaClO, $RuCl_3$ (0.5 mol%), MeCN; (c) NH_3/MeOH, Ti(O-Pr-i)$_4$, TMSCN; (d) (i) HCl (8 M), AcOH, (ii) H_2O.*

desired *trans*-epoxide **41** by exploiting the efficient directing effect of the free hydroxyl group. $RuCl_3$-mediated oxidation of **43**, followed by desilylation gave (+)-**29** in 95% yield (see Scheme 3.6).

The key intermediate (+)-**29** was converted to the second key intermediate **45** through ketalization with the bis-*O*-TMS ester of (*S*,*S*)-hydrobenzoin to form **44** (quantitative yield), and the subsequent $RuCl_3$-mediated oxidation (93% yield). Strecker reaction of **45** gave aminonitrile **46** with a very good diastereomer ratio (13.1 : 1 by HPLC analysis) as highly crystalline product. Thus, the desirable diastereomer of **46** was isolated as single product from the reaction mixture in 80–85% yield. Hydrolysis of ketal-aminonitrile **46** proceeded under much milder conditions than those for dithioketal-hydantoin **32** (see Scheme 4), and (+)-**7b** was isolated without utilizing ion-exchange column chromatography (94% yield) (see Scheme 3.7) [45]. Thus, this optimized process was able to avoid cumbersome handling, resolution using chiral HPLC and ion-exchange column chromatography, which were included in the original synthesis of (+)-**7b** (see Scheme 3.5).

3.4.4 Pharmacology and Pharmacokinetics of (+)-6a (MGS0008), (–)-7a (MGS0022) and (+)-7b (MGS0028)

3.4.4.1 In vitro *Pharmacology of Compounds 6 and 7*

The *in vitro* pharmacological data of fluorine-containing conformationally restricted glutamate analogues **6** and **7** are summarized in Table 3.1 [15]. Optically active (+)-**6a** (MGS0008), bearing a fluorine atom at the C-3 position of **4** (LY354740), is a potent and selective agonist for mGluR2/3. Compound (+)-**6a** exhibited a high agonist activity for mGluR2 (EC_{50} = 29.4 nM) and mGluR3 (EC_{50} = 45.4 nM) as well as a high binding affinity to mGluR2 (K_i = 47.7 nM) and mGluR3 (K_i = 65.9 nM), but did not exhibit significant

Table 3.1 In vitro *and* in vivo *pharmacological data of mGluR2/3 ligands* ***4***, ***6***, ***7*** *and* ***12***

Compound	X^1	X^2	X^3	Y	Agonist activity[a]		Antagonist activity[b]		Binding affinity[c]	
					$EC_{50} \pm$ SEM (nM)		$IC_{50} \pm$ SEM (nM)		*Ki* ± SEM (nM)[c]	
					mGluR2	mGluR3	mGluR2	mGluR3	mGluR2	mGluR3
4 (LY354740)	H	H	H	CH_2	18.3 ± 1.6	62.8 ± 12	–	–	23.4 ± 7.1	53.5 ± 13
(±)-**6a**	F	H	H	CH_2	67.7 ± 9.3	–	–	–	–	–
(+)-**6a** (MGS0008)	F	H	H	CH_2	29.4 ± 3.3	45.4 ± 8.4	>100 000	>100 000	47.7 ± 17	65.9 ± 7.1
(–)-**6a**	F	H	H	CH_2	2 640 ± 290	–	36 200	–	–	–
(±)-**6b**	H	F	H	CH_2	>100 000	–	17 100	–	–	–
(±)-**6c**	F	F	H	CH_2	>100 000	–	36 200	–	–	–
(±)-**7a**	H	H	F	CH_2	34.2 ± 6.3	–	–	–	–	–
(+)-**7a**	H	H	F	CH_2	1 120 ± 200	–	–	–	–	–
(–)-**7a** (MGS0022)	H	H	F	CH_2	16.6 ± 5.6	80.9 ± 31	>100 000	>100 000	22.5 ± 7.3	41.7 ± 7.1
(±)-**7b**	H	H	F	CO	1.26 ± 0.2	–	–	–	–	–
(+)-**7b** (MGS0028)	H	H	F	CO	0.570 ± 0.10	2.07 ± 0.40	>100 000	>100 000	3.30 ± 0.31	3.62 ± 1.6
(–)-**7b**	H	H	F	CO	94.7 ± 5.6	–	–	–	–	–
(±)-**7c**	H	H	Me	CH_2	>100 000	–	>100 000	–	–	–
(+)-**12a**	H	OH	H	CH_2	–	–	–	–	52[d]	89[d]
(±)-**12b**	H	Me	H	CH_2	–	–	1750 ± 620[e]	9830 ± 370[e]	983 ± 35[f]	146 ± 56[f]

–: not determined.

EC_{50}: 50% effective concentration.

IC_{50}: 50 inhibitory concentration.

K_i: inhibition constant.

[a] Compounds (+)-**6a**, (–)-**7a** and (+)-**7b** exhibited no significant agonist activities for rat mGluR1a, mGluR4, mGluR6, and mGluR7 expressed in CHO cells (ED_{50} > 100 000 nM).

[b] Compounds (+)-**6a**, (–)-**7a** and (+)-**7b** exhibited no significant antagonist activities for rat mGluR1a, mGluR4, mGluR6 and mGluR7 expressed in CHO cells (ED_{50} > 100 000 nM).

[c] Binding affinities for mGluR2 and mGluR3 were determined by binding study utilizing [^{3}H]-MGS0008 in rat mGluR-expressing cells.

[d] Displacement of [^{3}H]LY354740 binding in rat brain [36].

[e] cAMP responses in human mGluR-expressing cells [37].

[f] Displacement of [^{3}H]LY341495 binding to human mGluR-expressing cells [37].

agonist or antagonist activities for mGluR1a, mGluR4, mGluR5, mGluR6 or mGluR7, which was similar to compound **4** (LY354740). The agonist activity of (+)-**6a** was found to be highly stereospecific, since its enantiomer (−)-**6a** showed approximately 90-fold lower agonist activity (EC_{50} = 2,640 nM) than (+)-**6a** for mGluR2. Racemic (±)-**6b**, diastereomer of (±)-**6a** at the C-3 position, did not exhibit significant agonist (EC_{50} > 100 000 nM) or antagonist (EC_{50} = 17 100 nM) activities for mGluR2. The conspicuous difference in activities between (±)-**6a** and (±)-**6b** does not seem to be ascribable to the steric effect of fluorine incorporation, because mGluR2/3 antagonists **12a** (Ro 65–3479) or **12b** , bearing a hydroxyl group or a methyl group at the C-3α position of **4**, exhibited high or moderate binding affinities to mGluR2 (K_i = 52 nM or 983 nM) and mGluR3 (K_i = 89 nM or 146 nM) [36, 37]. These results suggest that the decrease in the activity of (±)-**6b** for mGluR2 is due to the large electronegativity of the incorporated fluorine atom, which influences the electron density of the amino and/or carboxyl groups at the C-2 position, depending on its stereochemistry. Similarly, the difluoro analogue (±)-**6c** did not exhibit appreciable agonist (EC_{50} = > 100 000 nM) or antagonist (EC_{50} = 36 200 nM) activities for mGluR2. This lack of activity may be due to the α-fluorine atom of (±)-**6c**, since (±)-**6b** with a fluorine atom at the C-3α position exhibited neither agonist nor antagonist activities for mGluR2, as mentioned above.

Racemic (±)-**7a**, bearing a fluorine at the C-6 position of (±)-**4**, exhibited a strong mGluR2 agonist activity (EC_{50} = 34.2 nM), which was also found to be highly stereospecific. The optically active (−)-**7a** (MGS0022) exhibited an approximately 67-fold higher agonist activity (EC_{50} = 16.6 nM) than its enantiomer (+)-**7a** (EC_{50} = 1 120 nM). Furthermore, compound (−)-**7a** exhibited a high agonist activity for mGluR3 (EC_{50} = 80.9 nM), but did not for mGluR4, mGluR6, mGluR7, mGluR1a, or mGluR5. No significant antagonist activity of (−)-**7a** was observed for mGluR1a and mGluRs2–7. In contrast, (±)-**7c**, bearing a methyl group at the C-6 position of (±)-**4**, exhibited neither agonist nor antagonist activities for mGluR2 (EC_{50} > 100 000 nM). The dramatic difference in activity between (±)-**7a** and (±)-**7c** may be ascribed to the difference in stereoelectronic properties between fluorine and a methyl group.

Interestingly, the replacement of the methylene group at the C-4 position of **7a** with a carbonyl group has substantially enhanced the agonist activity, that is, the resulting (+)-**7b** (MGS0028) is one of the best known agonists for mGluR2/3 to date. Compound (+)-**7b** demonstrated a potent and stereospecific agonist activity for mGluR2 (EC_{50} = 0.57 nM) and mGluR3 (EC_{50} = 2.07 nM), but neither showed agonistic effect on mGluR1a or mGluR4–mGluR7 or antagonistic effect on mGluR1a or mGluR2–mGluR7. The agonist activity of (+)-**7b** for mGluR2 was approximately 165-fold higher than that of its enantiomer (−)-**7b** (EC_{50} = 94.7 nM). The greatly enhanced agonist activity of (+)-**7b** as compared with (−)-**7a** may be ascribed to a conformational change caused by the replacement of the C-4 methylene moiety to a carbonyl group, especially the relative positions of the three key functional groups (i.e., one amino group and two carboxylic acids), as well as the stereoelectronic effects of the carbonyl group introduced.

3.4.4.2 Behavioral Pharmacology of (+)-6a (MGS0008) and (+)-7b (MGS0028)

The antipsychotic-like effects of (+)-**6a** (MGS0008) and (+)-**7b** (MGS0028) on laboratory animals are shown in Table 3.2 [15, 51]. It was recently found that phencyclidine

Table 3.2 *Antipsychotic-like effects of (+)-**6a** (MGS0008) and (+)-**7b** (MGS0028) on laboratory animals*

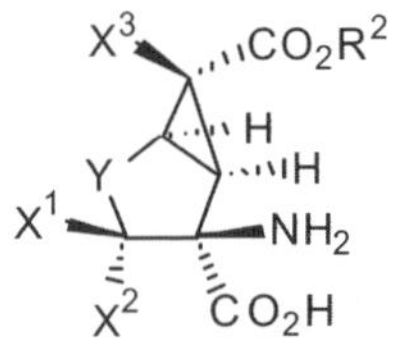

Compound	X^1	X^2	X^3	Y	PCP-induced hyperactivity	PCP-induced head-weaving behavior	Conditioned avoidance
					ED_{50} (mg/kg)	ED_{50} (μg/kg)	ED_{50} (mg/kg)
4 (LY354740)	H	H	H	CH_2	>100	3000	>30
(+)-**6a** (MGS0008)	F	H	H	CH_2	5.1	260	6.55
(+)-**7b** (MGS0028)	H	H	F	CO	0.30	0.090	1.67

ED_{50}: 50% effective dose.

(PCP)-induced head-weaving behavior in rats was inhibited by the intraperitoneal administration of **4** (LY354740) [17]. The oral administration of **4** (LY354740) also inhibited PCP-induced head-weaving behavior ($ED_{50} = 3.0$ mg/kg), but did not affect PCP-induced hyperactivity ($ED_{50} > 100$ mg/kg) or conditioned avoidance responses ($ED_{50} > 30$ mg/kg) in rats. In contrast, the oral administration of (+)-**6a** (MGS0008) ($ED_{50} = 0.26$ mg/kg) inhibited PCP-induced head-weaving behavior in rats more effectively (11-fold) than **4**. Furthermore, (+)-**6a** antagonized PCP-induced hyperactivity ($ED_{50} = 5.1$ mg/kg) and impaired conditioned avoidance responses ($ED_{50} = 6.55$ mg/kg) in rats. These results indicate that the PCP-induced head-weaving behavior is a sensitive method for screening mGluR2/3 agonists and the introduction of fluorine to **4** has clearly increased the oral activity. It has been reported that the oral bioavailability of **4** in rats is low ($F = 10\%$), apparently because of inefficient drug transfer across the intestinal epithelial membrane [52]. The enhanced oral activity of (+)-**6** can be attributed to the increase in oral bioavailability and BBB penetration as a result of the introduction of fluorine to the drug molecule, which would enhance lipophilicity as well as bring in unique properties associated with fluorine.

The oral administration of (+)-**7b** (MGS0028) very strongly inhibited PCP-induced head-weaving behavior in rats ($ED_{50} = 0.090$ μg/kg), that is, (+)-**7b** was much more effective than **4** (LY354740) and (+)-**6a** (MGS0008). Furthermore, the oral administration of (+)-**7b** strongly antagonized PCP-induced hyperactivity in rats ($ED_{50} = 0.30$ mg/kg) and impaired the conditioned avoidance responses ($ED_{50} = 1.67$ mg/kg) in rats. Excellent antipsychotic-like effects of (+)-**6a** (MGS0008) and (+)-**7b** (MGS0028) as mGluR2/3 agonists are very encouraging for their potential use in the treatment of schizophrenia.

3.4.4.3 Pharmacokinetics of (+)-7b (MGS0028)

The metabolism and disposition of (+)-**7b** (MGS0028) in three preclinical species (rats, dogs, and monkeys) are summarized in Table 3.3 [53]. In rats, (+)-**7b** (MGS0028) was

Table 3.3 *Pharmacokinetic parameters of (+)-**7b** (MGS0028) in Sprague-Dawley rats, rhesus monkeys, and beagle dogs*

Parameter[a]	Sprague-Dawley rats			Rhesus monkeys			Beagle dogs	
	3 mg/kg i.v.	1 mg/kg p.o.	10 mg/kg p.o.	1 mg/kg i.v.	5 mg/kg p.o.	0.3 mg/kg i.v.	0.1 mg/kg p.o.	0.3 mg/kg p.o.
C_{max} (μM)	NA[b]	1.0 ± 0.2	6.0 ± 1.9	NA[b]	1.0 ± 0.5	NA[b]	1.5 ± 0.8	4.0 ± 1.0
t_{max} (h)	NA[b]	0.8 ± 0.2	3.0 ± 1.4	NA[b]	2.3 ± 1.5	NA[b]	0.4 ± 0.1	0.5 ± 0.0
AUC_{0-24h} (μM h)	14.3 ± 0.8	3.4 ± 0.5	35.8 ± 3.8	10.2 ± 2.6	10.4 ± 5.0	8.2 ± 1.1	1.7 ± 0.9	5.7 ± 0.3
Effective $t_{1/2}$ (h)	1.1	NA[b]	NA[b]	1.1	NA[b]	0.9	NA[b]	NA[b]
CL_p (ml/min/kg)	16.2 ± 0.9	NA[b]	NA[b]	8.0 ± 1.8	NA[b]	2.8 ± 0.4	NA[b]	NA[b]
V_d (L/kg)	0.5 ± 0.2	NA[b]	NA[b]	0.6 ± 0.1	NA[b]	0.1 ± 0.01	NA[b]	NA[b]
F (%)	NA[b]	70.7	75.3	NA[b]	20.3[c]	NA[b]	63.4	69
$AUC_{Metabolite}/AUC_{parent}$	0.8	0.4	0.3	0.3	0.9	0.0	0.0	0.0

[a] C_{max}: maximum plasma or tissue concentration, t_{max}: time to reach maximum plasma concentration, AUC_{0-24}: area under the concentration–time curve from time 0 to 24 hours, Effective $t_{1/2}$: effective elimination half-life, CL_p: plasma clearance, V_d: volume of distribution, F: bioavailability.
[b] Not applicable.
[c] Some monkeys experienced emesis, and this may be partially responsible for low bioavailability.

widely distributed and was primarily excreted in the urine as the parent compound or as a single reductive metabolite, (1*R*,2*S*,4*R*,5*S*,6*S*)-2-amino-6-fluoro-4-hydroxybicyclo[3.1.0] hexane-2,6-dicarboxylic acid. Compound (+)-**7b** showed a low brain-to-plasma ratio at efficacious doses in rats and was eliminated more slowly in rat brain than in plasma. Plasma concentration was proportional to the drug dosage (1 mg/kg and 10 mg/kg) in rats, with good bioavailability (70.7% and 75.3%, respectively). This pharmacokinetic profile supported the oral antipsychotic-like effects of (+)-**7b** in rats. In dogs, oral bioavailabilities of (+)-**7b** were 63.4% and 69.0% at 0.1 and 0.3 mg/kg dosages, respectively; and no reductive metabolite was detected. In monkeys, however, the bioavailability was only 20.3% and a reductive metabolite was found at a relatively high level in the plasma. *In vitro* metabolic studies of (+)-**7b** in liver subcellular fractions (microsomes and cytosol) showed the presence of the reductive metabolite in specimens from rats, monkeys, and humans, but not in specimens from dogs. The metabolism of (+)-**7b** was not detected in liver microsomes from any of the examined species. Similar to the *in vivo* result, (+)-**7b** was metabolized to the reductive metabolite in the cytosol in a stereospecific manner. The order of *in vitro* metabolite formation (monkey >> rat ~ human >> dog) was consistent with the *in vivo* results in rats, dogs, and monkeys.

3.5 mGluR2/3 Antagonist 14a (MGS0039) and Its Prodrug 15a (MGS0210)

The original synthesis, pharmacology, and pharmacokinetics of compound **14** and its prodrug **15** are discussed in this section [25, 34].

3.5.1 Synthesis of 14a (MGS0039), Its Analogue, and Prodrugs of 14a (MGS0039)

The syntheses of mGluR2/3 antagonists, **13a**, **13b**, and **14**, as well as prodrug **15** (a prodrug of **14a**, MGS0039) from (−)-**25** (see Scheme 3.3) via key intermediates, **50** and **52**, are illustrated in Schemes 3.8 3.9 [25, 39].

First, the synthesis of mGluR2/3 antagonists, **13a**, **13b**, is discussed (see Scheme 3.8) [25]. Ketone (−)-**25** was converted to the corresponding enol triflate by reacting with *N*-p henylbis(trifluoromethanesulfonimide) and LHMDS, which was then subjected to the Pd-catalyzed carboalkoxylation with an alcohol (R^1OH) at room temperature and atmospheric pressure of carbon monoxide to give α,β-unsaturated ester **47** (71% (Et) and 33% (Bn) yields). Compound **47** was converted to key intermediate **50** using the protocol reported by Shao [54]. Thus, the stereoselective dihydroxylation of **47** with OsO_4 and *N*-methylmorphine-*N*-oxide (NMO) afforded **48** (91% (Et and Bn) yields) as single product, wherein OsO_4 reacted exclusively with the olefin moiety of **47** from the opposite face (*exo* face) of the fused cyclopropane ring. Then, **48** was converted to 2,3-cyclic sulfites with $SOCl_2$, followed by oxidization to cyclic sulfate **49** (93% (Et and Bn) yields). The regio- and stereoselective nucleophilic ring-opening of **49** with NaN_3, followed by hydrolysis gave key intermediate **50** (91% (Et) and 89% (Bn) yields).

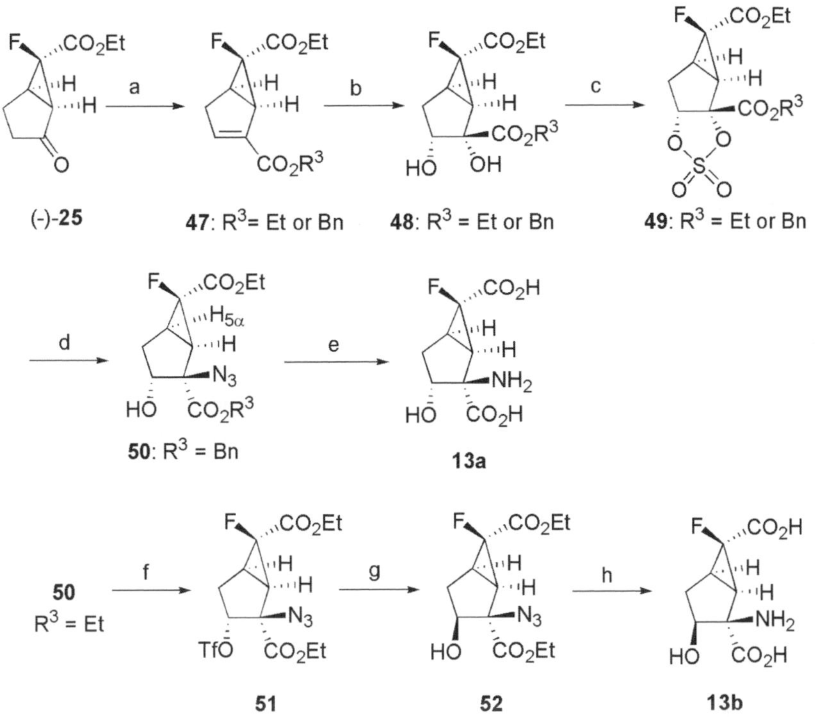

(-)-**25** **47**: R3= Et or Bn **48**: R3 = Et or Bn **49**: R3 = Et or Bn **50**: R3 = Bn **13a** **50** R3 = Et **51** **52** **13b**

Scheme 3.8 *Synthesis of* ***13a*** *and* ***13b****.* Reagents and conditions: *(a) (i) LHMDS, TfNPh, THF, (ii) CO, $Pd(OAc)_2$, $(i\text{-}Pr)_2NEt$, PPh_3, EtOH or $PhCH_2OH$, DMF; (b) OsO_4, NMO, MeCN; (c) (i) $SOCl_2$, Et_3N, CH_2Cl_2, (ii) $NaIO_4$, $RuCl_3$, H_2O, CCl_4, MeCN; (d) (i) NaN_3, DMF–H_2O, (ii) 20 % H_2SO_4; (e) (i) 10 % Pd/C, H_2, AcOH, H_2O, (ii) 10% HCl. (f) Tf_2O, pyridine, CH_2Cl_2; (g) KNO_2, 18-crown-6, DMF, rt; (h) (i) Me_3P, THF, H_2O, rt or 10% Pd/C, H_2, AcOH, H_2O (ii) LiOH, THF, H_2O.*

Compound **13a** was obtained from key intermediate **50** through reduction (hydrogenation) of the azido group to an amino group as well as deprotection (hydrogenolysis) of the benzyl ester over 10% Pd on carbon as the catalyst, followed by hydrolysis of the ethyl ester moiety with 10% HCl (79% yield) [25].

Compound **13b** was synthesized in four steps via key intermediate **52**, in which the configuration of the hydroxyl group at the C-3 position of **50** was inverted. Compound **50** was reacted with trifluoromethanesulfonyl anhydride and pyridine to give triflate **51** (96% yield), which was then reacted with KNO_2 in the presence of 18-Crown-6, followed by post-treatment of the resulting nitrous ester with water to afford key intermediate **52** (80% yield) [55, 56]. Reduction of the azide moiety of **52** by trimethylphosphine or hydrogenation on Pd on carbon, followed by hydrolysis of the ester moieties gave **13b** (48% yield).

Etherification of the hydroxyl group at the C-3 position of **50** through benzylation using benzyl trichloroacetimidate under acidic conditions [57, 58] or an alkyl triflate under basic conditions gave **53** (42% (Et) and 17–90% (Bn) yields). Azide **53** was reduced to

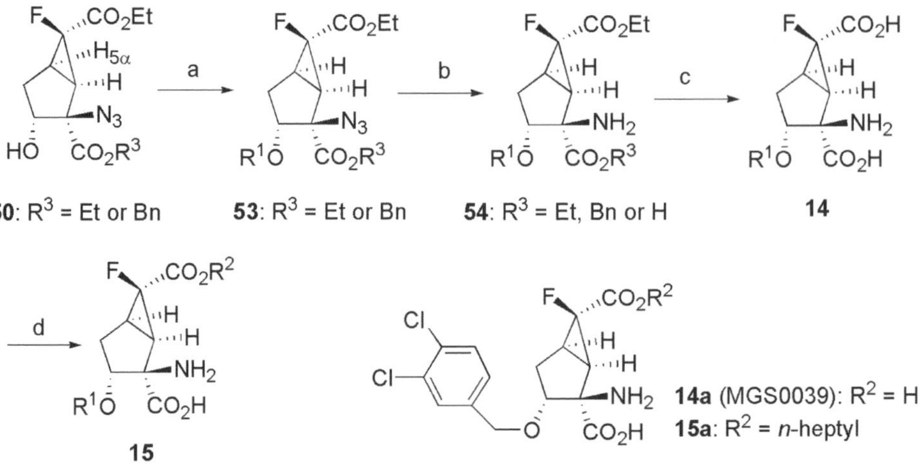
F
CO_2Et
$H_{5\alpha}$
H
N_3
HO
CO_2R^3
a
b
c
d
R^1O
NH_2
CO_2H
50: R^3 = Et or Bn
53: R^3 = Et or Bn
54: R^3 = Et, Bn or H
14
CO_2R^2
15
Cl
14a (MGS0039): R^2 = H
15a: R^2 = *n*-heptyl

Scheme 3.9 *Synthesis of **14a** (MGS0039) and related compounds.* Reagents and conditions: *(a) $R^1OC(=NH)CCl_3$, TfOH, $CHCl_3$, cyclohexane or R^1OTf, 2,6-*tert*-butylpyridine; (b) Me_3P, THF, H_2O or 10% Pd/C, H_2, AcOH, H_2O; (c) LiOH, THF, H_2O or 10 % HCl; (d) R^2OH, $SOCl_2$.*

amine **54** by the Staudinger reaction [59, 60] or catalytic hydrogenation over Pd on carbon (89% (Et) and 59–81% (Bn) yields). Amino-diester **54** was hydrolyzed under basic (LiOH) or acidic (HCl) conditions to afford **14** (18–82% yield). Finally, alkyl ester prodrugs **15** were obtained by the esterification of **14a** (MGS0039) with an alcohol R^2OH (see Table 3.4 for R^2) and thionyl chloride (40–81% yield). The structure of **15a** is shown as an example.

3.5.2 Pharmacology and Pharmacokinetics of 14a (MGS0039) and Its Analogue

mGluR2/3 antagonists bearing a bicyclo[3.1.0]hexane skeleton have been reported [24, 25, 61]. Among them, 3-alkoxy-2-amino-6-fluorobicyclo[3.1.0]hexane-2,6-dicarboxylic acids **14**, especially **14a** (MGS0039), is one of the best known mGluR2/3 antagonists, based on their pharmacological [23–31] and pharmacokinetic profiles [25].

3.5.2.1 In Vitro *Pharmacology of Compound 14a (MGS0039) and Its Derivatives*

The *in vitro* pharmacological profiles of **14a** (MGS0039) and its analogue are summarized in Table 3.3 [24, 25, 61]. Optically active **13a**, bearing a hydroxyl group at the C-3α position of (–)-**7a** (MGS0022), exhibits binding affinities for mGluR2 (K_i = 32.9 nM) and mGluR3 (K_i = 67.1 nM), which are similar to those of (–)-**7a** (mGluR2, K_i = 22.5 nM; mGluR3, K_i = 41.7 nM) as shown in Tables 3.1 and 3.3. Interestingly, however, **13a** shows a moderate antagonist activity (IC_{50} = 476 nM), but no significant agonist activity (EC_{50} > 100 000 nM) for mGluR 2. This makes a sharp contrast with (–)-**7a** (MGS0022), which is a strong mGluR2 agonist (EC_{50} = 16.6 nM) as shown in Table 3.1. The (3*S*)-isomer (β-OH) **13b**, exhibits a 3-fold lower affinity for mGluR2 than its (3*R*)- isomer

Table 3.4 In vitro *pharmacological data for optically active mGluR2/3 antagonists* ***11**, **12**, **13** and **14***

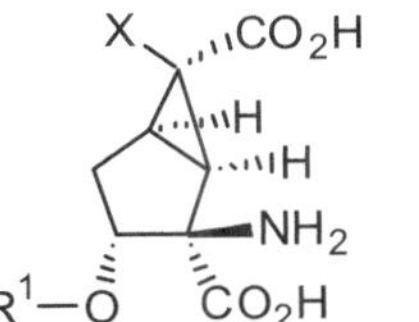

Compound	X	R[1]	Binding affinity[a]			Antagonist activity[b]		Agonist activity[c]	
			K_i (nM)			IC_{50} ± SEM (nM)		EC_{50} ± SEM (nM)	
			mGluR2	mGluR3	mGluR7	mGluR2	mGluR3	mGluR2	mGluR3
11			3.13	3.04	110	23.2 ± 8.75	14.2 ± 4.34	>100 000	>100 000
12a	H	H	52[d]	89[d]	–	–	–	–	–
13a	F	H	32.9	67.1	–	476 ± 134	–	>100 000	–
13b	F	H (β-OH)	105	–	–	803	–	–	–
14a (MGS0039)	F	3,4-Cl_2-$PhCH_2$	2.38	4.46	664	20.0 ± 3.67	24.0 ± 3.54	>100 000	>100 000
14b	F	Me	39.2	88.1	–	229 ± 77.3	–	–	–
14c	F	*n*-Pr	5.17	–	–	–	–	–	–
14d	F	$PhCH_2$	7.14	15.9	–	131 ± 44.9	–	–	–
14e	F	4-Cl-$PhCH_2$	3.17	4.77	–	29.1 ± 8.11	–	–	–
14f	F	3,4-F_2-$PhCH_2$	2.27	3.00	–	40.8 ± 12.6	–	–	–
14g	F	Ph_2CH	2.58	3.93	–	24.4 ± 6.53	–	–	–
14h	F	2-$NapCH_2$	2.53	5.43	–	22.7 ± 7.06	–	–	–
14i	H	3,4-Cl_2-$PhCH_2$	2.51	–	–	34.2	–	>100 000	–

–: not determined.
EC_{50}: 50% effective concentration.
IC_{50}: 50 inhibitory concentration.
K_i: inhibition constant.
[a] Binding affinities for mGluR2 and mGluR3, and affinities for mGluR7 were determined by binding study utilizing [^{3}H]MGS0008 and [^{3}H]-**11**, respectively.
[b] Compounds **14a** (MGS0039) and **11** exhibited no significant antagonist activities for mGluR1 (IC_{50} = 93 300 ± 14 600 and 8990 ± 907 nM), mGluR4 (IC_{50} = 1740 ± 1080 and 2650 ± 521 nM), mGluR5 (IC_{50} = 117 000 ± 38 600 and 11 400 ± 2700 nM), and mGluR6 (IC_{50} = 2060 ± 1270 and 1140 ± 378 nM) expressed in CHO cells, respectively.
[c] Compound **14a** (MGS0039) exhibited no significant agonist activities for mGluR1, mGluR4, mGluR5, and mGluR6 expressed in CHO cells (ED_{50} > 100 000 nM).
[d] Displacement of [^{3}H]-LY354740 binding in rat brain [36].

(α-OH) **13a**. The result indicates that the binding affinity of **13** depends on the stereochemistry of the hydroxyl group at the C-3 position of the bicyclo[3.1.0]hexane ring and the (3*R*)-configuration (α-OH) is critical for a high affinity for mGluR2.

The introduction of a methoxy group in place of the hydroxyl group at the C-3 position of **13a**, which provides **14b**, does not change the binding affinity for mGluR2 (K_i = 39.2 nM) or mGluR3 (K_i = 88.1 nM) or the antagonist activity for mGluR2 (IC_{50} = 229 nM). However, the introduction of a larger substituent as R^1, such as *n*-propyl (**14c**) and benzyl (**14d**) groups, resulted in binding affinities for mGluR2 several-fold higher than that of **14b**. 3-Benzyloxy derivative **14d** exhibits a moderate antagonist activity for mGluR2 (IC_{50} = 131 nM).

These findings indicate that the agonist/antagonist activities of 2-amino-6-fluorobicyclo[3.1.0]hexane-2,6-dicarboxylic acids, **13** and **14**, for mGluR2/3 are controlled by the C-3 substituent in a size-independent manner. This observation is supported by the results on nonfluorinated congeners, 3-alkoxy-2-aminobicyclo[3.1.0]hexane-2,6-dicarboxylic acids **14i** [61], **12a** (Ro 653479) [36], and **12b** [37].

Since **14d**, bearing a benzyl group at the C-3 position, showed promising binding affinity and antagonist activity for mGluR2/3, the optimization of **14d** by varying the substituents on the benzene ring and replacing the phenyl group with other aryl groups has been undertaken to find better mGluR2/3 antagonists. It has been found, to date, that 3,4-dichlorobenzyl derivative **14a** (MGS0039) is the best mGluR2/3 antagonist with regard to binding affinity, antagonist activity, selectivity, oral bioavailability, and BBB penetration.

Compound **14a** (MGS0039) exhibited a high affinity for mGluR3 (K_i = 4.46 nM) as well as mGluR2 (K_i = 2.38 nM). Furthermore, **14a** exhibited a lower affinity (K_i = 664 nM) for mGluR7 than the standard antagonist **11** (LY341495) (K_i = 110 nM). Compound **14a** exhibited potent antagonist activities for both mGluR2 and mGluR3 (IC_{50} = 20.0 nM and 24.0 nM, respectively), and much weaker antagonist activities for mGluR4 (IC_{50} = 1740 nM), mGluR6 (IC_{50} = 2060 nM), mGluR1 (IC_{50} = 93 300 nM) and mGluR5 (IC_{50} = 117 000 nM). In addition, **14a** did not exhibit significant agonist activities for mGluR2, mGluR3, mGluR4, mGluR6, mGluR1, and mGluR5 (EC_{50} > 100 000 nM). In contrast, **11** (LY341495) possesses an affinity for mGluR7 and mGluR8 [62]. Compound **14a** exhibited a 300-fold lower affinity for mGluR7 than that for mGluR2, while **11** exhibited a 35-fold lower affinity for mGluR7 than for mGluR2, as determined by [^{3}H]-**11** binding to recombinant mGluR7. Thus, **14a** may possess a greater specificity for mGluR2/3, although its effects on mGluR8 have yet to be determined. It should be noted that **14a** did not interact with other receptors and transporters, including *N*-methyl-D-aspartic acid (NMDA), α-amino-3-hydroxy-5-methylisoxazole-4-propionate (AMPA), and kainite receptors [24]. Moreover, in a preliminary experiment, **14a** did not inhibit glutamate transport through glutamate transporters, such as excitatory amino acid transporter (EAAT) 1, EAAT2, and EAAT3, even at 10 μM concentration [24]. These findings indicate that **14a** (MGS0039) is one of the most potent and selective antagonists for mGluR2/3 developed to date.

3.5.2.2 Behavioral Pharmacology of 14a (MGS0039)

Antidepressant-like and anxiolytic-like activities of **14a** (MGS0039) have been studied in experimental animal models [24, 29, 30]. Compound **14a** exhibited an antidepressant-like effect when evaluated using the forced swimming test in rats (lowest active dose 1 mg/kg, i.p.), the tail suspension test in mice (lowest active dose 1 mg/kg, i.p., both acutely and subchronically for 5 days) [24], and the learned helplessness test (escape failure) in rats (lowest active dose 10 mg/kg, i.p. for 7 days) [29]. In addition to these antidepressant-like effect, **14a** also showed anxiolytic-like activities when evaluated using the conditioned fear stress test in rats (lowest active dose 2 mg/kg, i.p.) [29], the marble-burying behavior test in mice (lowest active dose 3 mg/kg, i.p.) [30], and the stress-induced hyperthermia test in single housed mice (lowest active dose 1 mg/kg, i.p.) [31].

3.5.2.3 Pharmacokinetics of 14a (MGS0039) and its Derivatives

The pharmacokinetic profiles of **14a** (MGS0039) and **14e–14h**, selected from 3-alkoxy-2-amino-6-fluorobicyclo[3.1.0]hexane-2,6-dicarboxylic acids **14** as typical mGluR2/3 antagonists, are summarized in Table 3.5 [25]. As Table 3.5 shows, **14a** (MGS0039) exhibits the best pharmacokinetic parameters among these compounds. The oral administration of 3, 10, and 30 mg/kg of **14a** to fasting rats resulted in almost dose-dependent pharmacokinetic parameters (C_{max} = 214 ng/mL at 2.0 h, 932 ng/mL at 2.7 h and 2960 ng/mL at 3.3 h, $t_{1/2}$ = 2.15 h, 2.76 h and 2.77 h, AUC_{inf} = 1240 ng h/mL, 6260 ng h/mL and 19 300 ng h/mL, respectively). At doses of 3, 10, and 30 mg/kg, the ratios of C_{max} were 1.0, 4.4, and 13.9, respectively, while the ratios of AUC_{inf} were 1.0, 5.1, and 15.6, respectively. The mean maximum plasma level of **14a** (MGS0039) was 492.3 ng/mL at 6 h. After peaking, the plasma concentrations decreased with an estimated half-life of 2.3 h. The AUC_{inf} was 6813.0 ng h/mL [25].

The brain and plasma levels and pharmacokinetics parameters after oral administration were compared for compounds **14a** (MGC0039) and **14e–14h** (see Table 3.4) [25]. Again, **14a** exhibited the best BBB penetration among the compounds evaluated. The mean maximum cerebral level of **14a** was 13.22 ng/g at 6 h. After peaking, the cerebral concentrations decreased with an estimated half-life of 10.9 h. The cerebrum/plasma ratios of **14a** at 1, 3, 6, and 24 h were 0.01, 0.02, 0.03, and 1.99, respectively. The rate of elimination from the cerebrum was slower than that from the plasma.

Based on these pharmacokinetic data, the ability of **14a** to penetrate the BBB appears to be acceptable or even better than that of other known mGluR2/3 antagonists, but the oral bioavailability of **14a** might be insufficient for its development as a drug for the treatment of depression and/or anxiety.

3.5.3 Pharmacokinetics of 14a (MGS0039) Prodrugs

Various prodrugs of **14a** (MGS0039) have been examined and reported. [39, 40] The strategy for the development of prodrugs of **14a** is summarized in Figure 3.3. The synthesized prodrugs were initially evaluated in human liver S9 fractions. In this study, the prodrugs that were efficiently transformed to their active form, **14a**, were selected for further evaluation in monkey S9 and rat S9 fractions. The compounds selected in the S9 studies were then further evaluated using *in vivo* pharmacokinetic studies in monkeys and rats, and preclinical candidates were selected on the basis of the transformation of the prodrug to the active form of **14a** as well as its *in vivo* pharmacokinetics profile in monkeys.

Table 3.5 *Brain and plasma levels and pharmacokinetics parameters of mGluR2/3 antagonists* ***14a****,* ***14e****,* ***14f****,* ***14g****, and* ***14h*** *after peroral dosing to rats at dose of 10 mg/kg*

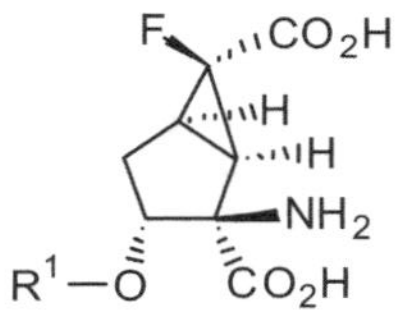

Compound	R^1	Tissue	Plasma / brain concentrations (ng/mL or g)[a]			Parameters[b] in plasma and brain			
			1 h	6 h	24 h	t_{max} (mg/kg)	C_{max} (ng/mL)	$t_{1/2}$ Lambda z (h)	$AUC_{inf\ (predicted)}$ (h ng/mL)
14a (MGS039)	3,4-Cl_2-$PhCH_2$	Plasma	364 ± 273	492 ± 344	2.10 ± 0.20	6.0	492	2.3	6810
		Brain	3.88 ± 2.46	13.2 ± 6.75	4.22 ± 1.30	6.0	13.2	10.9	269
14e	4-Cl-$PhCH_2$	Plasma	176 ± 39.3	98.2 ± 7.7	0.00 ± 0.00	3.0	225	2.5	1330
		Brain	2.51 ± 0.20	3.39 ± 1.08	0.00 ± 0.00	3.0	3.66	NA[c]	NA[c]
14f	3,4-F_2-$PhCH_2$	Plasma	213 ± 11.4	41.1 ± 9.8	0.80 ± 0.70	1.0	213	2.8	1210
		Brain	3.06 ± 0.34	2.40 ± 0.40	1.39 ± 1.21	3.0	3.11	19.5	88.7
14g	Ph_2CH	Plasma	61.3 ± 24.0	23.1 ± 11.5	0.20 ± 0.10	1.0	61.3	2.8	431
		Brain	0.00 ± 0.00	2.04 ± 0.60	0.51 ± 0.89	6.0	2.04	9	32.6
14h	2-$NapCH_2$	Plasma	121 ± 9.0	210 ± 11.1	2.20 ± 1.20	3.0	252	2.9	3040
		Brain	1.42 ± 0.21	8.72 ± 6.64	0.25 ± 0.02	6.0	8.72	7.1	132

[a] Results are expressed as the mean ± SD, $n = 3$.

[b] t_{max}: time to reach maximum plasma or brain concentration; C_{max}: maximum plasma or brain concentration; $t_{1/2}$ Lambda z: the terminal elimination half-life; AUC_{inf}: area under the concentration–time curve from time 0 to infinite time.

[c] Not applicable.

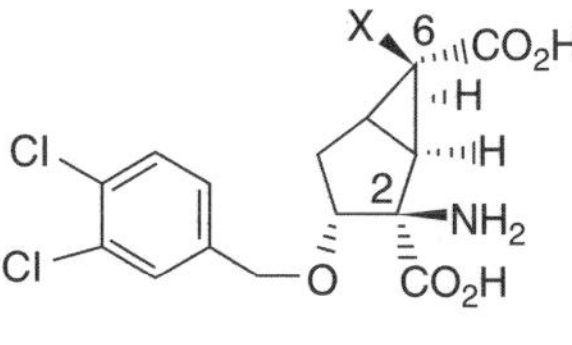

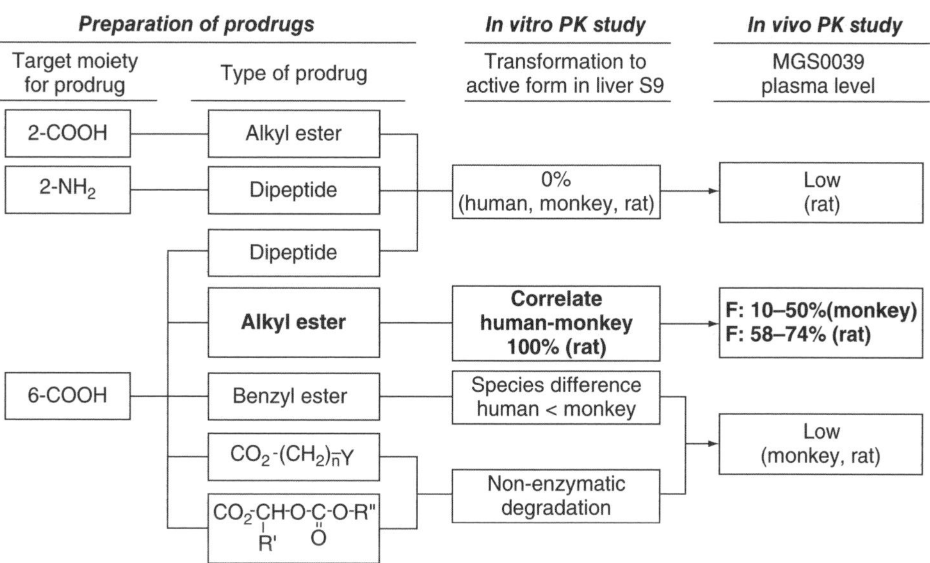

Figure 3.3 *Summary of strategy for discovering a prodrug of* ***14a*** *(MGS0039) (X = F).*

To prepare the prodrug for **14a**, the functional groups in **14a** – two carboxylic acids at the C-2 and C-6 positions and an amino group at the C-2 position – were exploited. It was found that the C-2 ester was not transformed to the active form **14a** in human, monkey, or rat liver S9 fractions. A dipeptide prodrug of **4** (LY354740) with natural amino acids, particularly with alanine at the 2-NH_2 position, has been reported as an effective prodrug [52]. However, in the case of **14a**, dipeptides, formed through coupling of the C-2 carboxylic acid with natural amino acids, such as leucine, or coupling of the C-2 amino group with natural amino acids, such as alanine, were surprisingly stable in human, monkey and rat liver S9 fractions. Thus, these dipeptide derivatives were not effective prodrugs of **14a**. Regarding the formation of prodrugs with modifications at the C-6 carboxylic acid, it was found that substituted alkyl esters, such as morpholinoethyl ester and alkoxycarbonyloxymethyl ester, were too unstable under nonenzymatic conditions to be used as prodrugs. The transformation of benzyl ester to **14a** revealed a large difference between humans and monkeys, in that transformation in a human liver S9 fraction was lower than that in a monkey liver S9 fraction. Finally, C-6 alkyl esters showed good to excellent transformation to **14a** in S9 fractions from both human and monkey.

The metabolic stability (transformed percentage to **14a**) of typical C-6 alkyl ester prodrugs of **14a** in liver S9 fractions from rats, monkeys, and humans as well as their pharmacokinetics parameters in rats and monkeys are summarized in Table 3.6. Linear

Table 3.6 *Transformation (%) from prodrug **15** to active substance **14a** and **14i** by liver S9 fractions from rats, monkeys and humans, pharmacokinetics parameters for active substances, **14a** and **14i**, after oral dosing (10 mg/kg) of **14a**, **14i**, and prodrug **15** to rats and monkeys, and antidepressant-like activity of **14a** and its prodrug **15a** (MGS0210)*

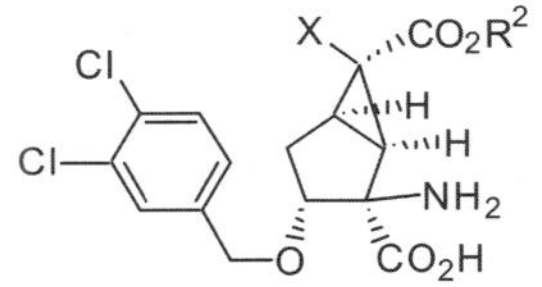

Compound	X	R^2	Metabolic stability			Pharmacokinetic parameter[a] of active form after oral administration of prodrug								*In vivo* pharmacology
			Transformed % to active form in liver S9 fractions			Rat				Monkey				Forced swimming test in rats /Tail suspension test in mice
			Rat	Monkey	Human	Dose (μmol/kg)	t_{max} (h)	C_{max} (μM)	*F* (%)	Dose (μmol/kg)	t_{max} (h)	C_{max} (μM)	*F* (%)	Lowest active dose (mg/kg)
14a (MGS0039)	F	H	NA[b]	NA[b]	0.2	26.4	2.7 ± 1.2	2.5 ± 0.6	10.9	26.4	1.3 ± 0.6	0.8 ± 0.5	12.6	1.0 (i.p.)/1.0 (i.p.)
14i	H	H	NA[b]	NA[b]	NA[b]	27.8	4.7 ± 3.1	0.4 ± 0.1	3.6	NA[b]	NA[b]	NA[b]	NA[b]	NA[b]
15a (MGS0210)	F	*n*-Hep	98.0	42.6	76.9	21.0	1.3 ± 0.6	12.9 ± 0.9	73.0	21.0	3.7 ± 3.8	4.0 ± 1.6	38.6	3.0 (p.o.)/3.0 (p.o.)
15b	F	Me	86.6	26.4	44.2	23.3	1.0 ± 0.0	28.9 ± 7.2	70.6	23.3	3.3 ± 1.2	1.3 ± 0.6	16.9	NA[b]
15c	F	Et	95.9	15.8	28.2	22.6	1.0 ± 0.1	20.7 ± 1.3	66.6	22.6	2.0 ± 0.0	1.4 ± 0.2	10.3	NA[b]

15d	F	n-Pr	100	39.2	39.2	NA[b]	NA[b]	NA[b]	NA[b]	NA[b]	NA[b]	NA[b]	NA[b]	NA[b]
15e	F	*n*-Bu	100	64.2	52.1	21.2	1.3 ± 0.6	17.0 ± 0.5	54.9	21.2	2.7 ± 1.2	4.0 ± 1.0	29.7	NA[b]
15f	F	*n*-Pen	100	68.4	43.9	22.3	1.3 ± 0.6	23.9 ± 4.0	59.1	22.3	3.3 ± 1.2	3.3 ± 0.7	32.9	NA[b]
15g	F	*n*-Hex	99.0	54.0	65.3	21.6	1.0 ± 0.0	20.8 ± 5.2	46.5	21.6	2.3 ± 1.5	3.3 ± 1.4	30.9	NA[b]
15h	F	*n*-Oct	94.9	43.2	68.8	20.4	1.0 ± 0.0	17.7 ± 4.2	43.6	20.4	1.3 ± 0.6	2.3 ± 0.6	25.2	NA[b]
15i	F	*n*-Dec	NA[b]	54.1	55.0	NA[b]	NA[b]	NA[b]	NA[b]	NA[b]	NA[b]	NA[b]	NA[b]	NA[b]
15j	F	*i*-Pr	100	5.9	5.4	NA[b]	NA[b]	NA[b]	NA[b]	NA[b]	NA[b]	NA[b]	NA[b]	NA[b]
15k	F	*i*-Bu	100	57.8	31.1	*21.2*	0.8 ± 0.3	24.1 ± 6.6	74.1	21.2	4.7 ± 3.1	3.8 ± 1.0	50.7	NA[b]
15l	F	3-Me-Bu	93.1	72.8	56.4	22.3	1.3 ± 0.6	17.8 ± 5.4	44.7	22.3	2.0 ± 0.0	5.0 ± 1.8	44.5	NA[b]
15m	F	4-Me-Pen	99.1	66.4	44.2	NA[b]	NA[b]	NA[b]	NA[b]	NA[b]	NA[b]	NA[b]	NA[b]	NA[b]
15n	F	5-Me-Hex	94.6	43.4	71.1	21.0	0.5 ± 0.0	17.3 ± 1.3	39.4	21.0	4.3 ± 3.5	2.7 ± 0.3	44.0	NA[b]
15o	F	6-Me-Hep	90.6	33.3	74.9	21.6	0.7 ± 0.3	15.0 ± 1.2	36.3	21.6	2.0 ± 0.0	2.8 ± 0.6	21.8	NA[b]
15p	F	cyclo-Hex	100	19.4	12.3	NA[b]	NA[b]	NA[b]	NA[b]	NA[b]	NA[b]	NA[b]	NA[b]	NA[b]
15q	F	cyclo-Hex-Me	99.1	69.0	10.0	NA[b]	NA[b]	NA[b]	NA[b]	NA[b]	NA[b]	NA[b]	NA[b]	NA[b]
15r	H	Et	17.8	NA[b]	NA[b]	23.5	2.0 ± 0.0	3.7 ± 0.2	20.2	NA[b]	NA[b]	NA[b]	NA[b]	NA[b]

[a] t_{max}: time to reach maximum plasma concentration; C_{max}: maximum plasma concentration; *F*: oral bioavailability.
[b] Not applicable.

alkyl (**15a–15i**), branched alkyl (**15j–15o**) and cycloalkyl (**15p**, **15q**) esters showed efficient transformation to **14a** in rat liver S9 fractions. However, lower linear alkyl (**15b–15d**), lower branched alkyl (**15j**, **15k**) and cycloalkyl (**15p**, **15q**) esters were too stable in human liver S9 fractions to be used as prodrugs. In contrast, long alkyl esters, especially *n*-heptyl ester **15a**, showed highly efficient transformation to **14a** in human liver S9 fractions, and also at an acceptable level in monkey liver S9 fractions.

Orally administered **14a** (MGS0039) exhibited a low bioavailability in rats ($F = 10.0\%$) and monkeys ($F = 12.6\%$). In contrast, orally administered *n*-heptyl ester prodrug **15a** (MGS0210) showed a much higher bioavailability of the active form, **14a**, in rats ($F = 73.0\%$) and monkeys ($F = 38.6\%$). Thus, prodrug **15a** is expected to have a good bioavailability to transport **14a** in humans based on the efficiency of transformation to **14a** in the S9 fractions of rats (98.0%), monkeys (42.6%), and humans (76.9%). Orally administered **15a** exhibited antidepressant-like effects when evaluated using the forced swimming test in rats (lowest active dose 3 mg/kg) and the tail suspension test in mice (lowest active dose 3 mg/kg) (see Table 3.6) [24].

3.6 Comparison of Fluorinated Glutamate Analogues with the Corresponding Hydrocarbon Analogues

As described above, the introduction of fluorine to conformationally restricted glutamate analogue **4** (LY354740) at a suitable position with appropriate stereochemistry did not affect the *in vitro* efficacy (binding affinity and functional activity). Thus, (+)-**6a** (MGS0008) and (–)-**7a** (MGS0022) exhibited almost the same binding affinities and agonist activities as **4** (see Table 3.1). However, oral activity was dramatically changed by the introduction of fluorine. The oral administration of **4** inhibited the PCP-induced head-weaving behavior at high doses, but did not significantly affect PCP-induced hyperactivity or conditioned avoidance responses in rats. In contrast, the oral administration of (+)-**6a** strongly (11-fold increase) inhibited PCP-induced head-weaving behavior and significantly affected both PCP-induced hyperactivity and conditioned avoidance responses in rats. No direct comparison of the oral pharmacokinetics between (+)-**6a** and **4** has been reported, but it seems reasonable to think that the low oral pharmacokinetic profile of **4** [52] has been significantly improved by the introduction of fluorine.

Concerning mGluR2/3 antagonists, fluorine-containing **14a** (MGS0039) and the corresponding hydrocarbon congener **14i** exhibited the same level affinities ($K_i = 2.38$ nM and 2.51 nM, respectively) and antagonist activities ($IC_{50} = 20.0$ nM and 34.2 nM, respectively) for mGluR2 (see Table 3.4). However, orally administered **14a** in rats ($F = 10.9\%$) showed a markedly higher bioavailability than that of **14i** ($F = 3.6\%$) (see Table 3.6). In addition to the improvement in the pharmacokinetic profile, fluorine incorporation is useful for the design of prodrugs. Ethyl ester prodrug **15c**, bearing fluorine at the C-6 carbon, exhibited a much higher transformation rate (95.9%) to carboxylic acid **14a** in rat liver S9 fractions than did the corresponding **15r** (17.8% to **14i**) (see Table 3.6). This finding suggests that the enzymatic reactivity of the ester moiety for enzymatic hydrolysis is enhanced by the introduction of fluorine to the α-carbon of the carboxylic acid moiety. Reflecting the high metabolic reactivity in rat liver S9 fractions, orally administered ethyl ester **15c**, a prodrug of **14a**, showed much higher bioavailability ($F = 66.6\%$ in the form of **14a**) than **15r** ($F = 20.2\%$ in the form of **14i**) (see Table 3.6), and the plasma level of **15c** was much lower ($C_{max} = 4.0\,\mu M$ at 0.7 h) than that of **14a** ($C_{max} = 20.7\,\mu M$ at 1.0 h). In contrast, the plasma level ($C_{max} = 4.8\,\mu M$ at 0.5 h) of **15r** was as low as that ($C_{max} = 3.7\,\mu M$ at 2.0 h) of **14i** transformed from **15r**. These findings indicated that prodrug **15c** was more effectively absorbed and more rapidly transformed to **14a** than prodrug **15r** of **14i** was, in oral administration.

3.7 Conclusion

The low oral activity of mGluR2/3 agonist **4** (LY354740) with regard to its antipsychotic-like effects in laboratory animal models was improved by the incorporation of fluorine into the bicyclo[3.1.0]hexane ring of **4** without changing the potency of the compound's affinity and agonist activities for mGluR2/3 as represented by (+)-**6a** (MGS0008). Furthermore, mGluR2/3 antagonist **14a** (MGS0039), derived from mGluR2/3 agonist (−)-**7a** (MGS0022) by the introduction of a 3,4-dichlorobenzyloxy group at the C-3 position, showed a higher bioavailability than the corresponding hydrocarbon congener **14i**, with no change in the potency of binding affinity and antagonist activity for mGluR2. The low oral bioavailability of **14a** was clearly improved by C-6 alkyl ester prodrug **15a**, which was discovered by exploiting a high enzymatic reactivity of the α-fluorocarboxylic acid moiety of **14**a. Thus, the incorporation of fluorine would provide a useful strategy for improving oral pharmacokinetic profiles, including those of prodrugs, in drug discovery and development.

References

1. Colinridge, G. L. and Lester, R. A. (1989) Excitatory amino acid receptors in the vertebrate central nervous system. *Pharmacological Reviews*, **40**, 143–210.
2. Monaghan, D. T., Bridges, R. J. and Cotman, C. W. (1989) The excitatory acid receptors: their classes, pharmacology, and distinct properties in the function of the central nervous system. *Annual Review of Pharmacology and Toxicology*, **29**, 365–402.
3. Nakanishi, S. (1989) Molecular diversity of glutamate receptors and implications for brain function. *Science*, **258**, 597–603.
4. Pin, J. P. and Duvoisin, R. (1992) The metabotropic glutamate receptors: structure and functions. *Journal of Neurochemistry*, **58**, 1184–1186.
5. Schoepp, D. D. and Conn, P. J. (1993) Metabotropic glutamate receptors in brain function and pathology. *Trends in Pharmacological Sciences*, **14**, 13–20.
6. Bockaert, J., Pin, J. and Fagni, L. (1993) Metabotropic glutamate receptors: an original family of G protein-coupled receptors. *Fundamental & Clinical Pharmacology*, **7**, 473–485.
7. Nakanishi, S. and Masu, M. (1994) Molecular diversity and functions of glutamate receptors. *Annual Review of Biophysics and Biomolecular Structure*, **23**, 319–348.
8. Hollmann, M. and Heinemann, S. (1994) Cloned glutamate receptors. *Annual Reviews of Neuroscience*, **17**, 31–108

9. Conn, P. J. and Pin, J-P. (1997) Pharmacology and functions of metabotropic glutamate. receptors. *Annual Review of Pharmacology and Toxicology*, **37**, 205–237.
10. Schoepp, D. D., Johnson, B. G. and Monn, J. A. (1992) Inhibition of cyclic AMP formation by a selective metabotropic glutamate receptor agonist. *Journal of Neurochemistry*, **58**, 1184–1186.
11. Nakajima, Y., Iwakabe, H., Akazawa, C., *et al.* (1993) Molecular characterization of a novel retinal metabotropic glutamate receptor mGluR6 with a high agonist selectivity for L-2-amino-4-phosphonobutyrate. *Journal of Biological Chemistry*, **268**, 11868–11873.
12. Meldrum, B. S. (2000) Glutamate as a neurotransmitter in the brain: review of physiology and pathology. *Journal of Nutrition*, **130**, 1007S–1015S.
13. Rorick-Kehn, L. M., Johnson, B. G., Knitowski, K. M., *et al.* (2007) *In vivo* pharmacological characterization of the structurally novel, potent, selective mGlu2/3 receptor agonist LY404039 in animal models of psychiatric disorders. *Psychopharmacology*, **193**, 121–136.
14. Rorick-Kehn, L. M., Johnson, B. G., Burkey, J. L., *et al.* (2007) Pharmacological and pharmacokinetic properties of a structurally-novel, potent, selective mGlu2/3 receptor agonist: in vitro characterization of LY404039. *Journal of Pharmacology and Experimental Therapeutics*, **321**, 308–317.
15. Nakazato, A., Kumagai, T., Sakagami, K., *et al.* (2000) Synthesis, SARs, and pharmacological characterization of 2-amino-3 or 6-fluorobicyclo[3.1.0]hexane-2,6-dicarboxylic acid derivatives as potent, selective, and orally active group II metabotropic glutamate receptor agonist. *Journal of Medicinal Chemistry*, **43**, 4893–4909.
16. Cartmell, J., Monn, J. A. and Schoepp, D. D. (2000) Attenuation of specific PCP-evoked behaviors by the potent mGluR2/3 receptor agonist, LY379268 and comparison with the atypical antipsychotic, clozapine. *Psychopharmacology*, **148**, 423–429.
17. Moghaddam, B. and Adams, B. W. (1998) Reversal of phencyclidine effects by a group II metabotropic glutamate receptor agonist in rats. *Science*, **281**, 1349–1352.
18. Grillon, C., Cordova, J., Levine, L. R. and Morgan, C. A. 3rd. (2003) Anxiolytic effects of a novel group II metabotropic glutamate receptor agonist (LY354740) in the fear-potentiated startle paradigm in humans. *Psychopharmacology*, **168**, 446–454.
19. Tizzano, J. P., Griffey, K. and Schoepp, D. D. (2002) The anxiolytic action of mGlu2/3 receptor agonist, LY35(4740) in the fear-potentiated startle model in rats in mechanistically distinct from diazepam. *Pharmacology, Biochemistry, and Behavior*, **73**, 367–374.
20. Helton, D. R., Tizzano, J. P., Monn, J. A., *et al.* (1998) Anxiolytic and side-effect profile of LY354740: a potent, highly selective, orally active agonist for group II metabotropic glutamate receptors. *Journal of Pharmacology and Experimental Therapeutics*, **284**, 651–660.
21. Monn, J. A., Valli, M. J., Massey, S. M., *et al.* (1997) Design, synthesis, and pharmacological characterization of (+)-2-aminobicyclo[3.1.0]hexane-2,6-dicarboxylic acid (LY354740): a potent, selective, and orally active group 2 metabotropic glutamate receptor agonist possessing anticonvulsant and anxiolytic properties. *Journal of Medicinal Chemistry*, **40**, 528–537.
22. Levine, L., Gaydos, B., Sheehan, D., *et al.* (2002) The mGlu2/3 receptor agonist, LY35(4740) reduces panic anxiety induced by a CO_2 challenge in patients diagnosed with panic disorder. *Neuropharmacology*, **43**, 294.
23. Kawashima, N., Karasawa, J., Shimazaki, T., *et al.* (2005) Neuropharmacological profiles of antagonists of group II metabotropic glutamate receptors. *Neuroscience Letters*, **378**, 131–134.
24. Chaki, S., Yoshikawa, R., Hirota, S., *et al.* (2004) MGS0039: a potent and selective group II metabotropic glutamate receptor antagonist with antidepressant-like activity. *Neuropharmacology*, **46**, 457–467.
25. Nakazato, A., Sakagami, K., Yasuhara, A., *et al.* (2004) Synthesis, in vitro pharmacology, structure-activity relationships and pharmacokinetics of 3-alkoxy-2-amino-6-fluorobicyclo[3.1.0]

hexane-2,6-dicarboxylic acid derivatives as potent and selective group II metabotropic glutamate receptor antagonists. *Journal of Medicinal Chemistry*, **47**, 4570–4587.

26. Karasawa, J., Shimazaki, T. and Chaki, S. (2006) A metabotropic glutamate 2/3 receptor antagonist, MGS(0039) increases extracellular dopamine levels in the nucleus accumbens shell. *Neuroscience Letters*, **393**, 127–130.
27. Yoshimizu, T. and Chaki, S. (2004) Increased cell proliferation in the adult mouse hippocampus following chronic administration of group II metabotropic glutamate receptor antagonist, MGS0039. *Biochemical and Biophysical Research Commununications*, **315**, 493–496.
28. Karasawa, J., Shimazaki, T., Kawashima, N. and Chaki, S. (2005) AMPA receptor stimulation mediates the antidepressantplike effect of a group II metabotropic glutamate receptor antagonist. *Brain Res.* **1042** (1), 92– 98.
29. Yoshimizu, T., Shimazaki, T., Ito, A. and Chaki, S. (2006) An mGluR2/3 antagonist, MGS(0039) exerts antidepressant and anxiolytic effects in behavioral models in rats. *Psychopharmacology*, **186**, 587–593.
30. Shimazaki, T., Iijima, M. and Chaki, S. (2004) Anxiolytic-like activity of MGS(0039) a potent group II metabotropic glutamate receptor antagonist, in a marble-burying behavior test. *European Journal of Pharmacology*, **501**, 121–125.
31. Iijima, M., Shimazaki, T., Ito, A. and Chaki, S. (2007) Effects of metabotropic glutamate 2/3 receptor antagonists in the stress-induced hyperthermia test in singly housed mice. *Psychopharmacology*, **190**, 233–239.
32. Ornstein, P. L., Bleisch, T. J., Arnold, M. B., *et al.* (1998) 2-Substituted (2*SR*)-2-amino-2-((1*SR*,2*SR*)-2-carboxycycloprop-1-yl)glycines as potent and selective antagonists of group II metabotropic glutamate receptors. 1. Effects of alkyl, arylalkyl, and diarylalkyl substitution. *Journal of Medicinal Chemistry*, **41**, 346–357.
33. Ornstein, P. L., Bleisch, T. J., Arnold, M. B., *et al.* (1998) 2-Substituted (2*SR*)-2-amino-2-((1*SR*,2*SR*)-2-carboxycycloprop-1-yl)glycines as potent and selective antagonists of group II metabotropic glutamate receptors. 2. Effects of aromatic substitution, pharmacological characterization, and bioavailability. *Journal of Medicinal Chemistry*, **41**, 358–378.
34. Jane, D. E., Jones, P. L., Pook, P. C., Tse, H. W. and Watkins, J. C. (1994) Actions of two new antagonists showing selectivity for different subtypes of metabotropic glutamate receptor in the neonatal rat spinal cord. *British Journal of Pharmacology*, **112**, 809–816.
35. Pelliciari, R., Marinozzi, M., Natalini, B., *et al.* (1996) Synthesis and pharmacological characterization of all sixteen stereoisomers of 2-(2′-carboxy-3′-phenylcyclopropyl)glycine. Focus on (2*S*,1′*S*,2′*S*,3′*S*)-2-(2′-carboxy-3′-phenylcyclopropyl)glycine, a novel and selective group II metabotropic glutamate receptor antagonist. *Journal of Medicinal Chemistry*, **39**, 2259–2269.
36. (a) Knoflach, F., Woltering, T., Adam, G., *et al.* (2001) Pharmacological property of native metabotropic glutamate receptors in freshly dissociated Golgi cells of the rat cerebellum. *Neuropharmacology*, **40**, 163–169. (b) Kew, J. N. C., Ducarre, J.-M., Woltering, T., *et al.* (1999) Frequency-dependent presynaptic autoinhibition by group II metabotropic glutamate receptors at the medial perforant path input to the dentate gyrus. *29th Annuual Meeting of the Society for Neurosciience*, October 23–28, 1999, Miami, Society for Neuroscience, Washington D.C., Abst. 494.10. (c) Adam, G., Huguenin-Virchaux, P. N., *et al.* (2000). 2-Amino-bicyclo[3.1.0]hexane-2,6-dicarboxylic acid derivatives and a process for the preparation thereof. U. S. Patent US6107342.
37. Dominguez, C., Prieto, L., Valli, M. J., *et al.* (2005) Methyl substitution of 2-aminobicyclo[3.1 .0]hexane 2,6-dicarboxylate (LY354740) determines functional activity at metabotropic glutamate receptors: identification of a subtype selective mGlu2 receptor agonist. *Journal of Medicinal Chemistry*, **48**, 3605–3612.
38. Welch, J. T. (1987) Advances in the preparation of biologically active organofluorine compounds. *Tetrahedron*, **43**, 3123–3197.

39. Yasuhara, A., Nakamura, M., Sakagami, K., *et al.* (2006) Prodrugs of 3-(3,4-dichlorobenzyloxy)-2-amino-6-fluorobicyclo[3.1.0]hexane-2,6-dicarboxylic acid (MGS0039): A potent and orally active group II mGluR antagonist with antidepressant-like potential. *Bioorganic & Medicinal Chemistry Letters*, **14**, 4193–4207.
40. Nakamura, M., Kawakita, Y., Yasuhara, A., *et al.* (2006) In vitro and in vivo evaluation of the metabolism and bioavailability of ester prodrugs of MGS0039 (3-(3,4-dichlorobenzyloxy)-2-amino-6-fluorobicyclo[3.1.0]hexane-2,6-dicarboxylic acid), a potent metabotropic glutamate receptor antagonist. *Drug Metabolism and Disposition*, **34**, 369–374.
41. Sakagami, K., Kumagai, T., Taguchi, T. and Nakazato, A. (2007) Scalable synthesis of (+)-2-amino-3-fluorobicyclo[3.1.0]hexane-2,6-dicarboxylic acid as a potent and selective group II metabotropic glutamate receptor agonist. *Chemical & Pharmaceutical Bulletin*, **55**, 37–43.
42. Wong, A., Welch, C. J., Kuethe, J. T., *et al.* (2004) Reactive resin facilitated preparation of an enantiopure fluorobicycloketone. *Organic & Biomolecular Chemistry*, **2**, 168–174.
43. Zhang, F., Song, Z. J., Tschaen, D. and Volante, R. P. (2004) Enantioselective preparation of ring-fused 1-fluorocyclopropane-1-carboxylate derivatives: Ene route to mGluR 2 receptor agonist MGS0028. *Organic Letters*, **6**, 3775–3777.
44. Yoshikawa, N., Tan, L., Yasuda, N., *et al.* (2004) Enantioselective syntheses of bicyclo[3.1.0]hexane carboxylic acid derivatives by intramolecular cyclopropanation. *Tetrahedron Letters*, **45**, 7261–7264.
45. Tan, L., Yasuda, N., Yoshikawa, N., *et al.* (2005) Stereoselective syntheses of highly functionalized bicycle[3.1.0]hexanes: A general methodology for the synthesis of potent and selective mGluR2/3 agonists. *Journal of Organic Chemistry*, **70**, 8027–8034.
46. Differding, E. and Ofner, H. (1991) N-Fluorobenzenesulfonimide: a practical reagent for electrophilic fluorinations. *Synlett*, 7–189.
47. Lal, G. S., Pez, G. P. and Syvret, R. G. (1996) Electrophilic NF fluorinating agents. *Chemical Reviews* **96**, 1737–1755.
48. Ohshima, E., Takatsuto, S., Ikekawa, N. and Deluca, H. F. (1984) Synthesis of 1α-fluorovitamine D_3. *Chemical & Pharmaceutical Bulletin*, **32**, 3518–3524.
49. Charles, R. G. (1957) Copper (II) and nickel (II) *N*-(*n*-alkyl)salicylaldimine chelates. *Journal of Organic Chemistry*, **22**, 677–679.
50. Pellicciari, R., Marinozzi, M., Natalini, B., *et al.* (1996) Synthesis and pharmacological characterization of all sixteen stereoisomers of 2-(2′-carboxy-3′-phenylcyclopropyl)glycine, a novel and selective group II metabotropic glutamate receptors antagonist. *Journal of Medicinal Chemistry*, **39**, 2259–2269.
51. Takamori, K., Hirota, S., Chaki, S. and Tanaka, M. (2003) Atipsychotic action of selective group II metabotropic glutamate receptor agonist MGS0008 and MGS0028 on conditioned avoidance responses in the rat. *Life Sciences*, **73**, 1721–1728.
52. Bueno, A. B., Collado, I., de Dios, A., *et al.* (2005) Dipeptides as effective prodrugs of the unnatural amino acid (+)-2-aminobicyclo[3.1.0]hexane-2,6-dicarboxylic Acid (LY354740), a selective group II metabotropic glutamate receptor agonist. *Journal of Medicinal Chemistry*, **48**, 5305–5320.
53. James, J. K., Nakamura, M., Nakazato, A., *et al.* (2005) Metabolism and disposition of a potent group II metabotropic glutamate receptor agonist, in rats, dogs and monkeys. *Drug Metabolism and Disposition*, **33**, 1373–1381.
54. Shao, H. and Goodman, M. (1996) An enantionmeric synthesis of *allo*-threonines and β-hydroxyvalines. *Journal of Organic Chemistry*, **61**, 2582–2583.
55. Radüchel, B. (1980) Inversion of configuration of secondary alcohols, in particular in the steroid and prostaglandin series. *Synthesis*, 292–295.
56. Jauch, J. (2001) A short synthesis of Mniopetal F. *European Journal of Organic Chemistry*, 473–476.

57. Wessel, H, -P., Iversen, T. and Bundle, D. R. (1985) Acid-catalysed benzylation and allylation by alkyl trichloroacetimidates. *Journal of the Chemical Society. Perkin Transactions* **1**, 2247–2250.
58. Widmer, U. (1987) A convenient benzylation procedure for β-hydoxy esters. *Synthesis*, 568–570.
59. Vaultier, M., Knouzi, N., Carrié, R. (1983) Reaction D'azines en amines primaries par une methode generale utilisant la reaction de Staudinger. *Tetrahedron Letters*, **24**, 763–764.
60. Gololobov, Y. G., Zhmurova, I. N. and Kasukhin, L. F. (1981) Sixty years of Staudinger reaction. *Tetrahedron*, **37**, 437–472.
61. Yasuhara, A., Sakagami, K., Yoshikawa, R., *et al.* (2006) Synthesis, in vitro pharmacology, and structure–activity relationships of 2-aminobicyclo[3.1.0]hexane-2,6-dicarboxylic acid derivatives as mGluR2 antagonists. *Bioorganic & Medicinal Chemistry*, **14**, 3405–3420.
62. Wright, R. A., Arnold, M. B., Wheeler, W. J., *et al.* (2000) Binding of [^{3}H](2S,1′S,2′S)-2-(9-xanthylmethyl)-2-(2′-carboxycyclopropyl)glycine ([^{3}H]LY341495) to cell membranes expressing recombinant human group III metabotropic glutamate receptor subtypes. *Naunyn-Schmiedeberg's Archives of Pharmacology*, **362**, 546–554.

4

Fluorinated Inhibitors of Matrix Metalloproteinases

Roberta Sinisi, Monika Jagodzinska, Gabriele Candiani, Florent Huguenot, Monica Sani, Alessandro Volonterio, Raffaella Maffezzoni, and Matteo Zanda

4.1 Introduction

Matrix metalloproteinases are a family of highly homologous Zn(II)-endopeptidases that collectively cleave a large number of the constituents of the extracellular matrix [1–8]. More than 20 human MMPs are known such as (i) the collagenases (MMP-1, MMP-8, MMP-13), which can degrade fibrillar collagens that are the major components of bone and cartilage; (ii) the gelatinases (MMP-2 and MMP-9), whose main substrates are denatured collagens (gelatins); and (iii) the stromelysins (MMP-3, MMP-10, MMP-11), which have a broad spectrum of matrix components as substrates except for those of collagenases.

MMPs play a pivotal role in a number of physiological processes, such as degradation of extracellular matrix and connective tissue remodeling. Some of them, e.g., MMP-9, are required for bone remodeling, wound healing, angiogenic revascularization of ischemic tissues, and so on, but are also implicated in a number of pathological processes in humans, such as cancer cell invasion, metastasis (especially MMP-2 and MMP-9), inflammatory and autoimmune diseases, arthritis (in particular MMP-1). Moreover it has been shown [5–7a] that several MMPs are overexpressed in various neoplasias, particularly in early growth and establishment of the tumors. Selective inhibition of MMPs might therefore represent an attractive strategy for therapeutic intervention. A number of rationally designed MMP inhibitors have shown some promise in the treatment of pathologies in which MMPs are involved in (see above). However, most of these, such as broad-spectrum MMP

Fluorine in Medicinal Chemistry and Chemical Biology Edited by Iwao Ojima

inhibitors (e.g., marimastat and batimastat, as well as selective inhibitors such as Trocade (cipemastat) (MMP-1)) have performed poorly in clinical trials [7b]. The failure of these MMP inhibitors has been largely due to toxicity (particularly musculoskeletal toxicity in the case of broad-spectrum inhibitors) and failure to show expected results (in the case of Trocade, promising results in rabbit arthritis models were not replicated in human trials). The reasons behind the largely disappointing clinical results of MMP inhibitors are unclear, especially in light of their activity in animal models. This disappointing situation initially dampened interest and enthusiasm for MMPs, but it is now becoming increasingly accepted that the impressive diversity in both substrates and functions of MMPs requires a much deeper knowledge of the physiological functions of these proteinases, which was not available at the time that the initial MMP inhibitor clinical trials were designed. This is expected to lead to a better understanding of the complex roles of these multifunctional enzymes in human pathology and, hopefully, to the design of improved MMP-inhibition strategies for therapeutic intervention.

In 2002, our group undertook a research project aimed at studying the "fluorine effect" in peptides and identifying selective fluorinated inhibitors of MMPs [9]. In fact, incorporation of fluorine into organic molecules is an effective strategy for improving and modifying their biological activity [10, 11]. In particular, the trifluoromethyl group occupies a prominent position in medicinal chemistry as a substituent with peculiar properties. It is highly hydrophobic, electron-rich, and sterically demanding. Moreover, it can provide high *in vivo* stability and features good mimicry (in terms of biological results) of several naturally occurring residues such as methyl, isopropyl, isobutyl, phenyl, and so on. However, we were also interested in studying the effect of the incorporation of other fluoroalkyl moieties (CHF_2, CH_2F, $CClF_2$, CF_2CF_3, etc.)having different degrees of electron-density, polarization, hydrophobicity, and steric demands, to investigate their effect on the MMP/inhibitor affinity, which depends on the inhibitor conformation as well as non-covalent ligand–receptor interactions such as hydrophobic contacts, Van der Waals interactions, and hydrogen bonding.

4.2 α-CF_3-Malic Hydroxamate Inhibitors

Some years ago, a new family of potent peptidomimetic hydroxamate inhibitors (such as **A**, see Figure 4.1) of MMP-1, MMP-3, and MMP-9 was described bearing a quaternary α-methylcarbinol moiety at the P1 position, and several different R^1 groups at P1′ [12, 13]. Interestingly, the other stereoisomers, including the epimers at the quaternary carbinol function, showed much lower activity, as the hydroxamic binding function was moved away from the catalytic Zn^{2+} center. The crystal structure of the inhibitor **A** ($R = CH_3$) complexed with MMP-3 revealed several interesting features, including the presence of a hydrogen bond between the quaternary hydroxyl (H-bond donor) of **A** and the Glu-202 residue of the MMP-3 active site.

We therefore decided to explore the effect of the replacement of the quaternary α-methyl group in A with a CF_3 group, in the hope of (i) increasing the affinity of the α- CF_3 malic inhibitors with MMPs by reinforcing the α-OH hydrogen bonding with increased acidity of the carbinol function bearing the electron-withdrawingα-CF_3 group and (ii)

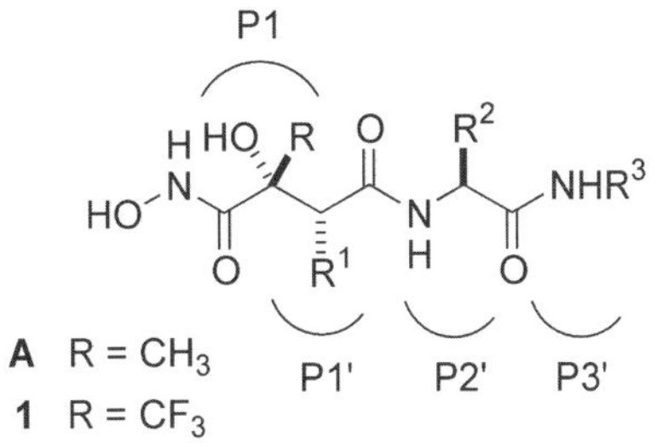

Figure 4.1 *Jacobsen's potent peptidomimetic hydroxamate inhibitors.*

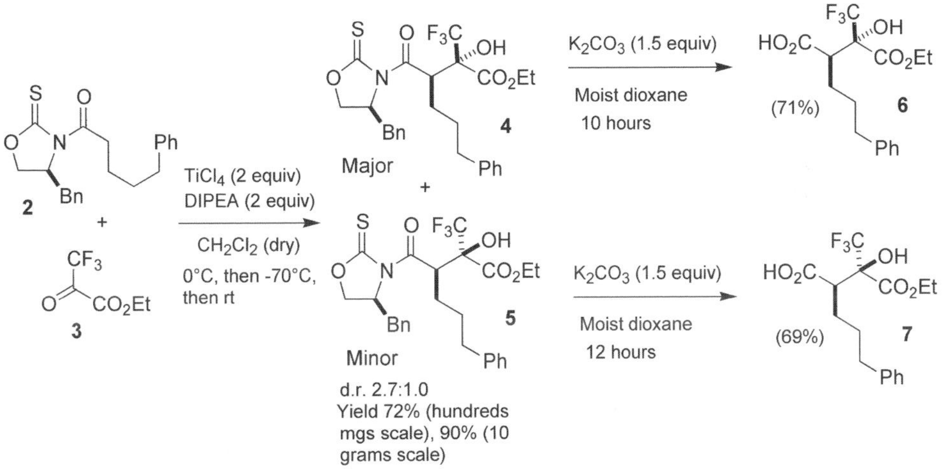

Scheme 4.1 *Aldol reaction to form the α-CF_3-malic framework and cleavage of the oxazolidin-2-thione auxiliary with K_2CO_3 in moist dioxane.*

improving the selectivity in favor of MMP-3 and MMP-9 through the increased stereo-electronic demands of the CF_3 group [14, 15].

The $TiCl_4$-catalyzed reaction of *N*-acyloxazolidin-2-thione 2 (see Scheme 4.1) with ethyl trifluoropyruvate **3** afforded two diastereomeric adducts **4** and **5**, out of four possibilities, in low diastereomeric ratio. The reaction features a favorable scale-up effect, affording ~70% yield on a hundred-milligram scale, and 90% on a ten-gram scale. Several alternative conditions were explored, but no improvement in diastereocontrol was achieved.

Cleavage of the oxazolidin-2-thione auxiliary was found to be considerably more challenging than expected, mainly due to α-epimerization. After considerable experimentation it was found that solid K_2CO_3 in moist dioxane (rt, 10–12 h), was able to produce directly the key carboxylic acid intermediates **6** and **7** (see Scheme 4.1), from **4** and **5**, respectively, in satisfactory yields and with only minor α-epimerization (2% for **6**, 9% for **7**).

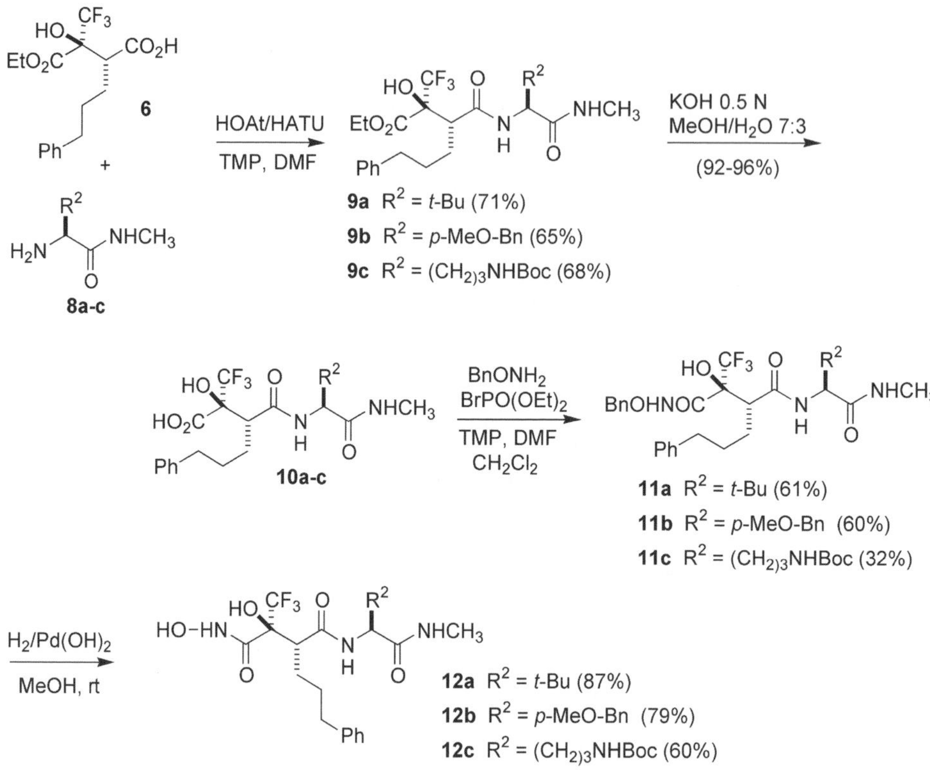

Scheme 4.2 *Synthesis of the peptidomimetics* ***12a–c*** *from the major diastereomer* ***6****.*

Coupling of the acid **6** with α-amino acid amides **8a–c** using HOAt/HATU occurred in good yields (see Scheme 4.2) [16]. The resulting peptidomimetic esters **9a–c** were submitted to basic hydrolysis, affording the acids **10a–c** in high yields. The subsequent coupling of 10a-c with *O*-Bn hydroxylamine proved to be extremely challenging, owing to the low reactivity and high steric hindrance of the carboxylic group bound to the quaternary α-CF_3 carbinol center. A number of "conventional" coupling agents for peptides [17] were tested, but the target *O*-Bn hydroxamates 11a-c were not obtained. Eventually, we found that freshly prepared $BrPO(OEt)_2$ was able to promote the coupling in reasonable yields (32–61%) [18, 19]. With **11a–c** in hand, we carried out the final *O*-Bn hydrogenolysis, which afforded the hydroxamates **12a–c** in good yields.

Since **12a–c** do not have the correct stereochemistry with respect to **A**, we decided to synthesize at least one exact analogue in order to have a complete set of biological data on the effect of the introduction of the CF_3 group. However, a new synthetic protocol had to be developed *ex novo*, because the minor diastereomer **7** (see Scheme 4.3) exhibited a dramatically different reactivity in the key-steps of the synthesis. First of all, the coupling of 7 and **8a** with HOAt/HATU gave rise to the formation of substantial amounts of the β-lactone **13**, which had to be processed separately, as well as the expected coupling product

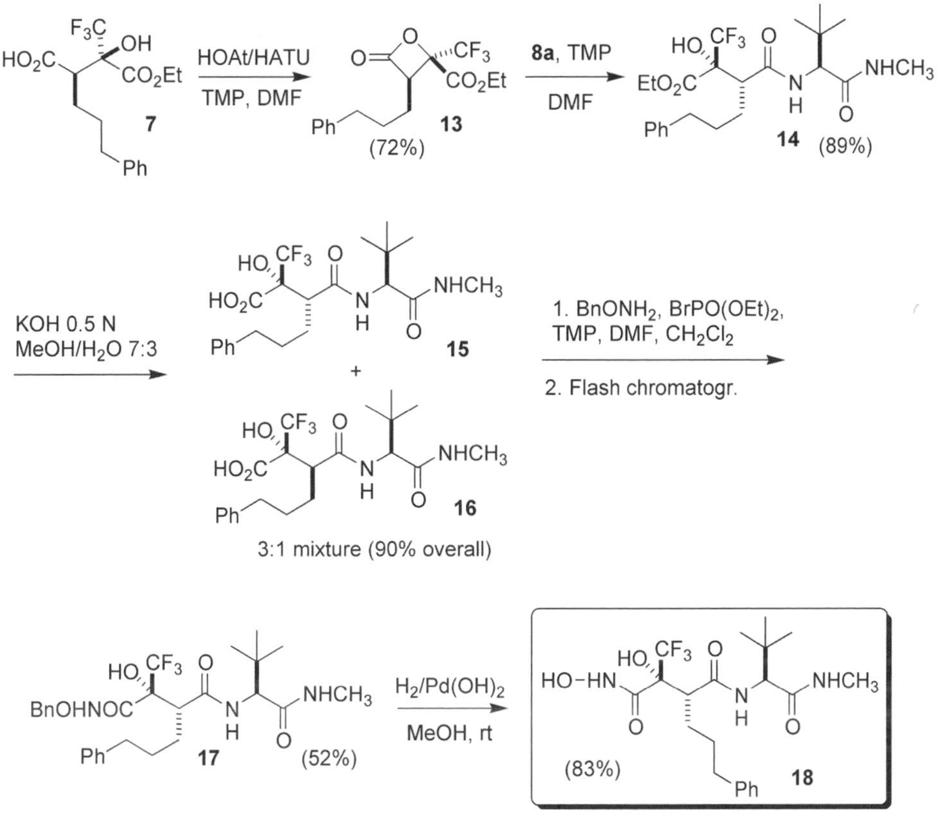

Scheme 4.3 *Synthesis of the target peptidomimetic* ***18*** *from the minor diastereomer* ***7****.*

14. Thus, the intermediate **13** was first prepared (72%), purified by short flash chromatography (FC), and then reacted with **8a** to afford the desired molecule **14** in high yields [20].

Saponification of the ester **14** proceeded effectively, but disappointingly a partial epimerization of the [$Ph(CH_2)_3$]-stereocenter occurred, affording a 3 : 1 mixture of diastereomers **15** and **16** under optimized conditions. Epimers **15** and **16**, which are difficult to separate by FC, were subjected together to coupling with $BnONH_2$ and the resulting diastereomeric *O*-Bn hydroxamates were separated by FC, affording pure **17** (52%), which was hydrogenated to the target free hydroxamate **18** in 83% yield.

The hydroxamates **12a–c** and **18** were tested for their ability to inhibit MMP-2 and MMP-9 activity using zymographic analysis. The IC_{50} values (μM) portrayed in Table 4.1 show that diastereomers **12a–c** displayed low inhibitory activity in line with the parent CH_3 compounds. Disappointingly, **18** showed a much lower activity than the exact CH_3-analogue **A** that was reported to be a low-nanomolar inhibitor of MMP-9. It is also worth noting that **12a** and **18** showed little selectivity, whereas **12b** and **12c** showed a better affinity for MMP-9 than for MMP-2.

Table 4.1 *IC_{50} values (μM) of the target CF_3-hydroxamates*

Compound	IC_{50} (μM)	
	MMP-2	MMP-9
12a	156	121
12b	407	84
12c	722	23
18	23	15

To understand the reasons for this unfavorable "fluorine-effect," we performed a molecular modeling study that allowed us to identify two concurrent reasons for the reduced activity of the fluorinated inhibitors: (i) reduced coordinating strength of the hydroxamate group close to the CF_3, and (i) the need for the fluorinated molecule to adopt, within the binding site, a conformation that does not coincide with its minimum-energy conformation in solution. Assuming additivity of these effects, we estimated that the overall binding energy of the fluorinated inhibitor **18** to the active site is reduced by approximately 11.3 kJ/mol compared with the original one (**A**). This result, at room temperature, of the decrease in the binding constant by two orders of magnitude is roughly in line with the experimental observation.

4.3 α-Trifluoromethyl-α-amino-β-sulfone Hydroxamate Inhibitors

In order to further probe the importance of the reduced coordinating strength of the zinc(II)-binding hydroxamate group and assess the compatibility of a CF_3 group in the α-position to the hydroxamic function, we undertook a study to investigate to the effect of a CF_3 group positioned as the R^1 substituent in structures **19** (see Figure 4.2). These were analogues of molecules **B** (see Figure 4.2) that were recently reported by Becker *et al.* [21] to be potent inhibitors of MMP-2, MMP-9, and MMP-13 [22]. Remarkably, these molecules exhibited limited inhibition of MMP-1, an enzyme thought to be responsible for the musculoskeletal side-effect observed clinically with the broad-spectrum MMP inhibitor marimastat [21, 23]. Although a large number of different alkyl and alkylaryl residues were well tolerated as nitrogen substituents R^2, only inhibitors **B** bearing R^1 = H, CH_3 or Ph were reported.

In the synthesis of hydroxamic acid **19a**, having a free quaternary amino group (see Scheme 4.4), the intermediate sulfone **21** was synthesized by Pd-catalyzed reaction of phenol with *p*-bromo derivative **20** [24]. Lithiation of **20**, followed by nucleophilic addition to the *N*-Cbz imine of trifluoropyruvate **22** [25] afforded the α-CF_3 α-amino acid derivative **23** in fair yields. Basic hydrolysis of the ester function gave the carboxylic acid **24**, which was submitted to condensation with *O*-Bn-hydroxylamine, affording hydroxamate **25**. The subsequent hydrogenolysis of **25** afforded the target molecule **19a**.

N-Alkylated analogues **19b–d** (see Scheme 4.5) were prepared using a modified procedure. To this end, sulfenyl diaryl ether **27** was prepared from phenol and **26** using an Ullmann-type reaction [26], and was then oxidized to sulfoxide **28**. Lithiation and

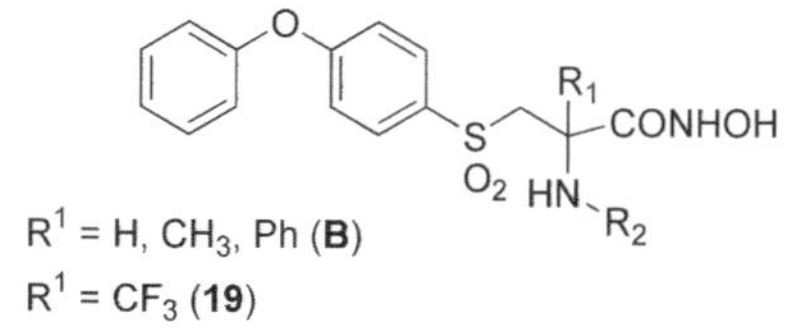

R^1 = H, CH_3, Ph (**B**)
R^1 = CF_3 (**19**)

Figure 4.2 *α-Amino hydroxamic MMP inhibitors.*

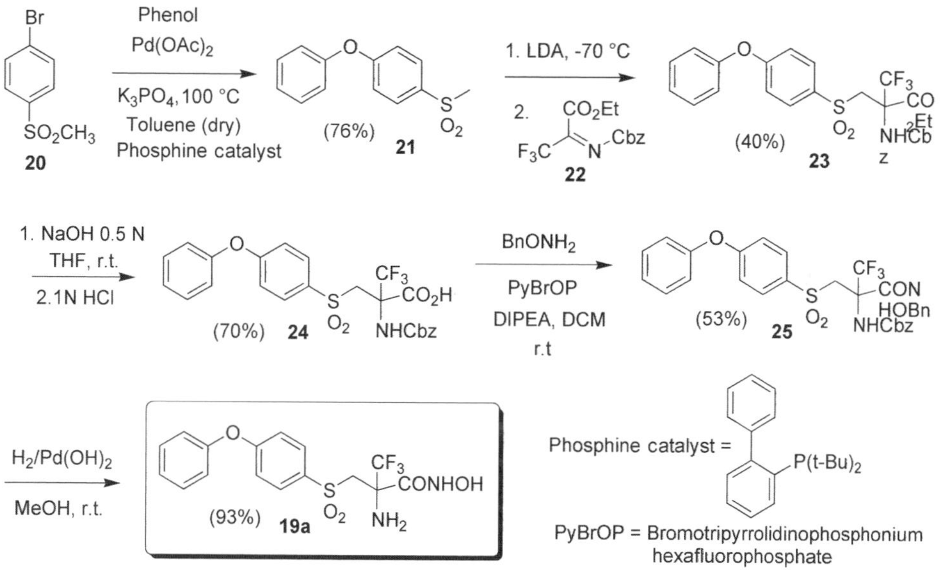

Phosphine catalyst =
PyBrOP = Bromotripyrrolidinophosphonium hexafluorophosphate

Scheme 4.4 *Synthesis of **19a**.*

Mannich-type reaction with **22** afforded a nearly equimolar mixture of sulfoxide diastereomers **29**, which were deoxygenated to racemic sulfide **30** using the Oae/Drabowicz protocol [27]. *N*-Alkylation occurred in good to excellent yields, affording the corresponding sulfides **31b–d**. Because of the presence of the sulfide functionality, which could interfere with a Pd-catalyzed hydrogenolysis, the Cbz group was cleaved with HBr [28], affording secondary amines **32b–d** in nearly quantitative yields. Ester saponification was readily performed, affording the carboxylic acids **33b–d** in good to excellent yields. Coupling of **33b–d** with *O*-Bn-hydroxylamine afforded the sulfenyl hydroxamates **34b–d**, which were oxidized to sulfones **35b–d**. The target hydroxamic acids **19b–d** were obtained in fair yields by hydrogenolysis with the Pearlman catalyst.

Enzyme inhibition assays on **19a–d** were performed on the catalytic domains of MMP-1, MMP-3, and MMP-9. the results are summarized in Table 4.2. As Table 4.2 shows, primary α-amino hydroxamate **19a** is the most potent compound, but it is worth noting that **19a–d** are all nanomolar inhibitors of MMP-3 and MMP-9. Even more importantly, **19a** showed excellent selectivity for MMP-9 as compared with that for MMP-1

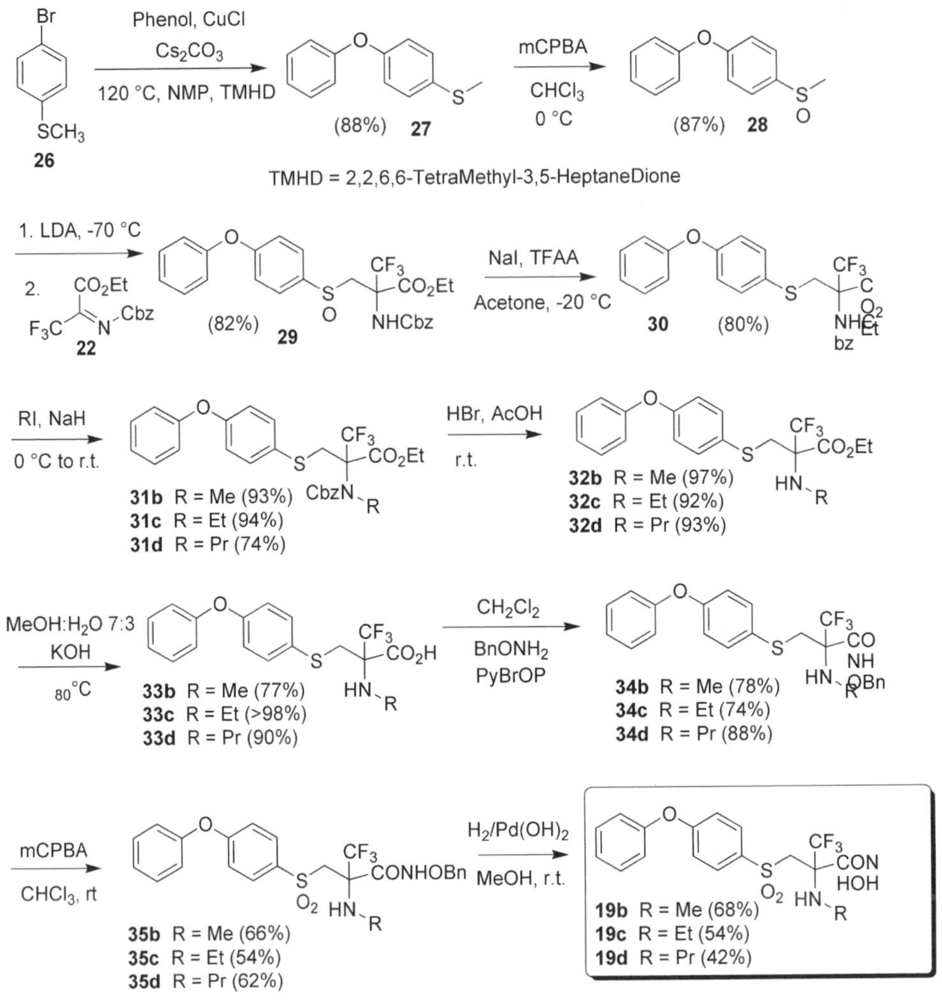

Scheme 4.5 *Synthesis of **19b–d**.*

Table 4.2 *Effect of the compounds **19a–d** on the proteolytic activity of different MMPs*

Compound	IC_{50}/MMP-3 (nM)	IC_{50}/MMP-9 (nM)	IC_{50}/MMP-1 (nM)
19a	14	1	>5000
19b	32	~20	n.a.
19c	28	63	n.a.
19d	53	59	n.a.

n.a.: not available.

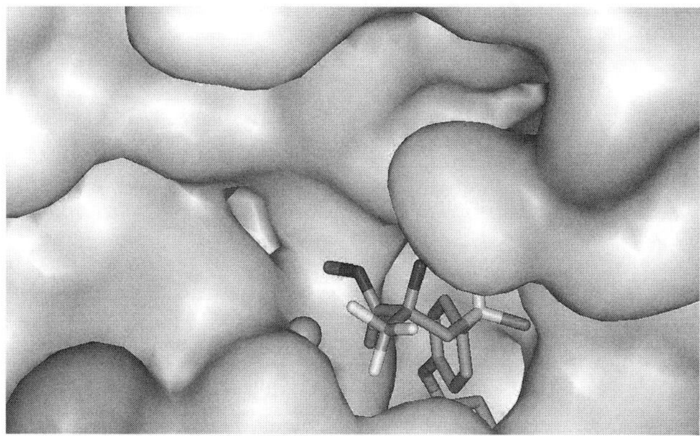

Figure 4.3 *(R)-**19a** in the MMP-9 active site (purple = Zn(II), cyan = F, red = O, blue = N, yellow = S). See color plate 4.3.*

(>5000-fold). These results show that a CF_3 group can be successfully used as a substituent in MMPs inhibitors, and is very well tolerated by the enzymes. This also suggests that the weak potency of compound **18** may be mainly due to a conformational change induced by the CF_3 group that lowers the affinity for the MMP active site.

Interestingly, the x-ray crystallographic structure of the complex of **19a** with the truncated catalytic domain of MMP-9 showed that the (*R*)-enantiomer binds preferentially, whereas the CF_3 group does not make significant interactions with the active-site residues of the protease, that is, it is essentially exposed to water (see Figure 3.3) [29]. This finding is in contrast with a previous crystallographic structure of a bis-CF_3-pepstatin complexed with Plasmepsin II [30], in which one CF_3 group was well accommodated into a pocket of the active site and was involved in relevant hydrophobic interactions.

4.4 A Nanomolar CF_3-Barbituric acid Inhibitor Selective for MMP-9

Gelatinases A and B (MMP-2 and MMP-9, respectively) play a pivotal role in a number of physiological processes. Selective inhibition of MMP-9 might represent an attractive strategy for therapeutic intervention. However, the active sites of MMP-2 and MMP-9 are closely related from the structural point of view, and hence selective inhibition of the latter is a challenging endeavor.

Barbiturates are potent and selective inhibitors of MMPs, sparing MMP-1 [31–35]. Compound **C** [36] (see Figure 4.4), a rather potent inhibitor of MMP-9 ($IC_{50} = 20\,nM$), was resynthesized in our laboratory and subjected to further enzyme inhibition assays, which confirmed the previous results as well as poor selectivity against MMP-2 ($IC_{50} = 43\,nM$, see Table 4.3), and very good selectivity against MMP-1 ($IC_{50} = 3.89 \times 10^5\,nM$). These results are in line with other barbiturates as MMP-9 inhibitors, which invariably exhibit excellent selectivity against MMP-1 but poor selectivity, if any, against MMP-2.

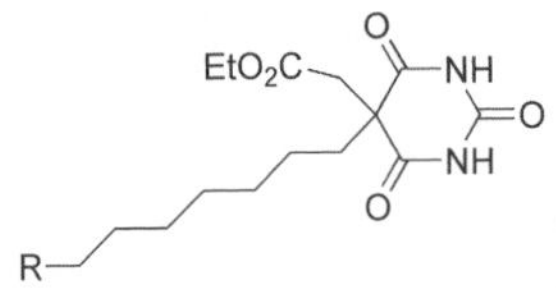

Figure 4.4 *The target fluorinated barbiturate (**36**) and its non-fluorinated analogue (C).*

Table 4.3 *IC_{50} values (nM) for the barbiturate inhibition of different MMPs*

Substrate	IC_{50} (nM)			
	MMP-1	MMP-2	MMP-3	MMP-9
36	$>10^6$	10.7	74.2	0.179
C	3.89×10^5	43.0	n.a.	20.0

A possible strategy to obtain barbiturate inhibitors selective for MMP-9 is represented by the fine-tuning of the P1′ substituent of the inhibitor into the "tunnel-like" hydrophobic S1′ cavity of the enzymes. Recent findings show that in MMP-2 the S1′ subsite is deeper than that in MMP-9, owing to the presence of the Arg-424 residue instead of smaller Thr-424 in MMP-2, which seems to partially obstruct S1′ [29]. We thought that probing the bottom part of the S1′ cavity of MMP-9 by placing a suitable P1′ substituent on the inhibitor might lead to some selectivity in favor of MMP-9 compared with MMP-2.

We therefore decided to replace the terminal methyl group of the 5-octyl chain of C, which was identified as the P1′ substituent (the 5′-CH_2CO_2Et group is the P2′ residue), with a CF_3 group (see **36**, Figure 4.4). In fact, the CF_3 group is bulkier and more hydrophobic than the CH_3, thus possibly leading to a better and deeper fit into the bottom part of the S1′ cavity of MMP-9, with little or no expected effect on the activity for MMP-2.

To synthesize the target barbiturate **36** we identified 1-bromo-8,8,8-trifluorooctane **43** (see Scheme 4.6) as the key building-block. This molecule is known, and was previously obtained only by fluorination of 8-bromooctanoic acid with SF_4 [37]. Unfortunately, the use of such an aggressive fluorinating agent requires specific experimental apparatus and presents considerable safety hazards, which are difficult to address in an ordinary synthetic laboratory. We therefore developed several alternative routes to **43** based on user-friendly protocols as well as the use of a cheap and commercially available source of fluorine, such as trifluoroacetic esters. One of these novel approaches is shown in Scheme 4.6.

Commercially available 6-bromohexan-1-ol **37** was *O*-benzylated to **38** and converted to the corresponding Grignard reagent, which was reacted with 0.25 equivalents of CF_3CO_2Et using an old but efficient methodology [38]. The Grignard acts first as a nucleophile and then as a reducing agent, converting the intermediate trifluoroketone into trifluorocarbinol **39**. Barton–McCombie radical deoxygenation of the methyl xanthate **40** occurred effectively, affording Bn-ether **41**. Compound **41** was hydrogenolyzed to the primary alcohol **42**, which was easily converted to the target **43**.

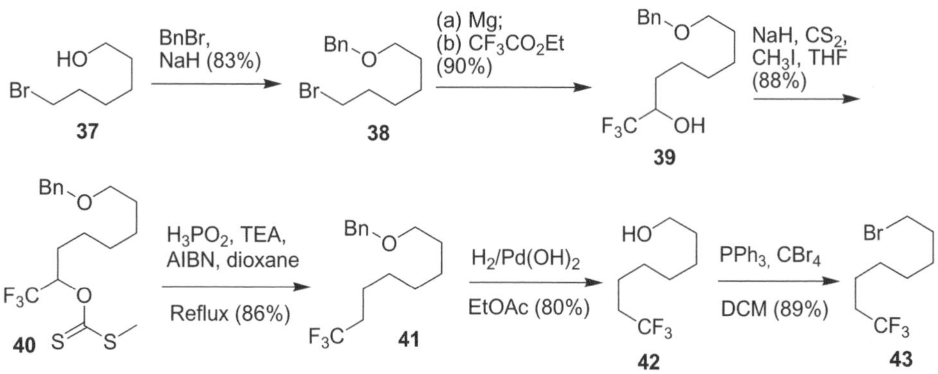

Scheme 4.6 *Preparation of the key intermediate **43**.*

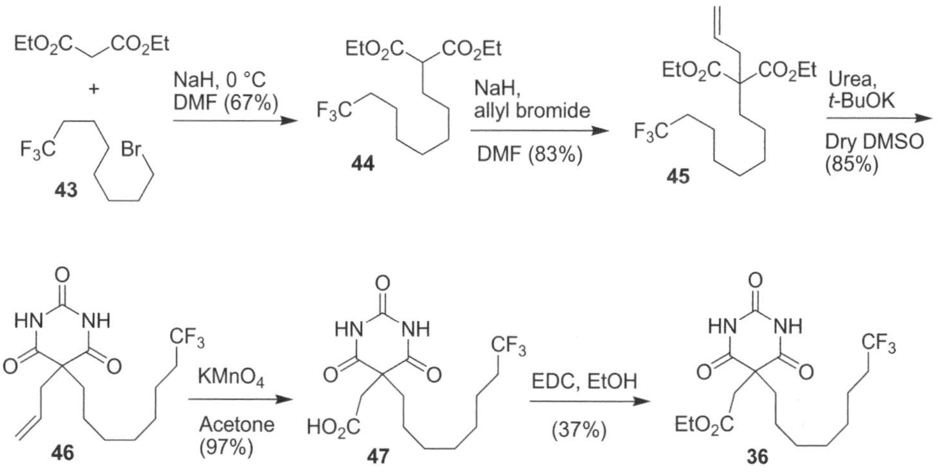

Scheme 4.7 *Completion of the synthesis of CF_3-barbiturate **36**.*

The synthesis of barbiturate **36** (see Scheme 4.7) was performed next. The sodium enolate of diethyl malonate was reacted with **43**, providing **44**, which was converted to 2,2-disubstituted malonate **45** by reaction with allyl bromide. Reaction of **45** with urea in the presence of *t*-BuOK as base afforded the barbiturate **46**, which was submitted to oxidative one-carbon demolition by the action of $KMnO_4$ to give carboxylic acid **47**. Compound **47** was esterified with ethanol to give the target **36**, but unfortunately in modest yields.

Enzyme inhibition assays conducted on a set of commercially available MMPs (see Table 4.3) showed that **36** is considerably more potent and more selective than the parent unfluorinated barbiturate **C**. More specifically, **36** is 100-fold more potent than **C** for MMP-9 inhibition and has an MMP-2/MMP-9 selectivity factor of 60, whereas the selectivity factor for **C** is only 2.

Although this strategy clearly needs further validation, fine tuning of the interactions between key functions of the ligand with the protease active site by introduction of fluoroalkyl groups, such as CF_3, seems to be a promising strategy to optimize the potency and increase the selectivity of an inhibitor.

4.5 β-Fluoroalkyl-β-sulfonylhydroxamic Acids

We next decided to study the effect on MMP-inhibitory potency of a fluoroalkyl group installed in a more distant position from the hydroxamic acid group, in order to better understand the unique stereoelectronic properties of fluoralkyl groups in a purely aliphatic position [39]. For this purpose we chose as a model system a structurally simple class of hydroxamic acid inhibitors bearing an arylsulfone moiety at the β-position such as **D** (see Figure 4.5), which showed nanomolar inhibitory potency for MMP-2, MMP-3, and MMP-13 [40–42].

In molecules **D**, the R side-chain was found to be critical for potency, but it could also dramatically influence the enzyme selectivity profile of the inhibition. Compounds **D** bearing large hydrophobic groups R (such as alkyl, cycloalkyl and arylalkyl groups) showed low-nanomolar, and even subnanomolar affinity for MMP-2, MMP-3, and MMP-13, and excellent selectivity against MMP-1.

4,4,4-Trifluorocrotonic acid **49** [43] (see Scheme 4.8) was used as the starting material for the synthesis of **48a**. The thia-Michael addition of **50** to **49** occurred in reasonable yields, affording carboxylic acid **51**. Coupling of **51** with *O*-Bn hydroxylamine gave *O*-Bn hydroxamate **52** in satisfactory yield, and the subsequent oxidation to the sulfone **53** took place in nearly quantitative yield. Hydrogenolysis of **53** with Pearlman's catalyst afforded the target racemic hydroxamic acid **48a** in good overall yields. An analogous reaction sequence from 4,4-difluorocrotonic acid afforded difluoro-hydroxamic acid **48b** (see Figure 4.5).

Pentafluoroethylhydroxamic acid **48c** was obtained through a different procedure (see Scheme 4.9). β-Hydroxy ester **54** was converted into triflate **55**, which was subjected to S_N2 reaction by thioanisol to give sulfide **56**. Acid hydrolysis provided the carboxylic acid

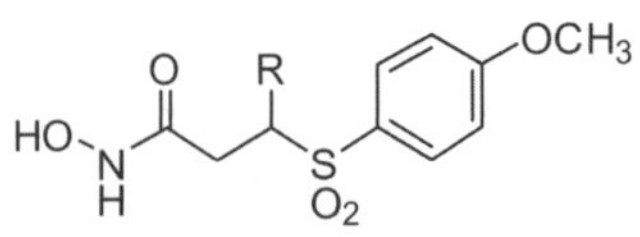

Figure 4.5 *β-Sulfonylhydroxamic acid inhibitors of MMPs.*

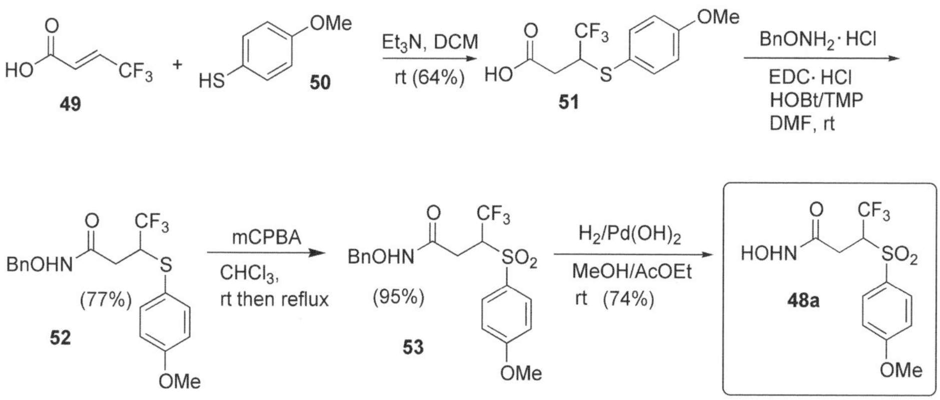

Scheme 4.8 *Synthesis of **48a**.*

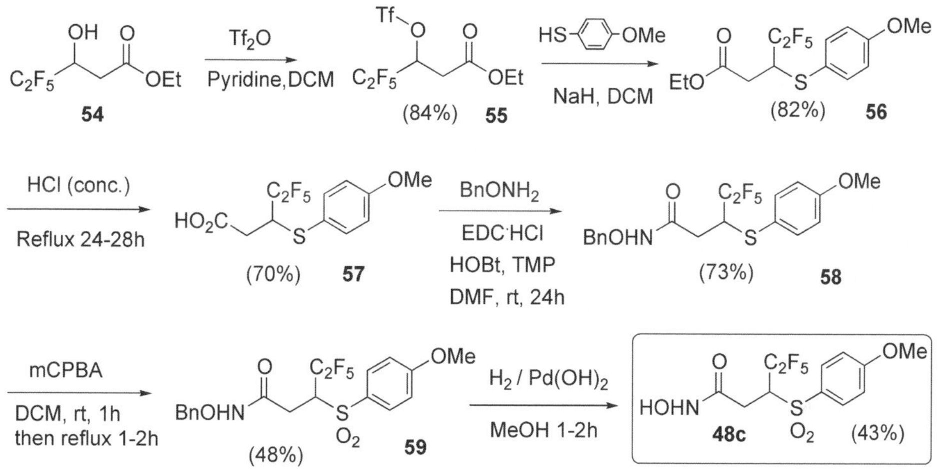

Scheme 4.9 *Synthesis of **48c**.*

57, which was coupled with *O*-Bn-hydroxylamine to give the protected hydroxamate **58**. The sulfide group of the latter was then oxidized to sulfone **59**. Finally, the Bn group was hydrogenolyzed to give the target hydroxamic acid **48c**.

The chlorodifluoro-analogue of **48c** (R = $CClF_2$; see Figure 5.5) was synthesized through the same methodology, but surprisingly this compound was found to be unstable at room temperature, thus precluding the enzyme inhibition assay.

The CF_3-compound **48a** showed a single-digit nanomolar IC_{50} for MMP-3 (see Table 4.4). Modest selectivity was observed against MMP-9 and MMP-2 (about 10-fold less potency), while **48a** showed very good selectivity against MMP-1 (about 1000-fold less potency). Interestingly, the pure enantiomers of **48a** (synthesized independently) and the racemic compound showed nearly identical inhibitory potency. This could be ascribed to

Table 4.4 *IC_{50} values (nM) for the inhibition of different MMPs by β-fluoroalkylhydroxamic acids* ***48a–c***

Substrate	IC_{50} (nM)			
	MMP-1	MMP-2	MMP-3	MMP-9
48a	4.0×10^3	78	8.0	52
48b	1.5×10^4	734	2	6
48c	947	32	93	1.7×10^3

an easy interchange of the position of the CF_3 and sulfone moieties in two different enzyme pockets, most likely S_1' and S_2'. Alternatively, one could hypothesize that enantiopure **48a** underwent racemization at some stage during the enzyme inhibition assays.

Diifluoro compound **48b** was even more potent than **48a**, showing a rather impressive inhibitory activity for both MMP-3 and MMP-9, and much better selectivity against both MMP-2 (>100-fold less potent) and MMP-1 (~10^4-fold less potent). Introduction of a C_2F_5 group (**48c**), however, had a surprising effect on the inhibitory activity: this substitution caused a dramatic drop in activity for MMP-9 (1000-fold less potent) and, to lesser extent, for MMP-3 (50-fold less potent) but brought about a higher potency for MMP-2 (200-fold more potent).

In summary, the results on fluoroalkylhydroxamic acids **19** and **48**, as well as those on the CF_3-barbiturate **36** show that (i) a fluoroalkyl group can be successfully used as a substituent in protease inhibitors [44–48] and is very well tolerated by the enzymes, and (ii) an electron-withdrawing CF_3 group at the α-position to the hydroxamic acid group brings about little effect on the zinc chelating capability of the latter. On the other hand, replacement of an α-methyl by a CF_3 group in malic hydroxamic acid inhibitor **18** (see Figure 4.1), was responsible for a dramatic loss of inhibitory potency. The final outcome of incorporation of a fluoroalkyl group is strongly dependent on the whole structure of the inhibitor, and the effect of the fluoroalkyl group on the conformation is the most critical factor in determining the biological activity of the molecule.

Abbreviations

Cbz	carbobenzyloxy
DCC	dicyclohexylcarbodiimide
DCM	dichloromethane
DIC	diisopropylcarbodiimide
DIPEA	diisopropylethylamine
DMAP	4-(*N*,*N*-dimethylamino)pyridine
DMF	*N*,*N*-dimethylformamide
EDC	1-(3-dimethylaminopropyl)-3-ethylcarbodiimide hydrochloride
HATU	*O*-(7-azabenzotriazol-1-yl)-*N*,*N*,*N*′,*N*′-tetramethyluronium hexafluorophosphate

HOAt	1-hydroxy-7-azabenzotriazole
HOBt	1-hydroxybenzotriazole
LDA	lithium diisopropylamide
PyBroP	bromotripyrrolidinophosphonium hexafluorophosphate
TEA	triethylamine
TFA	trifluoroacetic acid
TMP	*sym*-collidine (2,4,6-trimethylpyridine)

Acknowledgment

We thank the European Commission (IHP Network grant "FLUOR MMPI" HPRN-CT-2002-00181), MIUR (Cofin 2004, Project "Polipeptidi Bioattivi e Nanostrutturati"), Politecnico di Milano, and C.N.R. for economic support. We thank Professor Wolfram Bode (Max Planck Institute, Martinsried, Germany) for very useful discussions and suggestions.

References

1. Jacobsen, F. E., Lewis, J. A. and Cohen, S. M. (2007) The design of inhibitors for medicinally relevant metalloproteins. *ChemMedChem*, **2**, 152–171.
2. Verma, R. P. and Hansch, C. (2007) Matrix metalloproteinases (MMPs): chemical–biological functions and (Q)SARs. *Bioorganic & Medicinal Chemistry*, **15**, 2223–2268.
3. Pirard, B. (2007) Insight into the structural determinants for selective inhibition of matrix metalloproteinases. *Drug Discovery Today*, **12**, 640–646.
4. Nuti, E., Tuccinardi, T. and Rossello, A. (2007) Matrix metalloproteinase inhibitors: new challenges in the era of post broad-spectrum inhibitors. *Current Pharmaceutical Design*, **13**, 2087–2100.
5. Whittaker, M., Floyd, C. D., Brown, P. and Gearing, A. J. H. (1999) Design and therapeutic application of matrix metalloproteinase inhibitors. *Chemical Reviews*, **99**, 2735–2776.
6. Bode, W. and Huber, R. (2000) Structural basis of the endoproteinase-protein inhibitor interaction. *Biochimica et Biophysica Acta*, **1477**, 241–252.
7. (a) Giavazzi, R. and Taraboletti, G. (2001) Preclinical development of metalloprotease inhibitors in cancer therapy. *Critical Reviews in Oncology/Hematology*, **37**, 53–60; (b) Coussens, L. M., Fingleton, B. and Matrisian, L. M. (2002) Matrix metalloproteinase inhibitors and cancer – trials and tribulations. *Science*, **295**, 2387–2392.
8. Hu, J., Van den Steen, P. E., Sang, Q.-X. A. and Opdenakker, G. (2007) Matrix metalloproteinase inhibitors as therapy for inflammatory and vascular diseases. *Nature Reviews Drug Discovery*, **6**, 480–498.
9. Zanda, M. (2004) Trifluoromethyl group: an effective xenobiotic function for peptide backbone modification. *New Journal of Chemistry*, **28**, 1401–1411.
10. Müller, K., Faeh, C. and Diederich, F. (2007) Fluorine in pharmaceuticals: looking beyond intuition. *Science*, **317**, 1881–1886.
11. Banks, R. E., Smart, B. E. and Tatlow, J. C. (1994) *Organofluorine Chemistry: Principles and Commercial Applications*, Plenum Press, New York.

12. Jacobson, I. C., Reddy, P. G., Wasserman, Z. R., *et al.* (1998) Structure-based design and synthesis of a series of hydroxamic acids with a quaternary-hydroxy group in P1 as inhibitors of matrix metalloproteinases. *Bioorganic & Medicinal Chemistry Letters*, **8**, 837–842.
13. Jacobson, I. C. and Reddy, G. P. (1996) Asymmetric reactions of chiral imide enolates with α-keto esters. *Tetrahedron Letters*, **37**, 8263–8266.
14. Sani, M., Belotti, D., Giavazzi, R., *et al.* (2004) Synthesis and evaluation of stereopure α-trifluoromethyl-malic hydroxamates as inhibitors of matrix metalloproteinases. *Tetrahedron Letters*, **45**, 1611–1615.
15. Moreno, M., Sani, M., Raos, G., *et al.* (2006) Stereochemically pure α-trifluoromethyl-malic hydroxamates: synthesis and evaluation as inhibitors of matrix metalloproteinases. *Tetrahedron*, **62**, 10171–10181.
16. Carpino, L. A., El-Faham, A. and Albericio, F. (1995) Efficiency in peptide coupling: 1-hydroxy-7-azabenzotriazole *vs* 3,4-dihydro-3-hydroxy-4-oxo-1,2,3-benzotriazine. *Journal of Organic Chemistry*, **60**, 3561–3564 and references therein.
17. Humphrey, J. M. and Chamberlin, A. R. (1997) Chemical synthesis of natural product peptides: coupling methods for the incorporation of noncoded amino acids into peptides. *Chemical Reviews*, **97**, 2243–2266.
18. Gorecka, A., Leplawy, M., Zabrocki, J. and Zwierzak, A. (1978) Diethyl phosphorobromidate – an effective new peptide-forming agent. *Synthesis*, 474–476.
19. Goldwhite, H. and Saunders, B. C. (1955) Esters containing phosphorus. Part XIII. Dialkyl phosphorobromidates. *Journal of the Chemical Society*, 3549–3564.
20. Nelson, S. G., Spencer, K. L., Cheung, W. S. and Mamie, S. J. (2002) Divergent reaction pathways in amine additions to β-lactone electrophiles. An application to β-peptide synthesis. *Tetrahedron*, **58**, 7081–7091.
21. Becker, D. P., DeCrescenzo, G., Freskos, J., *et al.* (2001) α-Alkyl-α-amino-β-sulphone hydroxamates as potent MMP inhibitors that spare MMP-1. *Bioorganic & Medicinal Chemistry Letters*, **11**, 2723–2725.
22. Sinisi, R., Sani, M., Candiani, G., *et al.* (2005) Synthesis of α-trifluoromethyl-α-amino-β-sulfone hydroxamates: novel nanomolar inhibitors of matrix metalloproteinases. *Tetrahedron Letters*, **46**, 6515–6518.
23. Hockerman, S. L., Becker, D. P., Bedell, L. J., *et al.* (2003) U.S. Patent 6,583,299, 2003; Chemical Abstracts 2000, **134**, 29702.
24. Aranyos, A., Old, D. W., Kiyomori, A., *et al.* (1999) Novel electron-rich bulky phosphine ligands facilitate the palladium-catalyzed preparation of diaryl ethers. *Journal of the American Chemical Society*, **121**, 4369–4378.
25. Bravo, P., Capelli, S., Meille, S. V., *et al.* (1994) Synthesis of optically pure (*R*)- and (*S*)-α-trifluoromethyl-alanine. *Tetrahedron: Asymmetry*, **5**, 2009–2018.
26. Buck, E., Song, Z. J., Tschaen, D., *et al.* (2002) Ullmann diaryl ether synthesis: rate acceleration by 2,2,6,6-tetramethylheptane-3,5-dione. *Organic Letters*, **4**, 1623–1626.
27. Drabowicz, J. and Oae, S. (1977) Mild reductions of sulfoxides with trifluoroacetic anhydride/sodium iodide system. *Synthesis*, 404–405.
28. Cowart, M., Kowaluk, E. A., Daanen, J. F., *et al.* (1998) Nitroaromatic amino acids as inhibitors of neuronal nitric oxide synthase. *Journal of Medicinal Chemistry* **41**, 2636–2642.
29. Tochowicz, A., Maskos, K., Huber, R., *et al.* (2007) Crystal structures of MMP-9 complexes with five synthetic inhibitors: contribution of the flexible Arg424 side chain to selectivity. *Journal of Molecular Biology*, **371**, 989–1006.
30. Binkert, C., Frigerio, M., Jones, A., *et al.* (2006) Replacement of isobutyl by trifluoromethyl in pepstatin A selectively affects inhibition of aspartic proteinases. *ChemBioChem*, **7**, 181–186.

31. Brandstetter, H., Grams, F., Glitz, D., *et al.* (2001) The 1.8-Å crystal structure of a matrix metalloproteinase 8-barbiturate inhibitor complex reveals a previously unobserved mechanism for collagenase substrate Recognition *Journal of Biological Chemistry*, **276**, 17405–17412; corrigendum ibid., **276**, 31474.
32. Foley, L. H., Palermo, R., Dunten, P. and Wang, P. (2001) Novel 5,5-disubstitutedpyrimidine-2,4,6-triones as selective MMP inhibitors. *Bioorganic & Medicinal Chemistry Letters*, **11**, 969–972.
33. Dunten, P., Kammlott, U., Crowther, R., *et al.* (2001) X-ray structure of a novel matrix metalloproteinase inhibitor complexed to stromelysin. *Protein Science*, **10**, 923–926.
34. Breyholz, H.-J., Schäfers, M., Wagner, S., *et al.* (2005) C-5-disubstituted barbiturates as potential molecular probes for noninvasive matrix metalloproteinase imaging. *Journal of Medicinal Chemistry*, **48**, 3400–3409.
35. Sheppeck, J. E., II, Gilmore, J. L., Tebben, A., *et al.* (2007) Hydantoins, triazolones, and imidazolones as selective non-hydroxamate inhibitors of tumor necrosis factor-α converting enzyme (TACE). *Bioorganic & Medicinal Chemistry Letters*, **17**, 2769–2774.
36. Kim, S.-H., Pudzianowski, A. T., Leavitt, K. J., *et al.* (2005) Structure-based design of potent and selective inhibitors of collagenase-3 (MMP-13). *Bioorganic & Medicinal Chemistry Letters*, **15**, 1101–1106.
37. Stoll, G. H., Voges, R., Gerok, W. and Kurz, G. (1991) Synthesis of a metabolically stable modified long-chain fatty acid salt and its photolabile derivative. *Journal of Lipid Research*, **32**, 843–857.
38. Campbell, K. N., Knobloch, J. O. and Campbell, B. K. (1950) The preparation and some reactions of 1,1,1-trifluoro-2-alkenes. *Journal of the American Chemical Society*, **72**, 4380–4384.
39. Sani, M., Candiani, G., Pecker, F., Malpezzi, L. and Zanda M. (2005) Novel highly potent, structurally simple γ-trifluoromethyl γ-sulfone hydroxamate inhibitor of stromelysin-1 (MMP-3). *Tetrahedron Letters*, **46**, 2393–2396
40. Groneberg, R. D., Burns, C. J., Morrissette, M. M., *et al.* (1999) Dual inhibition of phosphodiesterase 4 and matrix metalloproteinases by an (arylsulfonyl)hydroxamic acid template. *Journal of Medicinal Chemistry*, **42**, 541–544.
41. Salvino, J. M., Mathew, R., Kiesow, T., *et al.* (2000) Solid-phase synthesis of an arylsulfone hydroxamate library. *Bioorganic & Medicinal Chemistry Letters*, **10**, 1637–1640.
42. Freskos, J. N., Mischke, B. V., DeCrescenzo, G. A., *et al.* (1999) Discovery of a novel series of selective MMP inhibitors: Identification of the γ-sulfone-thiols. *Bioorganic & Medicinal Chemistry Letters*, **9**, 943–948.
43. Jagodzinska, M., Huguenot, F. and Zanda, M. (2007) Studies on a three-step preparation of β-fluoroalkyl acrylates from fluoroacetic esters. *Tetrahedron*, **63**, 2042–2046.
44. Volonterio, A., Bravo, P. and Zanda, M. (2000) Synthesis of partially modified retro and retro-inverso ψNHCH(CF3).-peptides. *Organic Letters*, **2**, 1827–1830.
45. Molteni, M., Volonterio, A. and Zanda, M. (2003) Stereocontrolled synthesis of ψ[CH(CF3)NH]Gly-peptides. *Organic Letters*, **5**, 3887–3890.
46. Black, W. C., Bayly, C. I., Davis, D. E., *et al.* (2005) Trifluoroethylamines as amide isosteres in inhibitors of cathepsin K. *Bioorganic & Medicinal Chemistry Letters*, **15**, 4741–4744.
47. Li, C. S., Deschenes, D., Desmarais, S., *et al.* (2006) Identification of a potent and selective non-basic cathepsin K inhibitor. *Bioorganic & Medicinal Chemistry Letters* **16**, 1985–1989.
48. Sani, M., Volonterio, A. and Zanda, M. (2007) The trifluoroethylamine function as peptide bond replacement. *ChemMedChem*, **2**, 1693–1700.

5

Fluoro-Taxoid Anticancer Agents

Antonella Pepe, Larisa Kuznetsova, Liang Sun, and Iwao Ojima

5.1 Introduction

The importance of fluorine in bioorganic and medicinal chemistry has been demonstrated by a large number of fluorinated compounds approved by the FDA for medical use [1, 2]. According to our most up-to-date survey, 138 fluorine-containing drugs have received FDA approval for human diseases (of which 23 have been discontinued from the market, however), while 33 are currently in use for veterinary applications (see the Appendix). These statistics make fluorine the "second-favorite heteroatom" after nitrogen in drug design [3]. Small atomic radius, high electronegativity, nuclear spin of ½, and low polarizability of the C–F bond are among the special properties that render fluorine so attractive. Those atomic properties translate widely into equally appealing attributes of fluoroorganic compounds. Higher metabolic stability, often increased binding to target molecules, and increased lipophilicity and membrane permeability are some of the properties associated with the replacement of a C–H or C–O bond with a C–F bond in biologically active compounds. Because of the recognized value of fluorine, it is now a common practice in drug discovery to study fluoro-analogues of lead compounds under development [4]. Although medicinal chemists have long introduced fluorine into bioactive molecules on the basis of experience and intuition, it is only recently that experimental and computational studies have been conducted to better understand how the introduction of fluorine into small drug molecules results in higher binding affinities and selectivity [5]. An understanding of how the replacement of H with F affects the electronic nature and conformation of small molecules is crucial for predicting the interaction of fluoroorganic molecules with proteins and enzymes. As a result, the rational design of fluoroorganic molecules will lead to the generation of new and effective biochemical tools. In addition, ^{19}F NMR has found

Fluorine in Medicinal Chemistry and Chemical Biology Edited by Iwao Ojima

numerous applications to molecular imaging and promoted the development of molecular probes for imaging. The sensitivity of ^{19}F NMR spectroscopy, along with large ^{19}F–^{1}H coupling constants and the virtual absence of ^{19}F in living tissues, makes incorporation of fluorine into bioactive compounds a particularly powerful tool for the investigation of biological processes [6, 7].

This chapter gives an account of our research on the use of fluorine, exploiting the unique nature of this element, in the medicinal chemistry and chemical biology of taxoid anticancer agents (i.e., "Taxol-like" compounds).

5.2 Paclitaxel, Taxoids, and Second-generation Taxoids

Paclitaxel (Taxol®) and its semisynthetic analogue docetaxel (see Figure 5.1) are two of the most important chemotherapeutic drugs, currently used for the treatment of advanced ovarian cancer, metastatic breast cancer, melanoma, non-small-cell lung cancer, and Karposi's sarcoma [8, 9]. More recently, these drugs have been used for the treatment of neck, prostate, and cervical cancers [8, 9].

The mechanism of action of paclitaxel involves its binding to the β-subunit of α,β-tubulin dimer, accelerating the formation of microtubules. The resulting paclitaxel-bound microtubules are much more stable and less dynamic than the natural GTP-bound micro-

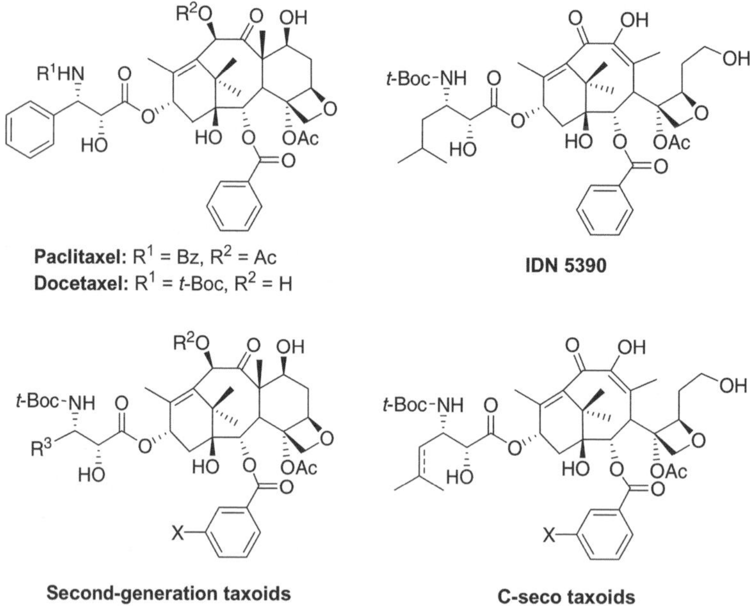

Figure 5.1 Structures of paclitaxel, docetaxel, second-generation taxoids, and C-seco-taxoids.

tubules, with a growth rate higher than the disassembly rate. The unnatural growth and stabilization of microtubules causes the arrest of the cell division cycle, mainly at the G_2/M stage, activating a cell-signaling cascade that induces apoptosis [10, 11].

Although paclitaxel and docetaxel possess potent antitumor activity, chemotherapy with these drugs encounters many undesirable side-effects as well as drug resistance [8, 12–14]. It is therefore important to develop new taxoid anticancer drugs as well as efficacious drug delivery systems with fewer side-effects, superior pharmacological properties, and improved activity against various classes of tumors, especially against drug-resistant cancers.

It has been shown that a primary mechanism of drug resistance is the overexpression of ABC transporters, for example, P-glycoprotein (Pgp), an integral membrane glycoprotein that acts as a drug-efflux pump to maintain the intracellular concentration of drugs below therapeutically active level [15]. In the course of our extensive studies on the design, synthesis, and structure–activity relationships (SARs) of taxoid anticancer agents, we discovered second-generation taxoids that possess one order of magnitude better activity against drug-sensitive cell lines and more than two orders of magnitude better activity against drug-resistant cell lines [16]. Several examples are shown in Table 5.1 (see Figure 5.1 for structures).

It was found that *meta*-substitution of the benzoyl group at the C-2 position dramatically increases the cytotoxicity of taxoids against drug-resistant cell lines [17]. This class of the second-generation taxoids is three orders of magnitude more potent than paclitaxel and docetaxel against drug-resistant cancer cell lines expressing multidrug resistance (MDR) phenotype [17] (see Table 5.1).

Because of the aforementioned advantages of introducing fluorine into biologically active molecules, we synthesized fluorine-containing paclitaxel and docetaxel analogues to investigate the effects of fluorine–incorporation on the cytotoxicity and the blockage of known metabolic pathways. Our earlier studies have been reported in several publications [18–21]. Thus, in this chapter, we describe the synthesis and biological evaluation of second-generation fluoro-taxoids bearing a difluoromethyl, trifluoromethyl, or difluorovinyl group at the C-3′ position.

Although the overexpression of Pgp and other ABC transporters is the main cause of multidrug resistance to paclitaxel and other hydrophobic anticancer drugs, it is not the only mechanism of drug resistance. Overexpression of specific tubulin isotypes has recently received substantial attention in terms of paclitaxel resistance and has not yet been successfully addressed [22–25]. Different β-tubulin isotypes form microtubules with anomalous behaviors *in vitro* with regard to assembly, dynamics, conformation, and ligand binding [23, 24, 26, 27]. Microtubules with altered β-tubulin isotype compositions respond differently to paclitaxel [28]. Derry and co-workers reported that $\alpha\beta_{III}$ microtubules are more dynamic than the most common $\alpha\beta_{II}$ or $\alpha\beta_{IV}$ [28]; hence, higher concentrations of paclitaxel are required to interfere with their dynamics. Recently Ferlini has reported that the C-seco-taxoid IDN 5390 is up to 8-fold more active than paclitaxel against an inherently drug-resistant OVCAR3 ovarian cancer cell line and paclitaxel-resistant human ovarian adenocarcinoma cell lines A2780TC1 and A2780TC3 (cell lines overexpressing class III β-tubulin) [29–31]. Following observation of the remarkable effects of *meta*-substitution at the C-2 benzoate moiety, mentioned above, we synthesized a series of IDN 5390 analogues with C-2-benzoate modifications, including a 3-fluorobenzoyl group. The

Table 5.1 In vitro *cytotoxicity (IC_{50} nM)*[a] *of representative second-generation taxoids*

Taxoid	R^2	R^3	R^4	MCF7[b]	MCF7-R[c]	R/S[d]	LCC6-WT[b]	LCC6-MDR[e]	R/S[d]
Paclitaxel	Ph	Ac	H	1.7	300	176	3.1	346	112
Docetaxel	Ph	H	H	1.0	235	235	1.0	120	120
SB-T-1213	Me_2C=CH-	EtCO	H	0.18	4.0	22	–	–	
SB-T-1103	Me_2CHCH_2-	EtCO	H	0.35	5.1	21	–	–	
SB-T-121303	Me_2C=CH-	EtCO	MeO	0.36	0.33	0.92	1.0	0.9	0.90
SB-T-121304	Me_2C=CH-	EtCO	N_3	0.9	1.1	1.2	0.9	1.2	1.3
SB-T-11031	Me_2CHCH_2-	EtCO	F	–	–	–	0.4	2.4	6.0
SB-T-11033	Me_2CHCH_2-	EtCO	MeO	0.36	0.43	1.19	0.9	0.8	0.89

[a] The concentration of compound that inhibits 50% of the growth of a human tumor cell line after 72 h drug exposure.
[b] Human breast carcinoma.
[c] Multidrug-resistant human breast cancer cell line (currently renamed NCI/ADR).
[d] Drug-resistance factor.
[e] Multidrug-resistant human breast carcinoma.

synthesis and biological evaluation of these novel fluoro-C-seco-taxoids are described below.

5.3 Second-generation Taxoids and C-seco-Taxoids with Strategic Incorporation of Fluorine

5.3.1 Synthesis and Biological Evaluation of C-3′-Difluoromethyl- and C-3′-Trifluoromethyl Taxoids

A series of the second-generation C-3′-CF_2H and C-3′-CF_3 taxoids were synthesized by the β-lactam synthon method [32, 33]. The synthesis of enantiopure β-lactam **8**, bearing a CF_2H group at the C-4 position is illustrated in Scheme 5.1. The [2+2] ketene-imine cycloaddition of acetoxyketene, generated *in situ* from acetoxyacetyl chloride and triethylamine, with *N*-PMP-3-methylbut-2-enaldimine (PMP = *p*-methoxyphenyl) gave racemic β-lactam 1. Enzymatic optical resolution of **1** by PS Amano lipase at 50 °C gave enantiopure (3*R*,4*S*)-3-AcO-4-isobutenyl-β-lactam **2(+)** [34]. Protection of the 3-hydroxyl group as TIPS ether and ozonolysis of the double bond of 2(+) afforded 4-formyl-β-lactam **5**. Difluoromethylation of **5** with diethylaminosulfur trifluoride (DAST) [35], followed by deprotection of PMP by cerium ammonium nitrate (CAN) and introduction of *tert*-butoxycabonyl group to the β-lactam nitrogen yielded enantiopure (3*R*,4*R*)-1-*t*-Boc-3-TIPSO-4-CF_2H-β-lactam **8**.

In a similar manner, enantiopure 1-*t*-Boc-3-TIPSO-4-CF_3-β-lactam **14** was synthesized from benzyloxyacetyl chloride/triethylamine and *N*-PMP-CF_3-aldimine through [2+2] ketene-imine cycloaddition, enzymatic optical resolution (PS Amano lipase at

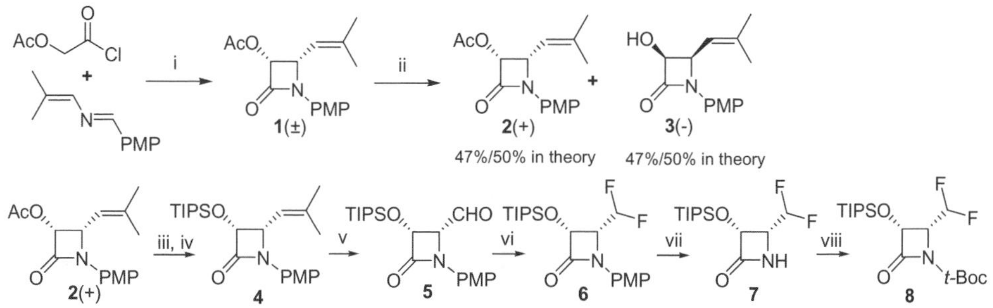

Scheme 5.1 *Synthesis of enantiopure C-4-difluoromethyl-β-lactam* ***8****. (i) Et_3N, CH_2Cl_2, −78 °C ~ r.t., 70%; (ii) PS-Amano, buffer pH 7.0, 10% CH_3CN, 50 °C; (iii) KOH, THF, 0 °C, 100%; (iv) TIPSCl, Et_3N, DMAP, CH_2Cl_2, 85%; (v) O_3, $MeOH/CH_2Cl_2$, −78 °C; Me_2S, 73%; (vi) DAST, CH_2Cl_2, 86%; (vii) CAN, H_2O/CH_3CN, −15 °C, 68%; (viii) t-Boc_2O, Et_3N, DMAP, CH_2Cl_2, 80%.*

(Source: *Reprinted with permission from Ojima, I.; Kuznetsova, L. V., and Sun, L. (2007)* Organofluorine Chemistry at the Biomedical Interface: A Case Study on Fluoro-Taxoid Anticancer Agents. *ACS Symposium Series 949,* Current Fluoroorganic Chemistry, *(2007) pp. 288–304. Copyright (2007) American Chemical Society.)*

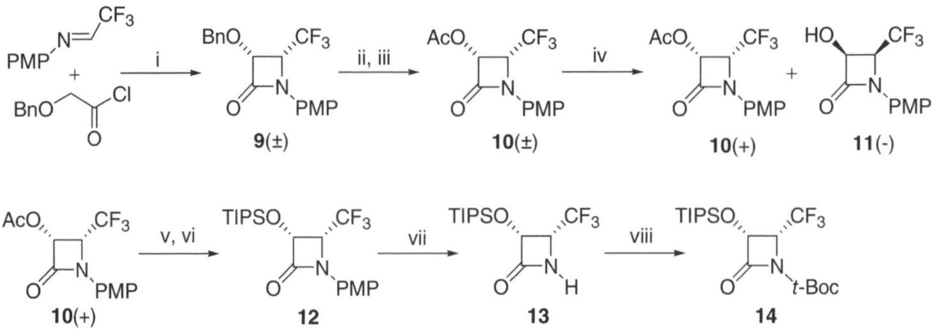

Scheme 5.2 Synthesis of enantiopure C-4-trifluoromethyl-β-lactam **14**. (i) Et_3N, CH_2Cl_2, 40 °C, 83%; (ii) H_2, Pd, MeOH, 45 °C, 98%; (iii) Ac_2O, DMAP, Py, CH_2Cl_2, 74%; (iv) PS-Amano, buffer pH 7, 10% CH_3CN, 0–5 °C; (v) KOH, THF, −5 °C, 100%; (vi) TIPSCl, Et_3N, CH_2Cl_2, 95%; (vii) CAN, CH_3CN/H_2O, −10 °C, 84%; (viii) t-Boc_2O, Et_3N, DMAP, CH_2Cl_2, 87%.

0–5 °C), deprotection–protection, and *t*-Boc carbamoylation as illustrated in Scheme 5.2 [34].

The Ojima–Holton coupling of 4-CF_2H- and 4-CF_3-β-lactams **8** and **14**, thus obtained, with 2,10-modified baccatins **15** was carried out at −40 °C in THF using LiHMDS as a base followed by deprotection of silicon protecting groups with HF/pyridine to give the corresponding second-generation 3′-CF_2H-taxoids (**18**) and 3′-CF_3-taxoids (19) in moderate to high overall yields (see Scheme 5.3) [20, 34, 36]. 2,10-Modified baccatins **15** were prepared by methods we have reported previously [17].

The second-generation fluoro-taxoids (**18** and **19**) were evaluated for their cytotoxicity *in vitro* against human breast cancer cell lines (MCF7-S and LCC6-WT), their corresponding drug-resistant cell lines (MCF7-R and LCC6-MDR), human non-small-cell lung (NSCL) cancer cell line (H460), and human colon cancer cell line (HT-29). The IC_{50} values were determined through 72 h exposure of the cancer cells to the fluoro-taxoids, following the procedure developed by Skehan *et al.* [37]. Results are summarized in Table 5.2 (3′-CF_2H-taxoids **18**) and Table 5.3 (3′-CF_3-taxoids **19**).

As Tables 5.2 and 5.3 show, all fluoro-taxoids **18** and **19** possess substantially higher potencies than those of paclitaxel and docetaxel, with cytotoxicity in single-digit nanomolar IC_{50} values against drug-sensitive MCF7-S, LCC6-WT, H460, and HT-29 cancer cell lines (except for a few cases). The cytotoxicity of the fluoro-taxoids **18** and **19** against multidrug-resistant MCF7-R and LCC6-MDR cell lines is more impressive: all fluoro-taxoids exhibit single-digit nanomolar IC_{50} values (except for a couple of cases) and are two orders of magnitude more potent than paclitaxel in average. These two series of fluoro-taxoids exhibit, in general, comparable cytotoxicity against all cancer cell lines examined.

However, the potency of 3′-CF_2H-taxoids **18** against MCF7-S and LCC6-WT appears to be higher and more uniform, with different substitution patterns, than that of 3′-CF_3-taxoids **19**, except for two cases (SB-T-12822-1, 0.19 nM, MCF7-S; SB-T-12824-1, 0.17 nM, MCF7-S). In contrast, compounds **19** exhibit more uniform potency against

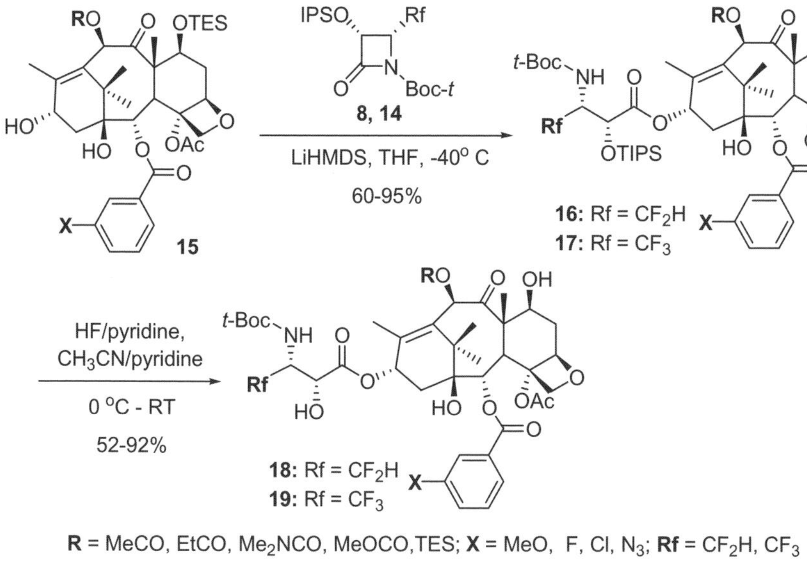

R = MeCO, EtCO, Me_2NCO, MeOCO,TES; **X** = MeO, F, Cl, N_3; **Rf** = CF_2H, CF_3

Scheme 5.3 *Synthesis of C-3′-CF_2-taxoids (**18**) and C-3′-CF_3-taxoids (**19**).*
(Source: *Reprinted with permission from Ojima, I.; Kuznetsova, L. V., and Sun, L. (2007)* Organofluorine Chemistry at the Biomedical Interface: A Case Study on Fluoro-Taxoid Anticancer Agents. *ACS Symposium Series 949,* Current Fluoroorganic Chemistry, *(2007) pp. 288–304. Copyright (2007) American Chemical Society.)*

multidrug-resistant MCF7-R and LCC6-MDR cell lines than do **18**. For fluoro-taxoids **18**, cytotoxicity against multidrug-resistant cell lines, MCF7-R and LCC6-MDR, depends on the nature of meta substituents of the C-2-benzoate moiety; the potency increases in the order F < MeO < Cl < N_3. In contrast, no clear trend is observed for fluoro-taxoids **19** against these multidrug-resistant cell lines. Among these fluoro-taxoids examined, SB-T-12842-4 (R = *n*-propanoyl; X = N_3) appears to be the most potent compound with a resistance factor (R/S ratio) of only 2.9–3.0 against two sets of human breast cancer cell lines.

5.4 Synthesis and Biological Evaluation of C-2-(3-Fluorobenzoyl)-C-seco-Taxoids

As mentioned above, paclitaxel and taxoids stabilize microtubules through their binding to the β-tubulin subunits, thereby blocking the cell mitosis, which leads to apoptosis [10, 11]. There are eight β-tubulin isotypes, of which class I β-tubulin is the major isotype in all mammalian tissues [38]. It has been shown that the overexpression of class III β-tubulin isotype causes paclitaxel drug resistance, which constitutes significant drug-resistance other than MDR [25]. Recently, a C-seco-taxoid, IDN5390 (see Figure 5.1), was reported to exhibit several times better potency than paclitaxel against drug-resistant ovarian cancer

Table 5.2 In vitro *cytotoxicity (IC_{50} nM)[a] of C-3′-CF_2H-taxoid (**18**)*

Taxoid	R	X	MCF7-S[b] (breast)	MCF7-R[c] (breast)	R/S[d]	LCC6-WT[b] (breast)	LCC6-MDR[e] (breast)	R/S[d]	H460[f] (lung)	HT-29[g] (colon)
Paclitaxel			1.7	300	176	3.1	346	112	4.9	3.6
Docetaxel			1.0	215	215	–	–	–	–	1.0
SB-T-12841-1	Ac	MeO	0.34	4.16	12	0.26	5.57	21	0.38	0.52
SB-T-12841-2	Ac	F	0.44	5.33	13	0.52	10.0	19	0.20	0.35
SB-T-12841-3	Ac	Cl	0.40	6.48	16	0.31	5.80	19	0.49	1.94
SB-T-12841-4	Ac	N_3	0.32	1.68	5.3	0.22	1.57	7.1	0.48	0.57
SB-T-12842-1	Et-CO	MeO	1.14	4.05	3.5	0.69	4.92	7.1	0.40	0.59
SB-T-12842-2	Et-CO	F	0.53	7.24	14	0.88	4.63	3.5	0.41	0.86
SB-T-12842-3	Et-CO	Cl	0.44	5.20	12	0.52	4.71	9.1	0.30	0.43
SB-T-12842-4	Et-CO	N_3	0.32	0.96	3.0	0.39	1.15	2.9	0.27	0.37
SB-T-12843-1	Me_2N-CO	MeO	0.45	4.51	10	0.69	7.06	10	0.40	0.43
SB-T-12843-2	Me_2N-CO	F	0.52	8.13	16	0.69	10.6	15	0.20	0.35
SB-T-12843-3	Me_2N-CO	Cl	0.31	2.96	9.5	0.21	3.87	18	0.36	0.58
SB-T-12843-4	Me_2N-CO	N_3	0.37	1.44	3.9	0.29	1.69	5.8	0.52	0.40
SB-T-12844-1	MeO-CO	MeO	0.81	6.59	8.1	1.03	10.2	9.9	0.30	0.44
SB-T-12844-2	MeO-CO	F	0.59	11.38	19	0.86	12.6	15	0.30	0.43
SB-T-12844-3	MeO-CO	Cl	0.26	2.08	8.0	0.13	1.82	14	0.25	0.29
SB-T-12844-4	MeO-CO	N_3	1.69	2.56	1.5	0.26	2.06	7.9	0.23	0.36

[a–e] See footnotes *a–e* of Table 5.1.
[f] Human non-small-cell lung carcinoma.
[g] Human caucasian colon adenocarcinoma.

Table 5.3 In vitro *cytotoxicity (IC_{50} nM)[a] of C-3′-CF_3-taxoids (**19**)*

Taxoid	R	X	MCF7-S[b] (breast)	MCF7-R[c] (breast)	R/S[d]	LCC6-WT[b] (breast)	LCC6-MDR[e] (breast)	R/S[d]	H460[f] (lung)	HT-29[g] (colon)
Paclitaxel			1.7	300	176	3.1	346	112	4.9	3.6
Docetaxel			1.0	215	215	...	...		...	1.0
SB-T-12821–1	Ac	MeO	0.32	8.8	28	0.33	3.99	12	0.38	0.69
SB-T-12821–2	Ac	F	0.45	5.58	13	0.38	5.93	16	0.49	1.11
SB-T-12821–3	Ac	Cl	0.40	5.04	13	0.22	4.96	23	0.5	0.85
SB-T-12821–4	Ac	N_3	0.47	3.85	8.2	1.18	4.00	3.4	0.20	0.50
SB-T-12822–1	Et-CO	MeO	0.19	2.16	11	0.45	4.24	9	0.41	0.54
SB-T-12822–2	Et-CO	F	0.68	3.78	5.6	0.82	4.27	5.2	0.59	1.15
SB-T-12822–3	Et-CO	Cl	0.34	3.28	9.6	0.39	2.54	6.5	0.63	1.11
SB-T-12822–4	Et-CO	N_3	0.38	1.61	4.2	1.09	2.56	2.3	0.20	0.40
SB-T-12823–1	Me_2NCO	MeO	0.57	1.84	3.2	0.28	4.48	16	0.35	0.68
SB-T-12823–2	Me_2NCO	F	0.32	2.64	8.3	0.32	5.57	17	0.5	0.76
SB-T-12823–3	Me_2NCO	Cl	0.12	1.02	8.5	0.27	2.55	9.4	0.42	0.45
SB-T-12823–4	Me_2NCO	N_3	0.47	2.61	5.6	1.27	3.52	2.8	0.30	0.50
SB-T-12824–1	MeOCO	MeO	0.17	2.88	17	0.27	3.99	15	0.38	0.53
SB-T-12824–2	MeOCO	F	0.31	4.88	16	0.39	5.81	15	0.61	0.85
SB-T-12824–3	MeOCO	Cl	0.65	4.72	7.3	0.29	5.08	18	0.43	0.68
SB-T-12824–4	MeOCO	N_3	0.47	2.92	6.2	1.09	4.00	3.7	0.20	0.40

[a–g]See footnotes of Table 5.2.

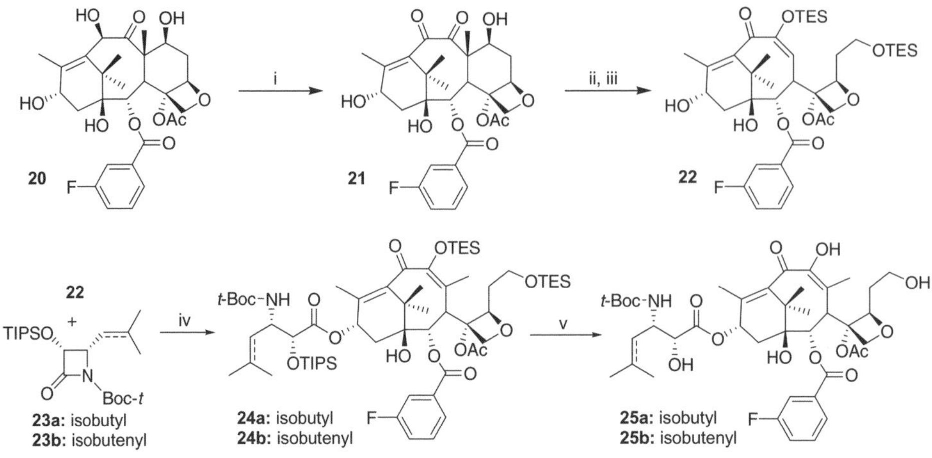

Scheme 5.4 *Synthesis of 2-(3-fluorobenzoyl)-C-seco-taxoids (**25**). (i) $Cu(OAc)_2$, MeOH, 86%; (ii) L-selectride, THF, −78°C 50–70%, (iii) methyl imidazole, TESCl, DMF, 0°C, 52; (iv) LiHMDS, THF, −40°C, 70–80%; (v) HF/pyridine, CH_3CN/pyridine, 0°C–RT, 52%–92%.*

cell lines overexpressing class III isotype [29]. Accordingly, we have performed a SAR study on IDN5390. As a part of this SAR study, we investigated two fluorine-containing analogues, SB-T-10104 (**25a**) and SB-T-10204 (**25b**) (see Scheme 5.4). Two C-seco-fluorotaxoids **25a** and **25b** were synthesized through the Ojima–Holton coupling of 7,9-di-TES-2-(3-fluorobenzoyl)-C-seco-baccatin (**22**) with β-lactams **23a** [39] and **23b** [40], respectively, under standard conditions, followed by deprotection with HF-pyridine (see Scheme 5.4). Di-TES-C-seco-baccatin **22** was prepared from 2-(3-fluorobenzoyl)-10-deacetylbaccatin **20** [17] using Appendino's protocol [41, 42] as follows: Baccatin **20** was oxidized with $Cu(OAc)_2$ and air to give the corresponding 10-oxo-baccatin **21**, which was then treated with L-selectride at −78 °C, followed by TES protection to afford di-TES-C-seco-baccatin **22** (see Scheme 5.4).

Novel C-seco-fluorotaxoids, **25a** and **25b**, were evaluated for their cytotoxicity against several human ovarian adenocarcinoma cell lines: A2780wt (drug-sensitive wild-type), A2780CIS, A2780TOP, A2780ADR (resistant to cisplatin, topotecan, and adriamycin/doxorubicin, respectively), and A2780TC1 and A2780TC3 (resistant to both paclitaxel and cyclosporine A). The drug resistance in the A2780ADR cell line is based on MDR, while that in the A2780TC1 and A2780TC3 cell lines is caused by the overexpression of class III β-tubulin subunit and other possible mutations. Thus, the activity of these two C-seco-fluorotaxoids is of particular interest. Results are shown in Table 5.4.

As Table 5.4 shows, SB-CST-10104 (**25a**) possesses remarkable potency against paclitaxel-resistant cell lines A2780TC1 and A2780TC3, especially the latter, that is, the most drug-resistant cell line for paclitaxel in this series. This C-seco-fluorotaxoid **25a** is 39 times more potent than paclitaxel against cell line A2780TC3. The resistance factor IC_{50} (A2780TC3)/IC_{50} (A2780 wt) for this cell line is 10 470 for paclitaxel, but it is only 41 for **25a**. For comparison, IDN5390 exhibits 8.0 times higher potency than paclitaxel

Table 5.4 In vitro *cytotoxicity* (IC_{50} *nM*)[a] *of C-seco-fluorotaxoids* (**24**)

C-seco-Taxoid	A2780wt[b]	A2780CIS[c]	A2780TOP[d]	A2780ADR[e]	A2780TC1[f]	A2780TC3[g]
Paclitaxel	1.7	2.2	7.2	1 239	10 027	17 800
IDN5390	17.4	16.8	27.5	2 617	2 060	2 237
SB-CST-10104 (25a)	11.1	11.8	12.8	3 726	1 497	460
SB-CST-10204 (25b)	6.1	4.9	6.9	2 218	4 454	745

[a] The concentration of compound that inhibits 50% of the growth of a human tumor cell line after 72 h drug exposure.
[b] Human ovarian carcinoma wild-type.
[c] Cisplatin-resistant A2780.
[d] Topotecan-resistant A2780.
[e] Adriamycin-resistant A2780.
[f,g] Clones derived from chronic exposition of A2780 to paclitaxel and cyclosporine.

with a resistance factor of 129 against the same cell line. This result is quite impressive taking into account the fact that the only structural difference between IDN5390 and **25a** is one fluorine substitution at the *meta* position of the C-2-benzoate moiety of the C-seco-taxoid molecule.

The C3′-substituents of C-seco-fluorotaxoids **25a** (3′-isobutyl) and **25b** (3′-isobutenyl) also show interesting effects on the potency, which is assumed to be related directly to their interaction with the class III β-tubulin. As Table 5.4 shows, **25b** exhibits higher potency than **25a** against A2780wt, A2780CIS, A2780TOP, and A2780ADR. However, the reversal of this SAR is observed against A2780TC1 and A2780TC3, in which the class III β-tubulin is overexpressed. Overall, it has been shown that the introduction of one fluorine to the C-2-benzoate moiety of C-seco-taxoid molecule substantially increases the potency against both paclitaxel-sensitive and paclitaxel-resistant human ovarian cancer cell lines.

5.5 Synthesis and Biological Evaluation of C-3′-Difluorovinyl-Taxoids

As described above, the introduction of isobutyl, isobutenyl, CF_2H, and CF_3 groups to the C3′-position of taxoids, replacing the phenyl group of paclitaxel and docetaxel, has led to the development of highly potent second-generation taxoids, especially against drug-resistant cancer cell lines expressing MDR phenotype. Our recent metabolism studies on 3′-isobutyl- and 3′-isobutenyl-taxoids has disclosed that the metabolism of second-generation taxoids (SB-T-1214, SB-T-1216, and SB-T-1103) is markedly different from that of docetaxel and paclitaxel [43]. These taxoids are metabolized (via hydroxylation) by CYP 3A4 of the cytochrome P450 family enzymes, primarily at the two allylic methyl groups of the C-3′-isobutenyl group and the methyne moiety of the 3′-isobutyl group (see Figure 5.2). This forms a sharp contrast with the known result that the *tert*-butyl group of the C-3′ *N-t*-Boc moiety is the single predominant metabolic site for docetaxel [44]. These unique metabolic profiles prompted us to design and synthesize 3′-difluorovinyl-taxoids, in order to block the allylic oxidation by CYP 3A4, which should enhance the metabolic stability and activity *in vivo*.

For the synthesis of a series of C-3′-difluorovinyl-taxoids **29**, novel (3*R*,4*S*)-1-*t*-Boc-3-TIPSO-4-difluorovinyl-β-lactam **28**(+) is the key component for the coupling with baccatins **15** (see Scheme 5.5). We prepared this β-lactam **28**(+) in three steps from 4-formyl-β-lactam **5**(+) (see Scheme 5.1) using the Wittig reaction of the formyl moiety with difluoromethylphosphorus ylide generated *in situ* from $(Me_2N)_3P/CF_2Br_2/Zn$ (see Scheme 5.5). The Ojima–Holton coupling reaction [45–47] of β-lactam **28**(+) with C-2-modified, C-10-modified or C-2,10-modified baccatins **15** (X = H, MeO, N_3) [17] and the subsequent removal of the silyl protecting groups gave the corresponding C-3′-difluorovinyl-taxoids **29** in good to excellent yields.

The cytotoxicities of the 3′-difluorovinyl-taxoids **29** were evaluated *in vitro* against MCF7-S, MCF7-R, HT-29 (human colon carcinoma), and PANC-1 (human pancreatic carcinoma) cell lines [37]. The results are summarized in Table 5.5.

As Table 5.5 shows, all difluorovinyl-taxoids **29** are exceedingly potent compared with paclitaxel. A clear effect of C-2-benzoate modification at the *meta* position (X = H

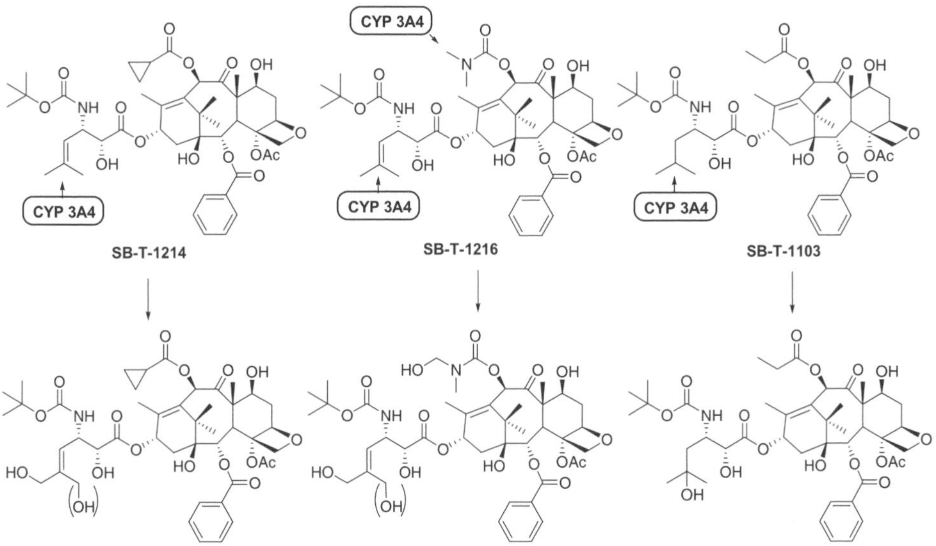

Figure 5.2 Primary sites of hydroxylation on the second-generation taxoids by P450 family enzymes.

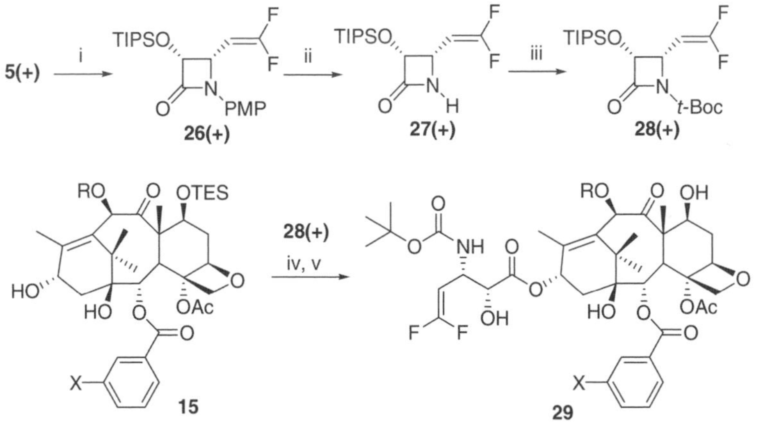

Scheme 5.5 Synthesis of C-3′-difluorovinyl-taxoids (**28**). (i) CBr_2F_2, HMPT, Zn, THF, 84%; (ii) CAN, $H_2O/CH3CN$, −15 °C, 92%; (iii) Boc_2O, Et_3N, DMAP, CH_2Cl_2, 96%; (iv) LiHMDS, THF, −40 °C; (v) HF/Py, Py/CH_3CN, overnight, 0 °C–RT, 57–91% (for two steps).

vs. X = MeO or N_3) is observed on the increase in potency against drug-sensitive and drug-resistant MCF7 cell lines (entries 2–5 vs. entries 6–11). Difluorovinyl-taxoids with 2,10-modifications (entries 6–11) have impressive potency, exhibiting IC_{50} values in the <100 pM range (78–92 pM), except for one case against MCF7-S (entry 7), and in the subnanomolar range (0.34–0.57 nM) against MCF7-R, which is 3 orders of magnitude

Table 5.5 In vitro *cytotoxicity (IC_{50} nM)[a] of 3′-difluorovinyl-taxoids(**29**)*

Entry	Taxoid	R	X	MCF7-S[b] (breast)	MCF7-R[c] (breast)	R/S	HT-29[d] (colon)	PANC-1[e] (pancreatic)
1	Paclitaxel			1.2	300	250	3.6	25.7
2	SB-T-12851	Ac	H	0.099	0.95	9.6	0.41	1.19
3	SB-T-12852	c-Pr-CO	H	0.12	6.0	50	0.85	5.85
4	SB-T-12853	Et-CO	H	0.12	1.2	10	0.34	0.65
5	SB-T-12854	Me_2N-CO	H	0.13	4.3	33	0.46	1.58
6	SB-T-12852-1	c-Pr-CO	MeO	0.092	0.48	5.2	–	–
7	SB-T-12853-1	Et-CO	MeO	0.34	0.57	1.7	–	–
8	SB-T-12855-1	MeO-CO	MeO	0.078	0.50	6.4	–	–
9	SB-T-12851-3	Ac	N_3	0.092	0.34	3.7	–	–
10	SB-T-12852-3	c-Pr-CO	N_3	0.092	0.45	4.9	–	–
11	SB-T-12855-3	MeO-CO	N_3	0.078	0.40	5.3	–	–

[a–d] See footnotes of Table 5.2.
[e] Human pancreatic carcinoma.

more potent than paclitaxel. The resistance factor for these taxoids is 1.7–6.4, while that for paclitaxel is 250. Difluorovinyl-taxoids with unmodified C-2-benzoate moiety (entries 2–5) also show highly enhanced potency against MCF7-S and MCF7-R as compared to paclitaxel. These taxoids exhibit impressive potency against HT-29 (human colon) and PANC-1 (human pancreatic) cancer cell lines as well. SB-T-12853 appears particularly promising against these gastrointestinal (GI) cancer cell lines. Although difluorovinyl-taxoids with 2,10-modifications (entries 6–11) have not yet been evaluated against HT-29 and PANC-1 cell lines, it is anticipated that these taxoids will exhibit remarkable potency against these GI cancer cell lines in a similar manner to that for MCF7-R.

5.6 Possible Bioactive Conformations of Fluoro-Taxoids

^{19}F NMR combined with advanced 2D spectroscopic methods provides a powerful tool for the study of dynamic conformational equilibria of fluorine-containing bioactive molecules. The wide dispersion of fluorine chemical shifts is particularly useful for the observation of conformers at low temperatures. We have successfully used fluorine-containing taxoids as probes for NMR analysis of the conformational dynamics of paclitaxel in conjunction with molecular modeling [48]. We have further applied the fluorine-probe protocol to solid-state magic angle spinning (SSMAS) ^{19}F NMR analysis with the radiofrequency driven dipolar recoupling (RFDR) method to measure the F–F distance in the microtubule-bound conformation of F_2-10-Ac-docetaxel (see Figure 5.3) [47]. Schaefer and co-workers used rotational echo double-resonance (REDOR) to investigate the structure of the microtubule-bound paclitaxel by determining the ^{19}F–^{13}C distances of a fluorine probe of paclitaxel (see Figure 5.3) [49]. These solid-state NMR studies have provided critical information on the bioactive conformation of paclitaxel and docetaxel.

Recently, we proposed a new bioactive conformation of paclitaxel, "REDOR-Taxol" [50], based on (i) the ^{19}F–^{13}C distances obtained by the REDOR experiment [49], (ii) the photoaffinity labeling of microtubules [51], (iii) the crystal structure (PDB code: 1TUB) of the Zn^{2+}-stabilized αβ-tubulin dimer model determined by cryo-electron microscopy (cryo-EM) [52], and (iv) molecular modeling (Monte Carlo; Macromodel) [50]. In this computational biology analysis, we first docked a paclitaxel-photoaffinity label molecule to the position identified by our photoaffinity labeling study and then optimized the

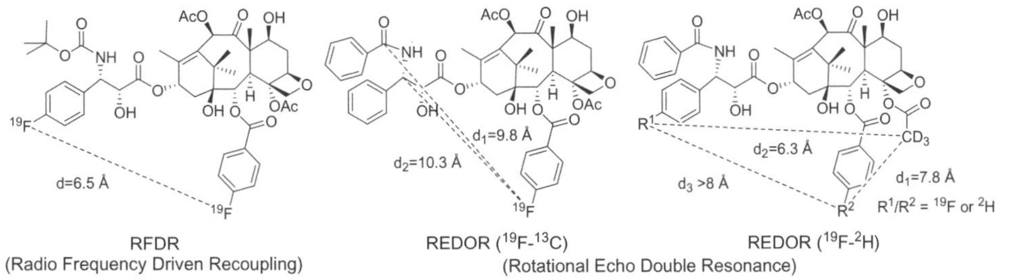

Figure 5.3 *Solid-state NMR studies on microtubule-bound fluoro-taxoid probes.*

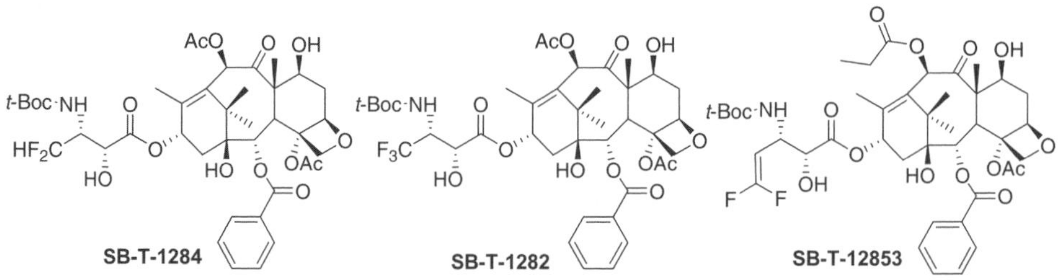

SB-T-1284 SB-T-1282 SB-T-12853

Figure 5.4 *Structures of fluoro-taxoids analyzed by molecular modeling.*

position with a free paclitaxel molecule in the binding space using the REDOR distances as filters [50].

More recently, three additional intramolecular distances of the key atoms in the microtubule-bound $^{19}F/^{2}H$-labeled paclitaxel were determined by the REDOR method (see Figure 5.3) [53]. It has also been shown that the optimized cryo-EM crystal structure of tubulin-bound paclitaxel (PDB code: 1JFF) [54] serves better for the computational structure analysis. Accordingly, we have optimized our REDOR-Taxol structure, using the 1JFF coordinates as the starting point, by means of molecular dynamics simulations (Macromodel, MMFF94) and energy minimization (InsightII 2000, CVFF) [55].

We applied the same computational protocol to investigate the microtubule-bound structures of the 3′-CF_2H-, 3′-CF_3-, and 3′-CF_2C=CH-taxoids, using the updated REDOR-Taxol [55] as the starting structure. Three fluoro-taxoids, SB-T-1284, SB-T-1282, and SB-T-12853 (see Figure 5.4), were docked into the binding pocket of paclitaxel in the β-tubulin subunit by superimposing the baccatin moiety with that of the REDOR-Taxol, and their energies were minimized (InsightII 2000, CVFF). The resulting computer-generated binding structures of three fluoro-taxoids are shown in Figure 5.5 (a, b, c).

As Figure 5.5 (a, b, c) shows, the baccatin moiety occupies virtually the same space in all cases, as expected. Each fluoro-taxoid fits comfortably in the binding pocket without any high-energy contacts with the protein. There is a very strong hydrogen bond between the C-2′-OH of a fluoro-taxoid and His227 of β-tubulin in all cases, which shares the same key feature with the REDOR-Taxol structure [50]. [*Note:* Our preliminary study on the tubulin-bound structures of these three fluoro-taxoids using the 1TUB coordinates [52] led to different structures in which the C-2′-OH had a hydrogen bond to Arg359 of β-tubulin [56]. However, the use of the updated REDOR-Taxol structure based on the 1JFF coordinates [54] unambiguously led to fluoro-taxoid structures bearing a strong hydrogen bond between the C-2′-OH and His227.]

The CF_2H and CF_3 moieties fill essentially the same space, as anticipated. However, the CF_2C=CH moiety occupies more extended hydrophobic space than the CF_2H and CF_3 moieties. It is likely that this additional hydrophobic interaction contributes substantially to the exceptional cytotoxicity of difluorovinyl-taxoids **29**. The overlay of SB-T-12853 with a representative second-generation taxoid, SB-T-1213 shows excellent fit, which may demonstrate that the difluorovinyl group mimics the isobutenyl group (see Figure 5.5, d). However, the difluorovinyl group is in between vinyl and isobutenyl groups in size, and two fluorine atoms may mimic two hydroxyl groups rather than two methyl groups

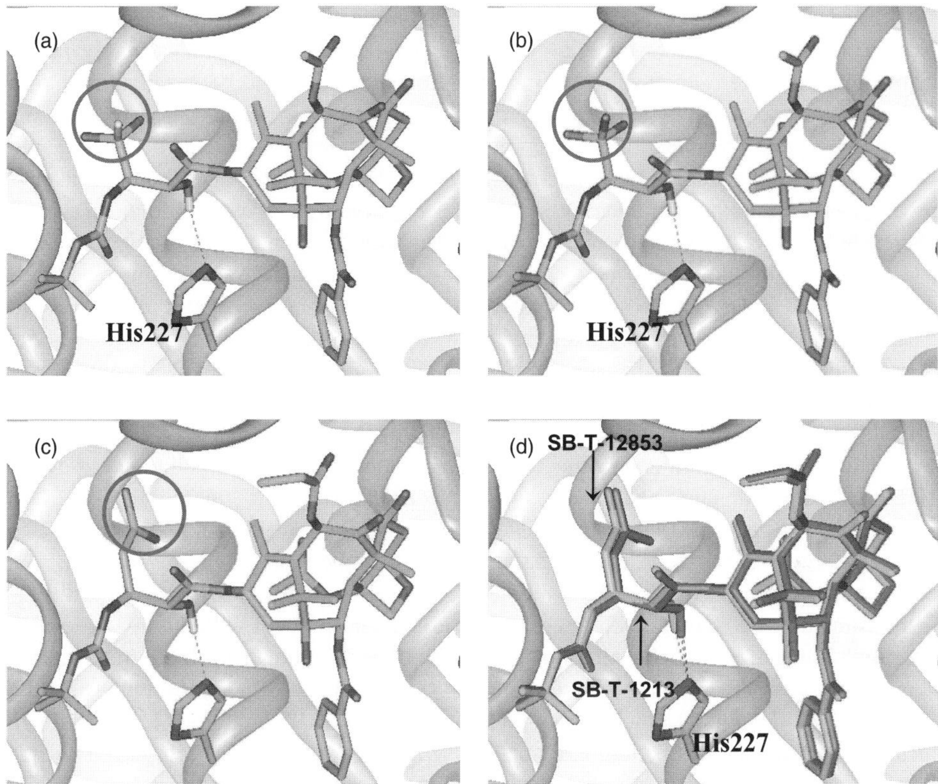

Figure 5.5 Computer-generated binding structures of fluoro-taxoids to β-tubulin: (a) SB-T-1284 (3′-CF_2H); (b) SB-T-1282 (3′-CF_3); (c) SB-T-12853 (3′-CF_2=CH); (d) Overlay of SB-T-12853 and SB-T-1213 (C3′-isobutenyl). See color plate 5.5.

electronically. Accordingly, the difluorovinyl group can be regarded as "magic vinyl" in drug design, similarly to "magic methyl" for the trifluoromethyl group, including its anticipated metabolic stability against P-450 family enzymes.

5.7 Use of Fluorine in Tumor-Targeting Anticancer Agents

Although current cancer chemotherapy is based on the premise that rapidly proliferating tumor cells are more likely to be killed by cytotoxic drugs, the difference in activity of cytotoxic drugs against tumor tissues and against primary tissues is relatively small. Consequently, the amount of an anticancer drug required to achieve clinically effective level of activity against the targeted tumor cells often causes severe damage to actively propagating non-malignant cells such as cells of the gastrointestinal tract and bone marrow, resulting in a variety of undesirable side-effects. Accordingly, it is very important to develop new chemotherapeutic agents with improved tumor specificity.

To address the tumor specificity issue, various tumor-targeting anticancer drug conjugates, consisting of a tumor-targeting molecule, a functional linker, and a highly potent cytotoxic drug, have been developed. Such a drug conjugate should have a high affinity to tumor cells through the tumor-targeting module, promoting efficient endocytosis of the whole conjugate. The functional linker moiety of the conjugate should be intelligently designed and engineered so that the conjugate is stable in the circulation but the linker is readily cleaved in tumors to release the potent cytotoxic drug efficiently. Monoclonal antibodies (mAbs), hyaluronic acid, folic acid, biotin, and somatostatin peptide mimic are among the tumor-targeting molecules most commonly used [57, 58]. For example, mAbs specifically bind to antigens that are overexpressed on the surface of tumor tissues or cells, distinguishable from normal tissues. Therefore, in principle, mAb–cytotoxic drug conjugates can be specifically delivered to the tumor and internalized via receptor-mediated endocytosis, releasing the cytotoxic drug [57, 59, 60]. Mylotarg® (gemtuzumab-ozogamicin) [61] was approved by the US FDA for the treatment of acute myelogenous leukemia (AML), providing the first mAb–drug immunoconjugate for the treatment of cancer in clinic use. Several other mAb–drug conjugates have reached human clinical trial stage [60, 62–65].

Our investigation in this area led to the development of novel mAb–taxoid conjugates as tumor-targeting anticancer agents that exhibited extremely promising results in human cancer xenografts in SCID (severe combined immunodeficiency) mice. The results clearly demonstrated tumor-specific delivery of a taxoid anticancer agent without any noticeable toxicity to the animals, curing all animals tested [66]. As the linker for these mAb–taxoid conjugates, we used a disulfide linker that was stable in blood circulation but efficiently cleaved by glutathione or other thiols in the tumor [*Note:* The glutathione level is known to be 1000 times higher in tumor tissues than in blood plasma [67]. However, in this first-generation of mAb-taxoid conjugates, the original taxoid molecule was not released because of the compromised modification of the taxoid molecule for attachment the disulfide linker. Accordingly, the cytotoxicity of the taxoid released in these conjugates was 8–10 times weaker than that of the parent taxoid [66].

To address this problem, we have been developing second-generation mechanism-based disulfide linkers. One of our approaches is the glutathione-triggered cascade drug release, forming a thiolactone as a side-product as illustrated in Scheme 5.6. This mechanism-based drug release concept has been nicely demonstrated in a model system by monitoring the reaction with ^{19}F NMR using fluorine-labeled compounds (see Figure 5.6) [68]. The strategic incorporation of a fluorine substituent at the *para* position to the disulfide linkage would direct the cleavage of this linkage by a thiol to generate the desirable thiophenolate or sulfhydrylphenyl species for thiolactonization. In addition, the incorporation of a fluorine substitution may increase the metabolic stability of the conjugate. This type of linkers is highly versatile and readily applicable to any tumor-targeting drug conjugates, including fluoro-taxoids.

Moreover, combination of a fluorine-containing linker and a fluoro-taxoid may provide fluorine probes for monitoring the internalization and drug release of these conjugates in the tumor cells and tissues by ^{19}F NMR as an alternative method to the use of fluorescence-labeled probes with confocal fluorescence microscopy [69].

Further applications of the strategic incorporation of fluorine into medicinally active substances as well as compounds useful as tools for biomedical research are actively under study in these laboratories.

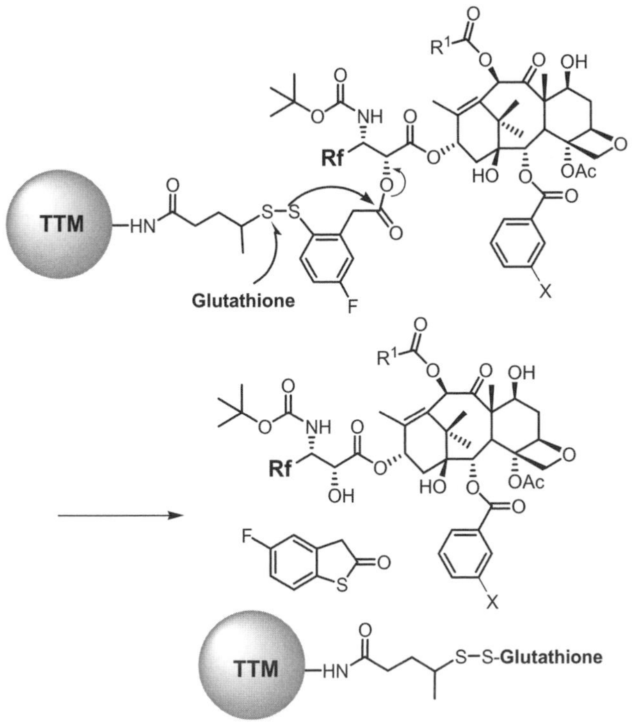

Scheme 5.6 Glutathione-triggered cascade drug release for tumor-targeting of fluoro-taxoid conjugates.

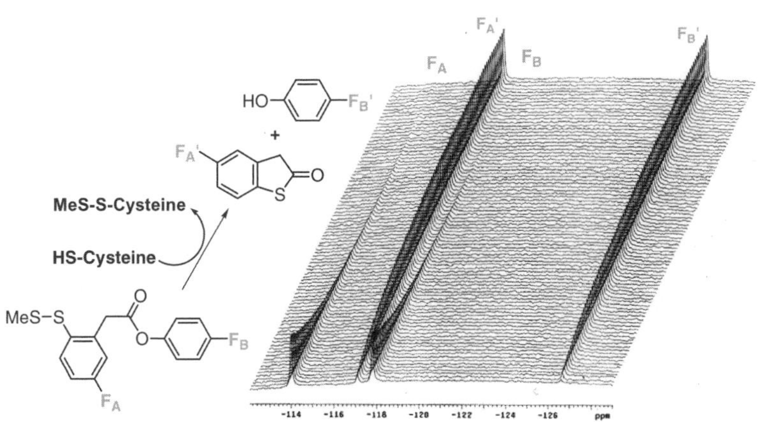

Figure 5.6 A model system for mechanism-based release.
(Source: *Reproduced with permission, from Ojima, I. Use of fluorine in the medicinal chemistry and chemical biology of bioactive compounds – a case study on fluorinated taxane anticancer agents.* ChemBioChem, *(2004) 5, 628–635. Copyright Wiley-VCH Verlag GmbH & Co. KGaA.)*

References

1. Begue, J.-P. and Bonnet-Delpon, D. (2006) Recent advances (1995–2005) in fluorinated pharmaceuticals based on natural products. *Journal of Fluorine Chemistry*, **127**, 992–1012.
2. Isanbor, C. and O'Hagan, D. (2006) Fluorine in medicinal chemistry: A review of anti-cancer agents. *Journal of Fluorine Chemistry*, **127**, 303–319.
3. Cottet, F., Marull, M., Lefebvre, O. and Schlosser, M. (2003) Recommendable routes to trifluoromethyl-substituted pyridine- and quinolinecarboxylic acids., *European Journal of Organic Chemistry*, 1559–1568.
4. Kirk, K. L. (2006) Fluorine in medicinal chemistry: Recent therapeutic applications of fluorinated small molecules *Journal of Fluorine Chemistry*, **127**, 1013–1029.
5. Müller, K., Faeh, C. and Diederich, F. (2007) Fluorine in pharmaceuticals: looking beyond intuition. *Science*, **317**, 1881–1886.
6. O'Hagan, D., Schaffrath, C., Cobb, S. L., *et al.* (2002) Biochemistry: Biosynthesis of an organofluorine molecule. *Nature*, **416**, 279–279.
7. Martino, R., Malet-Martino, M. and Gilard, V. (2000) Fluorine nuclear magnetic resonance, a privileged tool for metabolic studies of fluoropyrimidine drugs. *Current Drug Metabolism*, **1**, 271–303.
8. Rowinsky, E. K. (1997) The development and clinical utility of the taxane class of antimicrotubule chemotherapy agents. *Annual Review of Medicine*, **48**, 353–374.
9. FDA. (2004) Drug Approvals. http://www.fda.gov/cder/approval/t.htm.
10. Schiff, P. B., Fant, J. and Horwitz, S. B. (1979) Promotion of microtubule assembly in vitro by taxol. *Nature*, **277**, 665–667.
11. Jordan, M. A., Toso, R. J. and Wilson, L. (1993) Mechanism of mitotic block and inhibition of cell proliferation by taxol at low concentration. *Proceedings of the National Academy of Sciences of the U S A*, **90**, 9552–9556.
12. Georg, G. I., Chen, T. T., Ojima, I. and Vyas, D. M. (1995) *Taxane Anticancer Agents: Basic Science and Current Status*, American Chemical Society, Washington, D. C.
13. Suffness, M. (1995) *Taxol, Science and Applications*, CRC Press, New York.
14. Bristol-Myers, S. (2003) *Taxol (paclitaxel) Injection*. http://packageinserts.bms.com/pi/pi_taxol.pdf.
15. Gottesman, M. M., Fojo, T. and Bates, S. E. (2002) Multidrug resistance in cancer: role of ATP-dependent transporters. *Nature Reviews Cancer*, **2**, 48–58.
16. Ojima, I., Slater, J. C., Michaud, E., *et al.* (1996) Syntheses and structure-activity relationships of the second generation antitumor taxoids. Exceptional activity against drug-resistant cancer cells. *Journal of Medicinal Chemistry*, **39**, 3889–3896.
17. Ojima, I., Wang, T., Miller, *et al.* (1999) Synthesis and structure-activity relationships of new second-generation taxoids. *Bioorganic & Medicinal Chemistry Letters*, **9**, 3423–3428.
18. Ojima, I., Kuduk, S. D., Slater, J. C., *et al.* (1996) Syntheses of new fluorine-containing taxoids by means of β-lactam synthon method. *Tetrahedron*, **52**, 209–24.
19. Ojima, I., Kuduk, S. D., Slater, J. C., *et al.* (1996) Syntheses, biological activity, and conformational analysis of fluorine-containing taxoids, in *Biomedical Frontiers of Fluorine Chemistry*, ACS Symposium Series **639** (eds I. Ojima, J. R. McCarthy, J. T. and Welch.), American Chemical Society, Washington, D.C., pp. 228–243.
20. Ojima, I., Slater, J. C., Pera, P., *et al.* (1997) Synthesis and biological activity of novel 3′-trifluoromethyl taxoids. *Bioorganic & Medicinal Chemistry Letters*, **7**, 133–138.
21. Ojima, I., Inoue, T., Slater, J. C., *et al.* (1999) Synthesis of enantiopure F-containing taxoids and their use as anticancer agents as well as probes for biomedical problems, in *Asymmetric Fluoroorganic Chemistry: Synthesis, Application, and Future Directions*, ACS Symposium

Series **746** (ed. P. V. Ramachandran), American Chemical Society, Washington, D. C., pp 158–181.

22. Sullivan, K. F. (1988) Structure and utilization of tubulin isotypes. *Annual Review of Cell Biology*, **4**, 687–716.
23. Banerjee, A., Roach, M. C., Trcka, P. and Luduena, R. F. (1992) Preparation of a monoclonal antibody specific for the class IV isotype of beta-tubulin. Purification and assembly of alpha beta II, alpha beta III, and alpha beta IV tubulin dimers from bovine brain. *Journal of Biological Chemistry*, **267**, 5625–5630.
24. Panda, D., Miller, H. P., Banerjee, A., *et al.* (1994) Microtubule dynamics in vitro are regulated by the tubulin isotype composition. *Proceedings of the National Academy of Sciences of the USA*, **91**, 11358–11362.
25. Kavallaris, M., Kuo, D. Y. S., Burkhart, C. A., *et al.* (1997) Taxol-resistant epithelial ovarian tumors are associated with altered expression of specific beta-tubulin isotypes. *Journal of Clinical Investigation*, **100**, 1282–1293.
26. Sharma, J. and Luduena, R. F. (1994) Use of *N,N*(-polymethylenebis(iodoacetamide) derivatives as probes for the detection of conformational differences in tubulin isotypes. *Journal of Protein Chemistry*, **13**, 165–76.
27. Schwarz, P. M., Liggins, J. R. and Luduena, R. F. (1998) Beta-tubulin isotypes purified from bovine brain have different relative stabilities. *Biochemistry*, **37**, 4687–4692.
28. Derry, W. B., Wilson, L., Khan, I. A., *et al.* (1997) Taxol differentially modulates the dynamics of microtubules assembled from unfractionated and purified β-tubulin isotypes. *Biochemistry*, **36**, 3554–3562.
29. Ferlini, C., Raspaglio, G., Mozzetti, S., *et al.* (2005) The seco-taxane IDN5390 is able to target class III β-tubulin and to overcome paclitaxel resistance. *Cancer Research*, **65**, 2397–2405.
30. Haber, M., Burkhart, C. A., Regl, D. L., *et al.* (1995) Altered expression of Mbeta2, the class ii beta-tubulin isotype, in a murine J774.2 cell line with a high level of taxol resistance. *Journal of Biological Chemistry*, **270**, 31269–31275.
31. Kavallaris, M. (1999) Antisense oligonucleotides to class III beta-tubulin sensitize drug-resistant cells to Taxol. *British Journal of Cancer*, **80**, 1020–1025.
32. Ojima, I. (1995) Recent advances in β-lactam synthon method. *Accounts of Chemical Research*, **28**, 383–389.
33. Ojima, I. (1995) Asymmetric syntheses by means of β-lactam synthon method, in *Advances in Asymmetric Synthesis* (ed A. Hassner), JAI Press, Greenwich, pp 95–146.
34. Kuznetsova, L., Ungureanu, I. M., Pepe, A., *et al.* (2004) Trifluoromethyl- and difluoromethyl-β-lactams as useful building blocks for the synthesis of fluorinated amino acids, dipeptides, and fluoro-taxoids. *Journal of Fluorine Chemistry* **125**, 487–500.
35. Lin, S., Geng, X., Qu, C., *et al.* (2000) Synthesis of highly potent second-generation taxoids through effective kinetic resolution coupling of racemic β-lactams with baccatins. *Chirality* **12**, 431–441.
36. Ojima, I., Lin, S., Slater, J. C., *et al.* (2000) Syntheses and biological activity of C-3′-difluoromethyl-taxoids. *Bioorganic & Medicinal Chemistry*, **8**, 1619–1628.
37. Skehan, P., Storeng, R., Scudiero, D., *et al.* (1990) New colorimetric cytotoxicity assay for anticancer-drug screening. *Journal of the National Cancer Institute*, **82**, 1107–1112.
38. Huzil, J. T., Luduena, R. F. and Tuszynski, J. (2006) Comparative modelling of human β tubulin isotypes and implications for drug binding. *Nanotechnology* **17**, S90–S100.
39. Ojima, I., Slater, J. S., Kuduk, S. D., *et al.* (1997) Syntheses and structure-activity relationships of taxoids derived from 14β-hydroxy-10-deacetylbaccatin III. *Journal of Medicinal Chemistry*, **40**, 267–278.
40. Ojima, I., Sun, C. M. and Park, Y. H. (1994) New and efficient coupling method for the synthesis of peptides bearing norstatine residue and their analogs. *Journal of Organic Chemistry*, **59**, 1249–1250.

41. Appendino, G., Danieli, B., Jakupovic, J., *et al.* (1997) Synthesis and evaluation of C-seco paclitaxel analogues. *Tetrahedron Letters*, **38**, 4273–4276.
42. Appendino, G., Noncovich, A., Bettoni, P., *et al.* (2003) The reductive fragmentation of 7-hydroxy-9,10-dioxotaxoids. *European Journal of Organic Chemistry*, 4422–4431.
43. Gut, I., Ojima, I., Vaclavikova, R., *et al.* (2006) Metabolism of new generation taxanes in human, pig, minipig and rat liver microsomes. *Xenobiotica* **36**, 772–792.
44. Vuilhorgne, M., Gaillard, C., Sanderlink, G. J., *et al.* (1995) *Metabolism of Taxoid Drugs, in Taxane Anticancer Agents: Basic Science and Current Status*, ACS Symp. Ser. **583** (eds G. I. Georg, T. T. Chen, I. Ojima and D. M. Vyas), American Chemical Society, Washington, D. C., pp 98–110.
45. Ojima, I., Sun, C. M., Zucco, M., *et al.* (1993) A highly efficient route to taxotère by the β-lactam synthon method. *Tetrahedron Letters*, **34**, 4149–4152.
46. Ojima, I., Habus, I., Zhao, M., *et al.* (1992) New and efficient approaches to the semisynthesis of taxol and its C-13 side-chain analogs by means of beta-lactam synthon method. *Tetrahedron*, **48**, 6985–7012.
47. Ojima, I., Kuduk, S. D. and Chakravarty, S. (1998) Recent advances in the medicinal chemistry of taxoid anticancer agents, in *Advanced Medicinal Chemistry* (eds B. E. Maryanoff and A. B. Reitz,), JAI Press, Greenwich, CT, pp 69–124.
48. Ojima, I., Kuduk, S. D., Chakravarty, S., *et al.* (1997) A novel approach to the study of solution structures and dynamic behavior of paclitaxel and docetaxel using fluorine-containing analogs as probes. *Journal of the American Chemical Society*, **119**, 5519–5527.
49. Li, Y., Poliks, B., Cegelski, L., *et al.* (2000) Conformation of microtubule-bound paclitaxel determined by fluorescence spectroscopy and REDOR NMR. *Biochemistry*, **39**, 281–291.
50. Geney, R., Sun, L., Pera, P., *et al.* (2005) Use of the tubulin bound paclitaxel conformation for structure-based rational drug design, *Chemistry & Biology*, **12**, 339–348.
51. Rao, S., He, L., Chakravarty, S., *et al.* (1999) Characterization of the taxol binding site on the microtubule. Identification of Arg282 in β-Tubulin as the site of photoincorporation of a 7-benzophenone analogue of taxol. *Journal of Biological Chemistry*, **274**, 37990–37994.
52. Nogales, E., Wolf, S. G. and Downing, K. H. (1998) Structure of the αβ tubulin dimer by electron crystallography. *Nature* **391**, 199–203.
53. Paik, Y., Yang, C., Metaferia, B., *et al.* (2007) Rotational-echo double-resonance NMR distance measurements for the tubulin-bound paclitaxel conformation. *Journal of the American Chemical Society*, **129**, 361–370.
54. Lowe, J., Li, H., Downing, K. H. and Nogales, E. (2001) Refined structure of alpha beta-tubulin at 3.5 Å resolution. *Journal of Molecular Biology*, **313**, 1045–1057.
55. Sun, L., Geng, X., Geney, R., *et al.* (2008) Design, synthesis and biological evaluation of novel C14-C3′BzN–linked macrocyclic taxoids, *Journal of Organic Chemistry*, **73**, 9584–9593.
56. Ojima, I., Kuznetsova, L. V. and Sun, L. (2007) Organofluorine chemistry at the biomedical interface – a case study on fluoro-taxoid anticancer agents, in *Current Fluoroorganic Chemistry. New Synthetic Directions, Technologies, Materials and Biological Applications* (eds V. Soloshonok, K. Mikami, T. Yamazaki, J.T. Welch and J. Honek), American Chemical Society/Oxford University Press, Washington, D.C., pp 288–304.
57. Chen, J., Jaracz, S., Zhao, X., *et al.* (2005) Antibody-based toxin conjugates for cancer therapy. *Expert Opinion in Drug Delivery*, **2**, 873–890.
58. Jaracz, S., Chen, J., Kuznetsova, L. V. and Ojima, I. (2005) Recent advances in tumor-targeting anticancer drug conjugates. *Bioorganic & Medicinal Chemistry*, **13**, 5043–5054.
59. Chari, R. V. J. (1998) Targeted delivery of chemotherapeutics: tumor-activated prodrug therapy. *Advanced Drug Delivery Reviews*, **31**, 89–104.
60. Liu, C., Tadayoni, B. M., Bourret, L. A., *et al.* (1996) Eradication of large colon tumor xenografts by targeted delivery of maytansinoids. *Proceedings of the National Academy of Sciences of the U S A*, **93**, 8618–8623.

61. Hamann, P. R., Hinman, L. M., Hollander, I., *et al.* (2002) Gemtuzumab ozogamicin, a potent and selective anti-CD33 antibody-calicheamicin conjugate for treatment of acute myeloid leukemia. *Bioconjugate Chemistry*, **13**, 47–58.
62. Lam, L., Lam, C., Li, W. and Cao, Y. (2003) Recent advances in drug-antibody immunoconjugates for the treatment of cancer. *Drugs of the Future*, **28**, 905–910.
63. Saleh, M. N., LoBuglio, A. F. and Trail, P. A. (1998) Monoclonal antibody-based immunoconjugate therapy of cancer: studies with BR96-doxorubicin. *Basic and Clinical Oncology*, **15**, 397–416.
64. Chan, S. Y., Gordon, A. N., Coleman, R. E., *et al.* (2003) A phase 2 study of the cytotoxic immunoconjugate CMB-401 (hCTM01-calicheamicin) in patients with platinum-sensitive recurrent epithelial ovarian carcinoma. *Cancer Immunology, Immunotherapy* **52**, 243–248.
65. Gillespie, A. M., Broadhead, T. J., Chan, S. Y., *et al.* (2000) Phase I open study of the effects of ascending doses of the cytotoxic immunoconjugate CMB-401 (hCTMO1-calicheamicin) in patients with epithelial ovarian cancer. *Annals of Oncology*, **11**, 735–741.
66. Ojima, I., Geng, X., Wu, X., *et al.* (2002) Tumor-specific novel taxoid-monoclonal antibody conjugates. *Journal of Medicinal Chemistry*, **45**, 5620–5623.
67. Kigawa, J., Minagawa, Y., Kanamori, Y., *et al.* (1998) Glutathione concentration may be a useful predictor of antibody–cytotoxic agent conjugates for cancer therapy response to second-line chemotherapy in patients with ovarian cancer. *Cancer*, **82**, 697–702.
68. Ojima, I. (2004) Use of fluorine in the medicinal chemistry and chemical biology of bioactive compounds – a case study on fluorinated taxane anticancer agents. *ChemBioChem*, **5**, 628–635.
69. Ojima, I. (2008) Guided molecular missiles for tumor-targeting chemotherapy: case studies using the 2nd-generation taxoids as warheads. *Accounts of Chemical Research* **41**, 108–119.

6

Antimalarial Fluoroartemisinins: Increased Metabolic and Chemical Stability

Jean-Pierre Bégué and Danièle Bonnet-Delpon

6.1 Introduction

Malaria continues to be one of the most important infectious diseases in the world [1]. Of the four human malaria parasites, *Plasmodium falciparum* is the one overwhelmingly responsible for severe clinical malaria and death. A major cause of malaria morbidity and mortality is the current increasing resistance of malaria parasites, in particular *P. falciparum*, to the most prescribed antimalarial drugs (chloroquine, pyrimethamine, proguanil, halofantrine, etc.) [2]. Facing this alarming decline in the efficacy of antimalarial drugs, a broad consensus on the need to develop new antimalarial drugs is now well established [3]. Artemisinin **1** (see Figure 6.1), which was isolated from a plant, *Artemisia annua* L., and has been used to treat high fever over 2000 years in Chinese traditional medicine [4], has emerged as a desirable drug. Artemisinin is a sesquiterpene lactone that contains an endoperoxide bridge, as a part of a 1,2,4-trioxane core. This very unusual functional group in medicinal chemistry is an absolute requirement for the antimalarial activity. When the endoperoxide is lacking, as in the reduced desoxoartemisinin, the compound is devoid of antimalarial activity [5]. Consequently, the peroxide function has become a focus for experiments aimed at understanding the mode of action of artemisinin and other trioxanes. Numerous hypotheses have been proposed as possible mechanisms of action for artemisinin derivatives ("artemisinins"). They were supported by a large number of studies seeking an understanding at the molecular level [6]. However, the conclusion of these

Fluorine in Medicinal Chemistry and Chemical Biology Edited by Iwao Ojima

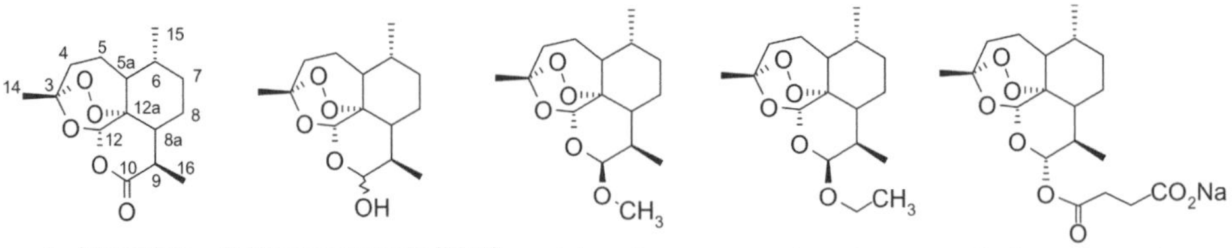

Figure 6.1 First-generation antimalarial artemisinins.

complementary but often contradictory experiments is that the *in vivo* activity is probably multifactorial: that is, artemisinins interacts with several targets [7].

The therapeutic value of artemisinin is limited by its low solubility in both oil and water. However, artemisinin provides a unique key molecular framework from which medicinal chemists can attempt to prepare more efficacious antimalarial drugs by retaining the pharmacologically essential 1,2,4-trioxane core [8].

The first-generation semisynthetic analogues, dihydroartemisinin **2**, artemether, arteether ("ethers of artemisinin") and sodium artesunate, were prepared in the 1980s (Figure 6.1) [5a, 9]. These artemisinin derivatives have made a substantial impact on the treatment of malaria [10–12]. They are fast-acting and potent antimalarial drugs against all *Plasmodium* parasites, in particular the multidrug-resistant *P. falciparum* [13]. They act at the early stages of development of parasites once the parasites have invaded red cells [14]. Artemisinins are gametocytocidal, but they do not kill the hepatic stages of parasites [15]. No clinical case of resistance to artemisinins has been reported so far [16].

These derivatives have poor oral bioavailability, but fortunately this is partially countered by their very high intrinsic activities. Furthermore, they all possess a short half-life in the body, which can lead to recrudescence of parasitemia [17]. According to Scheme 6.1, the human metabolism of an ether of artemisinin, such as artemether, involves first an oxidation in the liver by cytochrome P450 enzymes. This produces dihydroartemisinin (DHA) as the main metabolite, which retains the antimalarial activity [11]. DHA is then rapidly eliminated through phase II metabolism via generation of water-soluble conjugates such as α-DHA-β-glucuronides [18]. These two processes are responsible for the short plasma half-life of ethers of artemisinin [8]. A second important factor is the low stability of the acetal group under acidic conditions, such as those found in the stomach upon oral administration.

[*Note:* In artemisinin series, conventional designation of stereochemistry is used (e. g., α is below the plane, while β is above the plane), but the stereochemistry at C-10 is opposite to that used for glycosides.]

The use of these artemisinin derivatives in combination with an antimalarial drug with a longer half-life, as recommended by WHO, can not only delay emergence of resistance but also partially overcome the problem of short plasma half-life [19]. Nevertheless, chemically and metabolically more stable artemisinins would bring improvement in malaria therapy and bitherapy (or combination therapy) [20].

In order to improve the oral bioavailability, a huge number of compounds have been synthesized from artemisinin, which is a readily available naturally occurring compound [8]. Large-scale isolation and purification of artemisinin from wild or cultivated *Artemisia*

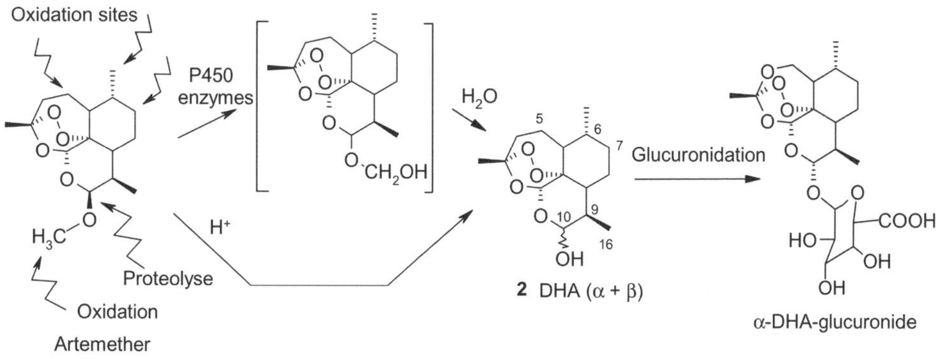

Scheme 6.1 *Metabolism of artemether in humans.*

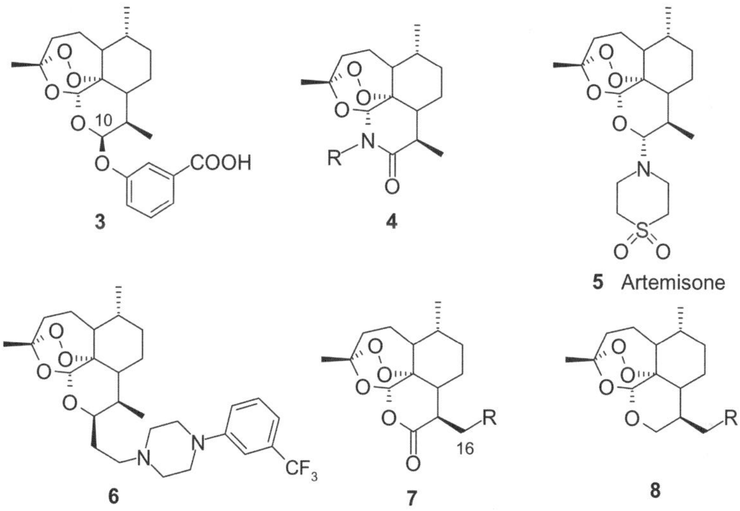

Figure 6.2 *Main structural variations of the artemisinin skeleton for antimalarial drug discovery.*

annua are rather easy (more than 10 tonnes/year is produced in south-east Asia). The two main objectives of medicinal chemists have been to obtain more stable derivatives and to increase water solubility by introducing polar or ionizable functional groups.

Numerous chemical modifications of artemisinin have been extensively investigated and reviewed [8, 20a, 21, 22]. The most effective approaches include the design of oxidation-resistant ethers of DHA **3** (see Figure 6.2), of 10-aza analogues **5** [23, 24], and of 10-carba analogues **6** [8, 20a, 21, 22, 25]. From these various "robust" artemisinins, current efforts are focused on incorporating functional groups in order to increase water solubility and to decrease log *P*. Good examples are artemisone **5**, developed by Bayer and MMV [26], and the piperazine derivative **6**, a potent *in vitro* and *in vivo* antimalarial [8, 27].

All structural modulations described above are concerned with the C-10 site. Artemisinin derivatives functionalized at C-16 (**7**, **8**) are more difficult to prepare, and the natural precursors, artemisitene and artemisinic acid, are less readily available than artemisinin [18c, 28, 29].

6.2 Fluoroartemisinins: Control of Metabolism

To prevent oxidative metabolism and/or to increase hydrolytic stability, ten years ago we initiated our approach to designing a metabolically more stable artemisinin using fluorine substitution. This approach is now classical in medicinal chemistry and has been validated by numerous examples [30, 31]. The decreased rate of metabolism as a result of fluorine substitution is a consequence of the intrinsic properties of the fluorine atom – the electronic structure and strong electronegativity of fluorine impart great chemical inertia to the C–F bond, in particular toward oxidation. Furthermore, the electronegativity of fluorine confers a strong electron-withdrawing character on fluoroalkyl substituents. This disfavors development of positive charge on the α-carbon, and hence the generation of cationic species involved in hydrolytic processes [30, 32].

We present here an overview of our approaches to and results on new potent antimalarial fluoro-artemisinins. The effects of fluorine substitution on chemical reactivity make the synthesis of fluoro-artemisinins not always straightforward. These specific synthetic challenges will be highlighted.

6.2.1 Fluoroalkyl Ethers of Dihydroartemisinin

6.2.1.1 Synthesis

As indicated in Scheme 6.1, *in vivo* or in liver homogenates, ethers of DHA undergo rapid hydroxylation by cytochrome P450 enzymes to generate a hemiacetal, which decomposes to produce DHA **2** and an aldehyde [18c, 33]. A feasible approach to prolonging the half-life of DHA ethers is to design poorer substrates for cytochrome P450 by introduction of a fluorinated substituent (R^f) at the α-methylene carbon of the alkoxy group (see Scheme 6.2). A slower rate of oxidative dealkylation would be expected since it has been

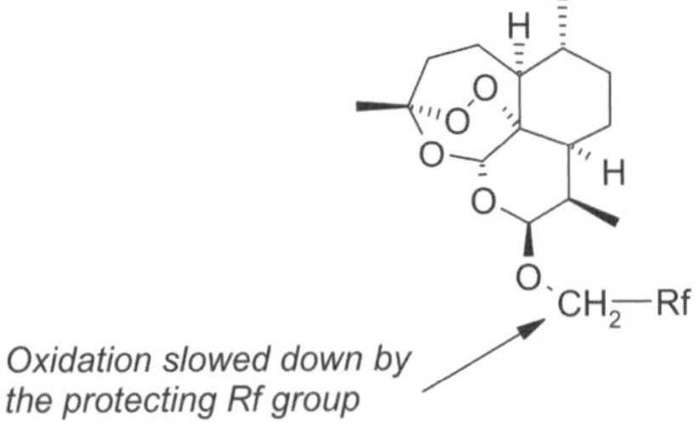

H
O
O
O
O
H
O
CH_2—Rf
Oxidation slowed down by
the protecting Rf group

Scheme 6.2 *Design of fluoroalkyl ethers of artemisinin.*

demonstrated that the protection against oxidative processes provided by a fluoroalkyl group is often extended to adjacent CH or CH_2 groups [30, 34–37].

Ethers of DHA are usually prepared by treatment of DHA **2** ($\alpha:\beta \sim 50:50$) with an appropriate alcohol in the presence of BF_3 and Et_2O. The stereochemistry of this reaction has been intensively investigated and discussed [38], in particular by Haynes *et al.* [18c, 24, 39]. This process is less efficient with fluoro-alcohols (R^fOH), because they are poor nucleophiles. For instance, the reaction of trifluoroethanol with DHA [40], in the presence of BF_3 and Et_2O in ether, yielded ether **9a** (67%), and dehydrodeoxoartemisinin **10** (20%) resulting from the competing deprotonation of the intermediate oxonium ion **11** (see Scheme 6.3) [27, 30]. However, a secondary fluoro-alcohol and pentafluorophenol did not react at all (see Table 6.1).

The Mitsunobu procedure was reported to be efficient in the case of fluoroalkyl alcohols [42]. Unlike nonfluorinated alcohols, the acidic fluoro-alcohols [43] are efficiently deprotonated by the PPh_3-DEAD adduct. This facilitates the displacement of the oxyphosphonium leaving group by an alkoxide. With the use of this reaction, a range of ethers **9** were prepared from fluoro-alcohols in good (from primary alcohols) to moderate yields (secondary alcohols) with high β-stereoselectivity (see Scheme 6.3 and Table 6.1).

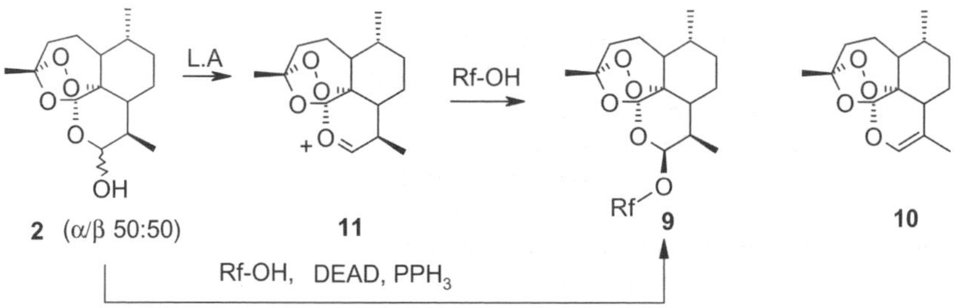

2 (α/β 50:50) —L.A→ **11** —Rf-OH→ **9**; **10**

2 —Rf-OH, DEAD, PPH_3→ **9**

Scheme 6.3 *Etherification reaction of dihydroartemisinin [40; T. Van Nhu* et al.*, unpublished results on "Preclinical development experimental pharmacology of fluorinated artemisinin derivatives", 2005]. DEAD: diethyl azodicarboxylate.*

Table 6.1 *Preparation of DHA fluoroalkyl ethers* **9**

Alcohol R^f-OH	Method	**9** Yields (%)	β/α
a CF_3CH_2OH	$BF_3,Et_2O/Et_2O$	67	94:4
	Mitsunobu	74	93:7
b $CF_3CF_2CH_2OH$	$BF_3,Et_2O/Et_2O$.	43	97:3
	Mitsunobu	80	93:3
c $CF_3CHOHCF_3$	$BF_3,Et_2O/Et_2O$	0	
	Mitsunobu	60	100:0
d C_6F_5OH	$BF_3,Et_2O/Et_2O$	0	
	Mitsunobu	50	94:6

6.2.1.2 Antimalarial Activity

The IC_{50} values for the fluoroalkyl ethers **9**, evaluated on *P. falciparum* (W-2 strain) (27 nM < IC_{50} < 72 nM), are higher than that of artemether (IC_{50} = 6 nM), wherein only a small influence of the structure of fluorinated chains is observed [40]. However, ethers **9** were very active *in vivo* (Peters test, *P. berghei*, intraperitoneal administration for 4 days at 35.5 µmol/kg), especially ethers **9b** and **9c**. All animals were cured before day 42 (D-42). Compound **9b** has been investigated more in detail [41].[1] High activity was also found after subcutaneous administration (ED_{50} = 1 mg/kg, ED_{90} = 1.5 mg/kg; cf. artesunate: ED_{50} = 2.8 mg/kg, ED_{90} = 5.4 mg/kg) or oral administration (ED_{50} = 2.8 mg/kg, ED_{90} = 4.3 mg/kg; cf. artesunate: ED_{50} = 10.5mg /kg, ED_{90} = 15.3 mg/kg). However, its water solubility is very low (log P = 6.1, water solubility at pH 7.4 < 5µg /mL). Thus, an appropriate formulation was required.

6.2.2 Fluorinated Analogues of Dihydroartemisinin and Artemether

Substantial effects on both stability and metabolism might be expected from the presence of a trifluoromethyl substituent at C-10 in DHA and its ethers. The following effects were anticipated:

- Glucuronidation of CF_3-DHA **12** should be greatly slowed relative to DHA, because the electron-withdrawing character of the trifluoromethyl group can affect the two possible glucuronidation processes. The nucleophilicity of the hydroxyl in CF_3-DHA **12** is strongly decreased. Consequently, **12** is expected to be a poorer glycosyl acceptor, regardless of its configuration at C-10. Alternatively, an intermediate oxonium ion **14** at C-10, which is a potential glycosyl donor, should be generated from **12** only with high activation energy (see Scheme 6.4).
- *In vivo* stability of the trifluoromethyl analogue **13a** of artemether is anticipated to be increased in two ways: first, the trifluoromethyl group could decrease the oxidation of the methoxy group [30, 34–37]; second, the intermediate oxonium ion **14** should be much more difficult to generate in acidic medium [30].

To validate these hypotheses, the synthesis of the trifluoromethyl analogues of DHA and artemether was investigated.

6.2.2.1 Chemistry

The hemiketal **12** was easily prepared in high yield (80%) from artemisinin by treatment with trifluoromethyl trimethylsilane ($TMSCF_3$) in the presence of tetrabutylammonium fluoride trihydrate (TBAF, $3H_2O$). Complete desilylation occurred after addition of water (see Scheme 6.5). The reaction was stereoselective, and the α configuration of the CF_3 group was unambiguously determined (β-**12**) [40, 44]. However, this configuration at C-10 is not the result of an α approach of the CF_3 but of a thermodynamic equilibrium of the

[1] For pharmacokinetic studies experiments were performed with an intravenous formulation (0.1 M Captisol™ in water) (solubility: 4600 µg/mL) and an oral standard suspension vehicle (SSV) formulation (0.5 % w/v CMC, 0.5 % v/v benzyl alcohol, 0.4 % v/v Tween 80 in 0.9 % w/v NaCl) (solubility: 585 µg/mL): K. A. McIntosh, W. N. Charman *et al.*, unpublished results, 2002.

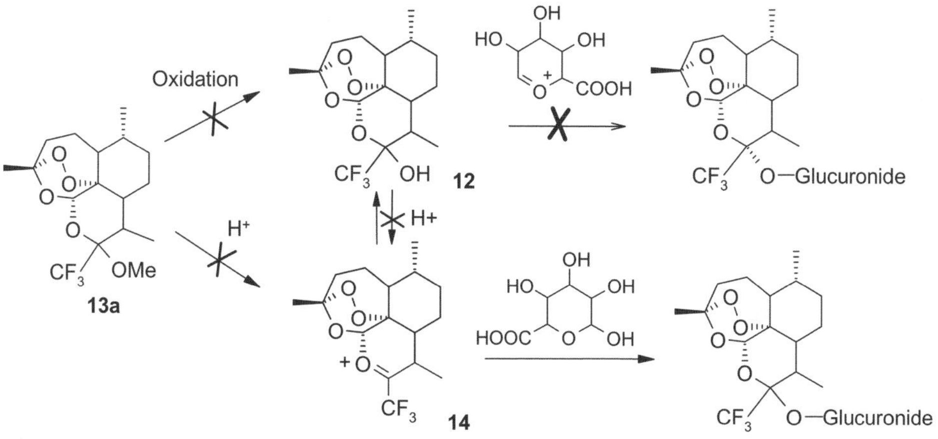

Scheme 6.4 *Expected effect of CF_3-substitution on the stability of artemether and DHA.*

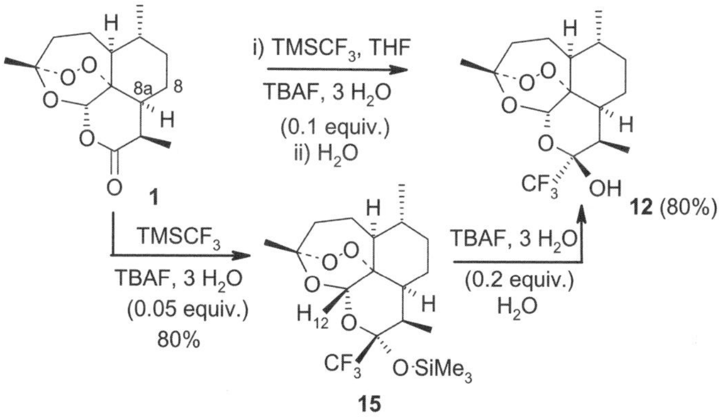

Scheme 6.5 *Reaction of $TMSCF_3$ with artemisinin. TBAF: tributylammonium fluoride.*

kinetic (β-CF_3, α-hydroxy) product into β-**12.** The primary product of the reaction, silyloxy ketal **15**, is β-CF_3. This indicates that, in the reaction of artemisinin with TMSCF3, there is a kinetic preference for a β-addition such as in the reaction of activated DHA involving an oxonium ion [18c, 24, 39]. The axial CF_3 group, which is a bulky substituent [45], probably suffers from a repulsive interaction with both the C-8a–C-8 axial bond and the methyl substituent, and thus favors the equilitrium towards β-**12** [46], while in DHA **2** the α : β equilibrium ratio is about 50 : 50.

We then investigated routes to the CF_3 analogue of artemether. All attempts at etherification or acylation of the hydroxyl moiety of hemiketal **12** failed. Starting hemiketal **12** was recovered under various conditions generally used for the functionalization of DHA [47]. This can be ascribed to the poor reactivity of the β-epimer of DHA wherein the axial hydroxyl experiences a 1,3-diaxial interaction with the C-8–C-8a bond [18c]. This steric

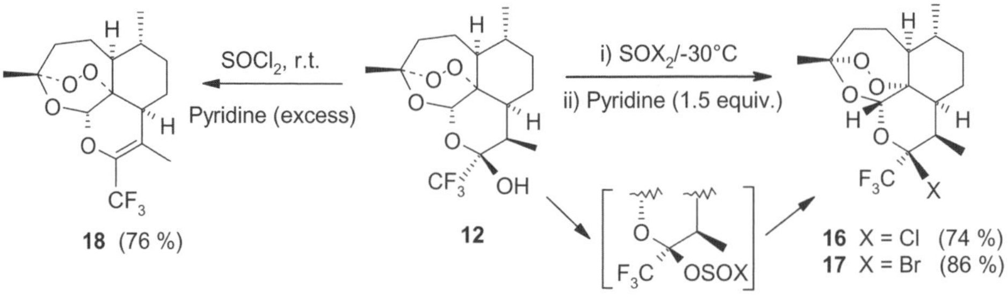

Scheme 6.6 *Preparation of 10-halogeno-10-deoxoartemisinins.*

hindrance, coupled with the low nucleophilicity of a tertiary trifluoromethyl hydroxyl, raises the activation energy of alkylation or acylation. A substitution reaction involving displacement of the hydroxyl group of **12** is not an effective alternative. In general, the substitution of α-trifluoromethyl hydroxyl groups through S_N1 is as difficult as that through S_N2 processes. The strong electron-withdrawing effect of the CF_3 group strengthens the C–O bond and destabilizes the intermediary carbocation in the S_N1 process [48]. Despite the presence of an alkoxy substituent at C-10, the hemiketal **12** remained unreacted when it was subjected to S_N1-type conditions (BF_3,Et_2O/MeOH). In an S_N2 process, the combination of steric and electronic repulsive effects of fluorine atoms on the incoming nucleophile decreases the reaction rate [48a,b]. Indeed, the Mitsunobu reaction, a typical S_N2 process, failed for the attempted coupling of **12** with benzoic acid [47].

Finally, halogenation was tried and found successful. Thus, chloride **16** and bromide **17** were prepared from **12** in good yields through reaction with thionyl chloride or thionyl bromide at −30 °C, using 1.5 equivalents of pyridine (see Scheme 6.6). These reaction conditions are very important for obtaining **16** and **17**: when the same reactions were performed using pyridine as solvent at *room temperature*, pseudo-glycal **18** was obtained in good yield in place of halogenation products. Halides **16** and **17** were stereoselectively obtained with β configuration at C-10 as in the starting CF_3-DHA. Thus, the substitution proceeds with retention of configuration through SNi mechanism [49]. These 10-halogeno compounds **16** and **17** are, a priori, better substrates than the hydroxyl counterpart (CF_3-DHA) for further substitution reactions.

The nucleophilic substitution of bromide **17** with methanol used as solvent was investigated in the presence of silver salts. The ether **13a** was obtained as a mixture of two diastereoisomers, accompanied by a large amount of glycal **18** (see Scheme 6.7) [47].

Taking into account the lack of chemoselectivity and stereoselectivity of these silver ion-mediated reactions, it is reasonable to postulate an S_N1-type mechanism for this process, leading to the formation of oxonium ion **14**, despite the electron-withdrawing effect of the CF_3 group (see Scheme 6.8). The stereochemical outcome should be ascribed to the steric or electronic preference of methanol addition to two possible diastereo-faces of the trifluoromethyl oxonium ion **14**, in which the interactions are probably different from those in nonfluorinated DHA derivatives. In order to disfavor the formation of oxonium **14**, the reaction was performed without silver salt in MeOH. While the reaction rate did not decrease significantly, only 23% of glycal **18** was obtained. Furthermore, the

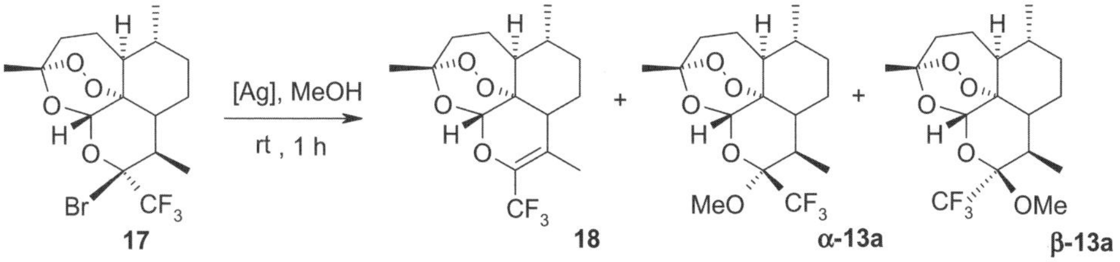

Scheme 6.7 *Substitution of bromide **17** with MeOH in the presence of silver salt.*

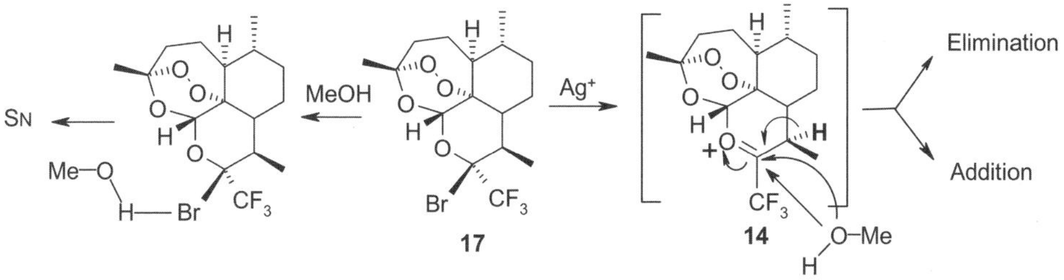

Scheme 6.8 *Both activation pathways of substitution reaction of bromide **17**.*

substitution leading to the formation of **13a** occurred with more than 80% retention of configuration ($\alpha/\beta = 17/83$). These results can be rationalized by assuming an activation of bromide **17** by methanol through a hydrogen bond ($\alpha_{MeOH} = 0.98$) [50].

An improvement required was to decrease the amount of alcohol used for the substitution. Reaction of bromide **17** in CH_2Cl_2 at room temperature with MeOH (10 equivalents) was slow (72% conversion after 15 h), but highly diastereoselective ($\alpha/\beta = 8/92$) and chemoselective (only 3% of glycal **18**). For further optimization, 1,1,1,3,3,3-hexafluoroisopropanol (HFIP) was used as additive [47]. HFIP possesses low nucleophilicity, high ionizing power, high hydrogen bond donor ability ($\alpha = 1.96$) and a strong ability to solvate anions [43b, 51, 52]. These properties have been exploited in solvolysis reactions [51] and various other reactions [43b, 53].

Under optimized experimental conditions (CH_2Cl_2, MeOH (10 equivalents), HFIP (5 equivalents)) the reaction was complete at room temperature in less than 5 h. Moreover, the reaction was still chemoselective (less than 5% of the elimination product **18**), and completely stereoselective (the α stereoisomer was not detected) (see Scheme 6.9) [47]. The nucleophile is delivered on the same β-face as the leaving group. Similar results from DHA glycosyl donor have been rationalized by the formation of the half-chair oxonium ion, and a stereoelectronically preferred axial addition with reactive nucleophiles [18c]. Compared to the reaction of **17** with the stronger Ag^+ electrophilic assistance (see Schemes 6.7 and 6.8), the low ratio of elimination product **16** and the high β-diastereoselectivity suggest a mechanism different from an addition on the oxonium ion **14**. It is also clear

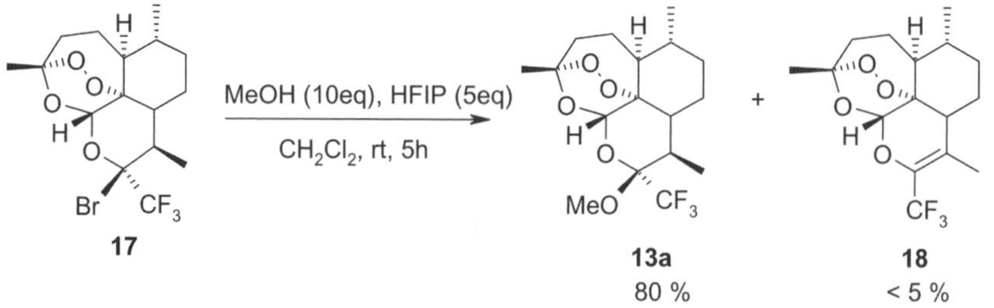

Scheme 6.9 *Preparation of **13a**.*

Table 6.2 *Antimalarial activities of compound **12** (CF_3-DHA) and **13a** (CF_3-artemether) compared to reference artemisinins*

	In vitro: IC_{50} (nM)			Subcutaneous (*P. berghei*) (mg/kg)		Oral (*P. berghei*) (mg/kg)		I.p. (*P. berghei*) (mg/kg)	
	D6	W2	FCB1	ED_{50}	ED_{90}	ED_{50}	ED_{90}	ED_{50}	ED_{90}
12	2.6	0.9		0.7	1.8	4.3	13.0		
Na artesunate	–	5.4		2.8	10.4	5.4	15.3		
Artemether			3.5					2.5	8.5
13a			0.8					1.25	6.4

that in **17** the side opposite to the bromide is hindered: the bulky trifluoromethyl moiety and the cyclic oxygen lone pairs offer a more repulsive face than C-8–C-8a and C–H-12 bonds.

6.2.2.2 Biological Properties

6.2.2.2.1 Antimarial Activity The 10-*R*-(trifluoromethyl)dihydroartemisinin **12** is highly active against D6 and W2 drug-resistant strains of *P. falciparum* (D6, $IC_{50} = 2.6\,nM$; W2, $IC_{50} = 0.9\,nM$), and more potent than artesunate (see Table 6.2). Moreover, it is active against wild isolates from African patients (with Senegalese isolates, $IC_{50} = 3.3\,nM$) [54]. Compound **12** is also more active *in vivo* (subcutaneous or oral administration) than sodium artesunate in mice infected with murine *P. berghei* [40]. All mice survived until day 42 in the Peters test.

In vitro activity on *P. falciparum* (FCB1 strain) of the trifluoromethyl analogue **13a** of artemether is higher than that of artemether itself (0.8 nM for CF_3-artemether) (see Table 6.2) [55]. *In vivo* ED_{50} shows that **13a** is about twice as active as artemether. More importantly, i.p. administration of the 10-CF_3 analogue of artemether to mice infected with *P. berghei* (NK173) completely cleared parasitemia from the end of treatment to day-25. This clearance has never been observed previously with artemether itself when used at the same concentration in our experiments (35.5 μmol/kg).

6.2.2.2.2 Stability Chemical and metabolic stabilities have been evaluated. A quantitative evaluation of the hydrolytic stability brought to the ketal function by the introduction

Table 6.3 *Half-lives at pH 2 and 37°C and* in vivo

	Acidic conditions (pH 2, 37°C) $t_{1/2}$ (h)	Plasma half-life $t_{1/2}$ (min)
DHA	17	23
CF_3-DHA **12**	760	86
Artemether	11	52
CF_3-artemether **13a**	660	–

of a trifluoromethyl group at C-10 showed that the trifluoromethyl analogues of DHA and artemether were dramatically more stable under acid conditions (pH 2) than DHA and artemether (see Table 6.3) [47, 55]. This confirms the hypothesis about the effect of the fluorine substitution on the protection toward proteolysis. Presence of CF_3 disfavors the formation of the oxocarbenium ion **14** [48, 55]. This higher stability under acidic conditions should increase their half-lives in the stomach and consequently improve oral bioavailability.

The plasma half-life of CF_3-DHA **12** ($t_{1/2}$ = 86 min), after intravenous (i.v, 10 mg/kg) or oral (50 mg/kg) administration to rats is also higher than that of DHA ($t_{1/2}$ = 23 min.) or artemether ($t_{1/2}$ = 52 min.) (see Table 6.3) [56]. Compound **12** possesses a high oral bioavailability (28%) compared with that of artemether (1.4%), probably due to a good compromise between a convenient log *P* (4.36) and a fairly good solubility (see footnote 1 earlier).

6.2.2.2.3 ***Toxicity*** The toxicity CF_3-DHA **12** has been investigated further. It displays low toxicity in rodents. The LD_{50} in mice orally treated was 820 mg/kg, while the ED_{90} is only 13 mg/kg. Subacute toxicity testing was done in rabbits (orally at 20 mg/kg once daily for 28 days), with no effect on the body weight of animals. The electrocardiographic index was the same in treated and nontreated groups. Hematological and biochemical (serum glutamic oxaloacetic transaminase (SGOT) and serum glutamic pyruvic transaminase (SGPT), etc.) parameters and histopathological examinations showed that liver and kidney microstructures were normal in the group treated [41, 57]. *In vitro* neurotoxicity evaluated on neuroblastoma cultures in the presence of liver microsomes showed that the neurotoxicity of **12** is half that of artemether. *In vivo*, the speed of establishing reflex, the speed of extinguishing reflex for searching food in a maze (mice), and conditional reflex (rats) were only slightly different from those of the nontreated group. Toxicity studies performed on monkeys did not show undesirable effects [41].

The CF_3-dihydroartemisinin **12** is chemically stable and is easily prepared on large scale, in one reaction step with purification by crystallization [56]. It is therefore considered to be a good candidate for future development.

6.3 Toward Water-Soluble Fluoroartemisinins

The clear effect on metabolic stability induced by the introduction of a fluoroalkyl substituent at C-10 of artemisinins has been demonstrated. This provided robust derivatives that therefore exhibited better antimalarial activity than reference artemisinins. Our further

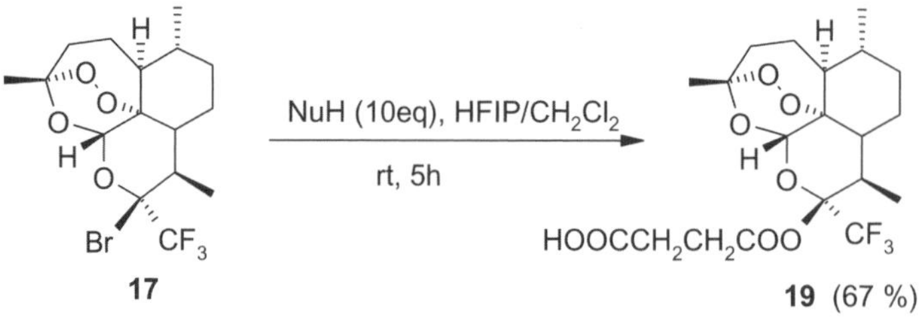

Scheme 6.10 *Preparation of the CF_3 analogue of artesunate.*

important objective was to introduce polar or ionizable functions into fluoroartemisinins to increase the water solubility. Several strategies have been investigated.

6.3.1 CF_3 Analogue of Artesunate

Conditions set up for the preparation of 10-CF_3-artemether **13a** by substitution of the bromide **17** in the presence of HFIP offered the possibility to introduce other nucleophiles. Various alcohols provided corresponding 10β-alkoxy-10α-trifluoromethyl deoxoartemisinins in good yields (73–89%), accompanied by only a small amount of glycal **18** (2–6%). These efficient conditions were used to prepare the 10-CF_3 analogue of artesunate. The carboxylate generated *in situ* from succinic acid and triethylamine reacted with **17**, in the presence of HFIP as a co-solvent (1 : 1), giving rise to the trifluoromethyl analogue **19** of β-artesunate in 67% yield (see Scheme 6.10).

Surprisingly, this trifluoromethyl analogue of artesunate was not highly efficacious *in vivo*. While it protected mice against *P. berghei* until day 7 (3.9% parasitemia), a rapid increase of parasitemia then appeared, thus exhibiting a profile similar to that of artemether

6.3.2 Reaction of Difluoroenoxysilanes with Artemisinin Derivatives

Another promising approach to combining the effect of fluoroalkyl substituent and water solubility was the introduction of a difluoroketone motif as a possible precursor of corresponding acids at C-10. For this purpose we investigated the chemistry of difluoroenoxysilanes, reported as a route to difluoro-C-glycosides [58]. Ziffer previously reported that DHA easily reacted with enoxysilanes under Lewis acid catalysis [59].

Aryl difluoroenoxysilanes **20** were prepared by Mg-promoted defluorination of trifluoromethyl ketones, according to Uneyama's procedure [60]. Their reaction with DHA acetate appeared to be much more critical than any other Lewis acid-catalyzed reaction and the setup was very troublesome. Each difluoroenoxysilane required specific reaction conditions (e.g., Lewis acid and rate of addition) (see Scheme 6.11 and Table 6.4) [61]. For instance, the best conditions found for the preparation of the difluoroketone **21a** ($SnCl_4$, 0.4 equivalents) provided a completely rearranged artemisinin skeleton when applied to the enoxysilane **20b**. Furthermore, the stereochemical outcome of reactions was unusual: difluoroketones **21a–c** all possess the *epi*-artemisinin configuration at C-9 (9α-Me). The mechanism of this epimerization at C-9 is not fully understood, but probably

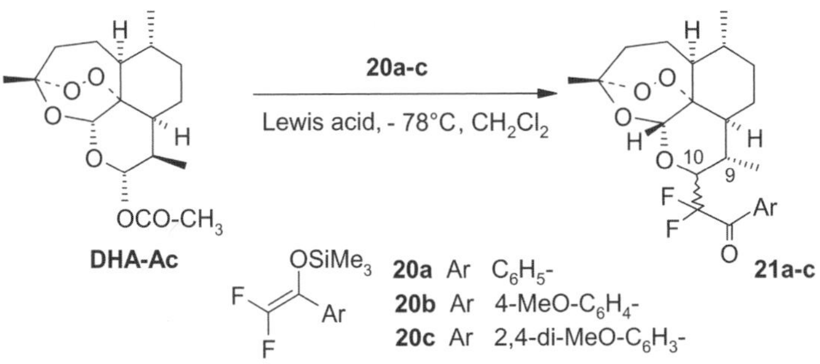

Scheme 6.11 *Reaction of DHA acetate with enoxysilanes **20a–c** in the presence of Lewis acid.*

Table 6.4 *Reaction of DHA acetate with enoxysilanes **20a–c** in presence of Lewis acid*

Enoxy silanes	Lewis acid (equiv.)	Products	Isolated yields	(β/α at C-10)
20a	$SnCl_4$ (0.4)	**21a** (9α-Me)	66%	100:0
20b	BF_3,Et_2O (0.2)	**21b** (9α-Me)	73%	100:0
20c	$SnCl_4$ (0.4)	**21c** (9α-Me)[a]	33%	70:30

[a] Accompanied by glycal **18**.

involves a deprotonation–reprotonation of glycal **10** and oxonium ions **11** (see Scheme 6.3). Such epimerization has been reported in rare cases for Lewis acid-catalyzed reactions with DHA or DHA acetate, but always as a minor process [40, 62].

The subtle chemistry of difluoroenoxysilanes has not been fruitful in our search for new antimalarial drug candidates. All our attempts to convert ketones **21** into acids failed (Baeyer–Villiger reaction, oxidation of the aryl moiety). Furthermore, the configuration at C-9 in artemisinins is known to be crucial for the antimalarial activity [5]. Ketones **21** are much less active *in vitro* and *in vivo* than artemether.

6.3.3 Functionalization at C-16 of 10-trifluoromethyl-anhydroartemisinin

Another approach to introduce a polar function into fluoroartemisinins was to exploit the allylic site in glycal **18.** However, the standard methods (P_2O_5 or $BF_3{\cdot}Et_2O$) [38, 63] reported for the nonfluorinated parent compound **10** from DHA failed for CF_3-DHA **12**, because of the great stability of a CF_3-substituted hydroxyl under acidic conditions [48, 64]. The glycal **18** could be selectively prepared in good yields by treatment of **12** with thionyl chloride and pyridine in large excess (see Scheme 6.6) [44].

Before starting any structural modulation, we first evaluated the effect of the C-9–C-10 unsaturation on antimalarial activity and ascertained that the presence of a CF_3 group at C-10 could disfavor the protonation leading to oxonium **14**, and hence the further glucuronidation (see Scheme 6.4). The antimalarial activities of glycals **10** and **18** have been assessed and compared (see Figure 6.3) [58]. Whereas *in vitro* activities on *P. falciparum*

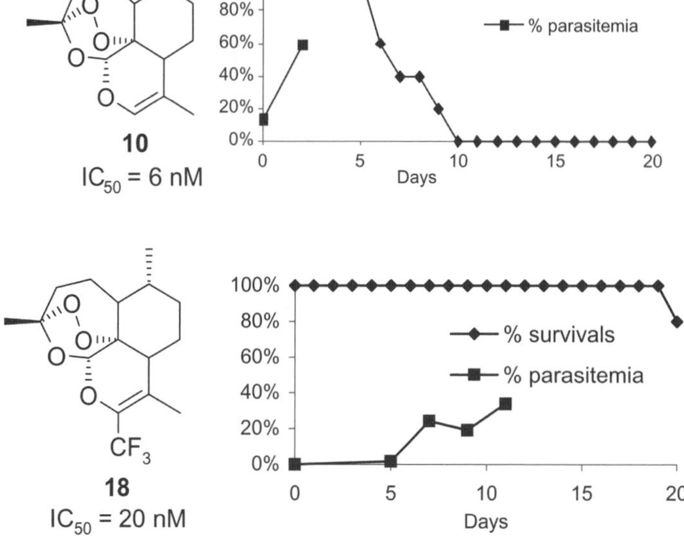

Figure 6.3 *Effect of fluorine substitution at C-10 on antimalarial activity of glycols.*

are similar, *in vivo* activities are remarkably different. With glycal **10**, there is no clearance of parasitemia at the end of the i.p. treatment of mice infected with *P. berghei NK 173* (35.5 μmol/kg according to the Peters test), and no survival at D-10. With **18**, all mice survived until D-20. This is a clear evidence of the protection against hydrolytic processes conferred by a CF_3 substituent. Glycal **18** could thus be used as precursor for the preparation of new antimalarials functionalized at C-16.

6.3.3.1 Allylic Bromination

Surprisingly, the allylic radical bromination of the nonfluorinated glycal **10** had not been reported, although the allyl bromide could obviously provide a shorter route to 16-substituted derivatives than the previously described approaches from artemisitene or artemisinic acid [28, 29, 59a, 65, 66]. Conversely, reactivity of **10** toward electrophilic brominating agents was well documented. Dibromides [44, 67] and bromohydrins [68, 69] have been used for the introduction of ionizable functions in artemisinin at C-10 and C-9.

Allylic bromination of glycals **10** and **18** was investigated under the usual Wohl–Ziegler conditions with *N*-bromosuccinimide (NBS/CCl_4/reflux) in the presence of azobisisobutyronitrile (AIBN) as initiator. Although both bromides were easily formed in this reaction, the nonfluorinated one appeared to be unstable and could not be isolated [70]. The parent 10-CF_3-16-bromo derivative **22** was also obtained in good yield (72%) (see Scheme 6.12). The compound could be purified by crystallization and stored for several weeks at 0 °C. Clearly, the electron-withdrawing character of the CF_3 group makes the allyl bromide less labile.

The presence of an initiator (AIBN) was not necessary in the bromination reaction of glycal **18** and yield was even improved to 90% when the reaction was performed without

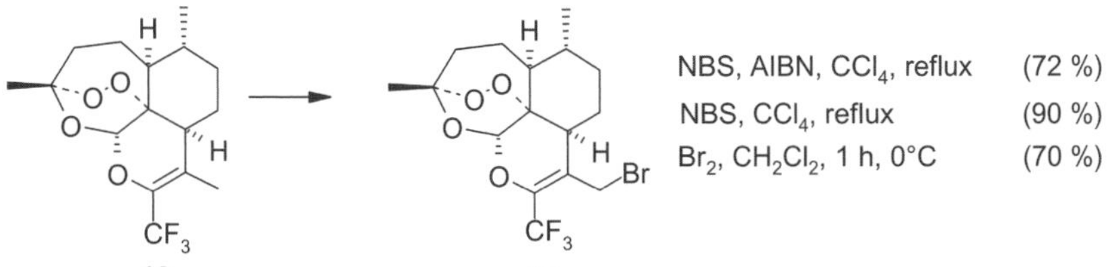

18 → 22

NBS, AIBN, CCl_4, reflux	(72 %)
NBS, CCl_4, reflux	(90 %)
Br_2, CH_2Cl_2, 1 h, 0°C	(70 %)

Scheme 6.12 *Allylic bromination reactions of glycal **18**.*

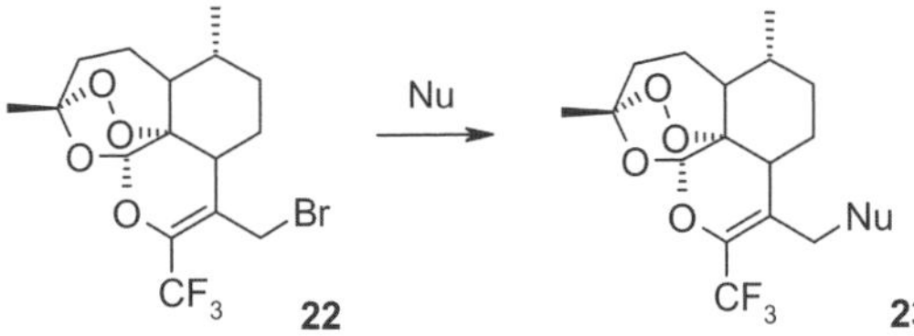

22 —Nu→ 23

Nu = Amines, alcoholates, malonate, etc...

Scheme 6.13 *Reactions of bromide **22** with* N-, O-, *and* C-*nucleophiles.*

AIBN. The ease of the allylic bromination of **18** has been attributed to the presence of the endoperoxide, which can initiate the radical reaction [44]. Similar results were obtained with Br_2 without initiator, whereas Br_2 reacts with glycal **10** through an electrophilic addition on the double bond [44, 67].

*6.3.3.2 Reactions of the Allyl Bromide **22** with Nucleophiles*

Trifluoromethyl allyl bromide **22** reacted selectively with amines and alcoholates, leading to 16-substituted compounds **23** (S_N2-type products), with no trace of the allylic rearrangement product (see Scheme 6.13 and Table 6.5) [57]. However, conditions were also found to provide predominantly the allylic rearrangement [71]. Excellent yields in **23** were obtained with secondary amines. With primary amines, polyalkylation could be avoided by using dilute conditions. Since reductive conditions to convert an azide into a primary amine are not compatible with the endoperoxide bridge [72], the primary amine at C-16 was prepared using NH_3 as nucleophile, in a very large excess to avoid polyalkylation. Sodium alkoxides, generated *in situ* with sodium hydride, also reacted well to generate corresponding ethers in good yield, but addition of a catalytic amount of KI was required (see Table 6.6). Dimethyl sodium malonate reacted with bromide **22**, leading to the diester **23q** in excellent yield (90%). A large series of new compounds **23a–r** were obtained.

*6.3.3.3 Biological Activity of Products **23***

Most of the 16-functionalized fluoroartemisinins exhibited strong antimalarial activity *in vitro* (see Table 6.7) and *in vivo* (see Table 6.8) [46, 57]. Only the diester **23q** and the diacid **23r** were found to be inactive against *Plasmodium falciparum*. The more promising

Table 6.5 *Reactions of bromide* ***22*** *with* N-*nucleophiles*

N-Nucleophile[a]	Equiv.	Time (h)	23	Yield
Morpholine	4	6	**a**	90%
Piperazine ethanol	4	6	**b**	87%
$EtNH_2$	10	4	**c**	85%
$MeNH_2$	10	3	**d**	98%
$NH_2CH_2CH_2NH_2$	10	3	**e**	95%
NaN_3[b]	1.5	1	**f**	95%
NH_3[c]		4	**g**	77%

[a] Reactions performed in THF.
[b] Reaction performed in DMSO.
[c] Large excess of NH_3 in a mixture NH_3/THF 1:1.

Table 6.6 *Reactions of bromide* ***22*** *with O-and C-nucleophiles*

Nucleophile[a]	Equiv.	NaH (equiv.)	Solvent	Time (h)	23	Yield[b]
EtOH	20	3	THF	18	**h**	98%
BnOH	1.5	3	THF	18	**i**	97%
$MeOCH_2CH_2OH$	3	2	DMSO	1	**j**	96%
CH_2=CH-CH_2OH	3	2.5	DMSO	2	**k**	81%
$HOCH_2CH_2OH$	4	1.5	DMSO	2	**l**	69%
$CH_2(COOMe)_2$	1.5	1.8	THF	5	**q**	90%

[a] In presence of KI (0.1 equiv.).
[b] Isolated yield.

Table 6.7 In vitro *activities of compounds* ***23*** *and artemether on the chloroquine-resistant* Plasmodium falciparum *FcB1 and W2 clones and calculated log*P *values (Clog*P*)*

	Nu	IC_{50} (nM)[a]	Clog *P*[b]
	Artemether	3.5±1.2[c]	2.92
23a	–Morpholino–	3.1±0.5[c]	3.89
23b	–Piperazinoethanol	15.2±6.7[d]	2.43
23c	–NHEt	13.4 ±4.5[d]	3.40
23d	–NHMe	9.2±2.4[c]	2.87
23e	$-NHCH_2CH_2NH_2$	1.2±0.7[c]	2.50
23f	$-N_3$	10.0±7.6[c]	4.97
23g	$-NH_2$	4.4±0.4[c]	3.13
23h	$-NHSO_2CH_3$	20.0±6.3[c]	3.07
23i	$-NHSO_2C_6H_4CH_3$	19.1±3.5[c]	5.25
23h	–OEt	25.0±7[d]	3.73
23i	–OBn	20±5[d]	5.72
23j	$-OCH_2CH_2OMe$	2.7±0.4[c]	3.26
23k	$-OCH_2CH=CH_2$	6.0±1.7[c]	4.43
23l	$-OCH_2CH_2OH$	2.4±0.4[c]	3.09
23m	$-OCOCH_3$	1.7±0.5[c]	3.97
23n	–OH	7.5±0.8[c]	3.05
23p	$-OCH_2CHOHCH_2OH$	3.7±0.5[c]	2.27
23q	$-CH(COOMe)_2$	>1000[d]	3.79
23r	$-CH(COOH)_2$	>1000[d]	3.00

[a] Mean±standard deviation was calculated for $n=3$ experiments.
[b] LogP values were calculated using the program available from http://www.daylight.com/cgihyphen;bin/contrib/pcmodels.cgi.
[c] FcB1 and [d] W2 clones.

candidates for oral treatment are the amines **23a–c**, which are more active than sodium artesunate, as shown by ED_{50} and ED_{90} values (see Table 6.8).

Pharmacokinetic studies performed on compound **23b** exhibited a significantly better bioavailability in rats than artemether (see Tables 6.9 and 6.10) [56]. C_{max} and T_{max} values

Table 6.8 *In vivo data for amines **23a–c**, ether **23h** and Na artesunate (P. berghei N.) subcutaneous (s.c.) and oral (p.o.) administration*

	Route	ED_{50} (mg/kg)	ED_{90} (mg/kg)	Reduction of parasitemia at D-4 at 10 mg/kg (%)
Na artesunate	s.c.	2.8	10.5	90
Na artesunate	p.o.	5.4	15.3	–
23a	s.c.	<10	<10	98.1
23a	p.o.	<10	<10	93.3
23b	s.c.	<10	<10	100
23b	p.o.	<10	<10	100
23c	s.c.	<10	<10	98.1
23c	p.o.	<10	<10	96
23h	s.c.	<10	<10	100
23h	p.o.	>10	nd	15.6

Table 6.9 *Log P/D values and equilibrium solubilities of **23b** and artemether, after incubation at 25 °C for 72 h*

	Artemether	**23b**
Log P	3.36	3.52
Log D (pH 7.4)	–	3.46
pK_a	–	6.59
Solubility		
I.v. formulation (μg/mL)[a]	>3000	>3000
Oral formulation (μg/mL)[b]	256	675.9
PBS (pH 7.4) (μg/mL)	63.4	234.5

[a] I.v. formulation is 0.1 M Captisol™ in water for artemether and 10% ethanol–0.1 M Captisol™, pH 3 for **23b**.
[b] Oral formulation contains 0.5% w/v carboxymethyl cellulose, 0.5% v/v benzyl alcohol, 0.4% v/v Tween 80 in 0.9% w/v NaCl for artemether, and 0.5% w/v hydroxypropylmethyl cellulose, 0.5% v/v benzyl alcohol, 0.4% v/v Tween 80 in 0.9% w/v NaCl for **23b**.

Table 6.10 *Pharmacokinetic data for artemether and **23b***

	Artemether[a]	**23b**
C_{max} (ng/mL)	168.4±50.2	607.9
t_{max} (min)	20.4±7.6	60
$t_{1/2}$ (min)[b]	52.2±5.8	55.3
Plasma clearance ([mL/min]/kg)	114.1±20.6	148.7
Volume of distribution (L/kg)[b]	8.4±1.7	11.8
Oral bioavailability (%)	1.4±0.6	34.6

[a] Artemether pharmacokinetic parameters from previous studies [57].
[b] After 10 mg/kg (i.v. administration).

for **23b**, determined from the oral dosing studies, were 607.9 ng/mL and 60 min, respectively, compared with 168.4±50.2 ng/mL and 20.4 ± 7.6 min, respectively, for artemether. At a 50 mg/kg oral dose, oral bioavailability was 34.6%, a significant improvement over the 1.4% observed for artemether. However, the piperazine ethanol derivative **23b** exhibited a faster metabolic clearance than the hemiketal **12**. Current experiments are in progress to exploit this new series by optimization of the polar substituent at C-16.

6.4 Conclusion

With the increasing resistance of *Plasmodium falciparum* toward most available antimalarial drugs, artemisinin derivatives constitute the greatest hope in the fight against malaria. First-generation artemisinin derivatives feature in all fixed combination therapies clinically available. Drawbacks stemming from their instability and short plasma half-life prompted medicinal chemists to design and prepare more efficacious derivatives while retaining the 1,2,4-trioxane pharmacophore. With this aim, the concept of "fluorine substitution" appeared to be useful to obtain an improved pharmacological profile. By taking into account the main sites of oxidative and hydrolytic degradation, various series of fluorinated derivatives of artemisinin have been designed, synthesized, and evaluated against *Plasmodium.*

The uncommon structural artemisinin framework, and the well-known difficulty in selectively introducing a fluorinated motif into a molecule, faced us with numerous chemical challenges. The evaluation of the effects of fluorine on physicochemical and antimalarial properties is reported. These data and preclinical data of lead compounds are encouraging, with strong and prolonged antimalarial activity of fluoro-artemisinins.

Acknowledgments

We thank Benoit Crousse, Ahmed Abouabdellah, Truong Thi Thanh Nga, Fabienne Grellepois, Fatima Chorki, Guillaume Magueur, and Constance Chollet for their contributions and Michèle Ourévitch for NMR experiments. The authors are also grateful to Jean-Charles Gantier, Florence Chrétien, E. Dumarquez (Parasitology, Châtenay-Malabry), Bruno Pradines, Véronique Sinou, Daniel Parzy (Institut de Médecine Tropicale, SSA, Le Pharo, France), Philippe Grellier, Sébastien Charneau, (Museum, Paris, France), Kylie A. McIntosh, William N. Charman (Monash University, Australia), Vu Dinh Hoang, G. Diem Pham, Nguyen Van Hung, (Institute of Chemistry, VAST, Hanoi, Vietnam), Le Dinh Cong, T. N. Hai, P. D. Binh, Truong Van Nhu, Doan Hanh Nhan (NIMPE, Hanoi, Vietnam) and Dr. Marie-Annick Mouries, (WHO DDR/TDR). Work at BIOCIS on fluoroartemisinins was funded by CNRS, University Paris-South, PAL+ program from Research and Technology French Ministery, the Délégation Générale de l′Armement and the GDR Parasitology (CNRS).

References

1. (a) Breman, J. G. (2001) The ears of the hippopotamus: Manifestations, determinants, and estimates of the malaria burden. *Am. J. Trop. Med. Hyg.*, **64** (Suppl.), 1–11; (b) White, N., Nosten J. F., Looareesuwan S., *et al.* (1999) Averting a malaria disaster. *Lancet*, **353**, 1965–1967.
2. Olliaro, P. L. and Bloland, P. B. (2001) Clinical and public health implicationsof antimalarial drug resistance, in *Antimalarial Chemotherapy: Mechanisms of Action, Resistance, and New Directions in Drug Discovery* (ed. P. J. Rosenthal), Humana Press, Totowa, NJ, pp. 65–83.
3. Ridley, R. G. (2002) Medical need, scientific opportunity and the drive for antimalarial drugs. *Nature*, **415**, 686–693.
4. Haynes, R. C., Vonwiller, S. C. (1997) From qinghao, marvelous herb of antiquity, to the antimalarial trioxane qinghaosu and some remarkable new chemistry. *Acc. Chem. Res.*, **30**, 73–79.
5. (a) Klayman, D. L. (1985) Qinghaosu (artemisinin). An antimalarial drug from China. *Science*, **228**, 1049–1055; (b) Jefford, C. W., Vicente, M. G. H., Jacquiern, Y., *et al.* (1996) The deoxygenation and isomerization of artemisinin and artemether and their relevance to antimalarial action. *Helv. Chim. Acta*, **79**, 1475–1487; (c) Jefford, C. W., Burger, U., Millasson-Schmidt P., *et al.* (2000) Epiartemisinin, a Remarkably poor antimalarial: implications for the mode of action. *Helv. Chim. Acta*, **83**, 1239–1246.
6. (a) ONeill, P. M., Rawe, S. L., Borstnik, K., *et al.* (2005) Enantiomeric 1,2,4-trioxanes display equivalent in vitro antimalarial activity versus *P. falciparum* malaria parasites: implications for the molecular mechanism of action of the artemisinins. *ChemBiochem*, **6**, 2048–2054; (b) Laurent, S.A-L., Robert, A. and Meunier, B. (2005) C10-modified artemisinin derivatives: efficient heme-alkylating agents. *Angew. Chem. Int. Ed.*, **44**, 2060–2063; (c) Stocks, P. A., Bray, P. G., Barton, V. E., *et al.* (2007) Evidence for a common non-heme chelatable-iron-dependent activation mechanism for semisynthetic and synthetic endoperoxide antimalarial drugs. *Angew. Chem.*, **46**, 6278–6283; and references cited therein.
7. (a) Ridley, R. G. (2003) To kill a parasite. *Nature*, **424**, 887–889; (b) Mercereau-Puijalon, O. and Fandeur, T. (2003) Antimalarial activity of artemisinins: identification of a novel target? *The Lancet*, **362**, 2035–2036.
8. O'Neill, P. and Posner, G. H., (2004) Medicinal chemistry perspective on artemisinin and related endoperoxides. *J. Med. Chem.*, **47**, 2945–2964.
9. China Cooperative Research Group on Qinghaosu and its derivatives as antimalarials (1982) Antimalarial efficacy and mode of action of qinghaosu and its derivatives in experimental models. *J. Trad. Chin. Med.*, **2**, 9–16.
10. Dhingra, V., Rao, K. V. and Narasu, M. L. (2000) Current status of artemisinin and of its derivatives as antimalarial drug. *Life Sci.*, **66**, 279–300.
11. Van Agtmael, M. A., Eggelte, T. A. and Van Boxtel, C. J. (1999) Artemisinin drugs in the treatment of malaria: from medicinal herb to registered medication. *Trends Pharmacol. Sci.*, **20**, 199–204.
12. Hien, T. T. (1996) An overview of the clinical use of artemisinin and its derivatives in the treatment of *P. falciparum* malaria in Vietnam. *N. Engl. J. Med.*, **335**, 76–83.
13. White, N. J. (1997) Assessment of the pharmacodynamic properties of antimalarial drugs in vivo *Antimicrob. Agents Chemother.*, **41**, 1413–1422.
14. Kumar, N. and Zheng, H. (1990) Stage-specific gametocytocidal effect in vitro of the antimalaria drug qinghaosu on *Plasmodium falciparum. Parasitol. Res.*, **76**, 214–218.
15. Price, R. N., Nosten, F., Luxemburger, C. *et al.* (1996) Effects of artemisinin derivatives on malaria transmissibility. *Lancet*, **347** (90016), 1654–1658.

16. (a) Malaria Unit, Division of Control of Tropical Disease, WHO. (1998) The use of artemisinin and its derivatives as antimalarial drugs. Report of a Joint CTD/DMP/TDR Informal Consultation, WHO, Geneva. (b) Krishna, S., Woodrow, C. J., Staines, H. M., *et al.* (2006) Re-evaluation of how artemisinins work in light of emerging evidence of in vitro resistance. *Trends Mol. Med.*, **12**, 200–205
17. Khanh, N. X., de Vries, P. J., Dang Ha, L., *et al.* (1999) Declining concentrations of dihydroartemisinin in plasma during 5-day oral treatment with artesunate for falciparum malaria, *Antimicrob. Agents Chemother.*, **43**, 690–692.
18. (a) Ilett, K. F., Ethell, B. T., Maggs, J. L., *et al.* (2002) Glucuronidation of dihydroartemisin in vivo and by human liver microsomes and expressed UDP-glucuronosyltransferases. *Drug Metab. Dispos.*, **30**, 1005–1012; (b) O'Neill, P. M., Scheinmann, F., Stachulski, A. V., *et al.* (2001) Efficient preparations of the β-glucuronides of dihydroartemisinin and structural confirmation of the human glucuronide metabolite. *J. Med. Chem.*, **44**, 1467–1470. (c) Haynes, R. K. (2006) From artemisinin to new artemisinin antimalarials: biosynthesis, extraction, old and new derivatives, stereochemistry and medicinal chemistry requirements, *Current Top. Med. Chem*, **6**, 509–537.
19. (a) Word Health Organization Antimalarial Drug Combination Therapy. (2001) Report of a WHO Technical Consultation. WHO/CDS/RBM/2001.35, WHO, Geneva. (b) Duffy, P. E. and Mutabingwa, T. K. (2004) Drug combinations for malaria: Time to ACT. *Lancet*, **363**, 3–4.
20. (a) Bégué, J. P. and Bonnet-Delpon, D. (2005) The future of antimalarials: artemisinins and synthetic endoperoxides. *Drug of the Future*, **30**, 509–522. (b) Haynes, R. K., Chan, H-W., Lung, C-M., *et al.* (2007) Artesunate and dihydroartemisinin (DHA): unusual decomposition products formed under mild conditions and comments on the fitness of dha as an antimalarial drug. *Chem Med Chem*, **2**, 1448–1463.
21. Jung, M., Lee, K., Kim, H. and Park M. (2004) Recent advances in artemisinin and its derivatives as antimalarial and antitumor agents. *Current Med. Chem.*, **11**, 1265–1284.
22. Haynes, R. K. (2001) Artemisinin and derivatives: The future for malaria treatment. *Current Opin. Infect. Dis.*, **14**, 719–726.
23. Torok, D. S., Ziffer, H., Meshnick, S. R., *et al.* (1995) Syntheses and antimalarial activities of *N*-substituted 11-azaartemisinins. *J. Med. Chem.*, **38**, 5045–5050.
24. Haynes, R. K., Chan, H. W., Ho, W. Y., *et al.* (2005) Convenient access both to highly antimalaria-active 10-arylaminoartemisinins, and to 10-alkyl ethers including artemether, arteether, and artelinate. *ChemBiochem.*, **6**, 659–667.
25. Haynes, R. K., Chan, H. W., Cheung, M. K., *et al.* (2003) Stereoselective preparation of 10α- and 10β-aryl derivatives of dihydroartemisinin. *Eur. J. Org. Chem.*, 2098–2114.
26. (a) Haynes, R. K. (2003) Preparation of antiparasitic artemisinin derivatives. *PCT Int. Appl.*, WO 2003-EP1839 2003224 [*Chem. Abstr.*, **139**, 261439]; (b) Haynes, R. K., Fugmann, B., Stetter, J., *et al.* (2006) Artemisone – a highly active antimalarial drug of the artemisinin class. *Angew. Chem.*, **45**, 2082–2088.
27. Hindley, S., Ward, S. A., Storr, R. C., *et al.* (2002) Mechanism-based design of parasite-targeted artemisinin derivatives: Synthesis and antimalarial activity of new diamine containing analogues. *J. Med. Chem.*, **45**, 1052–1063.
28. (a) El-Feraly, F. S., Ayalp, A., Al-Yahya, M. A., *et al.* (1990) Conversion of artemisinin to artemisitene. *J. Nat. Prod.*, **53**, 66–71. (b) Paitayatat, S., Tarnchompoo, B., Thebtaranonth, Y. and Yuthavong, Y. (1997) Correlation of antimalarial activity of artemisinin derivatives with binding affinity with ferroprotoporphyrin IX. *J. Med. Chem.*, **40**, 633–638. (c) Ma, Y. Y., Weiss, E., Kyle, D. E. and Ziffer H. (2000) Acid-catalyzed Michael additions to artimisitene. *Bioorg. Med. Chem. Lett.*, **10**, 1601–1603.
29. Jung, M. A., Lee, K. and Jung, H. (2001) (+) deoxoartemisitene and first C-11 derivatives. *Tetrahedron Lett.*, **42**, 3997–4000.

30. (a) Bégué, J. P. and Bonnet-Delpon, D. (2005) *Chimie Bioorganique et Médicinale du Fluor*, EDP-Sciences/CNRS Editions, Paris, pp. 10–23, 88–95, 94.
31. Böhm, H. J., Banner, D., Bendels, S., *et al.* (2004) Fluorine in medicinal chemistry. *ChemBioChem.*, **5**, 637–645.
32. Smart, B. E. (1994) Aspects of organofluorine chemistry, in *Organofluorine Chemistry: Principles and Commercial Applications* (eds R. E. Bank, B. E. Smart, J. C. Tatlow), Plenum Press, New York, pp. 12–38.
33. (a) Lee, I. S. and Hufford, C. D. (1990) Metabolism of antimalarial sesquiterpene lactones. *Pharmacol. Ther.*, **48**, 345–355; (b) Leskovac, V. ad Theoharides, A. D. (1991) Hepatic metabolism of artemisinin drugs – I. Drug metabolism in rat liver microsomes. *Comp. Biochem. Physiol.*, **99c**, 383–390; (c) Leskovac, V. and Theoharides, A. D. (1991) Hepatic metabolism of artemisinin drugs – II. Metabolism of arteether in rat liver cytosol. *Comp. Biochem. Physiol.*, **99c**, 391–396.
34. (a) Legros, J., Crousse, B., Maruta, M., *et al.* (2002) Trifluoromethylcyclohexane as a new solvent? Limits of use. *Tetrahedron*, **58**, 4067–4070.
35. Irrupe, J., Casas, J. and Messeguer, A. (1993) Resistance of the 2,2,2-trifluoroethoxy aryl moiety to the cytochrome P-450m etabolism in rat liver microsomes. *Bioorg. Med. Chem. Lett.*, **3**, 179–182.
36. Bird, T. G. C., Fredericks, P. M., Jones, E. R. H. and Meakins, G. D. (1980) Microbiological hydroxylation. Part 23. Hydroxylations of fluoro-5α-androstanones by the fungi *Calonectria decora*, *Rhizopus nigricans*, and *Aspergillus ochraceus*. *J. Chem. Soc. Perkin Trans.* **1**, 750–755.
37. Haufe, G. and Wölker, D. (2003) Blocking fluorine substitution in biotransformation of nortricyclanyl *N*-phenylcarbamates with *Beauveria bassiana*. *Eur. J. Org. Chem.*, 2159–2175.
38. (a) Lin, A. J., Lee, M. and Klayman, D. L. (1989) Antimalarial activity of new water-soluble dihydroartemisinin derivatives. 2. Stereospecificity of the ether side chain. *J. Med. Chem.*, **32**, 1249–1252; (b) Lin, A. J., Klayman, D. L. and Milhous, W. K. (1987) Antimalarial activity of new water-soluble dihydroartemisinin derivatives. *J. Med. Chem.*, **30**, 2147–2150; (c) Lin, A. J., Li, L. Q., Andersen, S. L. and Klayman, D. L. (1992) Antimalarial activity of new dihydroartemisinin derivatives. 5. Sugar analogues. *J. Med. Chem.*, **35**, 1639–1642; (d) Lin, A. J. and Miller, R. E. (1995) Antimalarial activity of new dihydroartemisinin derivatives. 6. α-Alkylbenzylic ethers, *J. Med. Chem.*, **38**, 764–770.
39. Haynes, R. K., Chan, H-W., Cheung, M-K., *et al.* (2002) C-10 ester and ether derivatives of dihydroartemisinin – 10-α artesunate, preparation of authentic 10-β artesunate, and of other ester and ether derivatives bearing potential aromatic intercalating groups at C-10. *Eur. J. Org. Chem.*, 113–132.
40. Thanh Nga, T. T., Ménage, C., Bégué, J. P., *et al.* (1998) Synthesis and antimalarial activities of fluoroalkyl derivatives of dihydroartemisinin. *J. Med. Chem.*, **41**, 4101–4108.
41. Van Nhu, T., Hanh Nhan, D., Minh Thu, N. T., *et al.* (2005) Preliminary study on sub-acute toxicity of trifluoromethyl hydroartemisinin (BB101) in monkeys. *Rev. Pharm*, **1**, 12–16.
42. (a) Falck, J. R., Yu, J. and Cho, H. S. (1994) A convenient synthesis of unsymmetric polyfluoroethers. *Tetrahedron Lett.*, **35**, 5997–6000; (b) Cho, H. S., Yu, J. and Falck, J. R. (1994) Preparation and scope of a remarkably robust primary alcohol protective group, *J. Am. Chem. Soc.*, **116**, 8354–8355; (c) Sebesta, D. P., O'Rourke, S. S. and Pieken, W. A. (1996) Facile preparation of perfluoro-*tert*-butyl ethers by the Mitsunobu reaction, *J. Org. Chem.*, **61**, 361–362.
43. (a) Henne, A. L. and Francis, W. C. (1953) Acidity and infrared absorption of fluorinated alcohols. *J. Am. Chem. Soc.*, **75**, 991–992; (b) Bégué, J. P., Bonnet-Delpon, D. and Crousse, B. (2004) Fluorinated alcohols: a new medium for selective and clean reaction. *Synlett*, 18–24.
44. Grellepois, F., Chorki, F., Crousse, B. and Ourévitch, M. (2002) Anhydrodihydroartemisinin and its 10-trifluoromethyl analogue: access to novel D-ring-contracted artemisinin trifluoromethyl ketones. *J. Org. Chem.*, **67**, 1253–1260.

45. (a) Bott, G., Field, L. D. and Sternhell, S. (1980) Steric effects. A study of a rationally designed system. *J. Am. Chem. Soc.*, **102**, 5618–5626; (b) Nagai, T., Nishioka, G., Koyama, M., *et al.* (1991) *Chem. Pharm. Bull.*, **39**, 233–235.
46. Bégué, J. P. and Bonnet-Delpon, D. (2007) Fluoroartemisinins: metabolically more stable antimalarial artemisinin derivatives. *ChemMedChem*, **2**, 608–624.
47. Magueur, G., Crousse, B., Bonnet-Delpon, D. and Bégué, J. P. (2003) Preparation of 10-trifluoromethyl artemether and artesunate. influence of hexafluoropropan-2-ol on substitution reaction. *J. Org. Chem.*, **68**, 9763–9766.
48. (a) Allen, A. D. and Tidwell, T. T., (1989) Fluorine-substituted carbocations, in *Advances in Carbocation Chemistry* (ed. X. Creary), JAI Press Inc.: Greenwich, CT, pp 1–44; (b) Creary, X. (1991) Electronegatively substituted carbocations. *Chem. Rev.*, **91**, 1625–1676; (c) Chennoufi, A., Malézieux, B., Gruselle, M., *et al.* (2000) Cobalt-induced C-N and C-C bond formation via metal-stabilized -CF_3 carbenium ion. *Org. Lett.*, **2**, 807–809.
49. Hoang, V. D., Chorki, F., Grellepois, F., *et al.* (2002) First synthesis of 10-(trifluoromethyl) deoxoartemisinin. *Org. Lett.*, **4**, 757–759.
50. Reichardt, C. (2003) *Solvents and Solvent Effects in Organic Chemistry*, 3rd edn., Wiley-VCH Verlag GmbH, Weinheim.
51. Allard, B., Casadevall, A., Casadevall, E. and Largeau, C. (1980) The acid-enhancing effect of HFIP as solvent. Carbocation rearrangemebt during solvolysis of 2-adamantyl tosylate in anhydrous HFIP. *Nouv. J. Chim.*, **4**, 539–545.
52. Eberson, L., Hartshorn, M. P., Persson, O. and Radner, F. (1996) Making radical cations live longer. *Chem. Commun.*, 2105–2112.
53. See for instance: (a) Ichikawa, J., Miyazaki, S., Fujiwara, M. and Minami, T. (1995) Fluorine-directed Nazarov cyclizations: a controlled synthesis of cross-conjugated 2-cyclopenten-1-ones. *J. Org. Chem.*, **60**, 2320–232; (b) Takita, R., Oshima, T. and Shibasaki, M. (2002) Highly enantioselective catalytic Michael reaction of α-substituted malonates using La-linked-BINOL complex in the presence of HFIP (1,1,1,3,3,3-hexafluoroisopropanol). *Tetrahedron Lett.*, **43**, 4661–4665; (c) Cooper, M. S., Heaney, H., Newbold, A. J. and Sanderson, W. R. (1990) Oxidation reactions using urea-hydrogen peroxide; a safe alternative to anhydrous hydrogen peroxide. *Synlett* 533–535; (d) Legros, J., Crousse, B., Bonnet-Delpon, D. and Bégué, J. P. (2002) Urea-hydrogen peroxide/hexafluoro-2-propanol: an efficient system for a catalytic epoxidation reaction without a metal. *Eur. J. Org. Chem.*, 3290–3293; (e) Das, U., Crousse, B., Kesavan, V., *et al.* (2000) Facile ring opening of oxiranes with aromatic amines in fluoro alcohols. *J. Org. Chem.*, **65**, 6749–6751.
54. Abouabdellah, A., Bégué, J. P., Bonnet-Delpon, D., *et al.* (1996) Synthesis and in vivo antimalarial activity of 12a-trifluoromethyl-hydroartemisinin. *Bioorg. Med. Chem. Lett.*, **6**, 2717–2720.
55. Magueur, G., Crousse, B., Charneau, S., *et al.* (2004) Fluoroartemisinin: trifluoromethyl analogue of artemether and artesunate. *J. Med. Chem.*, **47**, 2694–2699.
56. Grellepois, F., Chorki, F., Ourévitch, M., *et al.* (2004) Orally active antimalarials: hydrolytically stable derivatives of 10-trifluoromethyl anhydrodihydroartemisinin. *J. Med. Chem.*, **47**, 1423–1433.
57. Binh, P. D., Le Dinh Cong, D. H., Nhan, T. V., *et al.* (2002) The effect of 10α-trifluoromethyl hydroartemisinin on *Plasmodium berghei* infection and its toxicity in experimental animals. *Trans. Royal Soc. Trop. Med. Hyg.*, **96**, 677–683.
58. Berber, H., Brigaud, T., Lefebvre, O., *et al.* (2001) Reactions of difluoroenoxysilanes with glycosyl donors: synthesis of difluoro-C-glycosides and difluoro-C-disaccharides. *Chem. Eur. J.*, **7**, 903–909.
59. (a) Pu, Y. M. and Ziffer, H. (1995) Synthesis and antimalarial activities of 12β-allyldeoxoartemisinin and its derivatives. *J. Med. Chem.*, **38**, 613–616; (b) Ma, J., Katz, E.,

Kyle, D. E. and Ziffer, H. (2000) Syntheses and antimalarial activities of 10-substituted deoxoartemisinins. *J. Med. Chem.*, **43**, 4228–4232.
60. Amii, H., Kobayashi, T., Hatamoto, Y. and Uneyama, K. (1999) Mg^0-promoted selective C–F bond cleavage of trifluoromethyl ketones: a convenient method for the synthesis of 2,2-difluoro enol silanes. *J. Chem. Soc. Chem. Commun.*, 1323–1324
61. (a) Chorki, F., Crousse, B., Brigaud, T., *et al.* (2001) C-10-fluorinated derivatives of dihydroartemisinin: difluoromethylene ketones. *Tetrahedron Lett.*, **42**, 1487–1490; (b) Chorki, F., Grellepois, F., Ourevitch, M., *et al.* (2001) Fluoro artemisinins: difluoromethylene ketones. *J. Org. Chem.*, **66**, 7858–7863.
62. Wang, D. Y., Wu, Y. L., Wu and Y., Li, Y. (2001) Further evidence for the participation of primary carbon-centered free radicals in the antimalarial action of the qinghaosu (artemisinin) series of compounds. *J. Chem. Soc. Perkin Trans. I*, 605–609.
63. Lin, A. J., Li, L., Klayman, D. L., *et al.* (1990) Antimalarial activity of new water-soluble dihydroartemisinin derivatives. 3. Aromatic amine analogues. *J. Med. Chem.*, **33**, 2610–2614.
64. Creary, X. (1991) Electronegatively substituted carbocations. *Chem. Rev.*, **91**, 1625–1678.
65. Ekthawatchai, S., Kamchonwongpaisan, S., Kongsaeree, P., *et al.* (2001) C-16 artemisinin derivatives and their antimalarial and cytotoxic activities: syntheses of artemisinin monomers, dimers, trimers, and tetramers by nucleophilic additions to artemisitene, *J. Med. Chem.*, **44**, 4688–4695.
66. (a) Jung, M., Bustos, D. A., El Sohly, H. N. and McChesney, J. D. (1990) A concise and stereoselective synthesis of (+)-12-*n*-butyldeoxoartemisinin. *Synlett.*, 743–744; (b) Jung, M., Bustos, D. A., El Sohly H. N. and McChesney J. D. (1991) A concise synthesis of 12-(3′-hydroxy-*n*-propyl)-deoxoartemisinin. *Bioorg. Med. Chem. Lett.*, **1**, 741–744; (c) Haynes, R. K., Vonwiller, S. C. (1992) Efficient preparation of novel qinghaosu (artemisinin) derivatives: conversion of qinghao (artemisinic) acid into deoxoqinghaosu derivatives and 5-carba-4-deoxoartesunic. *Synlett*, 481–483.
67. Grellepois, F., Bonnet-Delpon, D. and Bégué, J. P. (2001) Ring-contracted artemisinin derivatives: stereoselective reaction of anhydrodihydroartemisinin toward halogenating reagents. *Tetrahedron Lett.*, **42**, 2125–2127.
68. Venugopalan, B., Bapat, C. P., Karnik, P. J., *et al.* (1991) *Eur. Pat. EP.* 456,149; *Chem. Abstr.* (1992), **116**, 83708z.
69. Venugopalan, B., Bapat, C. P. and Karnik, P. J. (1994) Synthesis of a novel ring contracted artemisinin derivative. *Bioorg. Med. Chem. Lett.*, **4**, 751–752.
70. Grellepois, F., Chorki, F., Ourévitch, M., *et al.* (2002) Allylic bromination of anhydrodihydroartemisinin and of its 10-trifluoromethyl analogue: a new access to 16-substituted artemisinin derivatives. *Tetrahedron Lett.*, **43**, 7837–7840.
71. Chollet, C., Crousse, B., Ourévitch, M. and Bonnet-Delpon, D. (2006) S_N/S_N' competition: selective access to new 10-fluoro artemisinins. *J. Org. Chem.*, **71**, 3082–3083.
72. Scriven, E. F. V. and Turnbull, K. (1988) Azides: their preparation and synthetic uses. *Chem. Rev.*, **88**, 297–368.

7

Synthesis and Biological Activity of Fluorinated Nucleosides

Tokumi Maruyama, Masahiro Ikejiri, Kunisuke Izawa, and Tomoyuki Onishi

7.1 Introduction

There are several reasons for introduction of fluorine atom(s) into drug-candidate compounds. Most of these involve the unique properties of the fluoro group, such as the similarity of its Van der Waals radius to that of hydrogen, its high electronegativity, and the stability of the C–F bond. The small size of the fluoro group makes it possible to replace almost any C–H bond in organic compounds with a corresponding C–F bond. For instance, fluorinated nucleosides bearing C–F bonds in their sugar moiety would not cause significant steric hindrance while penetrating into a pocket of their target enzymes. The fluoro group is the most electronegative of all elements and has one of the highest ionization potentials. Therefore, the fluorine substituent in the sugar moiety brings about a strong electron-withdrawing effect on C–F bonds, which induces intramolecular electrostatic interactions that can change the conformation of the molecule. Since the C–F bond is stronger than the corresponding C–H bond, the C–F bond distance becomes short enough to protect its carbon center from various nucleophilic attacks. Moreover, the carbon center of the C–F bond resists oxidation due to the strong electronegative nature of fluorine. Thus, the fluorine substituent makes organic compounds chemically more stable. Because fluorinated compounds often exhibit potent biological activities, they are attracting considerable attention in the pharmaceutical field. In addition, the unique properties of the fluoro group also contribute to improving pharmacokinetic properties, such as ADME (absorption, distribution, metabolism, and excretion), and reducing undesired effects.

Fluorine in Medicinal Chemistry and Chemical Biology Edited by Iwao Ojima

Natural products bearing fluorine atoms are so rare that only five compounds have been discovered. Interestingly, a nucleoside analogue, nucleocidin, is one of these fluorinated naturally occurring compounds [1] (see Figure 7.1). Its discovery motivated organic chemists to synthesize fluorinated nucleosides. Additionally, the practical uses of 5-fluorouracil [2], gemcitabine [3], and clofarabine [4] as anticancer drugs have also attracted interest (see Figure 7.2). A fluorinated nucleoside, clevudine (L-FMAU, Levovir) [5], is expected to get FDA approval as an anti-hepatitis B virus agent in the near future. Since fluoronucleosides have been shown to be useful in clinical studies, novel antiviral fluoronucleosides such as FddA (**1**, lodenosine) [6], MIV-210 (a prodrug of FddG (**2**)) [7], and FLT (alovudine, MIV-310) [8], are currently under development (see Figure 7.3).

There are a number of excellent reviews on the synthesis of nucleoside analogues bearing fluorine atom(s) in their sugar moieties [9–14]. Therefore, in this chapter, we describe the syntheses of several fluorinated nucleoside analogues from common ribonucleosides such as adenosine, guanosine, and uridine, focusing on synthetic issues. As practical applications, possible process-scale syntheses of FddA (9-(2,3-dideoxy-2-fluoro-β-D-*threo*-pentofuranosyl)adenine) and FddG (2′,3′-dideoxy-3′-α-fluoroguanosine) are also described. There are good reviews on the conformational analysis of nucleosides [15], but in this chapter the relationship between the conformations and biological activities of fluorinated nucleosides is discussed.

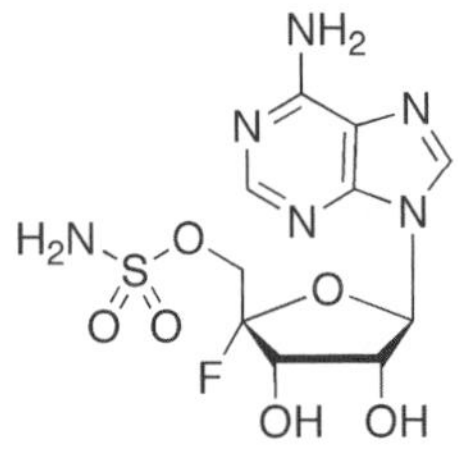

Figure 7.1 *Structure of nucleocidin.*

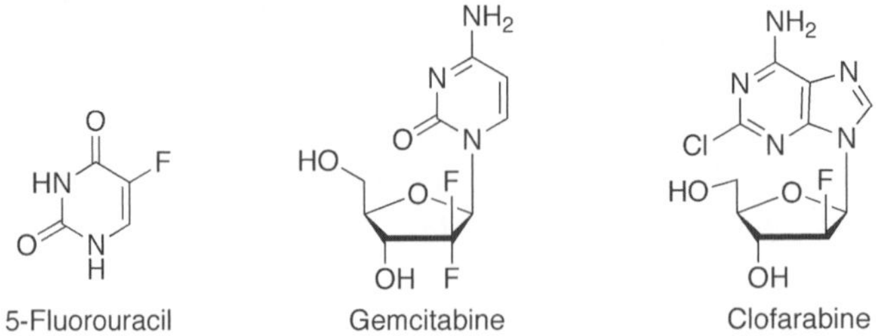

5-Fluorouracil
Gemcitabine
Clofarabine

Figure 7.2 *Structures of antitumor fluorinated nucleoside analogues.*

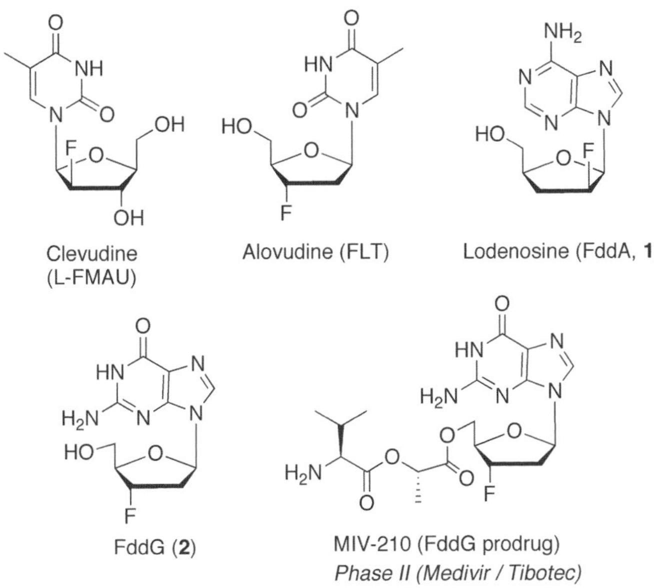

Figure 7.3 *Structures of antiviral fluorinated nucleoside analogues.*

7.2 Synthesis and Biological Activity of 2′-Deoxy-2′-α-fluororibonucleosides

This section describes the synthesis of 2′-deoxy-2′-α-fluoronucleosides using α-selective fluorination at C-2′ by several types of nucleophilic substitutions.

7.2.1 Synthesis of 2′-Deoxy-2′-α-fluoro Pyrimidine Nucleosides via Cleavage of Cyclonucleosides

In 1964, Fox and co-workers synthesized 2′-deoxy-2′-α-fluorouridine (**4**, X = H) via cleavage of 2,2′-*O*-cyclouridine (**3**) with hydrogen fluoride in 41–46% yield (see Scheme 7.1) [16a]. Compound **3** was readily prepared from uridine in one step. In general, a condensation method, in which sugars are subjected to fluorination prior to the introduction of nucleobases, has been used to prepare fluorinated nucleosides because the direct fluorination of nucleosides is difficult. There are thus several reasons why Fox's direct fluorination is successful. The conversion of the cleaved ether oxygen into the carbonyl group at the 2-position of uracil is one of the most important. The stability of pyrimidine nucleosides under severe reaction conditions (anhydrous HF, 100 °C) also makes this reaction possible. Thus, it is advisable to use 2,2′-*O*-cyclouridine to prepare 2′-α-substituted analogues of uridine; in fact, various substitutions (e.g., by other halogen and azide groups) have been reported [16a]. Interestingly, the cleavage of 2,3′-*O*-cyclouridine (**5**) with hydrogen fluoride in the presence of aluminum fluoride afforded the desired 3′-deoxy-3′-α-fluorouridine (**6**) in only 31% yield (path A); the major product was 2′-fluoro analogue **4** (X = H) (47%

yield). In contrast, in the reaction of the 2′,5′-di-*O*-trityl derivative of **5**, the desired 3′-deoxy-3′-α-fluorouridine analogue **6** was obtained in good yield as a sole product [17]. Based on these results, the formation of **4** from **5** can be explained by the isomerization of **5** to **3** through an epoxidation–recyclization pathway (path B) [16b].

7.2.2 Synthesis of 2′-Deoxy-2′-α-fluoro Purine Nucleosides via Nucleophilic Substitutions at the Carbon-2′ Position

The synthesis of 2′-deoxy-2′-α-fluoro purine nucleosides via cleavage of purine *O*-cyclonucleosides (see Scheme 7.2) has not been reported, presumably for the following reasons [18]: (i) the synthesis of purine *O*-cyclonucleosides is not so easy as that of pyrimidine *O*-cyclonucleosides; (ii) the glycosyl bond of purine nucleosides is rather unstable under severe reaction conditions such as the use of HF; and (iii) it is not easy to remove the

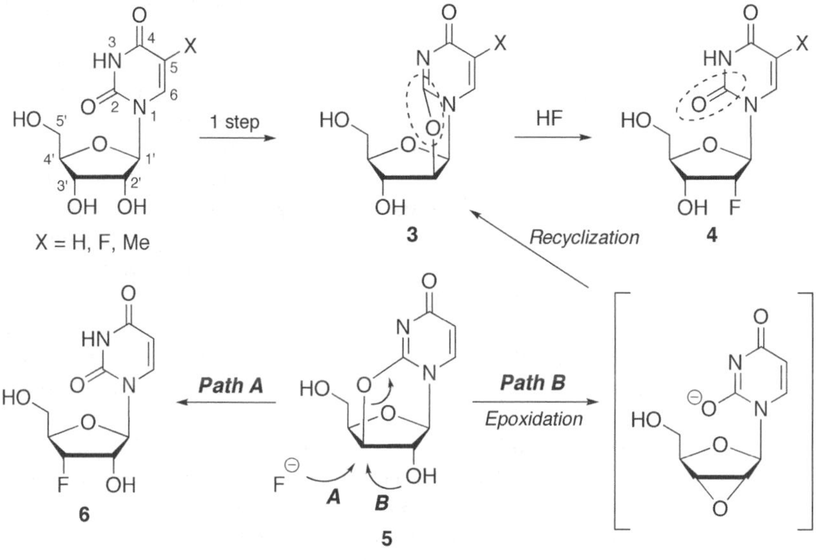

Scheme 7.1

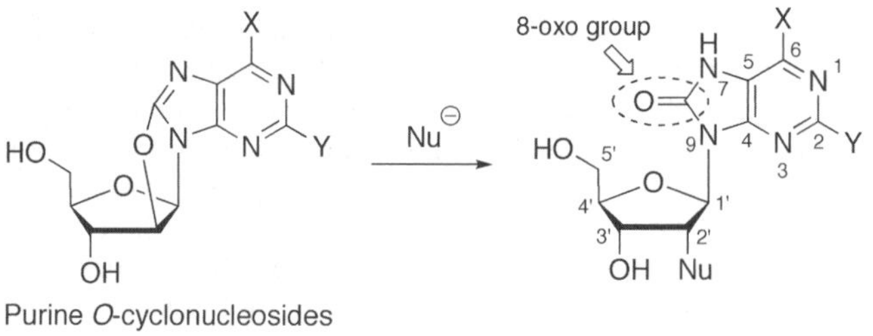

Scheme 7.2

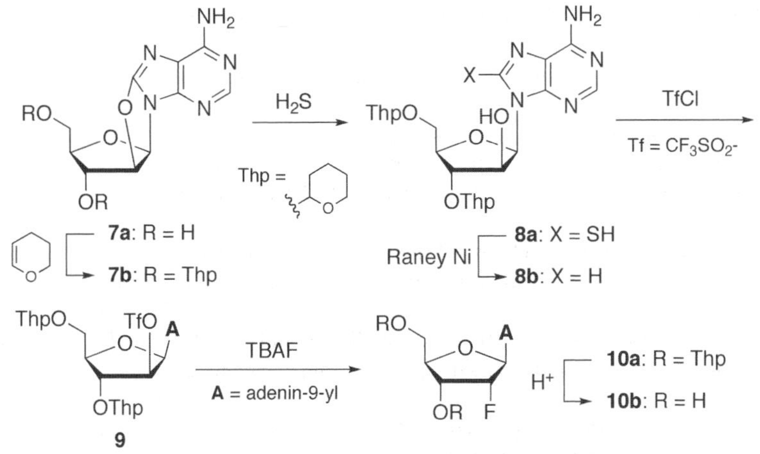

Scheme 7.3

resultant 8-oxo group even if the C-2′–*O* bond is successfully cleaved by nucleophilic substitution. Thus, another synthetic strategy for introducing fluorine at C-2′-α has been developed to achieve the synthesis of 2′-deoxy-2′-α-fluoro purine nucleosides. In this strategy, 3′,5′-di-*O*-protected arabanoside is initially prepared from 8,2′-*O*-cyclo purine nucleosides, and its 2′-hydroxyl group is then inverted by a nucleophilic (F^-) substitution to give the desired 2′-deoxy-2′-α-fluoronucleoside.

7.2.2.1 Synthesis of 2′-Deoxy-2′-α-fluoroadenosine via an 8,2′-O-Cycloadenosine Derivative

Ikehara and co-workers synthesized 2′-deoxy-2′-α-fluoroadenosine (**10b**) via 8,2′-anhydro-8-oxo-9-(β-D-arabinofuranosyl)adenine (8,2′-*O*-cycloadenosine, **7a**), as shown in Scheme 7.3. First, **7a** was efficiently prepared from 8-bromoadenosine in three steps using a regioselective 2′-*O*-tosylation as a key reaction [19, 20]. After both hydroxyl groups of **7a** were protected with a tetrahydro-2-pyranyl (Thp) group, the resultant **7b** was successively treated with H_2S and Raney nickel to afford arabinoside **8b** as a key intermediate [21]. Compound **8b** was converted to triflate **9**, which, upon subsequent substitution with tetrabutylammonium fluoride (TBAF), afforded the S_N2 product **10a** in 60% yield. Finally, removal of the Thp groups in **10a** gave 2′-deoxy-2′-α-fluoroadenosine (**10b**) in good yield [22]. Notably, its guanosine analogue, 2′-deoxy-2′-α-fluoroguanosine, was also synthesized by almost the same procedure [23], although the yield of the fluorination was reduced. As an alternative method for the preparation of **8b**, Ranganathan reported a synthetic route that started from a sugar derivative [24].

7.2.2.2 Synthesis of 2′-Deoxy-2′-α-fluoroadenosine Using a Protection Strategy with 3′- and 5′-Hydroxyl Groups

Various methods for protecting the hydroxyl groups of the ribofuranosyl moiety have been reported to date, and 2′,3′-*O*-isopropylidene and 5′-*O*-trityl groups have been used extensively in nucleoside chemistry [25a]. The synthesis of 3′,5′-di-*O*-acylnucleosides via selective deprotection of 2′,3′,5′-tri-*O*-acylnucleosides has also been reported by Ishido *et al.* [25b].

In 1978, Markiewicz and co-workers developed a novel diol-protecting group, that is, a tetraisopropyldisiloxan-1,3-diyl (TIDS) group, for regioselective protection of the 3′,5′-diol moiety [26]. We applied Markiewicz's strategy to the synthesis of base-modified analogues of 2′-deoxy-2′-α-fluoronucleosides, as shown in Scheme 7.4. 3′,5′-*O*-TIDS-protected 6-chloropurine riboside derivative **11a** was converted to triflate **11b**, which was subsequently transformed to 2′-β-acetate **12a** by S_N2 substitution with acetate ion. Since a TIDS group is unstable against fluoride ion, the TIDS of **12a** should be changed prior to fluorination at C-2′. Thus, **12a** was treated with TBAF to afford **12b**, whose hydroxyl groups were, in turn, protected again with Thp groups to provide **12c**. Subsequent hydrolysis of the 2′-acetate moiety gave 3′,5′-di-*O*-Thp arabinoside **13**, which is a substrate for fluorination. Compound **13** was treated with diethylaminosulfur trifluoride (DAST) to give **14a**, and further deprotection provided 2′-α-fluoride **14b**, which was subsequently treated with various nucleophiles to give the desired base-modified analogues (i.e., 6-substituted 2′-deoxy-2′-α-fluoroadenosine analogues) [27]. This synthetic method can also be applied to the synthesis of 2-substituted analogues. Thus, compound **15a**, which was prepared from 2-iodo-6-methoxypurine riboside by almost the same procedure, was converted to 2-iodoadenine analogue **15b**; its C-2 position was subjected to various nucleophilic sub-

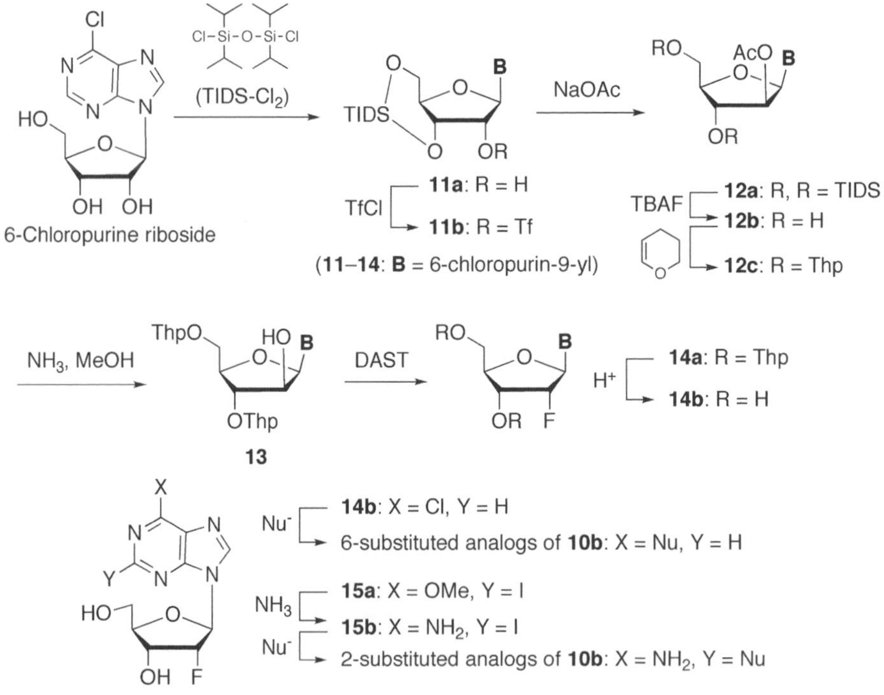

Scheme 7.4

stitutions to give a variety of 2-substituted 2′-deoxy-2′-α-fluoroadenosine analogues [28]. Unfortunately, these 2- or 6-substituted compounds did not show any significant anti-HIV activity [27, 28].

7.2.2.3 Synthesis of Various 2′-Deoxy-2′-α-fluororibosides

7.2.2.3.1 2′-Deoxy-2′-α-fluoropuromycin Puromycin (**20a**), a broad-spectrum antibiotic, inhibits protein biosynthesis at the translation stage by prematurely terminating a peptide chain (see Scheme 7.5) [29]. It is generally accepted that puromycin can bind to the A site of a ribosome, where the peptidyl chain of a peptidyl-tRNA is transferred to the amino group of the *p*-methoxyphenylalanine moiety of puromycin. However, the role of the 2′-hydroxyl group in this puromycin reaction is still unclear. Thus, we performed the synthesis and biological evaluation of 2′-deoxy-2′-α-fluoropuromycin (**20b**) to gain a better understanding. The 5′- and 3′-hydroxyl groups of 2′-*O*-acetate **12b** were modified with trityl and mesyl groups, respectively, to give **16b**, which was subsequently reacted with dimethylamine to afford 3′-*O*-mesyl arabinoside **17** (see Scheme 7.5). Treatment of **17** with sodium azide gave a mixture of 3′-azido-arabinoside **18** and its 2′-isomer (78% and 21% yields, respectively) via an epoxide intermediate. The major product **18** was

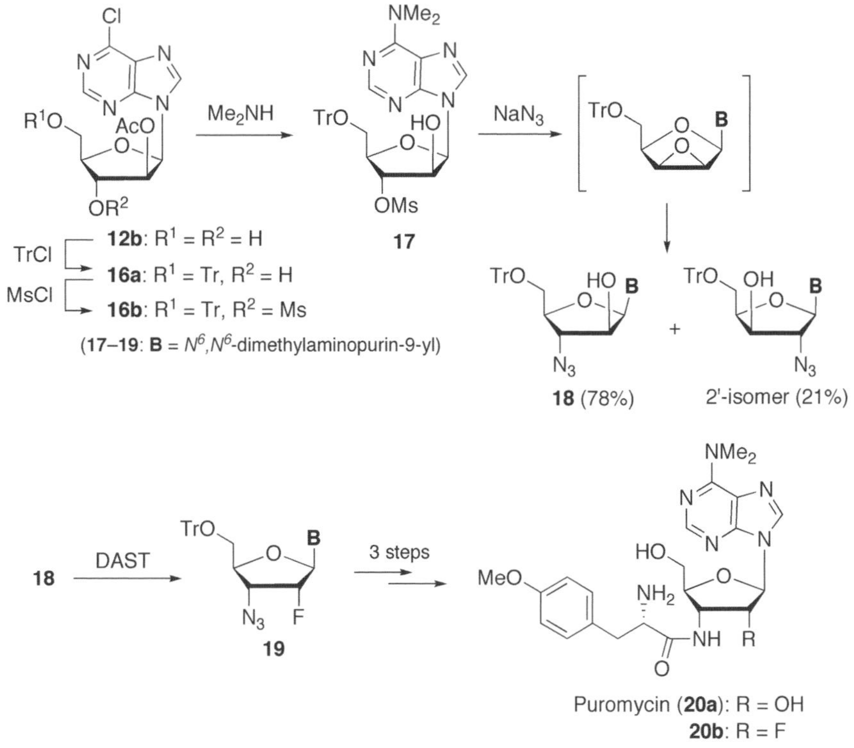

Scheme 7.5

treated with DAST to give **19**, which was converted to the desired **20b** in a three-step sequence (hydrogenation of the 3′-N_3 group, aminoacylation of the resultant 3′-NH_2 group, and removal of the protective groups) [30]. The biological activity of **20b** was evaluated. Interestingly, **20b** showed appreciable antitumor activity *in vitro* and also exhibited antibacterial potency that was comparable to that of puromycin. These activities are likely caused by the inhibition of protein biosynthesis.

It has been recognized that the 2′-hydroxyl group of puromycin is indispensable for its biological potency since 2′-deoxypuromycin showed no significant activity [31]. However, on the basis of our present result, it seems that the hydroxyl group at the 2′-position is not essential for the biological potency of puromycin. The coupling constant of ^{1}H NMR between 1′-H and 2′-H in **20b** ($J_{1'-2'}$) is nearly zero, which means that **20b** has the 3′-*endo* conformation (*N*-conformation), similar to that of ribonucleosides composing RNA. Therefore, the biological potency of puromycin does not seem to require the 2′-hydroxyl group itself, but the 3′-*endo* conformation of the sugar moiety, which can be formed by the presence of the hydroxyl or fluoro group at C-2′, is important.

7.2.2.3.2 2′FGpG Since some RNases have gained considerable attention because of their specific cytotoxicity toward cancer cells and their use as cancer chemotherapy drugs [32], there have been reported many x-ray crystallographic analyses of complexes of RNase with dinucleoside monophosphate seeking to better understand the structure–activity relationship (SAR) of such ribonucleases. Among them, dinucleoside monophosphates containing a 2′-deoxy-2′-α-fluoronucleoside are expected to be a good substrate for study because of their enzymatic resistance to RNases [33]. This resistance is likely due to the lack of the 2′-hydroxyl group, which is essential for the cleavage by RNase. Notably, it seems that dimer analogues composed by 2′-deoxy-2′-α-fluoronucleosides retain the 3′-*endo* conformation (RNA-type conformation). We reported the synthesis of a GpG analogue **24** (2′FGpG) containing 2′-deoxy-2′-α-fluoroguanosine (see Scheme 7.6). In our

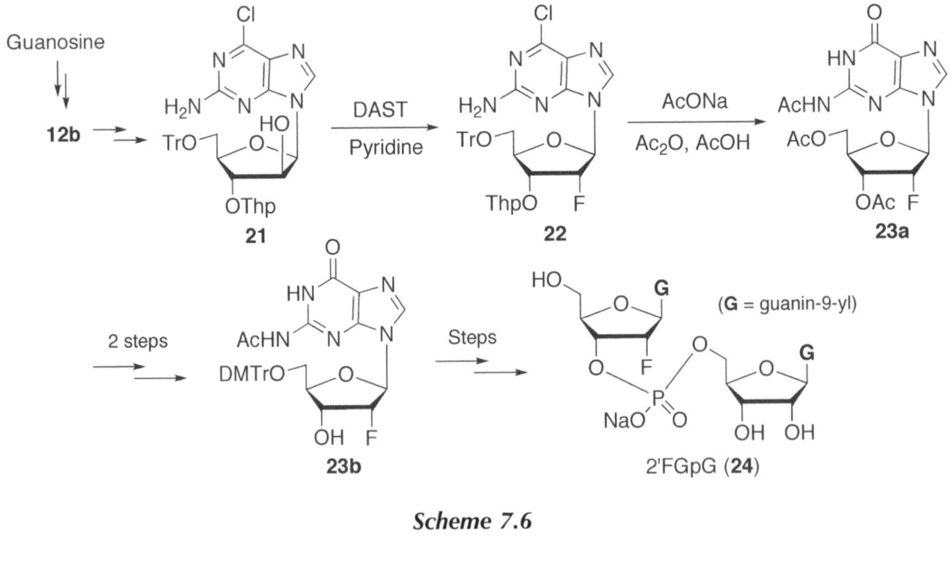

Guanosine → **12b** → **21** (DAST, Pyridine) → **22** (AcONa, Ac_2O, AcOH) → **23a**

2 steps → **23b** (Steps) → 2′FGpG (**24**)

(**G** = guanin-9-yl)

Scheme 7.6

synthesis, 2-amino-6-chloropurine riboside was used as a precursor for guanine, since previous reports revealed that attempted fluorination of the 2′-hydroxy group in arabinofuranosylguanine (ara-G) derivatives resulted in low yield [22, 34, 35]. Thus, 2-amino-6-chloropurine riboside, which was readily prepared from guanosine, was transformed to 2′-*O*-acetyl arabinoside (**12b'**, B = 2-amino-6-chloropurine) and further converted to **21** in three steps. Compound **21** was reacted with DAST in the presence of pyridine to afford 2′-α-fluoro compound **22** in 60% yield, which was treated with sodium acetate in a mixture of acetic acid and acetic anhydride to afford N^3,$O^{3'}$,$O^{5'}$-triacetyl-2′-deoxy-2′-α-fluoroguanosine (**23a**). Under these reaction conditions, the following transformations were performed in one pot: (a) hydrolysis of the 6-chloro moiety to the corresponding 6-oxo group (via 6-acetoxylation); (b) deprotection of the 3′- and 5′-hydroxyl groups; and (c) acetylation of the resulting hydroxyl groups and the 2-amino group. After the protecting groups of **23a** were changed to those of **23b** in two steps, compound **23b** was coupled with a guanosine derivative (3′-component) by a phosphoroamidate method to give the desired dimer **24** (2′FGpG) [36]. The 2′-deoxy-2′-α-fluoroguanosine moiety in **24** seems to have the 3′-*endo* conformation ($J_{1'-2'}$ = 3.0 Hz). Thus, compound **24** is expected to have a conformation similar to that of GpG.

7.2.2.3.3 N^3-Substituted 2′-Deoxy-2′-α-fluorouridine and Its Hypnotic Activity - Uridine and N^3-benzyluridine have been reported to show sleep-promoting activity [37] and hypnotic activity [38], respectively. We have been very interested in this topic and have studied the hypnotic activity of N^3-substituted 2′-deoxy-2′-α-fluorouridines to determine their SAR trends. For the synthesis of the key compound **4** (2′-deoxy-2′-α-fluorouridine), an alternative synthetic route via arabinoside was developed since a previously reported method [39] involving a ring-cleavage reaction of 2,2′-*O*-cyclouridine (**3**, X = H) required severe reaction conditions, such as exposure to HF (see Scheme 7.1). The new route is illustrated in Scheme 7.7. The hydroxyl groups of 2,2′-*O*-cyclouridine (**3**) were

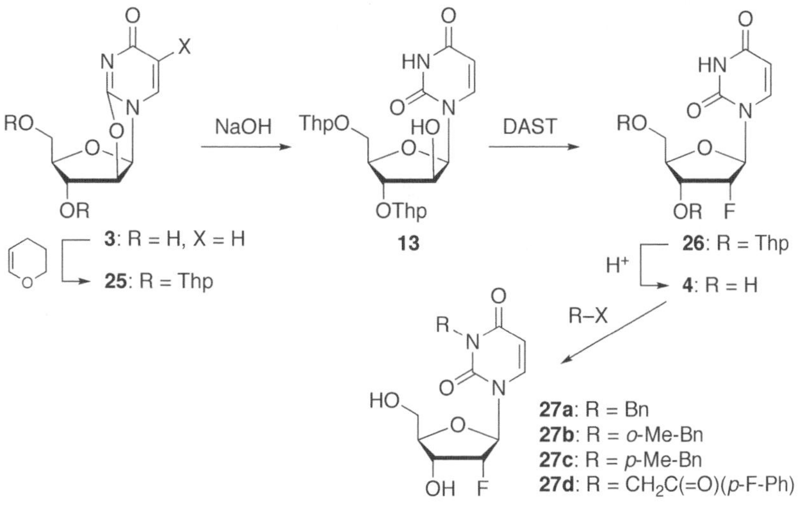

Scheme 7.7

protected with Thp groups to afford **25**, which was subsequently treated with sodium hydroxide to yield arabinoside **13**. Treatment of **13** with DAST in the presence of pyridine gave 2′-α-fluoro compound **26** in 57% yield, which was deprotected to afford 2′-deoxy-2′-α-fluorouridine (**4**). Compound **4** was reacted with various alkyl halides such as benzyl bromide and 2-chloro-4′-fluoroacetophenone to give the corresponding N^3-alkyl derivatives **27a–d** [39]. Although these derivatives exhibited hypnotic activity, they were weaker than the corresponding uridine derivatives. Since these derivatives are considered to have the 3′-*endo* conformation ($J_{1'-2'}$ = 3.5 Hz), the 2′-hydroxyl group seems to be important for the hypnotic activity.

7.3 Syntheses and Antiviral Activities of FddA and FddG

7.3.1 Synthetic Issues Regarding 2′-Deoxy-2′-β-fluoroarabinosides

To introduce a β-fluoro group at the C-2′ position of a ribonucleoside, one may first try an S_N2 reaction of a fluoride ion with a 2′-hydroxyl group activated as a leaving group, such as triflate. However, in most cases, its nucleobase (both pyrimidine and purine) prevents this reaction because (a) steric hindrance of the nucleobase interrupts the nucleophilic (F^-) attack from the top face, and (b) the nucleobase reacts intramolecularly with its own sugar moiety. Therefore, 2′-β-fluoro substitution is generally more difficult than the corresponding α-fluoro substitution.

7.3.1.1 Attempted synthesis of pyrimidine 2′-deoxy-2′-β-fluoroarabinosides

In the case of pyrimidine nucleosides, the S_N2 reaction of 2′-*O*-activated ribonucleoside (e.g., 2′-triflate) forms a 2,2′-*O*-cyclo bond prior to substitution of a nucleophile such as fluoride ion. Therefore, with the hope of suppressing this intramolecular substitution, the S_N2 reaction of 2,5′-*O*-cyclouridine derivative **28**, which places a triflate group at C-2′, was attempted (see Scheme 7.8). However, unexpectedly, a 5′-substituted 2,2′-*O*-cyclo derivative **29** was obtained [40]. The reaction mechanism to form **29** can be explained as follows. The nucleophilic attack at C-5′ of **28** forms a 5′-substituted intermediate, which causes an intramolecular substitution of the 2-oxo group at C-2′ to give **29**. In some cases, the introduction of an electron-withdrawing group at the N-3 position is effective for reducing the nucleophilicity of the 2-oxo group. Only one successful example has been reported by Matsuda and co-workers, in which an N^3-benzoyl derivative **30** was condensed with hydrogen azide using a Mitsunobu reaction to give 2′-α-azido analogue **31** [41]. However, no fluorination studies using the S_N2-type reactions like this have been reported. Therefore, the development of an effective method for suppressing the nucleophilicity of the 2-oxo group is still a challenging theme in the chemistry of pyrimidine nucleosides (the details were reviewed by Pankiewicz *et al.* [14].).

7.3.1.2 Synthetic Issues Regarding Purine 2′-deoxy-2′-β-fluoroarabinosides

The purine base also disturbs nucleophilic substitution at the sugar moiety, but the degree of this disruption is controllable. The nucleophilicity of the N-3 atom was reduced by the

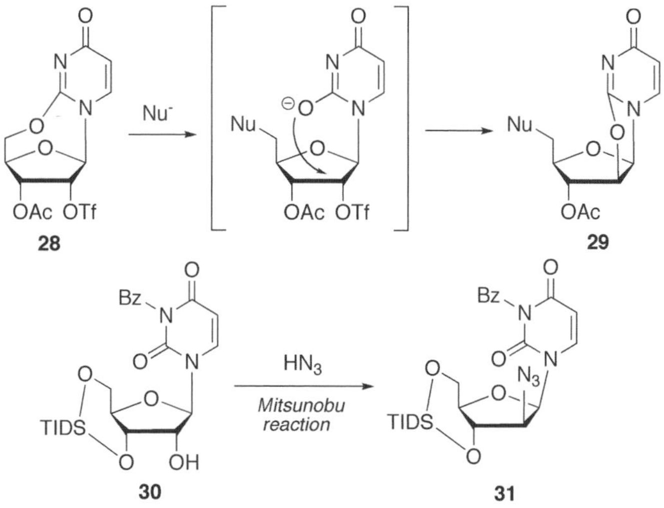

Scheme 7.8

introduction of electronegative group(s) to the purine base. In addition, the choice of the protective group(s) of sugar-OHs is important to control side-reactions.

7.3.1.2.1 Participation of Purine N-3 Atom In the case of purine ribonucleosides, the participation of the purine N-3 atom is a critical issue. For instance, the treatment of **32** with DAST furnished the desired 2′-β-analogue **34a** in only 30% yield because of an undesirable side-reaction that formed by-product **33** in 51% yield [42] (see Scheme 7.9). Compound **33** is probably formed as follows. The reaction of **32** with DAST gives the 2′-*O*-activated intermediate, which could receive not only the desired nucleophilic attack of fluoride ion, but also the intramolecular attack of the N-3 atom. The latter forms a new bond between N-3 and C-2′ of the sugar, which leads to an alternative attack of fluoride ion at the C-1′ position to yield the 1′-α-fluoro product. Therefore, reducing the nucleophilicity of the N-3 atom is effective for controlling this side-reaction. Pankiewicz and co-workers successfully synthesized the desired purine 2′-deoxy-2′-β-fluoroarabinosides **34b** in 63% yield by treating N^1-benzylinosine derivative **35** with DAST [42] (79% yield when tris(dimethylamino)sulfur(trimethylsilyl)difluoride (TASF) was used as another fluorinating agent) [43]. We performed this conversion with 6-chloropurine derivative **36**, whose N-3 atom was expected to show weaker nucleophilicity than that of **32**, and then the yield of **34c** was indeed improved to 87% [44]. Consequently, effective β-fluorination at C-2′ was achieved by the introduction of an electron-withdrawing group onto the base moiety.

7.3.1.2.2 Elimination Another problem in the synthesis of 2′-deoxy-2′-β-fluoroarabinosides is elimination. This issue seems to be influenced by the conformation of the sugar moiety. For instance, the reaction of **37** with TASF did not give a 2′-β-fluoro derivative, but rather two elimination products **38** and **39** [43] (see Scheme 7.10). The undesired eliminations are probably caused by the 3′-*endo* conformation of the sugar moiety in **37**, whose 3′-hydrogen and 2′-triflate are arranged in almost a *trans*-diaxial orientation that

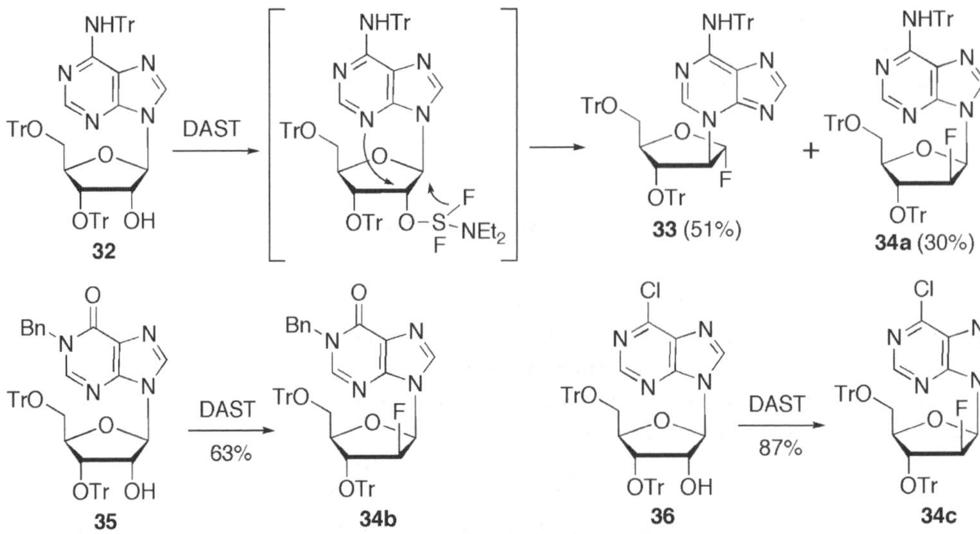

Scheme 7.9

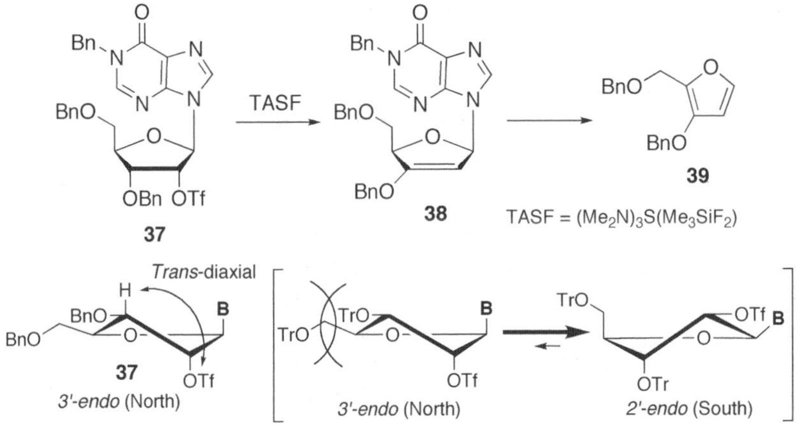

Scheme 7.10

favors elimination over substitution [42]. Ikehara and co-workers reported that the ratio of the 3′-*endo* conformer of 2′-α-substituted adenosines increased linearly with the electronegativity of the 2′-substituent [45]. Thus, **37** is thought to have the 3′-*endo* conformation because of the presence of electronegative 2′-triflate. To change the conformation to 2′-*endo*, which is unfavorable for elimination, the introduction of two trityl groups at the 3′- and 5′-hydroxyl groups is a reliable method, since the steric hindrance of the two trityl groups prevents 3′-*endo* formation and leads to puckering of the sugar to give the 2′-*endo* conformation [42–44]. This control of the conformation contributes to improving the

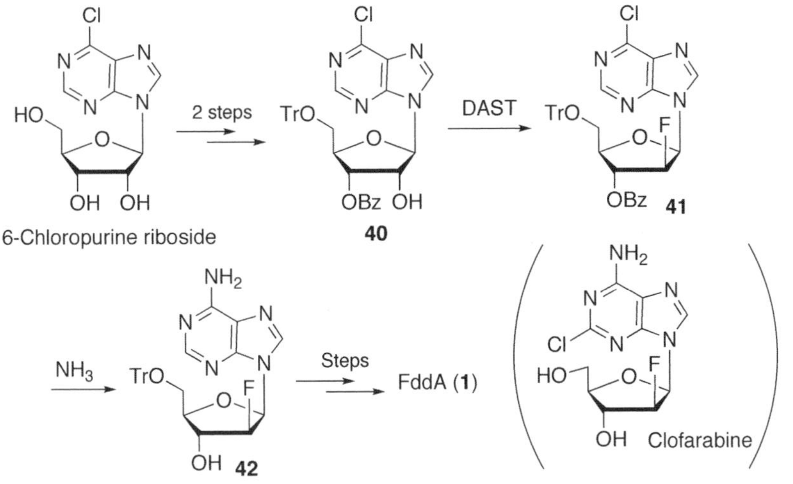

Scheme 7.11

product yield of fluorination (e.g., those of 3′,5′-di-*O*-trityl ethers **32**, **35**, and **36**; see Scheme 7.9).

However, there are some problems in the synthesis of such 3′,5′-di-*O*-trityl nucleosides, for example, a low yield (~20%) caused by the formation of a by-product 2′,5′-di-*O*-trityl derivative. The difficulty of removing the by-product from the desired 3′,5′-di-*O*-trityl derivative is another serious problem. Interestingly, we found that a 5′-*O*-trityl-3′-*O*-benzoyl derivative was also a good substrate for preparation of the 2′-deoxy-2′-β-fluoroarabinoside (see Scheme 7.11). Thus, 6-chloropurineriboside was treated with dibutyltin oxide and benzoyl chloride according to Moffatt's method to give a 3′-*O*-benzoate derivative, which was successively converted to 3′-*O*-benzoyl-5′-*O*-trityl derivative **40** by a conventional method. Treatment of **40** with DAST gave the 2′-β-fluoro analogue **41** in 78% yield [46]. It is believed that **40** gives the desired product **41** even though it is less sterically hindered than the corresponding 3′,5′-di-*O*-trityl derivative (and the leaving group at C-2′-α leads to puckering to the 3′-*endo* conformation), because the electron-withdrawing 3′-*O*-benzoyl group effectively leads to puckering to the 2′-*endo* conformation [15], and hence it reduces the competitive elimination reaction. Another merit of this protection system (i.e., 3′-OBz, 5′-OTr) is its generality, which enables the regioselective deprotection of each hydroxyl group. For instance, the regioselective deprotection of **41** with ammonia afforded a 3′-hydroxyl analogue **42**, which was used as a precursor of FddA through a Barton-deoxygenation reaction. Details are described in the Section 7.3.2. (see Scheme 7.15). This synthetic strategy should also be applicable to the synthesis of clofarabine (see Figure 7.2) with 2-chloroadenosine as a starting material.

7.3.1.2.3 FddA from 2′-Deoxy-2′-α-fluoroadenosine Until recently, the introduction of a fluoro group to the β-side of C-2′ has been quite difficult, as described earlier. Accordingly, a novel strategy that involves the inversion of C-2′ of 2′-deoxy-2′-α-fluororiboside to its 2′-β-fluoro epimer was developed by Marquez and co-workers (see Scheme 7.12). In their synthesis, N^6-benzoyl-2′-deoxy-2′-α-fluoroadenosine (**43a**) was converted to 3′-*O*-

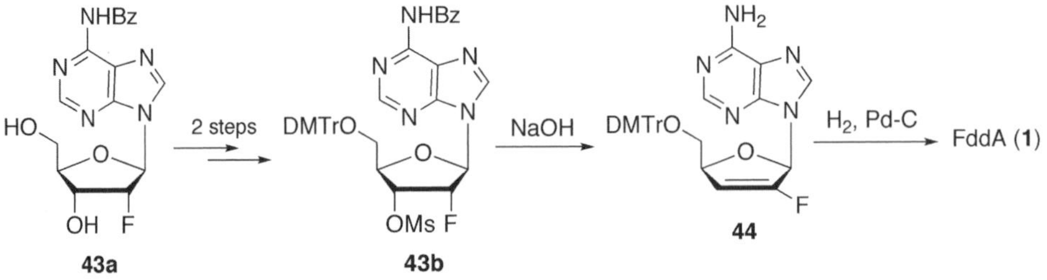

Scheme 7.12

mesylate **43b**, which was subsequently treated with sodium hydroxide to give 2′,3′-didehydro derivative **44** through elimination of methanesulfonic acid. Finally, hydrogenation of **44** afforded the desired compound, FddA (**1**) [47]. Although this synthetic route is quite elegant, it would not be suitable for a practical synthesis since five steps are required to prepare the intermediate **43a**, using an expensive starting compound, 9-(β-D-arabinofuranosyl)adenine (ara-A), and four additional steps are required to obtain FddA.

7.3.2 FddA (Lodenosine)

7.3.2.1 Antiviral Effect of FddA

FddA (**1**, lodenosine) is the 2′-β-fluoro-analogue of ddA, whose triphosphate is the active substance of didanosine, which is commonly used as an anti-HIV agent [48–51]. Didanosine (ddI) is first converted to ddA and then subsequent triphosphorylation gives the active metabolite ddA-TP. However, the dideoxynucleoside ddI is quite unstable toward acid, and therefore requires antacid agents to suppress decomposition in the stomach when given orally. This may cause a compliance issue for patients. On the other hand, the introduction of a fluorine atom to the 2′-β-position of the sugar moiety helps to improve the stability of dideoxynucleosides due to its strong electron-withdrawing property. FddA is sufficiently resistant to acid, and thus is suitable for oral administration without the need for antacid agents and eventually improves patient compliance. FddA has not exhibited cross-resistance to other dideoxynucleoside anti-HIV drugs such as 3′-azido-3′-deoxythymidine (AZT), ddI and ddC, and has shown synergistic activity with AZT [52, 53]. In addition, although certain 2′,3′-dideoxynucleosides penetrate the blood–brain barrier (BBB) rather poorly, FddA can be a brain-targeted prodrug or drug–carrier conjugate, since FddA is a good substrate for enzymes present in brain tissue such as adenosine deaminase (ADA) [54–58].

7.3.2.2 Industrial Synthesis of FddA

In general, industrial synthesis requires a good total yield. There are also several restrictions, such as (i) the raw materials and the reagents should be available in large quantities at reasonable costs; (ii) the reactions involved in the synthesis should be reasonably safe, without risks, such as explosion and contamination with toxic compounds; (iii) column chromatography should be avoided in separation and purification for economic reasons;

and (iv) the raw materials, reagents and solvents should be easy to dispose of. Even if the synthesis of a certain target compound becomes possible in the laboratory, it cannot be supplied in large quantities until these issues are resolved.

For example, while there have been many reports [49, 59–67] concerning the synthesis of lodenosine (**1**) there are drawbacks with each method from the viewpoint of industrialization. First, with the method via glycosylation, multiple steps are necessary to synthesize fluorinated sugar derivatives. In addition, the formation of α-anomer in glycosylation necessitates tedious separation and purification procedures, which generally give low yields. Therefore, a synthetic method that uses a nucleoside with only β-anomer as the starting material is desirable. However, as described in the Section 7.3.1, it is well known that fluorination of the nucleoside derivative directly at the 2′-β position is very difficult. In addition, tributyltin hydride is usually used as the reagent to convert the 3′-hydroxyl group into a deoxy compound, but its virulence and cost are problematic. Furthermore, in the case of deoxygenation after fluorination, there is a risk of losing the precious fluorine derivative produced by the difficult operation of fluorination. Thus, the development of an efficient deoxygenation method is needed.

In this section, we describe the industrial synthesis of FddA starting with 6-chloropurine riboside, which is readily available from inosine.

7.3.2.2.1 Method via F-ara-A (Route A) As described above (Section 7.3.1), we found that the 5′-*O*-Tr-3′-*O*-Bz derivative **40** can be fluorinated with DAST to give the desired 2′-β-fluorinated derivative **41** in 78% yield (see Scheme 7.11). Deprotection of the 3′-*O*-benzoyl group and displacement of the 6-chloro group of **41** could be achieved by treatment with ammonia in MeOH to afford **42** in 73% yield. Deoxygenation of the 3′-hydroxy group was achieved by the conventional method (see Scheme 7.13). Compound **42** was treated with phenyl chlorothionoformate to give **45** in 80% yield. The product was then treated with tris(trimethylsilyl)silane in the presence of 2,2′-azobis (isobutyronitrile) (AIBN) in toluene to give the deoxygenated compound **46** in 73% yield. Acid treatment of **46** gave the desired FddA (**1**) in 88% yield [46].

In this way, we obtained FddA via F-ara-A (**42**) in a fairly good overall yield from 6-chloropurine riboside. However, to establish a scalable process, each reaction step must be improved and further investigations are needed to optimize conditions for fluorination and deoxygenation on an industrial scale.

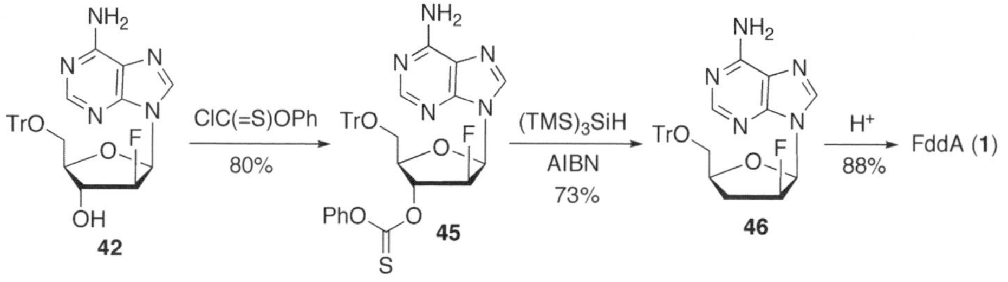

Scheme 7.13

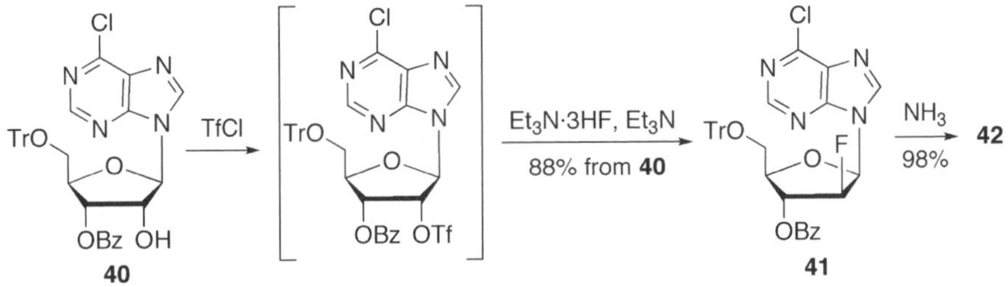

Scheme 7.14

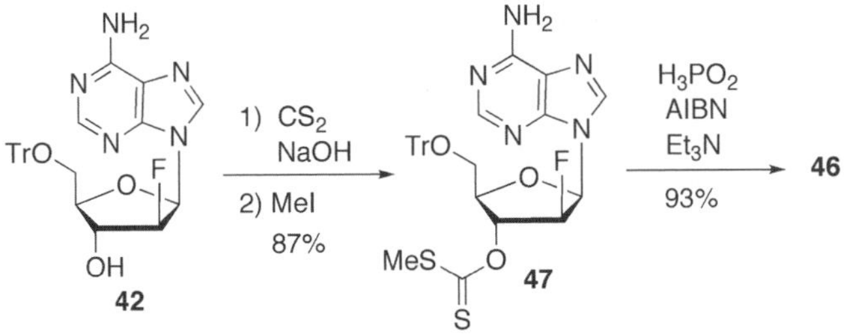

Scheme 7.15

Improvement in fluorination. DAST used for the laboratory-scale synthesis described above is not desirable for industrial synthesis in terms of availability and safety. Thus, we examined the fluorination of 2′-activated nucleoside with triethylamine trihydrogenfluoride. The triflate, which was quantitatively obtained from **40**, was reacted with 6 equivalents of $Et_3N \cdot 3HF$ and 3 equivalents of Et_3N in ethyl acetate. Fluorination proceeded very smoothly to give **41** in 88% yield [68]. To the best of our knowledge, this is the highest reported yield in the fluorination of a purine riboside at the 2′-position. Next, we treated compound **41** with ammonia to give **42** in almost quantitative yield by simultaneous 6-amination and 3′-benzoyl deprotection. (see Scheme 7.14).

Improvement in radical deoxygenation. As shown in Scheme 7.13, we achieved the effective radical deoxygenation of **45** using tris(trimethylsilyl)silane. However, when we examined the manufacturing cost in further detail, the silane used in the deoxygenation step was found to be relatively expensive. We therefore investigated another type of radical reduction for the xanthate **47** with hypophosphorous acid using the conditions reported by Barton *et al.* [69] and this proved to be the most efficient (see Scheme 7.15). The best yield (93%) was obtained with an excess amount of hypophosphorous acid and Et_3N [70]. In summary, we synthesized FddA from 6-chloropurine riboside in eight steps and 32.8% overall yield via route A.

7.3.2.2.2 Method via 6-Chloropurine 3′-deoxyriboside (Route B) Although the overall yield for the reaction sequence in the route A was the best reported thus far, we investigated other routes in an attempt to reduce the production cost. Thus, we investigated the synthesis of 6-chloropurine 3′-deoxyriboside.

The first synthesis of FddA (**1**) from 3′-deoxyadenosine derivative by Herdewijn *et al.* gave the fluorination product in only 10% overall yield after deprotection and purification [59]. Several years later, we also examined a similar reaction with the 5′-*O*-acetyl compound, but the yield was confirmed to be very poor [71]. We hypothesized that one possible reason for the low fluorination yield in the above reaction might be the nucleophilic participation of N-3 of the adenine ring, and that this might be overcome by using 6-chloropurine 3′-deoxyriboside as a starting material (see also Section 7.3.1.2.1).

7.3.2.2.3 Synthesis of 6-Chloropurine 3′-deoxyriboside We first investigated the practical synthesis of 6-chloropurine-3′-deoxyriboside starting with inosine, which was readily available in suitable quantities (see Scheme 7.16).

We used the acetoxybromination process in this synthesis, which was previously developed by us for the large-scale synthesis of ddA [72]. The reaction proceeded well and gave **48** in 80% overall yield from inosine. We then carried out the radical debromination of **48** using hypophosphorous acid as a reducing agent in the presence of a water-soluble radical initiator such as V-50 or VA-044 to give **49** in almost quantitative yield. When we used AIBN as a radical initiator, we observed a lower isolated yield due to a loss of product during purification to remove residual AIBN. 6-Chlorination and subsequent deacetylation of **50a** were carried out in a conventional manner, and the product **50b** was tritylated for the next fluorination step to give **50c** in 85% yield [73b]. This process for the synthesis of 6-chloro-3′-deoxyriboside gave an overall yield of 73% in five steps, and was shown to be scalable to 3000-liter vessels. After developing a practical process for the synthesis of 6-chloro-3′-deoxyriboside, we turned our attention to the fluorination of **50c**.

7.3.2.2.4 Fluorination of 6-Chloropurine 3′-deoxyriboside

Fluorination with DAST. To examine our hypothesis, we first attempted the fluorination of **50c** with DAST and obtained the product **51** in 43% yield (see Scheme 7.17) [73a]. In

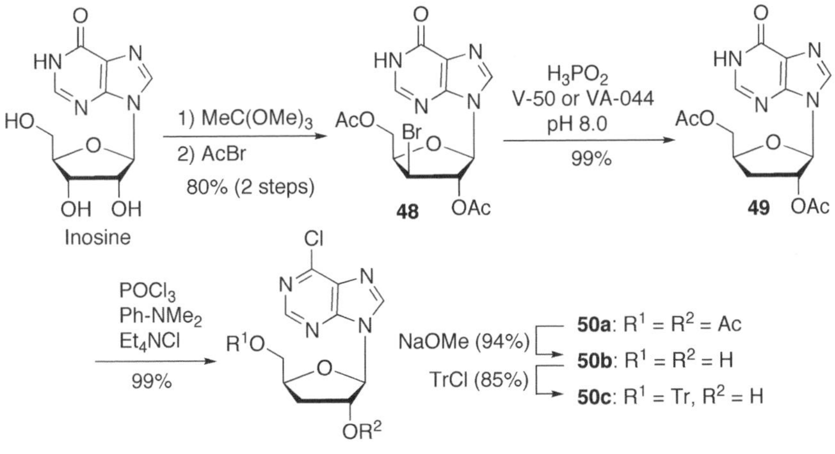

HO, Inosine, OH OH
1) $MeC(OMe)_3$
2) AcBr
80% (2 steps)
AcO, Br, **48**, OAc
H_3PO_2
V-50 or VA-044
pH 8.0
99%
AcO, **49**, OAc
$POCl_3$
$Ph\text{-}NMe_2$
Et_4NCl
99%
Cl, R^1O, OR^2
NaOMe (94%)
TrCl (85%)
50a: $R^1 = R^2 = Ac$
50b: $R^1 = R^2 = H$
50c: $R^1 = Tr$, $R^2 = H$

Scheme 7.16

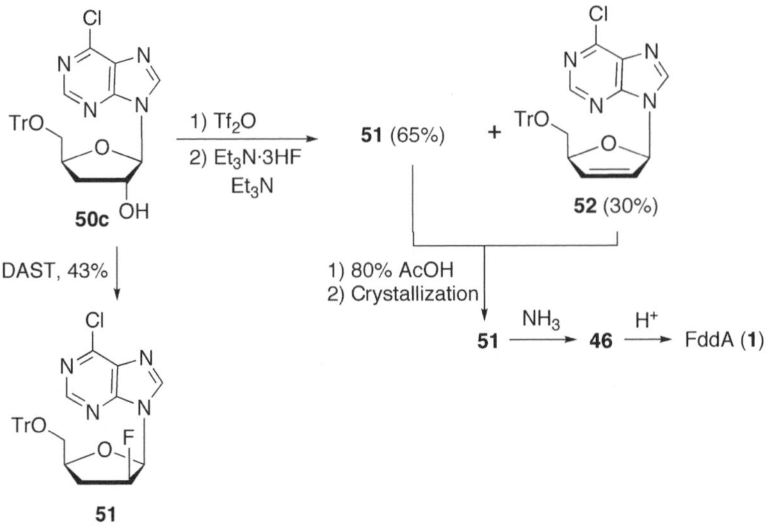

Scheme 7.17

comparison with the result reported for 3′-deoxyadenosine by Herdewijn, this result clearly supports our hypothesis that the electron-withdrawing effect of the 6-chloro group may prevent rearrangement of the purine moiety and therefore increase the reaction yield (see above). It is also conceivable that the N-3 nitrogen of the adenine ring may participate in the abstraction of a hydrogen atom at the 3′-β position. Replacement of an amino group of adenine with a chlorine atom may lead to decrease in the electron density at the C-2′ position, which would allow facile attack of the nucleophile.

Although we were able to improve the yield of fluorination with DAST, we examined other fluorination methods to develop a more economic and safer industrial process.

Fluorination with $Et_3N{\cdot}3HF$ after triflate formation. We applied the same fluorination method as that used in the synthesis of F-araA in Scheme 7.14. Compound **50c** was treated with $Et_3N{\cdot}3HF$ after triflate formation to give the fluorinated product **51** in 65% yield, which is much better than that with DAST (see Scheme 7.17). Although the fluorination yield was improved, the elimination product **52** was also formed in 30% yield, and this should be removed from the product. A 2′-β-fluorinated nucleoside such as **51** is generally very stable under acidic conditions. Thus, we were able to quantitatively recover the fluorinated product **51** as a pure form simply by treating the mixture of **51** and **52** with 80% acetic acid after the complete decomposition of **52** [73]. The fluorinated compound was then treated with ammonia followed by deprotection with hydrochloric acid to give FddA (**1**) in good yield. Thus, we have developed a practical process for the synthesis of FddA that does not require any chromatographic purification and corrosive reagents.

Fluorination with perfluoroalkanesulfonyl fluorides. Since triflate formation is an expensive procedure, we examined the use of an inexpensive perfluoroalkanesulfonyl fluoride such as perfluoro-1-butanesulfonyl fluoride (nonafluoro-1-butanesulfonyl fluoride, $C_4F_9SO_2F$, NfF) or perfluoro-1-octanesulfonyl fluoride ($C_8F_{17}SO_2F$, OctF) for fluorination,

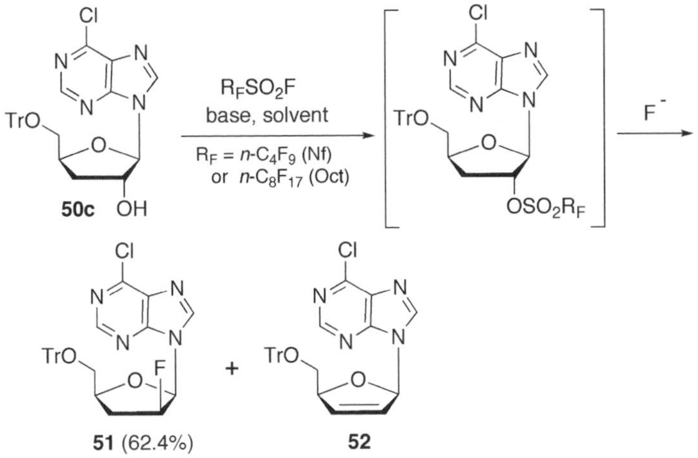

Scheme 7.18

and compared the yields with those using other fluorination agents [74] (see Scheme 7.18). These fluorination reagents are commercially available in suitable quantities and are known to be stable and less corrosive. Among the bases we examined, *N,N*-dimethylcyclohexylamine (DMCHA) seems to be a good base for fluorination. With 2 equivalents of NfF and 2 equivalents of DMCHA as a base, HPLC analysis indicated a 62.4% yield after aqueous work-up. The yield of **51** from **50**c with NfF was much greater than that using DAST and almost the same as that in the previous two-step method involving trifluoromethanesulfonylation and fluorination with $Et_3N{\cdot}3HF$ (see Scheme 7.17).

Remaining issues. We developed a new route to the practical synthesis of FddA from inosine in nine steps and 36% overall yield. During the course of this study, we greatly improved the fluorination of 3′-deoxyriboside, which had been very difficult and the bottleneck in FddA synthesis. However, even with this process, formation of the elimination byproduct was inevitable. To further improve the yield, studies are still needed to fix the 3′-deoxyriboside to the 2′-*endo* conformation, which does not easily give the elimination product.

7.3.3 FddG

7.3.3.1 Antiviral Effect of FddG

It has been reported that FddG (**2**), the 3′-α-fluoro analogue of ddG, showed a potent anti-HIV [75, 76] and anti-HBV [77] activities *in vitro*, and therefore is a promising antiviral drug candidate for clinical use. The antiviral activity of FddG was evaluated in the duck hepatitis-B virus (DHBV) system, and it was found that FddG is a strong inhibitor of DHBV replication not only *in vitro* but also *in vivo* [78]. The mechanism by which FddG inhibits HBV replication was investigated, and the results suggested that FddG-triphosphate was most likely a competitive inhibitor of dGTP incorporation and a DNA chain terminator [79]. FddG similarly inhibited the replication of wild-type, lamivudine (3TC)-resistant, adefovir dipivoxil (PMEA)-resistant, and 3TC-plus-PMEA-resistant HBV mutants in Huh7 cells that had been transfected with different HBV constructs. This cross-resistance profile of FddG suggests that this compound could be envisaged as a new alternative drug for the treatment of chronic HBV carriers who have developed resistance to currently approved drug regimens, and that FddG may also be valuable for the design and evaluation of new combination therapies with other polymerase inhibitors for the treatment of chronic HBV infection to prevent or delay the emergence of drug-resistant mutants [79]. A phase II clinical trial is now underway with MIV-210, which is a prodrug of FddG [7].

7.3.3.2 Industrial Synthesis of FddG

Although there have been several reports on the synthesis of FddG via the coupling of a guanine base with a sugar moiety [80–82], these methods may not be suitable for industrial-scale synthesis due to the formation of α-anomer and rather lengthy reaction steps. In this section, several approaches to the direct transformation of guanosine into FddG are described [11]. This approach has the great advantage that fewer reaction steps are needed and purification is easy because no α-anomer is formed.

7.3.3.2.1 Synthesis of FddG with Inversion of the Configuration at the C-3′ Position of Guanosine The first synthesis of FddG (**2**) from guanosine was reported by Herdewijn *et al.* [76, 80] After the C-5′-OH group was protected with a benzolyl group using benzoic anhydride and triethylamine, compound **55** was subjected to fluorination using DAST followed by deprotection to give FddG (**2**) (see Scheme 7.19). Fluorination took place with inversion of configuration at the C-3′-position. Although the fluorination yield was rather low (35%), this method may be regarded as one of the most convenient and efficient

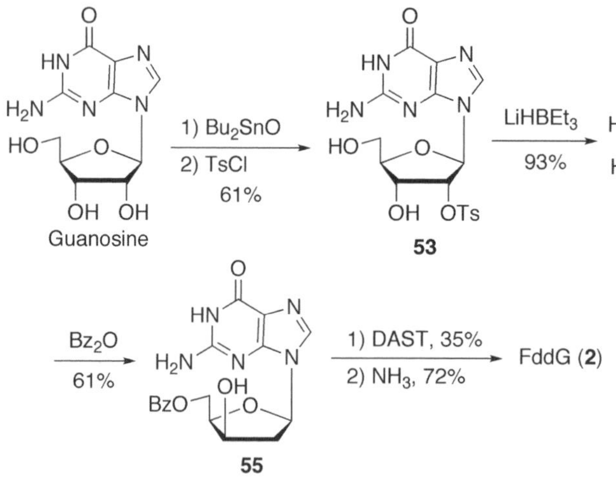

Scheme 7.19

approaches to the synthesis of FddG (**2**). A disadvantage of this method may be the use of Bu_2SnO, $LiHB(Et_3)$, and DAST, which are not appropriate for industrial-scale synthesis.

As mentioned above, we significantly improved the fluorination of 6-chloropurine-3′-deoxyriboside with DAST during the synthesis of FddA (**1**). We speculated that the introduction of a chlorine atom at the 6-position of the purine base might have a beneficial effect on the synthesis of FddG (**2**) as well. 2-Amino-6-chloropurine riboside analogues such as **56** would be readily transformed to FddG (**2**) by hydrolysis after fluorination of the sugar moiety, but **56** may also be transformed to 6-substituted FddG derivatives **59** (see Scheme 7.20).

As we expected, we confirmed that the fluorination yield with DAST increased to 60% from 35% in the case of guanine (**57**→**58a**) [83]. The fluorination product **58a** was then subjected to deprotection of the trityl group at the C-5′-position followed by treatment of **58b** with 2-mercaptoethanol under basic hydrolysis conditions to give FddG (**2**) in 67% yield (see Scheme 7.20).

Since 2-amino-6-chloropurine derivative **58b** is used as a synthetic intermediate, this method can be used not only for the synthesis of guanine nucleoside FddG (**2**), but also for the syntheses of 3′-α-fluoronucleosides with nucleic bases other than guanine. In fact, through conversion of the 6-position of **58b** by reduction or reaction with thiols, a series of 2-amino-6-substituted purine nucleosides **59a–c** can be synthesized. The anti-HBV (hepatitis-B virus) activities of these compounds were evaluated *in vitro* and 2-amino-6-chloropurine derivative **58b** was shown to have potent anti-HBV activity (EC_{50} = 10.4 μM), which is comparable to that of FddG (EC_{50} = 9 μM). In addition, with regard to the structure–activity relationship, 2-amino-6-arylthiopurine derivative **59c** showed more potent anti-HBV activity (EC_{50} = 3.6 μM) than FddG, while PMEA (anti-HBV drug used as a positive control) was 10-fold more potent than **59c** [83].

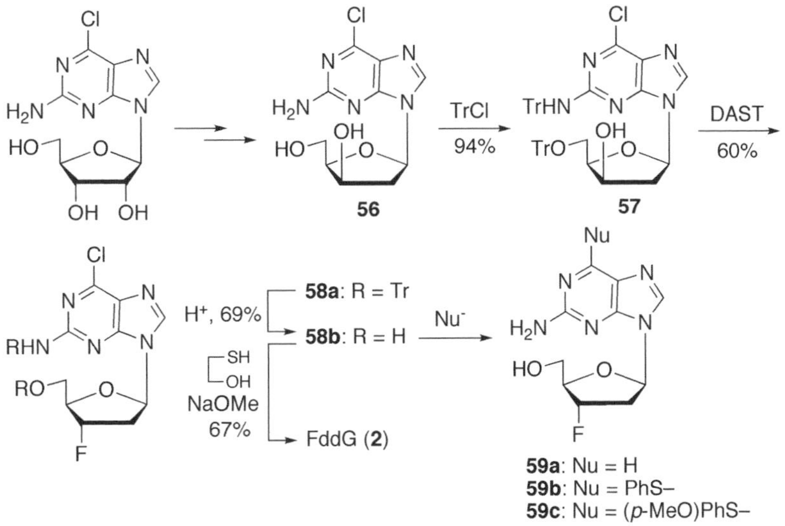

Scheme 7.20

7.3.3.2.2 Synthesis of FddG with Bromine Rearrangement during Fluorination We reported that the bromohydrin derivative **60b**, which was easily prepared from guanosine via **60a**, gave the C-3′-α-fluorinated compound **61** in 59% yield accompanied by the formation of a regioisomer (21%) under fluorination conditions (see Scheme 7.21) [84]. The reaction can be explained by considering the rearrangement of the bromine atom from the C-3′-β- to the C-2′-β-position via a bromonium ion intermediate on which the fluoride anion attacks from the α-side of the C-3′- or C-2′-position. The C-3′-α-fluorinated product **61** was converted to FddG (**2**) by radical debromination and deprotection.

We further investigated the fluorination conditions to improve the yield of C-3′-α-product **61**. We then found that the fluorination of compound **60b** with SF_4 furnished the C-3′-α-product **61** in 79% yield in an 3′-F:2′-F ratio of 7.8:1 [85]. This suggests that the diethylamino group of DAST may have a negative influence on the regioselectivity of the reaction. To our surprise, we also found that the reaction of compound **60b** with NfF in the presence of NEt_3 did not afford fluorination product **61**, but produced a 2′,3′-didehydro-2′,3′-dideoxyguanosine derivative in 62% yield. This method has advantages with respect to the number of reaction steps and high overall yield. However, one serious drawback is that this method uses SF_4-type reagents which require special equipment for handling.

7.3.3.2.3 Synthesis of FddG via Retentive Fluorination at the C-3′-α Position After concluding that the fluorination with bromine rearrangement described above might proceed via a bromonium ion intermediate, we considered that a similar reaction might take place via a sulfonium ion intermediate. We chose readily available 8,2′-anhydro-8-mercaptoguanosine as the starting material, anticipating that fluorination might proceed with the participation of a sulfur atom, which would facilitate the attack of fluoride ion to the C-3′-α- rather than the C-2′-α-position for steric reasons [86]. 8,2′-Anhydro-8-mercaptoguanosine was first converted to the $N^2,O^{5'}$-diacetyl derivative **62** which was subjected to fluorination with DAST (see Scheme 7.22). As anticipated, we obtained the

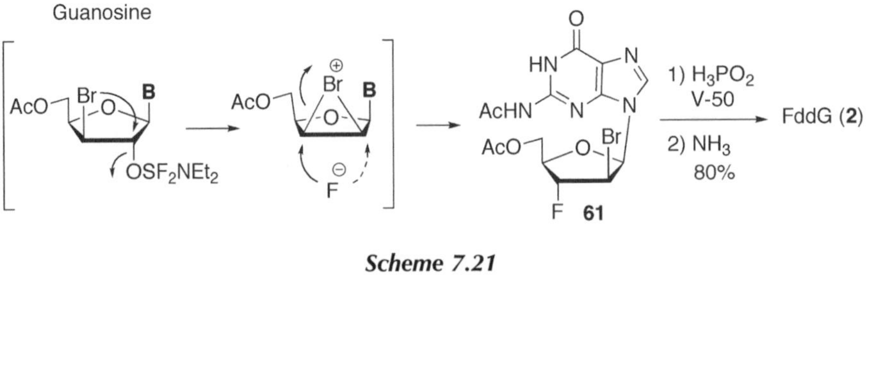

Scheme 7.21

C-3′-α-fluorinated product **63** in 47% yield without any formation of the C-2′-α-fluorinated product. In the case of fluorination with NfF, however, we found that elimination predominated (**63**:**64** = 1:3.5) (see Scheme 7.22).

Encouraged by these results, we further investigated the fluorination of ditrityl protected compound **65** using several fluorinating reagents (see Scheme 7.23). Although the reaction with DAST gave a complex mixture, the desired C-3′-α-product **66** was obtained in 63% yield with NfF/NEt_3, along with the elimination product. Next, we optimized the reaction conditions. When we used diisopropylethylamine (DIPEA) as the base, elimination was completely suppressed, albeit the yield was not improved. The best yield (91%) was achieved when we used an excess amount of NfF and DIPEA.

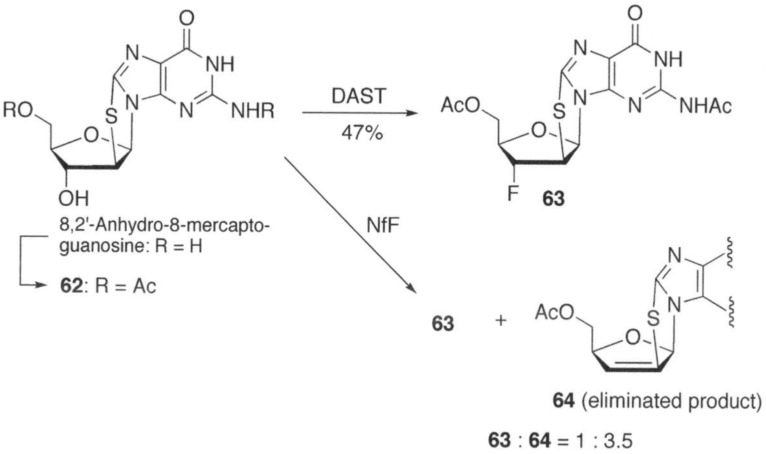

8,2'-Anhydro-8-mercapto-guanosine: R = H

62: R = Ac

DAST 47%

63

NfF

63 + **64** (eliminated product)

63 : **64** = 1 : 3.5

Scheme 7.22

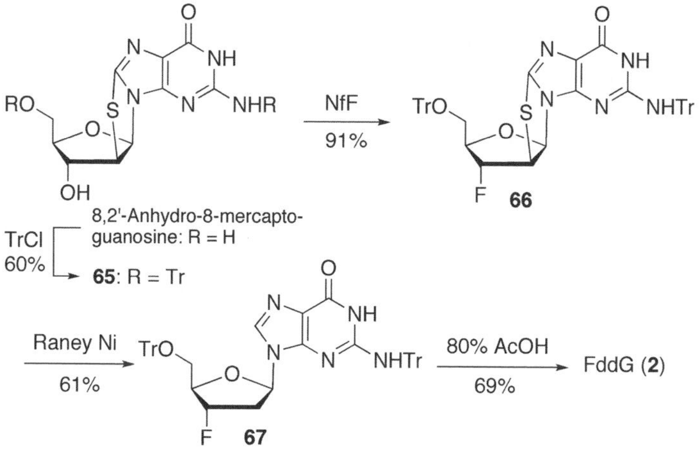

8,2'-Anhydro-8-mercapto-guanosine: R = H

TrCl 60%

65: R = Tr

NfF 91%

66

Raney Ni 61%

67

80% AcOH 69%

FddG (**2**)

Scheme 7.23

The fluorination product **66** was then treated with acetic acid followed by reductive desulfurization with Raney nickel in aqueous NaOH to give defluorination product ddG, but not FddG (**2**). Accordingly, we tried the desulfurization of compound **66** in various solvent systems prior to deprotection with the acid. After several unsuccessful attempts, we obtained the desired Tr_2-FddG **67** in 61% isolated yield when we carried out Raney-nickel reduction in toluene. This method is advantageous because fluorination can be achieved using the scalable reagent NfF.

In this section, synthetic approaches to FddG (**2**) with particular focus on industrially applicable fluorination methods are described. At the beginning of this study, these approaches which used a nucleoside as a starting material required SF_4 reagents such as DAST and morpholinosulfur trifluoride (MOST) for fluorination because other agents gave mainly elimination products. However, SF_4-type reagents are not desirable for industrial-scale synthesis because of their poor availability and inherent toxicity. To overcome these problems, we developed a new nucleoside fluorination method that involved the participation of neighboring groups. With this methodology, we achieved the fluorination of a guanosine derivative at the C-3′ position in good yield using readily available NfF. This would provide a useful stereoselective method for introducing a fluorine atom into the sugar moiety of nucleosides.

7.4 Biological Activity of Fluorinated Nucleosides and Correlations with Their Conformations

7.4.1 Sugar Ring Conformation of 2′-Deoxy-2′-α-fluoronucleosides

7.4.1.1 Properties of Modified RNA Containing 2′-Deoxy-2′-α-fluoronucleosides

In 1968, four years after the first synthesis of 2′-deoxy-2′-α-fluorouridine (**4**), Fox and co-workers reported that **4** adopts the 3′-*endo* conformation (N-form) more readily than the 2′-*endo* conformation (S-form) [87]. In the 1970s, some studies examined the relationships between the conformations of 2′-deoxy-2′-α-substituted nucleosides and the electronegativity of their 2′-substituents, and revealed that an increase in electronegativity led to a high proportion of the 3′-*endo* conformation. According to these studies, the proportion of the 3′-*endo* conformer of **4** is approximately 85% [88]. With regard to 2′-substituted purine nucleosides [89], Uesugi and co-workers also reported that 2′-deoxy-2′-α-fluoroadenosine (**10b**) favors the 3′-*endo* conformation, and this trend remained unchanged even when **10b** was contained in dinucleoside monophosphate [90]. Although ribonucleoside generally favors the 3′-*endo* conformation, 2′-deoxy-2′-α-fluoro analogues increase this tendency more than ribonucleosides. However, despite the difference in sugar puckering, the torsion angles of the C-4′–C-5′ bond and the C-5′–O-5′ bond in the 2′-deoxy-2′-α-fluoro analogues are analogous to those of ribonucleosides. Thus, a nucleic acid that contains the 2′-deoxy-2′-α-fluoronucleoside forms a structure similar to RNA, in which the proportion of the 3′-*endo* conformation is higher than that in RNA.

This property naturally influences their biological activities. For example, it has been reported that poly(2′-deoxy-2′-α-fluoroadenylic acid) (**69a**) is as good a template as poly(adenylic acid) (**68a**) for some reverse transcriptases and DNA polymerase [91]

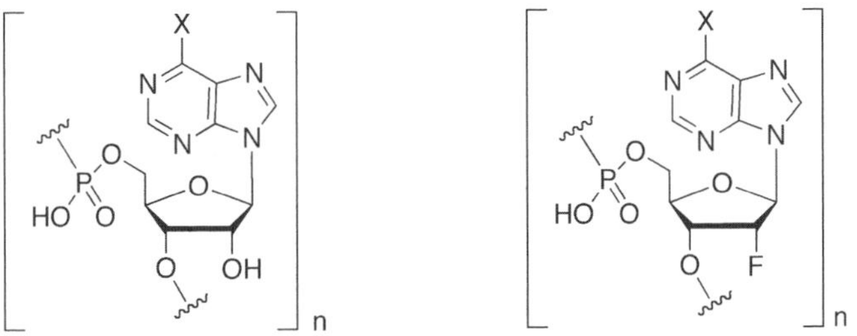

68a: X = NH_2; poly(adenylic acid)
68b: X = OH; poly(inosinic acid)

69a: X = NH_2; poly(2'-deoxy-2'-α-fluoro-adenylic acid)
69b: X = OH; poly(2'-deoxy-2'-α-fluoro-inosinic acid)

Figure 7.4 *Structures of polynucleotides and their analogues.*

(Figure 7.4). Likewise, poly(2′-deoxy-2′-α-fluoroinosinic acid) (**69b**) and poly(2′-deoxy-2′-α-fluorocytidylic acid) are effective templates for reverse transcriptases [92]. Notably, **69a** also functions as mRNA in protein-synthesizing systems *in vitro*, in which [^{14}C]lysine is incorporated into polypeptides [93]. Moreover, modified mRNA that contains 2′-deoxy-2′-α-fluoroadenosine (**10b**) can give rise to the corresponding protein (luciferase). Interestingly, modified RNA that contains 2′-deoxy-2′-α-fluorouridine (**4**) does not act as mRNA [94]. A similar trend has been observed in a study on the induction of interferon by double-helical RNA (poly I · poly C). The complex of poly(2′-deoxy-2′-α-fluoroinosinic acid) (**69b**) · poly C shows higher interferon-inducing activity than that of poly I · poly C, while the complex of poly I · poly(2′-deoxy-2′-α-fluorocytidylic acid) is not an effective interferon inducer [95].

7.4.1.2 Properties of 2′-Deoxy-2′-α-fluoronucleosides

Although the relationship between conformation and biological activity is not well documented at the nucleoside level, several 2′-deoxy-2′-α-fluoronucleosides are known to have biological activity, such as antiviral activity. Tuttle and co-workers reported that several 2′-deoxy-2′-α-fluoropurine nucleosides exhibited antiviral potency against influenza A and B viruses. The 2-amino group of the purine moiety is critical [96a] for this activity, and the antiviral activity is much better than those of amantadine and ribavirin [96b]. Notably, 2′-deoxy-2′-α-fluoro-2′-*C*-methylcytidine shows increased inhibitory activity in the hepatitis C virus (HCV) replicon assay compared to 2′-*C*-methylcytidine, and low cellular toxicity [97]. Influenza virus and HCV are classified as RNA viruses, which are a major target of current chemotherapy. However, the development of effective drugs against these virus infections is still slow; further investigation of 2′-deoxy-2′-α-fluoronucleosides is expected to accelerate the development of such drugs.

Another interesting biological activity of 2′-deoxy-2′-α-fluoropuromycin has been observed. Unlike other nucleoside analogues, puromycin inhibits protein biosynthesis without phosphorylation of the 5′-hydroxyl group. As mentioned above (Section 7.2.2.3.1), 2′-deoxy-α-2′-fluoropuromycin (**20b**) showed antitumor and antibacterial activity comparable to that of puromycin. Since 2′-deoxypuromycin, which favors the 2′-*endo* conformation, showed no biological activity, the 3′-*endo* conformation is thought to be essential for its activity.

In general, the 3′-*endo* conformation of fluoronucleoside is essential for antiviral activity against RNA viruses although the anti-HIV activity is an exception (the details of anti-HIV activity are addressed in the next section).

7.4.2 Sugar Ring Conformations of FddA and FddG and Their Anti-HIV Activities

Mu and co-workers reported that the introduction of a fluoro group to the C-2′-α position of 2′,3′-dideoxyadenosine 5′-triphosphate (ddATP) led to the 3′-*endo* conformation like a nucleoside congener, while introduction to its β position led to the 2′-*endo* conformation [98]. In contrast, the 2′-*endo* conformation was predominant when the fluoro group was introduced to the C-3′-α position of 2′,3′-dideoxynucleosides [99]. Further conformational analysis of alovudine (3′-deoxy-3′-fluorothymidine, FLT) (see Figure 7.3), a potent anti-HIV agent, has revealed that approximately 90% of FLT is fixed to the 2′-*endo* conformation in aqueous solution [100]. Another conformational analysis of four monofluorinated analogues of 2′,3′-dideoxyuridine (i.e., 2′-α-F, 2′-β-F, 3′-α-F, and 3′-β-F analogues, see Figure 7.5) also revealed that the 2′-up and 3′-down analogues favor the 2′-*endo* conformation [101]. This preference can be explained by a *gauche* effect, in which a more electronegative substituent favors an axial orientation [15].

We now discuss the relationship between these conformations and anti-HIV activity. In 1989, Roey and co-workers reported the correlation between the sugar ring conforma-

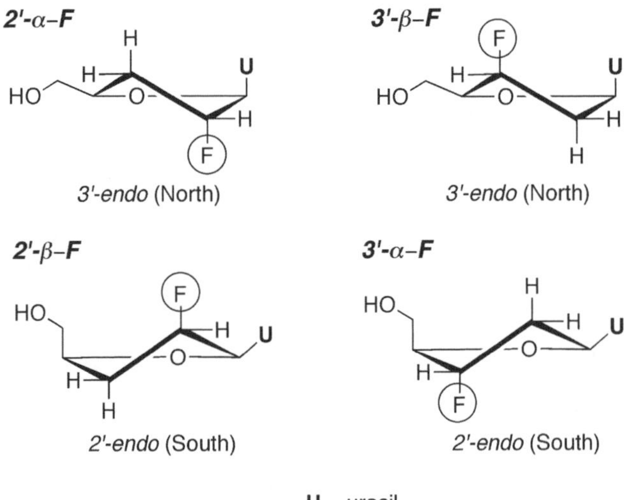

Figure 7.5 Sugar-puckering of four monofluorinated analogues of 2′,3′-dideoxyuridine.

tion and the anti-HIV activity of some nucleoside analogues. They discovered that the 2′-*endo* conformation is preferable for anti-HIV activity [102]. In the 2′-*endo* conformation, the C-5′ moiety takes an axial orientation, which directs its 5′-hydroxyl group to an advantageous position for the kinase-mediated phosphorylation process, or to a favorable position for its affinity with reverse transcriptase (RT). Mu and co-workers reported a molecular dynamics simulation of ternary complexes of an HIV-1-RT, a double-strand DNA, and 2′,3′-dideoxy-2′-fluoroadenosine 5′-triphosphate (FddATP), and discussed the activity of FddATPs (2′-α- or 2′-β-forms) as inhibitors [98]. Based on the preference of the RT to bind incoming dNTP (or ddNTP), a 3′-*endo* conformer (i.e., αFddATP) is believed to fit the polymerase site better than βFddATP. Interestingly, however, their study has revealed that Tyr115 of the RT, which appears to function as a steric gate against the incoming dNTP, prevents the effective binding of αFddATP, while the 2′-*endo* conformer, βFddATP, can pass the gate without hindrance to the RT. Consequently, the 2′-*endo* conformer, βFddATP, shows stronger anti-HIV activity, although this conformation shows a worse fit to the polymerase site of HIV-RT.

As mentioned above, nucleoside analogues have to clear many hurdles to exhibit biological activity. For example, those hurdles include phosphorylation by kinase, further conversion to triphosphate, penetration to the active site of the target enzyme through a steric gate, and strong affinity to the active site of the enzyme. Several dideoxynucleosides like ddI and ddC, which have been approved as anti-HIV agents, are flexible enough to clear all of these hurdles and show good anti-HIV activity. However, the introduction of a fluoro group to the sugar moiety of nucleoside brings about rigid 2′-*endo* or 3′-*endo* conformation, and such a fixed conformation makes it difficult to clear all of these hurdles in some cases. If these hurdles can be overcome, the many advantages of fluoronucleoside, such as chemical and pharmacological stability, could lead to stronger biological activity.

7.5 Conclusion

Since the discovery of nucleocidin, a large number of fluorinated nucleosides have been synthesized and have contributed to the development of biologically active agents. The practical use of gemcitabine as an anticancer agent and also the upcoming use of clevudine as an anti-HBV agent reflect this development. Early in this chapter, we described the syntheses and biological activities of 2′-deoxy-2′-α-fluoronucleosides and their oligomers. Such fundamental studies led to the use of fluorinated nucleosides in other fields, such as antisense and ribozymes, due to the chemical and biological stability of fluorinated nucleosides. Next, we described the synthesis of a potent antiviral agent, FddA, with special focus on synthetic issues (e.g., the difficulty of β-side-selective fluorination) and its process-scale synthesis. The possible process-scale synthesis of FddG was also described. The relationships between the sugar ring conformations and the biological activities of several fluorinated nucleosides were also discussed. We hope that this chapter provides a better understanding of the usefulness of fluorinated nucleosides as well as valuable references for their syntheses.

References

1. Morton, G. O., Lancaster, J. E., Van Lear, G. E., *et al.* (1969) Structure of nucleocidin. III. Revised structure. *J. Am. Chem. Soc.*, **91**, 1535–1537.
2. Heidelberger, C., Chaudhuri, N. K., Danneberg, P., *et al.* (1957) Fluorinated pyrimidines, a new class of tumour-inhibitory compounds. *Nature*, **179**(4561), 663–666.
3. Hertel, L. W., Kroin, J. S., Misner, J. W. and Tustin, J. M. (1988) Synthesis of 2-deoxy-2,2-difluoro-D-ribose and 2-deoxy-2,2-difluoro-D-ribofuranosyl nucleosides. *J. Org. Chem.*, **53**, 2406–2409.
4. Bonate, P. L., Arthaud, L., Cantrell, W. R., Jr., *et al.* (2006) Discovery and development of clofarabine: a nucleoside analogue for treating cancer. *Nat. Rev. Drug Discov.*, **5**, 855–863.
5. Korba, B. E., Furman, P. A. and Otto, M. J. (2006) Clevudine: a potent inhibitor of hepatitis B virus in vitro and in vivo. *Expert Rev. Anti-Infect. Ther.*, **4**, 549–561.
6. Billich, A. (1999) Lodenosine. *Curr. Opin. Anti-Infect. Invest. Drugs*, **1**, 179–185.
7. De Clercq, E. (2005) Emerging anti-HIV drugs. *Expert Opin. Emerg. Drugs*, **10**, 241–274.
8. Rusconi, S. (2003) Alovudine (Medivir). *Curr. Opin. Investig. Drugs (Thomson Current Drugs)* **4**, 219–223.
9. Takamatsu, S., Maruyama, T. and Izawa, K. (2005) Development of an industrial process for synthesizing lodenosine (FddA). *Yuki Gosei Kagaku Kyokaishi*, **63**, 864–878.
10. Izawa, K., Takamatsu, S., Katayama, S., *et al.* (2003) An industrial process for synthesizing lodenosine (FddA). *Nucleosides, Nucleotides Nucleic Acids*, **22**, 507–517.
11. Izawa, K., Torii, T., Onishi, T. and Maruyama, T. (2007) The synthesis of an antiviral fluorinated purine nucleoside: 3′-α-fluoro-2′,3′-dideoxyguanosine. *Current Fluoroorganic Chemistry, ACS Symposium Series* **949**, pp. 363–378.
12. Herdewijn, P., Van Aerschot, A. and Kerremans, L. (1989) Synthesis of nucleosides fluorinated in the sugar moiety. The application of diethylaminosulfur trifluoride to the synthesis of fluorinated nucleosides. *Nucleosides Nucleotides*, **8**, 65–96.
13. Pankeiwicz, K. W. (2000) Fluorinated nucleosides. *Carbohydr. Res.*, **327**, 87–105.
14. Pankeiwicz, K. W. and Watanabe, K. A. (1993) Synthesis of 2′-β-fluoro-substituted nucleosides by a direct approach. *J. Fluorine Chem.*, **64**, 15–36.
15. Saenger, W. (1984) *Pronciples of Nucleic Acid Structure*. Springer-Verlag, New York.
16. (a) Codington, J. F., Doerr, I. L. and Fox, J. J. (1964) Nucleosides. XVIII. Synthesis of 2′-fluorothymidine, 2′-fluorodeoxyuridine, and other 2′-halogeno-2′-deoxy nucleosides. *J. Org. Chem.*, **29**, 558–564. (b) Kowollik, G., Gaertner, K. and Langen, P. (1975) Nucleosides of fluorocarbohydrates. XIII. Synthesis of 3′-deoxy-3′-fluorouridine. *J. Carbohydr., Nucleosides, Nucleotides*, **2**, 191–195.
17. Misra, H. K., Gati, W. P., Knaus, E. E. and Wiebe, L. I. (1984) Reaction of 1-(2′,3′-epoxy-β-D-lyxofuranosyl)uracil with hydrogen fluoride. The unexpected formation of 1-(3′-fluoro-3′-deoxy-β-D-ribofuranosyl)uracil. *J. Heterocyclic Chem.*, **21**, 773–775.
18. (a) Ikehara, M. and Maruyama, T. (1976) Studies of nucleosides and nucleotides. LXIX. Purine cyclonucleosides. (30). Elimination of the 8-oxy function of purine nucleosides. *Chem. Pharm. Bull.*, **24**, 565–569. (b) Ikehara, M., Maruyama, T., Miki, H. and Takatsuka, Y. (1977) Studies of nucleosides and nucleotides. LXXXV. Purine cyclonucleosides. (35). Synthesis of purine nucleosides having 2′-azido and 2′-amino functions by cleavage of purine cyclonucleosides. *Chem. Pharm. Bull.*, **25**, 754–760.
19. Wagner, J., Verheyden, J. P. H. and Moffatt, J. G. (1974) Preparation and synthetic utility of some organotin derivatives of nucleosides. *J. Org. Chem.*, **39**, 24–30.
20. Ikehara, M. and Maruyama, T. (1975) Studies of nucleosides and nucleotides – LXV: Purine cyclonucleosides-26 a versatile method for the synthesis of purine *O*-cyclo-nucleosides. The

first synthesis of 8,2′-anhydro-8-oxy 9-β-D-arabinofuranosylguanine. *Tetrahedron*, **31**, 1369–1372.

21. Ikehara, M., Maruyama, T. and Miki, H. (1976) A new method for the synthesis of 2′-substituted purine nucleosides. Total synthesis of an antibiotic 2′-amino-2′-deoxyguanosine. *Tetrahedron Lett.*, **17**, 4485–4488.
22. Ikehara, M. and Miki, H. (1978) Studies of Nucleosides and Nucleotides. LXXXII. Cyclonucleosides. (39). Synthesis and Properties of 2′-Halogeno-2′-deoxyadenosines. *Chem. Pharm. Bull.*, **26**, 2449–2453.
23. Ikehara, M. and Imura, J. (1981) Studies on nucleosides and nucleotides. LXXXVII. Purine cyclonucleosides. XLII. Synthesis of 2′-deoxy-2′-fluoroguanosine. *Chem. Pharm. Bull.*, **29**, 1034–1038.
24. Ranganathan, R. (1977) Modification of the 2′-position of purine nucleosides: syntheses of 2′-α-substituted-2′-deoxyadenosine analogs. *Tetrahedron Lett.*, **18**, 1291–1294.
25. (a) McOmie, J. F. W. (1973) *Protective Groups in Organic Chemistry*. Plenum Press, New York. (b) Nishino, S., Takamura, H. and Ishido, Y. (1986) Regioselective protection of carbohydrate derivatives. Part 20. Simple, efficient 2′-*O*-deacylation of fully acylated purine and pyrimidine ribonucleosides through *tert*-butoxide. *Tetrahedron*, **42**, 1995–2004.
26. (a) Markiewicz, W. T. and Wiewiorowski, M. (1978) A new type of silyl protecting groups in nucleoside chemistry. *Nucleic Acids Res., Spec. Publ.*, **4**, 185–188. (b) Markiewicz, W. T. (1975) Tetraisopropyldisiloxane-1,3-diyl, a group for simultaneous protection of 3′- and 5′-hydroxy functions of nucleosides. *J. Chem. Res., Synopses*, 24–25.
27. Maruyama, T., Utzumi, K., Sato, Y. and Richman, D. D. (1994) Synthesis and anti-HIV activity of 6-substituted purine 2′-deoxy-2′-fluororibosides. *Nucleosides Nucleotides*, **13**, 527–537.
28. Maruyama, T., Utzumi, K., Sato, Y. and Richman, D. D. (1994) Synthesis and anti-HIV activity of 2-substituted 2′-deoxy-2′-fluoroadenosines. *Nucleosides Nucleotides*, **13**, 1219–1230.
29. Suhadolnik, R. J. (1979) *Nucleosides as Biological Probes*. John Wiley and Sons, Inc., New York, pp. 96–102.
30. Maruyama, T., Utsumi, K., Tomioka, H., *et al.* (1995) Synthesis, antiviral, antibacterial and antitumor cell activities of 2′-deoxy-2′-fluoropuromycin. *Chem. Pharm. Bull.*, **43**, 955–959.
31. Koizumi, F., Oritani, T. and Yamashita, K. (1990) Synthesis and antimicrobial activity of 2′-deoxypuromycin. *Agric. Biol. Chem.*, **54**, 3093–3097.
32. (a) Ledoux, L. and Baltus, E. (1954) The effect of ribonuclease on the cells of the Ehrlich carcinoma. *Experientia*, **10**, 500–501. (b) Youle, R. J. and D'Alessio, G. (1997) *Ribonucleases: Structures and Functions*. Academic Press, New York, pp. 491–514.
33. Nonaka, T., Nakamura, K. T., Uesugi, S., *et al.* (1993) Crystal structure of ribonuclease Ms (as a ribonuclease T1 homolog) complexed with a guanylyl-3′,5′-cytidine analog. *Biochemistry*, **32**, 11825–11837.
34. Ikehara, M. and Imura, J. (1981) Studies on nucleosides and nucleotides. LXXXIX. Purine cyclonucleosides. (43). Synthesis and properties of 2′-halogeno-2′-deoxyguanosines. *Chem. Pharm. Bull.*, **29**, 3281–3285.
35. Benseler, F., Williams, D. M. and Eckstein, F. (1992) Synthesis of suitably-protected phosphoramidites of 2′-fluoro-2′-deoxyguanosine and 2′-amino-2′-deoxyguanosine for incorporation into oligoribonucleotides. *Nucleosides, Nucleotides, Nucleic Acids*, **11**, 1333–1351.
36. Maruyama, T., Kozai, S., Nakamura, K. and Irie, M. (2002) Synthesis of 2′-deoxy-2′-fluoroguanyl-(3′,5′)-guanosine. *Nucleosides, Nucleotides, Nucleic Acids*, **21**, 765–774.
37. Honda, K., Komoda, Y., Nishida, S., *et al.* (1984) Uridine as an active component of sleep-promoting substance: its effects on nocturnal sleep in rats. *Neurosci. Res.*, **1**, 243–252.

38. Yamamoto, I., Kimura, T., Takeoka, Y., *et al.* (1987) *N*-Substituted oxopyrimidines and nucleosides: structure–activity relationship for hypnotic activity as CNS depressant. *J. Med. Chem.*, **30**, 2227–2231.
39. Sato, Y., Utsumi, K., Maruyama, T., *et al.* (1994) Synthesis and hypnotic and anti-human immunodeficiency virus-1 activities of N^3-Substituted 2′-Deoxy-2′-fluorouridines. *Chem. Pharm. Bull.*, **42**, 595–598.
40. Pankiewicz, K. W. and Watanabe, K. A. (1987) Nucleosides. CXLIII. Synthesis of 5′-deoxy-5′-substituted-2,2′-anhydro-1-(β-D-arabinofuranosyl)uracils. A new 2,5′- to 2,2′-anhydronucleoside transformation. Studies directed toward the synthesis of 2′-deoxy-2′-substituted arabino nucleosides. (4). *Chem. Pharm. Bull.*, **35**, 4494–4497.
41. Matsuda, A., Yasuoka, J. and Ueda, T. (1989) A new method for synthesizing the antineoplastic nucleosides 1-(2-azido-2-deoxy-β-D-arabinofuranosyl)cytosine (Cytarazid) and 1-(2-amino-2-deoxy-β-D-arabinofuranosyl)cytosine (Cytaramin) from uridine. *Chem. Pharm. Bull.*, **37**, 1659–1661.
42. Pankiewicz, K. W., Krzeminski, J., Ciszewski, L. A., *et al.* (1992) A synthesis of 9-(2-deoxy-2-fluoro-β -D-arabinofuranosyl)adenine and -hypoxanthine. An effect of C3′-endo to C2′-endo conformational shift on the reaction course of 2′-hydroxyl group with DAST. *J. Org. Chem.*, **57**, 553–559.
43. Krzeminski, J., Nawrot, B., Pankiewicz, K. W. and Watanabe, K. A. (1991) Synthesis of 9-(2-deoxy-2-fluoro-β-D-arabinofuranosyl)hypoxanthine. The first direct introduction of a 2′-β-fluoro substituent in preformed purine nucleosides. Studies directed toward the synthesis of 2′-deoxy-2′-substituted arabinonucleosides. 8. *Nucleosides Nucleotides*, **10**, 781–798.
44. Maruyama, T., Sato, Y., Oto, Y., *et al.* (1996) Synthesis and antiviral activity of 6-chloropurine arabinoside and its 2′-deoxy-2′-fluoro derivative. *Chem. Pharm. Bull.*, **44**, 2331–2334.
45. Uesugi, S., Miki, H., Ikehara, M., *et al.* (1979) A linear relationship between electronegativity of 2′-substituents and conformation of adenine nucleosides. *Tetrahedron Lett.*, **20**, 4073–4076.
46. Maruyama, T., Takamatsu, S., Kozai, S., *et al.* (1999) Synthesis of 9-(2-deoxy-2-fluoro-β-D-arabinofuranosyl)adenine bearing a selectively removable protecting group. *Chem. Pharm. Bull.*, **47**, 966–970.
47. Siddiqui, M. A., Driscoll, J. S. and Marquez, V. E. (1998) A new synthetic approach to the clinically useful, anti-HIV-active nucleoside, 9-(2,3-dideoxy-2-fluoro-β-D-*threo*-pentofuranosyl)adenine (β-FddA). Introduction of a 2′-β-fluoro substituent via inversion of a readily obtainable 2′-α-fluoro isomer. *Tetrahedron Lett.*, **39**, 1657–1660.
48. Marquez, V. E., Tseng, C. K.-H., Kelley, J. A., *et al.* (1987) 2′,3′-Dideoxy-2′-fluoro-ara-A. An acid-stable purine nucleoside active against human immunodeficiency virus (HIV). *Biochem. Pharmacol.*, **36**, 2719–2722.
49. Marquez, V. E., Tseng, C. K.-H., Mitsuya, H., *et al.* (1990) Acid-stable 2′-fluoro purine dideoxynucleosides as active agents against HIV. *J. Med. Chem.*, **33**, 978–985.
50. Ruxrungtham, K., Boone, E. B., Ford, H., Jr., *et al.* (1996) Potent activity of 2′-β-fluoro-2′,3′-dideoxyadenosine against human immunodeficiency virus type 1 infection in hu-PBL-SCID mice. *Antimicrob. Agents Chemother.*, **40**, 2369–2374.
51. Graul, A., Silvestre, J. and Castaner, J. (1998) Lodenosine: anti-HIV (reverse transcriptase inhibitor). *Drugs Future*, **23**, 1176–1189.
52. Driscoll, J. S., Mayers, D. L., Bader, J. P., *et al.* (1997) 2′-Fluoro-2′,3′-dideoxyarabinosyladenine (F-ddA): activity against drug-resistant human immunodeficiency virus strains and clades A-E. *Antivir. Chem. Chemother.*, **8**, 107–111.
53. Tanaka, M, Srinivas, R. V., Ueno, T., *et al.* (1997) *In vitro* induction of human immunodeficiency virus type 1 variants resistant to 2′-β-fluoro-2′,3′-dideoxyadenosine. *Antimicrob. Agents Chemother.*, **41**, 1313–1318.

54. Singhal, D., Morgan, M. E. and Anderson, B. D. (1997) Role of brain tissue localized purine metabolizing enzymes in the central nervous system delivery of anti-HIV agents 2′-β-fluoro-2′,3′-dideoxyinosine and 2′-β-fluoro-2′,3′-dideoxy-adenosine in rats. *Pharmaceutical Res.*, **14**, 786–792.
55. Driscoll, J. S., Siddiqui, M. A., Ford, H., Jr., *et al.* (1996) Lipophilic, acid-stable, adenosine deaminase-activated anti-hiv prodrugs for central nervous system delivery. 3. 6-Amino prodrugs of 2′-β-fluoro-2′,3′-dideoxyinosine. *J. Med. Chem.*, **39**, 1619–1625.
56. Johnson, M. D. and Anderson, B. D. (1996) Localization of purine metabolizing enzymes in bovine brain microvessel endothelial cells: an enzymic blood–brain barrier for dideoxynucleosides? *Pharm. Res.*, **13**, 1881–1886.
57. Ford, H., Jr., Siddiqui, M., Driscoll, J. S., *et al.* (1995) Lipophilic, acid-stable, adenosine deaminase-activated anti-HIV prodrugs for central nervous system delivery. 2. 6-Halo-and 6-alkoxy prodrugs of 2′-β-fluoro-2′,3′-dideoxyinosine. *J. Med. Chem.*, **38**, 1189–1195.
58. Johnson, M. D., Chen, J. and Anderson, B. D. (2002) Investigation of the mechanism of enhancement of central nervous system delivery of 2′-β-fluoro-2′,3′-dideoxyinosine via a blood–brain barrier adenosine deaminase-activated prodrug. *Drug Metab. Dispos.*, **30**, 191–198.
59. Herdewijn, P., Pauwels, R., Baba, M., *et al.* (1987) Synthesis and anti-HIV activity of various 2′-and 3′-substituted 2′,3′-dideoxyadenosines: a structure–activity analysis. *J. Med. Chem.*, **20**, 2131–2137.
60. Marquez, V. E. (1989) Design, synthesis, and antiviral activity of nucleoside and nucleotide analogs. *Nucleotide Analogues Antiviral Agents, ACS Symposium Series* **401**, pp. 140–155.
61. Wysocki, R. J., Jr., Siddiqui, M. A., Barchi, J. J., Jr., *et al.* (1991) A more expedient approach to the synthesis of anti-HIV-active 2,3-dideoxy-2-fluoro-β-D-threo-pentofuranosyl nucleosides. *Synthesis*, 1005–1008.
62. Siddiqui, M. A., Marquez, V. E., Driscoll, J. S. and Barchi, J. J., Jr. (1994) A diastereo-selective synthesis of (*S,S*)-α-fluoro-2,2-dimethyl-1,3-dioxolane-4-propanoic acid methyl ester, a key intermediate for the preparation of anti-HIV effective fluorodideoxy nucleosides. *Tetrahedron Lett.*, **35**, 3263–3266.
63. Siddiqui, M. A., Driscoll, J. S., Abushanab, E., *et al.* (2000) The "β-fluorine effect" in the non-metal hydride radical deoxygenation of fluorine-containing nucleoside xanthates. *Nucleosides, Nucleotides, Nucleic Acids*, **19**, 1–12.
64. Caille, J.-C., Miel, H., Armstrong, P. and McKervey, M. A. (2004) A new synthetic approach to 2,3-dideoxy-2-fluoro-β-D-threo-pentofuranose, the fluorofuranose unit of the anti-HIV-active nucleoside, β-FddA. *Tetrahedron Lett.*, **45**, 863–865.
65. Shanmuganathan, K., Koudriakova, T., Nampalli, S., *et al.* (1994) Enhanced brain delivery of an anti-HIV nucleoside 2′-F-ara-ddI by xanthine oxidase mediated biotransformation. *J. Med. Chem.*, **37**, 821–827.
66. Jin, F., Wang, D., Confalone, P. N., *et al.* (2001) (2*R*,3*S*,5*S*)-2-Acetoxy-3-fluoro-5-(*p*-toluoyloxymethyl) tetrahydrofuran: a key intermediate for the practical synthesis of 9-(2,3-dideoxy-2-fluoro-β-D-*threo*-pentofuranosyl)adenine (FddA). *Tetrahedron Lett.*, **42**, 4787–4789.
67. Choudhury, A., Jin, F., Wang, D., *et al.* (2003) A concise synthesis of anti-viral agent F-ddA, starting from (*S*)-dihydro-5-(hydroxymethyl)-2(3*H*)-furanone. *Tetrahedron Lett.*, **44**, 247–250.
68. Takamatsu, S., Maruyama, T., Katayama, S., *et al.* (2001) Improved synthesis of 9-(2,3-dideoxy-2-fluoro-β-D-*threo*-pentofuranosyl)adenine (FddA) using triethylamine trihydrofluoride. *Tetrahedron Lett.*, **42**, 2321–2324.
69. Barton, D. H. R., Jang, D. O. and Jaszberenyi, J. C. (1993) The invention of radical reactions. Part 32. Radical deoxygenations, dehalogenations, and deaminations with dialkyl phosphites and hypophosphorous acid as hydrogen sources. *J. Org. Chem.*, **58**, 6838–6842.

70. Takamatsu, S., Katayama, S., Hirose, N., *et al.* (2001) Radical deoxygenation and dehalogenation of nucleoside derivatives with hypophosphorous acid and dialkyl phosphites. *Tetrahedron Lett.*, **42**, 7605–7608.
71. Shiragami, H., Tanaka, Y., Uchida, Y., *et al.* (1992) A novel method for the synthesis of ddA and F-ddA via regioselective 2′-*O*-deacetylation of 9-(2,5-di-*O*-acetyl-3-bromo-3-deoxy-β-D-xylofuranosyl)adenine. *Nucleosides Nucleotides*, **11**, 391–400.
72. Shiragami, H., Amino, Y., Honda, Y., *et al.* (1996) Synthesis of 2′,3′-dideoxypurinenucleosides via the palladium catalyzed reduction of 9-(2,5-di-*O*-acetyl-3-bromo-3-deoxy-β-D-xylofuranosyl)purine derivatives. *Nucleosides Nucleotides*, **15**, 31–45.
73. (a) Takamatsu, S., Maruyama, T., Katayama, S., *et al.* (2001) Practical synthesis of 9-(2,3-dideoxy-2-fluoro-β-D-*threo*-pentofuranosyl) adenine (FddA) via a purine 3′-deoxynucleoside. *Tetrahedron Lett.*, **42**, 2325–2328. (b) Takamatsu, S., Maruyama, T., Katayama, S., *et al.* (2001) Synthesis of 9-(2,3-dideoxy-2-fluoro-β-D-*threo*-pentofuranosyl)adenine (FddA) via a purine 3′-deoxynucleoside. *J. Org. Chem.*, **66**, 7469–7477.
74. Takamatsu, S., Katayama, S., Hirose, N., *et al.* (2002) Convenient synthesis of fluorinated nucleosides with perfluoroalkanesulfonyl fluorides. *Nucleosides, Nucleotides, Nucleic Acids*, **21**, 849–861.
75. Balzarini, J., Baba, M., Pauwels, R., *et al.* (1988) Potent and selective activity of 3′-azido-2,6-diaminopurine-2′,3′-dideoxyriboside, 3′-fluoro-2,6-diaminopurine-2′,3′-dideoxyriboside, and 3′-fluoro-2′,3′-dideoxyguanosine against human immunodeficiency virus. *Mol. Pharmacol.*, **33**, 243–249.
76. Herdewijn, P., Balzarini, J., Baba, M., *et al.* (1988) Synthesis and anti-HIV activity of different sugar-modified pyrimidine and purine nucleosides. *J. Med. Chem.*, **31**, 2040–2048.
77. Schröder, I., Holmgren, B., Öberg, M. and Löfgren, B. (1998) Inhibition of human and duck hepatitis B virus by 2′,3′-dideoxy-3′-fluoroguanosine *in vitro*. *Antivir. Res.*, **37**, 57–66.
78. Hafkemeyer, P., Keppler-Hafkemeyer, A., al Haya, M. A., *et al.* (1996) Inhibition of duck hepatitis B virus replication by 2′,3′-dideoxy-3′-fluoro- guanosine *in vitro* and *in vivo*. *Antimicrob. Agents Chemother.* **40**, 792–794.
79. Jacquard, A.-C., Brunelle, M.-N., Pichoud, C., *et al.* (2006) *In vitro* characterization of the anti-hepatitis B virus activity and cross-resistance profile of 2′,3′-dideoxy-3′-fluoroguanosine. *Antimicrob. Agents Chemother.*, **50**, 955–961.
80. Marchand, A. Mathé, C., Imbach, J.-L. and Gosselin, G. (2000) Synthesis and antiviral evaluation of unnatural β-L-enantiomers of 3′-fluoro- and 3′-azido-2′,3′-dideoxy- guanosine derivatives. *Nucleosides, Nucleotides Nucleic Acids*, **19**, 205–217.
81. Chun, B. K., Schinazi, R. F., Cheng, Y. C. and Chu, C. K. (2000) Synthesis of 2′,3′-dideoxy-3′-fluoro-L-ribonucleosides as potential antiviral agents from D-sorbitol. *Carbohydr. Res.*, **328**, 49–59.
82. (a) Komatsu, H., Awano, H., Tanikawa, H., *et al.* (2001) Large-scale manufacturing of all four 2′-deoxynucleosides via novel strategies including a chemo-enzymatic process. *Nucleosides, Nucleotides Nucleic Acids*, **20**, 1291–1293. (b) Komatsu, H. and Araki, T. (2003) Chemo-enzymatic synthesis of 2′,3′-dideoxy-3′- fluoro-β-D-guanosine via 2,3-dideoxy-3-fluoro-α-D-ribose 1-phosphate. *Tetrahedron Lett.*, **44**, 2899–2901.
83. Torii, T., Onishi, T., Izawa, K., *et al.* (2006) Synthesis of 6-arylthio analogs of 2′,3′-dideoxy-3′-fluoroguanosine and their effect against hepatitis B virus replication. *Nucleosides, Nucleotides Nucleic Acids*, **25**, 655–665.
84. Torii, T., Onishi, T., Tanji, S. and Izawa, K. (2005) Synthesis study of 3′-α-fluoro -2′,3′-dideoxyguanosine. *Nucleosides, Nucleotides Nucleic Acids*, **24**, 1051–1054.
85. Ishii, A., Otsuka, T., Kume, K., *et al.* (2006) Regioselective preparation of 2-halo-2,3-dideoxy-3-fluoro-β-D-arabinofuranosyl nucleosides and their dehalogenation products. *JP Patent* 2006022009.

86. Torii, T., Onishi, T., Izawa, K. and Maruyama, T. (2006) A concise synthesis of 3′-α-fluoro-2′,3′-dideoxyguanosine (FddG) via 3′-α-selective fluorination of 8,2′-thioanhydronucleoside. *Tetrahedron Lett.*, **47**, 6139–6141.
87. Cushley, R. J., Codington, J. F. and Fox, J. J. (1968) Nucleosides. XLIX. Nuclear magnetic resonance studies of 2′-and 3′-halogeno nucleosides. The conformations of 2′-deoxy-2′-fluorouridine and 3′-deoxy-3′-fluoro-β-D-arabinofuranosyluracil. *Can. J. Chem.*, **46**, 1131–1140.
88. Guschlbauer, W. and Krzysztof, J. (1980) Nucleoside conformation is determined by the electronegativity of the sugar substituent. *Nucleic Acids Res.*, **8**, 1421–1433.
89. (a) Hakoshima, T., Omori, H., Tomita, K., *et al.* (1981) The crystal and molecular structure of 2′-deoxy-2′-fluoroinosine monohydrate. *Nucleic Acids Res.*, **9**, 711–729. (b) Uesugi, S., Miki, H. and Ikehara, M. (1981) Studies on nucleosides and nucleotides. LXXXVIII. Purine cyclonucleosides. XLIII. ^{13}C NMR spectra of 2′-substituted 2′-deoxyadenosines. Substituent effects on the chemical shifts in the furanose ring system. *Chem. Pharm. Bull.*, **29**, 2199–2204.
90. Uesugi, S., Takatsuka, Y., Ikehara, M., *et al.* (1981) Synthesis and characterization of the dinucleoside monophosphates containing 2′-fluoro-2′-deoxyadenosine. *Biochemistry*, **20**, 3056–3062.
91. Chandra, P., Demirhan, I. and De Clercq, E. (1981) A study of antitemplate inhibition of mammalian, bacterial and viral DNA polymerases by 2- and 2′-substituted derivatives of polyadenylic acid. *Cancer Lett.*, **12**, 181–193.
92. (a) Fukui, T., De Clerq, E., Kakiuchi, N. and Ikehara, M. (1982) Template activity of poly(2′-fluoro-2′-deoxyinosinic acid) for murine leukemia virus reverse transcriptase. *Cancer Lett.*, **16**, 129–135. (b) Aoyama, H., Sarih-Cottin, L., Tarrago-Litvak, L., *et al.* (1985) 2′-Fluoro-2′-deoxypolynucleotides as templates and inhibitors for RNA- and DNA-dependent DNA polymerases. *Biochim. Biophys. Acta*, **824**, 225–232.
93. Fukui, T., Kakiuchi, N. and Ikehara, M. (1982) Protein synthesis using poly(2′-halogeno-2′-deoxyadenylic acids) as messenger. *Biochim. Biophys. Acta*, **697**, 174–177.
94. Aurup, H., Siebert, A., Benseler, F., *et al.* (1994) Translation of 2′-modified mRNA *in vitro* and *in vivo*. *Nucleic Acids Res.*, **22**, 4963–4968.
95. De Clercq, E., Stollar, B. D., Hobbs, J., *et al.* (1980) Interferon induction by two 2′-modified double-helical RNAs, poly(2′-nuoro-2′-deoxymosimc acid) · poly(cytidylic acid) and poly(2′-chloro-2′-deoxyinosinic acid) · poly(cytidylic acid). *Eur. J. Biochem.*, **107**, 279–288.
96. (a) Tuttle, J. V., Tisdale, M. and Krenitsky, T. A. (1993) Purine 2′-deoxy-2′-fluororibosides as antiinfluenza virus agents. *J. Med. Chem.*, **36**, 119–125. (b) Tisdale, M., Appleyard, G., Tuttle, *et al.* (1993) Inhibition of influenza A and B viruses by 2′-deoxy-2′-fluororibosides. *Antiviral Chem. Chemother.*, **4**, 281–287.
97. (a) Clark, J. L., Hollecker, L., Mason, J. C., *et al.* (2005) Design, synthesis, and antiviral activity of 2′-deoxy-2′-fluoro-2′-*C*-methylcytidine, a potent inhibitor of hepatitis C virus replication. *J. Med. Chem.*, **48**, 5504–5508. (b) Murakami, E., Bao, H., Ramesh, M., *et al.* (2007) Mechanism of activation of β-D-2′-deoxy-2′-fluoro-2′-*C*-methylcytidine and inhibition of hepatitis C virus NS5B RNA polymerase. *Antimicrob. Agents Chemother.* **51**, 503–509.
98. Mu, L., Sarafianos, S. G., Nicklaus, M. C., *et al.* (2000) Interactions of conformationally biased north and south 2′-fluoro-2′,3′-dideoxynucleoside 5′-triphosphates with the active site of HIV-1 reverse transcriptase. *Biochemistry*, **39**, 11205–11215.
99. Khripach, N. B., Pupeiko, N. E., Skorynin, I. Yu. and Borisov, E. V. (1991) Conformational characteristics of 3′-deoxy-3′-fluoro- and 2′,3′-dideoxy-3′-fluororibonucleosides. *Bioorganicheskaya Khimiya*, **17**, 1521–1525.
100. Plavec, J., Koole, L. H., Sandström, A. and Chattopadhyaya, J. (1991) Structural studies of anti-HIV 3′-α-fluorothymidine and 3′-α-azidothymidine by 500 MHz ^{1}H-NMR spectroscopy and molecular mechanics (MM2) calculations. *Tetrahedron*, **47**, 7363–7376.

101. Barchi, J. J., Jr, Jeong, L.-S., Siddiqui, M. A. and Marquez, V. E. (1997) Conformational analysis of the complete series of 2′ and 3′ monofluorinated dideoxyuridines. *J. Biochem. Biophys. Methods*, **34**, 11–29.
102. Van Roey, P., Salerno, J. M., Chu, C. K. and Schinazi, R. F. (1989) Correlation between preferred sugar ring conformation and activity of nucleoside analogues against human immunodeficiency virus. *Proc. Natl. Acad. Sci. USA*, **86**, 3929–3933.

Synthetic Methods for Medicinal Chemistry and Chemical Biology

8

Synthesis of *gem*-Difluoromethylenated Nucleosides via *gem*-Difluoromethylene-containing Building Blocks

Wei-Dong Meng and Feng-Ling Qing

8.1 Introduction

Nucleosides have a very important place in medicinal chemistry as the structural basis for the development of antiviral and antitumor agents. Fluorine-containing nucleosides have attracted special attention because of their many unique properties [1]. Fluorine has been suggested to be an isopolar and isosteric substituent for oxygen [2]. Since 1-(2-deoxy-2,2-difluoro-β-D-*arabino*-furanosyl)cytosine (gemcitabine, gemzar or dFdC) was approved by FDA for treatment of inoperable pancreatic cancer and of 5-fluorouracil-resistant pancreatic cancer [3], considerable efforts have been focused on the synthesis and biological evaluation of the nucleosides containing CF_2 groups at the sugar moiety. The introduction of a *gem*-difluoromethylene group on the sugar moiety normally can be achieved by direct difluorination of keto groups at the sugar moiety or by using *gem*-difluoromethylene-containing building blocks. Here we describe recent advances in the development of *gem*-difluoromethylene-containing building blocks and their applications to the syntheses of nucleosides bearing a *gem*-difluoromethylene group at the sugar moiety in our laboratory.

Fluorine in Medicinal Chemistry and Chemical Biology Edited by Iwao Ojima

8.2 Synthesis of *gem*-Difluoromethylenated Nucleosides via *gem*-Difluorohomoallyl Alcohols

With inspiration from extensive studies on the synthesis and biological activities of 2′-deoxy-2′,2′-difluoronucleosides, 3′-deoxy-3′,3′-difluoro-D-*arabino*-furanosyl nucleosides were stereoselectively synthesized, featuring a novel and efficient strategy to prepare 3-deoxy-3,3-difluoro-D-*arabino*-furanose **7** via chiral *gem*-difluorohomoallyl alcohol **2** [4]. In the presence of indium powder, the reaction of (*R*)-glyceraldehyde acetonide **1** with 3-bromo-3,3-difluoropropene gave difluorohomoallyl alcohol **2** in 90% yield as a mixture of two diastereoisomers in a ratio of 7.7 : 1. Without separation, homoallyl alcohol **2** was subjected to benzylation. When 1.6 equivalents of NaH were used, benzylated compound **3** was obtained as a 5.7 : 1 mixture of *anti*- and *syn*-isomers in 93% yield. The C-3 position was slightly epimerized under such basic conditions. Accordingly, with the use of 0.8 equivalents of NaH, a kinetic resolution benzylation was achieved to give **3** with the *anti*/*syn* ratio of 21.8 : 1. The osmium-catalyzed dihydroxylation of compound **3** gave diols **4** and **5** as a 1 : 1 mixture in 95% yield. Compound **3** was subjected to Sharpless asymmetric dihydroxylation using $(DHQ)_2PYR$ as the ligand to give diol **5** together with diol **4** in a ratio of 4.4 : 1. These two diols were easily separated by column chromatography. Selective benzoylation of the primary hydroxyl group of diol **5** gave benzoate **6** in 90% yield, which was converted to furanose **7** by acidic hydrolysis of the isopropylidene group, followed by the oxidative scission of the resulting diol with sodium periodate. The stereochemistry at the C-2 position of furanose **7** was confirmed to have the *arabino* configuration. Thus, the stereoselective synthesis of 3-deoxy-3,3-difluoro-D-*arabino*-furanose **7** was achieved in five steps from chiral aldehyde **1** (see Scheme 8.1). Furanose **7** was then subjected to acetylation with acetic anhydride to give acetate **8** practically as the single product in α-form. The acetate **8** was coupled with silylated N^4-benzoylcytosine in refluxing acetonitrile to give the protected nucleosides **9a** and **9b** as a 1 : 1 mixture in 70% yield. These two isomers were separated by column chromatography. The protecting groups of **9a** and **9b** were removed by the standard hydrogenolysis procedures. Thus, the α-anomer **9a** gave difluoromethylenated nucleoside **10a** accompanied by the over-reduction product **11**, while the β-anomer **9b** gave the corresponding difluorinated nucleoside **10b** smoothly (see Scheme 8.1).

On the other hand, diol **4** was used to synthesize L-β-3′-deoxy-3′,3′-difluoronucleosides (see Scheme 8.2) [5]. Selective benzoylation of the primary hydroxyl group in diol **4** gave benzoate **12**, which was converted to furanose **13** through hydrolysis with acetic acid and oxidation with sodium periodate, followed by cyclization. Furanose **13** is a 1 : 1 mixture of two diastereoisomers based on the ^{19}F NMR analysis. Without separation of diastereomers, *O*-acetylation of furanose **13** by acetic anhydride afforded the β-anomer of acetate **14** almost exclusively through effective neighboring-group participation. According to the proposed mechanism by Vorbrüggen [6], the glycosylation reaction of a furanose with an acyloxy group at the C-2′ position should involve an oxonium intermediate, which would favor the attack of a nucleoside from the opposite face of the C-2′ acyloxy group. Thus, the benzyloxy group at the C-2′ position of **14** was converted to a benzoyloxy group to promote the formation of the desired β anomer in glycosylation reaction. Acetate **15** was thus obtained through debenzylation of **14** followed by benzoylation of the resulting

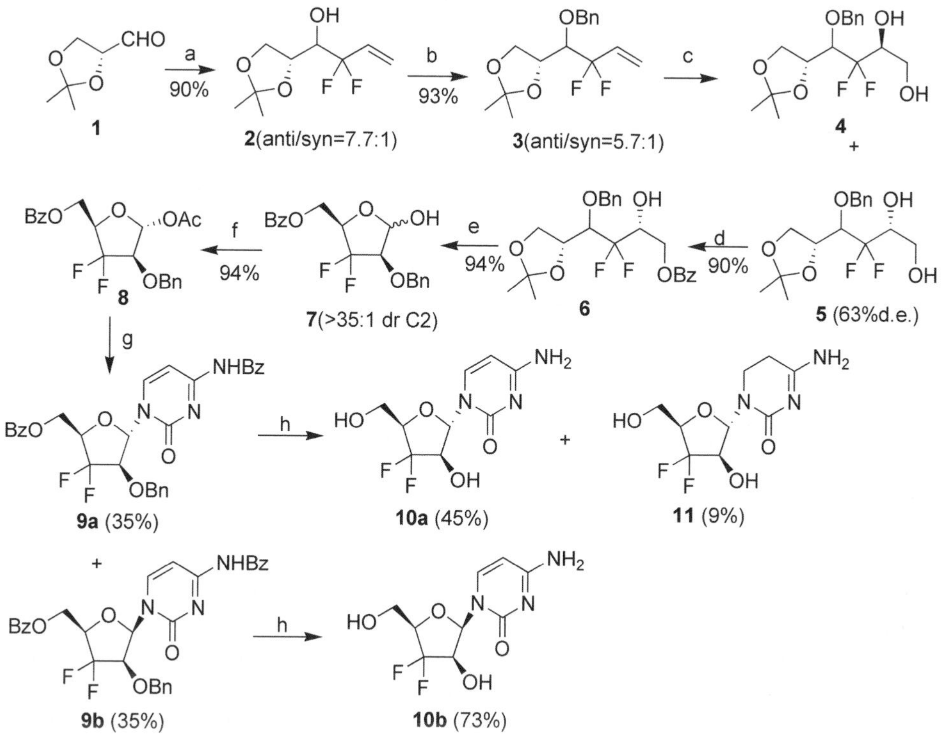

Scheme 8.1 *Reagents and conditions: (a) CH_2=$CHCF_2Br$, In powder; (b) NaH (1.6eq.), BnBr, TBAI, THF; (c) $(DHQ)_2PYR$, $K_2OsO_2(OH)_4$, $K_3Fe(CN)_6$, K_2CO_3; (d) BzCl, Py, CH_2Cl_2, −78 °C; (e) (i)75% HOAc, 50 °C; (ii) $NaIO_4$, acetone/H_2O; (f) Ac_2O, DMAP; (g) N^4-benzoylcytosine, N,O-bis(trimethylsilyl)acetamide, TMSOTf, CH_3CN, 0–80 °C; (h) (i) Sat. NH_3/MeOH, r.t.; (ii) $Pd(OH)_2$/C, H_2, MeOH/cyclohexane, r.t.*

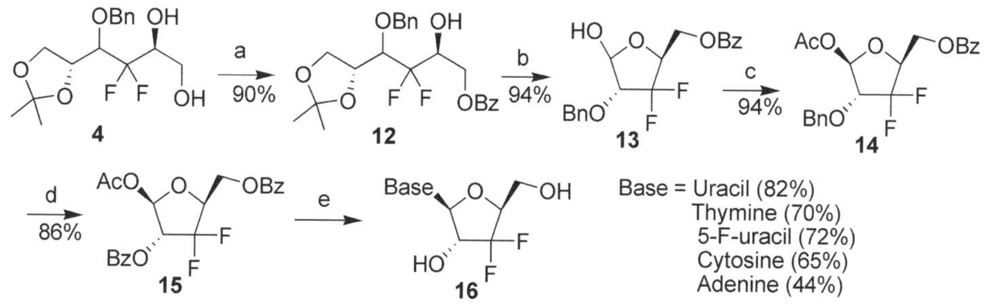

Scheme 8.2 *Reagents and conditions: (a) BzCl, Py, CH_2Cl_2; (b) (i) 75% HOAc, 50 °C; (ii) $NaIO_4$, acetone/H_2O, r.t.; (c) Ac_2O, DMAP, CH_2Cl_2; (d) (i) $NaBrO_3$, $Na_2S_2O_4$, EtOAc/H_2O; (ii) Bz_2O, DMAP, Et_3N, CH_2Cl_2; (e) (i) N,O-bis(trimethylsilyl)acetamide, pyrimidine, or 6-chloropurine, TMSOTf, CH_3CN; (ii) NH_3, MeOH.*

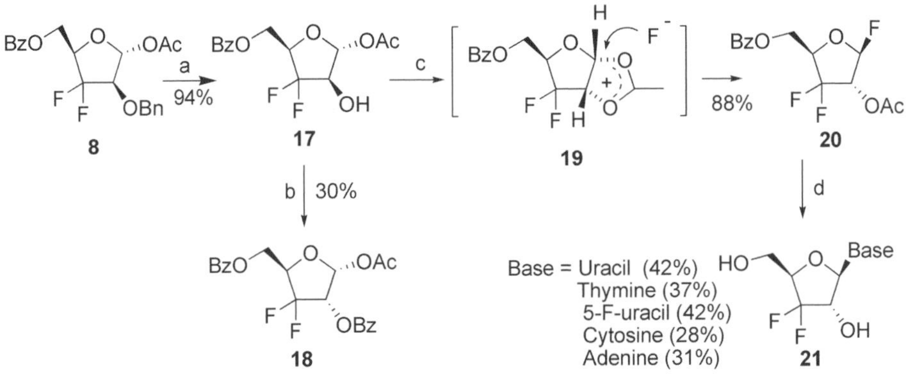

Scheme 8.3 *Reagents and conditions: (a) $NaBrO_3$, $Na_2S_2O_4$, EtOAc/H_2O; (b) (i) Tf_2O, pyridine, CH_2Cl_2; (ii) $PhCO_2Na$, toluene, 60 °C; (c) Deoxo-Fluor, toluene; (d) (i)* N,O*-bis(trimethylsilyl)acetamide, pyrimidine, or 6-chloropurine, TMSOTf, CH_3CN; (ii) NH_3, MeOH.*

alcohol. Acetate **15** was then coupled with various persilylated pyrimidines and 6-chloropurine in refluxing acetonitrile using Vorbrüggen conditions [7] to give L-nucleosides **16**. As expected, the coupling reactions of **16** with pyrimidine bases afforded only β-anomers through efficient neighboring-group participation. In the case of 6-chloropurine, the reaction afforded a separable mixture of β- and α-anomers in a ratio of 4.4 : 1 (see Scheme 8.2).

As described above, furanose **8** was used to synthesize D-*arabino* nucleosides, **10a** and **10b** via **9a** and **9b** in nonstereoselective manner (see Scheme 8.1). In order to synthesize D-β-3′-deoxy-3′,3′-difluoronucleosides **21**, it was necessary to invert the configuration at the C-2′ position (see Scheme 8.3). Thus, furanose **8** was debenzylated with $NaBrO_3$/$Na_2S_2O_4$ to give furanose **17**, which was treated with trifluoromethanesulfonic anhydride to give the corresponding triflate, followed by reaction with sodium benzoate to afford benzoate **18**. Compound **18** was obtained with inversion of configuration at C-2′, but the yield was low (30%). Fortunately, however, we serendipitously found that the reaction of compound **17** with bis(2-methoxyethyl)aminosulfur trifluoride (Deoxo-Fluor) did not proceed through the expected fluoride substitution of the C-2′-hydroxyl group, but gave acetate **20,** in which the configuration at C-2′ was inverted during the reaction, in high yield. It is very likely that the reaction involves oxonium ion intermediate **19**, and the subsequent nucleophilic attack of fluoride ion from the β-face. The coupling of acetate **20** with various persilylated pyrimidines and 6-chloropurine afforded the β-anomers of D-β-3′-deoxy-3′,3′-difluoronucleosides **21** (see Scheme 8.3) [5].

Direct difluorination of the keto groups of carbohydrates can yield the corresponding difluoromethylenated sugar moieties. However, the reaction could be complicated by neighboring group participation, group migration and elimination reactions [8]. To circumvent anticipated complications, we developed an efficient synthetic route to 3-deoxy-3,3-difluoro-D-ribohexose **26** from difluoro-diol **4** (see Scheme 8.4) [9]. The two hydroxyl groups of **4** were protected as *tert*-butyldimethylsilyl ether and the acetonide moiety of **4** was removed with $SnCl_2$ to give diol **22**. Selective benzoylation of the primary hydroxyl

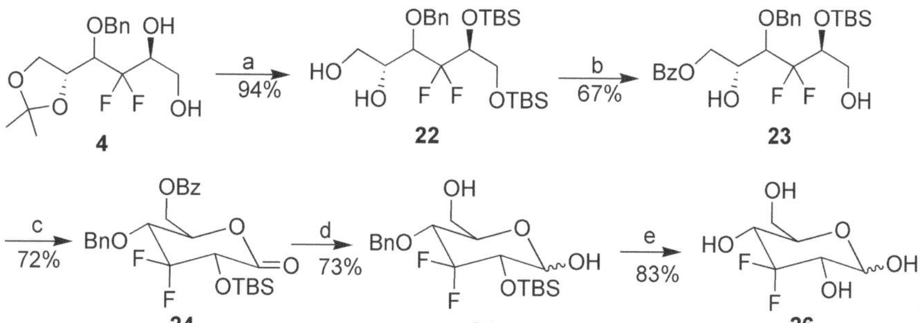

Scheme 8.4 Reagents and conditions: (a) (i) TBSCl, DMAP, imidazole; (ii) $SnCl_2$; (b) (i) BzCl, Py; (ii) TsOH, H_2O; (c) TCCA, TEMPO; (d) DIBAL-H; (e) (i) 4 M HCl; (ii) H_2, Pd/C.

group of **22**, followed by selective removal of the TBS group on the primary hydroxyl group gave 1,5-diol **23**. The reaction of **23** with 2 equivalents of trichloroisocyanuric acid in the presence of a catalytic amount of TEMPO afforded lactone **24** as the single product in 72% yield. The reduction of **24** with DIBAL-H gave lactol **25** as a 4 : 1 mixture of anomers. After deprotection of silyl and benzyl groups, difluoro-D-ribohexose **26** was obtained (4 : 1 mixture of anomers), which can be used for the synthesis of difluorinated pyranosyl nucleosides (see Scheme 8.4).

Nucleosides containing sulfur atoms instead of lactol oxygen have attracted special attentions for their potent biological activities [10]. Difluorofuranose **7** was used to synthesize difluoro-thionucleosides (see Scheme 8.5) [4]. Ring opening of **7** with $NaBH_4$ afforded diol **27** quantitatively, which was converted to thiofuranose **28** in three steps. Benzoylation of thiofuranose **28** gave **29** in quantitative yield. Oxidation of **29** with *m*-CPBA afforded sulfoxide **30**. Surprisingly, the coupling of **30** with a silylated cytosine gave nucleoside **31** (47%), in which the based was attached to the C-1′ position, along with the β-elimination product **32** (22%). Thiofuranose **28** was deoxylated to give thiofuranose **34**, which was treated in the same manner as that for **29** to give the deoxythionucleoside **35** (38%) accompanied by the elimination product **36.** It is apparent that the regiochemistry of these Pummerer reactions was determined by the acidity of the α-proton at C-1′ enhanced by the strongly electron-withdrawing C-2′-difluoromethylene group of thiofuranose **28** (see Scheme 8.5).

In view of the undesirable regiochemistry mentioned above, compound **6** was used to synthesize L-β-3′-deoxy-3′,3′-difluoro-4′-thionucleosides **41** (see Scheme 8.6) [11]. Triflation of **6**, followed by treatment with 5.4 equivalents of AcSH/CsF furnished thioacetate **38** in 86% yield. Hydrolysis of **38** with TFA, followed by oxidation with $NaIO_4$ gave an aldehyde, which was treated *in situ* with acidic methanol to afford thiofuranose **39** as the sole product in 64% yield in a stereospecific manner. Thiofuranose **39** was converted to acetate **40** only in the β-form, which was coupled with persilylated pyrimidine under the standard conditions to give thionucleosides **41**, which are the thio-containing analogues of gemcitabine (see Scheme 8.6).

gem-Difluorohomoallyl alcohol **2** was used to synthesize 2′,3′-dideoxy-6′,6′-difluoro-3′-thionucleosides **46** and **47** (see Scheme 8.7) [12]. Compound **2** was converted to its

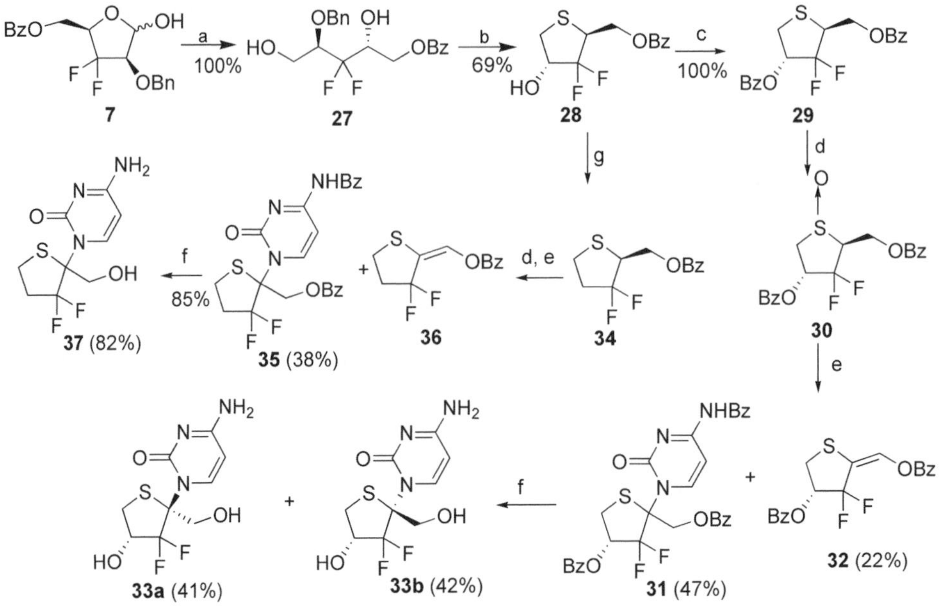

Scheme 8.5 Reagents and conditions: (a) $NaBH_4$, MeOH, 0 °C; (b) (i) MsCl, pyridine; (ii) Na_2S, DMF; (iii) BCl_3, MeOH; (c) Bz_2O, Et_3N, DMAP; (d). m-CPBA, CH_2Cl_2, −40 °C; (e) silylated N-benzoylcytosine, TMSOTf, DCE; (f). sat. NH_3/MeOH, r.t.; (g) (i) PhOC(S)Cl, DMAP, CH_3CN; (ii) Bu_3SnH, AIBN, toluene, 80 °C.

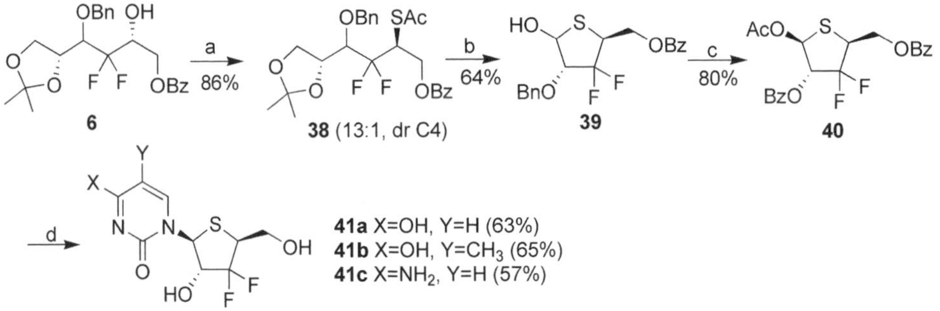

Scheme 8.6 Reagents and conditions: (a) (i)Tf_2O, Py; (ii) AcSH (5.4 eq.), CsF (5.4 eq.), DMF; (b) (i) TFA; (ii) $NaIO_4$, MeOH; (iii) 1 M HCl/MeOH; (c) (i) Ac_2O, DMAP; (ii) $NaBrO_3$, $Na_2S_2O_4$; (iii) Bz_2O, DMAP; (d) (i) N,O-BSA, pyrimidine; (ii) NH_3, MeOH.

triflate, which was reacted with sodium azide, followed by reduction to give amine **42** as the *syn*-isomer. The Boc protection of the amine group of **42** was carried out by using 5 equivalents of Boc_2O in order to suppress the formation of the dimeric urea by-product. The osmium-mediated dihydroxylation of **42** gave a 1 : 1 mixture of aminodiol **43** and its *syn*-isomer. After separation, **43** was converted to thiofuranose **44** in 36% yield from **42** through manipulations similar to those used for the preparation of **29**. Deprotection of the

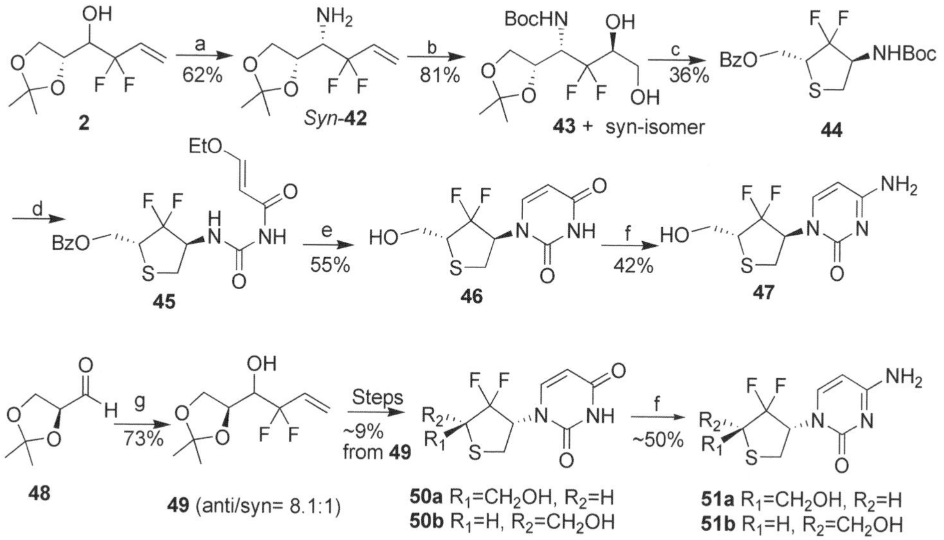

Scheme 8.7 *Reagents and conditions: (a) (i) Tf_2O, pyridine; (ii) NaN_3, DMF; (iii) PPh_3, THF, (iv) H_2O; (b) (i) Boc_2O; (ii) OsO_4, NMMO; (c) (i) AcOH; (ii) $NaIO_4$; (iii) $NaBH_4$/MeOH; (iv) MsCl, pyridine; (v) Na_2S, DMF; (d) (i) TFA; (ii) 3-ethoxy-2-propenoyl isocyanate; (e) (i) 2 M H_2SO_4; (ii) sat. NH_3/MeOH; (f) (i) Ac_2O, DMAP; (ii) TPSCl, DMAP, Et_3N; (iii) conc. $NH_3{\cdot}H_2O$; (g) In powder, $CH_2{=}CHCF_2Br$, DMF.*

Boc group of **44**, followed by the condensation of the resultant amine with 3-ethoxy-2-propenoyl isocyanate, afforded urea **45**. The cyclization of **45** with sulfuric acid and the subsequent removal of the benzoyl group gave thionucleoside **46**. Nucleoside **46** was further converted to thionucleoside **47**. Meanwhile, *gem*-difluorohomoallyl alcohol **49**, which was derived from (*S*)-glyceraldehyde acetonide **48** and 3-bromo-3,3-difluoropropene, was converted to thionucleosides **50a** and **50b** by applying the same strategy [13]. Thionucleosides **50a** and **50b** were further converted to thionucleosides **51a** and **51b** (see Scheme 8.7).

8.3 Synthesis of *gem*-Difluoromethylenated Azanucleosides via Difluoromethylenated L-Proline Derivatives

Azanucleosides, containing a nitrogen atom instead of the oxygen atom on the sugar ring, possess unique biological properties as the nitrogen atom not only has the heteroatom effect but also can bind strongly to certain DNA repair enzymes [14]. During the course of our study on fluorinated amino acids, a versatile procedure was developed for the preparation of 2′,3′-dideoxy-2′-difluoromethylazanucleosides (see Scheme 8.8) [15]. Naturally occurring *trans*-4-hydroxy-L-proline was converted to difluoro-olefin **52** in 30% yield via Swern oxidation, followed by difluoromethylenation of the resultant ketone with CF_2Br_2/Zn/HMPT2. Olefin **52** was reduced to give **53** as a 7 : 1 *cis*/*trans* mixture via

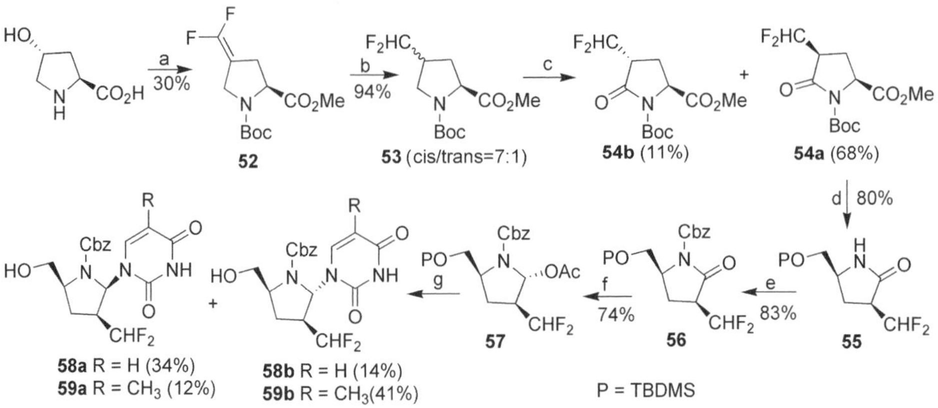

Scheme 8.8 *Reagents and conditions: (a) (i) $SOCl_2$, MeOH; (ii) Boc_2O, Et_3N, DMAP, CH_2Cl_2; (iii) Swern oxidation; (iv) CF_2Br_2/Zn/HMPT; (b) Pd/C, H_2, EtOH, r.t., 1 atm; (c) $RuO_2{\cdot}xH_2O$, $NaIO_4$, EtOAc, H_2O, r.t.; (d) (i) TFA, CH_2Cl_2; (ii) $NaBH_4$, MeOH, −78 °C–0 °C; (iii) TBDMSCl, imidazole, DMAP, CH_2Cl_2; (e) LHMDS, CbzCl, THF, −78 °C; (f) (i) $LiBEt_3H$, THF, −78 °C; (ii) Ac_2O, DMAP, Py; (g) (i) silylated uracil or silylated thymine, TMSOTf, MeCN; (ii) TBAF, THF.*

catalytic hydrogenation. After oxidation of the 5-methylene group of **53**, pyrrolidone **54a** was obtained in 68% yield along with its *trans*-isomer **54b**. After separation, **54a** was converted to **55** by reduction and silyl protection. Protection of the 1-amino group of **55** with a benzyloxycarbonyl group gave **56**, which was converted to acetate **57** by reduction and acetylation. Acetate **57** was then coupled with silylated uracil or thymine to furnish azanucleosides, **58a/b** or **59a/b**, as a mixture of α- and β-anomers (see Scheme 8.8). The absolute configuration of compound **59b** was confirmed by x-ray crystallographic analysis.

Another strategy to synthesize 3′-difluoromethylated azanucleosides **72–75**, starting from *trans*-4-hydroxy-L-proline, was also developed (see Scheme 8.9) [16]. Diol **61** was obtained in a straightforward manner by dihydroxylation of **60**, which was derived from *trans*-4-hydroxy-L-proline. Mono-benzoylation of diol **61** gave **62** as the main product (70%) along with compound **63** (17%). Compound **62** was then treated with Dess–Martin reagent to afford ketone **64**, which was converted to difluoromethylene-pyrrolidine **65** by reacting with CF_2Br_2/Zn/HMPT. Catalytic hydrogenation of **65** gave a separable mixture of **66** and **67**. Compound **66** was the main product when 10%Pd/C was used, whereas compound **67** was the major product with $Pd(OH)_2$/C. However, the attempted oxidation of the methylene group of **66** or **67** using $RuO_2{\cdot}xH_2O/NaIO_4$ failed to give the desired pyrrolidinones, probably due to the existence of *tert*-butyldiphenylsilyl group. Thus, the protecting groups of **66** and **67** were changed to *tert*-butyldimethylsilyl groups not only for the hydroxymethyl group at C-2, but also for the secondary hydroxyl group at C-4 to afford the corresponding **68** and **69**, respectively. Then, **68** and **69** were subjected to a series of transformations similar to those described for the synthesis of **58** and **59** from **53** (see Scheme 8.8) to give 3′-difluoromethylated azanucleosides **72**, **73**, **74** and **75** (see Scheme 8.9).

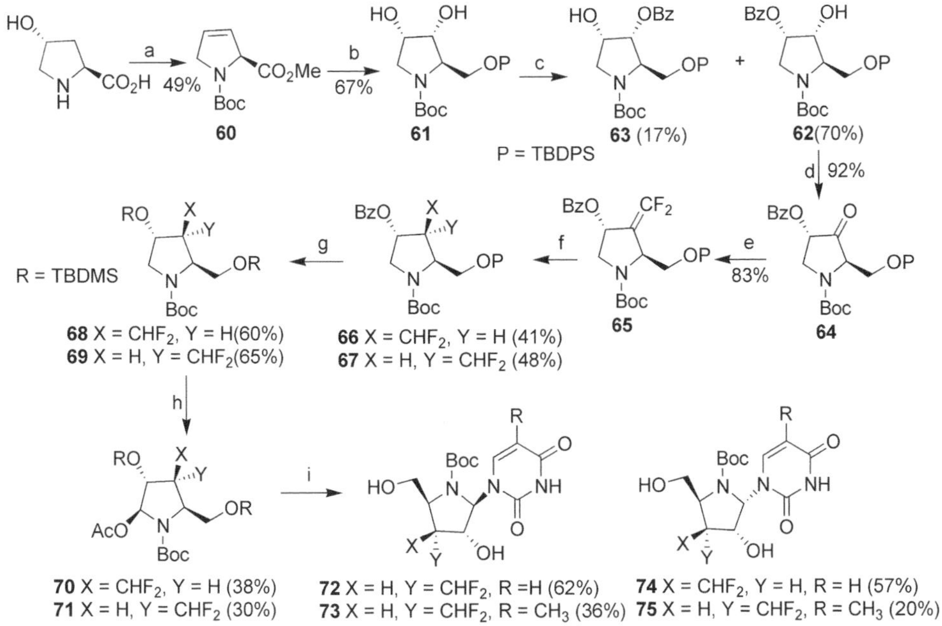

Scheme 8.9 *Reagents and conditions: (a) (i) $SOCl_2$, MeOH; (ii) Boc_2O, Et_3N, DMAP, CH_2Cl_2; (iii) MsCl, Et_3N, DMAP, CH_2Cl_2, r.t.; (iv) PhSeSePh, MeOH, reflux; v. H_2O_2, Py, r.t.; (b) (i) $LiAlH_4$, Et_2O, r.t.; (ii) TBDPSCl, imidazole, DMAP, CH_2Cl_2, r.t.; (iii) OsO_4, NMNO, acetone/H_2O, r.t.; (c) BzCl, Py, CH_2Cl_2, −10 °C, 24 h; (d) (i) BzCl, Py, CH_2Cl_2, −10 °C, 24 h; (ii) Dess–Martin oxidant, CH_2Cl_2, r.t.; (e) CF_2Br_2, HMPT, Zn, THF, reflux; (f) 10% Pd/C, 70 atm, or $Pd(OH)_2/C$, H_2, 80 atm; (g) (i) TBAF, THF, r.t.; (ii) TBDMSCl, imidazole, DMAP, CH_2Cl_2, r.t.; (iii) sat. NH_3/MeOH, r.t.; (iv) TBDMSCl, imidazole, DMAP, DMF; (h) (i) $RuO_2{\cdot}xH_2O$, $NaIO_4$, EtOAc, H_2O, r.t.; (ii) $LiBEt_3H$, THF, −78 °C; (iii) Ac_2O, CH_2Cl_2, Et_3N, DMAP, r.t.; (i) (i) silylated uracil or thymine, N,O-bis(trimethylsilyl)acetamide, TMSOTf, 0 °C to r.t.; (ii) TBAF, THF, r.t.*

8.4 Synthesis of *gem*-Difluoromethylenated Carbocyclic Nucleosides

In recent years, increasing attention has been paid to the structural modifications of carbocyclic nucleosides. Because of the absence of glycosidic linkage, carbocyclic nucleosides are chemically more stable and not vulnerable to phosphorylases that cleave the *N*-glycoside linkage in usual nucleosides [17]. The first synthesis of *gem*-difluoromethylenated carbocyclic nucleosides was reported by Borthwick *et al.* in 1990 via direct difluorination of a ketone moiety on a carbocyclic ring with DAST [18]. Recently, 2′,3′-dideoxy-6′,6′-difluorouracils, **85a** and **85b**, were synthesized using a new strategy in 14 steps starting from (*Z*)-2-butene-1,4-diol (**76**) (see Scheme 8.10) [19]. The new strategy included the construction of a carbocyclic ring via Reformatskii–Claisen rearrangement and ring-closing metathesis. (*Z*)-2-Butene-1,4-diol (**76**) was converted to chlorodifluoroacetate **77** in 69% yield, which was subjected to a silicon-induced Reformatskii–Claisen rearrangement to give difluoro ester **78**. Ester **78** was then transformed to Weinreb amide

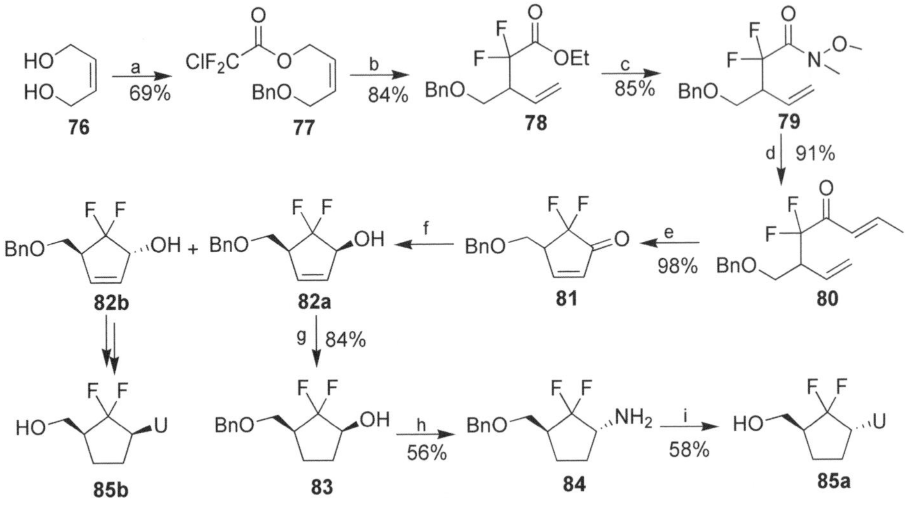

Scheme 8.10 *Reagents and conditions: (a) (i) NaH, DMF, BnBr, 0 °C; (ii) $ClCF_2CO_2H$, cat. H_2SO_4, toluene, 140 °C; (b) (i) zinc dust, TMSCl, CH_3CN, 100 °C; (ii) $ClCF_2CO_2H$, cat. H_2SO_4, toluene, 140 °C; (b) (i) zinc dust, TMSCl, CH_3CN, 100 °C; (ii) cat. H_2SO_4, EtOH, 40 °C; (c) N,O-dimethylhydroxylamine, $AlMe_3$, toluene; (d) (i) $CH_2{=}CHCH_2MgBr$, THF; (ii) Et_3N, THF; (e) Grubbs' II catalyst, toluene, 80 °C; (f) $CeCl_3{\cdot}7H_2O$, $NaBH_4$, 0 °C; (g) H_2, Pd black, benzene; (h) (i) Tf_2O, pyridine, CH_2Cl_2, −40 °C; (ii) NaN_3, DMF; (iii) H_2, cat. Pd black, benzene; (i) (i) (E)-EtOCH=CHCONCO, DMF, −25 °C; (ii) 2 N H_2SO_4, reflux; (iii) H_2, cat. Pd/C.*

79 and treated with allylmagnesium bromide to afford diene **80** via a double-bond isomerization. Diene **80** was subjected to ring-closing metathesis. However, when **80** was treated with the first-generation Grubbs catalyst, the reaction did not occur. The second-generation Grubbs catalyst was therefore employed for the reaction and the reaction proceeded smoothly to give cyclopentenone **81** in nearly quantitative yield. Luche reduction of **81** afforded a 2.9 : 1 separable mixture of unsaturated alcohols, **82a** and **82b**. Catalytic hydrogenation of **82a** gave alcohol **83**, which was transformed to α-2′,3′-dideoxy-6′,6′-difluorocarbocyclic uridine **85a**. The relative stereochemistry of **85a** was determined by x-ray crystallographic analysis. In the same manner, unsaturated alcohol **82b** was converted to β-anomer **85b**. Thus, racemic 2′,3′-dideoxy-6′,6′-difluorocarbocyclic uracils, **85a** and **85b**, were obtained in 14 steps from **76** in 7.6% and 1.5% overall yields, respectively (see Scheme 8.10).

8.5 Conclusion

gem-Difluoromethylenated nucleosides constitute important members of the extensively studied nucleoside analogues, which are an important class of candidates for new antiviral and antitumor agents. We have described here recent development in the syntheses of difluoromethylenated nucleosides in our laboratory. With the development of more practical difluoromethylenation methodologies and more knowledge of structure–activity relationships, more efficacious *gem*-difluoromethylenated nucleosides will be developed as novel antiviral and antitumor agents.

Acknowledgment

We are grateful to the National Natural Science Foundation of China, the Ministry of Education of China and Shanghai Municipal Scientific Committee for the financial support.

References

1. For recent fluorinated nucleosides reviews, see (a) Meng, W-D. and Qing F-L. (2006) Fluorinated nucleosides as antiviral and antitumor agents. *Current Topics in Medicinal Chemistry*, **6**, 1499–1528; (b) Qing, F-L. and Qiu, X-L. (2007) Synthesis of *gem*-difluoromethylated sugar nucleosides. In: *Current Fluoroorganic Chemistry*, ACS Symposium Series **949**, pp. 305–322 and references cited therein.
2. Blackburn, G. W., England, D. E. and Kolmann, F. (1981) Monofluoro- and difluoromethylenebisphosphonic acids: isopolar analogues of pyrophosphoric acid. *J. Chem. Soc., Chem. Commun.*, 930–932.
3. (a) Lee, D. (2003) Gemcitabine-based therapy in relapsed non-Hodgkin's lymphoma and cutaneous T-cell lymphoma. *Clin Lymphoma Myeloma*, **4**, 152–153; (b) Nobel, S. and Goa, K. L. (1997) Gemcitabine. A review of its pharmacology and clinical potential in non-small cell lung cancer and pancreatic cancer. *Drugs*, **54**, 447–472.
4. Zhang, X. G., Xia, H. R., Dong, X. C., *et al.* (2003) 3-Deoxy-3,3-difluoro-D-arabinofuranose: First stereoselective synthesis and application of *gem*-difluorinated sugar nucleosides. *J. Org. Chem.*, **68**, 9026–9033.
5. Xu, X. H., Qiu, X. L., Zhang, X. G. and Qing, F. L. (2006) Synthesis of L- and D-β-3′-deoxy-3′,3′-difluoronucleosides. *J. Org. Chem.*, **71**, 2820–2824.
6. Vorbrüggen, H. and Hofle, G. (1981) Nucleoside syntheses XXIII: On the mechanism of nucleoside synthesis. *Chem. Ber.*, **114**, 1256–1268.
7. Vorbrüggen, H, Krolikiewicz, K. and Bennua. B. (1981) Nucleoside syntheses XXII: Nucleoside synthesis with trimethylsilyl triflate and perchlorate as catalysts. *Chem. Ber.* **114**, 1234–1256.
8. (a) Barrena, M. I., Matheu, M. I. and Castillon, S. (1998) An improved synthesis of 4-*O*-benzoyl-2,2-difluorooleandrose from L-rhamnose. Factors determining the synthesis of 2,2-difluorocarbohydrates from 2-uloses. *J. Org. Chem.*, **63**, 2184–2188. (b) Aghmiz, M. L., Diaz, Y., Jana, G. H., *et al.* (2001) The reaction of pyranoside 2-uloses with DAST revised. Synthesis of 1-fluoro-ketofuranosyl fluorides and their reactivity with alcohols. *Tetrahedron*, **57**, 6733–6743.
9. Xu, X. H., You, Z. W., Zhang, X. G. and Qing, F. L. (2007) Synthesis of 3-deoxy-3,3-difluoro-D-ribohexose from *gem*-difluorohomoallyl alcohol. *J. Fluorine Chem.*, **128**, 535–539.
10. For thionucleosides reviews, see: Yokoyama, M. (2000) Synthesis and biological activity of thionucleosides. *Synthesis*, 1637–1655.
11. Zheng, F., Zhang, X. H., Qiu, X. L., *et al.* (2006) Synthesis of 1-β-3′-deoxy-3′,3′-difluoro-4′-thionucleosides. *Org. Lett.*, **8**, 6083–6086.

12. Wu, Y. Y., Zhang, X., Meng, W.-D. and Qing, F.-L. (2004) Synthesis of new 2′,3′-dideoxy-6′,6′-difluoro-3′-thionucleoside from *gem*-difluorohomoallyl alcohol. *Org. Lett.*, **6**, 3941–3944.
13. Yue, X., Wu, Y. Y. and Qing, F.-L. (2007) Synthesis of a series of novel 2′,3′-dideoxy-6′,6′-difluoro-3′-thionucleosides. *Tetrahedron*, **63**, 1560–1567.
14. For azanucleosides reviews, see: Yokoyama, M. and Momotake, A. (1999) Synthesis and biological activity of azanucleosides. *Synthesis*, 1541–1554.
15. Qiu, X. L. and Qing, F.-L. (2004) Synthesis of 2′,3′-dideoxy-2′-difluoromethyl azanucleosides. *Synthesis*, 334–340.
16. Qiu, X. L. and Qing, F.-L. (2005) Synthesis of 3′-dideoxy-3′-difluoromethyl azanucleosides from *trans*-4-hydroxy-l -proline. *J. Org. Chem.*, **70**, 3826–3837.
17. For carbocyclic nucleosides reviews, see: (a) Crimmins, M. T. (1998) New developments in the enantioselective synthesis of cyclopentyl carbocyclic nucleosides. *Tetrahedron*, **54**, 9229–9272. (b) Borthwick, A. D. and Biggadike, K. (1992) Synthesis of chiral carbocyclic nucleosides. *Tetrahedron*, **48**, 571–623. (c) Agrofoglio, L., Suhas, E., Farese, A., *et al.* (1994) Synthesis of carbocyclic nucleosides. *Tetrahedron*, **50**, 10611–10670. (d) Marquez, V. E. and Lim, M.-U. (1986) *Med. Res. Rev.*, Carbocyclic nucleosides **6**, 1–40. (e) Wu, Q. and Simons, C. (2004) Synthetic methodologies for C-nucleosides. *Synthesis*, 1533–1553. (f) Wang, J., Froeyen, M. and Herdewijn, P. (2004) Six-membered carbocyclic nucleosides. *Adv. Antivir. Drug Des.*, **4**. 119–145. (g) Rodriguez, J. B. and Comin, M. J. (2003) New progresses in the enantioselective synthesis and biological properties of carbocyclic nucleosides *Mini Rev. Med. Chem.*, **3**, 95–114.
18. Brothwick, A. D., Evans, D. N., Kirk, B. E., *et al.* (1990) Fluoro carbocyclic nucleosides: synthesis and antiviral activity of 2′-and 6′-fluoro carbocyclic pyrimidine nucleosides including carbocyclic 1-(2-deoxy-2-fluoro-β-d-arabinofuranosyl)-5-methyluracil and carbocyclic 1-(2′-deoxy-2-fluoro-β-d-arabinofuranosyl)5-iodouracil. *J. Med. Chem.*, **33**, 179–186.
19. Yang, Y. Y., Meng W.-D. and Qing, F.-L. (2004) Synthesis of 2′,3′-dideoxy-6′,6′-difluorocarbocyclic nucleosides. *Org. Lett.*, **6**, 4257–4259.

9

Recent Advances in the Syntheses of Fluorinated Amino Acids

Kenji Uneyama

9.1 Introduction

Fluorinated amino acids have been used as components of modified peptides and proteins in protein engineering [1] and have also found applications as potential enzyme inhibitors and antitumor and antibacterial agents [2–4]. To date, due to the potent biological activities of fluorinated α-amino acids, several synthetic strategies and methods have been the subjects of reviews and books [5, 6]. In this chapter synthetic strategies for fluorinated amino acids recently developed in the 2000s are summarized. Syntheses of fluorinated amino acids are classified into three groups on the basis of the strategies for creation of the chiral center: (1) enantioselective, (2) diastereoselective, and (3) racemic syntheses.

9.2 Enantioselective Synthesis

Enantioselective synthesis involves chemical modification via (section 9.2.1) introduction of a chiral center into achiral fluorinated building blocks, (9.2.2) chiral transposition, (9.2.3) introduction of fluorine functionality into nonfluorinated chiral building blocks, (9.2.4) modification of chiral fluorinated building blocks, and (9.2.5) enzymatic resolution of racemic fluorinated building blocks. The following sections summarize these five categories.

Fluorine in Medicinal Chemistry and Chemical Biology Edited by Iwao Ojima

9.2.1 Introduction of a Chiral Center into Achiral Fluorinated Building Blocks

9.2.1.1 Introduction of Chirality by Hydrogenation

Asymmetric hydrogenation of either a carbonyl or an imino group to a hydroxyl group or an amino group has frequently been employed for the introduction of chirality in amino acid syntheses. Corey's catecolborane–oxazaborolidine protocol enables transformation of difluoromethyl ketone **1** into alcohol **2** with excellent enantioselectivity. The reaction of diastereoselective amination of α-hydroxyaldehyde **3** with *N,N*-diallylamine and 2-furylboronic acid provides furyl amino alcohol **4** in good chemical yield along with excellent diastereoselectivity. This protocol is applicable for the preparation of amino acids and amino alcohols with a trifluoromethyl group by the combination of *N,N*-diallyl or *N,N*-dibenzyl amine and aromatic, heteroaromatic and alkenyl boronic acids [7]. The usual chemical transformations as shown in steps 5 to 8 in Scheme 9.1 lead to (2*S*,3*R*)-difluorothreonine **5** [8].

Palladium-catalyzed carboalkoxylation of imidoyl iodides **6** provides benzyl [9] and even *tert*-butyl[10] esters **7**. Asymmetric hydrogenation of the imino moiety of imono esters **7** in a $Pd(OCOCF_3)_2$ / (*R*)-BINAP / CF_3CH_2OH system gives enantio-enriched amino esters **8** in 85–91% ee (see Scheme 9.2) [11].The enantioselectivity achieved by the hydrogenation was much better than that by Corey's hydride reduction [12] and was employed for the syntheses of enantiomerically pure *N*-Boc-β,β-difluoroproline benzyl ester **9** (see Scheme 9.3)[13] and enantiomerically enriched *N*-Boc-β,β-difluoroglutamic acid benzyl ester [13].

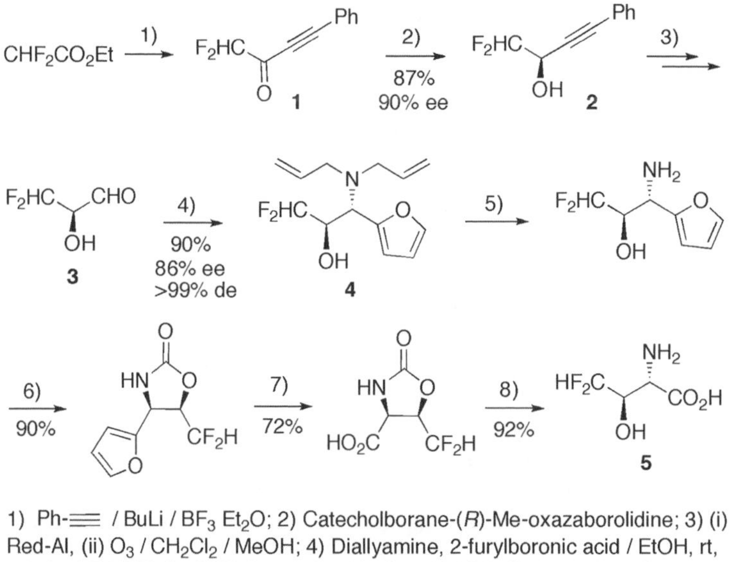

1) Ph-≡ / BuLi / BF_3 Et_2O; 2) Catecholborane-(*R*)-Me-oxazaborolidine; 3) (i) Red-Al, (ii) O_3 / CH_2Cl_2 / MeOH; 4) Diallyamine, 2-furylboronic acid / EtOH, rt, 1-2 d; 5) $Pd(PPh_3)_4$, *N,N'*-dimethylbarbituric acid / CH_2Cl_2; 6) (i) Boc_2O / dioxane, rt, (ii) NaH / DMF, rt; 7) MeOH / O_3, -78 °C; 8) 6 *N* HCl, reflux, 8 h.

Scheme 9.1

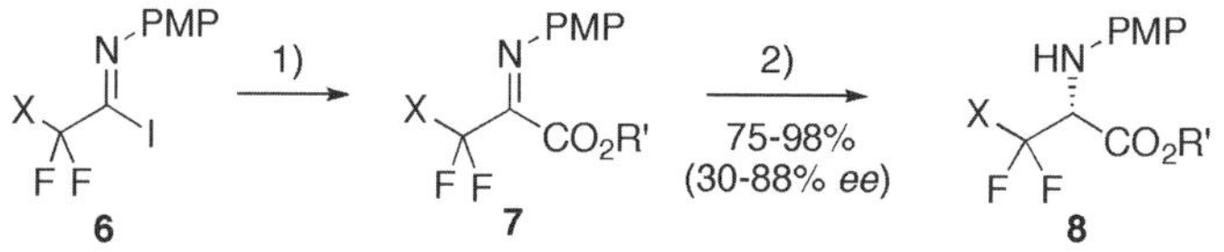

X = H, F, Cl, Br, C_2F_5
R' = Et, Bn, Bu^t

1) $Pd_2(dba)_3.CHCl_3$ (Pd: 0.10 eq.), CO (1 atm), R'OH, K_2CO_3, toluene or DMF, DMI, rt; 2) $Pd(OCOCF_3)_2$, (*R*)-BINAP, H_2 (100 atm), CF_3CH_2OH, rt, 24h.

Scheme 9.2

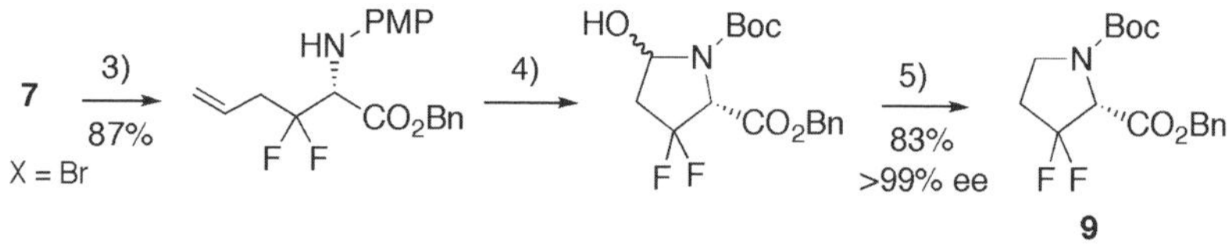

3) $AllylSnBu_3$, AIBN / toluene; 4) (i) CAN / MeCN-H_2O, (ii) $(Boc)_2O$ / $NaHCO_3$ (84%); (iii) O_3 / CH_2Cl_2 (90%); 5) (i) PPh_3Cl_2 / DMF, (ii) $RhCl(PPh_3)$-H_2

Scheme 9.3

Interestingly, the stereochemistry of the hydroxyl group on the imino nitrogen in oxime **10** affects the stereochemistry of hydroxylamine **11** (see Scheme 9.4). Thus, hydride reduction of either (*Z*)- or (*E*)-oximes **10** with hydroborane in the presence of chiral amino alcohols produces (*S*)- and (*R*)-*N*-benzyl oximes **11**, respectively, as shown in the table in Scheme 9.4, which were subsequently transformed to (*R*)-and (*S*)-trifluoroalanine **12** [14].

The carbon–carbon double bond of an enamine is also applicable for asymmetric hydrogenation leading to chiral amino acids. For example, hydrogenation of **13** by rhodium catalyst with ferrocenyl diphosphine **15** as a ligand was successful for the synthesis of methyl 3-amino-4-polyfluorophenylbutanoate **14** with excellent stereoselectivity (see Scheme 9.5) [15].

9.2.1.2 Asymmetric Dihydroxylation

Asymmetric dihydroxylation of trifluoromethylalkenes is also useful for construction of enantio-enriched trifluoromethylated diols usable for trifluoromethylated amino acids with chiral hydroxyl group. Thus, Sharpless AD reaction of **16** provides diol **17** with excellent enantioselectivity. Regioselective and stereospecific replacement of the sulfonate moiety in **18** with azide ion enables the introduction of nitrogen functionality. A series of well-known chemical transformation of **19** leads to 4,4,4-trifluorothreonine **20** (see Scheme 9.6) [16]. Dehydroxylative-hydrogenation of **21** by radical reaction via thiocarbonate and subsequent chemical transformation synthesize enantio-enriched (*S*)-2-amino-4,4,4-trifluorobutanoic acid **22** [16]. Both enantiomers of **20** and **22** were prepared in a similar manner from (2*R*,3*S*)-diol of **17**.

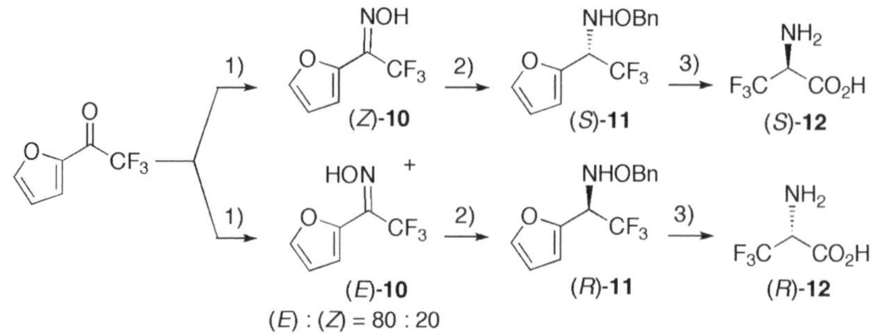

(*Z*)-**10** (*S*)-**11** (*S*)-**12**

(*E*)-**10** (*R*)-**11** (*R*)-**12**

(*E*) : (*Z*) = 80 : 20

1) NH_2OHHCl, AcONa / EtOH and then separation; 2) (i)BnBr, NaH / DMF, (ii) BH_3 , cat / THF, (iii) HCl / H_2O; 3) O_3 / MeOH, -78 °C

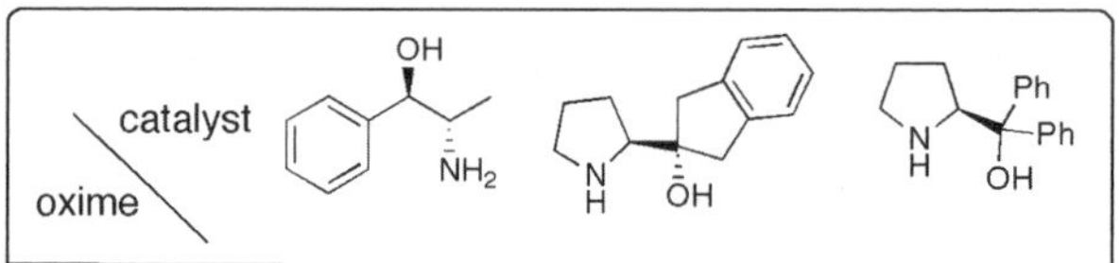

oxime \ catalyst	catalyst 1	catalyst 2	catalyst 3
(*Z*)	80 (76)(*S*)	73 (80)(*S*)	73 (88)(*S*)
(*E*)	74 (73)(*R*)	77 (82)(*R*)	74 (86)(*R*)

Yield (%) (% ee) (*absolute configuration*)

Scheme 9.4

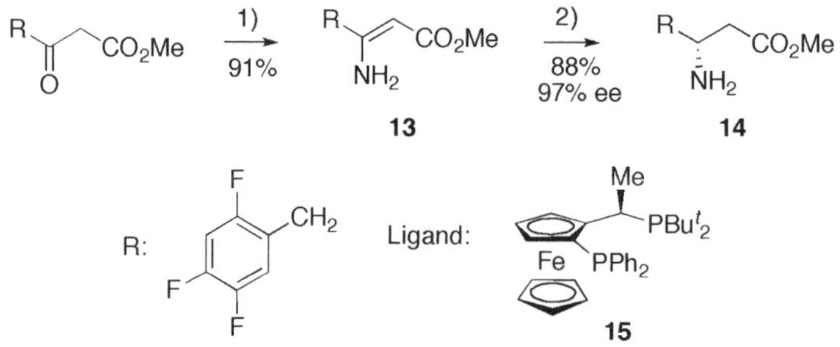

1) 91% **13** 2) 88% 97% ee **14**

R: Ligand: **15**

1) NH_4OAc / MeOH, reflux; 2) H_2, 0.1 mol% $[Rh(COD)Cl]_2$, ligand / MeOH

Scheme 9.5

9.2.1.3 Asymmetric Alkylation

Asymmetric alkylation of imines has been employed most frequently for the construction of chiral amino moieties involved in the syntheses of nitrogen heterocycles and amino acids [17]. This approach is also useful for fluorinated amino acid synthesis as shown in Scheme 9.7. Mannich reaction of enolate with imino ester **23** in the presence of L-proline gives α-amino esters **24** and **25** enantio-and diastereoselectively [18].

Asymmetric alkylation of *N*-protected glycine ester **26** under phase-transfer catalysis conditions is the well-known method for the syntheses of α-amino acids [19]. Scheme

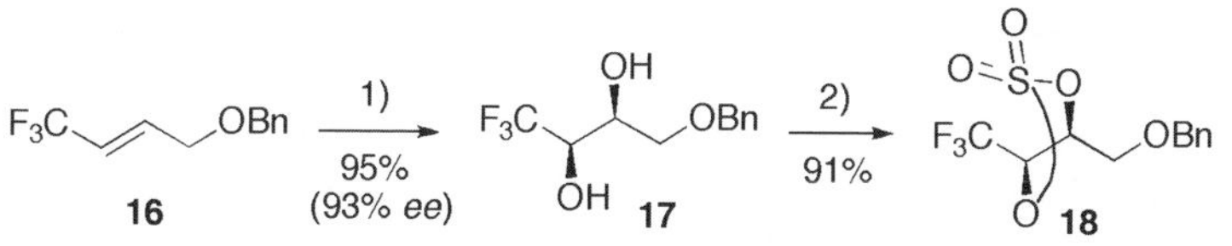

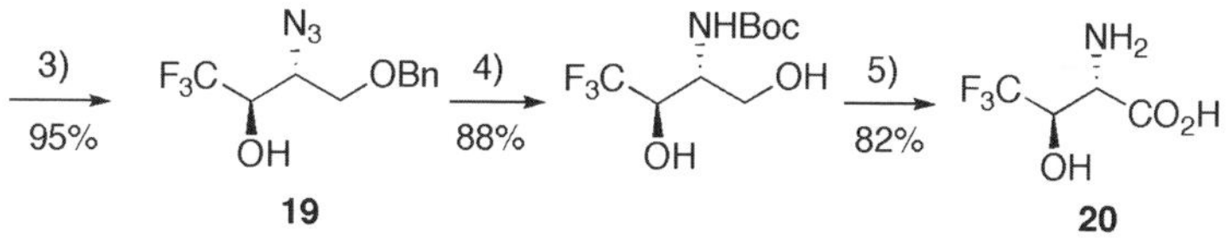

1) AD-mix-ß, $MeSO_2NH_2$ / H_2O-^{t}BuOH, rt, 4d.; 2) (i)$SOCl_2$, NEt_3, CH_2Cl_2, 0 °C; (ii) $NaIO_4$, $RuCl_3$ / MeCN-CCl_4, rt; 3) (i) NaN_3 / DMF, 80 °C, 4h; (ii) H_3O^+; 4) $Pd(OH)_2$, Boc_2O, H_2, THF, rt; 5) (i) Jones reagent; (ii) CF_3CO_2H, CH_2Cl_2, 0 °C;

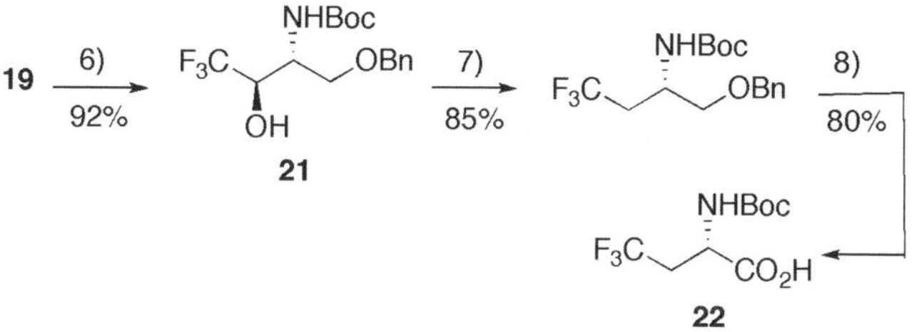

6) Pd/C, Boc_2O, H_2, THF, rt; 7) (i)PhOCSCl, DMAP, toluene; (ii) Bu_3SnH, AIBN, toluene; 8) (i) $Pd(OH)_2$, H_2, THF; (ii) Jones reagent.

Scheme 9.6

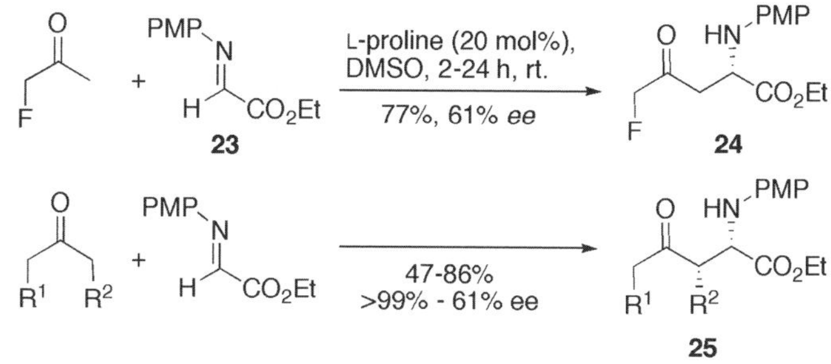

Scheme 9.7

9.8 shows two examples: one is synthesis of fluorinated phenylalanines in which benzylation to **26** proceeds in an excellent yield with almost perfect enantioselection under dimeric *Cinchona* alkaloid phase-transfer catalysis (α,α′-bis[*O*(9)-allylcinchonidinum]-*o*,*m*, or *p*-xylene) [20]; the other is S_N2' reaction of **26** with **29**, which provides **30** with moderate to good enantioselectivity [21]. Diastereoselective synthesis of fluorinated amino acids using 2-hydroxypinanone glycine Schiff base is described in section 9.2.

9.2.2 Chiral Transposition

Both enantiomers of α-phenethyl amines have most frequently been used as one of the easily available and economically feasible chiral auxiliaries for diastereoselective syntheses. Soloshonok developed a useful 1,3-proton shift methodology in which chirality is transposed concertedly from α-phenethyl amines to newly formed trifluoromethyl amines. This protocol is applicable for enantio-enriched amino acid synthesis, as shown in Scheme 9.9 [22]. The driving force for the proton shift is the thermodynamically lower stability of trifluoromethyl imine **32** than that of phenyl imine **33** [23]. Ketones and imines with a strongly electron-withdrawing α-substituent such as the trifluoromethyl group are, in general, unstable and are transformed to the corresponding hydrate as a stable form on exposure to water. The Schiff base of trifluoropyruvate **35** readily undergoes 1,3-proton

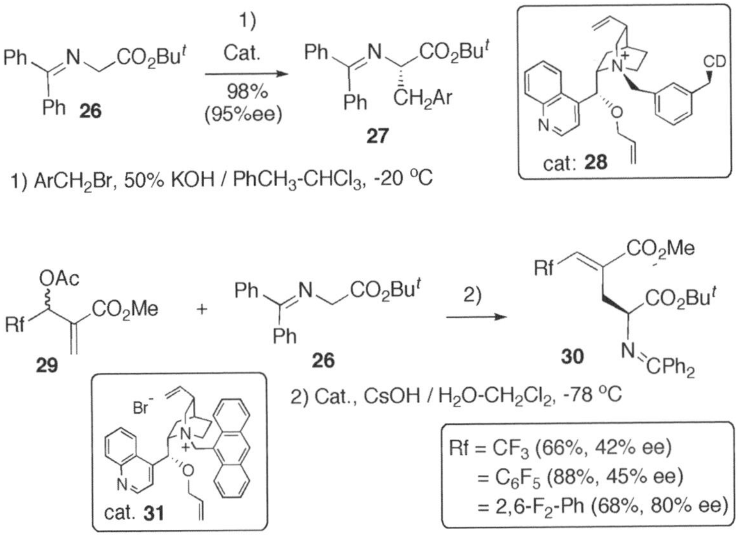

1) $ArCH_2Br$, 50% KOH / $PhCH_3$-$CHCl_3$, -20 °C

2) Cat., CsOH / H_2O-CH_2Cl_2, -78 °C

Rf = CF_3 (66%, 42% ee)
= C_6F_5 (88%, 45% ee)
= 2,6-F_2-Ph (68%, 80% ee)

Scheme 9.8

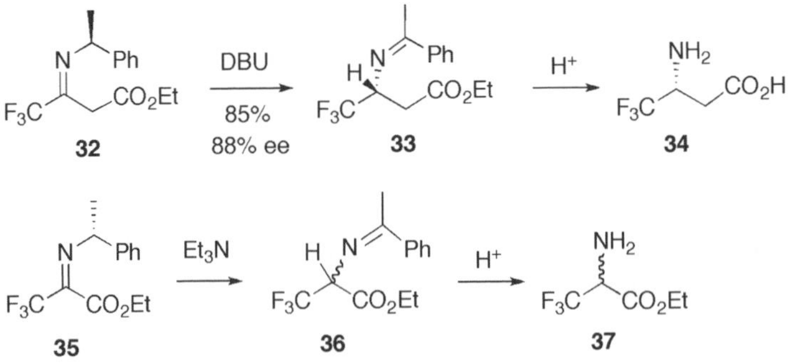

Scheme 9.9

shift with Et_3N, but trifluoroalanine derivative **36** was racemic. This undesired stereochemical outcome would arise from facile racemization of **36** under the reaction conditions due to the higher acidity of the methine proton of **36** than that of **33** [24].

9.2.3 Introduction of Fluorine Functionality into Nonfluorinated Chiral Building Blocks

9.2.3.1 Introduction of Chirality by Stereospecific Nucleophilic Substitution

Stereospecific nucleophilic substitution of a chiral *sec*-hydroxyl group is sometimes reliable when the hydroxyl group is preactivated by a strongly electron-withdrawing group. Some examples are shown in Schemes 9.10, 9.11, and 9.12. *sec*-Trifluoromethanesulfonate derived from **38** undergoes stereospecific S_N2 reaction with sodium azide, affording **40**, which is subsequently transformed to difluoroalanine derivative **41** (see Scheme 9.10) [25]. The carbomethoxy group of **42** is modified to difluoromethyl group of **43** via the formyl group so that **43** can be transformed into **44** in a similar manner [25].

The hydroxyl group of *sec*-alcohols is, in general, replaced with a fluorine atom stereospecifically (inversion) by reaction with DAST [26]. Thus, the hydroxyl group of **45** can be replaced stereospecifically with fluorine by morpho-DAST to give (4*S*)-fluoroproline ester **46** (see Scheme 9.11) [27]. Its enantiomer (4*R*)-fluoroproline, a mimic of (4*R*)-hydroxyproline, which controls the thermal stability of the collagen-like triple-helical structure [28], is also prepared in a similar manner (79%) [27].

Stereospecific nucleophilic substitution of the hydroxyl group of *sec*-trifluoromethylalcohol **47** with carbon nucleophiles has been a subject of active investigation; however, until now no successful result has been reported, although the S_N2 reaction of **47** with some heteroatom nucleophiles is known [29].

Konno *et al.* demonstrated that palladium catalysis accelerates the formal stereospecific replacement of the *sec*-MsO group in **48** with a carbon nucleophile generated from *N*-Boc-glycine **49** to give trifluoromethylated amino acid derivatives **50** (see Scheme 9.12) [30].

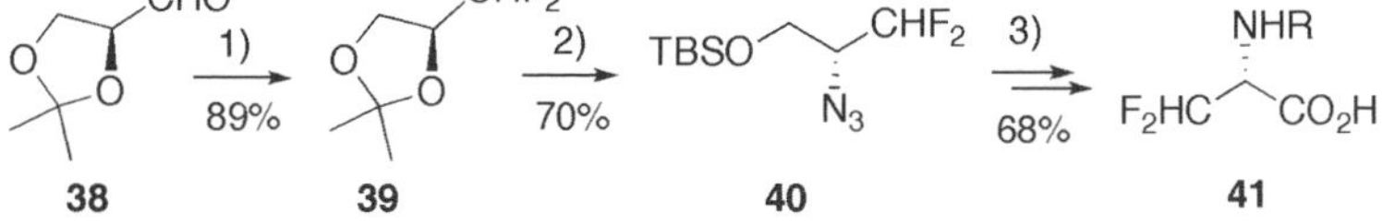

38 → (1) 89%) 39 → (2) 70%) 40 → (3) 68%) 41

1) DAST / CH_2Cl_2; 2) (i) Acid-hydrolysis, (ii) TBSCl, (iii) Tf_2O/Py, (iv) NaN_3 / DMF
3) (i) H_2 / $Pd(OH)_2$ / MeOH, (ii) FmocCl / $NaHCO_3$, (iii) HF/Py / THF, (iv) Jones oxidation

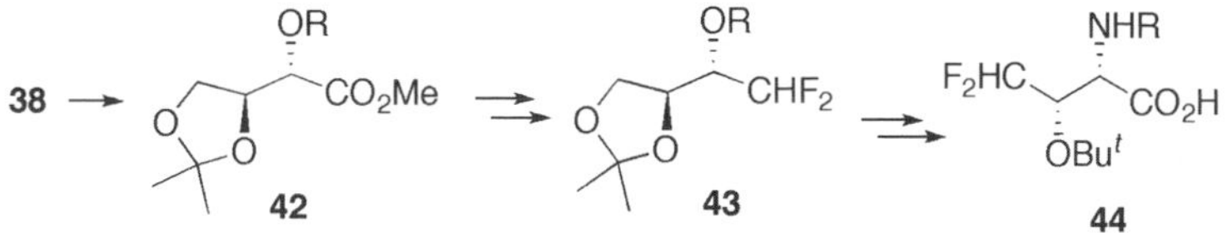

38 → 42 → 43 → 44

Scheme 9.10

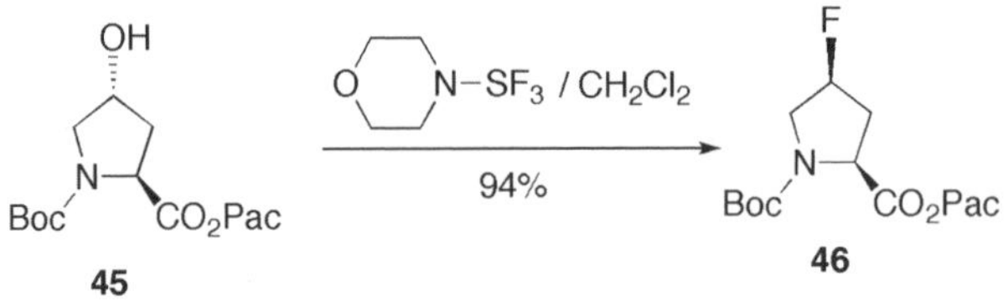

Scheme 9.11

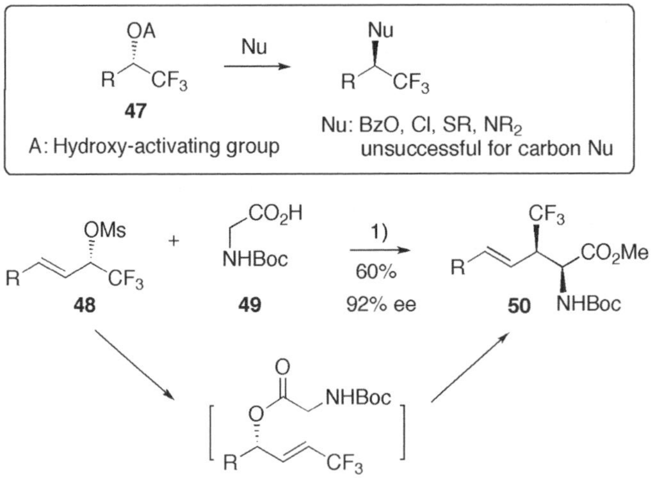

A: Hydroxy-activating group

Nu: BzO, Cl, SR, NR_2
unsuccessful for carbon Nu

1) (i) $Pd(PPh_3)_4$ / THF, 0 °C, (ii) *N*-Boc-glycine, Et_3N, (iii) $ZnCl_2$, LHMDS, reflux, (iv) Acid, (v) CH_2N_2

Scheme 9.12

9.2.3.2 Introduction of Fluorine Functionality into Nonfluorinated Chiral Building Blocks

(*R*)-Isopropylideneglyceraldehyde, Garner's aldehyde, (*S*)-serine, and L-proline have been most frequently employed as starting chiral building blocks. A fluorine functionality is introduced into the building blocks and the subsequent chemical modification of fluorinated chiral synthetic intermediates leads to syntheses of enantiomerically pure or enriched fluorinated amino acids as a final product. Examples of the syntheses from (*R*)-isopropylideneglyceraldehyde (**51**), Garner's aldehyde (**52**), (*S*)-serine (**53**), and proline are shown in Schemes 9.13–9.18.

A five-step chemical modification of **51** gives *N,O*-protected α-amino-α′-hydroxyketone **54**. Difluorination of the ketone **54** at the carbonyl carbon with morpho-DAST followed by conventional chemical modification results in the synthesis of β-amino-α,α-difluorocarboxylic acid **55** (see Scheme 9.13) [31]. Enantiomerically pure 5,5,5,5′,5′,5′-hexafluoroleucine **57** is efficiently synthesized from Garner's aldehyde **52** as shown in Scheme 9.14 [32]. Triphenylphosphine-induced reductive coupling of **52** with hexafluorothioacetone produces **56** in an excellent yield, which is conventionally transformed to

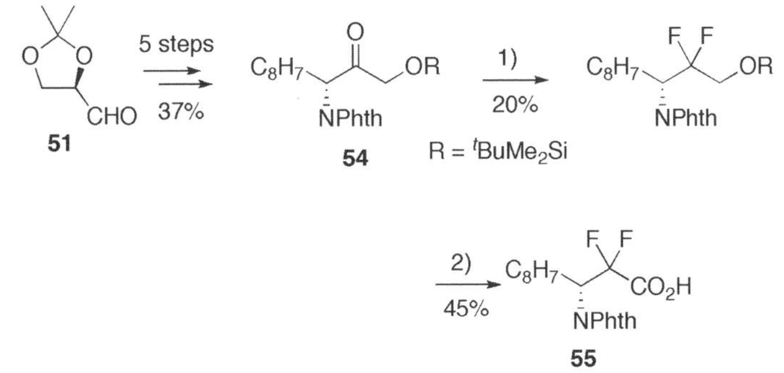

1) Morpho-DAST / CH_2Cl_2, 30 °C, 45 h; 2) (i) Dioxane-HCl-H_2O, 35-40 °C, 25 h, (ii) $NaIO_4$, $RuCl_3{\cdot}H_2O$ / CCl_4-MeCN-H_2O, rt, 24 h.

Scheme 9.13

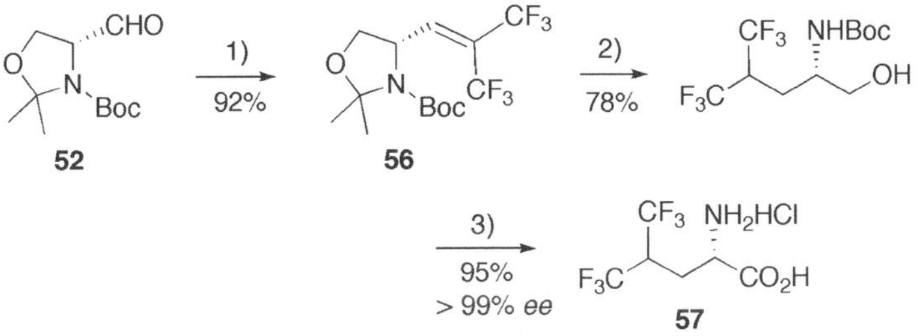

1) PPh_3, $[(CF_3)_2C]_2S_2$, Et_2O, -78 °C - rt, 3d; 2) (i) H_2, 10% Pd/C, THF, (ii) TsOH, MeOH, rt, 1d; 3) (i) PDC / DMF, 18h, (ii) 40% CF_3CO_2H/CH_2Cl_2 then HCl, 10min, rt.

Scheme 9.14

57. Reformatzky reaction of **52** followed by radical-initiated reductive hydrogenation of the hydroxyl group in **58** via a thiocarbamate intermediate provides **59**. The usual chemical modification of **59** produces enantiomerically pure 4,4-difluoroglutamine **60** (see Scheme 15) [33].

The hydroxyl group in serine **53** has a high potential for the various modifications in the synthesis of β-substituted α-amino acids. One example is shown in Scheme 9.16, in which the C–O bond on C-3 in serine **53** was at first converted into a C–I bond and then into a C–C bond. The overall transformation from serine **53** via **62** and **63** leads to the synthesis of hexafluoroleucine **64** [34]. L-Serine was also used for the synthesis of enantiomerically pure 4,4-difluoroglutamic acid **66** (see Scheme 9.17), where the carboxyl group in **53** was protected as an orthoester **65** and a difluoromethylene moiety was supplied from bromodifluoroacetate via Reformatzky reaction [35]. Interestingly, aldehyde **67** [36], readily prepared from L-serine, couples with 1,1-bis(*N*,*N*-dimethylamino)-2,2-difluoroethene **68** at room temperature to give adduct **69** as the sole stereoisomer. A bulky *N*-protecting group (PhFI) effectively controls approach of nucleophile **68** to the hindered carbonyl group of **67**.

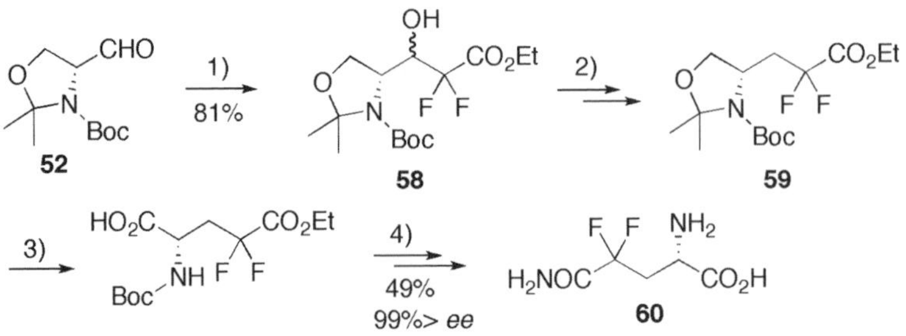

1) $BrCF_2CO_2Et$ / Zn / THF, us; 2) (i) $(imidazole)_2C{=}S$ / $ClCH_2CH_2Cl$, (ii) Et_3SiH / BPO, reflux; 3) cat.CrO_3, H_5IO_6 / MeCN, 0 °C -rt; 4) (i) aq NH_3, (ii) TFA / MeOH

Scheme 9.15

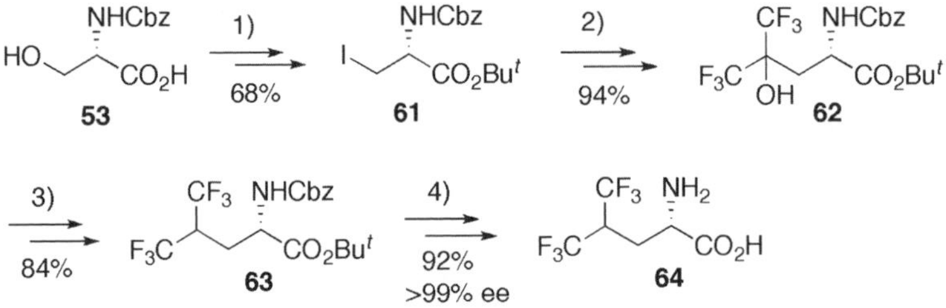

1) (i) K_2CO_3 / $BnNEt_3Cl$ / $^{t}BuBr$ / MeCN, 45-50 °C, (ii) $(PhO)_3PMeI$ / DMF; 2) (i) Zn / 1,2-dibromoethane / TMSCl / $CuBr{\cdot}SMe_2$ / DMF; (ii) Hexafluoroacetone; 3) (i)$ClCOCO_2Ph$, pyridine / toluene, (ii) Bu_3SnH, AIBN, toluene, 100 °C; (iii) KF-Celite, Et_2O; 4) (i) TFA/CH_2Cl_2, 0 °C - rt; (ii) H_2, 10% Pd/C / MeOH.

Scheme 9.16

Pyroglutaminol **70** is easily available from L-glutamic acid and is useful for amino acid synthesis as a functionalized chiral building block related to proline. Active methylene hydrogens are readily replaced with fluorine step-by-step via monofluoride by electrophilic fluorinating reagents such as *N*-fluorosulfone imines and *N*-fluoroammonium salts like Selectflor [37]. Thus, 4, 4-difluoroglutamic acid **71** was successfully synthesized from the readily available bicyclolactam **72** [38]. This synthesis involves electrophilic difluorination as a key step. Meanwhile, oxidation of 7**4** to **75** was found to be difficult due to the lower reactivity of the methylene group inactivated by the electron-withdrawing effect of the difluoromethylene group (see Scheme 9.18).

The commercially available lactam **76** is usable for amino acid synthesis as a chiral cyclic building block. 4-Fluoro- and 4,4-difluoro-3-aminocyclopentane carboxylic acids **79** and **81**, potential inhibitors of γ-aminobutanoic acid (GABA) aminotransferase, were synthesized as shown in Scheme 9.19 [39]. In this process, replacement of hydroxyl or carbonyl groups with fluorine was achieved by the use of DAST. Interestingly, the stereochemistry in substitution with DAST is retained [40], although it is by inversion in most cases [26]. Hydrolysis of **78** gave **80**.

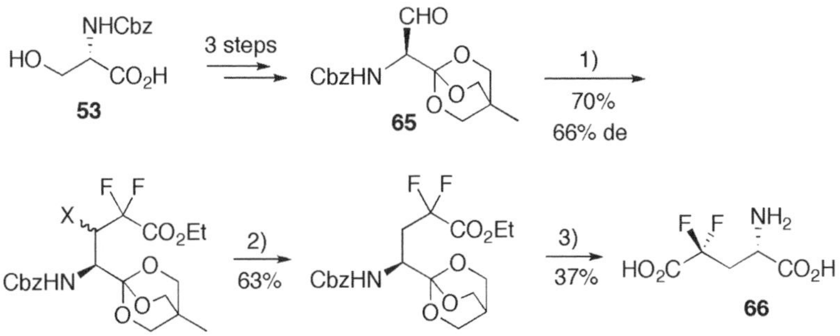

NHCbz HO CO$_2$H **53** 3 steps CHO CbzHN **65** 1) 70% 66% de

X CO$_2$Et CbzHN 2) 63% CO$_2$Et CbzHN 3) 37% HO$_2$C NH$_2$ CO$_2$H **66**

X=OH, OCS-$C_3H_3N_2$, H

1) $BrZnCF_2CO_2Et$, THF,rt.; 2) (i) $(C_3H_3N_2)C{=}S$, dry THF, rt, (ii) Et_3SiH, $(PhCO_2)_2$, benzene, reflux; 3) 6 *N* HCl, reflux

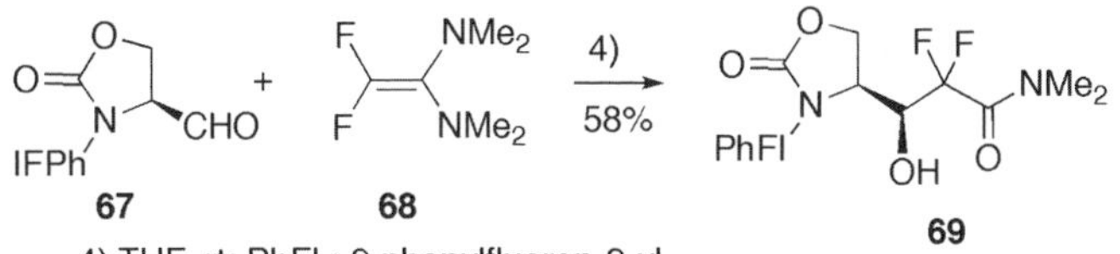

O N IFPh CHO **67** + F F NMe$_2$ NMe$_2$ **68** 4) 58% PhFl NMe$_2$ OH O **69**

4) THF, rt; PhFl : 9-phenylfluoren-9-yl

Scheme 9.17

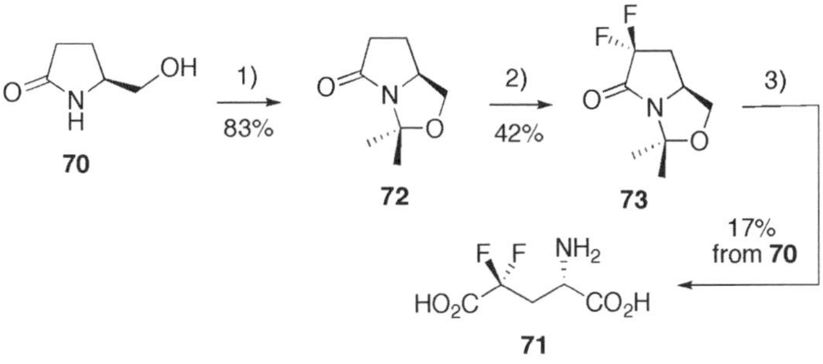

O N H OH **70** 1) 83% **72** 2) 42% **73** 3) 17% from **70** HO$_2$C NH$_2$ CO$_2$H **71**

1) DMP / TsOH / toluene, 80-90 °C; 2) LDA / NFSI / THF, -78 °C; 3) (i) AcOH / MeCN / H_2O, 90 °C, (ii) H_2CrO_4 / acetone, (iii) 6 *N* HCl, reflux

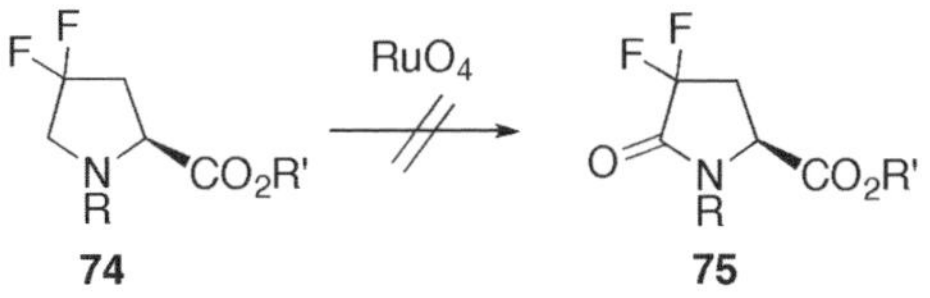

F F N R CO$_2$R' **74** RuO$_4$ F F O N R CO$_2$R' **75**

Scheme 9.18

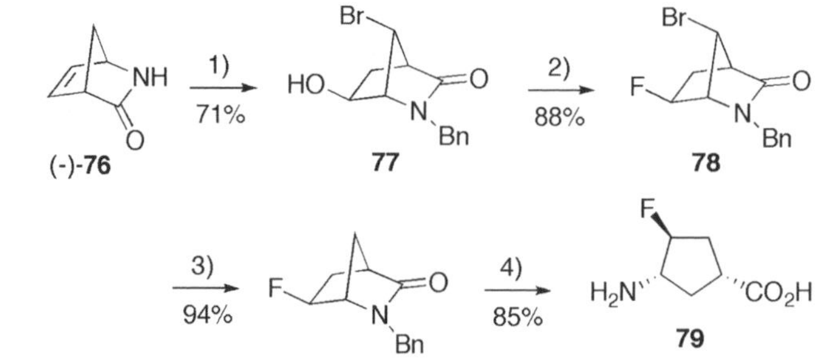

1) (i) BnBr, TMAI, KOH / DMSO, (ii) 1,3-dibromo-5,5-dimethylhydantoin / AcOH, (iii) K_2CO_3 / MeOH-H_2O; 2) DAST; 3) Bu_3SnH, AIBN / benzene: 4) (i) Na-^{t}BuOH / NH_3, (ii) 2M HCl

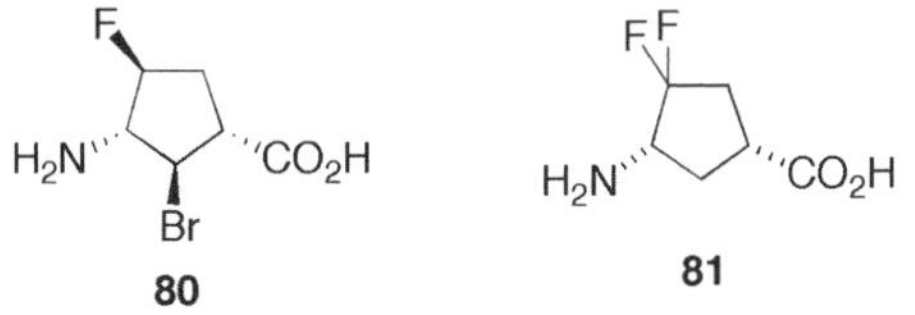

Scheme 9.19

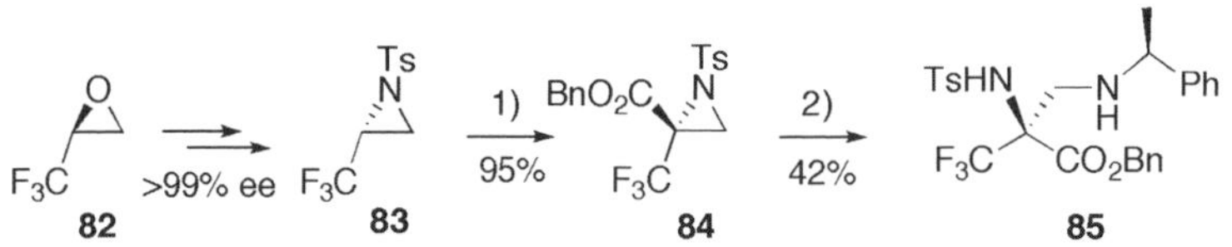

1) n-BuLi, $ClCO_2Bn$ / THF, -100 °C; 2) Phenethylamine / THF, 40 °C

Scheme 9.20

9.2.4 Modification of Chiral Fluorinated Building Block (*S*)-Trifluoropropene Oxide

(*S*)-Trifluoropropene oxide (75% ee) [41] is commercially available. Both enantiomers are readily prepared from racemic TFPO by Jacobsen's co-catalyzed enantioselective ring opening reaction [42]. The enantiomerically pure aziridine **83** is prepared from **82** by ring opening with amine and recyclization with inversion of stereochemistry [43]. Carbobenzyloxylation of **83** was achieved via lithiation followed by alkylation with benzyl chloroformate with complete retention of configuration [44]. Ring opening of **84** with (*S*)-1-phenethylamine provided **85** (see Scheme 9.20).

9.2.5 Enzymatic Resolution of Racemic Fluorinated Building Blocks

Both enzymatic esterification and hydrolysis are useful tools for resolution of racemic fluorinated building blocks. Among them, lipase-catalyzed reaction is reliable and most

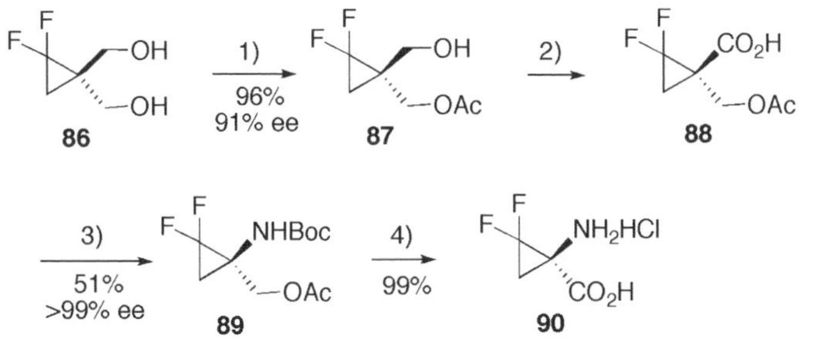

1) Lipase (Amano-PS); 2) CrO_3, H_2SO_4 / H_2O-acetone; 3) $(PhO)_2P(O)N_3$, ^tBuOH, Et_3N / benzene, reflux; 4) (i) K_2CO_3 / MeOH-H_2O, (ii) CrO_3, H_2SO_4 / H_2O-acetone, (iii) 3*N* HCl

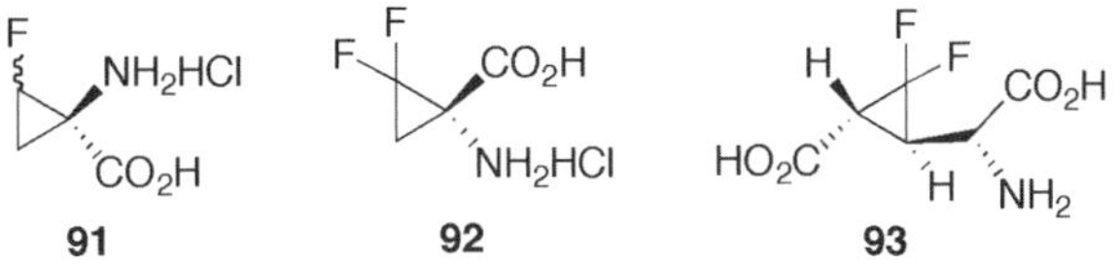

Scheme 9.21

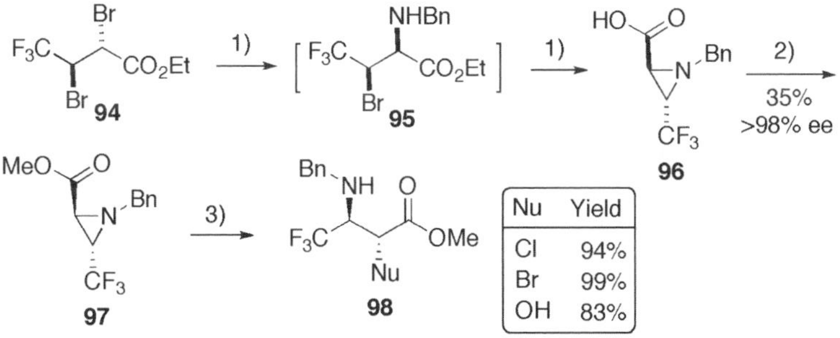

Nu	Yield
Cl	94%
Br	99%
OH	83%

1) $BnNH_2$; 2) *Candida antarctica* lipase; 3) Acid, Nu

Scheme 9.22

frequently used. Difluorocyclopropane diol **86** is resolved by lipase (Amano-PS)-catalyzed esterification to give **87** in excellent yield and enantioselectivity. Further conventional chemical transformation of **87** leads to enantiomerically pure difluorocyclopropane amino acid **90** (see Scheme 9.21) [45]. In relation to the biologically active 1-aminocyclopropane-1-carboxylic acid, several related fluorinated cyclopropane carboxylic acids **91–93** have been synthesized.

Starting from ethyl 4, 4, 4-trifluorocrotonate, racemic aziridine carboxylic acid **96** was prepared as shown in Scheme 9.22 [46] and was then subjected to lipase-catalyzed esterification. Methyl ester **97** was obtained in 35% yield with excellent enantiomeric purity. Acid-catalyzed ring opening of aziridine **97** proceeded regio- and stereoselectively, affording 2-substituted (2*R*,3*R*)- or (2*R*,3*S*)-3-amino-4,4,4-trifluorobutanoates **98** in high yields [47].

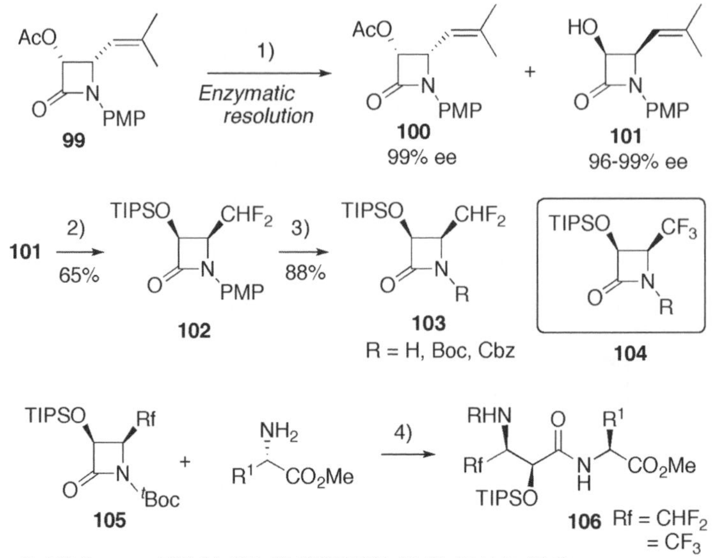

1) PS-Amano, 10% MeCN; 2) (i) TIPSCl, Et_3N, DMAP, (ii) O_3, MeOH-CH_2Cl_2, (iii) DAST / CH_2Cl_2; 3) (i) CAN / H_2O-MeCN, (ii) *N*-protection; 4) NMM / CH_2Cl_2, 25 °C, R = ^{t}Boc, Cbz, Ts, Yield; 50-100%

Scheme 9.23

Likewise, lipase-catalyzed hydrolysis of racemic lactam **99** gave both **100** and **101** in almost enantiomerically pure form (see Scheme 9.23). Conventional chemical conversion of the isobutenyl moiety of **101** to difluoromethyl and trifluoromethyl groups provided lactams **103** and **104**, which were further transformed to dipeptide **106** by ring-opening coupling with amino esters [48].

9.3 Diastereoselective Synthesis of Fluorinated Amino Acids

9.3.1 Chiral Auxiliary Approach

Easy availability, high diastereoselection, and removal under mild conditions are essential for auxiliaries that are feasible for synthetic organic chemistry. Very few chiral fluorinated building blocks are commercially available, so that highly diastereoselective fluorofunctionalization of nonfluorinated building blocks bearing a chiral auxiliary is useful for the synthesis of desired target fluorinated amino acids. Chiral auxiliaries often used for the synthesis of enantiomerically pure or enriched amino acids that appeared in references published since 2000 are summarized in Figure 9.1.

*9.3.1.1 (**S**)- and (**R**)-1-Phenylethylamines and Their Related Amines as Auxiliaries*

(*S*)- and (*R*)-1-Phenylethylamines and the related 2-hydroxy- and 2-methoxyamines are the most available and economically feasible auxiliaries and are frequently used as their

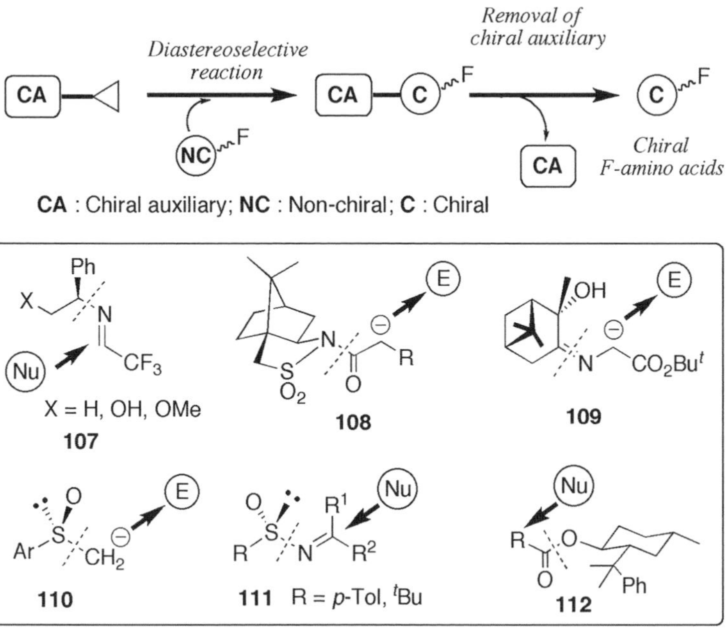

CA : Chiral auxiliary; **NC** : Non-chiral; **C** : Chiral

[a] ------ dot line indicates the point of removal of chiral auxiliary

Figure 9.1 *Chiral auxiliary approach. The dotted lines indicate the point of removal of the chiral auxiliary.*

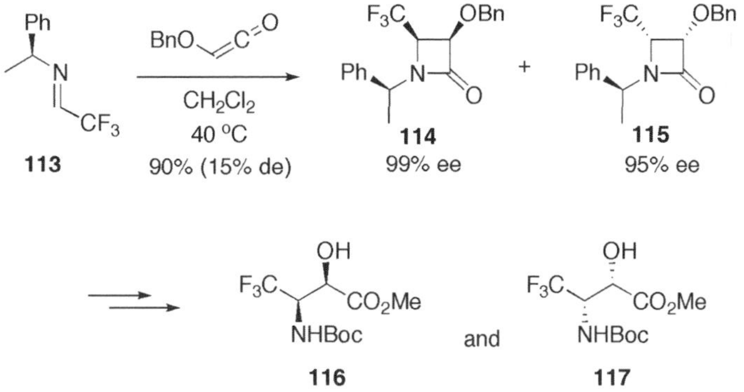

Scheme 9.24

aldimines and ketoimines bearing a fluorine-functionality. The corresponding aldimine **113** of trifluoroacetaldehyde undergoes cycloaddition or electrophilic addition with nucleophiles. Cycloaddition of **113** with ketene provides highly enantiomerically enriched trifluoromethylated lactams **114** and **115**, although diastereoselectivity is poor [49]. The lactams are transformed into *syn*-2-hydroxy-3-aminobutanoates **116** and **117**, respectively (see Scheme 9.24).

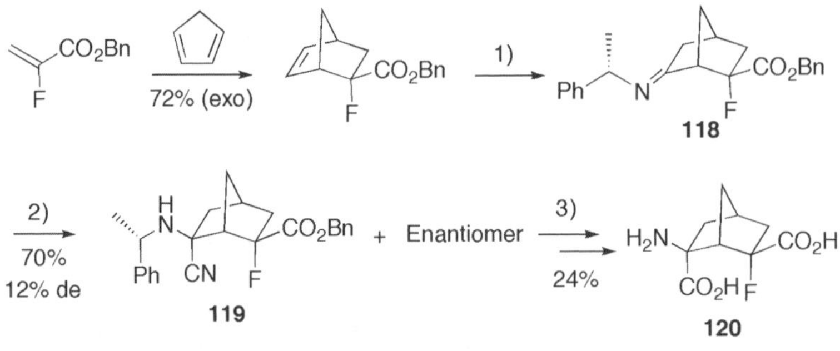

1) (i) BH_3-SMe_2, (ii) H_2O_2, NaOH, (iii) Dess-Martin oxidation, (iv) $TiCl_4$, (*S*)-PhCH(NH_2)Me; 2) TMSCN / $AlCl_3$; 3) (i) HCl / AcOH, 160 °C, (ii) CH_2N_2, (iii) Ti(OPr^i)$_4$ / BnOH, (iv) H_2 / Pd(OH)$_2$

Scheme 9.25

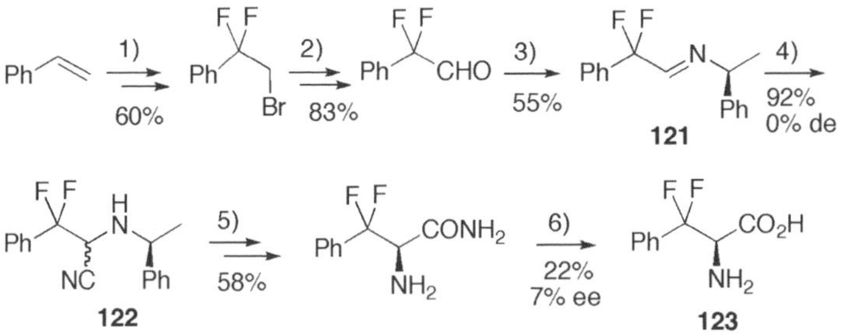

1) (i) NBS, Et_3N_3HF / CH_2Cl_2, (ii) ^{t}BuOK / THF, (iii) NBS, $Et_3N{\cdot}3HF$ / CH_2Cl_2; 2) (i) AcOK, 1,4,7,10,13,16-hexaoxacyclodecane / DMF, (ii) NaOH / EtOH, (iii) DMSO, $(COCl)_2$ / CH_2Cl_2; 3) (*S*)-1-phenylethylamine / toluene, reflux; 4) TMSCN, ZnI_2, 50 °C, 3 days; 5) (i) H_2SO_4, (ii) Separation of diastereomers, (iii) H_2, Pd/C / EtOH; 6) H_2SO_4 / H_2O, reflux, 4 h

Scheme 9.26

Likewise, imine **118** was transformd into **120**, a mimic of a conformationally rigid glutamic acid, via cyanide **119** (see Scheme 9.25) [50]. 1-Phenyl-1,1-difluoroacetaldehyde imine **121** was also used as a precursor of 3,3-difluorophenylalanine **123** via cyanide **122**, but no diastereoselectivity was observed in the cyanation step (see Scheme 9.26) [51].

The 2-methoxy group of 2-methoxy-1-phenylethylamine sometimes plays an important role in diastereoselection. Thus, imine **125** (X = OMe), which is prepared conventionally from imidoyl chloride **124**, undergoes allylation with allylzinc reagent in a diastereoselective manner (>98% de), affording **126** in a high yield. On the other hand, nonmethoxylated imine **125** (X = H) gives the product with a poor diastereoselectivity (40% de) (see Scheme 9.27). The methoxy group will chelate with the zinc atom tightly, making the transition state **127** favorable for the stereo-controlled allylation [52, 53]. However, the effect of the stereoselectivity enhancement by the methoxy group is not necessarily applicable for the stereochemical outcome of the Strecker-type cyanation of

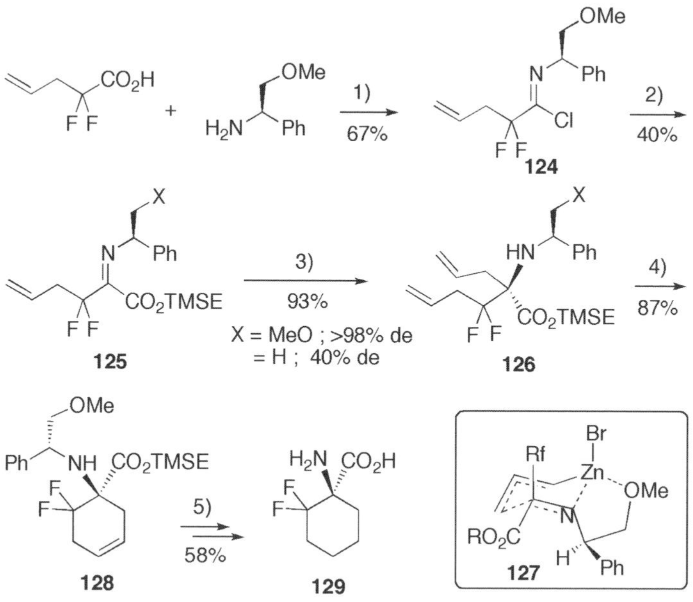

1) PPh_3, Et_3N, CCl_4, reflux; 2) (i) NaI /acetone, (ii) CO, TMSE-OH, K_2CO_3, $Pd(dba)_3$; 3) $CH_2{=}CHCH_2ZnBr$ / THF, -40 °C; 4) Grubbs RCM; 5) (i) H_2, $Pd(OH)_2$, (ii) TBAF / THF, rt. TMSE-OH : $Me_3SiCH_2CH_2OH$

Scheme 9.27

the Schiff base of *N*-(1-phenyl-2-methoxyethyl)amine with trifluoromethyl ketones since the reaction of the Schiff base with TMS-cyanide is strongly affected not only by substituent (X) but also by Lewis acid [54]. Grubbs' ring-closing metathesis (RCM) of **126** leads to cyclohexene skeleton **128**, which is finally converted to 1-amino-6,6-difluorocyclohexane-1-carboxylic acid **129**. Fluoride ion-promoted desilylation is useful for the deprotection of the trimethylsilylethyl group (TMSE) from 2-(trimethylsilyl)ethyl carboxylate **128** under mild conditions.

9.3.1.2 Chiral Oxazolidines as Auxiliaries

Oxazolidine **130** is a masked aldimine bearing a chiral *N*-(2-hydroxy-1-phenyl)ethyl moiety and is readily available from (*R*)-phenylglycinol. Mannich reaction of **130** with Reformatsky reagent of ethyl bromodifluoroacetate produces difluorolactam **131** in high diastereoselectivity, which is then transformed to enantio-enriched (*S*)-3-amino-2,2-difluoro-3-phenylpropanoic acid **133** (see Scheme 9.28) [55].

Oxazolidine **134** is a masked trifluoroacetaldehyde imine that generates *in situ* the corresponding imine **135** under Lewis acid catalysis conditions. Lewis acid-catalyzed reactions of **134** with TMS-cyanide and ketene silylacetal provide adducts **136** in high yields with good diastereoselectivities (see Scheme 9.29) [56]. Conventional chemical transformation of **137** produces 3-amino-4,4,4-trifluorobutanoic acid **138**. Similarly, trifluoroacetone oxazolidine **139** is used for the synthesis of 2-trifluoromethylalanine **142**

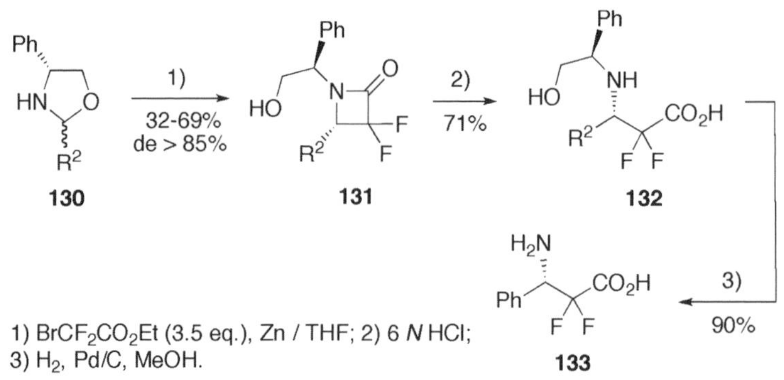

1) $BrCF_2CO_2Et$ (3.5 eq.), Zn / THF; 2) 6 *N* HCl;
3) H_2, Pd/C, MeOH.

Scheme 9.28

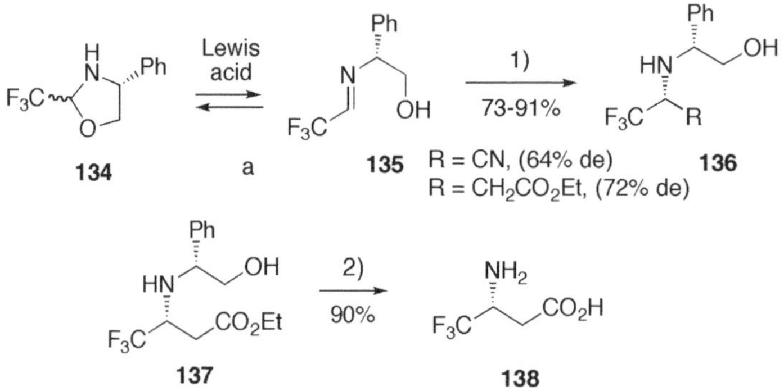

1) TMSCN or $H_2C{=}C(OTMS)OEt$, Lewis acid;
2) $Pb(OAc)_4$, CH_2Cl_2 / MeOH (2:1), then conc. HCl, reflux, 4h., H^+ resin.

Scheme 9.29

via **141**. The stereochemistry of C-2 in **139** does not affect diastereoselectivity in cyanation to ketimine **140** (see Scheme 9.30) [54].

Oxazolidine **144** obtained from amino alcohol **143** and ethyl trifluoropyruvate is also a synthetic intermediate for 2-amino-2-trifluoromethylpentanoic acid **145**. Lewis acid-catalyzed allylation of **144** with allyl silane occurs in excellent yield with a moderate stereoselectivity. Meanwhile, *O-tert*-butyldimethylsilyl-protected imine **146** gives better diastereoselectivity although yield is poor (see Scheme 9.31) [57].

A 1-phenylethylamino moiety is used for diastereomeric control not only in addition of nucleophiles to *N*-(1-phenylethyl)imines but also in diasteroselective Michael addition to α,β-unsaturated esters. Thus, lithium *N*-(1-phenylethyl)-*N*-benzylamide **148** is employed for a one-pot tandem Michael addition-fluorination reaction (see Scheme 9.32) [58]. The reaction provides *anti*-3-amino-2-fluoroesters **149** exclusively, whose diastereoselectivities (64–66% de) to the chiral carbon of the 1-phenylethyl-group are good enough.

9.3.1.3 (2R)-Bornane-10,2-sultam as an Auxiliary

Oppolzer's (2*R*)-bornane-10,2-sultam **150** is a good auxiliary that transmits its chirality to the α-carbon of an amide. The reaction of lithium enolate of **151** with the corresponding

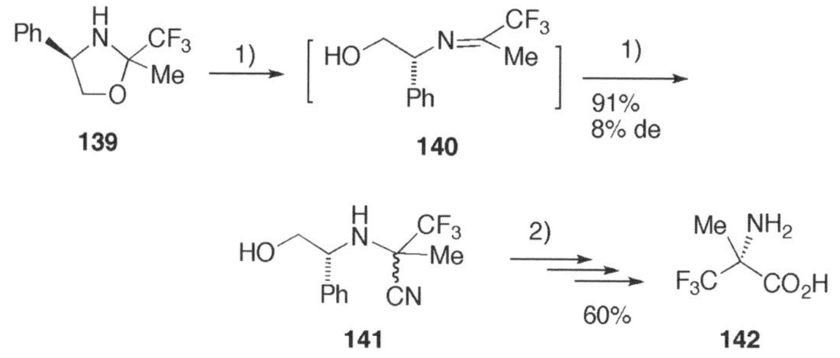

1) TMSCN, TMSOTf / CH_2Cl_2, 0 °C to rt; 2) (i) MeOH, HCl, MeOH, 0 °C, 6 h, (ii) H_2, $Pd(OH)_2$, MeOH, (iii) HCl AcOH, H_2O, 8h, reflux; Propylene oxide, 24 h, rt

Scheme 9.30

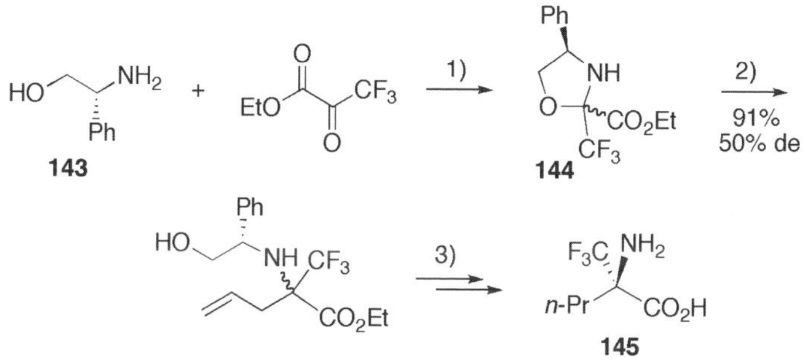

1) Py·*p*-TsOH, PhMe, rt, overnight, rt-reflux; 2) $H_2C{=}CHCH_2SiMe_3$, BF_3-Et_2O / CH_2Cl_2, rt- -15 °C; 3) (i) *p*-TsOH, PhMe, 24 h, rt-reflux, (ii) CF_3CO_2H, H_2, $Pd(OH)_2$, MeOH, overnight, rt, 4 bar, (iii) CF_3CO_2H, H_2, $Pd(OH)_2$, MeOH, 72 h, rt, 4 bar

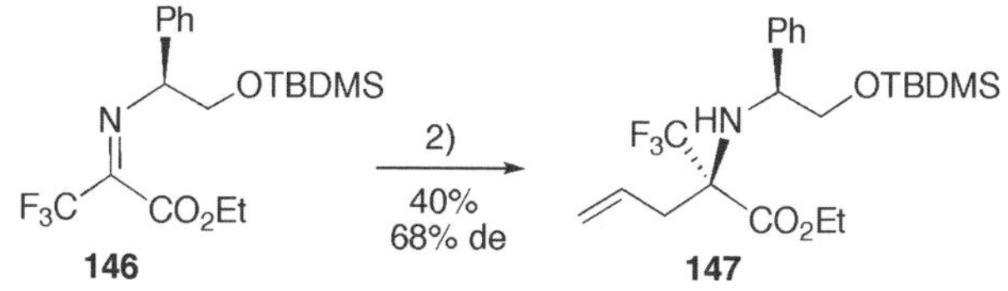

Scheme 9.31

propargyl bromide affords **152** with excellent diastereoselectivity (see Scheme 9.33) [59]. The subsequent reactions starting from **151** – hydrostannylation, iodine–tin exchange, and Pd-catalyzed Suzuki–Miyaura coupling with aryl borane of **151** – furnish a total synthesis of **153**, 5,5-diaryl-2-amino-4-pentenoate, a novel class of biologically active molecules targeted toward the recently cloned glycine reuptake transport system.

The radical perfluoroalkyl-iodination reaction of α,β-unsaturated lactam **154** proceeds diastereoselectively. The stereospecific azide formation and subsequent chemical transformation of the (*R*)-isomer result in the synthesis of a series of perfluoroalkylated α-amino acids **156** (see Scheme 9.34) [60].

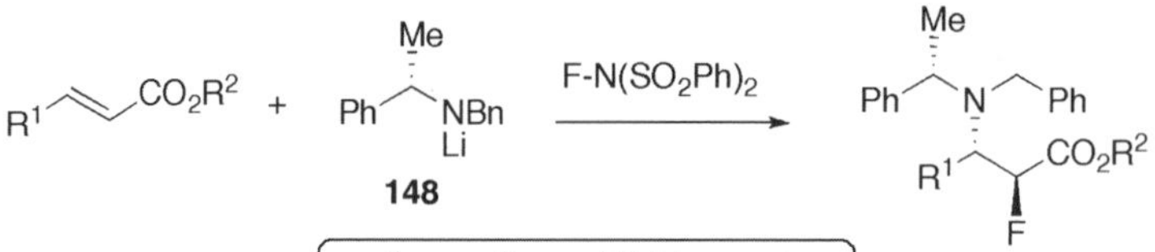

R^1	R^2	Yield	de
Ph	^{t}Bu	Quant	64%
Me	^{t}Bu	39%	66%

Scheme 9.32

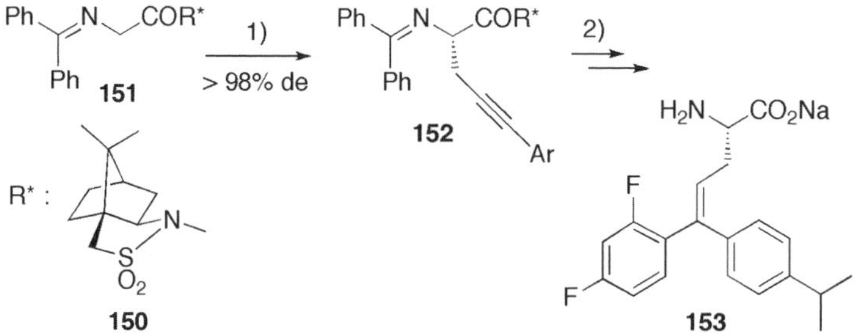

1) LDA / THF-HMPA, Ar-propargyl bromide, -78 °C; 2) (i) Bu_3SnH, $PdCl_2(PPh_3)_2$ / THF, (ii) I_2 / CH_2Cl_2, (iii) 2,4-diF_2-C_6H_3-$B(OH)_2$, $Pd(PPh_3)_4$, Na_2CO_3, DME, 110 °C, (iv) TFA / H_2O, (v) 1*N* NaOH / MeOH

Scheme 9.33

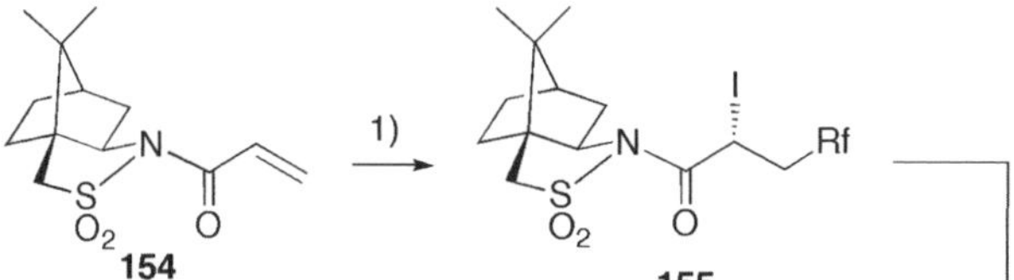

Rf	Yield of **155**	de
n-C_3F_7I	73%	62%
i-C_3F_7I	75%	66%
C_2F_5I	68%	58%
CF_3I	63%	60%

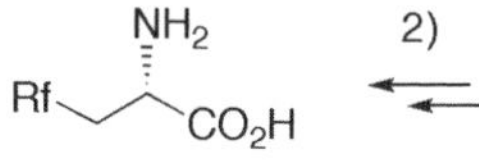

Rf = *n*-C_3F_7: 55%
= C_2F_5 : 55%

1) RfI, $Na_2S_2O_3$ / CH_2Cl_2, Hg lamp; 2) (i) Separation, (ii) NaN_3 / DMF, (iii) LiOH / H_2O/THF, (iv) H_2, Pd/C-AcOH

Scheme 9.34

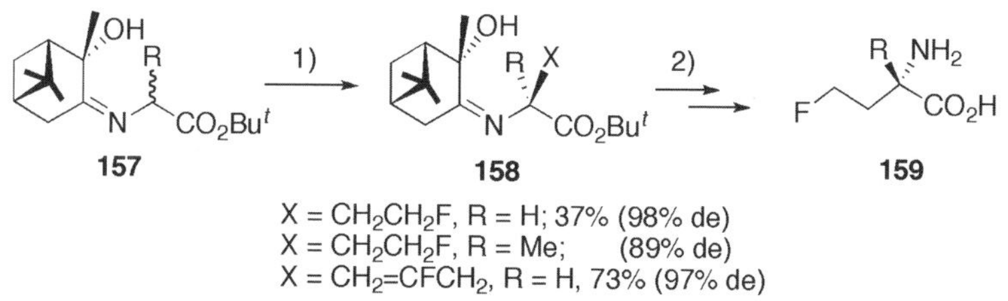

157 → 1) → 158 → 2) → 159

X = CH_2CH_2F, R = H; 37% (98% de)
X = CH_2CH_2F, R = Me; (89% de)
X = $CH_2{=}CFCH_2$, R = H, 73% (97% de)

1) LDA / THF, -78 °C → rt; 2) (i) 15% citric acid, (ii) HCl, propene oxide

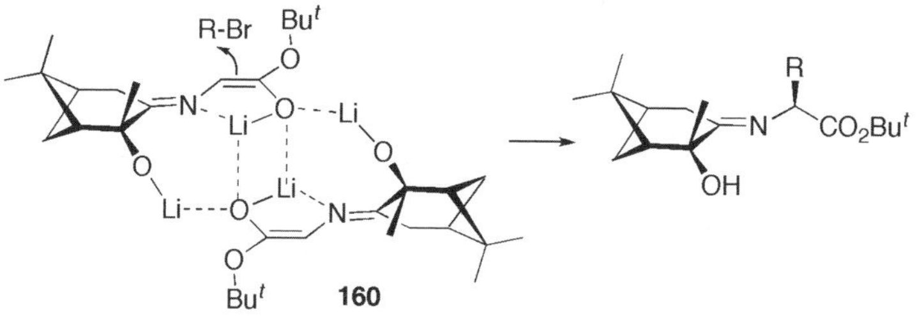

160

Scheme 9.35

161 → 1) 48% → 162 → 2) 82% (96% ee) → 163 → 3) → 164

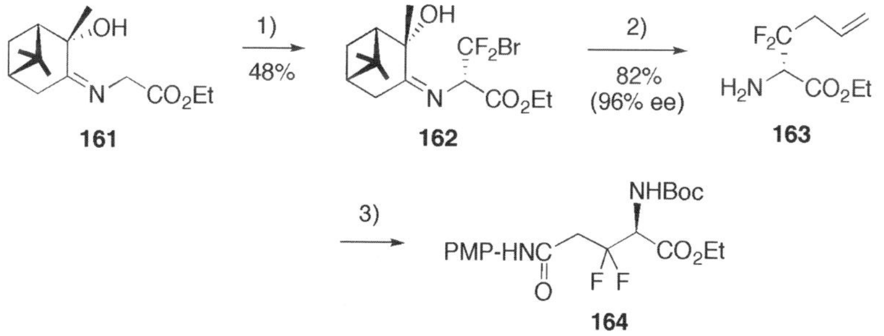

1) LTMP / CF_2Br_2 / TMSOTf / THF /, -78 °C, 5 h; 2) (i) $AllylSnBu_3$ / AIBN, (ii) 1*N* HCl;
3) (i) $(Boc)_2O$ / $NaHCO_3$, (ii) RuO_2 / $NaIO_4$, (iii) DCC / *p*-anisidine

Scheme 9.36

9.3.1.4 2-Hydroxy-3-Pinanone as an Auxiliary

Alkylation of the Schiff base of glycine with 2-hydroxy-3-pinanone proceeds in an extremely diastereoselective manner. Thus, fluoro-functionalization on the α-carbon of the Schiff base followed by hydrolysis provides fluorinated α-amino acids in a highly enantiomerically enriched form. 2-Fluoroethylation and 2-fluoroallylation of **157** (see Scheme 9.35) and bromodifluoromethylation of **161** (see Scheme 9.36) give the desired adducts **158** and **162**, respectively, with excellent diastereoselectivities. Lithium enolate dimer **160** has been proposed as a reactive intermediate for the stereocontrolled alkylation [61]. The adducts **158** and **162** were transformed to 4-fluoro-2-amino acids (>96% ee) **159** [61] and 3,3-difluoroglutamine **164** [62], respectively.

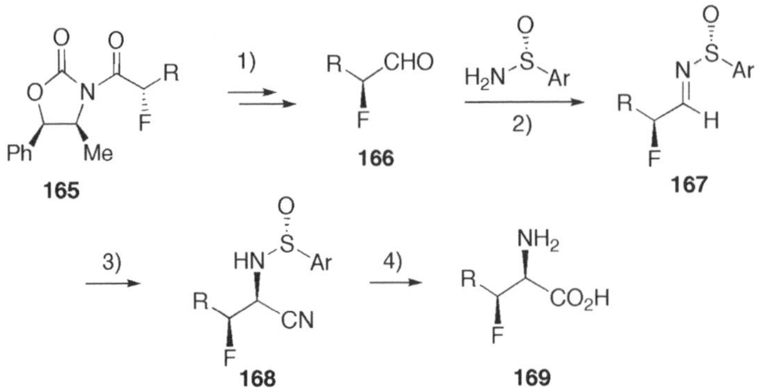

165 → 1) → 166 → $H_2N\text{-}S(O)\text{-}Ar$, 2) → 167 → 3) → 168 → 4) → 169

Step	1)	2)	3)	4)
	R = Ph, 72% (80% ee)	62%, (80% de)	78% (78% de)	69%
	R = Pr^i, 82% (95% ee)	70%, (98% de)	63% (96% de)	58%

1) (i) $LiBH_4$ / THF, 0 °C, (ii) Dess-Martin periodinate oxidation; 2) MS 4A / CH_2Cl_2, rt, 7 h; 3) Et_2AlCN / ^{i}PrOH; 4) 6 *N* HCl / propene oxide

Scheme 9.37

9.3.1.5 Oxazolidinone as an Auxiliary

4-Substituted 2-oxazolidones **165** are useful chiral auxiliaries for diastereoselective functionalization at the α-carbon of their amide carbonyl group. The α-fluoroaldehydes **166** were prepared by a series of reactions: electrophilic fluorination of the corresponding oxazolidinone sodium enolates with *N*-fluorobenzenesulfonimine; reductive removal of the auxiliary with $LiBH_4$; and Dess–Martin oxidation. The aldehydes are so unstable for isolation that they are converted with (*R*)-*p*-toluenesulfinamide to *p*-toluenesulfinimines **167**, which are isolable and satisfactorily enantio-enriched. Chiral sulfinimine-mediated diastereoselective Strecker cyanation with aluminum cyanide provided cyanides **168** in excellent diastereoselectivity, which were finally derived to 3-fluoroamino acids **169** (see Scheme 9.37) [63].

Diastereoslective Michael addition of the amino group in the polymer-supported amino ester **171** to 4,4,4-trifluorocrotonamide **170** is another application of 4-substituted 2-oxazolidone for fluoroamino acid synthesis (see Scheme 9.38) [64].

9.3.1.6 Chiral Sulfinamides as Auxiliaries

Both (*S*)-*p*-toluenesulfinamide **174** [65] and (*R*)-*tert*-butylsulfinamide **182** (see Scheme 9.41) [66] are used for amino acid synthesis. Diastereoselective alkylation to their imines is a key reaction for the creation of chiral amines.

Sulfimines **175** are obtained in excellent yield by Staudinger condensation of sulfinamide **174** with trifluoropyruvates. In contrast, and quite surprisingly, the Staudinger reaction of *tert*-butylsulfinamide **182** with the pyruvates under the same conditions does not work. Sulfimines **175** are alkylated with Grignard reagents in good yields but with poor to moderate diastereoselectivities (see Scheme 9.39) [67].

Meanwhile, Reformatsky reaction of sulfinimines **178** with bromodifluoroacetate provides the adducts in excellent diastereoselectivities (see Scheme 9.40) [68]. The ste-

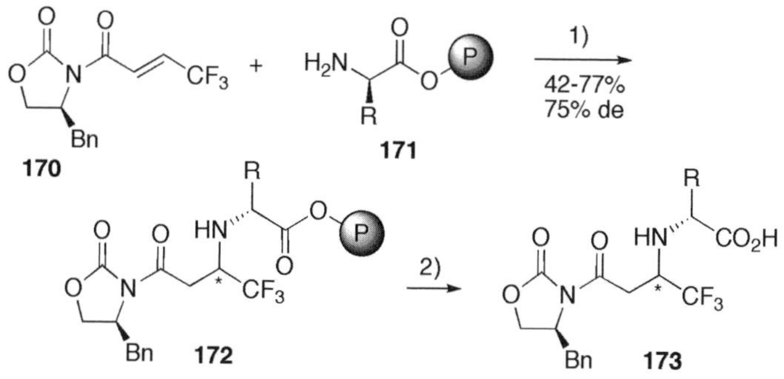

170 171 172 173

1) CH_2Cl_2; 2) TFA / CH_2Cl_2

Scheme 9.38

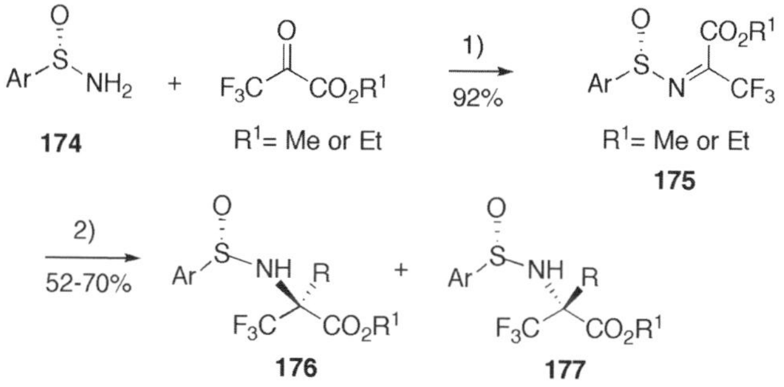

174 175 176 177

R^1= Me or Et

R	Yield (%)	**176**:**177**
Bn	68	30 : 70
Allyl	55	34 : 66
^{i}Bu	65	88 : 12
Et	70	73 : 27
Me	52	55 : 45

1) PPh_3, DEAD / THF, rt; 2) (i) RMgX / THF, -70 °C, (ii) NH_4Cl / H_2O

Scheme 9.39

reochemistry of (*S*) at C-3 and the high diastereoselectivity arise from the six-membered transition state shown in **181**. The same reaction with the corresponding nonfluorinated Reformatsky reagent (bromoacetate) provides 3-aminoester with (*R*)-configuration at C-3 for **179b** (R = aryl, 76%, 94% de), demonstrating that the fluorine atoms of Reformatsky reagent have no effect on the stereochemical outcome of the C–C bond formation. The acid-catalyzed hydrolysis of the sulfinamide **179a** produces enantiomerically enriched 3-amino-2,2-difluorocarboxylates **180** [68].

Likewise, *tert*-butylsulfinamide is also used for highly stereocontrolled syntheses of 3-amino-2,2-difluorocarboxylates **185** (see Scheme 9.41)[69] and 2-phenyl-3,3,3-trifluoroalanine **189** (see Scheme 9.42) [70]. Surprisingly, solvent is significantly effective for

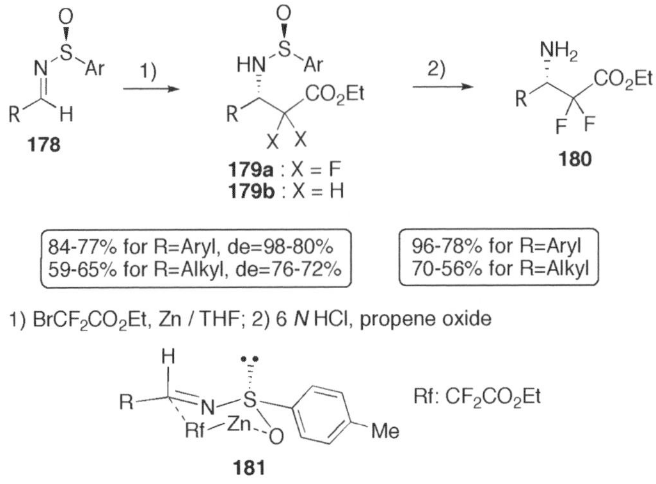

178 → 1) → 179a : X = F; 179b : X = H → 2) → 180

84-77% for R=Aryl, de=98-80%
59-65% for R=Alkyl, de=76-72%

96-78% for R=Aryl
70-56% for R=Alkyl

1) $BrCF_2CO_2Et$, Zn / THF; 2) 6 *N* HCl, propene oxide

Rf: CF_2CO_2Et

181

Scheme 9.40

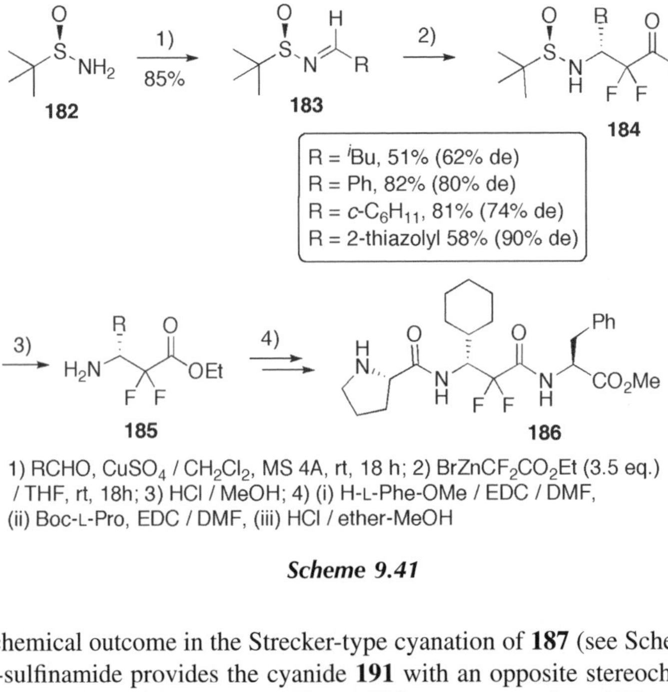

182 → 1) 85% → 183 → 2) → 184

R = ^{i}Bu, 51% (62% de)
R = Ph, 82% (80% de)
R = *c*-C_6H_{11}, 81% (74% de)
R = 2-thiazolyl 58% (90% de)

3) → 185 → 4) → 186

1) RCHO, $CuSO_4$ / CH_2Cl_2, MS 4A, rt, 18 h; 2) $BrZnCF_2CO_2Et$ (3.5 eq.) / THF, rt, 18h; 3) HCl / MeOH; 4) (i) H-L-Phe-OMe / EDC / DMF, (ii) Boc-L-Pro, EDC / DMF, (iii) HCl / ether-MeOH

Scheme 9.41

the stereochemical outcome in the Strecker-type cyanation of **187** (see Scheme 9.42). The use of (*R*)-sulfinamide provides the cyanide **191** with an opposite stereochemistry. Acid-catalyzed hydrolysis of sulfinamides **184** and **188** generates amines **185** and **189**, respectively. The difluoroamino acid **185** is incorporated into oligopeptide **186**.

9.3.1.7 Optically Active Aryl Methyl Sulfoxide

The optically active arylsulfinylmethylene moiety plays two important roles in asymmetric synthesis of optically active amino acids: one is as chiral auxiliary, and the other is as

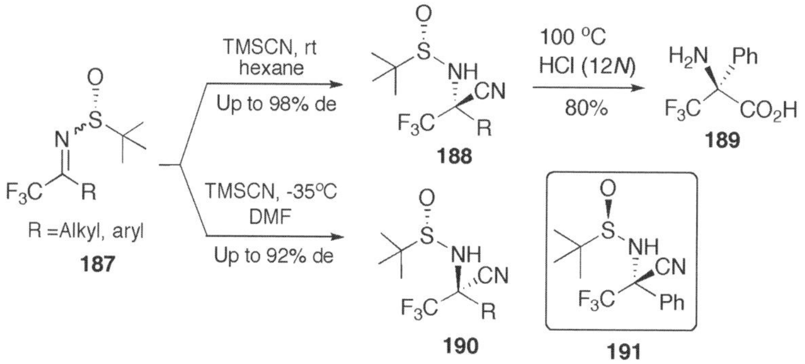

Scheme 9.42

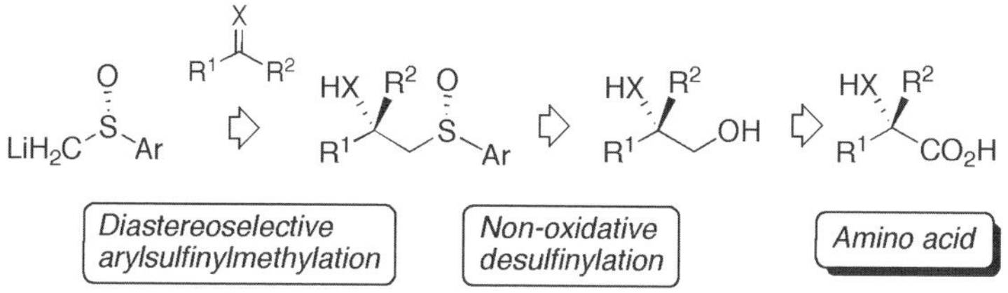

Scheme 9.43

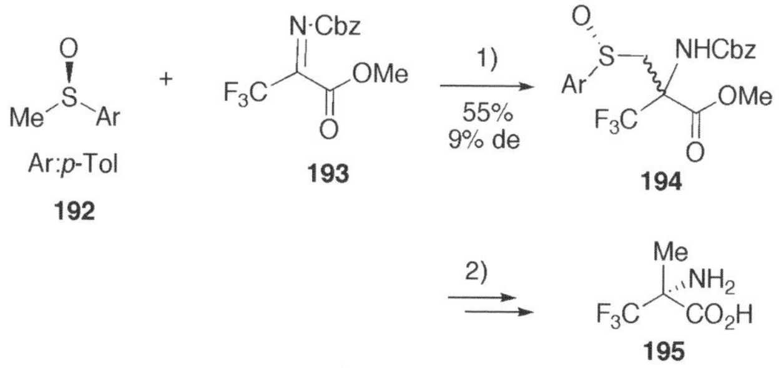

1) LDA/THF; 2) (i) $(CF_3CO)_2O$, NaI, Me_2CO, (ii) H_2 / Ni, EtOH; (iii) HCl / H_2O, (vi) Me_3SiI / MeCN

Scheme 9.44

synthon for the hydroxymethyl group. The arylsulfinylmethylene moiety is introduced to, for example, acyl halides, imines, and imidoyl halides by diastereoselective nucleophilic addition of arylsulfinylmethyl lithium. Then, nonoxidative desulfinylation followed by oxidation or conventional desulfinylation leads to the synthesis of enantio-enriched amino acids (see Scheme 9.43).

A typical example is shown in Scheme 9.44. Arylsulfinylmethyl lithium is introduced diastereoselectively to imine **193** to give **194**. Then, enantio-enriched

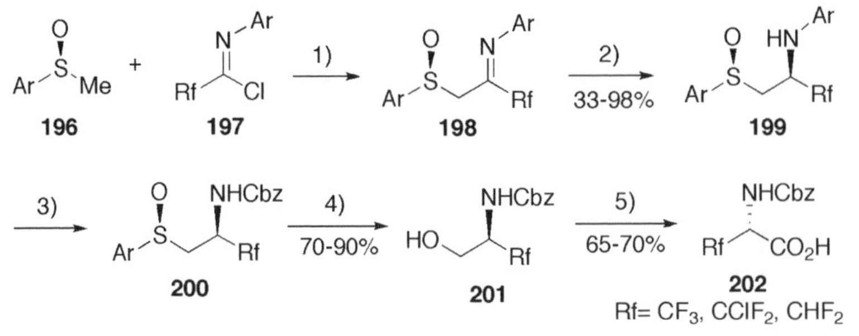

1) LDA (2.0 eq.), THF, -78 °C; 2) Bu_4NBH_4/ MeOH, -70 °C, 3) (i) CAN, MeCN/H_2O, rt, (ii) $ClCO_2Bn$ / dioxane, 50% aq. K_2CO_3; 4) (i) TFAA / MeCN, *s*-collidine, 0 °C, (ii) 10% K_2CO_3, (iii) $NaBH_4$, H_2O; 5) $RuO_2 \cdot xH_2O$ / $NaIO_4$, acetone / H_2O, rt.

Rf= CF_3, $CClF_2$, CHF_2

Scheme 9.45

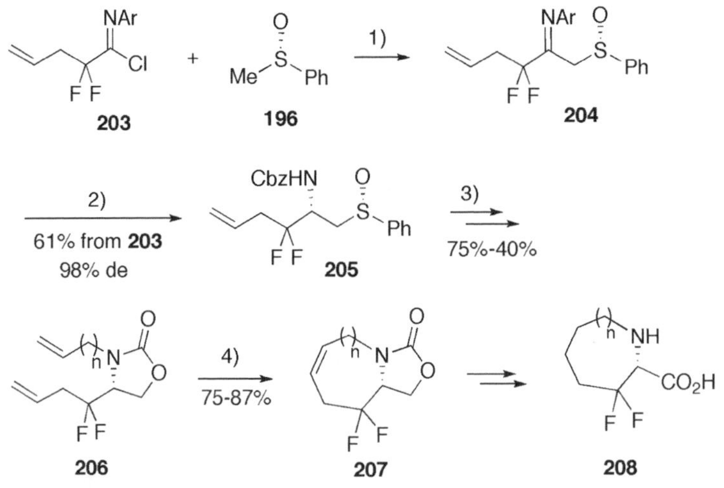

1) (i) LDA / THF, -78 °C; 2) (i) Bu_4NBH_4 THF-MeOH, -70 °C, (ii) CAN / MeCN-H_2O, (iii) $ClCO_2Bn$, K_2CO_3; 3) (i) TFA, (ii) K_2CO_3, (iii) CH_2=$CH(CH_2)_nBr$, NaH, DMF; 4) Grubbs' RCM

Scheme 9.46

2-trifluoromethylalanine **195** is synthesized, where arylsulfinyl group is transformed to a methyl group [71].

Another example is shown in Scheme 9.45, in which arylsulfinylmethyl lithium couples with polyfluoroalkylimidoyl chlorides **197**. Diastereoselective reduction of imine **198** to amine **199** followed by nonoxidative desulfinylative hydroxylation and oxidation of hydroxymethyl group provides trifluoro- and difluoroalanines **202** (see Scheme 9.45) [72]. The same protocol is applicable for the synthesis of 3,3-difluorocyclic amino acids **208** (see Scheme 9.46) [72, 73]. The borohydride reduction of sulfinylimine **204** proceeds with almost complete diastereoselectivity (>98% de). The conventional chemical modifi-

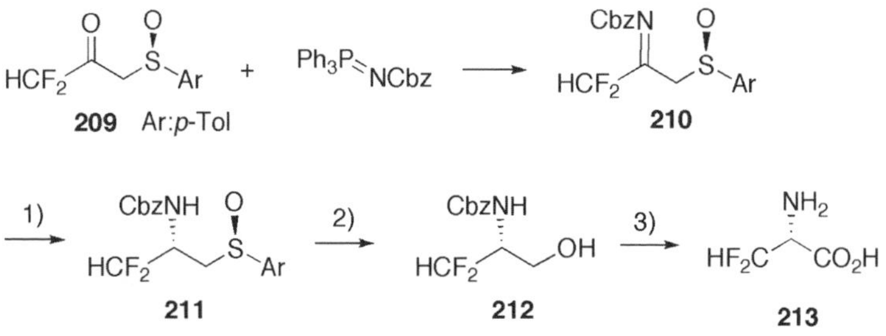

1) $NaBH_4$ / MeOH; 2) (i) *s*-Collidine, $(CF_3CO)_2O$ / MeCN; 3) $NaIO_4$, RuO_2 / $MeCN-H_2O$, (iii) H_2 / $Pd(OH)_2$ / EtOH

Scheme 9.47

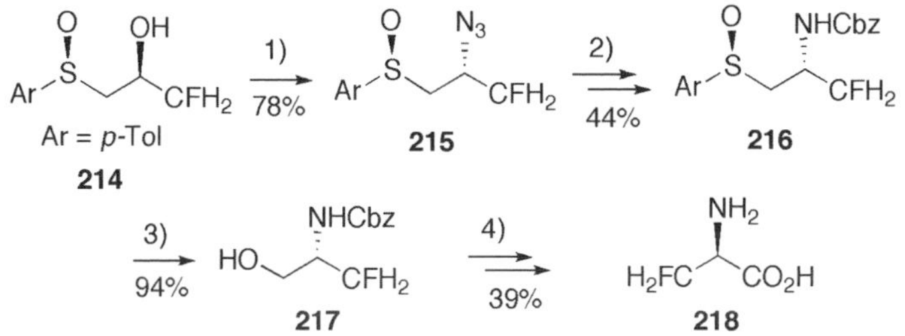

1) NaN_3, PPh_3, CBr_4 / DMF; 2) (i) $HS(CH_2)_3SH$, Et_3N / MeOH, (ii) $ClCO_2Bn$, K_2CO_3 / H_2O; 3) (i) TFAA, collidine / MeCN, (ii) K_2CO_3, $NaBH_4$ / H_2O; 4) (i) $NaIO_4$, RuO_2 / H_2O-acetone, (ii) Ra-Ni, H_2 / EtOH

Scheme 9.48

cation and Grubbs' RCM afford bicyclooxazolidone **207**, which is finally transformd to **208**. Starting from sulfinylketone **209**, enantiomerically pure difluoroalanine **213** is synthesized by the same sequence of reactions as shown in Scheme 9.47 [74]. Stereospecific nucleophilic substitution of the hydroxyl group in sulfinyl alcohol **214** [75] with azide ion and subsequent reduction of the azide group with thiol gave sulfinyl amine **216**. The non-oxidative desulfinylation followed by oxidation of the hydroxymethyl group afforded monofluoroalanine **218** (see Scheme 9.48) [76].

9.3.1.8 Menthyl Group as a Chiral Auxiliary

The menthyl group has often been employed as an easily available chiral auxiliary. 8-Phenylmenthyl 2-amino-3,3-difluorocyclopentenecarboxylate **222** prepared from **219** via **221** was hydrogenated diastereoselectively to form *cis*-2-aminocarboxylate **223** (see Scheme 9.49) [77]. The higher diastereoselectivity induced by the 8-phenylmenthyl group was also observed in ZnI_2-catalyzed $NaBH_4$ reduction of open-chain β-amino-α,β-unsaturated esters **224**. In contrast, the unsubstituted menthyl ester was reduced with no practical diastereoselectivity (see table in Scheme 9.49).

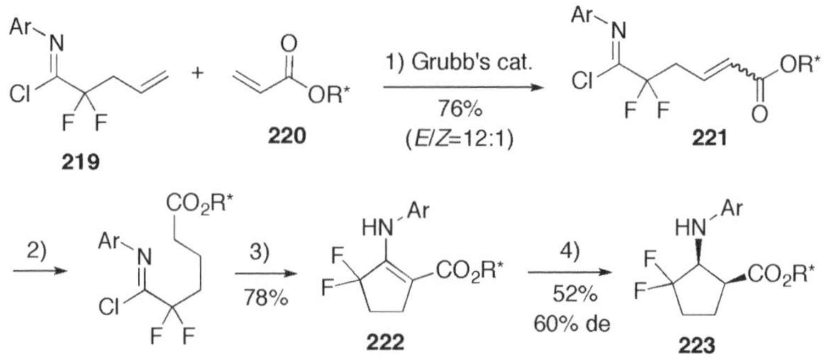

1) Grubb's cat. 76% (E/Z=12:1)

R* : (1*R*, 2*S*, 5*R*)-8-phenylmenthyl ; Ar = *p*-MeO-phenyl

1) [(IMesH_2)(PCy_3)Cl_2Ru=CHPh] (5 mol %); 2) H_2 / Pd / C; 3) LDA / THF; 4) HCO_2NH_4, Pd / C (10%) / EtOH, microwave, 100 °C / 45 min.

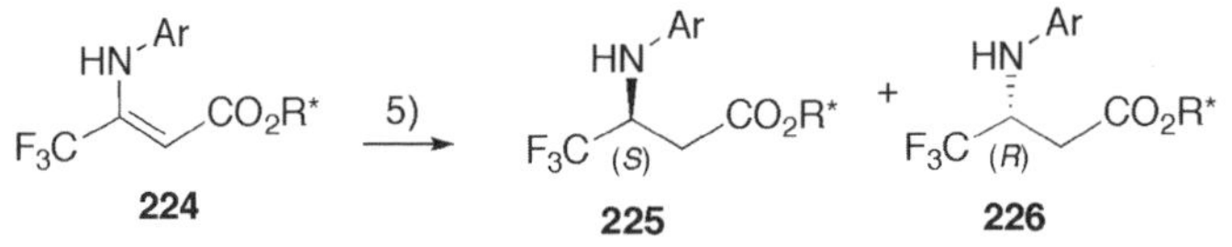

5) $NaBH_4$, ZnI_2 / CH_2Cl_2, rt

R*	Yield (%)	(*S* / *R*)
8-phenylmenthyl	85	80 / 20
Menthyl	93	55 / 45

Scheme 9.49

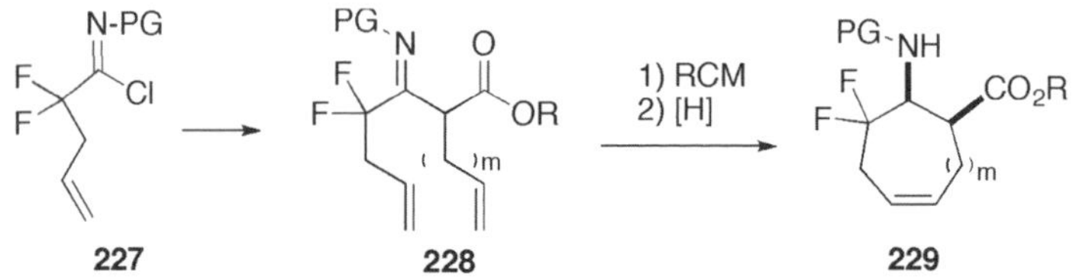

1) RCM 2) [H]

Scheme 9.50

A similar approach was employed for the synthesis of racemic $m = 1$ and higher members of cyclic β-amino esters **229** starting from **227** (see Scheme 9.50) [78].

The oxazoline moiety is a masked carboxyl group. A chiral oxazoline moiety attached to enamine **230** confers chirality to the double bond on reduction. Separation of the diastereomers followed by hydrolysis of the oxazoline moiety provide 3-amino-4,4-difluoro- or 3-amino-4,4,4-trifluorobutanoates **233** in enantiomerically pure forms (see Scheme 9.51) [79].

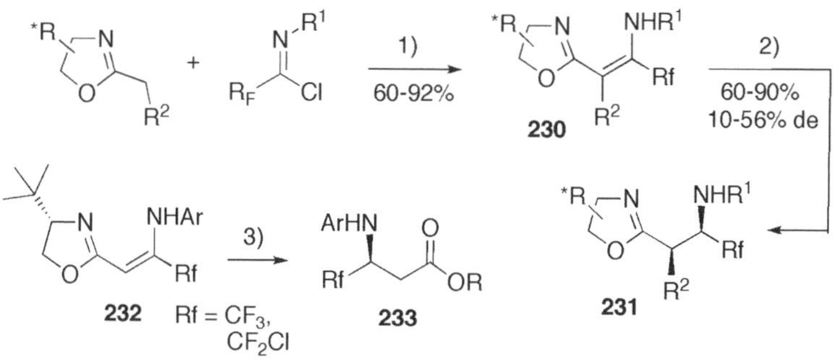

1) LDA / THF, -78 °C, 2-8 h; 2) $NaBH_4$, ZnI_2 / CH_2Cl_2, rt or H_2, Pd/C, MeOH, rt;
3) (i)1*N* HCl, heat, (ii) ROH / HCl.

Scheme 9.51

9.3.2 Building Blocks

Chiral fluorinated building blocks are in general less readily available, so that fluoro-functionalization of available chiral building blocks is one feasible approach for the asymmetric synthesis of fluorinated amino acids (see Fig. 9.2).

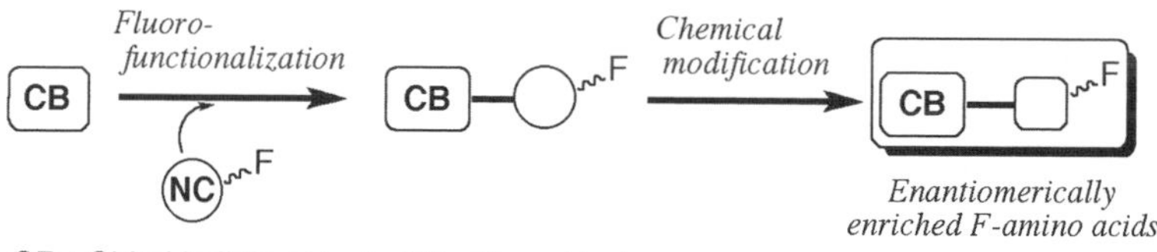

Figure 9.2 *Chiral building block approach.*

9.3.2.1 Garner's Aldehyde

Garner's aldehyde is readily available from serine. Synthetic processes so far reported for fluoroamino acids mostly consist of (1) fluoro-functionalization of the Garner's aldehyde at the formyl group, (2) acid-catalyzed hydrolytic opening of the oxazoline ring, and (3) oxidation of the hydroxymethyl group to a carboxyl group. Reaction of **234** with ethyl 2-fluoro-2-phosphonoacetate followed by diastereoselective hydrogenation of **235** provides monofluoroester **236**. A sequence of reactions – condensation of **236** with nitromethane, catalytic reduction of **237a**, reaction of **237b** with acetimidate, and transformation of the oxazoline ring to an α-aminocarboxyl moiety – synthesize 4-fluoro-L-lysine **238** (see Scheme 9.52) [80]. 4,4-Difluoroglutamine **242** is synthesized similarly (see Scheme 9.53) [81].

The aldehyde **234** is also applicable for construction of the proline skeleton. The oxazoline moiety of **234** was transferred to the N-$CHCO_2H$ moiety of the prolines **249**, the trifluoromethyl group of which was introduced by cross-coupling of bromide **244** with trifluoromethyl copper reagent (see Scheme 9.54) [82].

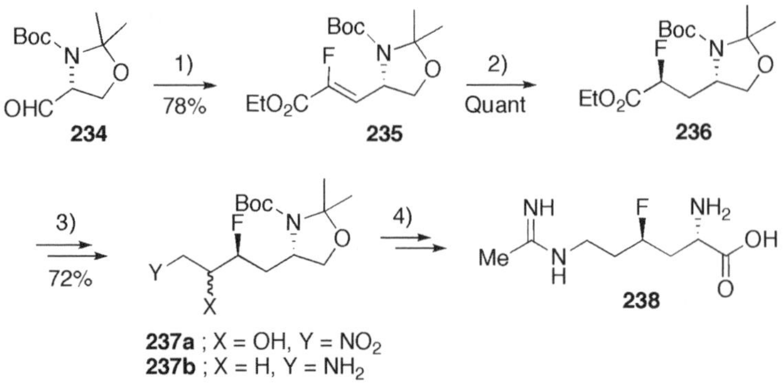

1) (i) $(EtO)_2P(O)CHFCO_2Et$, NaH, THF, -40 °C-rt; 2) H_2, 60 psi, Pd/C, EtOH; 3) (i) $NaBH_4$, MeOH, -50-0 °C, (ii) $MeNO_2$, Na_2CO_3, THF, rt, 2 d, (iii) Pd/C, MeOH, H_2, 5 psi, rt; 4) Ethyl acetimidate hydrochloride, NaOH / EtOH, rt.

237a ; X = OH, Y = NO_2
237b ; X = H, Y = NH_2

Scheme 9.52

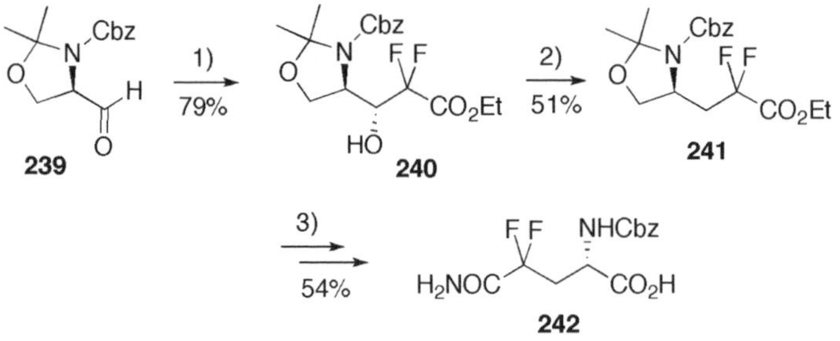

1) $BrCF_2CO_2Et$, Zn / THF; 2) (i)$(Im)_2C=S$ / THF, (ii) Et_3SiH, dioxane, BPO
3) (i) TsOH / EtOH, (ii) CrO_3, H_5IO_6, MeCN-H_2O, (iii) NH_4OH / MeOH

Scheme 9.53

9.3.2.2 (R)-2,3-O-Isopropylideneglyceraldehyde

(*R*)-2,3-*O*-Isopropylideneglyceraldehyde **250** (see Scheme 9.55) is also one of the most readily available chiral building blocks for amino acid synthesis. Here again, a similar sequence to that with Garner's aldehyde is used: fluoro-functionalization at the formyl group and oxidation to carboxyl group of either the secondary or the primary hydroxyl group involved in the masked glyceraldehydes **250**. The Mitsunobu protocol for transformation of a hydroxyl group to an amino group is used as a key reaction for synthesis of fluorinated amino acids [83].

In the synthesis of 3-fluoroamino acid **256** (Scheme 9.55), fluorination of alcohol **251** (R^2 = H) with morpho-DAST afforded the desired alcohol only in low yield (10–25%). The same reaction of trimethylsilyl ether **251** (R^2 = TMS) improved the yield (50%). The Mitsunobu amination of the secondary hydroxyl group in **254** successfully gives **255** in 89% yield. Starting from **250**, several fluorinated α- and β-amino acids **257–259** have been prepared [31].

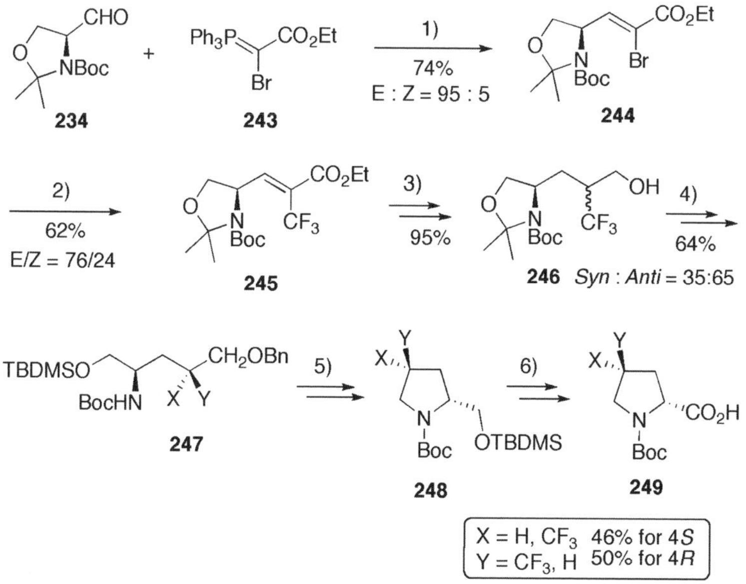

X = H, CF_3 46% for 4*S*
Y = CF_3, H 50% for 4*R*

1) CH_2Cl_2, reflux, 1 d; 2) $FSO_2CF_2CO_2Me$, CuI / DMF-HMPA; 3) (i) H_2, Raney-Ni, MeOH, rt, overnight, (ii) $LiAlH_4$, Et_2O, 0 °C; 4) (i) BnBr, NaH, TBAF, THF, rt, 5 h, (ii) 80% AcOH, 50 °C, overnight, (iii) TBDMSCl, imidazole / CH_2Cl_2, rt, 1h; 5) (i) 10% Pd/C, H_2, EtOH, rt, overnight, (ii) MsCl, Et_3N, CH_2Cl_2, rt, overnight, (iii) KHMDS / THF, 0 °C, 24 h; 6) (i) TBAF, THF, rt, 2 h, (ii) Jones reagent.

Scheme 9.54

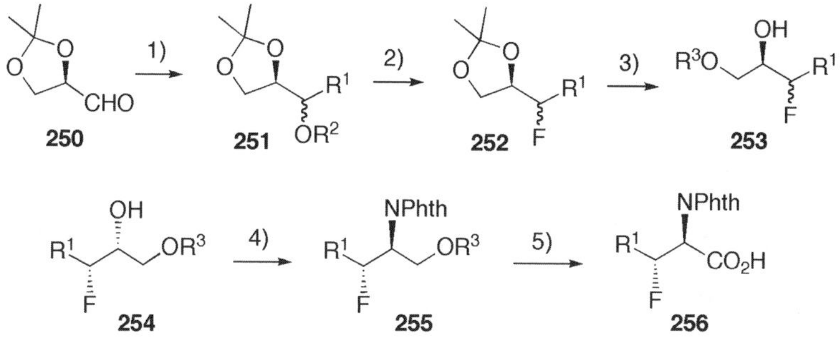

1) $C_8H_{17}MgBr$, 3*R*:3*S* = 2.5:1 ; 2) Morpho-DAST / CH_2Cl_2, 3*R*:3*S* = 5.7:1 ; 3) 4 *N* HCl-THF, then ^{t}BuMe$_2$SiCl/DMAP / CH_2Cl_2; 4) PhthNH / DEAD / PPh_3 / toluene; 5) $NaIO_4$ / RuO_2 $R^1 = C_8H_{17}$

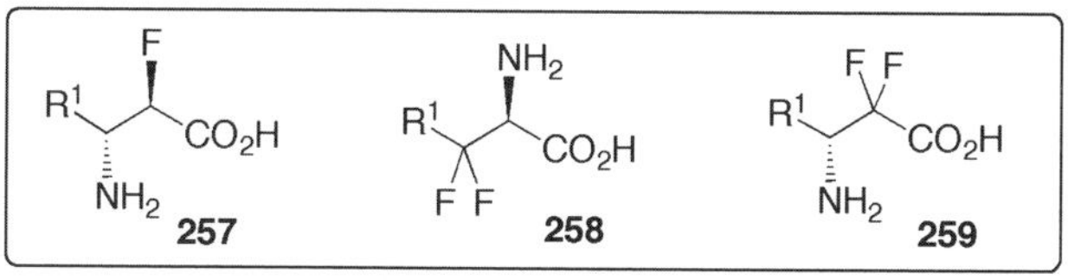

Scheme 9.55

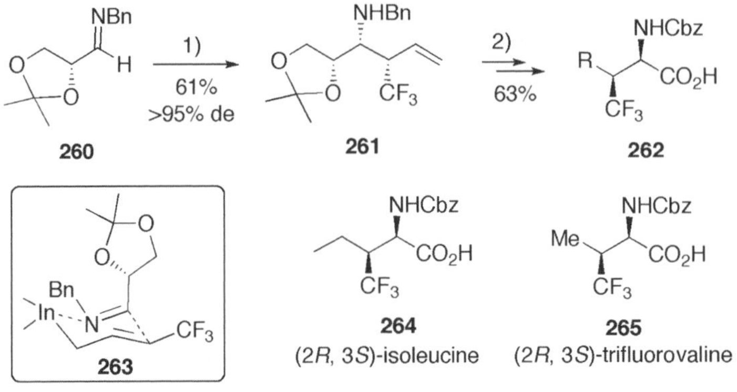

1) In, $CF_3CH{=}CHCH_2Br$ / DMF; 2) (i) Pd/C, H_2, (ii) CbzCl, $NaHCO_3$ (iii) TsOH, (iv) RuO_2, $NaIO_4$

Scheme 9.56

The benzyl imine **260** of aldehyde **250** is an excellent building block for the synthesis of enantiomerically pure trifluoromethylated isoleucine **264** and valine **265** (see Scheme 9.56). The In-mediated alkylation of imine **260** with 4,4,4-trifluorocrotyl bromide in DMF proceeds with excellent diastereoselectivity (>95% de), affording **261**. In contrast, poor diastereoselectivity (20% de) is obtained in the same In-mediated aldol reaction of aldehyde **250**. The transition state structure **263** is proposed to explain the exclusive stereocontrol [84].

9.3.2.3 Proline Derivatives

Diastereoselective fluoro-functionalization of proline derivatives produces fluorinated prolines. Both enantiomerically pure 4-trifluoromethyl and 4-difluoromethyl prolines **268** and **270** have been prepared (see Scheme 9.57) [85]. The key reaction for the stereo-controlled synthesis is diastereoselective Pd-catalyzed hydrogenation of **267** and **269**.

Nickel(II) complex **271** of the Schiff bases of glycine with *o*-[*N*-α-pycolylamine]be nzophenone plays two roles in stereocontrolled C–C bond formation; an excellent chiral auxiliary and a building block. Its diastereoselective Michael addition to **272** occurs cleanly at the α-position of the glycine moiety, affording **273** almost quantitatively. The adduct **273** is hydrolyzed by acid-catalysis to give **274**, which is then cyclized to pyroglutamic acid **275** [86]. During the two hydrolysis steps, both *o*-[*N*-α-pycolylamine]benzoph enone and (*S*)-5-phenyl-2-oxazolidone can be recovered for recycling (see Scheme 9.58).

9.3.3 Other Diastereoselective Syntheses

Other interesting diastereoselective syntheses of fluorinated amino acids are briefly summarized in this section.

Tandem alkene metathesis–intramolecular Michael addition of (*R*)-amine **276** proceeds diastereoselectively, providing pyrrolidines **278** and **279**. These pyrrolidines can be

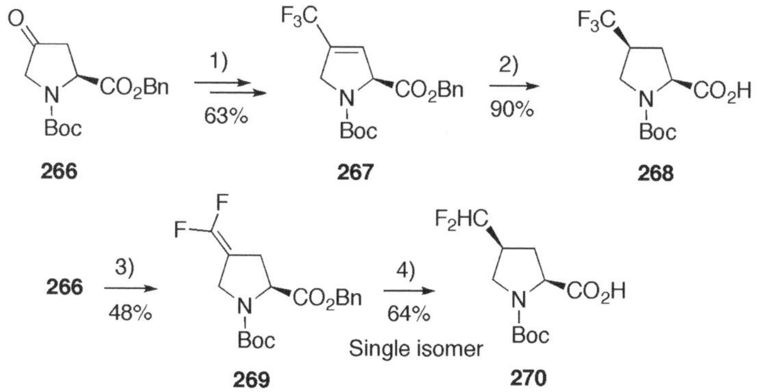

1) (i)CF_3SiMe_3, TBAF(cat.), rt, overnight; (ii) Sat. aq. NH_4Cl, rt, 15 min, then, TBAF, rt, 1 h, (iii) $SOCl_2$, pyridine, reflux, 20 min; 2) Pd/C, H_2, EtOH; 3) CF_2Br_2, Zn, HMPT, THF, reflux, 3.5 h ; 4) Pd/C, H_2, EtOH

Scheme 9.57

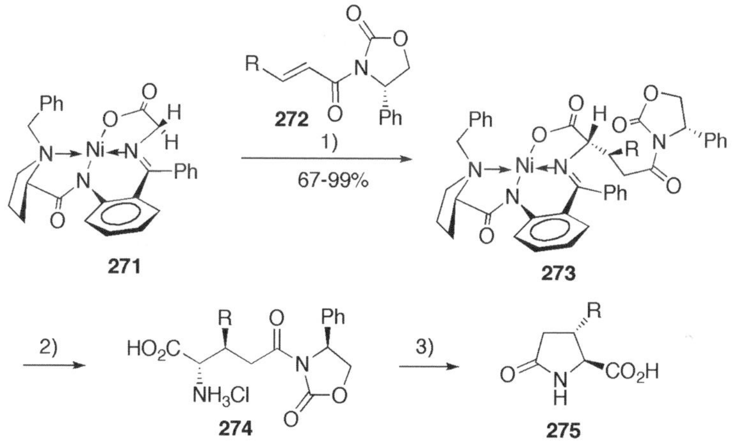

1) DBU / DMF, 30 min; 2) 2 *N* HCl / MeOH; 3) (i) Conc. NH_3, (ii) Dowex

Scheme 9.58

promising precursors of 5-CF_3-prolines and the related amino acids (see Scheme 9.59) [87].

Introduction of a trifluoroethyl group at the α-carbon of the nitro group of **280** by a sequence of reactions – (1,5-diazabicyclo[4.3.0]non-5-ene) (DBU)-catalyzed alkylation with trifluoroacetaldehyde hemiacetal in MeCN, dehydration with acetic anhydride–pyridine, and reduction of the C–C double bond with $NaBH_4$ – produces **281**. 6-Trifluoroethyl-L-lysine **282** can be prepared from **281** (see Scheme 9.60) [88].

Diastereoselective fluorination of enolate **283**, subsequent epoxidation of **284**, and separation of diastereomers provide **285**. Activation of epoxy-oxygen with Et_2AlCl and deprotonation from the CHF group in **285** with LDA induces stereospecific

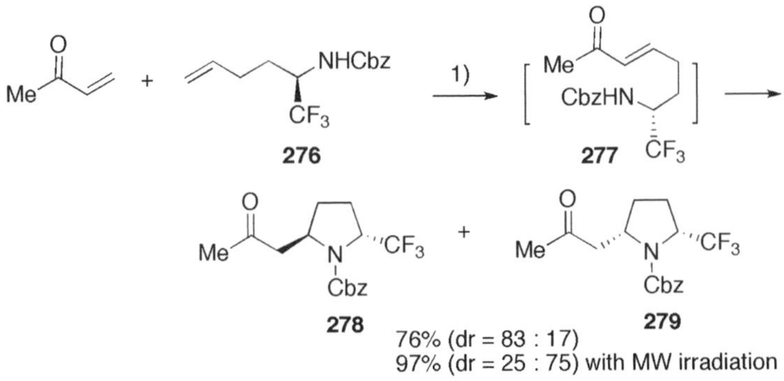

76% (dr = 83 : 17)
97% (dr = 25 : 75) with MW irradiation

1) Hoveyda-Grubbs' cat. (5 mol%), $BF_3 \cdot OEt_2$ (1 mol%) / DCM, 45 °C, 4 d. or MW

Scheme 9.59

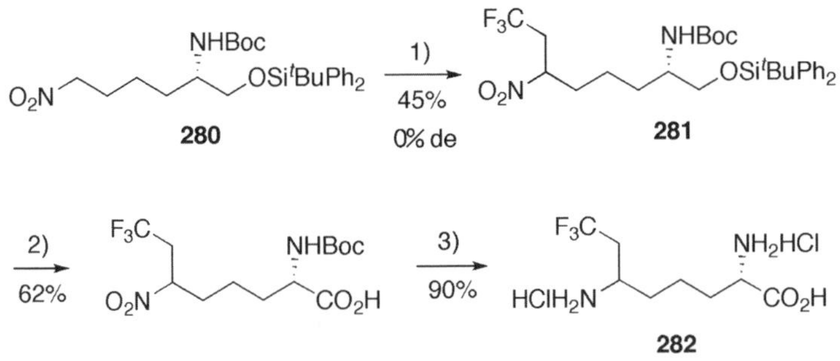

1) (i) $CF_3CH(OH)OEt$, DBU / MeCN, (ii) Ac_2O / py, (iii) $NaBH_4$ / THF-H_2O; 2) (i) TBAF / THF, (ii) PCC / DMF; 3) (i) $Pd(OH)_2$ / EtOH, 60 psi, (ii) HCl / dioxane-AcOH

Scheme 9.60

cyclopropanation, affording bicyclo product **286**, which is further transformed to bicyclic α-amino acid **287** (see Scheme 9.61) [89].

9.4 Racemic Amino Acids

In this section, recent advances in the syntheses of racemic fluorinated amino acids and the related peptides that involve conceptually new synthetic designs are briefly summarized.

Base-catalyzed deprotonation–alkylation of trifluoroalanine is not easy because α-trifluoromethylated carbanions readily undergo defluorination in general [90]. Therefore, α-alkylated trifluoroalanines have been prepared either by alkylation of imines obtained from trifluoropyruvates [54, 57]or by Strecker cyanation [56] to trifluoroketimines fol-

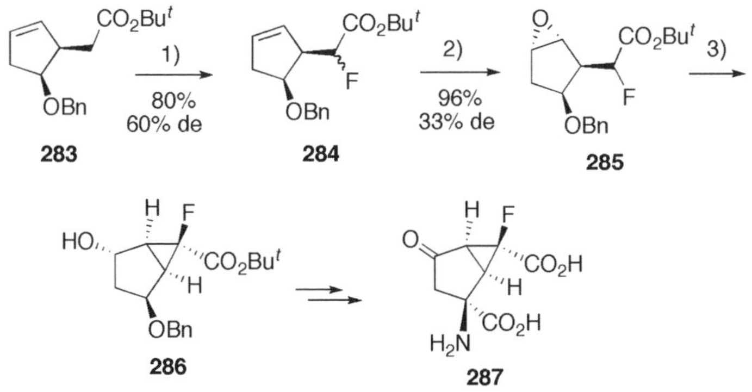

1) LDA, $(PhSO_2)_2NF$ / THF; 2) (i) *m*CPBA, Na_2HPO_4 / CH_2Cl_2, (ii) Chromatographic separation; 3) LDA, Et_2AlCl / THF

Scheme 9.61

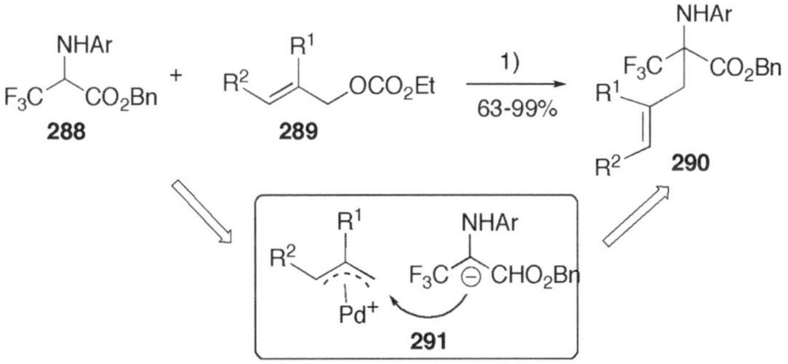

1) $Pd(dppe)_2$, MS 5A / THF, rt or reflux

Scheme 9.62

lowed by hydrolysis. π-Allyl palladium **291** is strongly electrophilic, so that it can trap the enolate of **288** to give α-allylated trifluoroalanine **290** in excellent yields (see Scheme 9.62) [91].

Stereospecific nucleophilic substitution of the hydroxyl group in *sec*-trifluoromethyl alcohols with nucleophiles is one of the promising approaches for synthesis of optically active trifluoromethylated tertiary carbon skeletons, since enantiomerically enriched *sec*-trifluoromethylalcohols are readily available [92]. However, the S_N2 reaction with carbon nucleophiles has never been developed, although the reactions with heteroatom nucleophiles are known [92]. Only the intramolecular stereospecific cyclopropanation has been employed for the syntheses of racemic trifluoromethylated coronamic acids **294** and **297**, as shown in Scheme 9.63 [93]. It is noteworthy that the stereochemistry for the cyclopropanation is governed by electrostatic repulsion between the trifluoromethyl group and the electronic charge on *ortho*-positions in the aryl group rather than by steric repulsion between the two substituents [94].

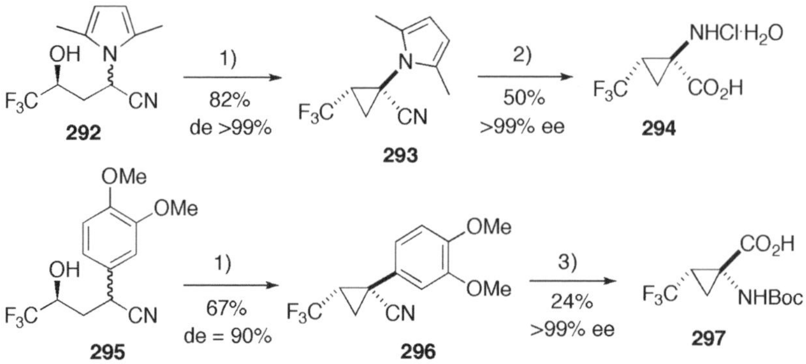

1) (i) *p*-TsCl, NaH, THF, rt, (ii) Recrystallization; 2) (i) $RuCl_3$ (cat.), $NaIO_4$ / $MeCN$-CCl_4-H_2O; (ii) aq. HCl; 3) (i) NaOH, H_2O_2, reflux; (ii) $Pb(OAc)_4$ / *t*-BuOH, reflux; (iii) $RuCl_3$ (cat.), $NaIO_4$ / $MeCN$-CCl_4-H_2O.

Scheme 9.63

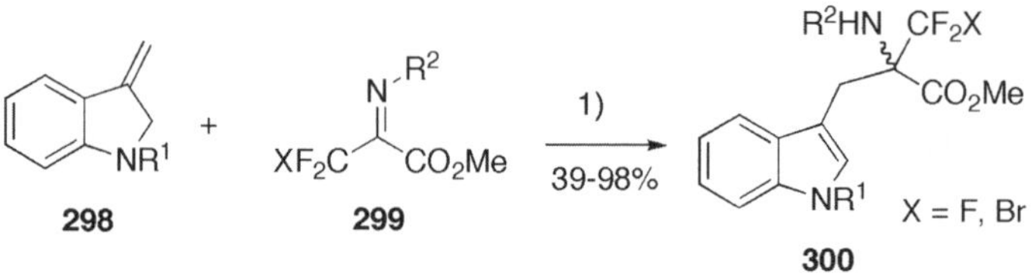

1) Benzene, rt, $R^1 = SO_2Ph$, $R^2 = SO_2Ar$, Boc

Scheme 9.64

Imines **299** of trifluoropyruvates are so strongly electrophilic that they undergo spontaneous ene-type carbon–carbon bond formation between **298** and **299** on their being mixed, as shown in Scheme 9.64. α-Trifluoromethylated tryptophan and its related amino esters **300** have been prepared (see Scheme 9.64) [95].

Copper-catalyzed thermal reaction of diazo-trifluoropyruvate **301** with aryl aldehydes in the presence of $SbBu_3$ provides alkenes **302**, which are transformed to racemic amino acids **304** (see Scheme 9.65) [96]. Antimony ylide **305** is proposed as an intermediate.

Dipeptides are mostly prepared by the condensation of two amino acids, in which a carboxyl group of one of the amino acids must be activated by condensation-promoting reagents. Noncondensation synthesis of trifluoroalanine dipeptides is shown in Scheme 9.66 [97]. Pentafluoroenamine **306** reacts with the amino group of amino esters in DMF at 0 °C within 1 hour, affording trifluoroalanine dipeptides in a one-pot synthesis. The overall reactions involve addition of amino ester to **306**, dehydrofluorination of **308**, and acid-catalyzed hydrolysis of imidoylfluorides **309**. Enamine **306** is readily prepared in 95% yield by magnesium-promoted defluorination of the corresponding hexafluoroacetone imine. Enamine **306** is strongly electrophilic and reacts spontaneously with even a weak nucleophile such as an amino ester due to the electron-withdrawing nature of both CF_2 and CF_3 groups [98] and is highly chemoselective since it accepts the amino group of

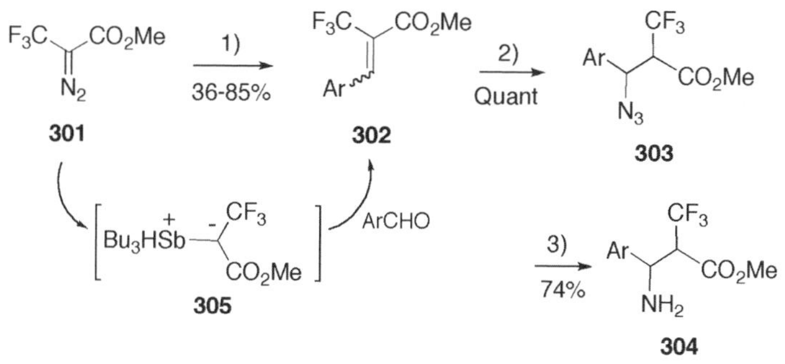

1) $SbBu_3$ / $Cu(acac)_2$, ArCHO; 2) HN_3, AcOH; 3) Pd/C / H_2, MeOH

Scheme 9.65

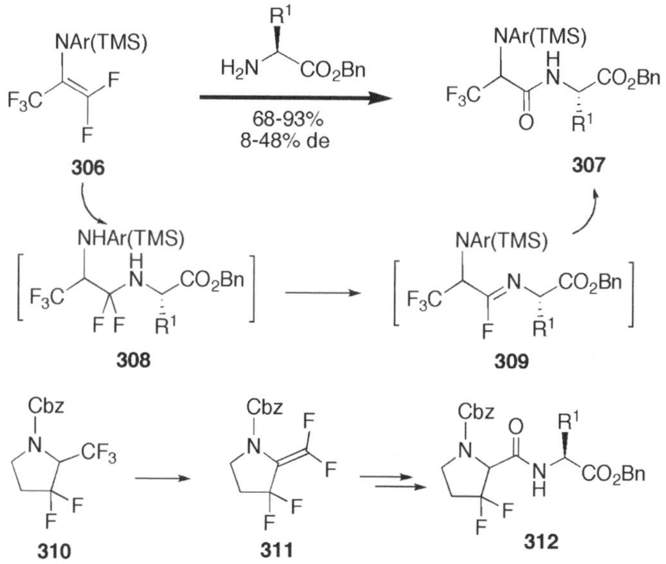

Scheme 9.66

serine and tyrosine without protecting the hydroxyl group. The same protocol is applicable for the synthesis of difluoroproline dipeptides of **312** from **310** [99].

9.5 Conclusions

Recent advances in the syntheses of fluorinated amino acids have been summarized on the basis of the strategies for creating chiral centers in target amino acids: (1) enantioselective, (2) diastereoselective, and (3) racemic syntheses. Remarkable progress in fluorinating reagents and fluoro-functionalization methodologies; the commercial availability of fluorinated building blocks; and easy accessibility of computer-based information on the chemistry and science of organofluorine compounds along with the well-collected and well-arranged reference books recently published [100] – all of these increase the potential for synthesis of desirable fluorinated amino acids.

References

1. (a) Babu, I. R., Hamill, E. K. and Kumar, K. (2004) A highly stereospecific and efficient synthesis of homopentafluorophenylalanine. *J. Org. Chem.*, **69**, 5468–5470; (b) Wang, P., Tang, Y. and Tirrell, D. A. Incorporation of Trifluoroisoleucine into proteins in vivo. (2003) *J. Am. Chem. Soc.*, **125**, 6900–6906.
2. (a) Soloshonok V. A. (ed.) (2005) Fluorine-containing amino acids and peptides: fluorinated synthons for life sciences. in *Fluorine-containing Synthons*, ACS Symposium Series **911**, American Chemical Society, Washington, D.C.; (b) Soloshonok, V. A., Mikami, K., Yamazaki, T., *et al.* (eds) (2006) New biological application, in *Current Fluoroorganic Chemistry*, ACS Symposium Series **949**, American Chemical Society, Washington, D.C.; (c) Filler, R. and Kobayashi, Y. (1982) *Biomedicinal Aspects of Fluorine Chemistry*, Kodansha, Ltd., Tokyo; (d) Ishikawa, N. (ed.) (1987) *Synthesis and Speciality of Organofluorine Compounds*, CMC, Tokyo; (e) Ishikawa, N. (ed.) (1990) *Biologically Active Organofluorine Compounds*, CMC: Tokyo; (f) Welch, J. T. and Eswarakrishnan, S. (1991) *Fluorine in Bioorganic Chemistry*, John Wiley & Sons, Inc., New York; (g) Ojima, I., McCarthy, J. R. and Welch, J. T. (eds) (1996) *Biomedical Frontiers of Fluorine Chemistry*, ACS Symposium Series **639**, American Chemical Society, Washington, D.C.
3. (a) Ondetti, M. A.; Cushman, D. W. and Rubin, B. (1977) Design of specific inhibitors of angiotensin-converting enzyme: new class of orally active antihypertensive agents. *Science*, **196**, 441–444; (b) Cushman, D. W., Cheung, H. S., Sabo, E. F. and Ondetti, M. A. (1977) Design of potent competitive inhibitors of angiotensin-converting enzyme. Carboxyalkanoyl and mercaptoalkanoyl amino acids. *Biochemistry*, **16**, 5484–5491; (c) Kollonitsch, J., Perkins, L. M. and Patchett, A. A., *et al.* (1978) Selective inhibitors of biosynthesis of aminergic neurotransmitters. *Nature*, **274**, 906–908. (d) Wyvratt, M. J., Tristram, E. W., Ikeler, T. J., *et al.* (1984) Reductive amination of ethyl 2-oxo-4-phenylbutanoate with L-alanyl-L-proline. Synthesis of enalapril maleate. *J. Org. Chem.*, **49**, 2816–2819.
4. Kukhar, V. P. and Soloshonok, V. A. (1995) *Fluorine-Containing Amino Acids: Synthesis and Properties*, John Wiley & Sons, Inc., New York.
5. Qiu, X.-L., Meng, W.-D. and Qing, F.-L. (2004) Synthesis of fluorinated amino acids. *Tetrahedron*, **60**, 6711–6743. This is the most up-to-date review on the synthesis of fluorinated amino acids.
6. (a) Liu, M. and Sibi, M. P. Recent advances in the stereoselective synthesis of β-amino acids. (2002) *Tetrahedron*, **58**, 7991–8035; (b) Ojima, I., Inoue, T. and Chakravarty, S. Enantiopure fluorine-containing taxoids: potent anticancer agents and versatile probes for biomedical problems. (1999) *J. Fluorine Chem.*, **97**, 3–10; (c) Uneyama, K. (1999) Asymmetric synthesis of fluoro-amino acids, in *Enantiocontrolled Synthesis of Fluoro-Organic Compounds* (ed. V. A. Soloshonok), John Wiley & Sons, Ltd, Chichester, pp. 3914–3918; (d) Uneyama, K., Katagiri, T. and Amii, H. New approaches to stereoselective synthesis of fluorinated amino acids. (2002) *J. Synth. Org. Chem. Jpn*, **60**, 1069–1075; (e) Welch, J. T. Tetrahedron report number 221, Advances in the Preparation of Biologically Active Organofluorine Compounds. (1987) *Tetrahedron*, **43**, 3123–3197.

7. (a) Petasis, N. A. and Zavialov, I. A. (1997) A new and practical synthesis of α-amino acids from alkenyl boronic acids. *J. Am. Chem. Soc.*, **119**, 445–446; (b) Petasis, N. A. and Zavialov, I. A. (1997) A new synthesis of α-arylglycines from aryl boronic acids. *Tetrahedron*, **53**, 16463–16470; (c) Petasis, N. A. and Zavialov, I. A. (1998) Highly stereocontrolled one-step synthesis of *anti*-β-amino alcohols from organoboronic acids, amines, and α-hydroxy aldehydes. *J. Am. Chem. Soc.*, **120**, 11798–11799; (d) Petasis, N. A. and Boral, A. (2001) One-step three-component reaction among organoboronic acids, amines and salicylaldehydes. *Tetrahedron Lett.*, **42**, 5395–5342; (e) Petasis, N. A. and Patal, Z. D. (2000) Synthesis of piperazinones and benzopiperazinones from 1,2-diamines and organoboronic acids. *Tetrahedron Lett.*, **41**, 9607–9611.
8. Prakash, G. K. S., Mandal, M., Schweizer, S., *et al.* (2002) Stereoselective synthesis of *anti*-α-(difluoromethyl)-β-amino alcohols by boronic acid based three-component condensation. Stereoselective preparation of (2*S*,3*R*)-difluorothreonine. *J. Org. Chem.*, **67**, 3718–3723.
9. Watanabe, H., Hashizume, Y. and Uneyama, K. (1992) Homologation of trifluoroacetimidoyl iodides by palladium-catalyzed carbonylation. An approach to α-amino perfluoroalkanoic acids. *Tetrahedron Lett.*, **33**, 4333–4336.
10. Amii, H., Kishikawa, Y., Kageyama, K. and Uneyama, K. (2000) Palladium-catalyzed *tert*-butoxycarbonylation of trifluoroacetimidoyl iodides. *J. Org. Chem.*, **65**, 34043–34408.
11. Abe, H., Amii, H. and Uneyama, K. (2001) Pd-catalyzed asymmetric hydrogenation of α-fluorinated iminoesters in fluorinated alcohol: a new and catalytic enantioselective synthesis of fluoro α-amino acid derivatives. *Org. Lett.*, **3**, 313–315.
12. Sakai, T., Yan, F., Kashino, S. and Uneyama, K. (1996) Asymmetric reduction of 2-(*N*-arylimino)-3,3,3-trifluoropropanoic acid esters leading to enantiomerically enriched 3,3,3-trifluoroalanine. *Tetrahedron*, **52**, 233–244.
13. Suzuki, A., Mae, M., Amii, H. and Uneyama, K. (2004) Catalytic route to the synthesis of optically active β,β-difluoroglutamic acid and β,β-difluoroproline derivatives. *J. Org. Chem.*, **69**, 5132–5134.
14. Demir, A. S., Sesenoglu, O. and Gercek-Arkin, Z. (2001) An asymmetric synthesis of both enantiomers of 2,2,2-trifluoro-1-furan-2-yl-ethylamine and 3,3,3-trifluoroalanine from 2,2,2-trifluoro-1-furan-2-yl-ethanone. *Tetrahedron Asymmetry*, **12**, 2309–2313.
15. Kubryk, M. and Hansen, K. B. (2006) Application of the asymmetric hydrogenation of enamines to the preparation of a beta-amino acid pharmacophore. *Tetrahedron Asymmetry*, **17**, 205–209.
16. Jiang, Z.-X, Qin, Y.-Y. and Qing, F.-L. (2003) Asymmetric synthesis of both enantiomers of *anti*-4,4,4-trifluorothreonine and 2-amino-4,4,4-trifluorobutanoic acid. *J. Org. Chem.*, **68**, 7544–7547.
17. Kobayashi, S. and Ishitani, H. (1999) Catalytic enantioselective addition to imines. *Chem. Rev.*, **99**, 1069–1094.
18. Córdova, A., Nots, W., Zhong, G., *et al.* (2002) A highly enantioselective amino acid-catalyzed route to functionalized α-amino acids. *J. Am. Chem. Soc.*, **124**, 1842–1843.
19. Reviews on phase-transfer catalysis for the syntheses of α-amino acids. (a) Ooi, T. and Maruoka, K. (2004) Asymmetric organocatalysis of structurally well-defined chiral quaternary ammonium fluorides. *Acc. Chem. Res.*, **37**, 526–533; (b) Maruoka, K. and Ooi, T. (2003) Enantioselective amino acid synthesis by chiral phase-transfer catalysis. *Chem. Rev.*, **103**, 3013–3028; (c) Ooi, T. and Maruoka, K. (2003) Enantioselective synthesis of α-amino acids by chiral phase-transfer catalysis. *Yuki Gosei Kagaku Kyokaishi (J. Synth. Org. Chem.)* **61**, 1195–1206.
20. Jew, S.-S., Jeong, B.-S., Yoo, M.-S., *et al.* (2001) Synthesis and application of dimeric Cinchona alkaloid phase-transfer catalysts: α,α′-bis[*O*(9)-allylcinchonidinium]-*o*, *m*, or *p*-xylene dibromide. *Chem. Commun.* 1244–1245.
21. Ramachandran, P. V., Madhi, S. and O'Donnell, M. J. (2007) Synthesis of fluorinated glutamic acid derivatives via vinylalumination. *J. Fluorine Chem.*, **128**, 78–83.

22. (a) Soloshonok, V. A., Ono, T. and Soloshonok, I. V. (1997) Enantioselective biomimetic transamination of β-keto carboxylic acid derivatives. An efficient asymmetric synthesis of β-(fluoroalkyl)-β-amino acids. *J. Org. Chem.*, **62**, 7538–7539; (b) Soloshonok, V. A., Ohkura, H. and Yasumoto, M. (2006) Operationally convenient asymmetric synthesis of (*S*)- and (*R*)-3-amino-4,4,4-trifluorobutanoic acid. *J. Fluorine Chem.*, **127**, 924–929; (c) Soloshonok, V. A., Ohkura, H. and Yasumoto, M. (2006) Operationally convenient asymmetric synthesis of (*S*)- and (*R*)-3-amino-4,4,4-trifluorobutanoic acid. *J. Fluorine Chem.*, **127**, 930–935.
23. Uneyama, K. (2006) Discussion on unique chemical phenomena of highly fluorinated carbonyl and imino compounds, *Organofluorine Chemistry*, Blackwell Publishing, Oxford, pp.31–36.
24. Soloshonok, V. A. and Kuklar, V. P. (1997) Biomimetic transamination of α-keto perfluorocarboxylic esters. An efficient preparative synthesis of β,β,β-trifluoroalanine. *Tetrahedron*, **53**, 8307–8314.
25. Li, G. and van der Donk, W. A. (2007) Efficient synthesis of suitably protected β-difluoroalanine and γ-difluorothreonine from L-ascorbic acid. *Org. Lett.*, **9**, 41–44.
26. For scope of fluorination with DAST, see ref. 23, pp. 233–239.
27. Doi, M., Nishi, Y., Kiritoshi, N., *et al.* (2002) Simple and efficient syntheses of Boc- and Fmoc-protected 4(*R*)- and 4(*S*)-fluoroproline solely from 4(*R*)-hydroxyproline. *Tetrahedron*, **58**, 8453–8459.
28. Raines, R. T. (2006) Hyperstable collagen based on 4-fluoroproline residues, in *Current Fluoroorganic Chemistry*, ACS Symposium Series **949**, American Chemical Society, Washington, D.C.
29. (a) Katagiri, T. and Uneyama, K. (2003) Stereospecific substitution at α-carbon to trifluoromethyl group: application to optically active fluorinated amino acid syntheses. *Chirality*, **15**, 4–9; (b) Katagiri, T., Uneyama, K. (2005) Functional group transformation at α-carbon to trifluoromethyl group, in *Fluorine-containing Synthons*, ACS Symposium Series **911**, American Chemical Society, Washington, D.C.
30. Konno, T., Daitoh, T., Ishihara, T. and Yamanaka, H. (2001) A novel and expedient synthesis of optically active fluoroalkylated amino acids via palladium-catalyzed allylic rearrangement and Ireland-Claisen rearrangement. *Tetrahedron Asymmetry*, **12**, 2743–2748.
31. Fokina, N. A., Kornilov, A. M., Kulik, I. B. and Kukhar, V. P. (2002) Towards optically pure mono- and difluorinated amino acids: common methodology based on (*R*)-2,3-*O*-isopropylideneglyceraldehyde. *Synthesis*, 2589–2596.
32. Xing, X., Fichera, A. and Kumar, K. (2001) A novel synthesis of enantiomerically pure 5,5,5,5′,5′,5′-hexafluoroleucine. *Org. Lett.*, **3**, 1285–1286.
33. Meffre, P., Dave, R. H., Leroy, J. and Badet, B. (2001) A concise synthesis of L-4,4-difluoroglutamine. *Tetrahedron Lett.*, **42**, 8625–8627.
34. Anderson, J. T., Toogood, P. L. and Marsh, E. N. G. (2002) A short and efficient synthesis of L-5,5,5,5′,5′,5′-hexafluoroleucine from *N*-Cbz-L-serine. *Org. Lett.*, **4**, 4281–4283.
35. Ding, Y., Wang, J., Abbound, K. A., *et al.* (2001) Synthesis of L-4,4-difluoroglutamic acid via nucleophilic addition to a chiral aldehyde. *J. Org. Chem.*, **66**, 6381–6388.
36. (a) Lubell, W. and Rapoport, H. (1989) Surrogates for chiral aminomalondialdehyde. Synthesis of *N*-(9-phenylfluoren-9-yl)serinal and *N*-(9-phenylfluoren-9-yl)vinylglycinal. *J. Org. Chem.*, **54**, 3824–3831; (b) Lubell, W. and Rapoport, H. (1987) Configurational stability of N-protected α-amino aldehydes. *J. Am. Chem. Soc.*, **109**, 236–239; (c) Lubell, W. D., Jamison, T. F. and Rapoport, H. (1990) *N*-(9-phenylfluoren-9-yl)-α-amino ketones and *N*-(9-phenylfluoren-9-yl)-α-amino aldehydes as chiral educts for the synthesis of optically pure 4-alkyl-3-hydroxy-2-amino acids. Synthesis of the C-9 amino acid MeBmt present in cyclosporin. *J. Org. Chem.*, **55**, 3511–3522.
37. Lal, G. S., Pez, G. P. and Syvret, R. G. (1996) Electrophilic N–F fluorinating agents. *Chem. Rev.*, **96**, 1737–1755.

38. Konas, D. W. and Coward, J. K. (1999) Synthesis of L-4,4-difluoroglutamic acid via electrophilic difluorination of a lactam. *Org. Lett.*, **1**, 2105–2107.
39. Qiu, J. and Silverman, R. B. (2000) A new class of conformationally rigid analogs of 4-amino-5-halopentanoic acids, potent inactivators of γ-aminobutyric acid aminotransferase. *J. Med. Chem.*, **43**, 706–720.
40. Very few examples of retention: (a) Kozikowski, A. P., Powis, G., Fauq, A. H., *et al.* (1994) Synthesis and biological activity of the D-3-deoxy-3-fluoro and D-3-chloro-3-deoxy analogs of phosphatidylinositol. *J. Org. Chem.*, **59**, 963–971; (b) Gharbaoui, T., Legraverend, M., Ludwing, O., *et al.* (1995) Synthesis of new carbocyclic oxetanocin analogs. *Tetrahedron*, **51**, 1641–1652.
41. Katagiri, T. and Uneyama, K. (2000) Chemistry of TFPO. *J. Fluorine Chem.*, **105**, 285–293.
42. Schaus, S. E., Brandes, B. D., Larrow, J. F., *et al.* (2002) Highly selective hydrolytic kinetic resolution of terminal epoxides catalyzed by chiral (salen)CoIII complexes. Practical synthesis of enantioenriched terminal epoxides and 1,2-diols. *J. Am. Chem. Soc.*, **124**, 1307–1315.
43. (a) Yamauchi, Y., Kawate, T., Katagiri, T. and Uneyama, K. (2003) Trifluoromethyl-stabilized optically active oxiranyl and aziridinyl anions for stereospecific syntheses of trifluoromethylated compounds. *Tetrahedron* **59**, 9839–9847; (b) Yamauchi, Y., Katagiri, T. and Uneyama, K. (2002) The first generation and stereospecific alkylation of α-trifluoromethyl oxiranyl anion. *Org. Lett.*, **4**, 173–176.
44. Yamauchi, Y., Kawate, T., Katagiri, T. and Uneyama, K. (2003) Generation and reactions of α-trifluoromethyl stabilized aziridinyl anion, a general synthetic precursor for stereospecific construction of α-amino-α-trifluoromethylated quaternary carbon. *Tetrahedron Lett.*, **44**, 6319–6322.
45. Kirihara, M., Kawasaki, M., Takuwa, T., *et al.* (2003) Efficient synthesis of (*R*)- and (*S*)-1-amino-2,2-difluorocyclopropanecarboxylic acid via lipase-catalyzed desymmetrization of prochiral precursors. *Tetrahedron Asymmetry*, **14**, 1753–1761.
46. Prati, F., Moretti, I., Forni, A., *et al.* (1998) Synthesis and stereodirected *N*-halogenation of *trans*-3-trifluoromethyl-2-methoxycarbonylaziridine. *J. Fluorine Chem.*, **89**, 177–181.
47. Davoli, P., Forni, A., Franciosi, C., *et al.* (1999) Stereoselective synthesis of fluorinated β-amino acids from ethyl *trans-N*-benzyl-3-trifluoromethylaziridine-2-carboxylate. *Tetrahedron Asymmetry*, **10**, 2361–2371.
48. Kuznetsova, L., Ungureanu, I. M., Pepe, A., *et al.* (2004) Trifluoromethyl- and difluoromethyl-β-lactams as useful building blocks for the synthesis of fluorinated amino acids, dipeptides, and fluoro-taxoids. *J. Fluorine Chem.*, **125**, 487–500.
49. Abouabdellah, A., Begue, J-P., Bonnet-Delpon, D. and Nga, T. T. T. (1997) Diastereoselective Synthesis of the Nonracemic Methyl *syn*-(3-Fluoroalkyl)isoserinates. *J. Org. Chem.*, **62**, 8826–8833.
50. Ito, H., Saito, A., Kakuuchi, A. and Taguchi, T. (1999) Synthesis of 2-fluoro analog of 6-aminonorbornane-**2**,6-dicarboxylic acid: a conformationally rigid glutamic acid derivative. *Tetrahedron*, **55**, 12741–12750.
51. Schlosser, M., Brugger, N., Schmidt, W. and Amrhein, N. (2004) β,β-Difluoro analogs of α-oxo-β-phenylpropionic acid and phenylalanine. *Tetrahedron*, **60**, 7731–7742.
52. Fustero, S., Sanchez-Rosello, M., Rodrigo, V., *et al.* (2006) Asymmetric synthesis of new β,β-difluorinated cyclic quaternary α-amino acid derivatives. *Org. Lett.*, **8**, 4129–4132.
53. Osipov, S. N., Artyushin, O. I., Kolomiets, A. F., *et al.* (2001) Synthesis of racemic cyclic amino acids; Synthesis of fluorine-containing cyclic α-amino acid and α-amino phosphonate derivatives by alkene metathesis. *Eur. J. Org. Chem.*, 3891–3897.
54. Huguenot, F. and Brigaud, T. (2006) Concise synthesis of enantiopure α-trifluoromethyl alanines, diamines, and amino alcohols via the Strecker-type reaction. *J. Org. Chem.*, **71**, 7075–7078.

55. Marcotte, S., Pannecoucke, X., Feasson, C. and Quirion, J.-C. (1999) Enantioselective synthesis of α,α-difluoro-β-amino acid and 3,3-difluoroazetidin-2-one via the Reformatskii-type reaction of ethyl bromodifluoroacetate with chiral 1,3-oxazolidines. *J. Org. Chem.*, **64**, 8461–8464.
56. (a) Lebouvier, N., Laroche, C., Huguenot, F. and Brigaud, T. (2002) Lewis acid activation of chiral 2-trifluoromethyl-1,3-oxazolidines. Application to the stereoselective synthesis of trifluoromethylated amines, α- and β-amino acids. *Tetrahedron Lett.*, **43**, 2827–2830; (b) Huguenot, F. and Brigaud, T. (2006) Convenient asymmetric synthesis of β-trifluoromethyl-β-amino acid, β-amino ketones, and γ-amino alcohols via Reformatsky and Mannich-type reactions from 2-trifluoromethyl-1,3-oxazolidines. *J. Org. Chem.*, **71**, 2159–2162.
57. Chaume, G., Van Severen, M-C., Marinkovic, S. and Brigaud, T. (2006) Straightforward synthesis of (*S*)- and (*R*)-α-trifluoromethyl proline from chiral oxazolidines derived from ethyl trifluoropyruvate. *Org. Lett.*, **8**, 6123–6126.
58. Andrews, P. C., Bhaskar, V., Bromfield, K. M., *et al.* (2004) Stereoselective synthesis of β-amino-α-fluoro esters via diastereoselective fluorination of enantiopure β-amino enolates. *Synlett*, **5**, 791–794.
59. Isaac, M., Slassi, A., Da Silva, K. and Xin T. (2001) Synthesis of chiral and geometrically defined 5,5-diaryl-2-amino-4-pentenoates: novel amino acid derivatives. *Tetrahedron Lett.*, **42**, 2957–2960.
60. Yajima, T. and Nagano, H. (2007) Photoinduced diastereoselective addition of perfluoroalkyl iodides to acrylic acid derivatives for the synthesis of fluorinated amino acids. *Org. Lett.*, **9**, 2513–2515.
61. Laue, K. W., Kroger, S., Wegelius, E. ad Haufe, G. (2000) Stereoselective synthesis of γ-fluorinated α-amino acids using 2-hydroxy-3-pinanone as an auxiliary. *Eur. J. Org. Chem.*, 3737–3743.
62. Katagiri, T., Handa, M., Matsukawa, Y., *et al.* (2001) Efficient synthesis of an optically pure β-bromo-β,β-difluoroalanine derivative, a general precursor for β,β-difluoroamino acids. *Tetrahedron Asymmetry*, **12**, 1303–1311.
63. Davis, F. A., Srirajan, V. and Titus, D. D. (1999) Efficient asymmetric synthesis of β-fluoro α-amino acids. *J. Org. Chem.*, **64**, 6931–6934.
64. (a) Volonterio, A., Bravo, P., Moussier, N. and Zanda, M. (2000) Solid-phase synthesis of partially-modified retro and retro-inverso-ψ[NHCH(CF_3)]-peptides. *Tetrahedron. Lett.*, **41**, 6517–6520; (b) Volonterio, A., Chiva, G., Fustero, S., *et al.* (2003) Stereocontrolled solid-phase synthesis of fluorinated partially-modified retropeptides via tandem aza-Michael/enolate-protonation. *Tetrahedron Lett.*, **44**, 7019–7022; (c) Volonterio, A., Bravo, P. and Zanda, M. (2000) Synthesis of partially modified retro and retroinverso-ψ [NHCH(CF_3)]-peptides. *Org. Lett.*, **2**, 1827–1830; (d) Volonterio, A., Bravo, P. and Zanda, M. (2001) Solution/solid-phase synthesis of partially modified retro-ψ[NHCH(CF_3)]-peptidyl hydroxamates. *Tetrahedron Lett.*, **42**, 3141–3144.
65. Davis, E. A., Reddy, R. E., Szewezyk, J. M., *et al.* (1997) Asymmetric synthesis and properties of sulfinimines (thiooxime *S*-oxides). *J. Org. Chem.*, **62**, 2555–2563.
66. (a) Cogan, D. A., Liu, G., Kim, K., *et al.* (1998) Catalytic asymmetric oxidation of *tert*-butyl disulfide. Synthesis of *tert*-butanesulfinamides, *tert*-butyl sulfoxides, and *tert*-butanesulfinimines. *J. Am. Chem. Soc.*, **120**, 8011–8018; (b) Backes B. J., Dragoli, D. R. and Ellman, J. A. (1999) Chiral *N*-acyl-*tert*-butanesulfinamides: the "safety-catch" principle applied to diastereoselective enolate alkylations. *J. Org. Chem.*, **64**, 5472–5478.
67. Asensio, A., Bravo, P., Crucianelli, M., *et al.* (2001) Synthesis of nonracemic α-trifluoromethyl α-amino acids from sulfinimines of trifluoropyruvate. *Eur. J. Org. Chem.*, 1449–1458.
68. (a) Soloshonok, V. A., Ohkura, H., Sorochinsky, A., *et al.* (2002) Convenient, large-scale asymmetric synthesis of β-aryl-substituted α,α-difluoro-β-amino acids. *Tetrahedron Lett.*, **43**,

5445–5448; (b) Sorochinsky, A., Voloshin, N., Markovsky, A., *et al.* (2003) Convenient asymmetric synthesis of β-substituted α,α-difluoro-β-amino acids via Reformatskii reaction between Davis' *N*-sulfinylimines and ethyl bromodifluoroacetate. *J. Org. Chem.*, **68**, 7448–7454.

69. Staas, D. D., Savage, K. L., Homnick, C. F., *et al.* (2002) Asymmetric synthesis of α,α-difluoro-β-amino acid derivatives from enantiomerically pure *N*-tert-butylsulfinimines. *J. Org. Chem.*, **67**, 8276–8279.
70. Wang, H., Zhao, X., Li, Y. and Lu, L. (2006) Solvent-controlled asymmetric Strecker reaction: stereoselective synthesis of α-trifluoromethylated α-amino acids. *Org. Lett.*, **8**, 1379–1381.
71. Bravo, P., Capelli, S., Meille, S. V., *et al.* (1994) Synthesis of optically pure (*R*)- and (*S*)-α-trifluoromethylalanine. *Tetrahedron Asymmetry*, **5**, 2009–2018.
72. Fustero, S., Navorro, A., Pina, B., *et al.* (2001) Enantioselective synthesis of fluorinated α-amino acids and derivatives in combination with ring-closing metathesis: intramolecular π-stacking interactions as a source of stereocontrol. *Org. Lett.*, **3**, 2621–2624.
73. Fustero, S., Sanz-Cervera, J. F., del Pozo, C. and Acena, J. L. (2006) Nitrogen-containing organofluorine compounds, in *Current Fluoroorganic Chemistry*, ACS Symposium Series **949**, American Chemical Society, Washington, D.C.
74. Arnone, A., Bravo, P., Capelli, S., *et al.* (1996) New versatile fluorinated chiral building blocks: synthesis and reactivity of optically pure α-(fluoroalkyl)-β-sulfinylenamines. *J. Org. Chem.*, **61**, 3375–3387.
75. Bravo, P. and Resnati, G. (1987) An efficient approach to enantiomerically pure fluorhydrins. *Tetrahedron Lett.*, **28**, 4865–4866.
76. Bravo, P., Cavicchio, G., Crucianelli, M., *et al.* (1997) Stereoselective synthesis of the antibacterial 3-fluoro-D-alanine. *Tetrahedron Asymmetry*, **8**, 2811–2815.
77. (a) Fustero, S., Sanchez-Rosello, M., Sanz-Cervera, J. F., *et al.* (2006) Asymmetric synthesis of fluorinated cyclic β-amino acid derivatives through cross metathesis. *Org. Lett.*, **8**, 4633–4636; (b) Fustero, S., Pina, B., Savalert, E., *et al.* (2002) New strategy for the stereoselective synthesis of fluorinated β-amino acids. *J. Org. Chem.*, **67**, 4667–4674.
78. Fustero, S., Bartolome, A., Sanz-Cervera, J. F., *et al.* (2003) Diastereoselective synthesis of fluorinated, seven-membered β-amino acid derivatives via ring-closing metathesis. *Org. Lett.*, **5**, 2523–2526.
79. Fustero, S., Salavert, E., Pina, B., *et al.* (2001) Novel strategy for the synthesis of fluorinated β-amino acid derivatives from Δ2-oxazolines. *Tetrahedron*, **57**, 6475–6486.
80. Hallinan, E. A, Kramer, S. W., Houdek, S. C., *et al.* (2003) 4-Fluorinated L-lysine analogs as selective i-NOS inhibitors: methodology for introducing fluorine into the lysine side chain. *Org. Biomol. Chem.*, **1**, 3527–3534.
81. Konas, D. W., Pankuch, J. and Coward, J. K. (2002) The synthesis of (2*S*)-4,4-difluoroglutamyl γ-peptides based on Garner's aldehyde and fluoro-Reformatskii chemistry. *Synthesis*, 2616–2626.
82. Qiu, X.-L and Qing, F. -L. (2002) Synthesis of Boc-protected *cis*- and *trans*-4-trifluoromethyl-D-prolines. *J. Chem. Soc., Perkin Trans.*, **1**, 2052–2057.
83. (a) Kolb, M., Barth, J., Heydt, J.-G. and Jung, M. J. (1987) Synthesis and evaluation of mono-, di-, and trifluoroethenyl-GABA derivatives as GABA-T inhibitors. *J. Med. Chem.*, **30**, 267–272; (b) Bey, P., Ducep, J. R. and Schirlin, D. (1984) Alkylation of malonates or Schiff base anions with dichlorofluoromethane as a route to *a*-chlorofluoromethyl or α-fluoromethyl α-amino acids. *Tetrahedron Lett.*, **25**, 5657–5660; (c) Bravo, P., Cavicchio, G., Crucianelli, M., *et al.* (1997) Stereoselective synthesis of the antibacterial 3-fluoro-D-alanine. *Tetrahedron Asymmetry*, **16**, 2811–2815.
84. Chen, Q, Qiu, X-L. and Qing, F-L. (2006) Indium-mediated diastereoselective allylation of D- and L-glyceraldimines with 4-bromo-1,1,1-trifluoro-2-butene: Highly stereoselective synthesis of 4,4,4-trifluoroisoleucines and 4,4,4-trifluorovaline. *J. Org. Chem.*, **71**, 3762–3767.

85. Qiu, X.-L. and Qing, F.-L. (2002) Practical synthesis of Boc-protected *cis*-4-trifluoromethyl and *cis*-4-difluoromethyl-L-prolines. *J. Org. Chem.*, **67**, 7162–7164.
86. Soloshonok, V. A., Cai, C. and Hruby, V. J. (2000) A practical asymmetric synthesis of enantiomerically pure 3-substituted pyroglutamic acids and related compounds. *Angew. Chem. Int. Ed.*, **39**, 2172–2175.
87. Fustero, S., Jimenez, D., Sanchez-Rosello, M. and del Pozo, C. (2007) Microwave-assisted tandem cross metathesis intramolecular aza-Michael reaction: an easy entry to cyclic β-amino carbonyl derivatives. *J. Am. Chem. Soc.*, **129**, 6700–6701.
88. Hallinan, E. A., Dorn, C. R., Moore, W. M., *et al.* (2003) The synthesis of 6-trifluoroethyl-L-lysine: a method to introduce functionality at C-6 of L-lysine. *Tetrahedron Lett.*, **44**, 7345–7347.
89. Yoshikawa, N., Tan, L., Yasuda, N., *et al.* (2004) Enantioselective syntheses of bicyclo[3.1.0]hexane carboxylic acid derivatives by intramolecular cyclopropanation. *Tetrahedron Lett.*, **45**, 7261–7264.
90. Uneyama, K., Katagiri, T. and Amii, H. (2008) The chemistry of α-trifluoromethylated carbanions is discussed. *Acc. Chem. Res.*, **41**, 817–829.
91. Konno, T., Kanda, M., Ishihara, T. and Yamanaka, H. (2005) A novel synthesis of fluorine-containing quaternary amino acid derivatives via palladium-catalyzed allylation reaction. *J. Fluorine Chem.*, **126**, 1517–1523.
92. Katagiri T. and Uneyama, K. (2003) Stereospecific substitution at α-carbon to trifluoromethyl group: application to optically active fluorinated amino acid syntheses. *Chirality*, **15**, 4–9.
93. Katagiri, T, Irie, M. and Uneyama, K. (2000) Syntheses of optically active trifluoronorcoronamic acids. *Org. Lett.*, **2**, 2423–2425.
94. Katagiri, T., Yamaji, S., Handa, M., *et al.* (2001) Diastereoselectivity controlled by electrostatic repulsion between the negative charge on a trifluoromethyl group and that on aromatic rings. *Chem. Commun.*, 2054–2055.
95. Osipov, S. N., Kova, N. M. K., Kolomiets, A. F., *et al.* (2001) α-Fluoromethyl tryptophans via imino ene reaction. *Synlett*, 1287–1289.
96. Pang, W., Zhu, S., Jiang, H. and Zhu, S. (2006) Transition metal-catalyzed formation of CF_3-substituted β-unsaturated alkene and the synthesis of α-trifluoromethyl-substituted β-amino ester. *Tetrahedron*, **62**, 11760–11765.
97. Guo, Y., Fujiwara, K. and Uneyama, K. (2006) A novel route to dipeptides via noncondensation of amino acids: 2-Aminoperfluoropropene as a synthon for trifluoroalanine dipeptides. *Org. Lett.*, **8**, 827–829.
98. The synthetic scope and mechanistic aspect of reactions of difluoromethylene compounds are summarized in ref. 23, pp 112–121.
99. Guo, Y., Fujiwara, K., Amii, H. and Uneyama, K. (2007) Selective defluorination approach to *N*-Cbz-3,3-difluoro-2-difluoromethylenepyrrolidine and its application to 3,3-difluoroproline dipeptide synthesis. *J. Org. Chem.*, **72**, 8523–8526.
100. (a) Hiyama, T. (2000) *Organofluorine Comounds*, Springer-Verlag, Berlin; (b) Chambers D. R. (2004) *Fluorine in Organic Chemistry*, Blackwell Publishing, Oxford; (c) Kirsh, P. (2004) *Modern Fluoroorganic Chemistry*, Wiley-VCH Verlag GmbH, Weinheim; (d) Uneyama, K. (2006) *Organofluorine Chemistry*, Blackwell Publishing, Oxford.

10

Fluorinated Moieties for Replacement of Amide and Peptide Bonds

Takeo Taguchi and Hikaru Yanai

10.1 Introduction

Chemical modifications of peptides to improve their biological activity, *in vivo* stability or bioavailability have been studied extensively [1, 2]. As one of the promising approaches, replacement of a specific amide bond in the target peptide molecule by an appropriate amide bond surrogate can not only bring about enhanced recognition by the target enzyme, which induces the parent peptide to exhibit biological responses, but also facilitate peptidase resistance, conformational control or increase of the lipophilicity of the target peptide (see Figure 10.1) [1, 2].

Isosteric amide replacements are commonly shown using the symbol Ψ[ ], where the Ψ indicates the absence of an amide bond and the structure that is replacing the amide is shown in the brackets. The alkene dipeptide isostere Ψ[CH=CH], which is conformationally fixed and enzymatically nonhydrolyzable (peptidase resistance), has been well studied because of its close similarity to the amide bond with regard to the steric demand, bond lengths, and bond angles, as described in Chapter 1 [3]. However, the low polarity of the alkene moiety compared with the amide bond, as well as the small dipole moment and the lack of ability to form hydrogen bonds are quite different characteristics of Ψ[CH=CH] as a replacement of the amide bond. Introduction of fluorine into the double bond remarkably changes the electronic nature of the double bond to one much closer to that of the amide bond, while still keeping steric similarity [4–7]. In addition to the fluoro-olefin dipeptide isostere Ψ[CF=CH], several types of fluorinated moieties such

Fluorine in Medicinal Chemistry and Chemical Biology Edited by Iwao Ojima

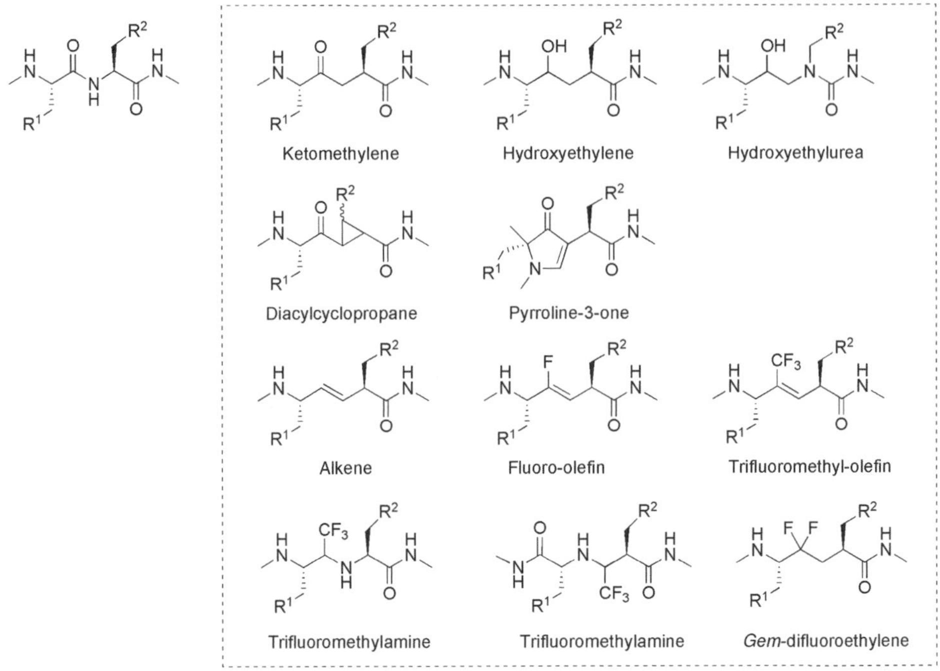

Figure 10.1 *Various peptide bond mimics.*

as trifluoromethyl-olefin Ψ[C(CF_3)=CH], trifluoromethylamine Ψ[CH(CF_3)-NH] and difluoroethylene Ψ[CF_2-CH_2] have been developed as dipeptide isosteres. In this chapter, recent achievements in fluoro-olefin Ψ[CF=CH] and trifluoromethyl-olefin Ψ[C(CF_3)=CH] dipeptide isosteres are reviewed. Developing bioorganic chemistry of fluoroarenes as replacement for nucleoside bases is also briefly described.

10.2 Fluoro-olefins Ψ[CF=CH] as Dipeptide Isosteres

10.2.1 Synthetic Methods

As described above, the fluoro-olefins are the most studied fluorinated peptide mimetics. For the synthesis of the fluoro-olefin Ψ[CF=CH] dipeptide isostere or the depsipeptide isostere (5-hydroxy derivative, see Figure 10.2), stereochemical control of the olefin configuration (either *E* or *Z*) and the relative stereochemistry of the two chiral centers at C-2 and C-5 (either *syn* or *anti*) is a major issue to be solved. Furthermore, the use of readily available starting material is also important. Since the pioneering work by Allmenginger *et al.* in the early 1990s [8], many research groups, as described in this section, have contributed to this development of facile synthetic methods for fluorinated dipeptide isosteres, in particular for optically pure forms, but this subject is still challenging. In this section are summarized (i) synthetic methods mainly focused on the olefination reactions of car-

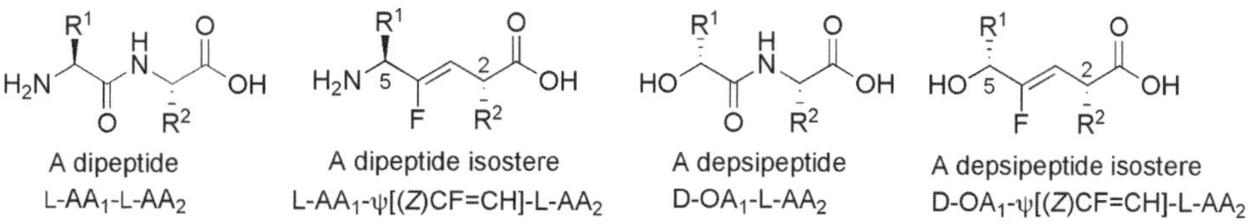

Figure 10.2 *Replacement of amide bonds by (Z)-fluoro-olefins.*

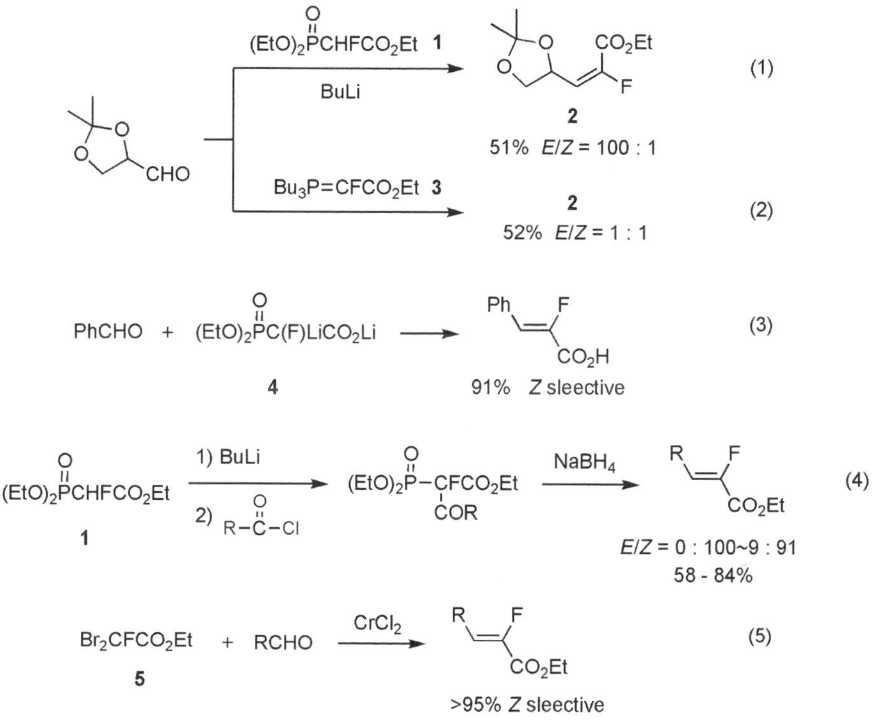

Scheme 10.1

bonyl compounds and related reactions, (ii) utilization of fluoroalkenyl building blocks, and (iii) application to the modifications of the target peptides of biological interest.

10.2.1.1 Utilization of Organophosphorus and Organosilyl Reagents

The Honor–Wadsworth–Emmons (HWE) reaction of α-fluoro-α-phosphonoacetate **1** with aldehydes or ketones, providing α-fluoro-α,β-unsaturated carbonyl compounds, is one of commonest methods for fluoro-olefins having suitable functionality for further elaboration to the target molecules (see equation 1, Scheme 10.1) [9]. The stereochemical outcome of this reaction with both aliphatic and aromatic aldehydes is generally excellent, giving rise to the (*E*)-isomer. In contrast, the Wittig reaction of phosphorane **3** ($Bu_3P{=}CFCO_2Et$) with aldehyde proceeds in nonstereoselective manner (see equation 2, Scheme 10.1). The reactions of the dianion of α-fluoro-α-phosphonoacetic acid **4**, with aromatic aldehydes give *Z*-isomers exclusively, while those with aliphatic aldehydes are nonstereoselective (see equation 3, Scheme 10.1) [10]. Sano and Nagao reported the *Z*-selective synthesis through acylation of **1** with acyl chloride, followed by $NaBH_4$ reduction of the ketone group (see equation 4, Scheme 10.1) [11]. Falck and Mioskowski reported a high yield and *Z*-selective synthesis of α-haloacrylates by the Cr(II)-mediated Reformatsky type reaction of trihaloacetate, including dibromofluoroacetate **5** for fluoro derivatives, with aldehydes (see equation 5, Scheme 10.1) [12].

Allmenginger reported the preparation of Phe-ψ[(*E*)-CF=CH]-Gly in racemic form via *E*-selective HWE reaction, yielding α-fluoroenoate **7**. Conversion of the (*E*)-aldehyde **8** to *N*-silylimine, followed by the addition of the Grignard reagent afforded amine **9** without isomerization of the double bond (see Scheme 10.2) [13].

The HWE reaction of $(EtO)_2P(O)CHFCO_2Et$ (**1**) with ketones proceeds in nonselective manner. For example, Augustyns reported the HWE reaction of **1** with cyclopentanone derivative **10** to give a mixture of α-fluoroenoate **11a** ($E/Z = 1.3{:}\ 1$), which, after separation of isomers, was converted to *N*-alkyl Gly-ψ[CF=C]-Pro-CN (**14**) for their SAR studies on dipeptidyl peptidase (DPP) inhibitors (see Scheme 10.3) [14]. In this synthesis, conversion of amide **12** to amine **13** was achieved by $LiAlH_4$ reduction of the imidoyl derivative formed by treating **12** with $POCl_3$. Without this pre-treatment, $LiAlH_4$ reduction of **12** gave **13** in very low yield because of several side-reactions (see Scheme 10.3).

For the preparation of Gly-ψ-[CF=CH]-Pro in relation to the study of cyclophilin A inhibitors, Welch and co-workers employed the Peterson reaction of α-fluoro-α-trimethylsilyl acetate (**15a,b**) with ketone **10**. *E/Z* selectivity was found to be influenced by the ester part of the acetate (see Scheme 10.4) [15]. The reaction of *tert*-butyl ester **15a** gave almost an equal amount of the isomers (**11b**, $E{:}Z = 1{:}1.1$), while moderate *E* selectivity was observed when trimethylphenyl ester **15b** was used (**11c**, $E{:}Z = 6{:}1$). Conversion of ester **Z-11b** to amino derivative **16** was achieved via the Mitsunobu reaction of phthalimide with the alcohol formed by the DIBAL-H reduction of **Z-11b**.

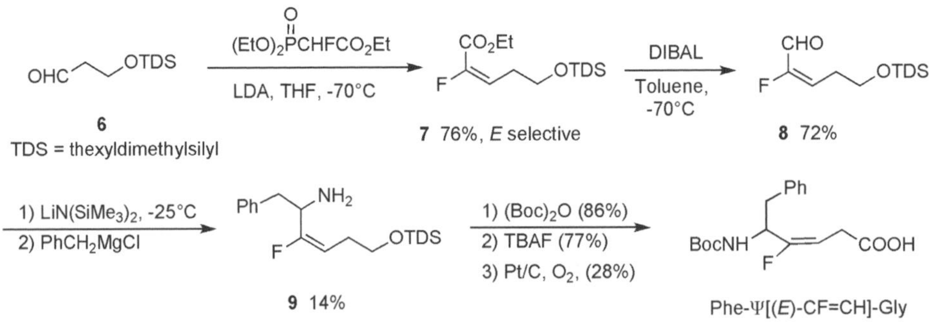

OHC–CH₂CH₂–OTDS (**6**); TDS = thexyldimethylsilyl → $(EtO)_2P(O)CHFCO_2Et$, LDA, THF, -70°C → **7** 76%, *E* selective → DIBAL, Toluene, -70°C → **8** 72% → 1) $LiN(SiMe_3)_2$, -25°C 2) $PhCH_2MgCl$ → **9** 14% → 1) $(Boc)_2O$ (86%) 2) TBAF (77%) 3) Pt/C, O_2, (28%) → Phe-Ψ[(*E*)-CF=CH]-Gly

Scheme 10.2

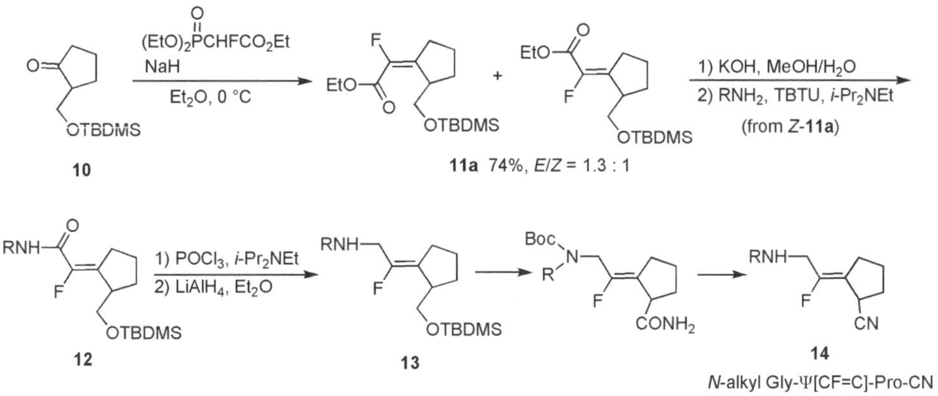

Scheme 10.3

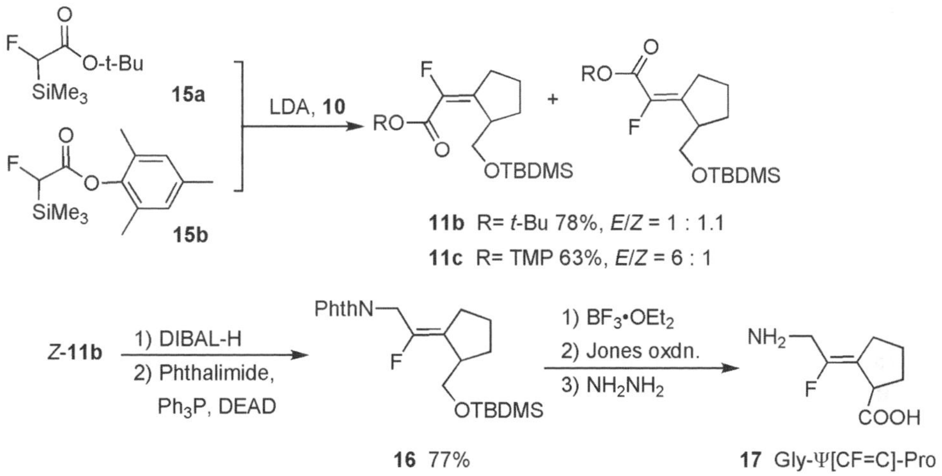

Scheme 10.4

10.2.1.2 Ring Opening Reaction of Oxygenated Chlorofluorocyclopropanes to (Z)-α-Fluoroenal

Under acidic conditions, α-fluoro-α,β-unsaturated aldehydes form the thermodynamically stable *Z*-isomer. For the preparation of (*Z*)-aldehyde **19**, acid hydrolysis of pyrane derivative **18**, whose olefin-configuration corresponds to the *E* configuration, was affected due to facile isomerization under the reaction conditions. Pyrane **18** was obtained by chlorofluorocarbene addition to dihydrofuran under phase-transfer conditions (see Scheme 10.5) [13]. (*Z*)-Aldehyde **19** was converted to Ph-ψ[(*Z*)-CF=CH]-Gly (**20**) using the same reactions for the preparation of the (*E*)-isomer as described above (see Scheme 10.2).

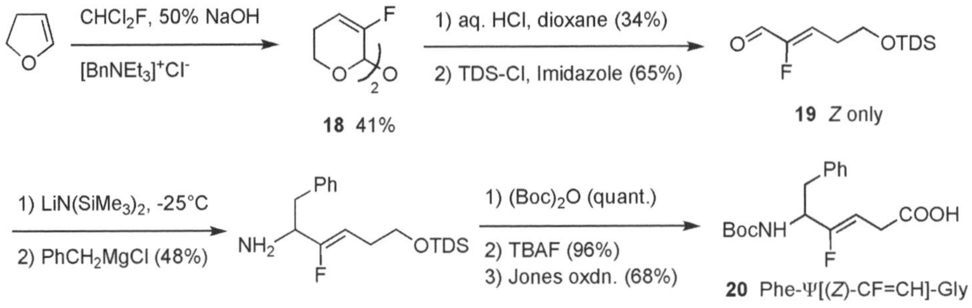

Scheme 10.5

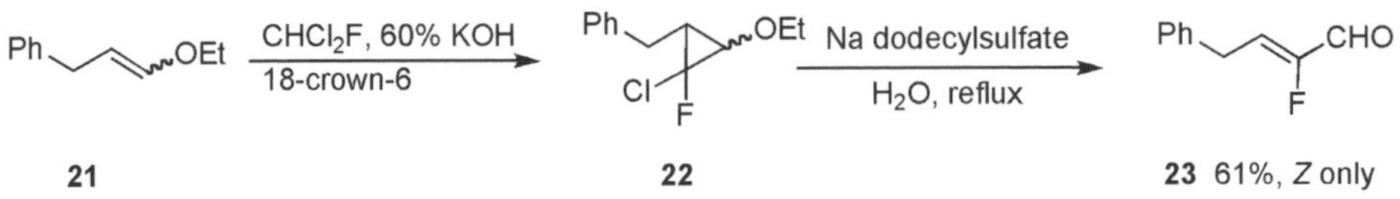

Scheme 10.6

Carbene addition to an *E*/*Z* mixture of vinyl ether **21**, followed by solvolysis of the resultant cyclopropane **22** afforded (*Z*)-α-fluoroenal **23** (Scheme 10.6) [16]. (*Z*)-α-Fluoroenal **23** was used for the preparation of Substance P analogues containing a Phe-Ψ[(*Z*)-CF=CH]-Gly dipeptide unit as described next.

10.2.1.3 Imidate [3,3]-Sigmatropic Rearrangement

Transposition of an allylic hydroxyl group into an amino group through [3,3]-sigmatropic rearrangement is well known as the Overmann rearrangement, which has been applied to the total synthesis of a variety of nitrogen-containing natural products. In general, the rearrangement proceeds via a six-membered chair-like transition state, resulting in a perfect chirality transfer of the stereogenic center at the allylic hydroxyl moiety. For the synthesis of a Phe-Ψ[(*Z*)-CF=CH]-Gly dipeptide unit in optically pure form to be applied to Substance P analogues, Allemendinger employed enantioselective aldol reactions of chiral enolates **24** and **25** with (*Z*)-α-fluoroenal **23**, giving rise to both enantiomers (*R*)-**26** and (*S*)-**26**, respectively, with high optical purities (see Scheme 10.7) [16].

Enantiomerically pure imidate **27** was subjected to thermal rearrangement at 140 °C to give the desired amide derivative **28** in good yield, retaining the optical purity of **26** (see Scheme 10.8). Jones oxidation of the primary alcohol to carboxylic acid afforded dipeptide isostere **29** in optically pure form, which was applied to the preparation of Substance P analogues **30** [16]. It should be noted that nonfluorinated analogues easily undergo isomerization of the double bond to form α,β-unsaturated amides, while the fluorinated double bond in these compounds and the corresponding intermediates is stable under basic and acidic conditions employed, reflecting the stabilizing effect of fluorine on the double bond [17].

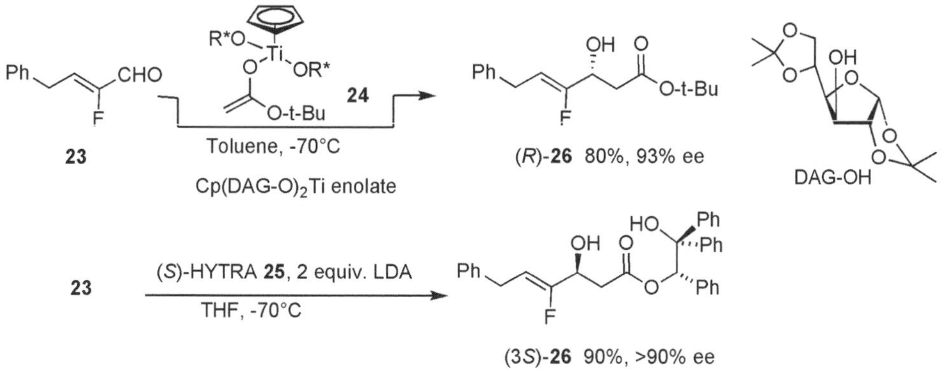

Scheme 10.7

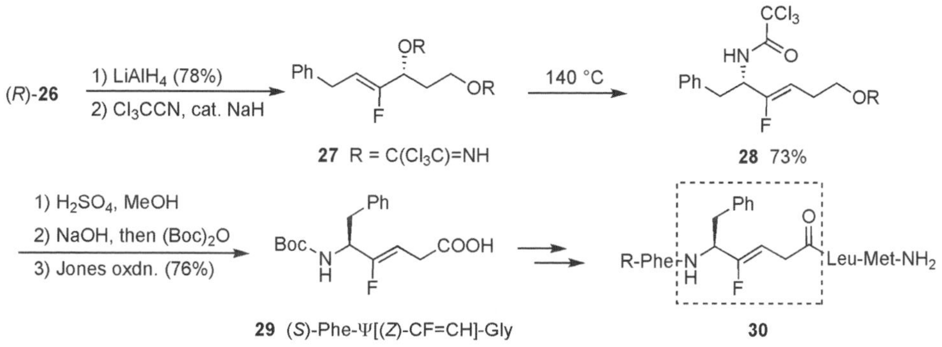

Scheme 10.8

For the synthesis of fluoro-olefin dipeptide isosteres corresponding to dipeptides consisting of two chiral amino acids in stereochemically pure forms with respect to both the configuration of the C–C double bond and the chiral centers at C-2 and C-5, we reported the imidate rearrangement route starting from optically pure aldehydes derived from, for example, commercially available ester **31** (see Scheme 10.9) [18]. Indium-mediated difluoroallylation of aldehyde **32** gave difluorohomoallyl alcohol **33** as a mixture of diastereomers, which were separated by column chromatography. In the presence of CuI (20 mol%), Grignard reaction of **32** with difluorohomoallyl alcohol **33** afforded allylic substitution product **34** in a highly *Z*-selective manner, wherein both aliphatic and aromatic Grignard reagents can be employed. Scheme 10.9 illustrates a typical example. The imidate rearrangement of *syn*-**34c** led to the formation of *syn*-**35c** and the subsequent conversion to the carboxylic acid proceeded smoothly to give Leu-Ψ[CF=CH]-Ala (*syn*-**36c**) without any loss in optical purity. This method has some generality since a variety of Grignard reagents can be used for the preparation of fluoroallyl alcohols **34** from **33**. Alternatively, the starting chiral aldehydes, such as **32**, could also be readily prepared by hydroxymethylation of carboxylic acid derivatives using a chiral auxiliary protocol [19].

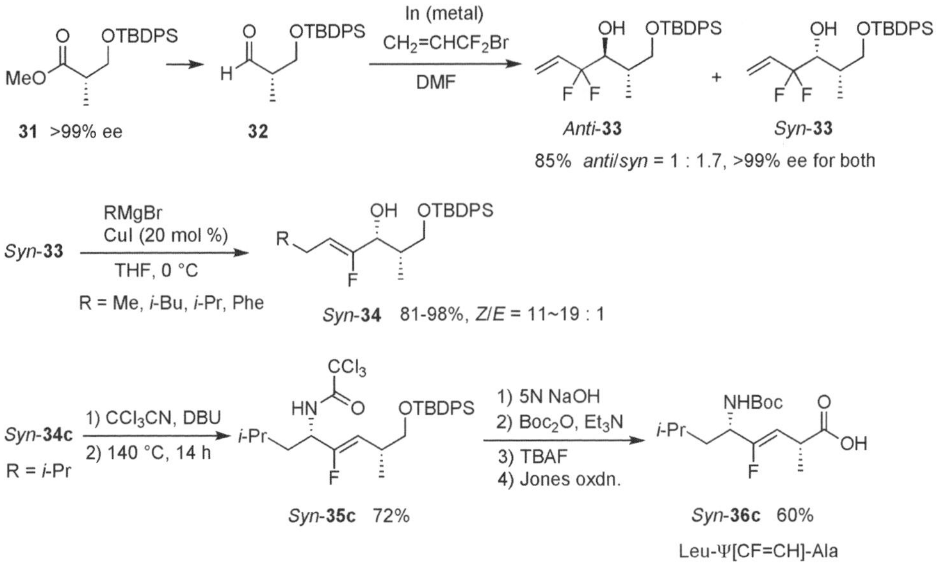

Scheme 10.9

Preparation of *N*-Ac-Glu-Ψ[CF=CH]-Ala was also reported using the imidate rearrangement protocol [20].

10.2.1.4 Aldolization of Fluorooxaloacetate

Bartlett and Otake employed aldol reaction of an in-situ-generated fluorooxaloacetate with a chiral aldehyde derived from **37** to prepare α-fluoroenoate **38**. Stereoselectivity was moderate ($Z\!:\!E = 2.3\!:\!1$), with preferential formation of the *Z*-isomer (see Scheme 10.10) [19]. Further elaboration led to the preparation of tripeptide analogues, Cbz-Gly-Ψ[(*Z*)-CF=CH]-Leu-Xaa (**39**).

10.2.1.5 Sulfoxide-elimination Reaction

Alkylation of the anion of α-fluoro-α-phenylsulfinylacetate **40** with an alkyl halide, followed by thermal *syn*-elimination of sulfinic acid, gave α-fluoroenoate **41** in *Z*-selective manner (see Scheme 10.11) [21].

10.2.1.6 Defluorinative Allylic Substitution and Related Reactions

Reaction of γ,γ-difluoro-α,β-enoate derivatives **42** with Me_2CuLi (5 equivalents), followed by aqueous work-up afford the reductive defluorination product **44** without forming the α-methylated product. It is postulated that enolate intermediate **43** reacts with alkyl halide smoothly and regioselectively to give the desired α-alkylated product **45** (see Scheme 10.12) [22, 23]. The reaction proceeded in an excellently *Z*-selective manner.

Otaka and Fujii demonstrated that alkyl transfer from Me_2CuLi can be promoted by air oxidation of intermediate **43** (see Scheme 10.13) [24].

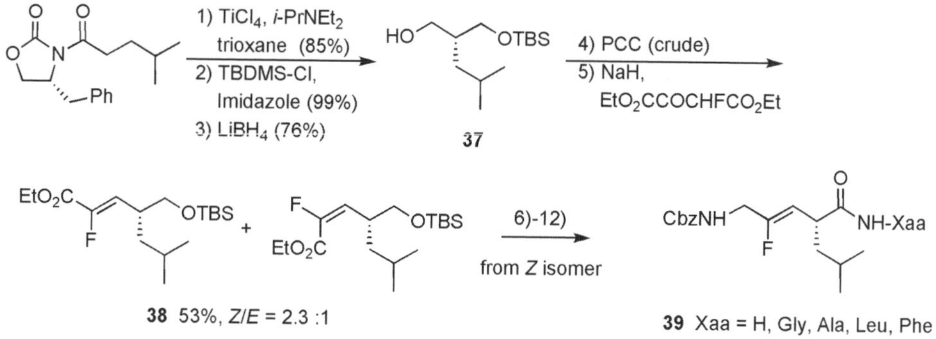

1) $TiCl_4$, *i*-$PrNEt_2$ trioxane (85%)
2) TBDMS-Cl, Imidazole (99%)
3) $LiBH_4$ (76%)

37

4) PCC (crude)
5) NaH, $EtO_2CCOCHFCO_2Et$

38 53%, *Z*/*E* = 2.3 :1

6)-12) from *Z* isomer

39 Xaa = H, Gly, Ala, Leu, Phe

Reagents: 6) NH_3, MeOH (89%), 7) $LiAlH_4$ 8) Cbz-Cl, Et_3N (67%), 9) TBAF (99%), 9) Jones Oxdn. 11) NH_3, EDC, HOBt (70%) or 11) Xaa-OMe, EDC, HOBt, *i*-Pr_2NEt (87-90%), 12) LiOH (quant.)

Scheme 10.10

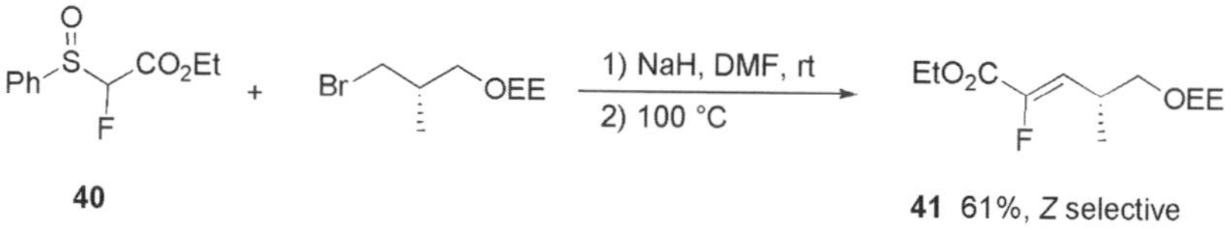

40 + Br–OEE

1) NaH, DMF, rt
2) 100 °C

41 61%, *Z* selective

Scheme 10.11

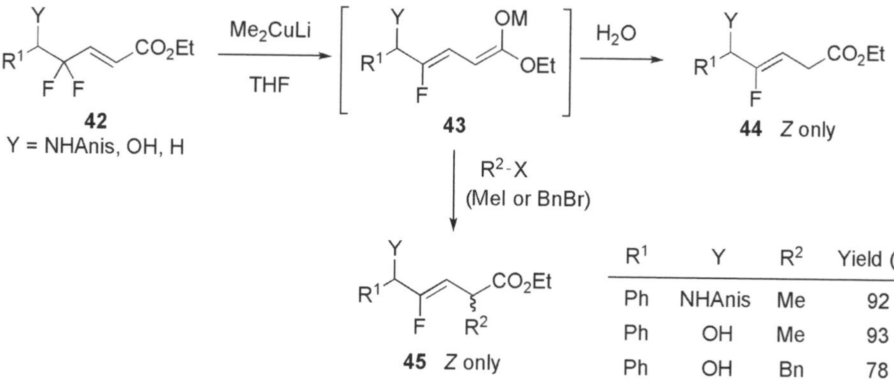

42
Y = NHAnis, OH, H

Me_2CuLi THF

43

H_2O

44 *Z* only

R^2-X (MeI or BnBr)

45 *Z* only

R^1	Y	R^2	Yield (%)
Ph	NHAnis	Me	92
Ph	OH	Me	93
Ph	OH	Bn	78
Ph	H	Me	95

Scheme 10.12

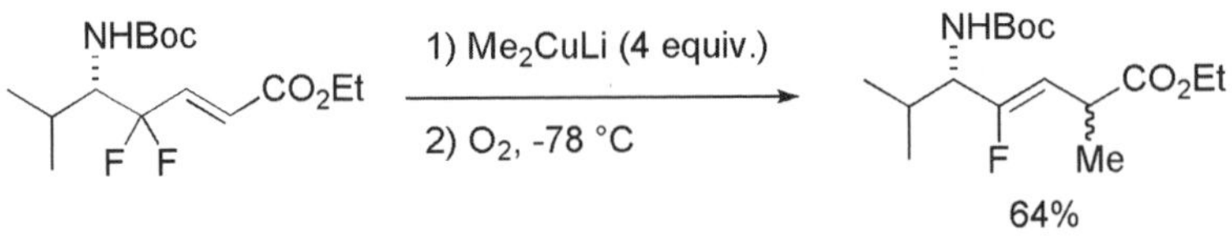

Scheme 10.13

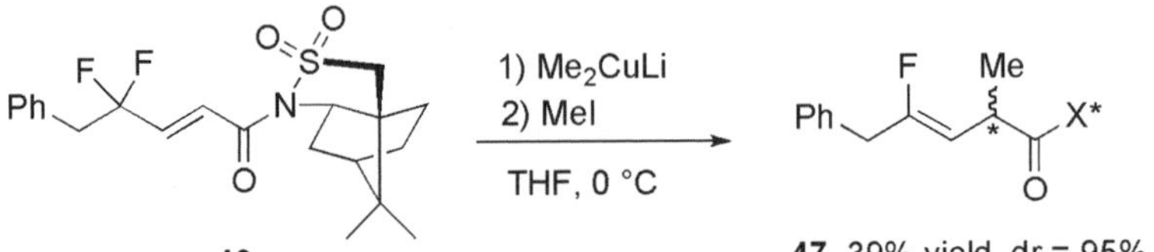

Scheme 10.14

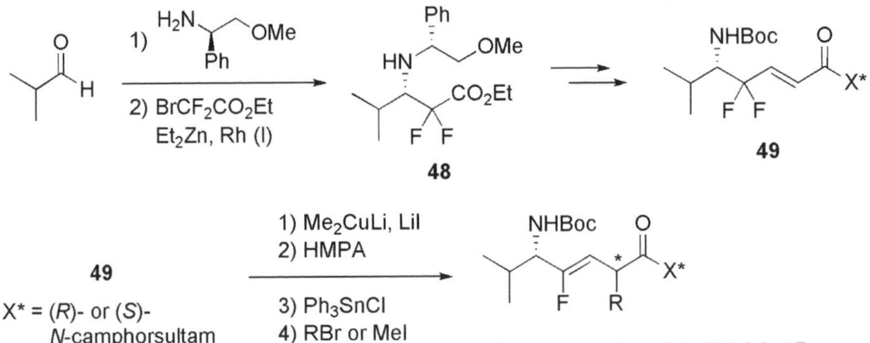

Scheme 10.15

Although Me_2CuLi-mediated reductive defluorination of **42**, followed by α-alkylation with alkyl halide, gave product **45** in high yield and in *Z*-selective manner, the relative stereochemistry at C-5 and C-2 was not controlled (see Scheme 10.12). For the stereocontrolled alkylation at C-2, a chiral auxiliary at the carboxyl moiety was used, and relatively high diastereoselectivity was realized in the case of camphorsultam derivative **46** (see Scheme 10.14) [23].

For the preparation of chiral nonracemic dipeptide isosteres, Otaka and Fujii also reported the use of camphorsultam derivative **49**, which was prepared in optically pure form through the Reformatsky reaction of $BrCF_2CO_2Et$ with a chiral imine (see Scheme 10.15) [25].

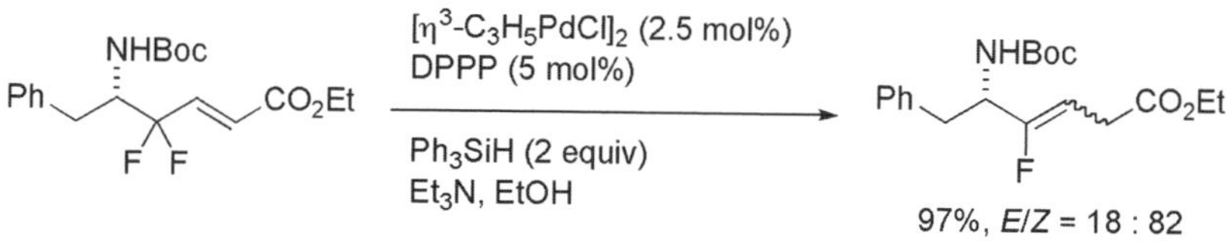

Scheme 10.16

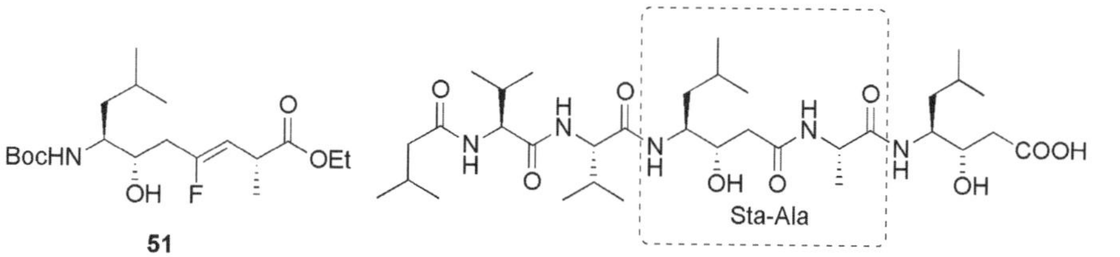

Figure 10.3 *Pepstatin and Sta-ψ[CF=CH]-Ala.*

A facile synthesis of fluoro-olefins through Pd(0)-catalyzed reductive defluorination of allylic *gem*-difluorides, in particular γ,γ-difluoro-α,β-enoates was recently reported (see Scheme 10.16) [26]. Stereoselectivities were moderate to high ($E:Z = 1:1$–$1:9$).

The reductive defluorination–alkylation method shown in Scheme 10.12 was successfully applied to the synthesis of *N*-Boc-Sta-Ψ[CF=CH]-Ala ethyl ester **51**, which mimics the central Sta-Ala unit of pepstatin, a naturally occurring inhibitor of aspartyl proteases including renin, pepsin, HIV-1 and HIV-2 proteases (see Figure 10.3) [27].

γ,γ-Difluoro-α,β-enoate **56**, the key substrate for the reductive defluorination–alkylation reaction, was synthesized in optically pure form using *N*-Boc leucinol **52** as the starting material (see Scheme 10.17). The Reformatsky reaction of bromodifluoroacetate with aldehyde **53**, having natural statin configuration, gave a 3 : 1 diastereomeric mixture of β-hydroxy ester **54**, which was subjected to radical deoxygenation reaction to give difluoro ester **55**. DIBAL reduction of **55**, followed by the HWE reaction using phosphonoacetate provided the key substrate **56**.

As expected, Me_2CuLi-mediated reductive defluorination of **56** and subsequent methylation with methyl iodide proceeded smoothly to give methylated product **57** in 96% yield. Regarding the stereoselectivity, a moderate *Z*-selectivity ($Z:E = 6.4:1$) was observed for the olefin moiety, but the C-2-methylation proceeded in nonselective manner. Fortunately, two diastereomers were readily separable by medium-pressure liquid chromatography (MPLC) as shown in Scheme 10.18. Structural assignments were made based on the x-ray crystallographic analysis of (2*S*)-**57**.

As described above, an organocopper reagent derived from alkyl lithium, such as Me_2CuLi, readily reacts with γ,γ-difluoro-α,β-enoate **42** to form a reductive defluorinated metal species **43**, exclusively (see Scheme 10.12). To facilitate the reactivity of fluorine as a leaving group, we also investigated the effects of alkylaluminum on this reaction as

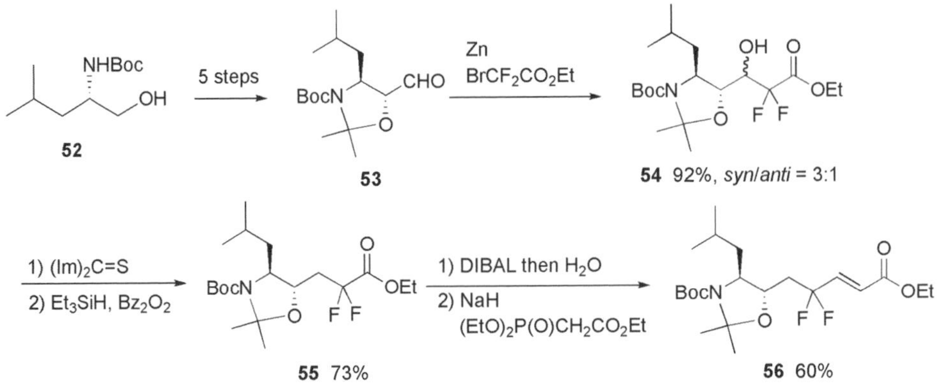

Scheme 10.17

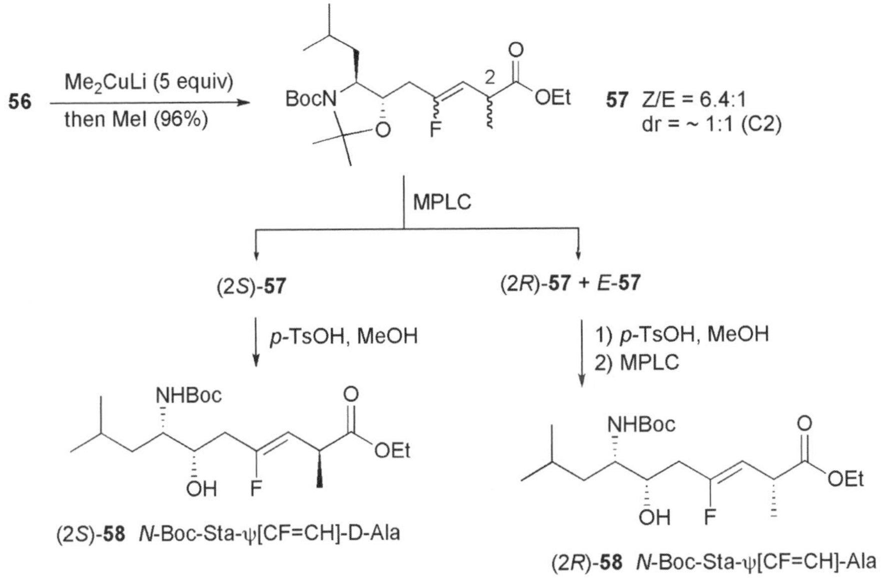

Scheme 10.18

an additive and/or alkylating reagent, since it is well documented that aluminum, in addition to its high oxygenophilicity (O-Al 511 ± 3 kJ/mol), has exceedingly high affinity toward fluorine (F-Al 663 ± 3 kJ/mol) [28]. We found that, in the presence of Cu(I), the reaction of γ,γ-difluoro-α,β-enoate **42b** having a free hydroxyl group at the δ-position with trialkylaluminum (5 equivalents) gave allylic alkylation product **44**, whereas γ,γ-difluoro-α,β-enoate having amino group **42a** or hydrogen derivative **42c** did not give the alkylated products (see Scheme 10.19) [22, 23]. With the hydroxyl substrate **44b**, the reaction proceeded in a *Z*-selective manner but without or with very low diastereoselectivity.

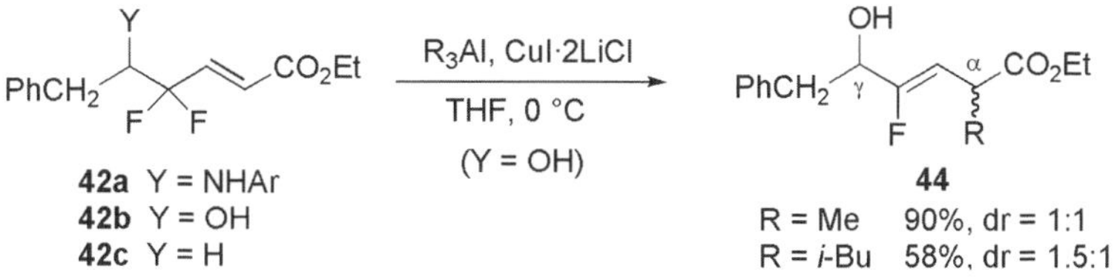

Scheme 10.19

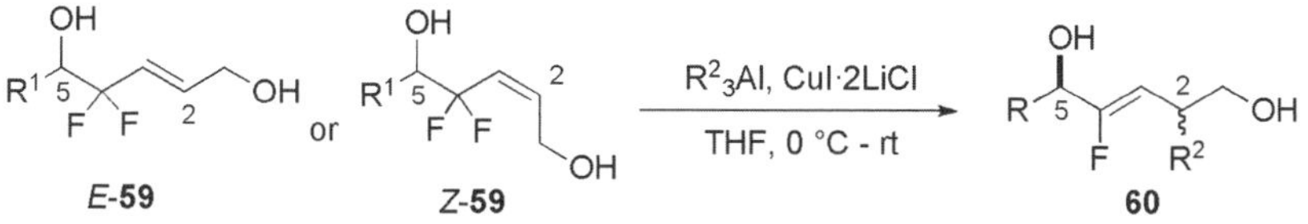

	R^1	$R^2{}_3Al$	**60** Yield(%)	*Z/E*	*Syn* : *anti*
E-**59a**	$PhCH_2$	Me_3Al	98	>95	>95 : 1
E-**59a**	$PhCH_2$	i-Bu_3Al	78	>95	>95 : 1
E-**59b**	Ph	Me_3Al	77	>95	>95 : 1
Z-**59b**	Ph	Me_3Al	53	>95	1 : 11

Scheme 10.20

Excellent diastereoselectivity was realized on using the substrate of allylic alcohol substrates **59** instead of α,β-unsaturated ester substrate **42** (Y = OH) mentioned above. Thus, in the presence of CuI·2LiCl, the reaction of the *E*-isomer of 5-hydroxy-4,4-difluoro-2-alken-1-ol (*E*)-**59** with trialkylaluminum proceeded smoothly to give the allylic substitution product **60** with complete 2,5-*syn*- and Z-selectivity (see Scheme 10.20) [29, 30]. With the *Z*-isomer substrate *Z*-**59**, reaction proceeded in a manner that favored the 2,5-*anti*-product, although the reactivity of the *Z*-isomer was clearly low compared with that of the *E*-isomer. It should be noted that the C-5 free hydroxyl functionality is crucial for the reaction to occur, since the lack of this OH group (i.e., protection as its ether or amine derivative in place of the OH group), resulted in no reaction under similar conditions, recovering the starting material.

This highly stereoselective reaction would provide an efficient method for the preparation of fluoroalkene dipeptide isosteres **62** and related molecules such as depsipeptide isosteres **61** (see Scheme 10.21) [29,30].

10.2.1.7 Cross-coupling Reaction of Fluorinated Alkenyl Bromides

In 2002, Burton and co-workers demonstrated that the *Z*- and *E*-isomers of bromofluoromethylene compound **63**, easily obtained by the Wittig reaction of carbonyl compound with Ph_3P-$CFBr_3$-Zn, were obtained through efficient kinetic separation using the palladium-catalyzed carboalkoxylation reaction (see Scheme 10.22) [31]. Since the *E*-isomer

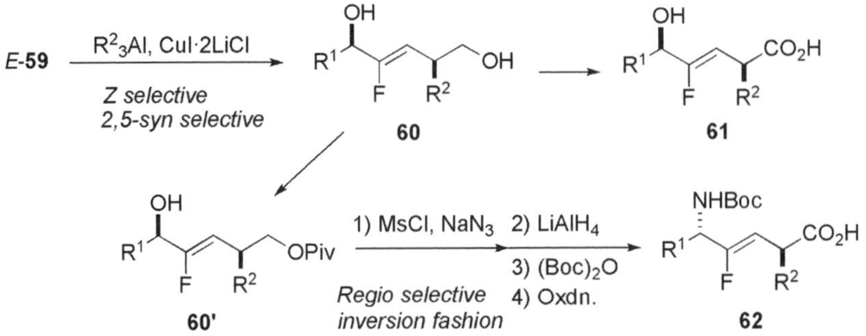

Scheme 10.21

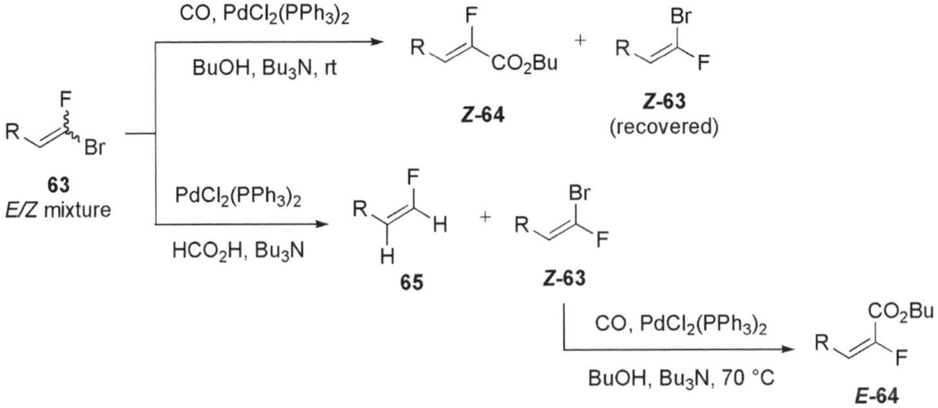

Scheme 10.22

of bromide **63** reacted much faster than the *Z* isomer, (*Z*)-α-fluoroenoate (*Z*)-**64** was preferentially obtained at low temperature. After a complete consumption of the *E*-isomer by Pd-catalyzed reduction with HCO_2H, the *Z*-isomer of bromide (*Z*)-**63** remaining was subjected to carboalkoxylation at 70 °C to give (*E*)-**64**.

Pannecoucke and co-workers developed a highly stereospecific synthesis of (*Z*)- and (*E*)-α-fluoroenones **66** through a kinetically controlled Negishi coupling reaction catalyzed by palladium (see Scheme 10.23) [32].

Further elaboration of both (*Z*)- and (*E*)-**66** led to the formation of fluoro-olefin ψ[CF=CH] dipeptide isosteres in enantiomerically pure form. For example, chiral imine (*Z*)-**68** prepared by Ti(IV)-promoted condensation of α-fluoroenone **66a** with Ellman's sulfonamide **67** was reduced in a highly diastereoselective manner to give the reductive amination product **69**, which was converted to the dipeptide isostere **70** via several steps (see Scheme 10.24) [33].

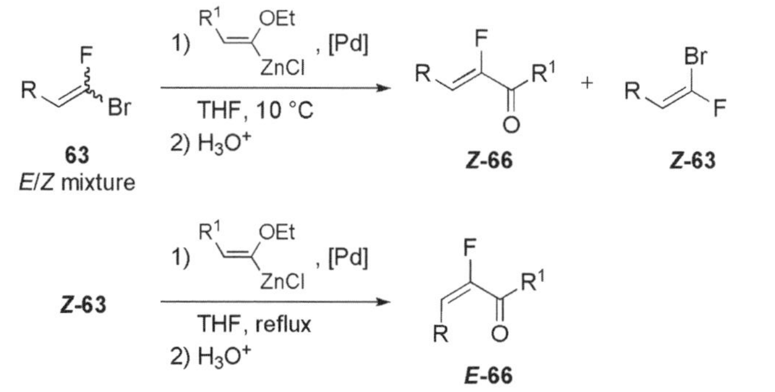

Scheme 10.23

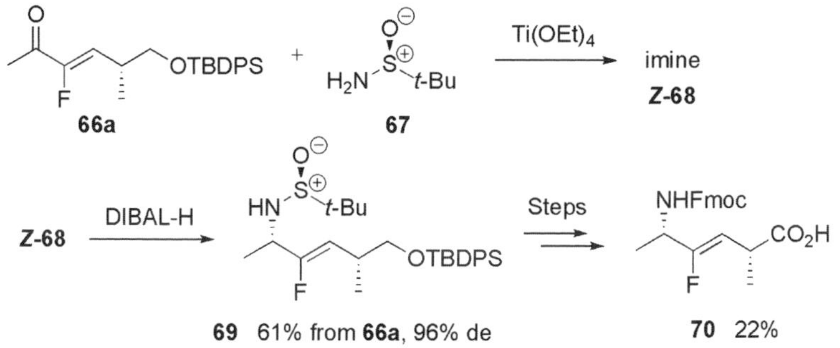

Scheme 10.24

10.2.2 Applications to Biologically Active Peptides and Related Studies

10.2.2.1 Substance P and Thermolysin

As an early example of the application of fluoro-olefin Ψ[CF=CH] dipeptide isosteres, Allmendinger *et al.* reported the preparation of neuropeptide substance P (SP) analogues containing the Phe-Ψ[(Z)-CF=CH]-Gly dipeptide unit [16]. In a receptor binding assay (see Table 10.1), SP analogue (*S*)-**30a** having the natural *S*-configuration in the Phe moiety was almost as active as SP itself (entry 1 vs. entry 6), while its diastereomer (*R*)-**30a**, the unnatural *R*-configured analogue, was 10 times less active than (*S*)-**30a** (entry 2). A similar difference was also observed in the binding affinities of the pyro-Glu-containing hexapeptide analogues (*S*)-**30b** and (*R*)-**30b** (entries 3 and 4). Furthermore, compared to the non-fluorinated olefinic analogue (*S*)-**30c**, the fluorinated analogue (*S*)-**30b** bound 10 times more strongly to the receptor (entry 3 vs. entry 5).

Bartlett and Otake experimentally verified the applicability of tripeptide analogues with the Cbz-Gly-Ψ[(Z)-CF=CH]-Leu-Xaa structure as ground-state analogue inhibitors of the zinc endopeptidase thermolysin (see Figure 10.4) [19]. These fluoro-olefin tripeptide

Table 10.1 *Receptor binding of Substance P and fluoroalkene dipeptide analogues*

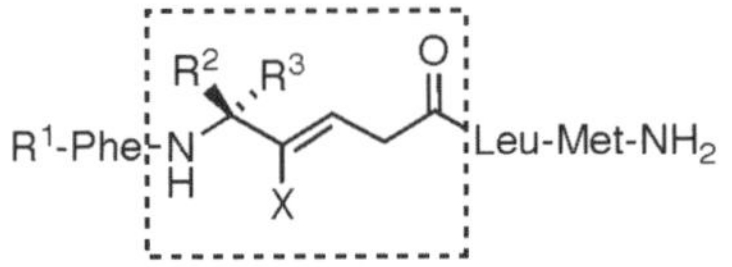

entry		X	R^1	R^2	R^3	IC_{50}
1	(*S*)-**30a**	F	Arg-Pro-Lys-Pro-Gln-Gln-Phe	$PhCH_2$	H	2 nM
2	(*R*)-**30a**	F	Arg-Pro-Lys-Pro-Gln-Gln-Phe	H	$PhCH_2$	20 nM
3	(*S*)-**30b**	F	Pyro-Glu-	$PhCH_2$	H	0.8 μM
4	(*R*)-**30b**	F	Pyro-Glu-	H	$PhCH_2$	10 μM
5	(*S*)-**30c**	H	Pyro-Glu-	$PhCH_2$	H	>10 μM
6	Substance P					1.3 nM

Figure 10.4 *Tripeptide analogues of the Cbz-Gly-Ψ[CF=CH]-Leu-Xaa structure.*

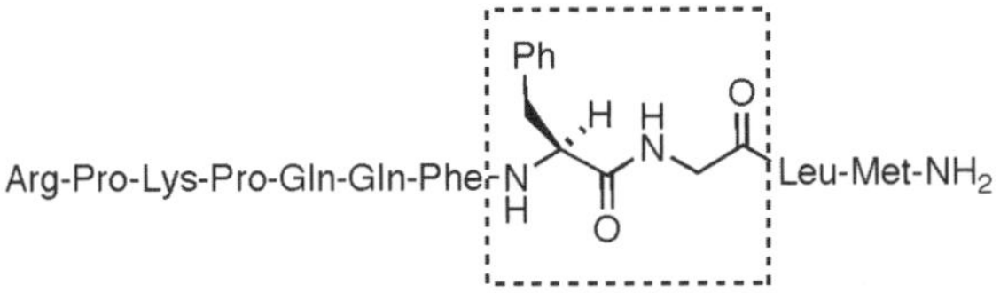

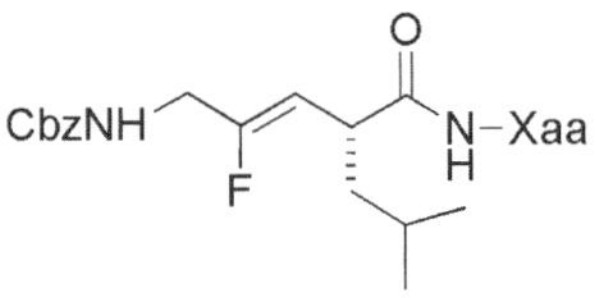

analogues **39** bind to thermolysin about one order of magnitude more tightly than the substrates, in which the binding mode of the fluoro-olefin inhibitor and substrate are not identical on the basis of K_i vs. K_m correlation,. Although these analogues are not particularly potent as enzyme inhibitors, they are useful as substrate models in structural studies.

10.2.2.2 Dipeptidyl Peptidase Inhibitors

Dipeptidyl peptidase IV (DPP IV, EC 3.4.14.5, CD26) is a serine protease cleaving off dipeptides from the amino terminus of peptides or proteins having proline or alanine at the penultimate position. Since prolylamides are known to play a critical role in peptide structure and function, and because of their high resistance toward nonspecific enzymatic hydrolysis, a few enzymes capable of cleaving this structural motif have attracted considerable attention. It has been shown that DPP IV inhibition can be used as a new tool for controlling type II diabetes [34]. Typically, these inhibitors possess a dipeptide skeleton with a free amine terminus and a pyrrolidine ring attached to an electrophilic site for Ser-OH scavenging such as a nitrile, a boronic acid, or a hydroxamate (see Figure 10.5).

Since the discovery of Xaa-(2-cyano)pyrrolidines (**71a**, X = CN, Figure 10.5) as potent and reversible inhibitors of DPP IV, the optimization of these "lead" structures in terms of activity and selectivity have been extensively investigated [35]. It was suggested that for strong enzyme inhibition, reversible interaction of the electrophilic nitrile group in the inhibitors and the Ser-OH group in the active site of the enzyme would play an important role. However, these nitrile-containing inhibitors **71a** suffer from the inactivation process to form cyclic amidines **74** and/or hydrolyzed product, diketopiperadine derivative **75**, due to a facile intramolecular cyclization of *N*-terminal amine to the nitrile group through conformational change from *trans*-amide to *cis*-amide as shown in Scheme 10.25.

To solve the inactivation issues mentioned above, the concept of conformationally restricted (*Z*)-fluoro-olefin dipeptide isosteres that mimic the active *trans* conformation of the DPP IV inhibitors was applied by Welch and co-workers to the preparation of inhibitors having Ala-Ψ[CF=CH]-Pro structure (**76** and **78**) and their inhibitory activities were evaluated (see Figure 10.6) [36, 37]. DPP IV inhibitory activities and the stability of the inhibitors **76** and **77**, in comparison with those of the model dipeptide Ala-Pro derivative **78** are summarized in Table 10.2. These fluoro-olefin analogues, **76** and **77**, showed better DPP IV inhibitory activity than that of **78**. In particular, *u*-**76** having the same relative stereochemistry as the natural dipeptide (L-Xaa-L-X′aa) configurations exhibited potent

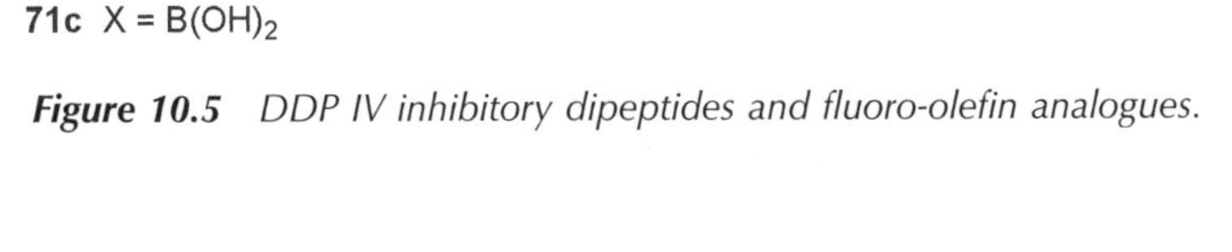

71a X = CN
71b X = CONHOCOAr
71c X = $B(OH)_2$

72a X = CN
72b X = CONHOCOAr

73a X = H
73b X = CN

Figure 10.5 *DDP IV inhibitory dipeptides and fluoro-olefin analogues.*

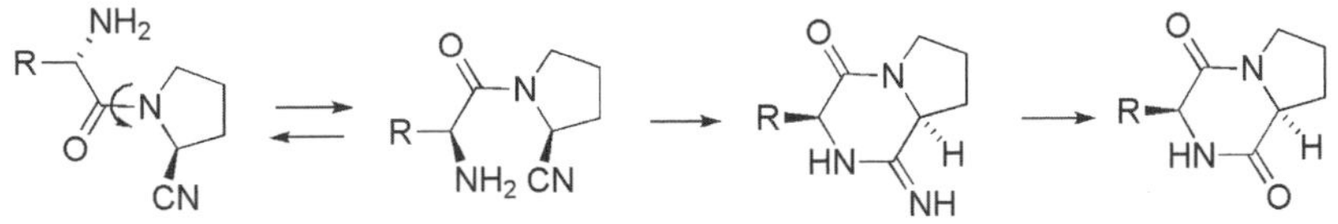

71a *trans*-amide ⇌ **71a** *cis*-amide → **74** → **75**

Scheme 10.25

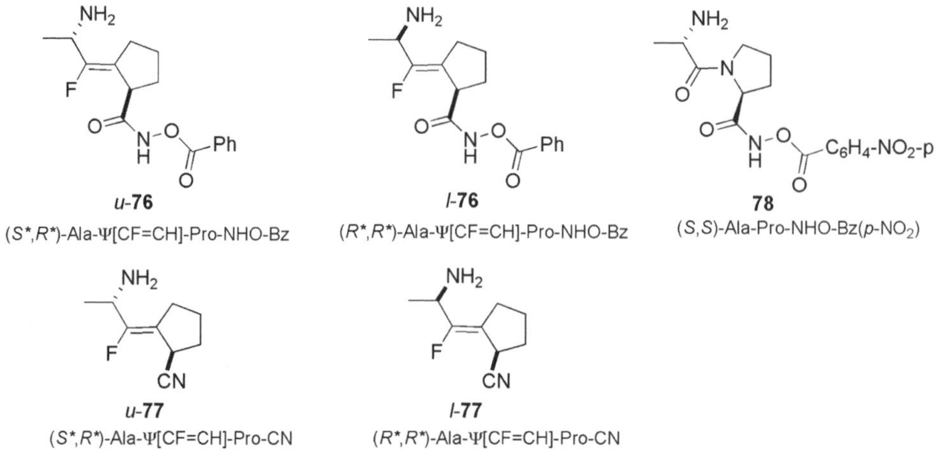

u-**76**
(S^*,R^*)-Ala-Ψ[CF=CH]-Pro-NHO-Bz

l-**76**
(R^*,R^*)-Ala-Ψ[CF=CH]-Pro-NHO-Bz

78
(S,S)-Ala-Pro-NHO-Bz(p-NO_2)

u-**77**
(S^*,R^*)-Ala-Ψ[CF=CH]-Pro-CN

l-**77**
(R^*,R^*)-Ala-Ψ[CF=CH]-Pro-CN

Figure 10.6 *Ala-Ψ[CF=CH]-Pro analogues as DDP IV inhibitors.*

Table 10.2 *Inhibition of DPPIV and stability of inhibitors*

inhibitor	Inhibition, %	([I], μM)	K_i, μM	$K_d \times 10^4$, min^{-1} [a]	$t_{1/2}$, h
u-**76**	42 100	(10) (250)	0.19	1.1	103
l-**76**	4 17	(10) (500)	14.4	ND	ND
u-**77**	16 50	(1) (10)	7.69	– [b]	
l-**77**	14 47	(1) (10)	6.03	– [b]	
78	29	(1,100)	30.0	13.0	8.8

[a] K_d: Decomposition rate constant
[b] No detectable degradation at pH 7.6 under buffered conditions.

activity ($K_i = 0.19\,\mu M$). Moreover, the stability of these fluoro-olefin-containing peptide mimetics was remarkable. For example, no detectable degradation of the cyano derivatives **77** was observed at pH 7.6 under buffered conditions. Although the effects of the replacement of the parent amide bond with the fluoro-olefin moiety in these molecules on inhibitory potency and stability might not be fully deduced from these limited examples, good affinity of these fluorine-modified peptide mimetics to the enzyme was experimentally verified.

Augustyns and co-workers reported a structure–activity relationship (SAR) study of fluoro-olefin analogues of *N*-substituted glycylpyrrolidines **79a**, glycylpiperidines **80a**, and glycyl-(2-cyano)pyrrolidines **81** as well as the corresponding olefin analogues **79b** and **80b** for their activities as DPP IV or DPP II inhibitors (Table 10.3) [38]. Except for **79c-1**, most of these compounds exhibited a strong preferential binding to DPP II. Recent crystallographic analyses of peptidic inhibitor–DPP IV complexes revealed that the P_2–P_1 amide

Table 10.3 *Inhibition of DPPIV and DPPII*

R	n = 1			n = 2		
	compd.	IC_{50} (μM) DPP IV	IC_{50} (μM) DPP II	Compd.	IC_{50} (μM) DPP IV	IC_{50} (μM) DPP II
1 (cyclohexyl)	**79a-1**	>1000	90 ± 1	**80a-1**	>500	500 ± 50
	79b-1	>1000	62 ± 12	**80b-1**	>1000	136 ± 14
	79c-1	148 ± 26	276 ± 72	**80c-1**	>1000	177 ± 54
2 (phenyl-CH2)	**79a-2**	>1000	38 ± 3	**80a-2**	>1000	34 ± 4
	79b-2	>1000	26 ± 3	**80b-2**	>1000	55 ± 4
	79c-2	1000 ± 10	500 ± 50	**80c-2**	>1000	397 ± 25
3 (Bn-N piperidinyl)	**79a-3**	>250	1.3 ± 0.2	**80a-3**	>250	1.0 ± 0.1
	79a-3	ND	ND	**80c-3**	>1000	3.1 ± 0.1

oxygen atom of inhibitors is involved in hydrogen bonding with the Asp710 and Arg125 residues of the enzyme. Assuming the importance of such a hydrogen bond, the observed low affinity of the fluoro-olefin isosteres, **79a** and **80a**, for DPP IV could be attributed to the fact that they are much weaker hydrogen-bond acceptors than an amide functionality. In contrast, the hydrogen-bond formation seems less critical for DPP II inhibition. Moreover, **81**, wherein a nitrile group was introduced to fluoro-olefin isostere **79a**, did not result in increased DPP IV or DPP II inhibitory activity, although these analogues showed substantially increased stability in solution over the parent amide compound.

10.2.2.3 Peptide-Nucleic Acid

Peptide-nucleic acid (PNA) is a DNA analogue based on a polyamide backbone. It is known that PNAs strongly bind to complementary DNA and RNA sequences, obeying the Watson–Crick hydrogen-bonding rules, by taking advantage of the lack of electrostatic repulsion between anionic ribose-phosphate moieties in the backbone [39]. The potential utility of PNAs as gene-therapy agents has been attracting much attention from medicinal chemists, although their poor cellular uptake, low solubility in water, and self-aggregation are remaining problems for clinical applications [40, 41]. To realize more specific and higher affinity to DNA and RNA, modification of PNA backbone has been extensively studied. Leumann and co-workers reported the synthesis and evaluation of fluoroalkene and nonfluorinated alkene isosteres of PNAs as conformationally locked mimics (see Figure 10.7) [42–45].

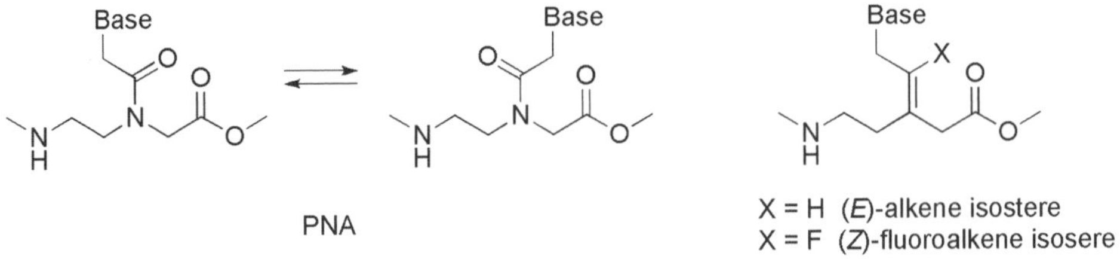

PNA

X = H (*E*)-alkene isostere
X = F (*Z*)-fluoroalkene isosere

Figure 10.7 *Rotameric forms of PNA and the structures of alkenic isosteres.*

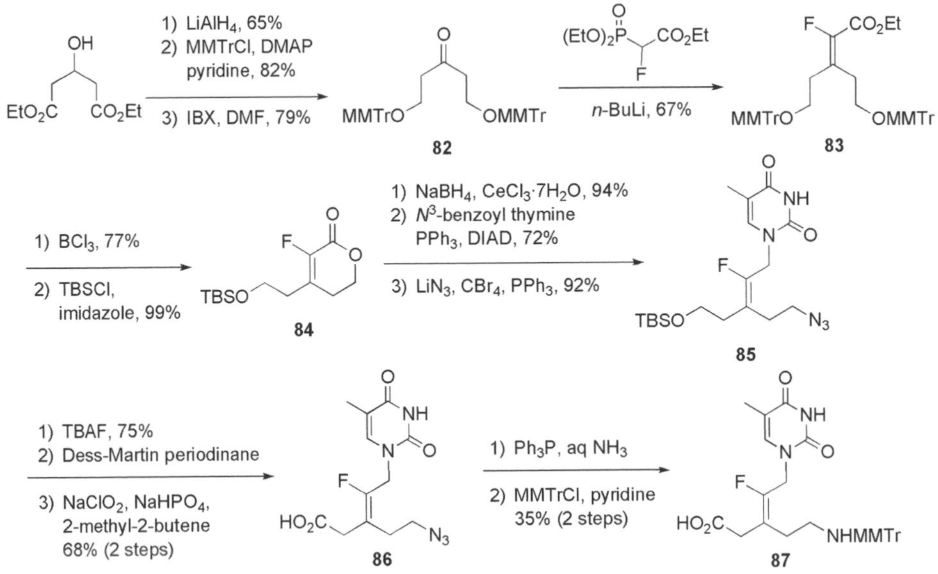

Scheme 10.26

The thymidyl-PNA monomer **87** having a fluoroalkene structure was obtained by multistep reactions using the Wittig reaction for the construction of the fluoroalkene functionality as shown in Scheme 10.26. The Wittig reaction of bis-MMTr-protected symmetric ketone **82**, derived from diethyl 3-hydroxyglutarate with $(EtO)_2P(O)CHFCO_2Et$, gave fluoroalkene **83** in 67% yield. The reaction of **83** with BCl_3 and the subsequent silylation gave fluorinated lactone **84**. The introduction of thymidyl group and amino functionality to **84** was achieved by the chemoselective Mitsunobu reaction of a diol, prepared by 1,2-reduction of **84**, with N^3-benzoyl thymine and Ph_3P/CBr_4-mediated S_N2-type substitution of the homoallylic alcohol moiety with LiN_3 to give azide **85**. Conversion of the azide group of **85** to an MMTr-amino acid functionality in five steps provided monomeric fluoroalkene isostere unit **87**. With the use of essentially the same method, a fluoroalkene isostere of an adenyl-PNA was also synthesized. By means of MMTr/acyl solid-phase peptide synthesis [46], these fluoroalkenic monomers were incorporated into the corre-

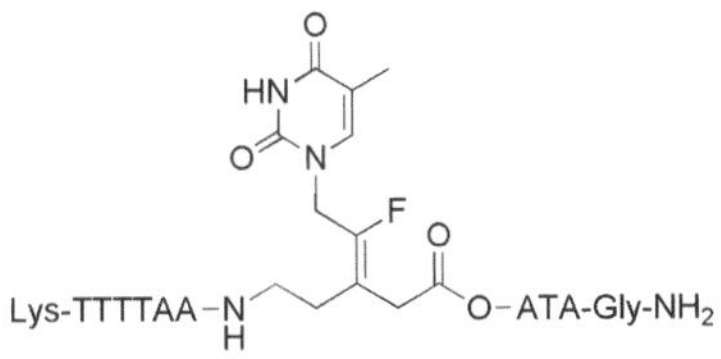

sequence	T_m (°C) d(TATATTAAAA) (antiparrallel)	T_m (°C) d(AAAATTATAT) (parallel)
Lys-TTTTAATATA-Gly-NH_2 (**88**)	33.2	11.6
Lys-TTTTAAt^{F}ATA-Gly-NH_2 (**89a**)	35.6	13.0
Lys-TTTt^{F}AATATA-Gly-NH_2 (**89b**)	27.0	8.0
Lys-TTt^{F}TAATATA-Gly-NH_2 (**89c**)	25.1	9.0
Lys-TTTTAAt^{E}ATA-Gly-NH_2 (**90**)	36.7	13.1
Lys-TTTTAAt^{Z}ATA-Gly-NH_2 (**91**)	28.0	6.1

T_m data are from UV-melting curves (260 nm) of duplex with antiparallel component .

Figure 10.8 *T_m data of PNA, alkenic PNA and fluoroalkenic PNA.*

sponding PNA sequence, which was equipped with lysine units at the N-termini to enhance water solubility and for easy purification.

The T_m value of fluoroalkene isostere **89a** with antiparallel DNA component was found to be slightly higher than that of the normal PNA oligomer **88** (ΔT_m = +2.4 °C) (see Figure 10.8). This result indicates better stabilization of the PNA/DNA duplex through introduction of a fluoroalkene unit at the central part of the PNA backbone. Nonfluorinated (*E*)-alkene isostere **90** [*note:* The (*E*)-alkene isostere has the same geometric configuration as the (*Z*)-fluoroalkene isostere] at the same location in the PNA sequence provided better stabilization of the PNA/DNA duplex with antiparallel DNA component than that of **88** (ΔT_m = +3.5 °C). In contrast, nonfluorinated (*Z*)-alkene isostere **91** resulted in significant destabilization of the duplex (ΔT_m = −5.2 °C). Thus, it is suggested that the Z-configuration of the double bond may cause a structural mismatch. Interestingly, the stability of the PNA/DNA duplexes of decamers **89**, containing single fluoroalkene modification with antiparallel DNA, was found to be strongly dependent on its position in the PNA sequence, with ΔT_m values ranging from +2.4 to −8.1 °C. Additionally, a similar tendency was observed for the PNA/DNA duplexes with parallel DNA.

10.2.2.4 Fluoroarenes as Nucleoside Base Mimics

The use of fluoroarene derivatives as structural mimetics of nucleobases has been studied extensively [47, 48]. In pioneering work in this field, Kool and co-workers reported the structural similarity between natural 2′-deoxythymidine (dT) and 1′,2′-dideoxy-1′-(2,4-difluoro-5-methylphenyl)-β-D-ribofuranoside **92** (dF) (see Figure 10.9) [49, 50].

As shown in Scheme 10.27, fluoroarene isostere **92** was prepared from 2,4-difluoro-5-bromotoluene and silyl-protected lactone **93** in three steps [51]. Thus, the coupling reaction of lactone **93** with an aryl Grignard regent, which was prepared from 2,4-difluoro-5-bromotoluene and magnesium in the presence of catalytic amount of iodine, and the subsequent reductive dehydroxylation of lactol intermediate by Et_3SiH resulted in the β-selective introduction of difluorotoluyl group. Further treatment with TBAF deprotected the silyl groups to give **92** in excellent yield. With the use of essentially the same procedure, dichloro- and dibromo-analogues were also prepared. To incorporate **92** into the DNA sequence, 5′-OH selective protection of **92** with 4,4′-dimethoxytrityl chloride (DMTr-Cl) and subsequent phosphorylation with 2-cyanoethyl *N,N*-diisopropylchlorophosphoramidite gave the fully protected nucleotide precursor **94**, which was subjected to

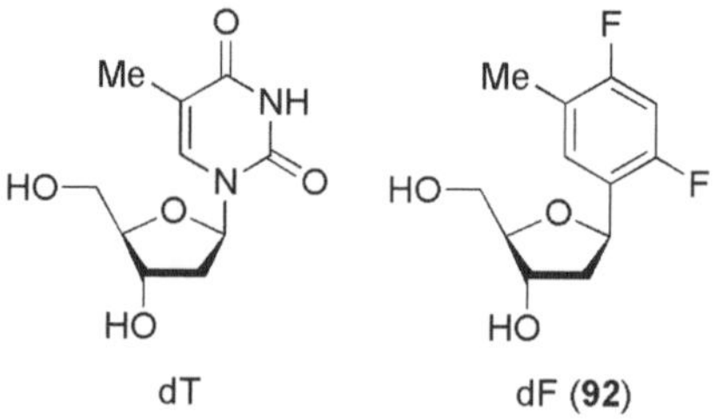

Figure 10.9 Natural 2′-deoxythymidine and 1′,2′-dideoxy-1′-(2,4-difluoro-5-methylphenyl)-β-D-ribofuranoside.

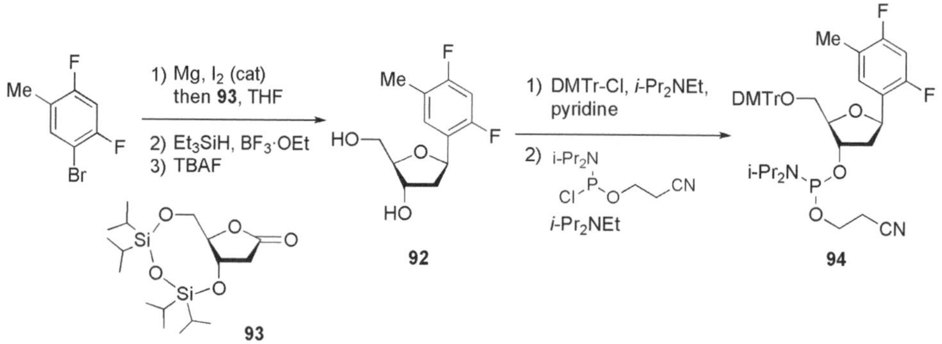

Scheme 10.27

oligonucleotide synthesis using an automated synthesizer based on phosphoramidite chemistry [52].

Fluoroarene isostere **92** was designed as a nonpolar nucleoside with the closest possible steric and structural similarity to the natural deoxythymidine, but without hydrogen-bonding ability. Despite the proposed ability of C–F groups in fluoroarene derivatives as weak hydrogen-bonding acceptors and aromatic C–H groups as weak hydrogen-bonding donors [53, 54], ^{1}H-NMR titration of 9-ethyladenine with 2,4-difluorotoluene in aqueous media did not show significant change of chemical shifts in 9-ethyladenine [50]. In general, it is known that the hydrogen-bonds between molecules in aqueous solution are weak due to the high dielectric constant of the medium as well as competitive hydrogen-bonding with the solvent itself. Therefore, this observation indicates the lack of hydrogen-bonding between the fluoroarene nucleoside isostere and the complementary natural nucleoside in aqueous media. On the other hand, Engeles and co-workers recently found intermolecular C–H···F–C hydrogen bonding in the crystalline form of 1′-deoxy-1′-(4-fluorophenyl)-β-D-ribofuranose **95a** (see Figure 10.10) [55]. The x-ray crystallographic analysis of **95a** (recrystallized from methanol) confirmed that the distance between the fluorine atom and the arylic hydrogen at the 3-position of another molecule (2.30 Å) is significantly shorter than the sum of the van der Waals radii of fluorine and hydrogen (2.55 Å). Unfortunately,

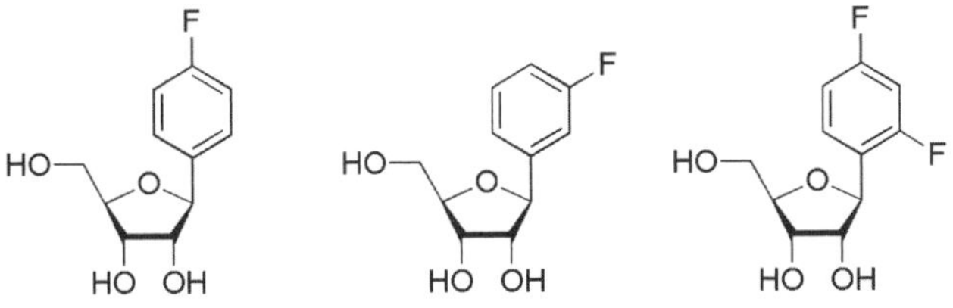

95a **95b** **95c**

Figure 10.10 *Fluoroarene nucleoside isosteres.*

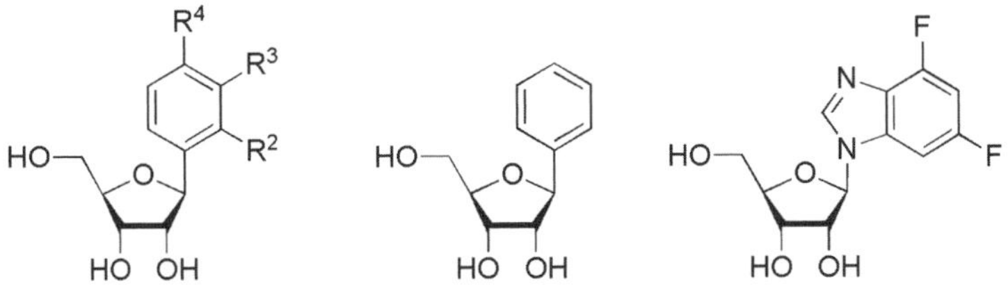

95a 4-F, **95b** 3-F, **95c** 2,4-F_2 **95d** **96**

X	Y = A		Y = C		Y = G		Y = U		Y = **95a**		Y = **95c**	
	T_m	$\Delta G°$	T_m	$\Delta G°$	T_m	$\Delta G°$	T_m	$\Delta G°$	T_m	$\Delta G°$	T_m	$\Delta G°$
U	37.8	11.9	30.4	9.8	38.6	11.9	30.1	9.7	-	-	-	-
95a	23.8	7.9	24.1	8.0	24.2	8.0	25.6	8.4	29.9	9.6	32.5	10.2
95b	24.7	8.2	25.0	8.2	25.0	8.2	25.7	8.4	-	-	-	-
95c	27.4	9.0	27.3	8.9	27.6	9.0	27.9	9.1	-	-	-	-
95d	23.0	7.7	22.6	7.6	23.5	7.9	23.1	7.7	-	-	-	-
96	28.4	9.2	28.7	9.2	29.4	9.5	29.4	9.3	31.3	10.0	34.6	11.2

T_m [°C]; $\Delta G°$ [kcal/mol] (T = 298 K); Errors: T_m: ±0.2 °C; $\Delta G°$: ±2%

Figure 10.11 *Thermodynamic properties of modified/unmodified duplex RNA (5′-CUU UUC XUU CUU paired with 3′-GAA AAG YAA GAA).*

in the cases of 3-fluorophenyl-β-D-ribofuranose **95b** and 2,4-difluorophenyl-β-D-ribofuranose **95c**, no H···F distance shorter than 2.55 Å was observed. This intermolecular C–H···F–C hydrogen bond also showed nearly linear arrangement with dihedral angle of 158 °. Interestingly, the H···F distance in a single crystal of **95a** recrystallized from water was found to be longer (2.38 Å) than in that recrystallized from methanol (2.30 Å) due to the incorporation of a water molecule into the unit cell: that is, a water molecule is placed between 2′-OH and 5′-OH, which makes the F···H distance longer.

Engeles and co-workers also reported the effect of these fluoroarene isosteres on the thermodynamic properties of the duplex RNA [55] (see Figure 10.11). In the 12-mer oligonucleotides (5′-CUU UUC XUU CUU paired with 3′-GAA AAG YAA GAA), the wobble base pair U·G showed the highest T_m (38.6 °C). Single modification of oligonucleotide sequence by the incorporation of nonfluorinated phenyl isostere **95d** significantly destabilized the RNA duplex structure, although fluorinated analogues **95a**, **95b** and **95c** improved the thermal stability of the duplex RNA compared with **95d**. In particular, the T_m values of singly modified duplex RNAs with 2,4-difluorophenyl isostere **95c** were notably higher than those with monofluorinated isosteres. 4,6-Difluorobenzoimidazole isostere **96** was also found to be an effective mimic of purine bases such as inosine. Surprisingly, the pairing of fluorinated nucleobase isosteres with a purine or pyrimidine nucleobase resulted in similar T_m values. Thus, these fluoroarene isosteres potentially act as a new class of universal base, which can pair with all natural bases without energy loss. The evaluation of *doubly* modified duplex RNAs also suggested a significant interaction between **96** and **95c** (T_m = 34.6 °C). It was also demonstrated by the exhaustive thermal analysis that the duplex RNA with **95c·96** base pair is 0.6 kcal/mol more stable than that of the calculated value. To explain this stabilization, Engeles and co-workers proposed a weak F···H hydrogen-bond between the fluorinated nucleotides [55].

Kool and co-workers also reported 1′,2′-dideoxy-1′-(4-fluoro-6-methylbenzoimidazolyl)-β-D-ribofuranose (**97**, dH) as a highly effective nonpolar isostere of 2′-deoxyguanosine (dG) (see Figure 10.12) [56]. In crystalline forms, the bond lengths and base shapes in these compounds are quite close, although the orientation of the base with respect to the furanose ring is different. That is, the natural dG adopts an *anti* conformation with a torsion angle χ of −122 °, while the base part of **97** is twisted by 58 ° relative to this and falls between typical *anti* and *syn* ranges with a measured torsion angle χ of −65 °. However, NMR studies in D_2O revealed that the conformation of **97** in solution was essentially identical to that of dG. For instance, the conformations of the deoxyribose parts of both nucleosides were classified to be 70% S for dG and 66% S for **97**. In general, the 2-deoxyribofuranoside structure preferentially occupies the *S*(2′-*endo*) conformation. Additionally, nuclear Overhauser effect (NOE) data of **97** in D_2O strongly indicate that nucleoside **97** is favored as the *anti*-conformer.

One of the most important aspects of 4-fluoro-6-methylbenzimidazole as a modified nucleobase is its strong stabilization of duplex DNA through a stacking effect. The stacking properties of **97** and other modified nucleosides can be measured by the dangling end thermal denaturation studies of DNA duplexes with self-complementary stands (dXCGC-

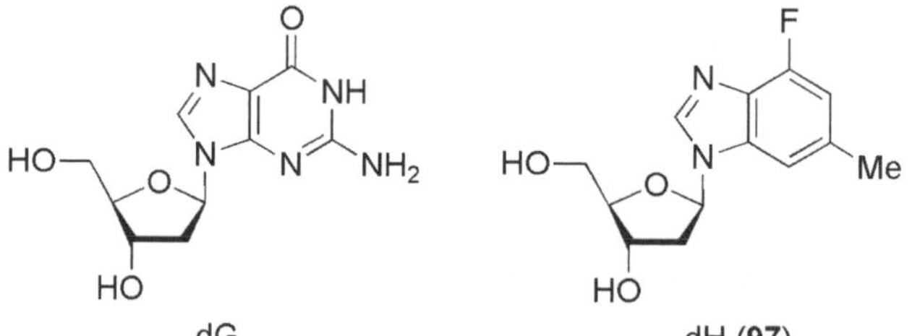

Figure 10.12 *2′-Deoxyguanosine and 1′,2′-dideoxy-1′-(4-fluoro-6-methylbenzoimidazolyl)-β-D-ribofuranose.*

GCG.) (see Figure 10.13a) [57]. Regarding the free energy of stacking for the natural/modified nucleobases, this study showed that the nucleobase mimic **97** stacked considerably more strongly than any of the four natural DNA bases. A possible stacking model was proposed as shown in Figure 10.13. In this model, the 4-fluoro-6-methylbenzimidazole ring nicely overlaps the cytosine base.

As an unique application of fluoroarene isosteres, Kool and co-workers recently reported that sets of 1′,2′-dideoxy-1′-(2,4-dihalo-5-methylphenyl)-β-D-ribofuranose (**98**) and 1′-deoxy-1′-(2,4-dihalo-5-methylphenyl)-β-D-ribofuranose (**99**) as thymidine or uridine analogues with gradually increasing size can be used as "molecular rulers" (see Figure 10.14) [56]. X-ray crystallographic analysis and various NMR studies in D_2O showed that the conformations of these isosteres in both crystalline state and solution closely resemble that of thymidine. Additionally, PM3 calculations of **98** and **99** suggested that change of the substituent X from hydrogen to a series of halogen atoms would increase the bond lengths at the 2,4-positions, corresponding to oxygens of thymidine, ranging from 1.09 Å to 1.97 Å (see Figure 10.14). Therefore, enzymatic evaluation using a set of these compounds should be able to elucidate the fidelity of substrate specificity for the target enzyme.

According to this concept, in 2005 Kool and Essingmann reported the utility of "molecular rulers" as probes for the active-site tightness of DNA polymerase [58, 59]. In kinetic studies of single nucleotide insertion in the steady state using a gel electorophoresis-based analysis, it was found that the efficiency of the replication of 28-mer/23-mer template-primer duplexes, having the sequence (5′-ACT GAT CTC CCT ATA GTG AGT CGT ATT A·5′-T AAT ACG ACT CAC TAT AGG GAG A) with variably sized nucleoside triphosphates **100** promoted by DNA polymerase I from *E. coli* (Klenow fragment, exonuclease deficient), was highly dependent on the size of compounds (see Figure 10.15). A comparison of the efficiency for nucleotide insertions, which was measured by V_{max}/K_M, shows that the Cl-substituted analogue dLTP, whose chloroarene ring is larger than natural thymine by 0.5 Å, is the most efficient compound in this series. In addition, the relative efficiency of dLTP is almost the same as that of natural dTTP within experimental error,

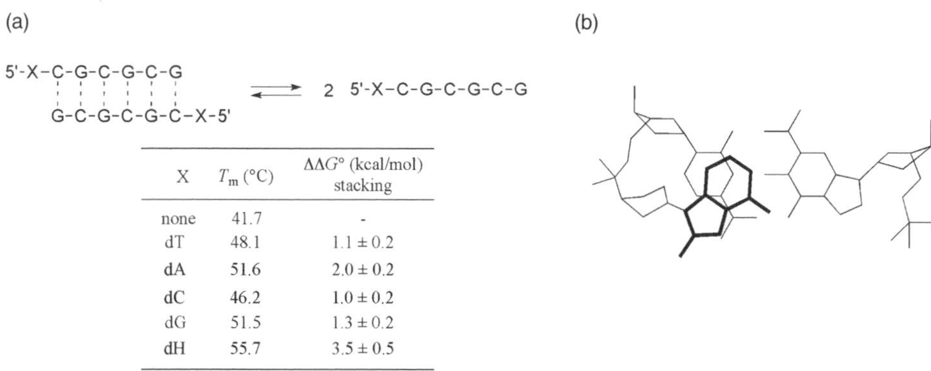

X	T_m (°C)	$\Delta\Delta G°$ (kcal/mol) stacking
none	41.7	-
dT	48.1	1.1 ± 0.2
dA	51.6	2.0 ± 0.2
dC	46.2	1.0 ± 0.2
dG	51.5	1.3 ± 0.2
dH	55.7	3.5 ± 0.5

Figure 10.13 *(a) The helix–coil equilibrium of 5′-XCGCGCG and dangling end thermal denaturation studies. (b) Possible 5′-end stacking geometry for the aromatic rings.*

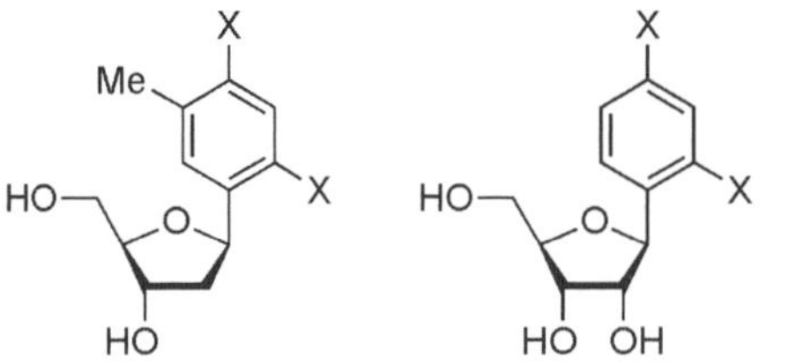

98 (X = H, F, Cl, Br, I)

99 (X = H, F, Cl, Br)

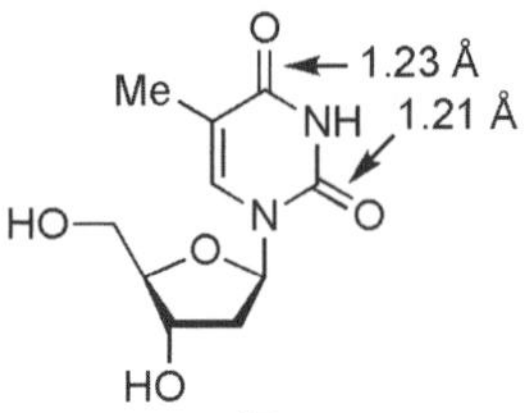

dT

	PM3-calculated bond length (Å)				
	X = H	X = F	X = Cl	X = Br	X = I
98	1.10	1.34	1.69	1.87	1.97
99	1.09	1.34	1.68	1.87	-

Figure 10.14 *PM3-calculated bond length of arene isosteres.*

dNTP	Efficiency (V_{max}/K_M)	Relative efficiency
dTTP	5.7 x 10^5	1
dHTP (X = H)	1.0 x 10^3	6.3 x 10^{-5}
dFTP (X = F)	1.0 x 10^5	6.3 x 10^{-3}
dLTP (X = Cl)	2.2 x 10^6	1.4 x 10^{-1}
dBTP (X = Br)	5.7 x 10^5	3.6 x 10^{-2}
dITP (X = I)	5.7 x 10^5	6.3 x 10^{-3}

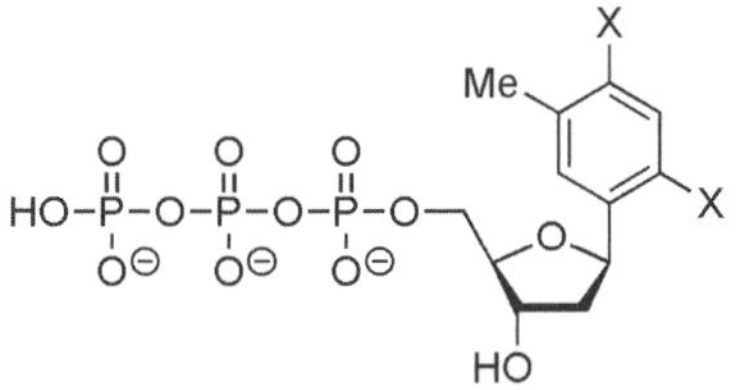

Modified dNTP **100**
(X = H, F, Cl, Br, I)

Figure 10.15 *Steady-state kinetic efficiencies for single-nucleotide insertion by DNA polymerase I with variably sized nucleoside triphosphate.*

despite the nonpolar nature of dLTP. When the molecule size exceeded that of the chlorinated analogue, the relative efficiency dropped markedly by a factor of 164 for the largest dITP in response to a subtle size increase of 0.35 Å over the optimum. This result suggests that the active site of DNA polymerase strictly recognizes the steric bulkiness and the shape of the substrate.

The "molecular ruler" methodology for determining the size of the active site of DNA polymerase was also applied to the investigations of other DNA polymerases, such as T7 DNA polymerase and Dpo4 polymerase [60, 61].

On the other hand, a set of 1′-deoxy-1′-(2,4-dihalo-5-methylphenyl)-β-D-ribofuranoses (**99**, rN) was used for the investigation of flexible active site of reverse transcriptase of HIV-1 virus (HIV-RT) (see Figure 10.16) [62]. Since HIV-RT is well known as a highly mutagenic polymerase, the understanding of its mutagenicity potentially provides a solution to the rapid development of drug resistance during treatment of AIDS. The efficiencies of the dATP incorporation to singly modified RNA templates by HIV-RT are shown in Figure 10.16. All uracil analogues **99** directed preferential incorporation of dATP. Regard-

Template	Efficiency (V_{max}/K_M)	Relative efficiency
rU	3.4 x 10^4	1
rH (X = H)	3.4 x 10^1	1.0 x 10^{-3}
rF (X = F)	1.2 x 10^3	3.5 x 10^{-2}
rL (X = Cl)	9.9 x 10^2	2.9 x 10^{-2}
rB (X = Br)	1.1x 10^3	3.2 x 10^{-2}

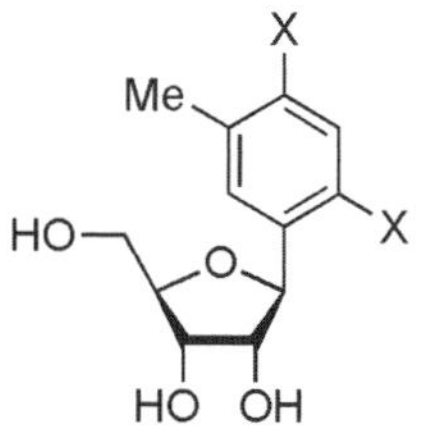

rN (**99**, X = H, F, Cl, Br)

Figure 10.16 *Steady-state kinetic efficiencies for single-nucleotide insertion by HIV-1 reverse transcriptase to duplex RNA templates [5′-r(ACU GXU CUC CCU AUA GUG AGU CGU AUU A),·55′-d(T AAT ACG ACT CAC TAT AGG GAG A)] with dATP.*

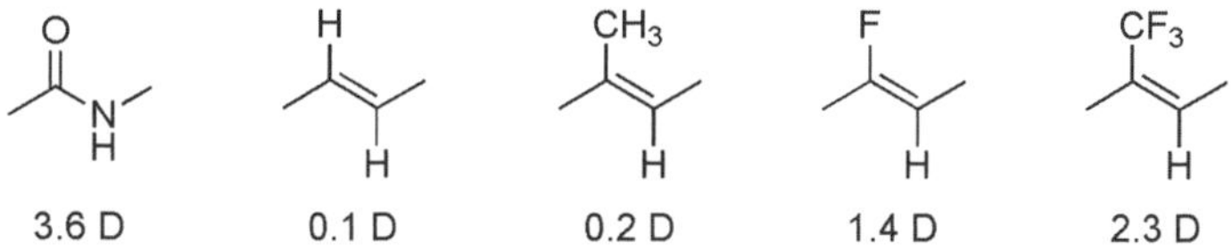

3.6 D 0.1 D 0.2 D 1.4 D 2.3 D

Figure 10.17 *Dipolar moments (debye) of amide and various alkene isosteres.*

ing the effect of size on the efficiency, the smallest analogue (X = H) was less tolerated than other analogues in the template, but there was otherwise little or no difference in dATP incorporation efficiencies. The results are remarkably different from those for the DNA polymerases, and clearly suggest the high structural flexibility of HIV-RT.

10.3 Trifluoromethylated Alkenes Ψ[CH(CF_3)=CH] as Dipeptide Isosteres

In 1998, Wipf and co-workers demonstrated that the replacement of a peptidic amide bond with a trifluoromethylated alkene mimics well the β-turn structure of the original peptide [7]. AM1-calculated dipolar moments of the amide bond and various alkene isosteres suggested the electrostatic similarity between the amide and the trifluoromethylalkene functionalities (see Figure 10.17) [7].

10.3.1 Synthetic Method

The synthesis of trifluoromethylalkene dipeptide isosteres was reported by Wipf and coworkers [7]. As shown in Scheme 10.28, the multistep synthesis of Ala-Ψ[C(CF_3)=CH]-Xaa type isosteres was achieved from 1,1,1-trichloro-2,2,2-trifluoroethane. The carboxylation of trifluorotrichloroethane via trifluorodichloroethylzinc intermediate, esterification with benzyl alcohol, and Reformatsky addition–elimination with acetaldehyde

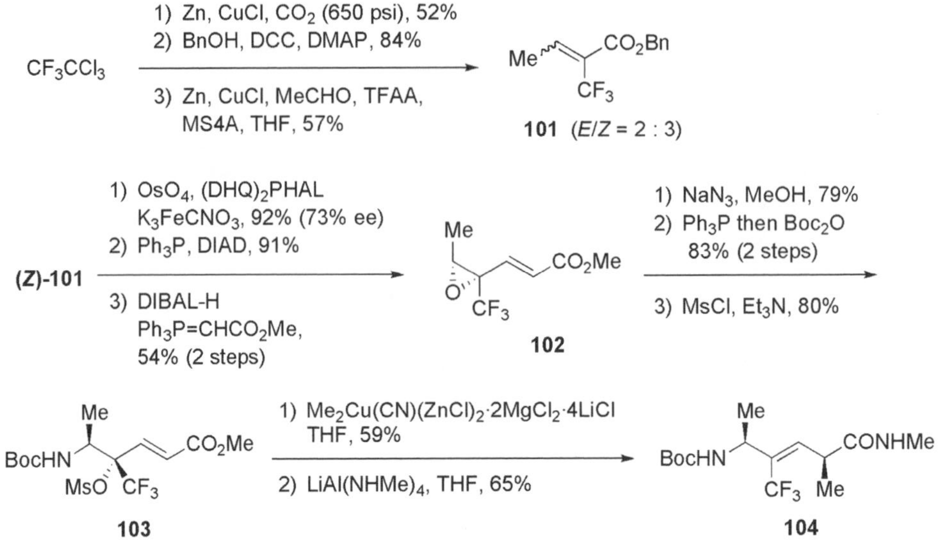

Scheme 10.28

yielded α-trifluoromethylated crotonate **101** as a 2:3 mixture of *E*- and *Z*-isomers. The Sharpless asymmetric dihydroxylation of (*Z*)-**101**, which was isolated by column chromatography on silica gel, afforded chiral diol in 73% ee. Conversion of this diol to an epoxide under Mitsunobu conditions, half-reduction of the ester moiety with DIBAL-H, and Wittig olefination gave trifluoromethylated epoxide **102**. Introduction of amino group through ring opening of epoxide **102**, followed by allylic methylation of the resultant γ-mesyloxy-α,β-enoate **103** with methylcuprate, and the final amidation gave Ala-Ψ[C(CF_3)=CH]-Ala **104**.

Later, Wipf and co-workers also reported the improved synthesis of trifluoromethylalkene isostere **110** through addition reaction of internal alkynes to *N*-phosphonylimines (see Scheme 10.29) [63]. As shown in Scheme 10.29, the reaction of *N*-phosphonylimine **104** with vinylzinc species, derived from alkynylstannane **103** through hydrozirconation, followed by transmetallation with Me_2Zn, gave vinylstannane **105** in 65% yield. The reaction of **105** with NIS (*N*-iodosuccinimide) afforded iodoalkene **106** in 63% yield. It should be noted that vinylstannane **105** was also converted to fluoroalkene **107** in moderate yield through reaction with XeF_2 and AgOTf [64]. The resulting iodoalkene **106** was transformed to naphthylamide **109** in a further four steps. Finally, the stereoselective introduction of a CF_3 group through reaction of **109** with $FSO_2CF_2CO_2Me$ [65] in the presence of Cu(I) thiophenecarboxylate (CuTc) gave trifluoromethylalkene isostere **110** in reasonable yield.

10.3.2 Conformational Analysis and Biological Aspects

One of the most important structural properties of trifluoromethylalkene isosteres is the reproduction of β-turn topography. As shown in Figure 10.18, natural β-turn structure can

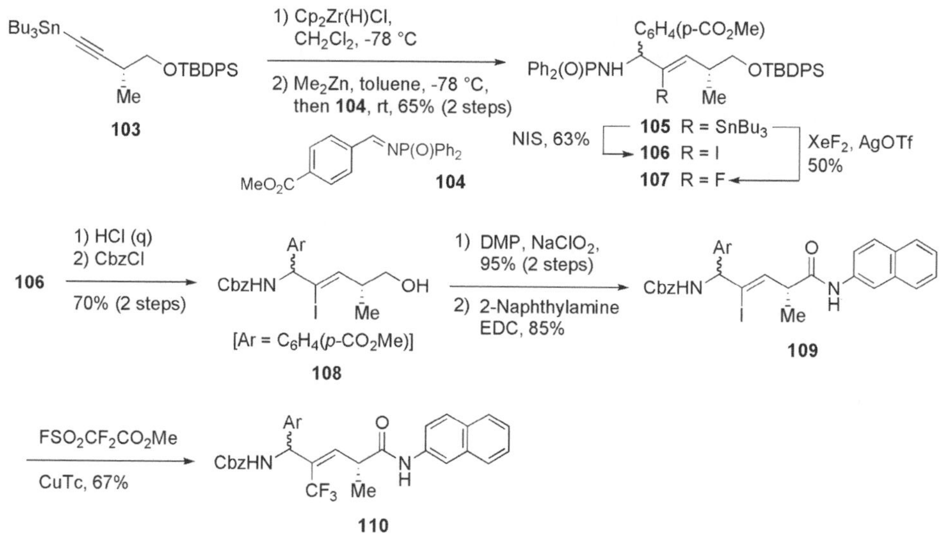

Scheme 10.29

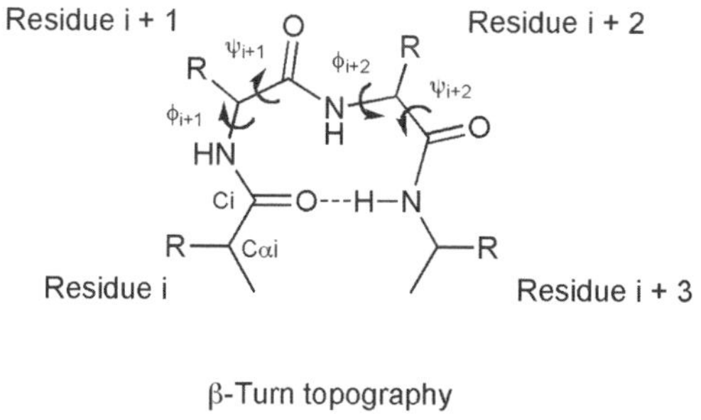

Turn type	ϕ_{i+1}	ψ_{i+1} (degrees)	ϕ_{i+2}	ψ_{i+2}
I	-60	-30	-90	0
I'	60	30	90	0
II	-60	120	80	0
II'	60	-120	-80	0
III	-60	-30	-60	-30
III'	60	30	60	30
IV	See caption			
V	-80	80	80	-80
V'	80	-80	-80	80
VI	See caption			
VII	See caption			

Figure 10.18 *Structure and classification of β-turns. Type IV β-turns are defined as those having two or more angles which differ by at least 40° from the definitions of β-turn types I, I', II, II', III and III'. Type VI β-turns have a cis-Pro at the position i+2. Type VII β-turns form a kink in the protein chain created by* $\psi_{i+1} \approx 180°$ *and* $|\phi_{i+2}| < 60°$ *or* $|\psi_{i+1}| < 60°$ *and* $\phi_{i+2} \approx 180°$.

be classified broadly into 11 types [66]. Compared with nonfluorinated alkene and methylalkene isosteres, the corresponding trifluoromethylalkene isosteres act as more effective β-turn promoters. Thus, the combination of $A^{1,3}$ and $A^{1,2}$ strains at the trifluoromethylated alkyne moiety leads to considerable restrictions in ϕ,ψ-dihedral angles in these peptides.

For instance, BocNH-L-Ala-Ψ[C(CF_3)=CH]-D-Ala-CONHMe takes solely type-II β-turn structure ($\phi_2 = -60°$, $\psi_2 = 120°$, $\phi_3 = 110°$, $\psi_3 = -23°$) in crystalline form (see Figure 10.19). In this case, the corresponding methylated alkene isostere also exists as a similar conformer in crystalline form, although the simple alkene isostere does not have a hydrogen-bond between the carbonyl oxygen at residue 1 and the amide proton at residue 4.

A similar observation was reported for the trifluoromethylalkene isostere of gramicidin S (GS), which is an antibiotic cyclodecapeptide (see Figure 10.20) [67]. In natural GS, the rigid and amphipathic antiparallel β-pleated sheet is held in place by two type-II′ β-turns at both D-Phe-Pro positions as well as four intramolecular hydrogen-bonds between the valine and leucine residues. Base on the x-ray crystallographic analysis, it is confirmed that trifluoromethylalkene isostere **102** adopts the pleated antiparallel β-sheet structure with two hydrogen bonds in solid state. The D-Phe-Pro unit has an ideal set of dihedral angles for the type II′ β-turn ($\phi_2 = 137°$, $\psi_2 = -95°$, $\phi_3 = -82°$, $\psi_3 = -5.7°$). Comparison of natural GS and Ψ[C(CF_3)=CH]$_2$GS in solution by variable temperature NMR and CD (circular dischroism) analyses also indicates the structural similarity of these two compounds. In contrast, CD spectra of nonfluorinated isostere Ψ[C(CH_3)=CH]$_2$GS **103** suggest the presence of disordered peptide conformation. Antibacterial activities of des-Cbz-**102** and des-Cbz-**103** also mimic that of the natural GS (MIC = 5–15 μg/mL).

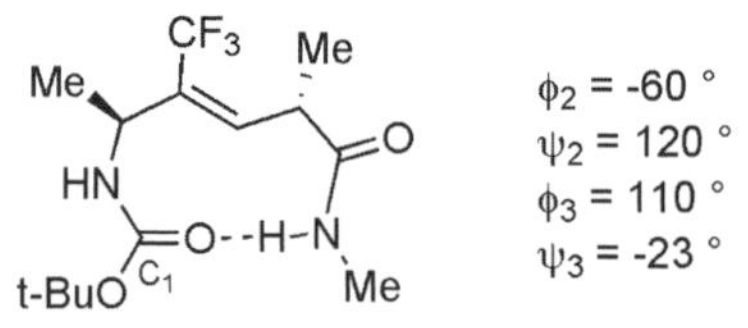

ϕ_2 = -60 °
ψ_2 = 120 °
ϕ_3 = 110 °
ψ_3 = -23 °

BocNH-Ala-ψ[C(CF_3)=CH]-D-Ala-CONHMe

Figure 10.19 *β-Turn structure of BocNH-L-Ala-ψ[C(CF$_3$)=CH]-D-Ala-CONHMe.*

Gramicidin S (GS) | Ψ[C(CF_3)=CH]$_2$GS **102** | Ψ[C(CH)$_3$=CH]$_2$GS 103

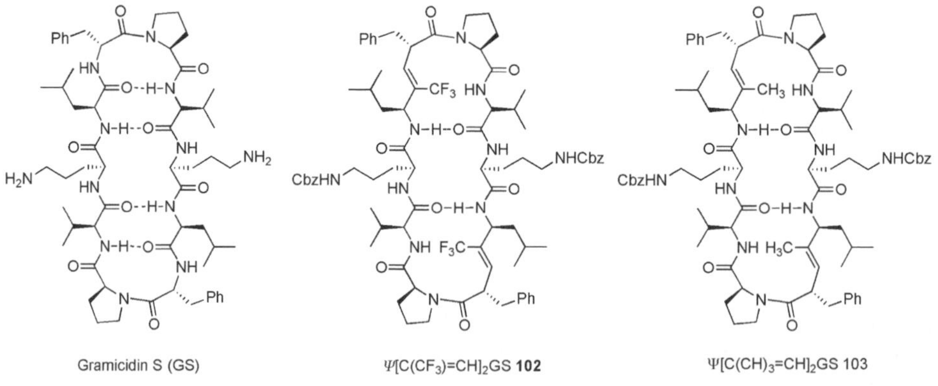

Figure 10.20 *Structures of gramicidin S, Ψ[C(CF$_3$)=CH]$_2$GS, and Ψ[C(CH$_3$)=CH]$_2$GS.*

10.4 Conclusion

Recent advances in the bioorganic and medicinal chemistry of fluoropeptide mimetics and related compounds are reviewed. The fluoro-olefin dipeptide isosteres Ψ[CF=CH] have been studied extensively. Although steady advances in synthetic methods should be noted, the development of more efficient and selective methods that tolerate various substitution patterns and produce products in enantiomerically pure forms with a short-step procedure is still required. Such development in synthetic methods will undoubtedly promote applications of fluoropeptide mimetics and related compounds to medicinal research. Weak or lack of hydrogen bond-forming ability of fluorinated isosteres as amide mimics has been suggested to explain the observed weak enzyme–substrate interactions or low biological activities. On the other hand, their enhanced *in vivo* stability and/or enhanced affinity to enzymes have also been experimentally verified. Fluorinated dipeptide isosteres and nucleobase mimics also provide new and useful tools for bioorganic chemistry research, as exemplified in their use in molecular recognition of oligonucleotides.

References

1. Gante, J. (1994) Peptidemimetics – tailored enzyme inhibitors. *Angew. Chem. Int. Ed.*, **33**, 1699–1720.
2. Kazmierski, W. M. (1999) *Peptidomimetics Protocol*, Humana Press. Totawa, NJ.
3. Hann, M. M., Sammes, P. G., Kennewell, P. D. and Taylor, J. B. (1982) On the double bond isostere of the peptide bond: preparation of an enkephalin analogue. *J. Chem. Soc. Perkin Trans. I*, 307–314.
4. Abraham, R. J., Ellison, S. L. R., Schonholzer, P. and Thomas, W. A. (1986) A theoretical and crystallographic study of the geometries and conformations of fluoro-olefin as peptide analogues. *Tetrahedron*, **42**, 2101–2110.
5. Cieplack, P., Kollmann, P. A. and Radomski, J. P. (1996) Molecular design of fluorine-containing peptide mimetics, in *Biomedicinal Frontiers of Fluorine Chemistry*, ACS Symposium Series **639**, American Chemical Society, Washington, D.C., pp. 143–156.
6. Welch, J. T., Lin, J., Boros, L. G., *et al.* (1996) Fluoo-olefin isosteres as peptidemimetics in *Biomedicinal Frontiers of Fluorine Chemistry*, ACS Symposium Series **639**,. American Chemical Society, Washington, D.C., pp. 129–142.
7. Wipf, P., Henninger, T. C. and Geib, S. J. (1998) Methyl- and (trifluoromethyl)alkene peptide isosteres: Synthesis and evaluation of their potential as β-turn promoters and peptide mimetics. *J. Org. Chem.*, **63**, 6088–6089.
8. Allmendinger, T., Felder, E. and Hungerbuehler, E. (1991) Fluoro-olefin dipeptide isosteres, in *Selective Fluorination in Organic and Bioorganic Chemistry*, ACS Symposium Series **456**, American Chemical Society, Washington, D.C., pp. 186–195.
9. Burton, D. J., Yang, Z.-Y. and Qiu, W. (1996) Fluorinated ylides and related compounds. *Chem. Rev.*, **96**, 1641–1715.
10. Grison, Ph. C. C. and Sauvetre, R. (1987) 2-Diethoxyphosphoryl alkanoic acid dianions IV. A direct route to 2-fluoro-2-alkenoic acids by the Horner synthesis. Application in the field of pyrethroids. *J. Organomet. Chem.*, **332**, 1–8.
11. Sano, S., Saito, K. and Nagao, Y. (2003) Tandem reduction-olefination for the stereoselective synthesis of (*Z*)-α-fluoro-α,β-unsaturated esters. *Tetrahedron Lett.*, **44**, 3987–3990.

12. Barrma, D. K., Kundeu, A., Zhang, H., *et al.* (2003) (*Z*)-α-Haloacrylates: an exceptionally stereoselective preparation via Cr(II)-mediated olefination of aldehydes with trihaloacetate. *J. Am. Chem. Soc.*, **125**, 3218–3219.
13. Allmendinger, T., Furet, P. and Hungerbruhler, E. (1990) Fluoroolefin dipeptide isosteres I. The synthesis of GlyΨ[CF=CH]Gly and racemic PheΨ[CF=CH]Gly. *Tetrahedron Lett.*, **31**, 7297–7300.
14. Van der Veken, P., Kertesz, I., Senten, K., *et al.* (2003) Synthesis of (*E*)- and (*Z*)-fluoro-olefin analogues of potent dipeptidyl peptidase IV inhibitors. *Tetrahedron Lett.*, **44**, 6231–6234.
15. Boros, L. G., De Corte, B., Gimi, R. H., *et al.* (1994) Fluoroolefin peptide isosteres – tools for controlling peptide conformations. *Tetrahedron Lett.*, **35**, 6033–6036.
16. Allmendinger, T., Felder, E. and Hungerbruhler, E. (1990) Fluoroolefin dipeptide isosteres II. Enantioselective synthesis of both antipodes of the Phe-Gly dipeptide mimic. *Tetrahedron Lett.*, **31**, 7301–7304.
17. Dolbier, Jr., W. R., Medinger, K. S., Greenberg, A. and Liebman, J. F. (1982) The thermodynamic effect of fluorine as a substituent. *Tetrahedron*, **38**, 2415–2420.
18. Watanabe, D., Koura, M., Saito, A., *et al.* (2005). A highly stereoselective synthesis of fluoroolefin dipeptide isostere. *125th Annual Meeting of the Pharmaceutical Society of Japan*, Tokyo, Japan. Pharmaceutical Society of Japan, Abstract 31-0205.
19. Bartlett, P. A. and Otake, A. (1995) Fluoroalkenes as peptide isosteres: ground state inhibitors of thermolysin. *J. Org. Chem.*, **60**, 3107–3111.
20. Lamy, C., Hofmann, J., Parrot-Lopez, H. and Goekjian, P. (2007) Synthesis of a fluoroalkene peptidemimetic precursor of *N*-acetyl-L-glutamyl-L-alanine. *Tetrahedron Lett.*, **48**, 6177–6180.
21. Allmendinger, T. (1991) Ethyl phenylsulfinyl fluoroacetate, a new and versatile reagent for the preparation of α-fluoro-α,β-unsaturated carboxylic acid esters. *Tetrahedron*, **47**, 4905–4914.
22. Okada, M., Nakamura, Y., Saito, A., *et al.* (2002) Synthesis of α-alkylated (Z)-γ-fluoro-β,γ-enoates through organocopper mediated reaction of γ,γ-difluoro-α,β-enoates: a different reactivity of R_3Al-Cu(I) and Me_2CuLi. *Chem. Lett.*, 28–29.
23. Nakamura, Y., Okada, M., Koura, M., *et al.* (2006) An efficient synthetic method for Z-fluoroalkene dipeptide isosteres: application to the synthesis of the dipeptide isostere of Sta-Ala. *J. Fluorine Chem.*, **127**, 627–636.
24. Otaka, A., Watanabe, H., Yukimasa, A., *et al.* (2001) New access to α-substituted (*Z*)-fluoroalkene dipeptide isosteres utilizing organocopper reagents under "reduction-oxidative alkylation (R-OA)" conditions. *Tetrahedron Lett.*, **42**, 5443–5446.
25. Narumi, T., Niida, A., Tomita, K., *et al.* (2006) A novel one-pot reaction involving organocopper-mediated reduction/transmetalation/asymmetric alkylation, leading to the diastereoselective synthesis of functionalized (*Z*)-fluoroalkene dipeptide isosteres. *Chem. Commun.*, 4720–4722.
26. Narumi, T., Tomita, K., Inokuchi, E., *et al.* (2007) Facile synthesis of fluoroalkenes by palladium-catalyzed reductive defluorination of allylic *gem*-difluorides. *Org. Lett.*, **9**, 3465–3468.
27. Babine, R. E. and Bender, S. L. (1997) Molecular recognition of protein–ligand complexes: applications to drug design. *Chem. Rev.*, **97**, 1359–1472.
28. Lide, D. R. (1997) *CRC Handbook of Chemistry and Physics*, 78th edn, CRC Press, New York.
29. Okada, M., Nakamura, Y., Saito, A., *et al.* (2002) Stereoselective construction of functionalized (*Z*)-fluoroalkenes directed to depsipeptide isosteres. *Tetrahedron Lett.*, **43**, 5845–5848.
30. Nakamura, Y., Okada, M., Sato, A., *et al.* (2005) Stereoselective synthesis of (*Z*)-fluoroalkenes directed to peptide isosteres: Cu(I) mediated reaction of trialkylaluminum with 4,4-difluoro-5-hydroxyallylic alcohol derivatives. *Tetrahedron*, **43**, 5741–5743.
31. Xu, J. and Burton, D. J. (2002) Kinetic separation methodology for the stereoselective synthesis of (*E*)- and (*Z*)-α-fluoro-α,β-unsaturated esters via the palladium-catalyzed carboalkoxylation of 1-bromo-1-fluoroalkenes. *Org. Lett.*, **4**, 831–833.

32. Dutheuil, G., Paturel, C., Lei, X., *et al.* (2006) First stereospecific synthesis of (*E*)- or (*Z*)-α-fluoroenones via a kinetically controlled Negishi coupling reaction. *J. Org. Chem.*, **71**, 4316–4319.
33. Dutheuil, G., Couve-Bonnaire, S. and Pannecoucke, X. (2007) Diastereomeric fluoroolefins as peptide bond mimics prepared by asymmetric reductive amination of α-fluoroenones. *Angew. Chem. Int. Ed.*, **46**, 1290–1292.
34. Augustyns, K., Bal, G., Thonus, G., *et al.* (1999) The unique properties of dipeptidyl-peptidase IV (DPP IV / CD26) and the therapeutic potential of DPP IV inhibitors. *Curr. Med. Chem.*, **6**, 311–327.
35. Ashworth, D. M., Atrash, B., Baker, G. R., *et al.* (1996) 2-Cyanopyrrolidines as potent, stable inhibitors of dipeptidyl peptidase IV. *Bioorg. Med. Chem. Lett.*, **6**, 1163–1166.
36. Lin, J., Toscano, P. J. and Welch, J. T. (1998) Inhibition of dipeptidyl peptidase IV (DPP IV) by fluoroolefin-containing *N*-peptidyl-*O*-hydroxylamine peptidemimetics. *Proc. Natl Acad. Sci. U. S. A.*, **95**, 14020–14024.
37. Zhao, K., Lim, D. S., Funaki, T. and Welch, J. T. (2003) Inhibition of dipeptidyl peptidase IV (DPP IV) by 2-(2-amino-1-fluoropropylidene)-cyanopentanecarbonitrile, a fluoroolefin containing peptidemimetic. *Bioorg. Med. Chem.*, **11**, 207–215.
38. Van der Veken, P., Senten, K., Kertesz, I., *et al.* (2005) Fluoro-olefins as peptidomimetic inhibitors of dipeptidyl peptidases. *J. Med. Chem.*, **48**, 1768–1780.
39. Egholm, M., Buchardt, O., Christensen, L., *et al.* (1993) PNA hybridizes to complementary oligonucleotides obeying the Watson–Crick hydrogen-bonding rules. *Nature*, **365**, 566–568.
40. Armitage, B. A. (2003) The impact of nucleic acid secondary structure on PNA hybridization. *Drug Discov. Today*, **8**, 222–228.
41. Koppelhus, U. and Nielsen, P. E. (2003) Cellular delivery of peptide nucleic acid (PNA) *Adv. Drug Del. Rev.*, **55**, 267–280.
42. Hollenstein, M., Gautschi, D. and Leumann, C. J. (2003) Fluorinated peptide nucleic acids. *Nucleosides Nucleotides Nucleic Acids*, **22**, 1191–1194.
43. Hollenstein, M. and Leumann, C. J. (2003) Synthesis and incorporation into PNA of fluorinated olefinic PNA (F-OPA) monomers. *Org. Lett.*, **5**, 1987–1990.
44. Hollenstein, M. and Leumann, C. J. (2005) Fluorinated olefinic peptide nucleic acid: synthesis and pairing properties with complementary DNA. *J. Org. Chem.*, **70**, 3205–3217.
45. Schutz, R., Cantin, M., Roberts, C., *et al.* (2000) Olefinic peptide nucleic acids (OPAs): new aspects of the molecular recognition of DNA by PNA. *Angew. Chem. Int. Ed.*, **39**, 1250–1253.
46. Will, D. W., Breipohl, G., Langner, D., *et al.* (1995) The synthesis of polyamide nucleic acids using a novel monomethoxytrityl protecting-group strategy. *Tetrahedron*, **51**, 12069–12082.
47. For a review, see: Kool, E. T., Morales, J. C. and Guckian, K. M. (2000) Mimicking the structure and function of DNA: insights into DNA stability and replication. *Angew. Chem. Int. Ed.*, **39**, 990–1009.
48. For a review, see: Kool, E. T. and Sintim, H. O. (2006) The difluorotoluene debate – a decade later. *Chem. Commun.*, 3665–3675.
49. Schweitzer, B. A. and Kool, E. T. (1994) Aromatic nonpolar nucleosides as hydrophobic isosteres of pyrimidine and purine nucleosides. *J. Org. Chem.*, **59**, 7238–7242.
50. Schweitzer, B. A. and Kool, E. T. (1995) Hydrophobic, non-hydrogen-bonding bases and base pairs in DNA. *J. Am. Chem. Soc.*, **117**, 1863–1872.
51. Kim, T. W. and Kool, E. T. (2004) A set of nonpolar thymidine nucleoside analogues with gradually increasing size. *Org. Lett.*, **6**, 3949–3952.
52. Kim, T. W. and Kool, E. T. (2005) A series of nonpolar thymidine analogues of increasing size: DNA base pairing and stacking properties. *J. Org. Chem.*, **70**, 2048–2053.
53. Dunitz, J. D. and Taylor, R. (1997) Organic fluorine hardly ever accepts hydrogen bonds. *Chem. Eur. J.*, **3**, 89–98.

54. Thalladi, V. R., Weiss, H.-C., Bläeser, D., *et al.* (1998) C–H···F interactions in the crystal structures of some fluorobenzenes. *J. Am. Chem. Soc.*, **120**, 8702–8710.
55. Parsch, J. and Engeles, J. W. (2002) C–F···H-C hydrogen bonds in ribonucleic acids. *J. Am. Chem. Soc.*, **124**, 5664–5672.
56. O'Neill, B. M., Ratto, J. E., Good, K. L., *et al.* (2002) A highly effective nonpolar isostere of deoxyguanosine: synthesis, structure, stacking, and base pairing. *J. Org. Chem.*, **67**, 5869–5875.
57. Guckian, K. M., Schweitzer, B. A., Ren, R. X.-F., *et al.* (2000) Factors contributing to aromatic stacking in water: evaluation in the context of DNA. *J. Am. Chem. Soc.*, **122**, 2213–2222.
58. Kim, T. W., Delaney, J., Essigmann, J. M. and Kool, E. T. (2005) Probing the active site tightness of DNA polymerase in subangstrom increments. *Proc. Natl. Acad. Sci. U. S. A.*, **102**, 15803–15808.
59. Sintim, H. O. and Kool, E. T. (2006) Remarkable sensitivity to DNA base shape in the DNA polymerase active site. *Angew. Chem. Int. Ed.*, **45**, 1974–1979.
60. Kim, T. W., Brieba, L. G., Ellenberger, T. and Kool, E. T. (2006) Functional evidence for a small and rigid active site in a high fidelity DNA polymerase: probing T7 DNA polymerase with variably sized base pairs. *J. Biol. Chem.*, **281**, 2289–2295.
61. Mizukami, S., Kim, T. W., Helquist, S. A. and Kool, E. T. (2006) Varying DNA base-pair size in subangstrom increments: evidence for a loose, not large, active site in low-fidelity Dpo4 polymerase. *Biochemistry* **45**, 2772–2778.
62. Silverman, A. P. and Kool, E. T. (2007) RNA probes of steric effects in active sites: high flexibility of HIV-1 reverse transcriptase. *J. Am. Chem. Soc.*, **129**, 10626–10627.
63. Wipf, P., Xiao, J. and Geib, S. J. (2005) Imine additions of internal alkynes for the synthesis of trisubstituted (*E*)-alkene and cyclopropane peptide isosteres. *Adv. Synth. Catal.*, **347**, 1605–1613.
64. Tius, M. A. and Kawakami, J. K. (1992) Vinyl fluorides from vinylstannanes. *Synth. Commun.*, **22**, 1461–1471.
65. Chen, Q.-Y. and Wu, S.-W. (1989) Methyl (fluorosulfonyl)difluoroacetate; a new trifluoromethylating agent. *J. Chem. Soc. Chem. Commun.*, 705–706.
66. Ball, J. B., Hughes, R. A., Alewood, P. F. and Andrews, P. R. (1993) β-Turn topography. *Tetrahedron*, **49**, 3467–3478.
67. Xiao, J., Weisblum, B. and Wipf, P. (2005) Electrostatic versus steric effects in peptidomimicry: Synthesis and secondary structure analysis of gramicidin S analogues with (*E*)-alkene peptide isosteres. *J. Am. Chem. Soc.*, **127**, 5742–5743.

11

Perfluorinated Heteroaromatic Systems as Scaffolds for Drug Discovery

David Armstrong, Matthew W. Cartwright, Emma L. Parks, Graham Pattison, Graham Sandford, Rachel Slater, Ian Wilson John A. Christopher, David D. Miller, Paul W. Smith, and Antonio Vong

11.1 Introduction

The drug discovery process within the pharmaceutical industry has undergone many significant changes over the past 40 years in order to decrease the high attrition rate of drug candidates in the highly resource intensive discovery and development phases [1–3]. Advances in technology brought about by the introduction of high-throughput screening (HTS) in 1989 now allow hundreds of thousands of compounds to be assessed for biological activity by a variety of *in vitro* assays within a very short time frame [4]. The subsequent introduction of combinatorial chemistry techniques [1] allowed the synthesis of compound collections comprising around a million different entities for biological screening purposes [5]. However, despite some early promise, the production of such large numbers of new molecules did not lead to the desired increase in the number of suitable candidates for hit-to-lead generation [6, 7].

Consequently, medicinal chemists have begun to pay closer attention to the structure, nature and function of molecules synthesized and screened as drug candidates in order to be able to recognize and, crucially, predict *drug-like* entities to aid the medicinal chemist in their choice of synthetic target molecules. Approaches to predict *drug-likeness* have been developed in recent years and include simple counting schemes, functional group methods and the analysis of the multidimensional "chemical space" occupied by drugs and neural network learning systems, and all these differing strategies have been discussed

Fluorine in Medicinal Chemistry and Chemical Biology Edited by Iwao Ojima

Table 11.1 *Lipinski parameters (RO5) for predicting "drug-likeness" of potential molecular targets*

Rule of 5 [10]	• Molecular weight ≤500 • The calculated log of the octanol/water partition coefficient, Clog $P \leq 5$ • Hydrogen-bond donors ≤5 • Hydrogen-bond acceptors (sum of N and O atoms) ≤10
Extensions [11]	• The polar surface area ≤140 Å^2 or sum of hydrogen-bond donors and acceptors ≤12 • Rotatable bonds ≤10

at length [5, 8]. In general, however, molecular "drug-like" properties can be defined as being a combination of favourable physiochemical (e.g. solubility, stability) and biological parameters (e.g. absorption, distribution, metabolism, elimination and toxicity; ADME-Tox) [2].

In a seminal publication that outlines one popular approach, published in 1997, Lipinski [1] analysed the physiochemical properties that are required for determining whether a molecule will be orally bioavailable, and a statistical analysis of existing successful drug systems was formalised as simple heuristic guidelines known as the "Rules of five" (RO5) (see Table 11.1). These "rules" were intended to be "conservative predictors" [9] to aid medicinal chemists to synthesize and select compounds that would possess suitable physiochemical drug-like properties and, therefore, help reduce compound attrition rates during clinical stages because, in principle, resources would not be utilized on the synthesis of molecules that would not have drug-like characteristics.

The research by the Pfizer group resulted in a paradigm shift of thinking for the medicinal chemist [12] and has been cited well over a thousand times in the literature [9]. Subsequently, similar analysis was carried out by scientists at other major pharmaceutical companies such as GSK [11], Boehringer Ingelheim [13], Astra Zeneca [14], Bayer [15] and Lilly [16], and very similar conclusions were formulated. From these further investigations, two notable extensions [11] to the RO5 (see Table 11.1) were established and are now widely used as additional predictive filters to identify suitable drug-like structures.

The RO5 have proved to be an extremely useful set of physiochemical parameters in the decision-making process in hit-to-lead generation and have been applied to the synthesis and analysis of many libraries of drug-like molecules that contain low-molecular-weight systems that possess diverse structural features [1]. If a particular compound has structural features that lie outside these guidelines, it is highly probable that solubility and permeability problems will occur in subsequent stages of drug development [9]. There are, however, exceptions to these rules [1], such as various medicinally important antibiotics, antifungals, vitamins and cardiac glycosides, and it is very important to note that RO5-compliant molecules are not, of course, automatically good drugs. Additionally, the RO5 do not give any indication as to the types of molecular structures that should be synthesized. They are merely a guideline as to whether a particular structural class of compound may possess drug-like features [9].

Consequently, it falls to the medicinal chemist to identify and design the structures of suitable drug-like entities using the RO5 as predictive guidelines, and subsequent con-

Table 11.2 *Features of "privileged structures"*

• Obeys RO5 (particularly MW ≤ 500) [3]
• Present in a large amount of natural products with varied biological activities
• Contains one or more rigid ring system and easily chemically modified to produce a diverse library range

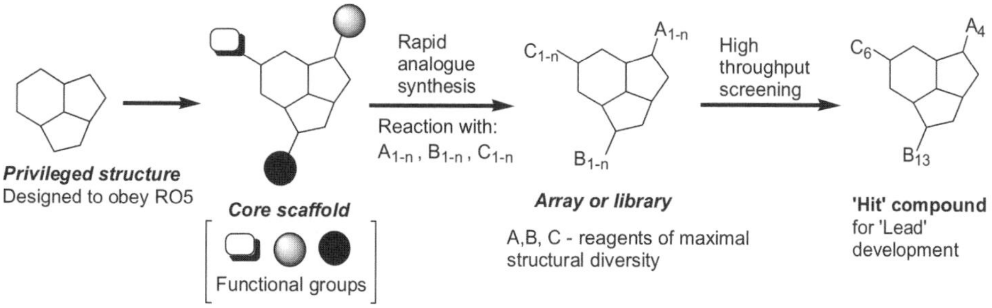

Figure 11.1 *Privileged structures, core scaffolds and RAS in the drug discovery process.*

struction of libraries or arrays of appropriate drug-like molecules has been increasingly based upon the concept of *privileged structures.* First proposed by Evans in 1988 [17], privileged structures can be described as molecules that possess versatile binding properties so that, through modification of various structural features, they are selective for a variety of different biological receptors. The parameters defined when choosing a privileged structure system can be summarized as in Table 11.2 [18].

Once a suitable privileged structure has been identified, rapid analogue synthesis (RAS) techniques are used to produce arrays of compounds containing the required privileged structure as the "heart" of the molecule and these methods can have a powerful impact within the drug discovery arena [19, 20]. Methodology that may be applied to RAS must be highly selective and high-yielding to allow the synthesis of libraries of privileged polyfunctional entities [21]. Most importantly, RAS requires the synthesis of "core scaffolds', that is, molecules that themselves have privileged structure features and possess many functional groups that allow rapid and selective functionalization, to give an array of analogues based upon the privileged core system. The design and synthesis process from identification of privileged structure to the RAS of many analogues of drug-like systems and identification of lead compounds by high-throughput screening is shown schematically in Figure 11.1.

It can reasonably be argued, therefore, that the key to the hit-to-lead generation process is the construction of a range of readily available "core scaffolds" possessing multiple functionality that may be processed by efficient, regioselective and stereoselective, versatile, short and high-yielding RAS methodology.

Based on the criteria outlined above, low-molecular-weight, polyfunctional heteroaromatic compounds should make excellent candidates for privileged structures because heteroaromatic structural subunits are present among many natural products and in about 70% of all pharmaceutical products that may be administered orally (some examples are shown in Figure 11.2).

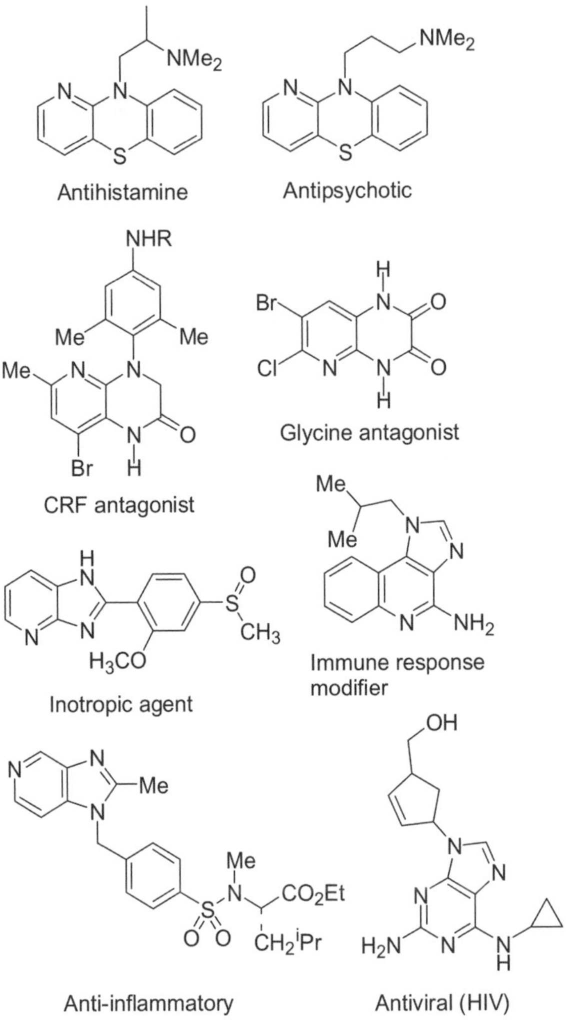

Figure 11.2 *Examples of biologically active heteroaromatic systems.*

Generally, rigid, polycyclic heteroaromatic systems, such as quinazolinones, are excellent privileged structures and many heterocyclic systems that have been assessed and developed as privileged structures have been thoroughly reviewed [3]. One key structural feature exhibited by heteroaromatic systems is their ability to contain a diverse range of substituents in a well-defined three-dimensional space around a core structural scaffold which, in comparison with their acyclic analogues, results in the constriction of conformational freedom [22]. In order to be more exploitable, heterocyclic "core scaffolds" must bear several reactive sites that may be readily functionalized in a controlled manner in high yield and with high regioselectivity. Unfortunately, scaffold and subsequent analogue synthesis of many polyfunctionalized heterocyclic systems is hampered by the inherent low reactivity and low regioselectivity of many aromatic heterocyclic systems [23, 24] and there exists, therefore, a continuing requirement in the life science industries for accessible, novel, heterocyclic scaffolds that bear multiple functionality and can be processed into systems that possess maximally diverse structural features.

In this chapter, we describe the use of highly fluorinated heteroaromatic systems as effective core scaffolds for the synthesis of an increasingly wide range of polyfunctional heterocyclic systems.

11.2 Perfluorinated Heterocycles as Core Scaffolds

Perfluoroheteroaromatic compounds [25, 26], in which all the hydrogen atom substituents of the heterocyclic ring have been replaced by fluorine atoms, were first synthesized in the 1960s by reaction of potassium fluoride with appropriate perchlorinated heteroaromatic precursors [25, 26] and a range of perfluorinated heteroaromatic systems may be accessed by halogen exchange (Halex) techniques as shown in Figure 11.3.

Perfluorinated heteroaromatic derivatives are highly susceptible towards nucleophilic attack owing to the presence of several highly electronegative fluorine atoms attached to the heteroaromatic ring; consequently, the chemistry of perfluoroheteroaromatic systems is dominated by nucleophilic aromatic substitution processes and new chemistry continues to emerge [25, 26]. Highly halogenated heteroaromatics are not merely of academic interest: a variety of such systems are commercially available from the usual research chemical suppliers and, indeed, some are prepared on the industrial scale for use in the fibre reactive dye industry and as intermediates for life-science products; for example 3,5-dichloro-2,4,6-trifluoropyridine is a precursor for a herbicide manufactured by the Dow company [27]. No special handling procedures are required for syntheses

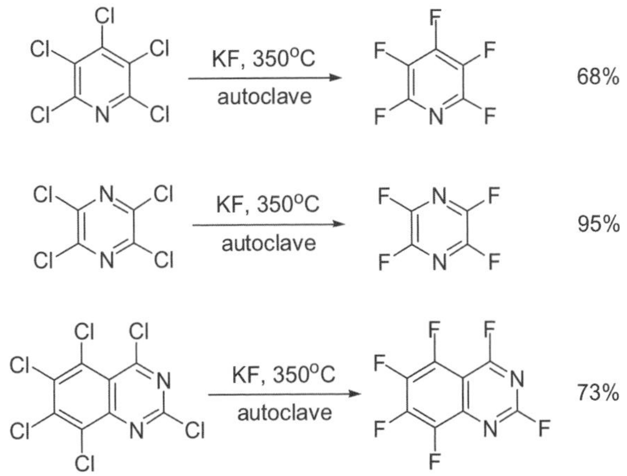

Figure 11.3 *Synthesis of perfluoroheteroaromatic systems [25, 26].*

involving perhaloheterocyclic derivatives; for example, pentafluoropyridine is a readily available, inexpensive, colourless liquid that has a boiling point of 100 °C. Consequently, their chemistry can be used by the general organic chemistry community and can be scaled up if required.

11.2.1 Pentafluropyridine as a Core Scaffold

Reactions between pentafluoropyridine **1** and a variety of nucleophiles have been established [25, 26] and are found to be regioselective in the vast majority of cases, giving products **2** arising from nucleophilic aromatic substitution of the fluorine atom located at the 4-position of **1** via the corresponding well-known Meisenheimer intermediate as shown in Figure 11.4, although a few exceptions to this general rule occur [28]. The regioselectivity of these processes is due to (i) the presence of the ring nitrogen, which activates *para* positions preferentially, and (ii) maximizing the number of activating fluorine atoms that are located *ortho* to the site of nucleophilic attack [25, 26].

Monosubstituted systems **2** are, of course, still activated towards nucleophilic attack and, indeed, all five fluorine atoms can be replaced upon reaction with an excess quantity of suitably reactive sulfur nucleophile species to give **3** [29] (see Figure 11.4), demonstrating the feasibility of using highly fluorinated heterocycles as scaffolds where the carbon–fluorine bond acts as the readily substituted functional group.

The order of nucleophilic attack for pentafluoropyridine **1** is established [25] to be $4 > 2 > 3$, for monosubstitution reactions involving a range of nucleophiles. The order of substitution for a succession of five nucleophilic substitution steps, where Nuc_1 is the first nucleophile, and Nuc_2 is the second, etc., can therefore be postulated to occur as outlined in Figure 11.4 to give **4** ultimately, although the outcome of successive reactions may depend upon the nature and effect of the substituent that is installed onto the pyridine ring in each nucleophilic substitution step. The wide range of functional nucleophiles available

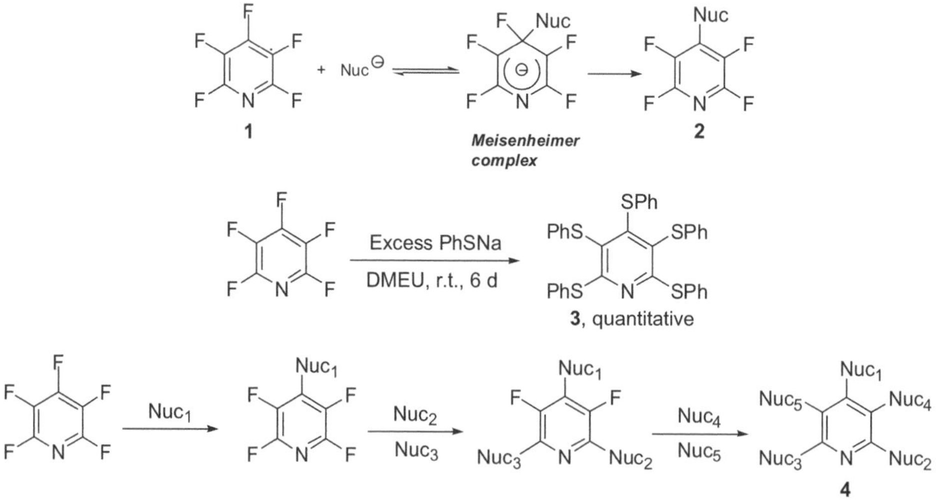

Figure 11.4 *Nucleophilic aromatic substitution of pentafluoropyridine [25, 26, 29].*

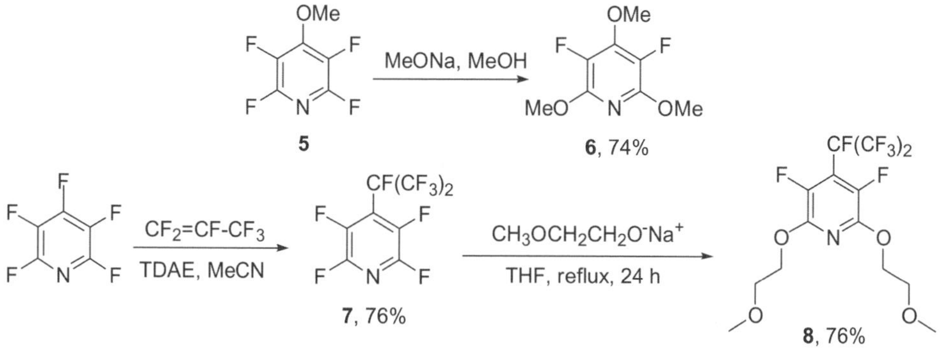

Figure 11.5 *Polysubstituted products from pentafluoropyridine [30, 32].*

(e.g. O-, N-, C-, S-centred) makes the theoretical number of highly functionalized pyridine ring systems that could be accessed by this methodology very large indeed if the sequence of nucleophilic aromatic substitution processes can be achieved in a controllable manner. Surprisingly, however, the number of sequential nucleophilic substitution processes using pentafluoropyridine as the starting material is very small despite the potential reactivity of these systems, but the few examples reported so far demonstrate the viability of the potential use of this substrate as a scaffold in the manner indicated in Figure 11.4. For example, as shown in Figure 11.5, heating 4-methoxytetrafluoropyridine **5** with an excess of sodium methoxide in methanol gave the 2,4,6-trimethoxypyridine derivative **6** [30]. In a controlled stepwise process, perfluoro-4-isopropylpyridine **7**, prepared by nucleophilic substitution of the 4-fluorine atom of pentafluoropyridine by perfluoroisopropyl anion (i.e. $Nuc_1 = (CF_3)_2CF^-$, Figure 11.4), generated *in situ* from hexafluoropropene and TDAE [31], subsequently gave trisubstituted products **8** (Nuc_2 and Nuc_3, Figure 11.4) by reaction with a range of oxygen-, nitrogen- and carbon-centred nucleophiles (see Figure 11.5) [32].

The reactivity profile established for pentafluoropyridine, where the 4, 2- and 6-positions are sequentially, regiospecifically substituted by a succession of oxygen-centred nucleophiles, has allowed medicinal chemists to use pentafluoropyridine as a core scaffold for the synthesis of small arrays of biologically active pyridine systems that fall within the Lipinski parameters (see Table 11.3).

The factor VIIa/TF (tissue factor) complex and Xa are proteins known to be involved in the blood coagulation cascade [33] and, as such, are validated targets in the search for novel antithrombotic drugs [34, 35]. Having established that a series of 2,6-diphenoxypyridines, including several 3,5-difluoro-4-methyldiaryloxypyridines derived from 4-methyltetrafluoropyridine, to be modest inhibitors of factor Xa, medicinal chemists synthesized a small library of 3,5-difluorotriaryloxypyridines **9** from pentafluoropyridine [36]. The 2,4,6-substitution pattern was obtained through sequential nucleophilic aromatic substitution by substituted phenols in typically high yields and polysubstitution could often be accomplished in a single reaction vessel. Systems derived from this series of 2,6-diphenoxypyridines served as a basis for creating a more potent system, leading to the potent FVIIa/TF inhibitor **10**, formed by reaction of a sequence of nitrogen, oxygen and oxygen-centred nucleophiles.

Table 11.3 *Biologically active polysubstituted systems synthesized from pentafluoropyridine*

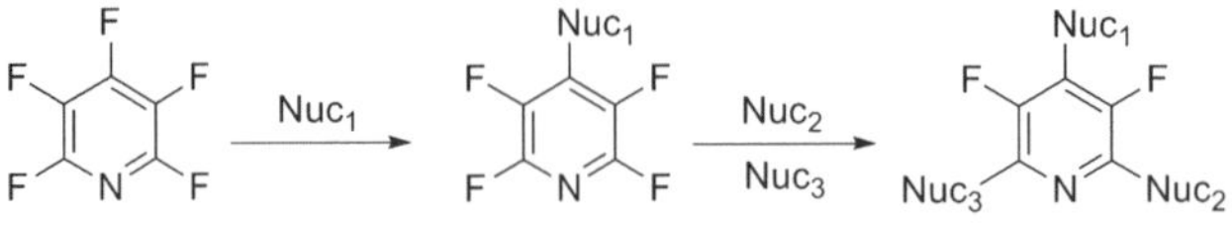

Nuc$_1$	Nuc$_2$	Nuc$_3$	Product
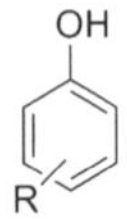	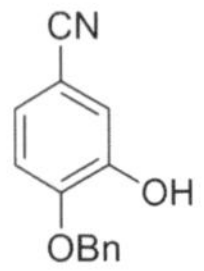	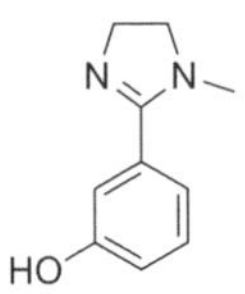	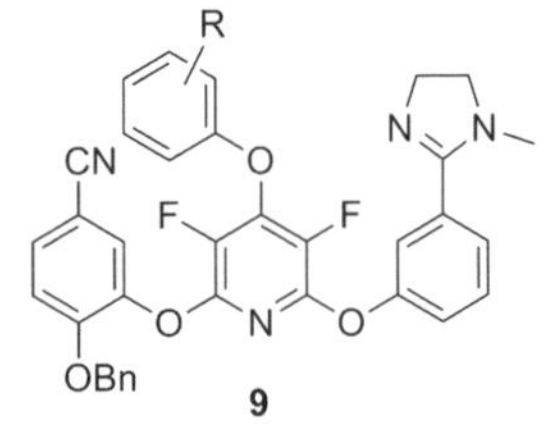 **9**
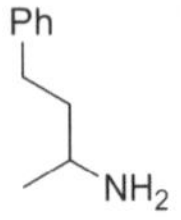	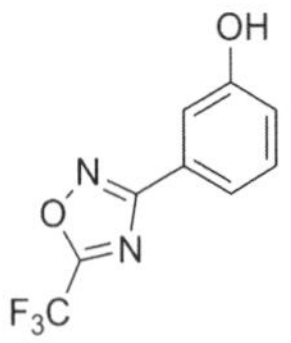	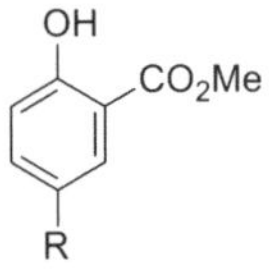	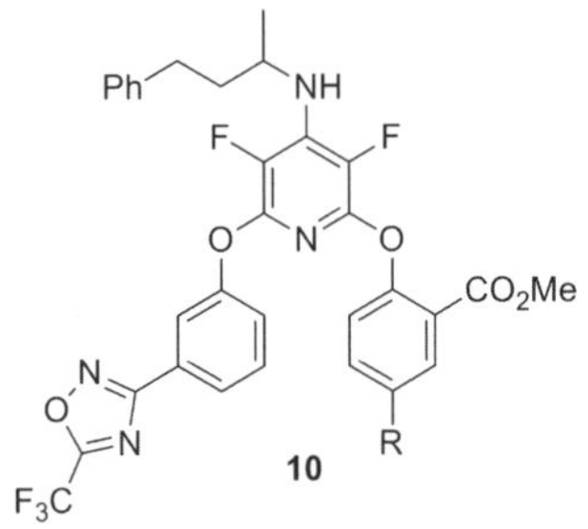 **10**
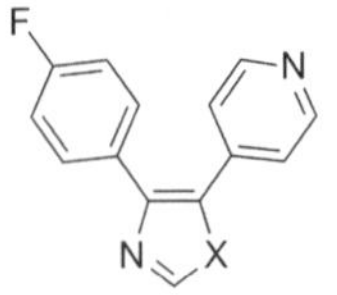	NH_3	NH_3	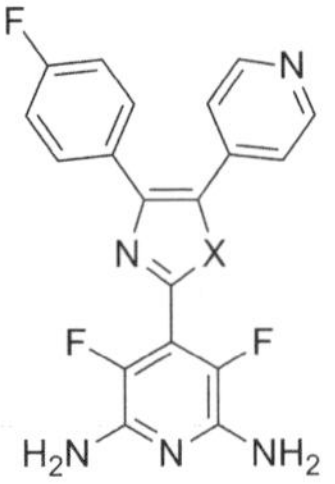 **11**

It is thought that inhibition of the p38 kinase protein could treat the underlying cause of chronic inflammatory diseases, and it is in this context that chemists in Switzerland prepared a diverse set of aryl-substituted pyridinylimidazoles **11** to achieve potentially high-affinity binding to the active site [37]. Deprotonation of the SEM (2-[trimethylsilyl] ethoxymethyl)-protected imidazole gave a carbanion that reacted as a nucleophile with pentafluoropyridine to give the expected 4-substituted pyridine. Bromination at the 4- and 5-positions of the imidazole was followed by regioselective Stille reaction to yield a pyridinylimidazole derivative, while subsequent Suzuki or Stille coupling at the remaining carbon–bromine bond was followed by deprotection of the imidazole. Finally, diamination at the 2- and 6-positions of the tetrafluoropyridine gave the desired biologically active 3,5-difluoropyridine system **11**.

11.2.2 Tetrafluoropyrimidine as a Core Scaffold

Perfluorinated diazines (pyrimidine, pyrazine and pyridazine) are typically 1000 times more reactive towards nucleophiles than is pentafluoropyridine and application of the sequential nucleophilic substitution methodology to reactions involving various diazine systems with a range of nucleophiles would, in principle, lead to the synthesis of many novel polyfunctional diazine derivatives. However, only a very limited number of reports concerning reactions of perfluorinated diazines with nucleophiles have been published [25, 26] and the use of tetrafluorodiazines as scaffolds has not been developed to any great extent. Several instances of reactions of tetrafluoropyrimidine **12** (Table 11.4) with a small range of nucleophiles have been reported [38] and, in all cases, nucleophilic substitution occurs selectively at the 4-position. A recent systematic study of reactions of the 4-aminopyrimidine systems (see Table 11.4) found that second and third substitution processes occurred selectively at the 6- and 2-positions, respectively, giving ready access to a small array of 5-fluoro trisubstituted pyrimidine derivatives **13** [39].

11.2.3 Perbromofluoropyridine Scaffolds

Clearly, the range of nucleophiles that is available, the functionality that could be installed (for example upon a pyridine or pyrimidine ring) and, of course, the functional groups on pendant substituents may, in principle, allow access to a great variety of polyfunctional pyridine analogues. However, despite this, the reactions outlined above are restricted to a sequence of nucleophilic substitution processes, thereby limiting the variety of structural arrays that can be synthesized from such perfluorinated core scaffolds. Consequently, related perhalogenated scaffolds that have more flexible functionality may be advantageous and, in this context, 2,4,6-tribromo-3,5-difluoropyridine **14**, synthesized by reaction of pentafluoropyridine with a mixture of hydrogen bromide and aluminium tribromide in an autoclave at 140 °C [40] was assessed as a potential polyfunctional scaffold system (see Figure 11.6). In a series of model reactions it was established that the bromofluoropyridine system reacts with "hard" nucleophiles (e.g. oxygen-centred nucleophiles) to give products **15** arising from selective replacement of fluorine, whereas "soft" nucleophiles (sulfur, nitrogen, etc.) selectively replace bromine to give product **16** [40]. The presence of carbon–bromine bonds on this scaffold allows Pd-catalysed Sonogashira [41] and Suzuki [42] coupling reactions to occur giving, for example, **17**, and selective debromo-lithiation at the 4-position followed by trapping of the lithiated pyridine species [43] by a variety of electrophiles, giving access to a wide range of polyfunctional pyridine systems **18** that could be utilised as scaffolds in their own right. Subsequently, a combination of nucleophilic aromatic substitution and Pd-catalysed Sonogashira reactions involving pentafluoropyridine as the core scaffold has enabled the synthesis of several pentasubstituted pyridine systems such as **19** and **20** [44] (see Figure 11.6).

Table 11.4 Tetrafluoropyrimidine as a core scaffold [39]

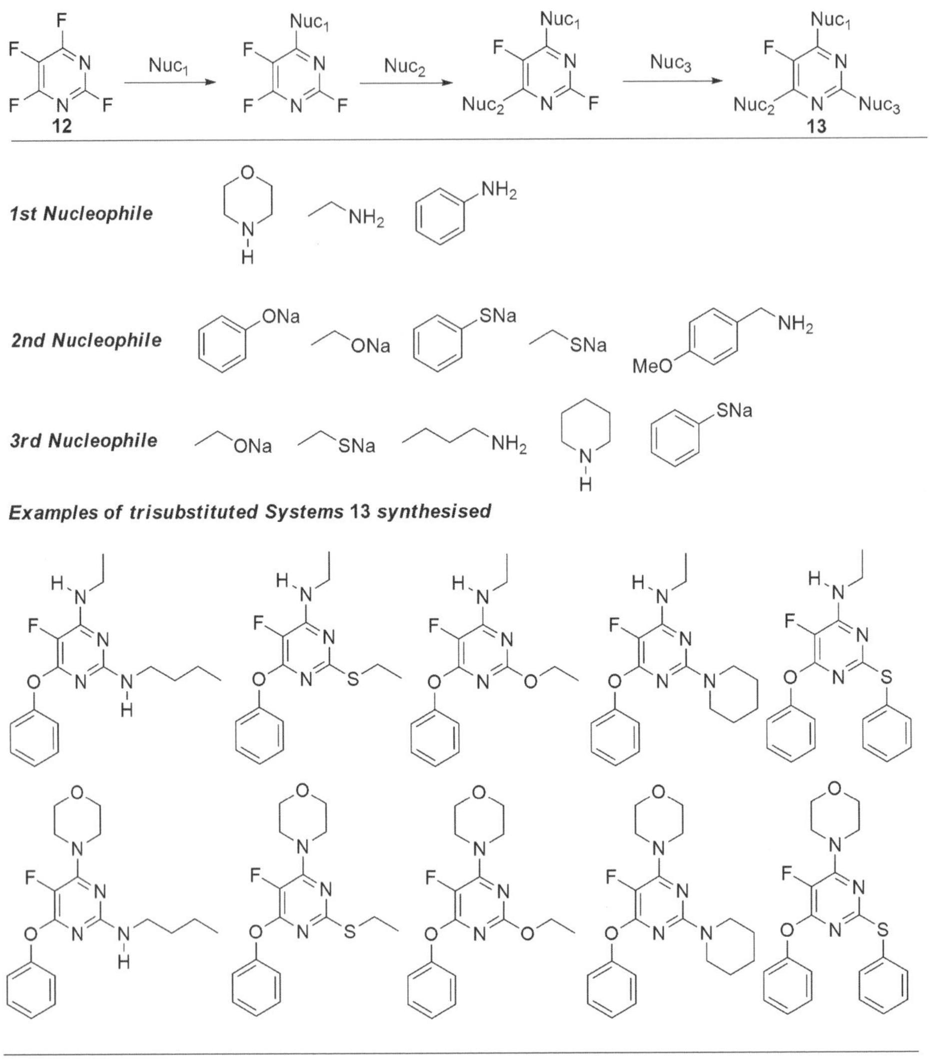

1st Nucleophile

2nd Nucleophile

3rd Nucleophile

Examples of trisubstituted Systems 13 synthesised

11.2.4 Polyfunctional Fluorinated [5,6]- and [6,6]-Bicyclic Heteroaromatic Scaffolds

Bicyclic nitrogen heterocyclic systems can have a range of very valuable biological activity and there are several examples of [5,6]- and [6,6]-ring-fused systems in which a ring-fused pyridine motif is a constituent part (see Figure 11.2). However, many bicyclic nitrogen-containing heterocycles remain surprisingly inaccessible [23] despite the relative simplicity of their molecular structures, and the chemistry of even the least complex heterocycles of this class remains largely unexploited. Inevitably, this provides great opportunities for the discovery of new small-molecule chemical entities that fall within

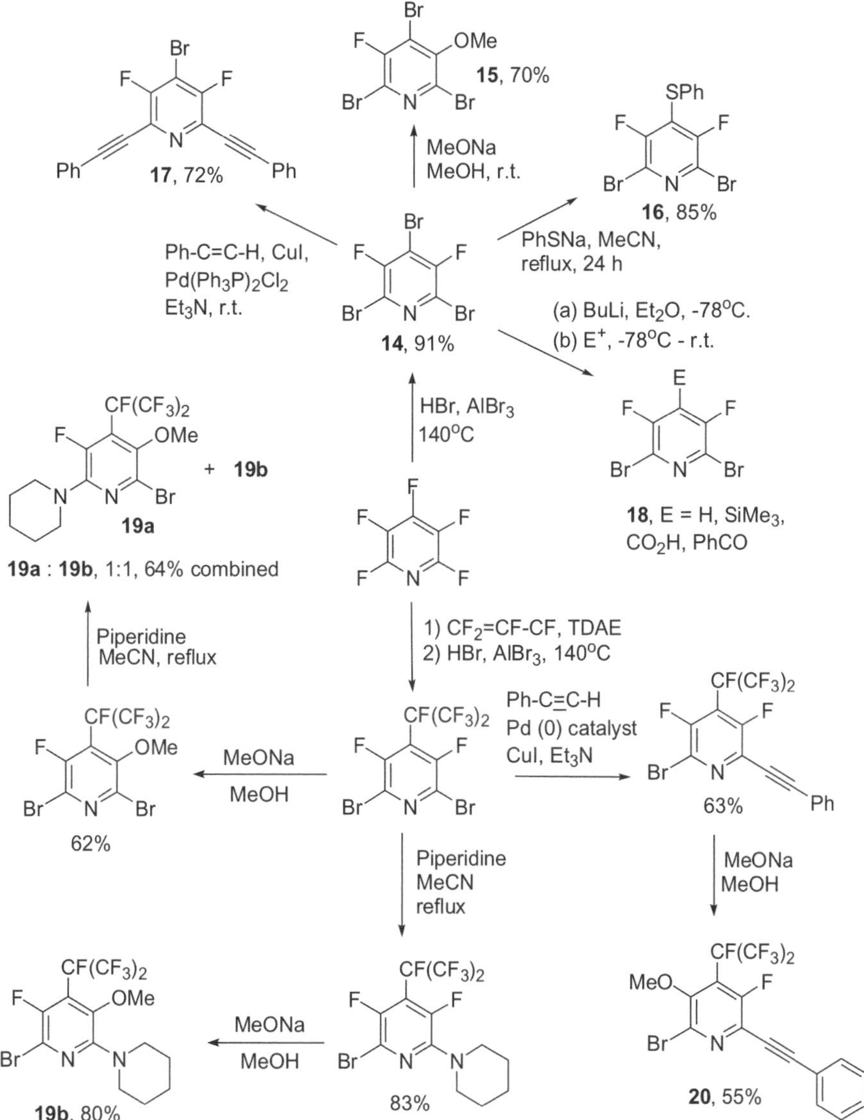

Figure 11.6 *Perfluorobromopyridine derivatives as core scaffolds [40–44].*

the RO5, if suitable polyfunctional, bicyclic, nitrogenated heterocyclic scaffolds can be reliably accessed.

In this context, a polyfluorinated imidazopyridine system **21** has been prepared by a multistep synthetic route (see Figure 11.7) beginning from 3-chlorotetrafluoropyridine **22**, via the formation and subsequent condensation of 3,4-diamino-2,5,6-trifluoropyridine with diethoxymethyl acetate [45]. Subsequent reaction of the ring-fused system with ammonia demonstrated the potential of such systems as polyfunctional scaffolds, if much shorter

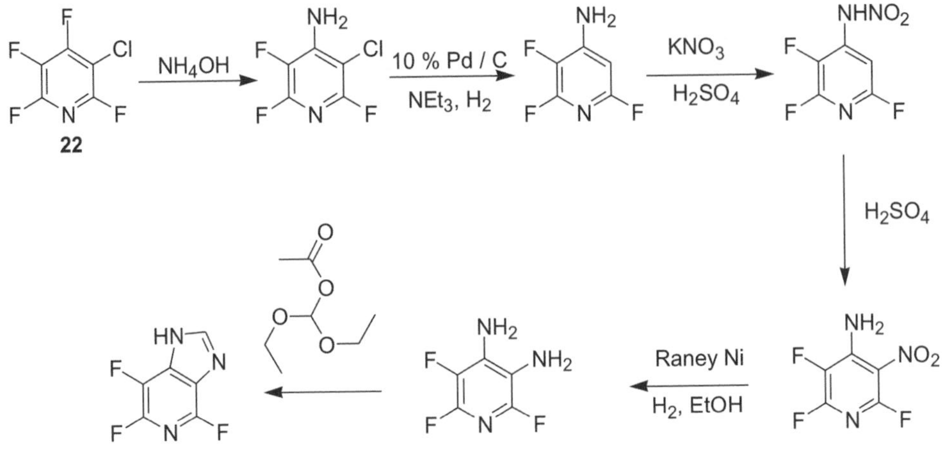

Figure 11.7 *Synthesis of imidazopyridine systems **21** from 3-chlorotetrafluoropyridine **22**.*

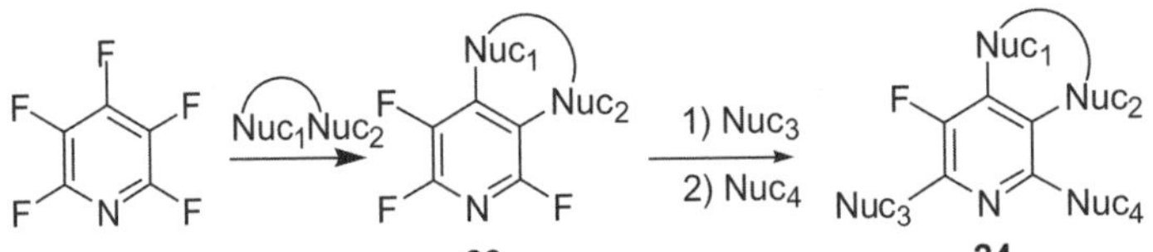

Figure 11.8 *Strategy for the synthesis of ring-fused heterocycles from pentafluoropyridine.*

synthetic sequences for the preparation of a range of these structural core scaffolds could be developed.

A convenient synthetic strategy for the synthesis of polyfluorinated polycyclic ring scaffolds involving reaction of pentafluoropyridine with bifunctional nucleophiles could, in principle, provide access to a wide range of polyfunctional systems (see Figure 11.8). Here, for example, substitution of the 4-position of the pentafluoropyridine scaffold would, in principle, be followed by attack at the adjacent 3-position owing to the geometric constraints of the system to give appropriate ring-fused systems **23**. The bicyclic scaffold possesses further sites that are activated towards nucleophilic attack and could, therefore, provide approaches to the synthesis of a wide variety of functional fused ring systems **24**, if the orientation of subsequent nucleophilic substitution processes could be controlled.

However, initial attempts to use this strategy to prepare azapurine derivatives, by reaction of pentafluoropyridine and either guanidine or thiourea, led to 4-aminopyridine **25** and bispyridyl derivatives **26** respectively via base-induced elimination processes [46] (see Figure 11.9).

In these cases, while nucleophilic substitution at the 4-position of pentafluoropyridine by guanidine could be achieved readily [46], attack at the less activated 3-position by the weak NH_2 nucleophile present on the guanidine moiety made cyclization a less favoured process than elimination. Consequently, in order to achieve cyclization by nucleophilic substitution in the second step, a more reactive second nucleophile is necessary.

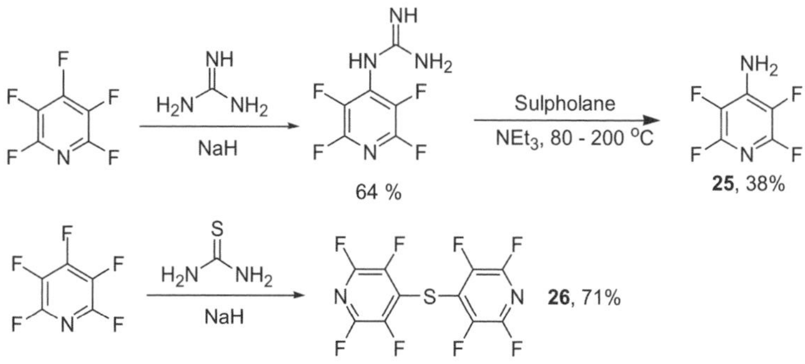

Figure 11.9 Reactions of pentafluoropyridine with urea and guanidine systems [46].

Pentafluoropyridine and several nucleophilic model difunctional secondary diamines, such as *N,N′*-dimethylethylene diamine **27a**, gave pyrido[3,4-*b*]pyrazine systems **28** in a two-step one-pot annelation reaction, upon reflux or microwave heating [47] (see Table 11.5). The chemistry of tetrahydropyrido[3,4-*b*]pyrazine systems is relatively undeveloped because of the difficult, low-yielding multistep syntheses, either from diaminopyridine [48] or chloroaminopyridine [49] precursors or by reduction of pyrido[3,4-*b*]pyrazine derivatives by metal hydrides [50–55], which are required to prepare even the simplest member of this heterocyclic class. Indeed, for the synthesis of appropriate multisubstituted scaffold systems based upon the tetrahydropyrido[3,4-*b*]pyrazine subunit, this difficult situation is magnified further. Less-nucleophilic primary diamine nucleophiles such as **27b** gave only noncyclized products **30** arising from substitution of the 4-position of pentafluoropyridine.

The pyrido[3,4-*b*]pyrazine scaffold reacted with a series of nucleophiles to give major products arising from substitution at the 7-position [47]. Although this process was not regiospecific, the alkoxylated scaffold could be isolated and purified by column chromatography from the minor isomer that was formed by substitution of the fluorine atom located at the 5-position.

The trifluorinated scaffold reacts preferentially at the 7-position with some product arising from competing substitution at the 5-position and this selectivity is due to the activating influence of ring nitrogen and maximizing of the number of activating fluorine atoms that are *ortho* to the site of attack (see Figure 11.10). The difluorinated scaffolds are still activated towards nucleophilic attack and a short series of fluoropyrido[3,4-*b*]pyrazines **29** has been synthesized in which the remaining fluorine atoms that are located *ortho* to pyridine nitrogen were displaced [47].

By a similar strategy, an imidazopyridine scaffold **32** was synthesized by reaction of pentafluoropyridine and benzamidine **31** [56] and, in this case, subsequent nucleophilic substitution occurs at the 5 position to give **33**, presumably because of interaction of the nucleophile and the imidazo ring NH bond which directs the incoming nucleophile to the less activated site adjacent to pyridine ring nitrogen.

This overall synthetic strategy (see Figure 11.8) was adapted to give a range of isomeric pyrido[2,3-*b*]pyrazines structures **34** by simply varying the order of reaction with appropriate mono- and difunctional nucleophiles [57, 58] (see Table 11.6). For example, reaction of sodium phenylsulfinate with pentafluoropyridine gives the

Table 11.5 Synthesis of pyrido[3,4-b]pyrazine and imidazopyridine scaffolds [47, 56]

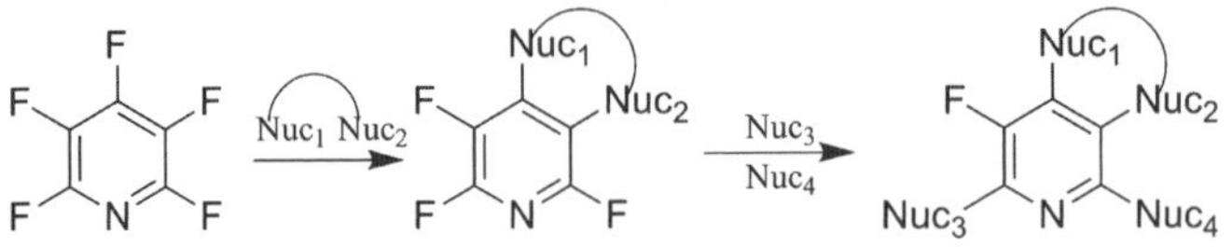

Nuc_1 Nuc_2	Core scaffold	Nuc_3	Nuc_4	Product
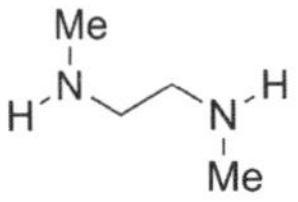 **27a**	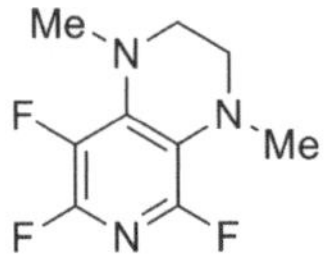 **28**, 97%	NaOMe	NaOEt	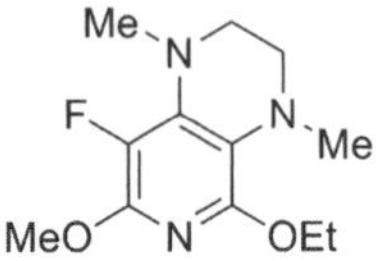 **29a**, 77%
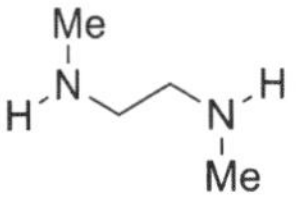 **27a**	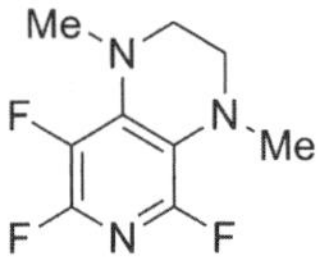 **28**, 97%	NaOMe	Et_2NH	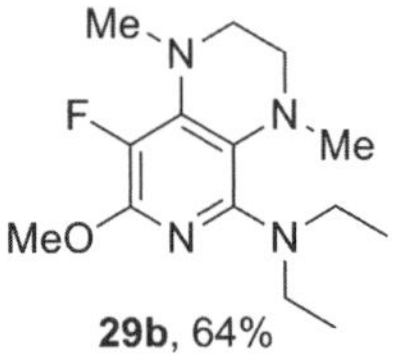 **29b**, 64%
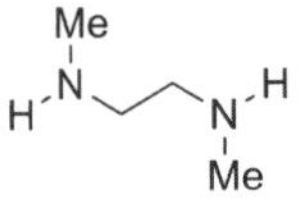 **27a**	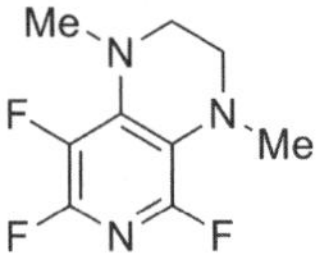 **28**, 97%	NaOMe	BuLi	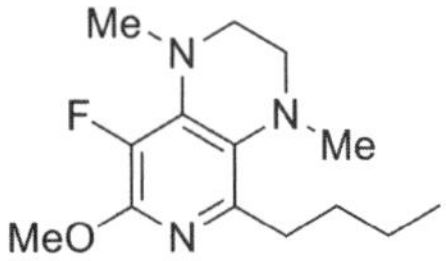 **29c**, 31%
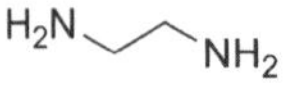 **27b**	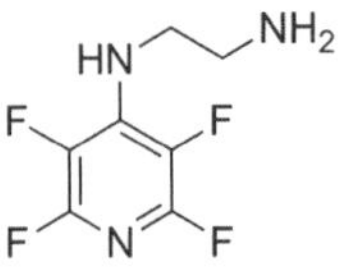 **30**, 75%	-	-	-
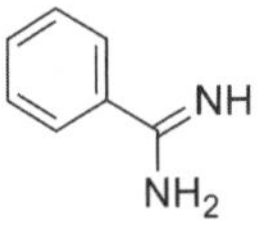 **31**	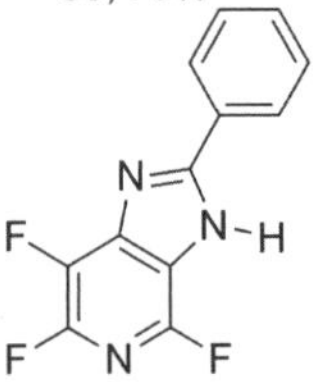 **32**, 91%	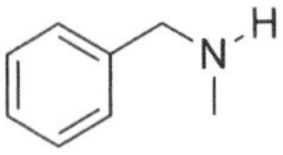		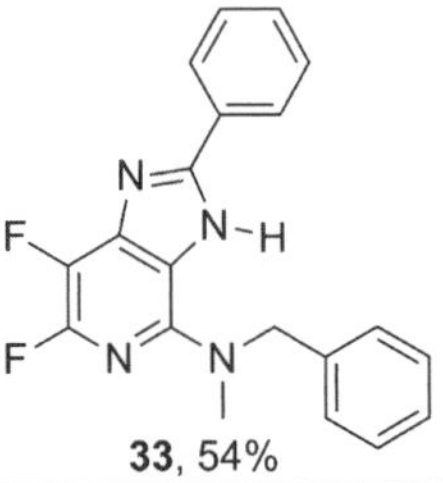 **33**, 54%

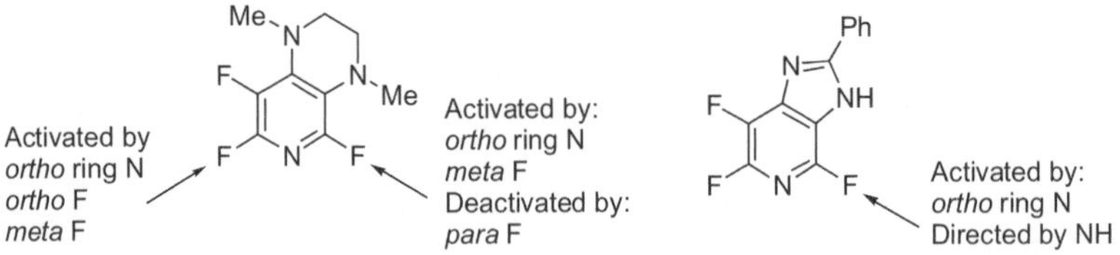

Activated by
ortho ring N
ortho F
meta F

Activated by:
ortho ring N
meta F
Deactivated by:
para F

Activated by:
ortho ring N
Directed by NH

Figure 11.10 *Regioselectivity of nucleophilic substitution of pyrido[3,4-b]pyrazine and imidazopyridine scaffolds [56].*

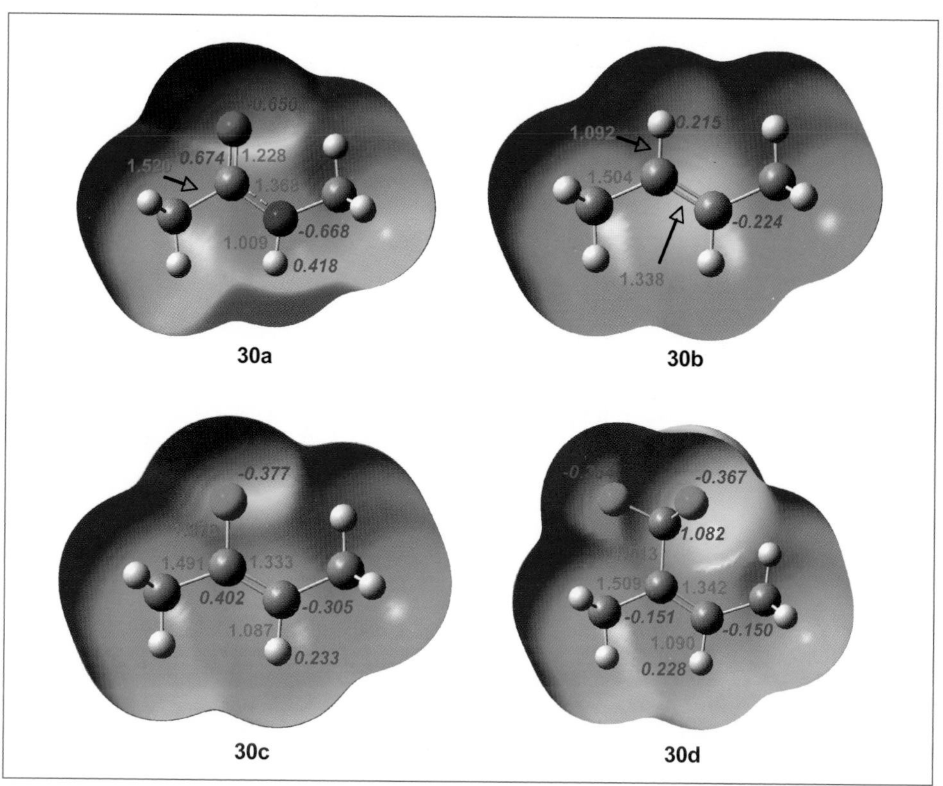

Plate 1.17 *Electrostatic potential of the model compounds **30a** to **30d** (color alteration from red to blue describes the shift of the electronically rich to deficient circumstance).*

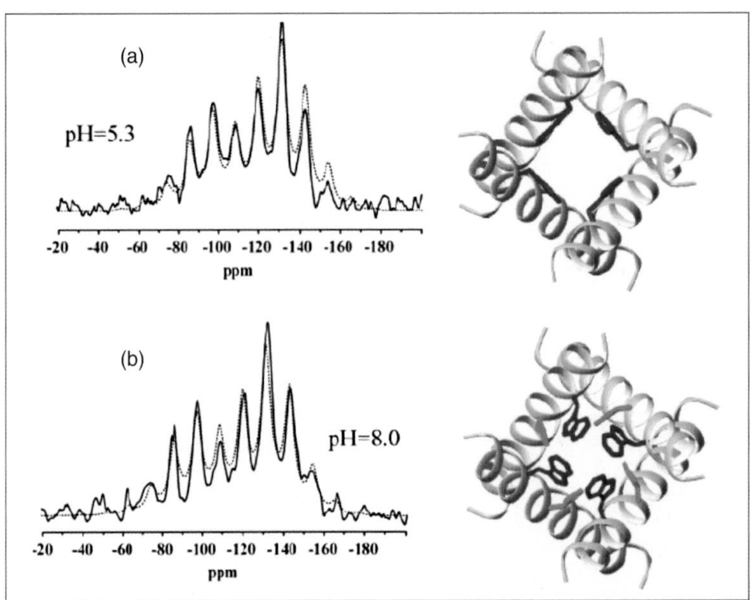

Plate 1.28 *Experimental (solid lines) and simulated (dashed lines) spin-echo spectra of 6F-Trp41-M2TMD at 6.5 kHz MAS at pH 5.3 (a) and pH 8.0 (b). Side-chain conformations (bottom view) of Trp41 (blue) and His37 (green) in the TM channel structure of the homo-tetrameric M2 protein are shown to the right side of the spectra. At pH 8.0, the structural parameters implicate an inactivated state, while at pH 5.3 the tryptophan conformation represents the activated state.*

(Source: *Reprinted with permission from Witter, R., Nozirov, F., Sternberg, U., Cross, T. A., Ulrich, A. S., and Fu, R. Solid-state 19F NMR spectroscopy reveals that Trp41 participates in the gating mechanism of the M2 proton channel of influenza A virus,* J. Am. Chem. Soc. *(2008)* **130**, *918–924. Copyright (2008) American Chemical Society.)*

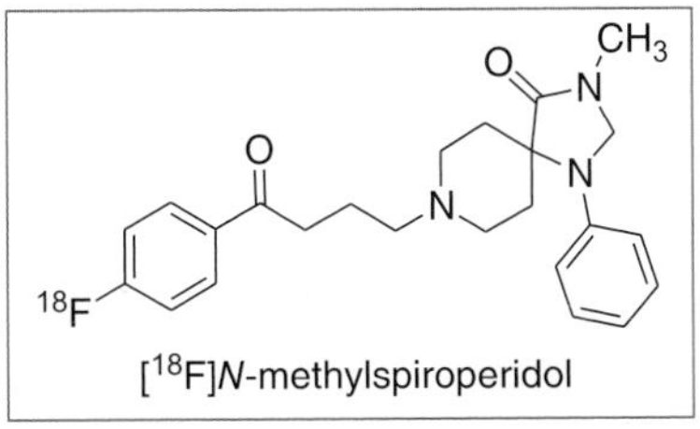

[^{18}F]*N*-methylspiroperidol

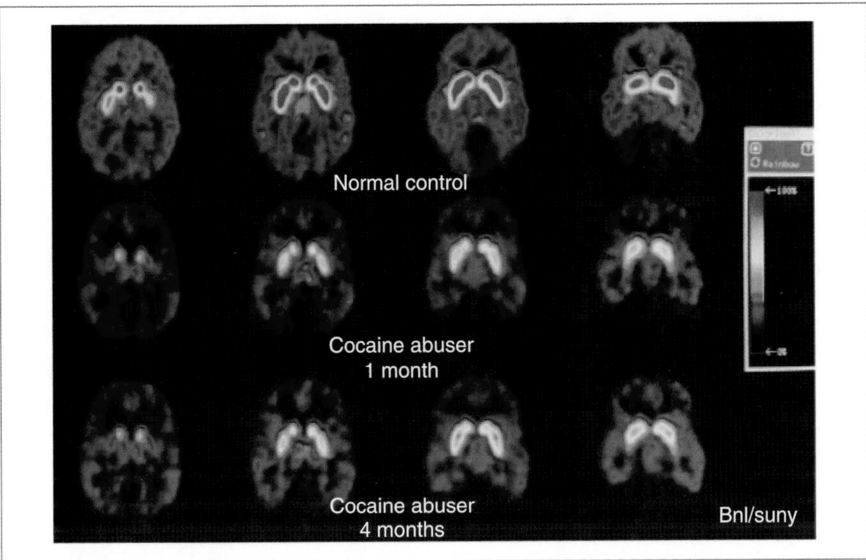

Plate 1.32 *[^{18}F]N-methylspiroperidol images in a normal control and in a cocaine abuser tested 1 month and 4 months after last cocaine use. The images correspond to the four sequential planes where the basal ganglia are located. The color scale has been normalized to the injected dose.*
(Source: *Reprinted with permission from Volkow, N. D., Fowler, J. S., Wang, G. J., Hitzemann, R., Logan, J., Schlyer, D. J., Dewey, S. L., and Wolf, A. P. Decreased dopamine D2 receptor availability is associated with reduced frontal metabolism in cocaine abusers,* Synapse *(1993)* **14**, *169-177. Reprinted with permission of Wiley-Liss Inc., a subsidiary of John Wiley & Sons, Inc. Copyright (1993) Wiley Interscience.)*

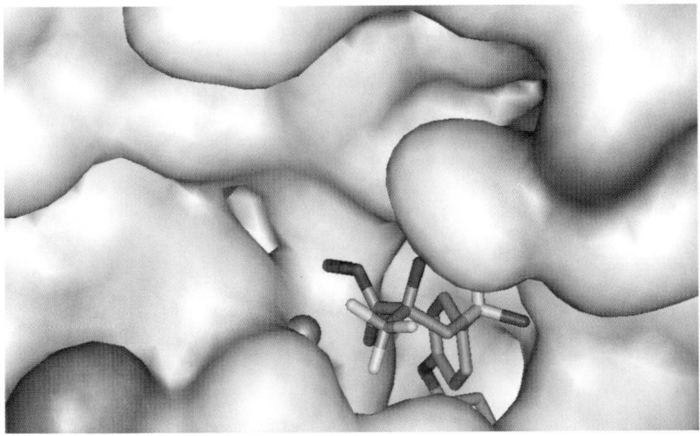

Plate 4.3 (R)-***19a*** *in the MMP-9 active site (purple = Zn(II), cyan = F, red = O, blue = N, yellow = S).*

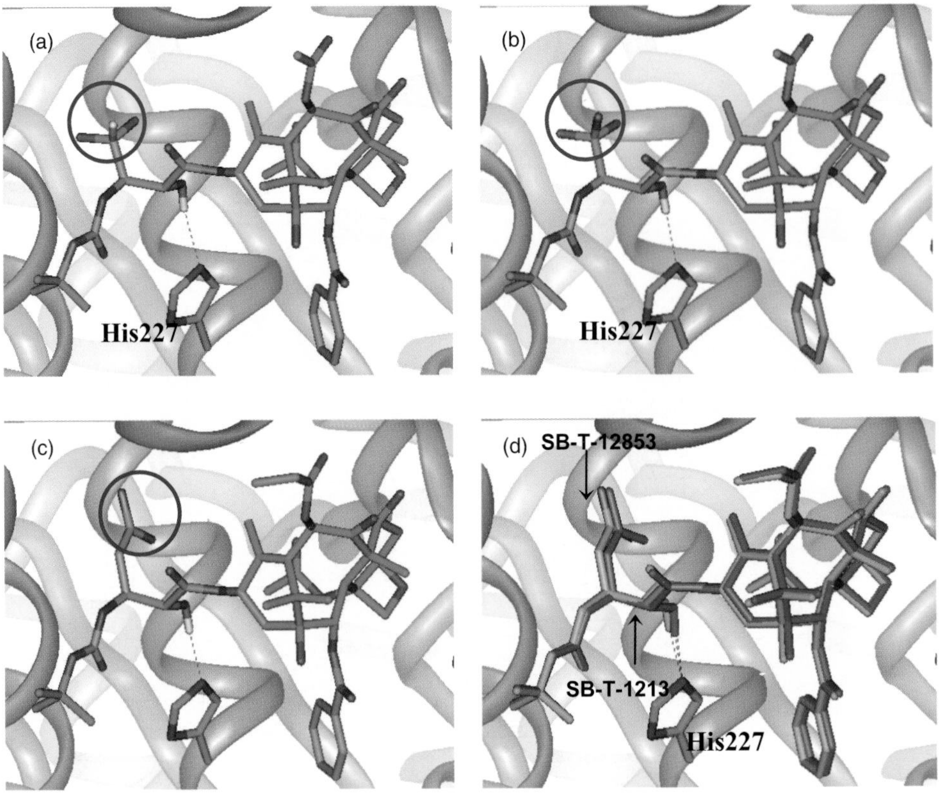

Plate 5.5 *Computer-generated binding structures of fluoro-taxoids to β-tubulin: (a) SB-T-1284 (3′-CF_2H); (b) SB-T-1282 (3′-CF_3); (c) SB-T-12853 (3′-CF_2=CH); (d) Overlay of SB-T-12853 and SB-T-1213 (C3′-isobutenyl).*

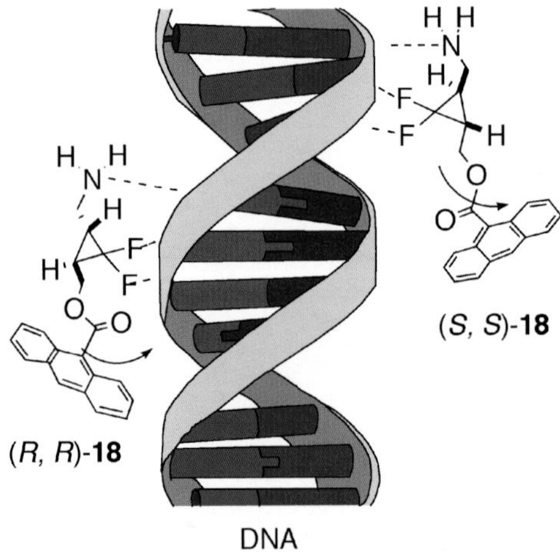

Plate 12.11 *Working hypothesis for the origin of difference in DNA-cleaving activities between (S,S)-***18** *and (R,R)-***18***. The (R,R)-isomer may bind to DNA more tightly than the (S,S)-isomer via intercalation or minor-groove binding.*

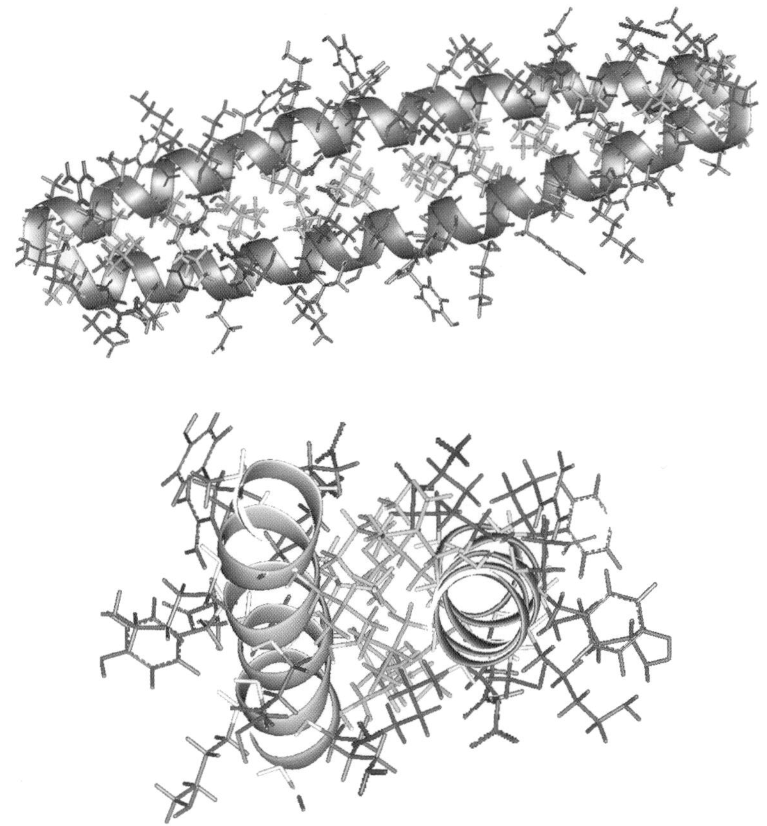

Plate 15.1 *Schematic representation of a modelled antiparallel coiled coil homodimer in both a side view (top) and a view along the superhelical axis (bottom). Hydrophobic side-chains (Leu) are represented in yellow, complementary salt bridges in red (Glu), and blue (Lys or Arg).*

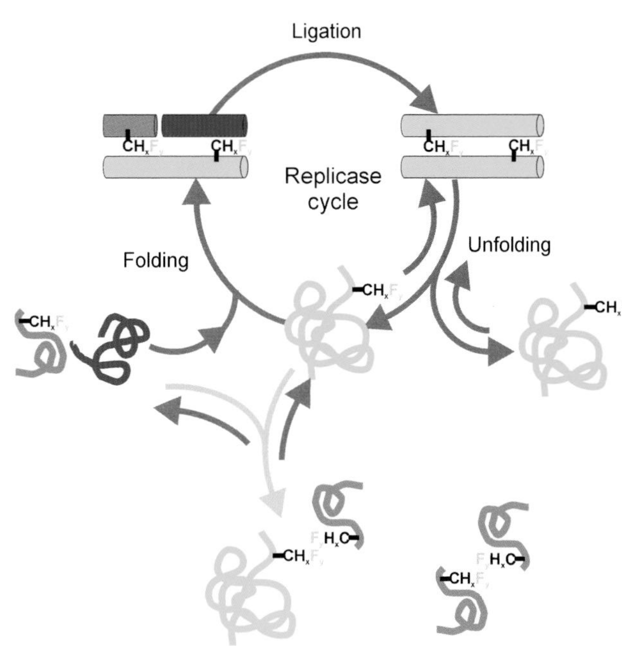

Plate 15.7 *The rate of replication depends on the association–dissociation equilibrium. The data for time-dependent turnover raise the question whether fluorous interactions interfere with the association of the electrophilic fragment with the template thereby inhibiting catalysis [9].*

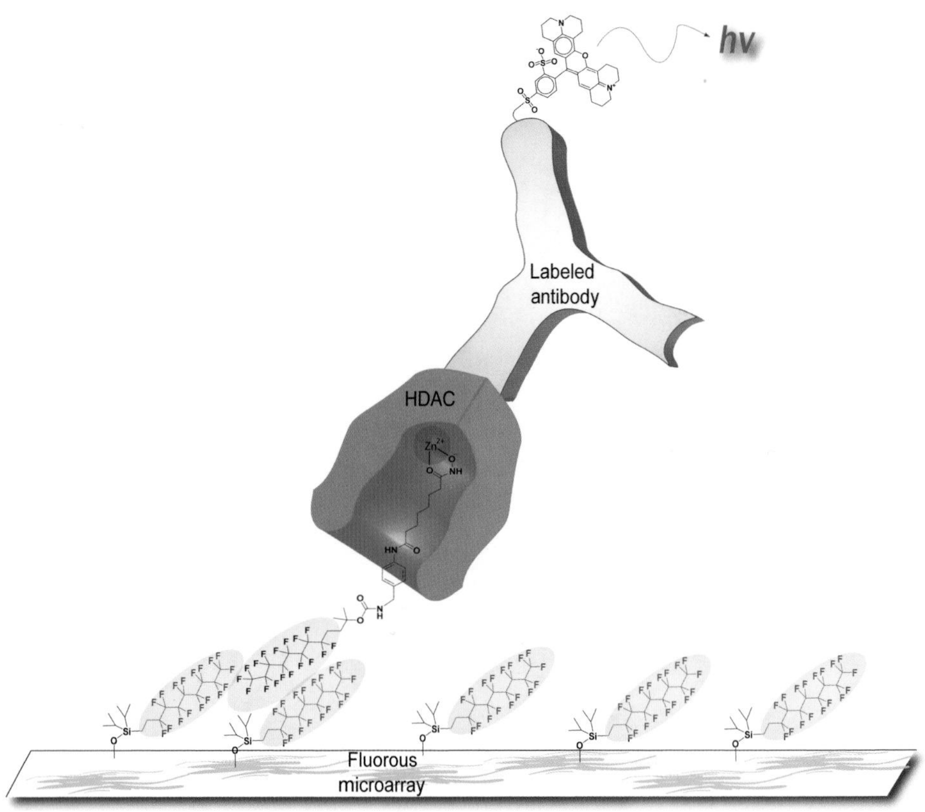

Plate 16.2 *Fluorous small-molecule microarrays. Small-molecule histone deacetylase (HDAC) binders are noncovalently immobilized onto a glass slide coated with fluorocarbon compounds. An antibody labeled with a fluorescent dye recognizes HDAC proteins.*
*(*Source: *Vegas, A. J., Brander, J. E., Tang, W. et al., Fluorous-based small-molecule microarrays for the discovery of histone deacetylase inhibitors, Angew. Chem. Int. Ed. (2007),* **46**, *7960–7964. Copyright Wiley-VCH Verlag GmbH & Co. KGaA. Reproduced with permission.)*

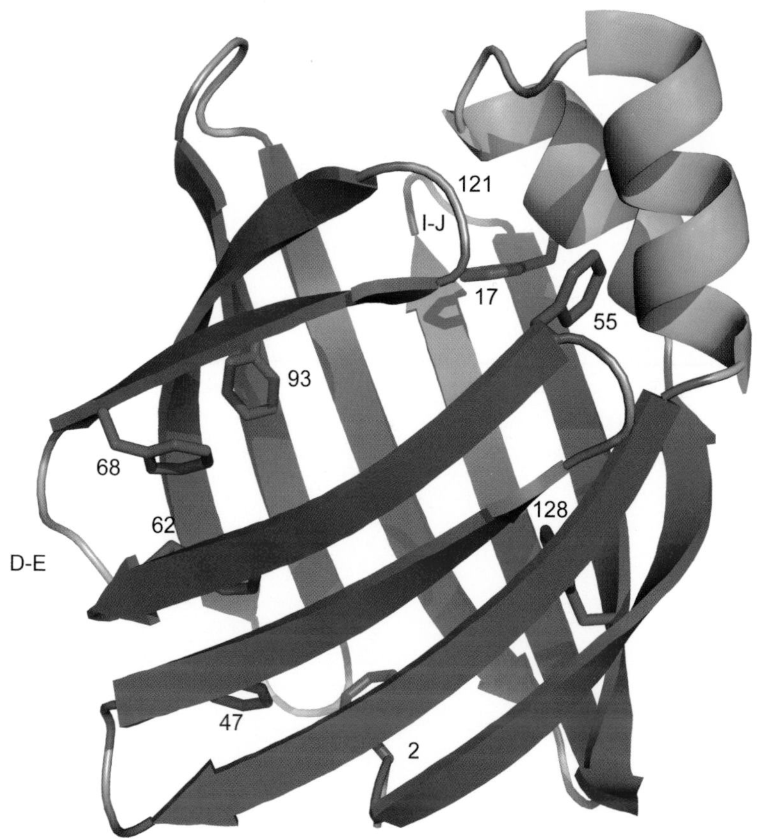

Plate 16.3 *The structure of apo-IFABP (PDB code: 1IFB). Eight Phe residues (shown in stick representation), the D–E and I–J regions, and location of G121 are indicated by labels. The structure was generated using MacPyMOL (DeLano Scientific LLC, Palo Alto, CA, U.S.A.).*

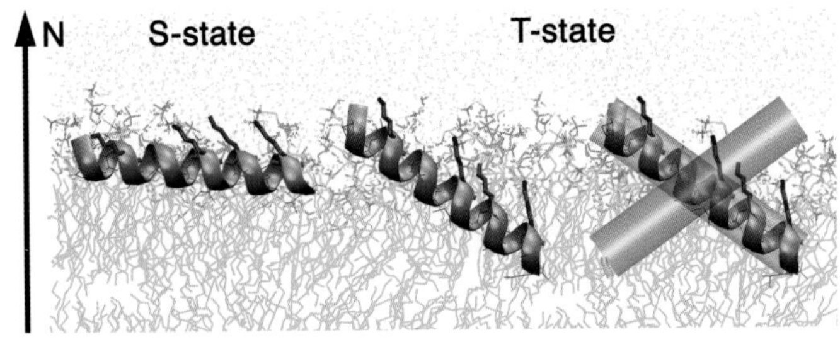

Plate 16.4 *Illustration of the S-state and T-state of the antimicrobial peptide PGLa in a DMPC membrane. At low concentrations, PGLa adopts an S-state. At high concentrations, PGLa assumes a tilted T-state. PGLa forms a dimer in the T-state (shown in purple).*
*(*Source: *Glaser, R. W., Sachse, C., Durr, U. H. N. et al., Concentration-dependent realignment of the antimicrobial peptide PGLa in lipid membranes observed by solid-state F-19-NMR, Biophys. J. (2005)* **88***, 3392–3397. Reproduced with permission from the Biophysical Society.)*

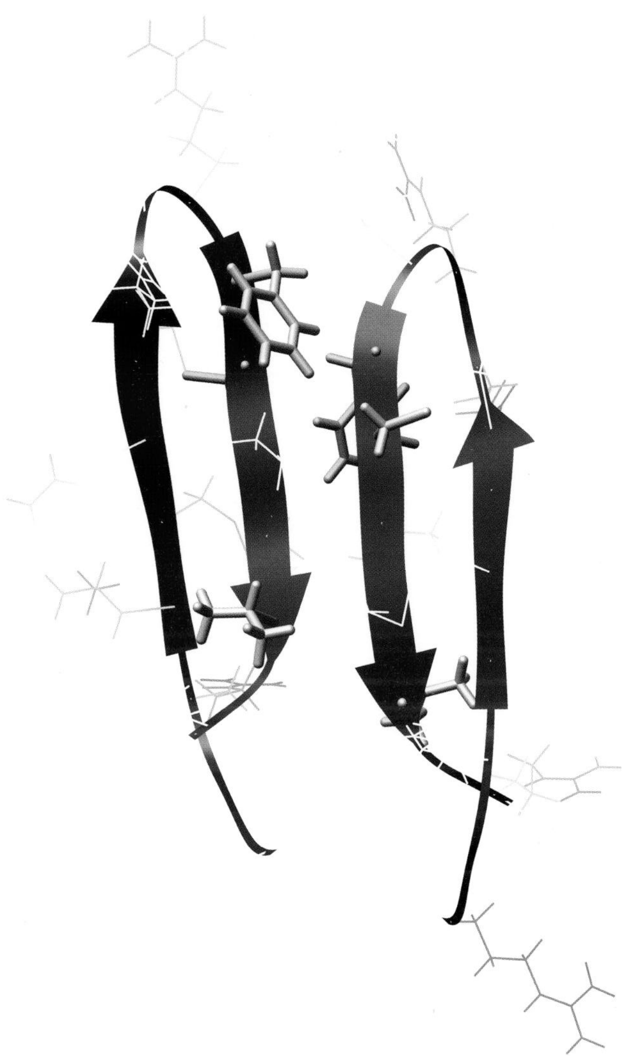

Plate 16.5 *The dimer structure of PG-1 (PDB code: 1ZY6) in POPC bilayers as determined by solid-state NMR. Isotopically labeled amino acids are shown in stick format; Phe12 was labeled with ^{19}F, Cys12 with ^{15}N and ^{13}C, and Val16 with ^{13}C. Inter- and intramolecular distances were used to determine the relative position of the two monomers.*

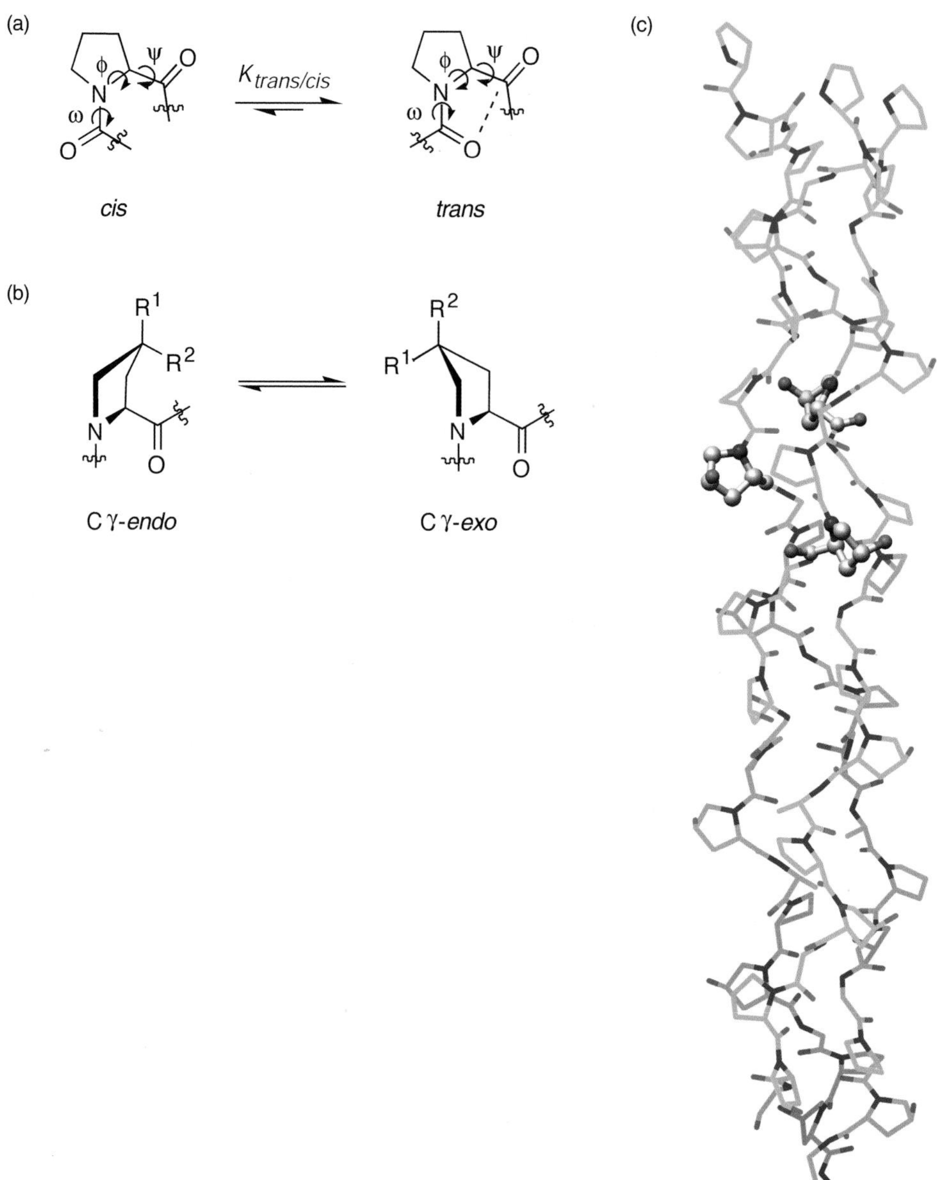

Plate 16.7 *Fluorinated prolines in collagen. (a) The trans/cis isomerization of amide bonds and main-chain angles of proline residues. The n → π* interaction, depicted by a dashed line, helps stabilize backbone dihedral angles. (b) Electron-withdrawing groups at Cγ influence the ring conformation of proline residues through the "gauche effect." The Cγ-endo pucker is favored when R^1 = H and R^2 = F, OH, or H. The Cγ-exo pucker is favored when R^1 = OH or F, and R^2 = H. Preorganization of the pyrrolidine ring contributes to the thermal stability of collagen and mimics. Flp: R^1 = H and R^2 = F; flp: R^1 = F and R^2 = H; Hyp: R^1 = H and R^2 = OH; hyp: R^1 = OH and R^2 = H. (c) Model structure of collagen (PDB code: 1CAG). Three hydroxyprolines at Yaa positions from each peptide chain are shown in ball-and-stick representation. Carbon = gray; Nitrogen = blue; Oxygen = red.*

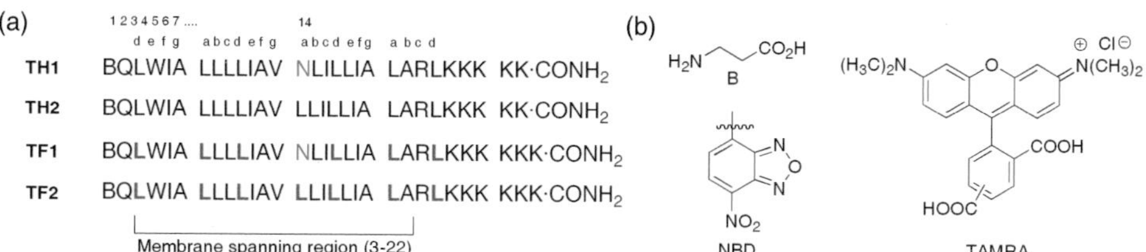

Plate 16.9 *(a) Sequences of transmembrane peptides TH1, TF1, TH2, and TF2. Asn (N) residues are assumed form interhelical hydrogen bonds in the core. L (green): hexafluoroleucine. (b) Structures of β-alanine, NBD (donor fluorophore), and TAMRA (acceptor fluorophore).*

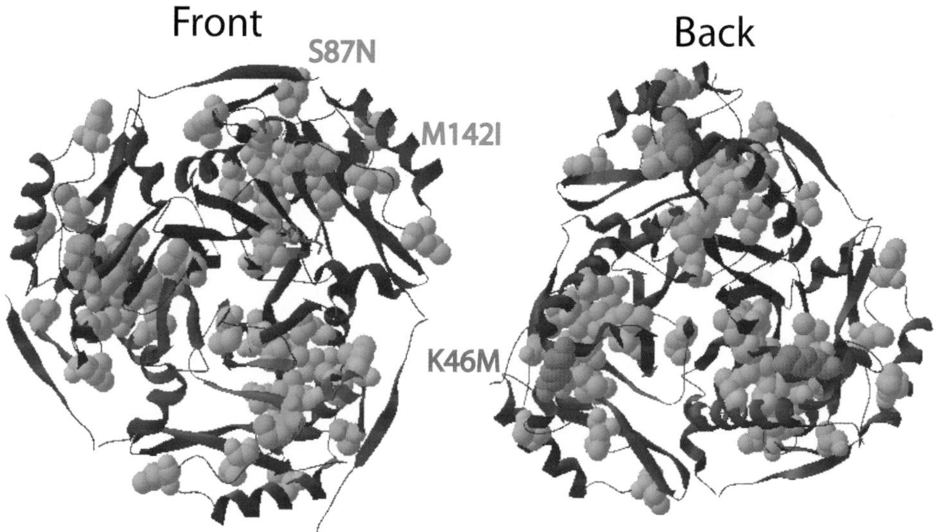

Plate 16.12 *Front and back views of the CAT trimer. The three stabilizing mutations in L2-A1 are S87N, M142I in orange, and K46M in pink. Trifluorolecuines/leucines are colored blue and chloramphenicol red.*
(Source: Montclare, J. K., Tirrell, D. A., Evolving proteins of novel composition. Angew. Chem. Int. Ed. (2006) **45**, *4518–4521. Copyright Wiley-VCH Verlag GmbH & Co. KGaA. Reproduced with permission.)*

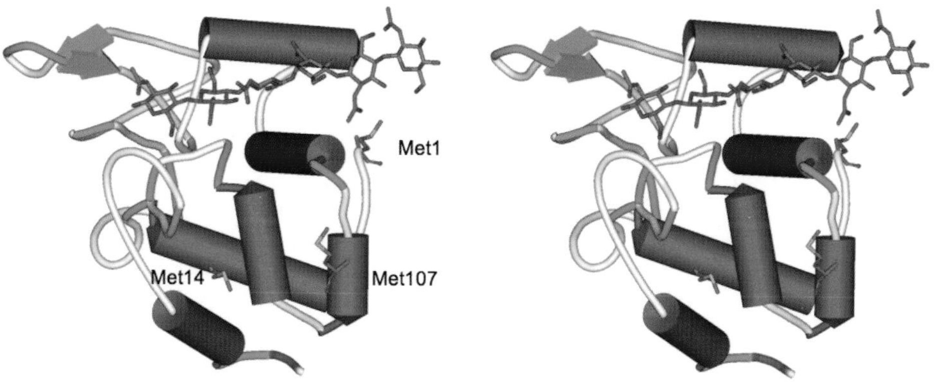

Plate 17.3 *Cartoon stereo diagram of the x-ray structure of wild-type bacteriophage lambda lysozyme as solved for the $(GlcNAc)_6$ complex and showing the position of the three methionine residues in the enzyme (PDB 1d9u) [32].*

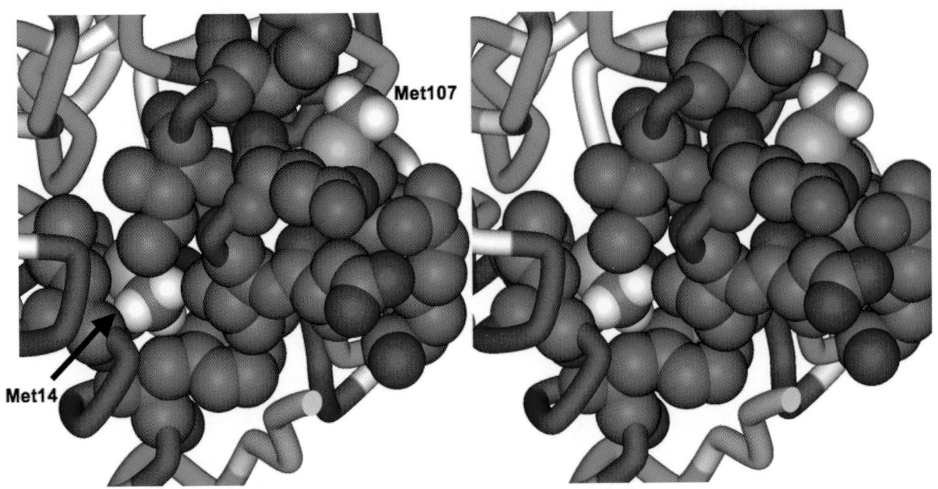

Plate 17.4 *Close-up of the detailed interactions between Met14 and adjacent residues and the location of Met107 (PDB 1d9u) [32].*

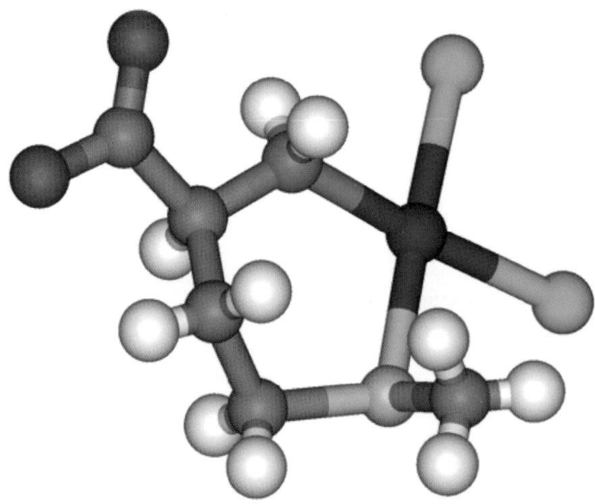

Plate 17.7 *Crystal structure of the methionine–$PtCl_2$ complex (coordinates from Wilson et al. [37]).*

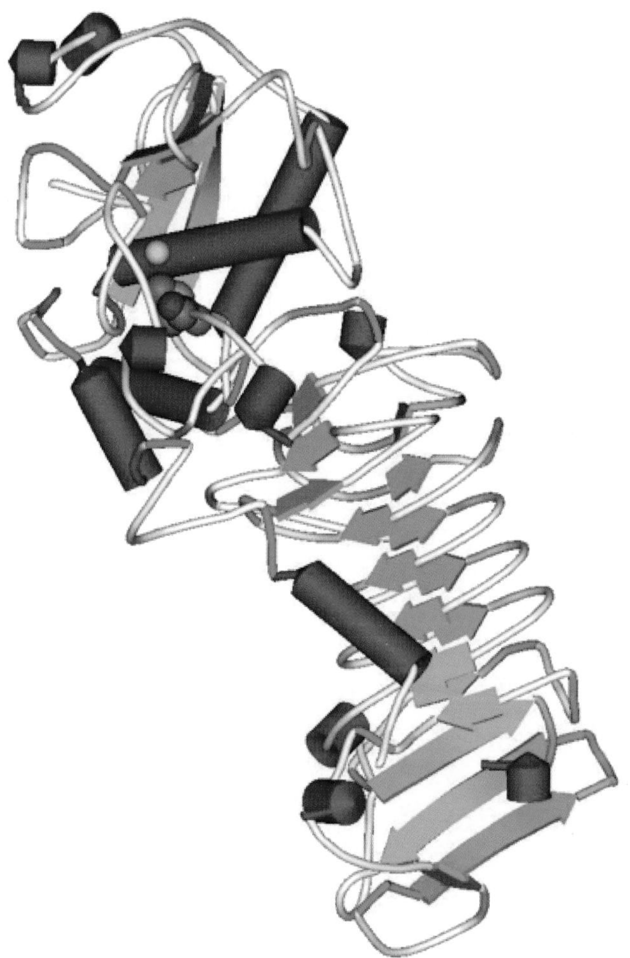

Plate 17.8 *Ribbon diagram of the* P. aeruginosa *alkaline protease showing the active site* Zn^{2+} *and the invariant methionine residue (PDB 1kap. [38]).*

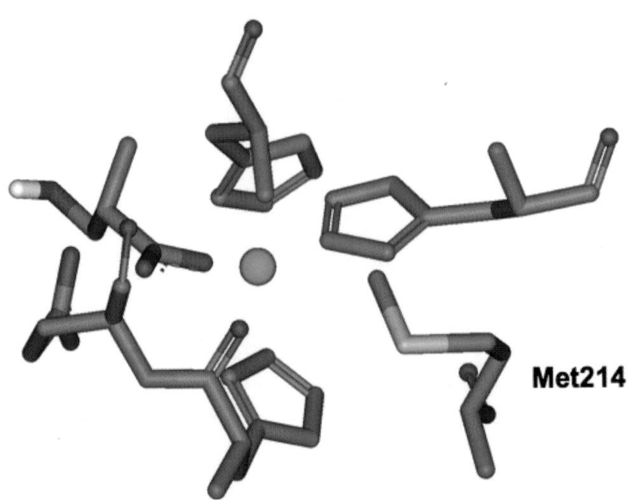

Plate 17.9 *Close-up of the active-site region of* P. aeruginosa *alkaline protease showing* Zn^{2+} *and protein ligands and the position of the thiomethyl group in close proximity to these side-chains (PDB 1kap. [38]).*

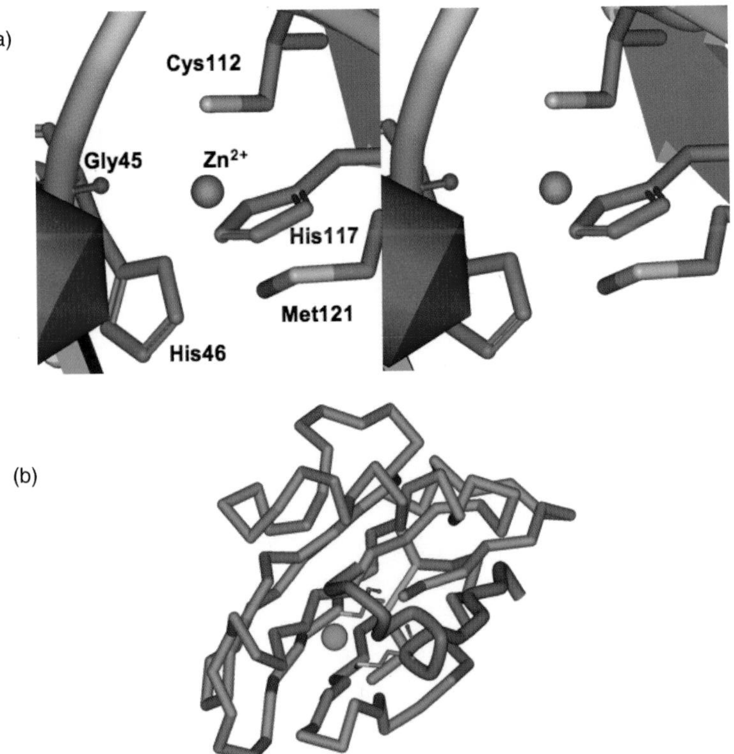

Plate 17.10 *(a) Stereo view of the active-site structure of* P. aeruginosa *azurin (PDB 4azu) [46]. (b) Tube diagram of the* P. aeruginosa *azurin showing the section of protein that was contributed by intein ligation.*

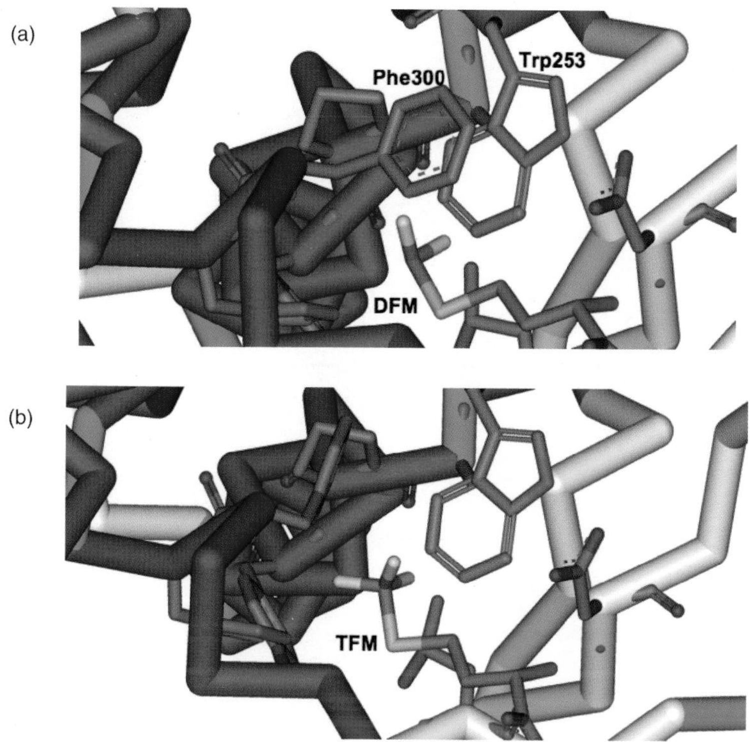

Plate 17.11 *X-ray structures of (a) DFM (PDB 1pfv) [48] and (b) TFM (PDB 1pfw) [48] complexes with Met-tRNA synthetase from* E. coli.

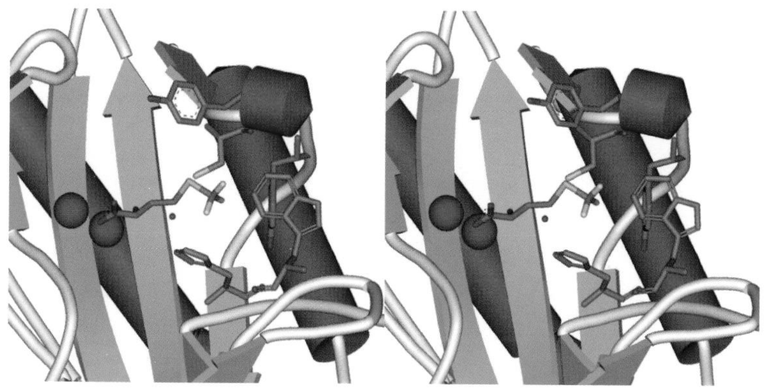

Plate 17.12 *Stereo view of the active-site interactions of TFM with* E. coli *methionine aminopeptidase (PDB 1c22) [49].*

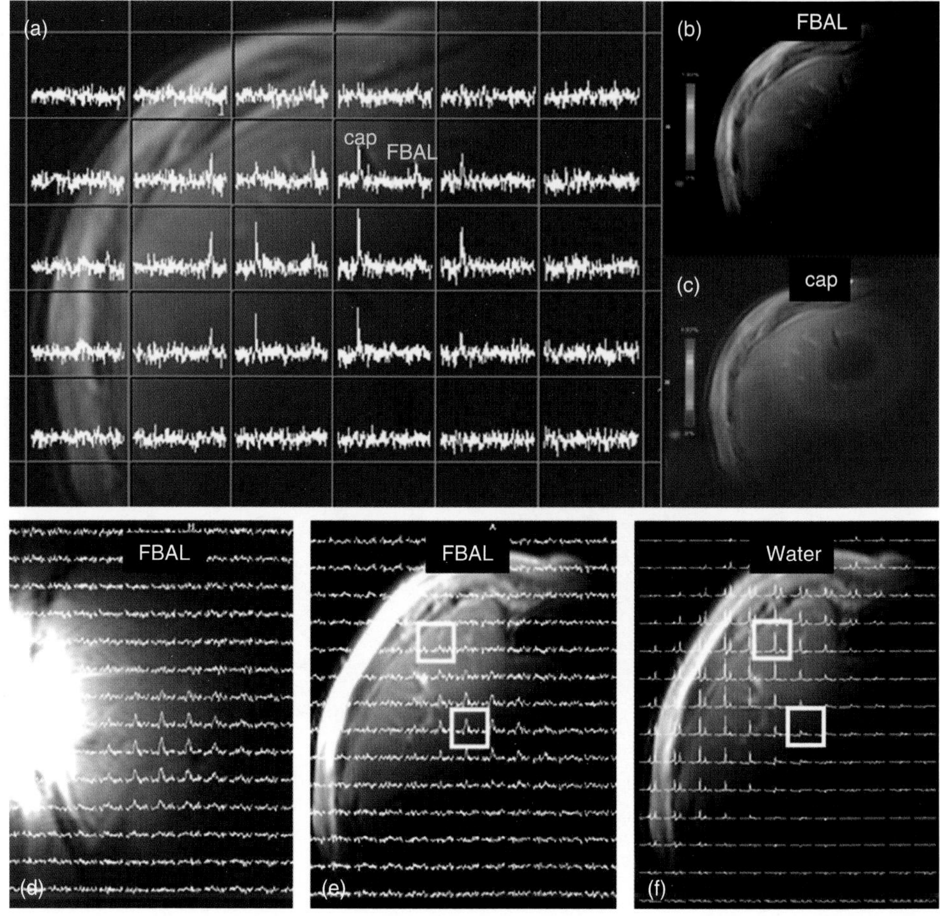

Plate 19.7 *Distribution of capecitabine and its metabolite FBAL in the liver of a patient treated with oral capecitabine at 3 T. (a) Spatially localized* ^{19}*F MR CSI spectra overlaid on the axial proton image of the liver acquired using the same surface coil. (b) and (c) Color depiction of distribution of FBAL in the axial plane and capecitabine in the coronal plane, respectively. (d) and (e) Distribution of FBAL in the coronal and axial planes, respectively, depicted by CSI spectra. (f) Distribution of water signal in the axial plane.*
*(*Source: *Klomp D, van Laarhoven H, Scheenen T, Kamm Y, Heerschap A., Quantitative* ^{19}*F MR spectroscopy at 3 T to detect heterogeneous capecitabine metabolism in human liver, NMR in Biomedicine (2007)* **20**, *485–492. Copyright (2007) John Wiley & Sons. Reprinted with permission.)*

Table 11.6 *Synthesis of pyrido[2,3-*b*]pyrazine scaffolds [57, 58]*

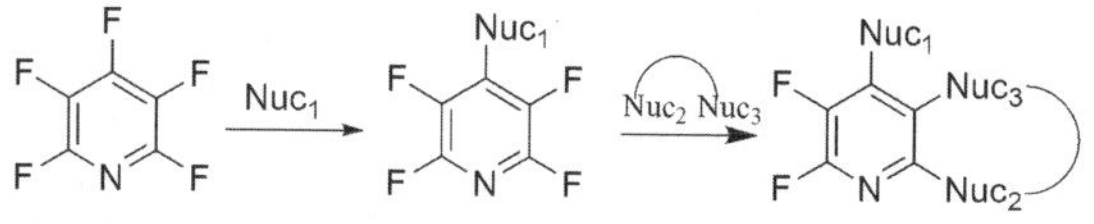

Nuc_1	Nuc_2 Nuc_3	Product
$PhSO_2Na$	Ph Ph —N N— H H	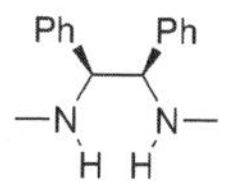 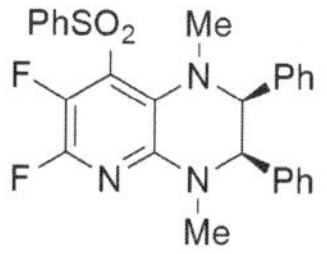 **34a**, 34%
$PhSO_2Na$	H H N N	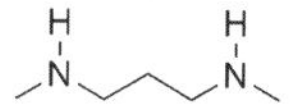 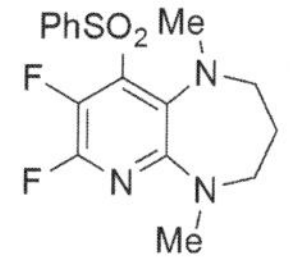 **34b**, 62%
$PhSO_2Na$	H_2N NH_2	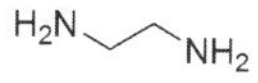 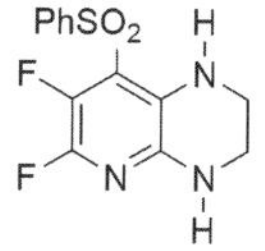 **34c**, 92%
$PhSO_2Na$	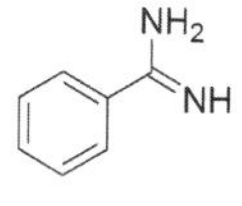 NH_2 NH	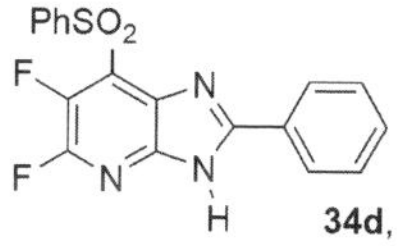 **34d**, 36%
H	H N N H	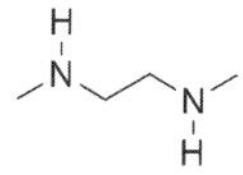 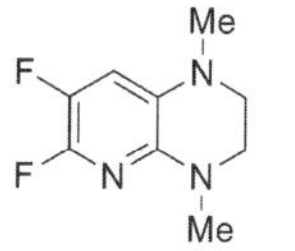 **34e**, 66%
Br	H N N H	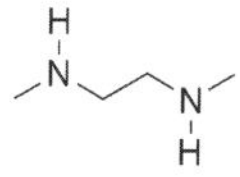 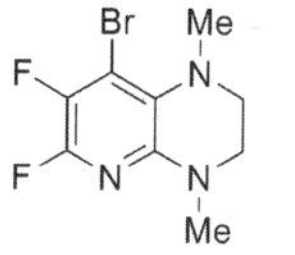 **34f**, 20%
CN	H N N H	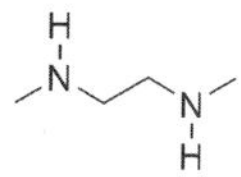 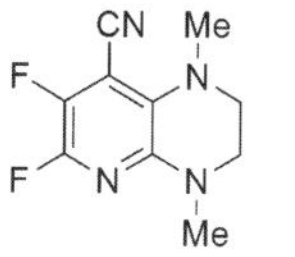 **34g**, 90%
OEt	H N N H	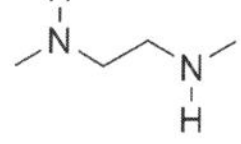 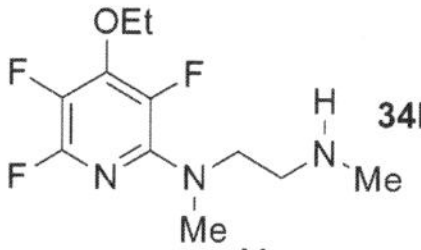 **34h**, 23%
NO_2	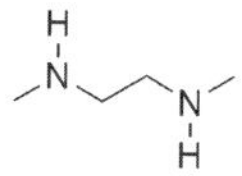 H N N H	 **34j**, 46%; 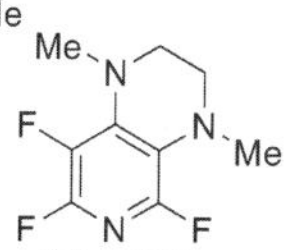 **34k**, 20%

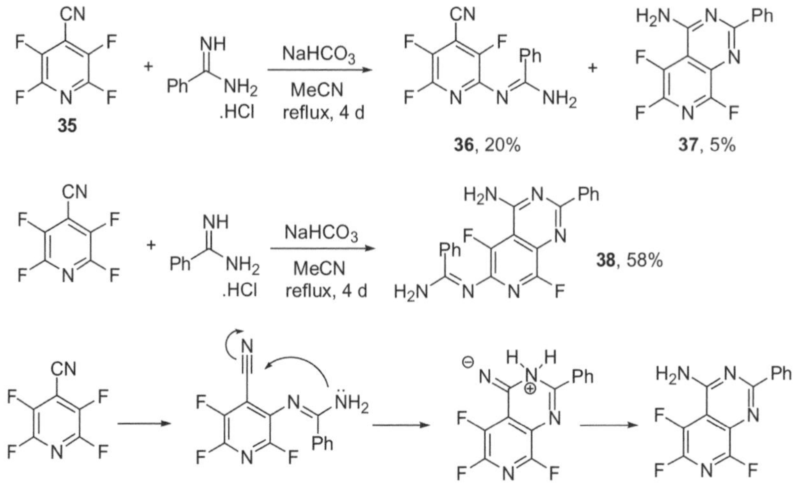

Figure 11.11 *Reaction of 4-cyanotetrafluoropyridine with amidines [56].*

4-phenylsulfonylpyridine derivative and, since the phenylsulfonyl group is strongly electron withdrawing, annelation by reaction with appropriate diamines proceeded readily to give pyrido[2,3-*b*]pyrazine scaffolds [57, 58]. In addition, 4-bromo- and 4-cyanotetrafluoropyridine systems gave the corresponding pyrido[2,3-*b*]pyrazines, **34f** and **34g** respectively, upon reaction with *N,N′*-dimethylethylene diamine [58].

However, not all tetrafluoropyridine derivatives were suitable substrates for the synthesis of pyrido[2,3-*b*]pyrazine scaffolds by analogous annelation reactions [58]. The 4-ethoxy and 4-dimethylaminotetrafluoropyridine systems gave only noncyclized products **34h** and **34i**, respectively, arising from substitution of the 2-position. However, the 4-nitro derivative led to the formation of a mixture of products **34j, k** arising from substitution of both the 2-fluorine and 4-nitro group, which is itself a very labile leaving group in nucleophilic aromatic substitution processes [58].

Additionally, 4-cyanotetrafluoropyridine **35** reacts with amidines [56] to give a mixture of products **36** and **37** arising from substitution of fluorine at both the 2- and 3-positions in a 1 : 1 ratio (see Figure 11.11). The product derived from 3-substitution and subsequent cyclization onto the cyano group, reflects the activating influence of the strongly electron-withdrawing cyano group on adjacent heteroaromatic carbon sites and the electrophilicity of the cyano group itself. An excess of the amidine leads to high yields of **38** [56].

Preliminary reactions involving scaffold **39** derived from 1,2-diaminoethane with nitrogen and sulfur nucleophiles led to products **40** arising from substitution of the phenylsulfonyl group, a very good leaving group that is located at an activated site *para* to pyridine nitrogen, and reaction of acetic anhydride gave selective acylation yielding **41** (see Figure 11.12), indicating some of the possibilities for functionalization of these bicyclic scaffolds [57].

In an application of this annelation strategy to a medicinal chemistry project, tricyclic 2-pyridone **42** was synthesized using pentafluoropyridine as the core scaffold and all but one of the fluorine atoms were displaced in a multistep process in which a carbon–oxygen centred difunctional nucleophile is used as the annelating reagent (see Figure 11.13).

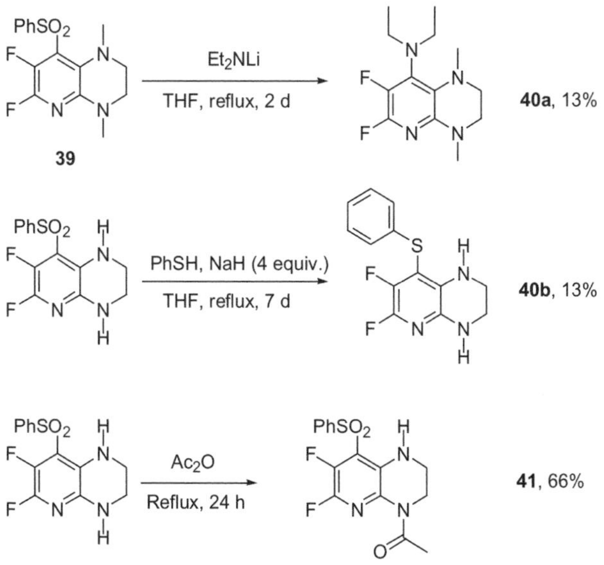

Figure 11.12 *Reactions of pyrido[2,3-*b*]pyrazine scaffolds [57].*

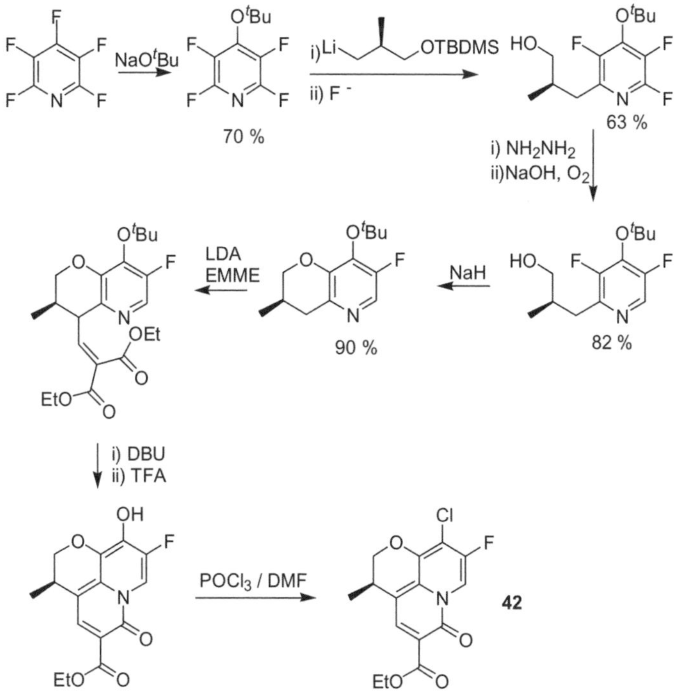

Figure 11.13 *Synthesis of 2-pyridone systems from pentafluoropyridine.*

Fluoroquinolones have proved to be one of the most successful classes of antibacterial agents both economically and clinically [59] and, as such, 2-pyridones, as bio-isosteres of quinolones, provide a valuable source of potential new therapeutic agents [60].

11.3 Summary

The drug discovery arena requires the availability of an ever-increasing, diverse range of polyfunctional heterocyclic core scaffolds based upon privileged structures with proven biological activity from which to synthesize arrays of structurally similar analogues that follow the Lipinski RO5 parameters with the aim of reducing the resources used in hit-to-lead progression. In this chapter, we have outlined how perfluorinated heteroaromatic systems such as pentafluoropyridine and tetrafluoropyrimidine can be utilized as very effective core scaffolds for the synthesis of a diverse array of polysubstituted pyridine, pyrimidine, [5,6]- and [6,6]-bicyclic fused-ring systems. Given the vast diversity of mono- and difunctional nucleophiles that are readily available from commercial suppliers, the core scaffolds and polyfunctional privileged structure systems that may be accessed from perfluorinated heteroaromatic starting materials offer, potentially, many great opportunities.

Acknowledgments

We thank GlaxoSmithKline and EPSRC for funding our research in this area.

References

1. Lipinski, C. A., Lombardo, F., Dominy, B. W. and Feeney P. J. (1997) Experimental and computational approaches to estimate solubility and permeability in drug discovery and development settings. *Adv. Drug Deliv. Rev.*, **23**, 3–25.
2. Sugiyama, Y. (2005) Druggability: selecting optimized drug candidates. *Drug Discov. Today*, **10**, 1577–1579.
3. Horton, D. A., Bourne, G. T. and Smythe, M. L. (2003) The combinatorial synthesis of bicyclic privileged structures or privileged substructures. *Chem. Rev.*, **103**, 893–930.
4. Patel, D. V. and Gordon, E. M. (1996) Applications of small-molecule combinatorial chemistry to drug discovery. *Drug Discov. Today*, **1**, 134–144.
5. Walters, W. P., Murcko, A. and Murcko, M. A. (1999) Recognizing molecules with drug-like properties. *Curr. Opin. Chem. Biol.*, **3**, 384–387.
6. Leach, A. R. and Hann, M. M. (2000) The *in silico* world of virtual libraries. *Drug Discov. Today*, **5**, 326–336.
7. Bailey, D. and Brown, D. (2001) High-throughput chemistry and structure-based design: survival of the smartest. *Drug Discov. Today*, **6**, 57–59.

8. Walters, W. P. and Murcko, M. A. (2002) Prediction of "drug-likeness". *Adv. Drug Deliv Rev.*, **54**, 255–271.
9. Lipinski, C. A. (2004) Lead- and drug-like compounds: the rule-of-five revolution. *Drug Discov. Today: Technologies*, **1**, 337–241.
10. Lipinski, C. A., Lombardo, F., Dominy, B. W. and Feeney, P. J. (2001) Experimental and computational approaches to estimate solubility and permeability in drug discovery and development settings. *Adv. Drug Deliv. Rev.*, **46**, 3–26.
11. Veber, D. F., Johnson, S. R., Cheng, H. Y., *et al.* (2002) Molecular properties that influence the oral bioavailability of drug candidates. *J. Med. Chem.*, **45**, 2615–2623.
12. Keller, T. H., Pichota, A. and Yin, Z. (2006) A practical view of "druggability". *Curr. Opin. Chem. Biol.*, **10**, 357–361.
13. Fichert, T., Yazdanian, M. and Proudfoot, J. R. (2003) A structure–permeability study of small drug-like molecules. *Bioorg. Med. Chem. Lett.*, **13**, 719–722.
14. Wenlock, M. C., Austin, R. P., Barton, P., *et al.* (2003) A comparison of physiochemical property profiles of development and marketed oral drugs. *J. Med. Chem.*, **46**, 1250–1256.
15. Muegge, I. (2003) Selection criteria for drug-like compounds. *Med. Res. Rev.*, **23**, 302–321.
16. Vieth, M., Siegel, M. G., Higgs, R. E., *et al.* (2004) Characteristic physical properties and structural fragments of marketed oral drugs. *J. Med. Chem.*, **47**, 224–232.
17. Evans, B. E., Rittle, K. E., Bock, M. G., *et al.* (1988) Methods for drug discovery – development of potent, selective, orally effective cholecystokinin antagonists. *J. Med. Chem.*, **31**, 2235–2246.
18. Nicolaou, K. C., Pfefferkorn, J. A., Roecker, A. J., *et al.* (2000) Natural product-like combinatorial libraries based on privileged structures. 1. General principles and solid-phase synthesis of benzopyrans. *J. Am. Chem. Soc.*, **122**, 9939–9953.
19. Mason, J. S., Morize, I., Menard, P. R., *et al.* (1999) New 4-point pharmacophore method for molecular similarity and diversity applications: overview of the method and applications, including a novel approach to the design of combinatorial libraries containing privileged substructures. *J. Med. Chem.*, **42**, 3251–3264.
20. DeSimone, R. W., Currie, K. S., Mitchell, S. A., *et al.* (2004) Privileged structures: applications in drug discovery. *Comb. Chem. High Throughput Screening*, **7**, 473–494.
21. Benmansour, H., Chambers, R. D., Hoskin, P. R. and Sandford, G. (2001) Multi-substituted heterocycles. *J. Fluorine Chem.*, **112**, 133–137.
22. Beck, B., Picard, A., Herdtweck, E. and Domling, A. (2004) Highly substituted pyrrolidinones and pyridones by 4-CR/2-CR sequence. *Org. Lett.*, **6**, 39–42.
23. Katritzky, A. R. and Rees, C. W. (1984) *Comprehensive Heterocyclic Chemistry*. Pergamon Press, Oxford.
24. Eicher, T. and Hauptmann, S. (1995) *The Chemistry of Heterocycles*. Thieme, Stuttgart.
25. Chambers, R. D. and Sargent, C. R. (1981) Polyfluoroheteroaromatic compounds. *Adv. Heterocycl. Chem.*, **28**, 1–7.
26. Brooke, G. M. (1997) The preparation and properties of polyfluoroaromatic and heteroaromatic compounds. *J. Fluorine Chem.*, **86**, 1–76.
27. Banks, R. E., Smart, B. E. and Tatlow, J. C. (1994) *Organofluorine Chemistry. Principles and Commercial Applications*. Plenum, New York.
28. Banks, R. E., Jondi, W. and Tipping, A. E. (1989) SNAr displacement of fluorine from pentafluoropyridine by sodium oximates: unprecedented substitution patterns. *J. Chem. Soc., Chem. Commun.*, 1268–1269.
29. Gilmore, C. J., MacNicol, D. D., Murphy, A. and Russell, M. A. (1984) Discovery and x-ray crystal structure of a new host compound: 1,3,4-tris(phenylthio)[1]benzothieno[3,2-c]pyridine. *Tetrahedron Lett.*, **25**, 4303–4306.

30. Banks, R. E., Burgess, J. E., Cheng, W. M. and Haszeldine, R. N. (1965) Nucleophilic substitution in pentafluoropyridine: the preparation and properties of some 4-substituted 2,3,5,6-tetrafluoropyridines. *J. Chem. Soc.*, 575–581.
31. Chambers, R. D., Gray, W. K. and Korn, S. R. (1995) Amines as initiators of fluoride ion catalysed reactions. *Tetrahedron*, **51**, 13167–13172.
32. Chambers, R. D., Hassan, M. A., Hoskin, P. R., *et al.* (2001) Reactions of perfluoroisopropylpyridine with oxygen, nitrogen and carbon nucleophiles. *J. Fluorine Chem.*, **111**, 135–146.
33. Davie, E. W., Fujikawa, K. and Kisiel, W. (1991) The coagulation cascade: initiation, maintenance, and regulation. *Biochemistry*, **30**, 10363–10370.
34. Kranjc, A., Kikelj, D. and Peterlin-Masic, L. (2005) Recent advances in the discovery of tissue factor/factor, VIIa inhibitors and dual inhibitors of factor, VIIa/Factor, Xa. *Curr. Pharm. Des.*, **11**, 4207–4227.
35. Walenga, J. M., Jeske, W. P., Hoppensteadt, D. and Fareed, J. (2003) Factor Xa inhibitor: today and beyond. *Curr. Opin. Investig. Drugs*, **4**, 272–281.
36. Ng, H. P., Buckman, B. O., Eagen, K. A., *et al.* (2002) Design, synthesis, and biological activity of novel factor Xa inhibitors: 4-aryloxy substituents of 2,6-diphenoxypyridines. *Bioorg. Med. Chem.*, **10**, 657–666.
37. Revesz, L., Di Padova, F. E., Buhl, T., *et al.* (2002) SAR of 2,6-diamino-3,5-difluoropyridinyl substituted heterocycles as novel p38 MAP kinase inhibitors. *Bioorg. Med. Chem. Lett.*, **12**, 2109–2112.
38. Banks, R. E., Field, D. S. and Haszeldine, R. N. (1967) Nucleophilic substitution in tetrafluoropyrimidine. *J. Chem. Soc. (C)*, 1822–1826.
39. Parks, E. L. and Sandford, G. (2007) Reactions of tetrafluoropyrimidine. Unpublished results.
40. Chambers, R. D., Hall, C. W., Hutchinson, J. and Millar, R. W. (1998) Fluorinated nitrogen heterocycles with unusual substitution patterns. *J. Chem. Soc., Perkin Trans* **1**, 1705–1713.
41. Benmansour, H., Chambers, R. D., Sandford, G., *et al.* (2007) Sonogashira reactions of 2,4,6-tribromo-3,5-difluoropyridine. *Arkivoc*, **(xi)**, 1–10.
42. Benmansour, H., Chambers, R. D., Sandford, G., *et al.* (2007) Suzuki reactions of 2,4,6-tribromo-3,5-difluoropyridine. *J. Fluorine Chem.*, **128**, 718–722.
43. Benmansour, H., Chambers, R. D., Sandford, G., *et al* (2001) Multi-functional heterocycles from bromofluoro pyridine derivatives. *J. Fluorine Chem.*, **112**, 349–355.
44. Chambers, R. D., Hoskin, P. R., Sandford, G., *et al.* (2001) Synthesis of multi-substituted pyridine derivatives from pentafluoropyridine. *J. Chem. Soc., Perkin Trans* **1**, 2788–2795.
45. Liu, M.-C., Luo, M.-Z., Mozdziesz, D. E., *et al.* (2001) Synthesis of halogen-substituted 3-deazaadenosine and 3-deazagunaosine analogues as potential antitumour/antiviral agents. *Nucleosides Nucleotides Nucleic Acids*, **20**, 1975–2000.
46. Coe, P. L., Rees, A. J. and Whittaker, J. (2001) Reactions of polyfluoropyridines with bidentate nucleophiles: attempts to prepare deazapurine analogues. *J. Fluorine Chem.*, **107**, 13–22.
47. Sandford, G., Slater, R., Yufit, D. S., *et al.* (2005) Tetrahydro-pyrido[3,4b]pyrazine scaffolds from pentafluoropyridine. *J. Org. Chem.*, **70**, 7208–7216.
48. Savelli, F. and Boido, A. (1992) Synthesis of new dipyridopyrazines. *J. Heterocycl. Chem.*, **29**, 529–533.
49. Couture, A. and Grandclaudon, P. (1991) A convenient synthetic methodology for the elaboration of the pyrido[2,3-*b*] or [3,4-*b*]-pyrazine and [1,4]-diazepine skeletons. *Synthesis*, **11**, 982–984.
50. DeSelms, R. C. and Mosher, H. S. (1960) Stereospecific $LiAlH_4$ reduction of 2,3-dimethylquinoxaline and related triazanaphthalenes. *J. Am. Chem. Soc.*, **82**, 3762–3765.

51. Archer, R. A. and Mosher, H. S. (1967) Stereochemistry and conformations of reduced quinoxalines, phenazines and pteridines. *J. Org. Chem.*, **32**, 1378–1381.
52. Boutte, D., Queguiner, G. and Pastour, P. (1971) Pyrido-[2,3-*b*]- and -[3,4-*b*]-pyrazines. *C. R. Acad Sci. Ser. C.*, **273**, 1529–1532.
53. Vinot, N. and Maitte, P. (1976) Study of the reaction of organomegnesium compounds with pyrido-[2,3-*b*]-pyrazines. *Bull. Chim. Soc. Fr.*, 251–254.
54. Cosmao, J. M., Collignon, N. and Queguiner, G. (1979) Alkylating reduction of some polynitrogen heteroaromatic compounds. *J. Heterocycl. Chem.*, **16**, 973–976.
55. Armand, J., Boulares, L., Bellec, C. and Pinson, J. (1988) Preparation, chemical and electrochemical reduction of pyrido[2,3-b]quinoxalines and pyrido[3,4-b]quinoxalines. *Can. J. Chem.*, **66**, 1500–1505.
56. Cartwright, M. W., Sandford, G., Yufit, D. S., *et al.* (2007) Imidazopyridine scaffolds from perfluorinated pyridine derivatives. *Tetrahedron*, **63**, 7027–7035.
57. Baron, A., Sandford, G., Slater, R., *et al.* (2005) Polyfunctional tetrahydro-pyrido[2,3b]pyrazine scaffolds from 4-phenylsulfonyl-tetrafluoropyridine. *J. Org. Chem.*, **70**, 9377–9381.
58. Sandford, G., Hargreaves, C. A., Slater, R., *et al.* (2007) Synthesis of tetrahydropyrido [2,3-b]pyrazine scaffolds from 2, 3, 5, 6-tetrafluoropyridine derivatives. *Tetrahedron*, **63**, 5204–5211.
59. Mitscher, L. A. (2005) Bacterial topoisomerase inhibitors: quinolone and pyridone antibacterial agents. *Chem. Rev.*, **105**, 559–592.
60. Li, Q., Mitscher, L. A. and Shen, L. L. (2000) The 2-pyridone antibacterial agents: bacterial topisomerase inhibitors. *Med. Res. Rev.*, **20**, 231–293.

12

gem-Difluorocyclopropanes as Key Building Blocks for Novel Biologically Active Molecules

Toshiyuki Itoh

12.1 Introduction

The cyclopropyl group is a structural element present in a wide range of naturally occurring biologically active compounds found in both plants and microorganisms [1]. Since substitution of two fluorine atoms on the cyclopropane ring is expected to alter both chemical reactivity and biological activity due to the strong electron-withdrawing nature of fluorine [2], *gem*-difluorocyclopropanes are expected to display unique biological and physical properties and a great deal of attention has been given to their chemistry and singular properties [3,4]. This chapter describes the chemistry of these *gem*-difluorocyclopropanes that have unique biological activities. In particular, we focus on the methodology for the synthesis of optically active *gem*-difluorocyclopropanes as well as our investigation into the novel biological activities of *gem*-difluorocyclopropane derivatives.

12.2 Biologically Active *gem*-Difluorocyclopropanes

The first example of biologically active *gem*-difluorocyclopropane derivative was reported in 1986 by Johnson and his co-workers [5a]. Halogen-substituted analogues of DDT-

Fluorine in Medicinal Chemistry and Chemical Biology Edited by Iwao Ojima

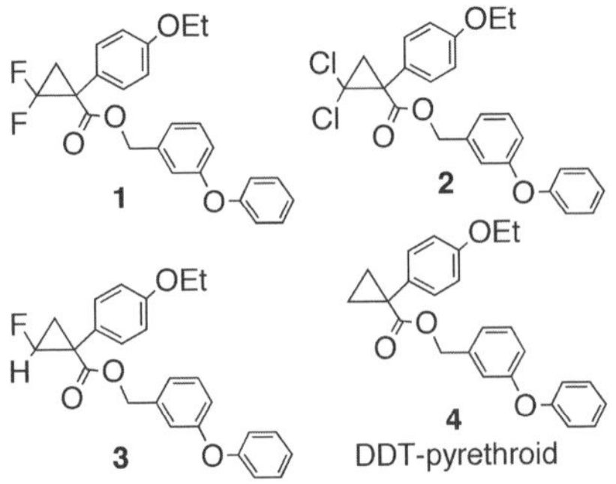

Compound	LD_{50}, mg/Kg L. cuprina	B. germanica
1	3.9	9.3
2	3.9	31.5
3	98.9	>200
4	>200	>200

W. M. Johnson (1986)[5a]

Figure 12.1 *Effect of substitution of fluorine at the C-2 position of the cyclopropyl ring on insecticidal activity.*

pyrethroid were synthesized and their insecticidal activity was examined. Subsequently, it was found that *gem*-difluorocyclopropane **1** possesses strong toxicity to some insects. It was revealed that the optimum insecticidal activity required two fluorine substituents at the C-2 position of the cyclopropane ring since chlorine- or monofluorine-substituted compounds, **2** and **3**, were less active than *gem*-difluorocyclopropane **1** (see Figure 12.1).

The next important examples of biologically active *gem*-difluorocyclopropane derivatives were reported in 1995–1997 from several groups independently [6, 8, 9]. Pfister and co-workers found a strong multidrug resistance (MDR) reversal activity of *gem*-difluorinated compound **5**, which is an analogue of the MDR reversal agent MS-073. When *anti*-(*R*)-**5** was used with the anticancer agent doxorubicin (1 μg/mL) against the CHRC-5 Chinese hamster cell line, expressing P-glycoprotein-mediated MDR, the cytotoxicity of doxorubicin was substantially increased (EC_{50} 20.7 ± 3.8 nM), though anti-(*R*)-**5** alone did not show any cytotoxicity [6]. *gem*-Difluorocyclopropane anti-(*R*)-**5** exhibited an effect ~2-fold stronger than the corresponding nonfluorinated compound, *anti*-H-**5**, and ~3-fold stronger than MS-073 (see Figure 12.2) [6]. In addition, *anti*-(*R*)-**5** was found to show excellent stability in acidic medium ($t_{1/2}$ = >72 h at pH 2.0 and 37 °C) compared with MS-073 ($t_{1/2}$ = ~15 min at pH 2.0 and 37 °C). Among four stereoisomers, *anti*-(*R*)-**5** possesses the most potent activity. It is suggested that the superior activity of anti-isomers over syn-isomers may be ascribed to efficient π-stacking of the rigid *anti*-methanobenzosuberyl group with the phenylalanine residue extending outwards in the α-helical backbone of *P*-glycoprotein [6]. Barnett and co-workers recently reported an improved synthesis of *anti*-(*R*)-**5** [7].

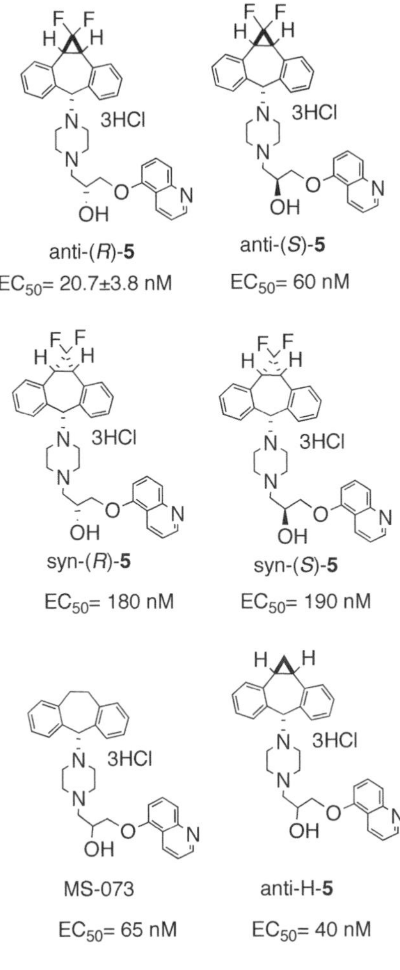

Figure 12.2 *Effect of the* gem-*difluorinated analogue of MS-073 on the proliferation of multidrug resistant CH^RC5 Chinese hamster cells in the presence of doxorubicin (1 μg/mL).*

Boger and Jenkins reported the synthesis of 9,9-difluoro-1,2,9,9*a*- tetrahydrocyclopropyl[*c*]benzo[*e*]indol-4-one (F_2CBI) and its trimethylindolcarbonyl derivatives such as F_2CBI-TMI (**6**) as a fluorinated analogue of CC-1065, duocarmycin SA, and duocarmycin A, which are potent antitumor antibiotics exhibiting their cytotoxicities based on sequence-selective DNA alkylation through the nucleophilic ring-opening of the cyclopropane moiety [8] (see Figure 12.3). It was found that the introduction of two fluorine atoms onto the cyclopropane ring brings about enhanced reactivity in the acid-catalyzed solvolysis (~500 times compared with that of nonfluorinated analogues) without altering the inherent regioselectivity. Single-crystal x-ray structure analysis of the difluoro

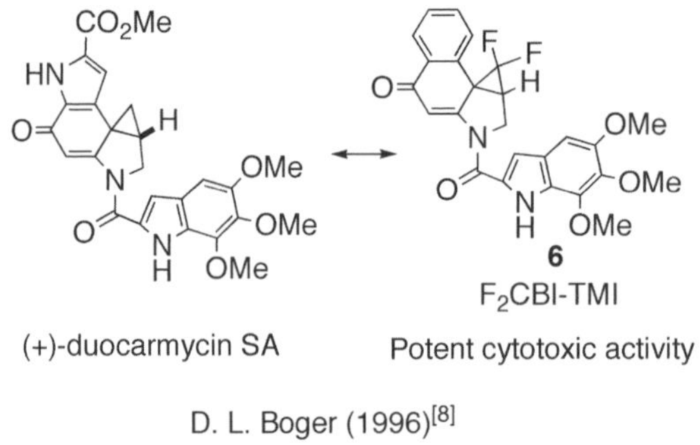

(+)-duocarmycin SA

6

F_2CBI-TMI

Potent cytotoxic activity

D. L. Boger (1996)[8]

Figure 12.3 *Biologically active* gem*-difluorocyclopropane F_2CBI as a duocarmycin analogue.*

analogues explained this enhanced reactivity and regioselectivity; that is, the cyclopropane C–CF_2–C bond angle was expanded and the carbon–carbon bond opposite to the difluoromethylene moiety was lengthened to induce additional strain energy. As a consequence, the stability of F_2CBI at pH 3 and 25 °C ($t_{1/2}$ = 4.2 h) was much lower than that of nonfluorinated counterpart, CBI, under the same conditions ($t_{1/2}$ = 930 h). The observed difference in the stability of the F_2CBI structure as compared to the CBI structure was reflected to the cytotoxicity of their derivatives. Thus, **6** (IC_{50} 36 nM) was ~1000 times less cytotoxic than CBI-TMI (IC_{50} 30 pM). Also, the introduction of the *gem*-difluorocyclopropane moiety showed no perceptible effect on the DNA alkylation selectivity with reduced efficiency. Thus, in this case, fluorinated analogues did not provide better biological activity than the nonfluorinated counterparts, although the introduction of fluorine induced substantial changes in chemical and biological properties.

Taguchi and co-workers reported the potential of *gem*-difluorocyclopropane as a source of biologically active compounds [9]. *gem*-Difluorinated methylenecyclopropane (F_2MCP) **7** was synthesized and its reactivity as a Michael acceptor was examined. The crossover reaction of F_2MCP **7** with 4-aminophenylthiol in the presence of the nonfluorinated methylenecyclopropane (MCP) **8** showed that F_2MCP **7** reacted exclusively with the thiol to give adduct **9** in excellent yield, while nonfluorinated MCP did not react with the thiol at all and **8** was recovered (see Figure 12.4) [9]. The result suggests that F_2MCP might act an irreversible inhibitor for certain enzymatic reactions such as acyl-CoA dehydrogenase-catalyzed reactions.

As described above, incorporation of fluorine into bioactive compounds is well recognized to be very effective means for the enhancement of the physiological properties of molecules [1–3]. Since fluorine is the most electronegative element, its substitution influences the acidity/basicity of neighboring groups and the lipophilicity of molecules (see Chapter 1). Due to the isoelectronic nature of fluorine to oxygen, a fluorine substituent can mimic a hydroxyl group in some biological responses [10]. Increased ring-strain energy might be another important characteristic of fluorine-substituted cyclopropanes [3a, 11]. Cyclopropane ring-opening reactions are important for the mechanism of cyclopropane-based enzyme inhibitors [12] and fluorine-substitution may enhance the

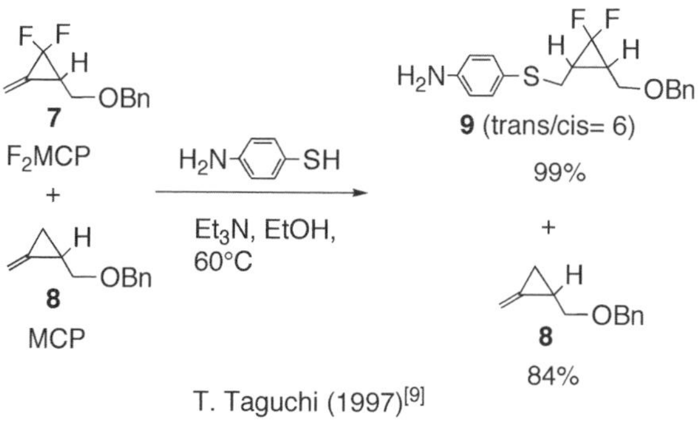

T. Taguchi (1997)[9]

Figure 12.4 *High reactivity of F_2MCP as a Michael acceptor.*

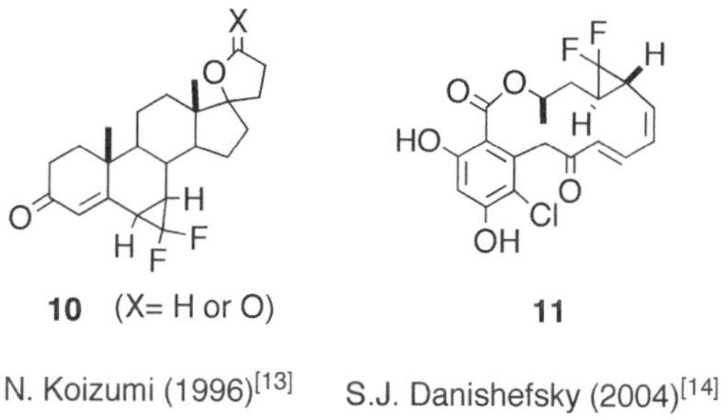

N. Koizumi (1996)[13] S.J. Danishefsky (2004)[14]

Figure 12.5 *Examples of biologically active* gem-*difluorocyclopropanes (complex molecules).*

inhibitory activity of these compounds. Examples are summarized in Figures 12.5–12.8.

Figure 12.5 shows two examples of *gem*-difluorocyclopropanes as a part of a rather complex natural product framework. Koizumi and co-workers synthesized *gem*-difluorocyclopropane **10** as an analogue of the antiandrogenic agent, KNP 215 [13] and examined its activity in terms of the suppressive effect on the androgen-stimulated weight gain. Compound **10** exhibited an activity similar to that of KNP 215. Danishefsky and co-workers reported the synthesis of the *gem*-difluorocyclopropane analogue of anticancer agent cycloproparadicicol **11**, but the potency of **11** was only ~1/185 of that of the parent compound [14].

Various biological activities have been reported for simple *gem*-difluorcyclopropane molecules (see Figure 12.6). Conformationally restricted analogues of glutamate, known as an excitatory neurotransmitter in mammals, have attracted much attention for identification of receptor subtype-specific agonists or antagonists (see Chapter 3). Introduction of a cyclopropane moiety to such biologically active substances is one of the key protocols in investigating an active conformer for the target receptor. It was anticipated that the introduction of fluorine would bring about little change in the steric size of the molecule but would enhance the acidity of the neighboring carbonyl group and decrease

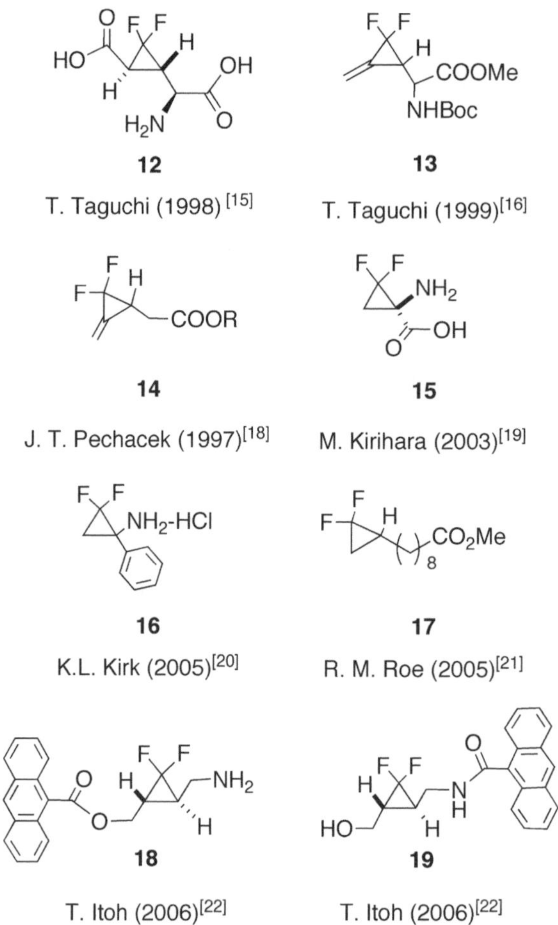

Figure 12.6 *Examples of biologically active* gem-*difluorocyclopropanes (simple molecules).*

the basicity of the amino group due to the strong electron-withdrawing nature of fluorine.

Taguchi and co-workers synthesized several biologically active *gem*-difluorocyclopropanes, such as metabotropic glutamate receptor agonist **12** [15] and methylene-*gem*-difluorocyclopropane **13** [16]. 2-(2-Carbohydroxy-3,3-difluorocyclopropyl)-glycines (F_2CCGs) were synthesized by means of a chiral auxiliary method (for F_2CCG-I **12** [15a, 15c]) as well as a chiral pool method (for all eight possible stereoisomers [15c]). Enhancement of acidity of the ω-carboxyl group by fluorine incorporation was confirmed by NMR titration experiments. Evaluation of the neuropharmacological activities of F_2CCGs in rat, compared with those of the corresponding cyclopropylglycine isomers, revealed that the (2*S*,1′*S*,2′*S*)-isomer, L-F_2CCG-I **12**, was a potent agonist for metabotropic glutamate receptors, which induced a priming effect on α-aminopimelate and L-glutamate [15c, 15d, 5e]. Methylenecyclopropylglycine (MCPG) and its metabolite, MCPF-CoA, are known to have inhibitory activity against enoyl-CoA hydratase (crotonase), responsible for fatty acid metabolism [17]. Four possible stereoisomers of methylene-*gem*-difluorocyclopropylglycines **13** (F_2MCPG) were synthesized [16], although their biological activities were not reported.

Barger and co-workers also synthesized F_2MCP derivatives **14** and evaluated their activities against the nematode pest *Meloidogyne incognite* [18]. Unfortunately, compounds **14** were inactive up to a concentration of 400 ppm, while the corresponding non-fluorinated compound (R = H) showed 100% inhibition of the microbe at only 7.8 ppm [18]. Kirihara *et al.* reported a highly efficient method for the synthesis of the optically active *gem*-difluorocyclopropane analogue **15** of 1-aminocyclopropane-1-carboxylic acid using a chemo-enzymatic strategy [19]. Kirk, Haufe, and their co-workers synthesized 2,2-difluoro-1-phenylcyclopropylamine (**16**) and evaluated its inhibitory activity for flavin-containing monoamine oxidase (MAO). Unfortunately, no inhibitory effect was observed and it was concluded that monofluorine substitution was essential to cause MAO inhibition [20]. Recently, Roe *et al.* reported an interesting biologically active simple *gem*-difluorocyclopropane **17**, which acted as a competitive inhibitor of the juvenile hormone-epoxide hydrolase [21].

We developed a synthetic methodology for preparing optically active *gem*-difluorocyclopropane building blocks using lipase technology [4]. Using this methodology, *gem*-difluorocyclopropanes **18** and **19**, bearing a 9-anthracenecarbonyl group, were synthesized. These compounds showed a strong DNA-cleavage property switched on by photoirradiation [22] (see Figure 12.6).

Nucleoside analogues have been recognized as potential chemotherapeutic agents and in fact several carbocyclic nucleosides exhibit potent anti-HIV activities. Accordingly, *gem*-difluorocyclopropane-containing nucleoside analogues and mimics have attracted substantial interest and various compounds have been synthesized (see Figures 12.7 and 12.8). Csuk pioneered this field and his group prepared a good number of *gem*-difluorocyclopropane-containing nucleoside mimics as represented by compounds **20–25** [23], some which were found to possess antitumor activity and anti-HIV activity. Although all compounds reported by Csuk's group are racemic, optically active compounds could be prepared using the enzymatic resolution technology developed by us [4] and by Kirihara's group [19].

Zemlicka and co-workers synthesized methylene-*gem*-difluorocyclopropane mimics of nucleoside and found that the *Z*-isomer of adenine-9-yl analogue **26** exhibited anticancer activity against leukemia and solid tumor cell lines *in vitro* (see Figure 12.8) [24]. Adenin-9-yl, 2-amino-6-chloropurin-9-yl, and guanin-9-yl analogues of **26** were synthesized and evaluated for their activities against HCMV, HSV-1, HSV-2, EBV, VZV, and HIV-1. However, only the *E*-isomer of adeninyl analogue **26** showed moderate antiviral activity against the Towne strain of HCMV propagated in human foreskin fibroblast (HFF) cells (EC_{50} 21 μM), which was noncyctotoxic at this concentration [24]. Recently, Robins and co-workers reported the synthesis of the *gem*-difluorocyclopropane analogues of nucleosides such as **27** and **28** (see Figure 12.8) [25], but no biological activities of these compounds have yet been disclosed.

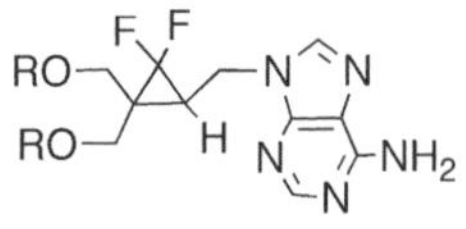

20
R. Csuk (1998)[23a]

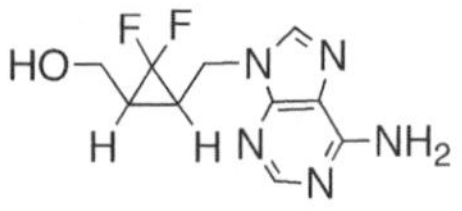

21
R. Csuk (2003)[23b]

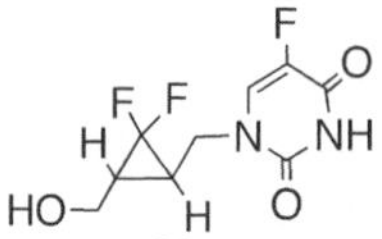

22
R. Csuk (2003)[23b]

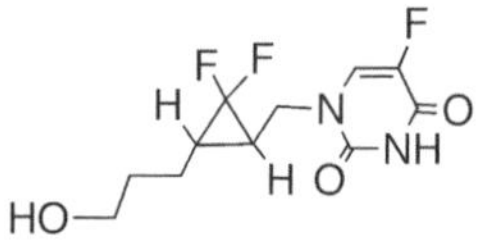

23
R. Csuk (2003)[23c]

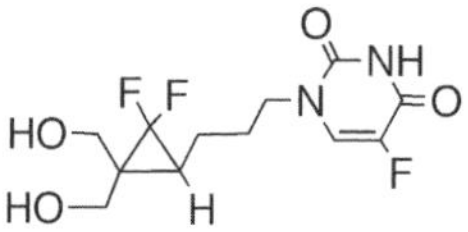

24
R. Csuk (2003)[23d]

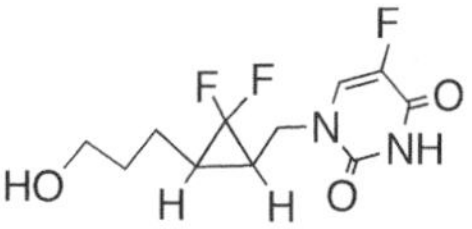

25
R. Csuk (2003)[23e]

Figure 12.7 *Nucleoside-type biologically active* gem-*difluorocyclopropanes prepared by the Csuk group.*

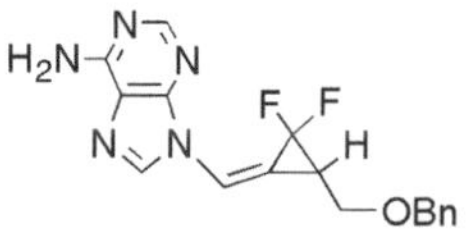

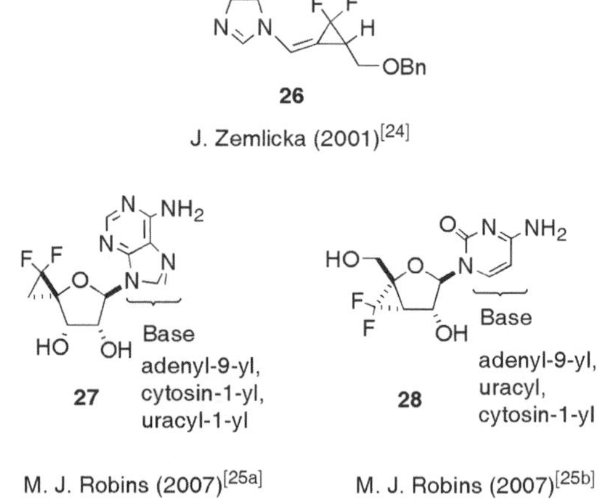

Figure 12.8 *Nucleoside-type biologically active* gem-*difluorocyclopropanes prepared by the Zemlick and Robins group.*

12.3 Synthesis of *gem*-Difluorocyclopropanes

Three synthetic methods for *gem*-difluorcyclopropane derivatives have been developed to date. Shibuya and Taguchi reported an efficient route to optically active *gem*-difluorocyclopropane derivative **30a** through diastereoselective Michael addition of Evans enolate **29a** to 4-bromo-4,4-difluorobut-2-enoate, followed by Et_3B-mediated radical coupling (see equation 1, Scheme 12.1) [15a]. Alternatively, for the synthesis of metabotropic glutamate receptor agonist F_2CCG-I, the Michael acceptor having the Evans oxazolidinone as the chiral auxiliary was reacted with *N*-diphenylmethyleneglycinate **29b** in DMF to give difluorocyclopropylglycine derivative **30b** with an excellent diastereoselectivity (>95% de) (see equation 2, Scheme 12.1) [15c].

The most versatile method for constructing *gem*-difluorocyclopropanes is the carbine-addition reaction to olefins, which includes two types as shown in Scheme 12.2 (type A and type B). Boger and Jenkins synthesized the *gem*-difluorocyclopropane moiety through intramolecular metal-carbenoid addition to 1,1-difluoroalkene group (type A). Thus, diazoquinone **31** was treated with 0.1–0.2 equivalents of $Rh_2(OAc)_4$ in toluene at 110 °C to give *gem*-difluorocyclopropane **32** in good yield [8] (see equation 3 in Scheme 12.2). The most widely used method for preparing *gem*-difluorocyclopropanes is the difluorocarbene addition to olefins (type B), which is applicable to large-scale preparation. Although numerous reactions for generating difluorocarbene have been reported, three major methods are shown under type B in Scheme 12.2 [26–29]: (a) pyrolysis of chlorodifluoroacetate [26] or trimethylsilyl fluorosulfonyldifluoroacetate (TFDA), known as Dolbier's reagent [27]; (b) generation from dibromodifluoromethane following Barton's

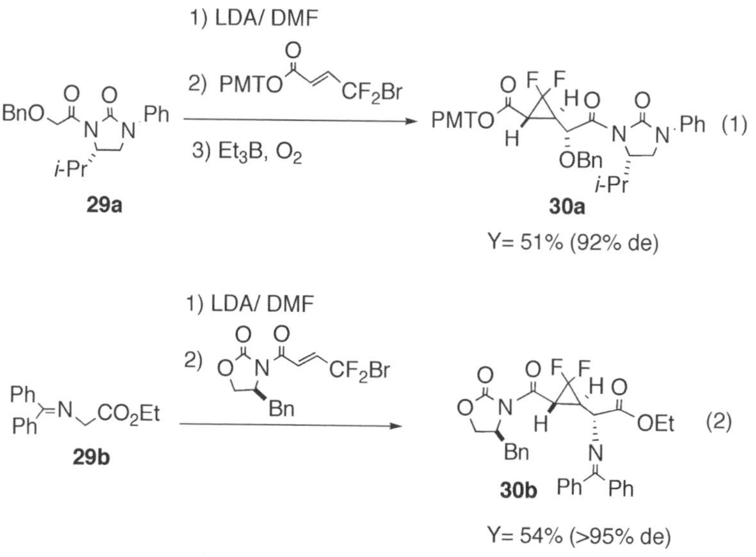

1) LDA/ DMF
2) PMTO(C=O)CH=CHCF$_2$Br
3) Et_3B, O_2

29a → **30a** (1)

Y= 51% (92% de)

1) LDA/ DMF
2) oxazolidinone-N-C(=O)CH=CHCF$_2$Br (Bn)

29b → **30b** (2)

Y= 54% (>95% de)

Scheme 12.1 *Synthesis of gem-difluorocyclopropane through diastereoselective Michael reaction followed by radical coupling.*

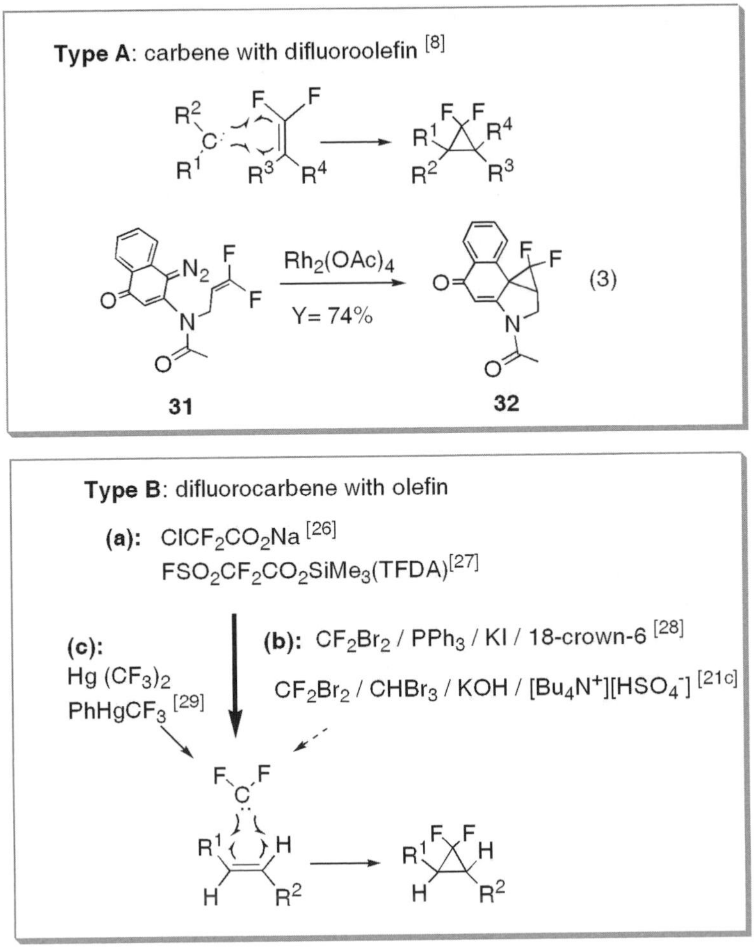

Type A: carbene with difluoroolefin [8]

Rh$_2$(OAc)$_4$

Y= 74%

(3)

31 **32**

Type B: difluorocarbene with olefin

(a): ClCF$_2$CO$_2$Na [26]
FSO$_2$CF$_2$CO$_2$SiMe$_3$(TFDA) [27]

(b): CF$_2$Br$_2$ / PPh$_3$ / KI / 18-crown-6 [28]
CF$_2$Br$_2$ / CHBr$_3$ / KOH / [Bu$_4$N$^+$][HSO$_4^-$] [21c]

(c):
Hg (CF$_3$)$_2$
PhHgCF$_3$ [29]

Scheme 12.2 *Preparation of* gem-*difluorocyclopropanes through addition of difluorocarbene to olefin.*

or Schlosser's procedure [28]; and (c) generation from trifluoromethylmercury compounds, known as Seyferth's reagents [29]. Of these methods, (b) and (c) can generate difluorocarbenes under mild conditions. Unfortunately, CF$_2$Br$_2$ and Seyferth's reagent are no longer commercially available because of their inherently hazardous properties. Therefore, ClCF$_2$COONa is currently the sole commercial source for difluorocarbene generation. Since this reaction requires harsh conditions (180–190 °C in diglyme), development of a new and efficient method is desirable for the generation of difluorocarbene under mild conditions.

Scheme 12.3 summarizes typical examples for the synthesis of *gem*-difluorocyclopropanes via difluorocarbene addition to olefins. Although high temperature is necessary to generate difluorocarbene by pyrolysis of ClCF$_2$COONa, addition of difluorocarbene to carbon–carbon double bonds proceeds in a stereospecific cis manner

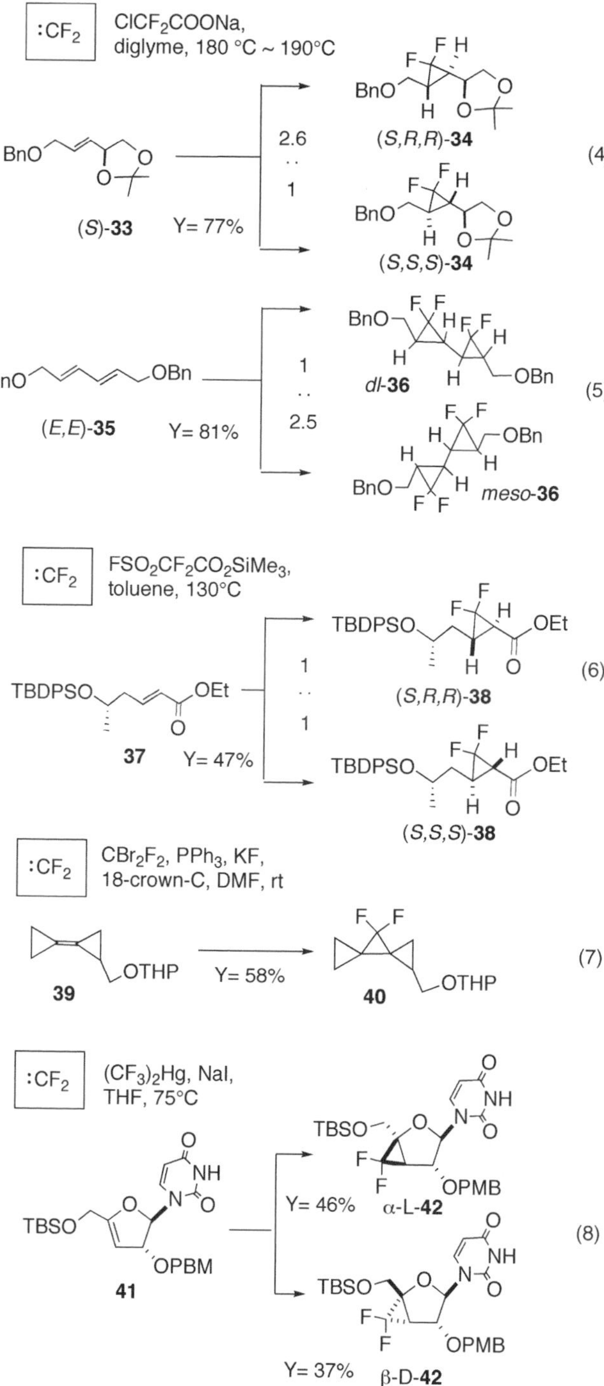

Scheme 12.3 *Typical examples of the synthesis of* gem-*difluorocyclopropanes using difluorocarbene addition to olefin.*

[19a]. Taguchi and co-workers prepared optically pure *gem*-difluorocyclopropane **34** through addition of difluorocarbene to optically active olefin (*S*)-**33**, followed by chromatographic separation of the resulting diastereomers, (*S,R,R*)-**34** and (*S,S,S*)-**34** (2.6 : 1, 77% yield) (see equation 4, Scheme 12.3) [15a]. We prepared novel bis-*gem*-difluorocyclopropanes **36** as a mixture of *dl*-**36** and *meso*-**36** (1 : 1.5, 81% yield) through difluorocarbene addition to (*E,E*)-diene **35** [4b] (see equation 5, Scheme 12.3). Generally, pyrolysis of trimethylsilyl fluorosulfonyldifluoroacetate (TFDA) proceeds under milder conditions (130 °C) as shown in equation 6, Scheme 12.3 [14]. However, TFDA is available only in a few countries at present. Difluorocarbene can be generated at room temperature under the Barton–Schlosser conditions (CF_2Br_2, PPh_3, KF, and 18-crown-6). Interesting spiro-*gem*-difluorocyclopropane **40** was synthesized by de Meijere and co-workers using this method (see equation 7, Scheme 12.3) [30]. To the best of our knowledge, this is the most efficient method for preparation of *gem*-difluorocyclopropanes. However, as mentioned above, it is impossible to obtain hazardous CF_2Br_2 commercially. For experienced researchers using appropriate safety precautions, toxic Seyferth's reagent is a useful source for preparing *gem*-difluorocyclopropanes. Robins and co-workers synthesized *gem*-difluorocyclopropane nucleoside **42** using this method because substrate **41** was not tolerant of high temperature (see equation 8, Scheme 12.3) [25].

12.4 Synthesis of Optically Active *gem*-Difluorocyclopropanes via Enzymatic Resolution

Enzymatic resolution has been successfully applied to the preparation of optically active *gem*-difluorocyclopropanes (see Scheme 12.4). We succeeded in the first optical resolution of racemic *gem*-difluorocyclopropane diacetate, *trans*-**43**, through lipase-catalyzed enantiomer-specific hydrolysis to give (*R,R*)-(−)-**44** with >99% ee (see equation 9, Scheme 12.4) [4a]. We also applied lipase-catalyzed optical resolution to an efficient preparation of monoacetate *cis*-**46** from prochiral diacetate *cis*-**45** (see equation 10, Scheme 12.4) [4a]. Kirihara *et al.* reported the successful desymmetrization of diacetate **47** by lipase-catalyzed enantiomer-selective hydrolysis to afford monoacetate (*R*)-**48**, which was further transformed to enantiopure amino acid **15** (see equation 11, Scheme 12.4) [19]. We demonstrated that the lipase-catalyzed enantiomer-specific hydrolysis was useful for bis-*gem*-difluorocyclopropane **49**. Thus, optically pure diacetate (*R,S,S,R*)-**49** and (*S,R,R,S*)-diol **50**, were obtained in good yields, while meso-**49** was converted to the single monoacetate enantiomer (*R,S,R,S*)-**51** via efficient desymmetrization (see equation 12, Scheme 12.4) [4b, 4e]. Since these mono- and bis-*gem*-difluorocyclopropanes have two hydroxymethyl groups to modify, a variety of compounds can be prepared using them as building blocks [4, 22].

de Meijere and co-workers reported the synthesis of enantiopure 7,7-difluorodispiro [2.0.2.1]heptylmethanols (**54**) via difluorocyclopropanation of methylenespirobiscyclopropane (1*S*,3*R*)-**53** (>99% ee), which was obtained by lipase-catalyzed enantiomer-specific transesterification of (±)-**52**, followed by separation of two diastereomers (see equation 13, Scheme 12.4) [30].

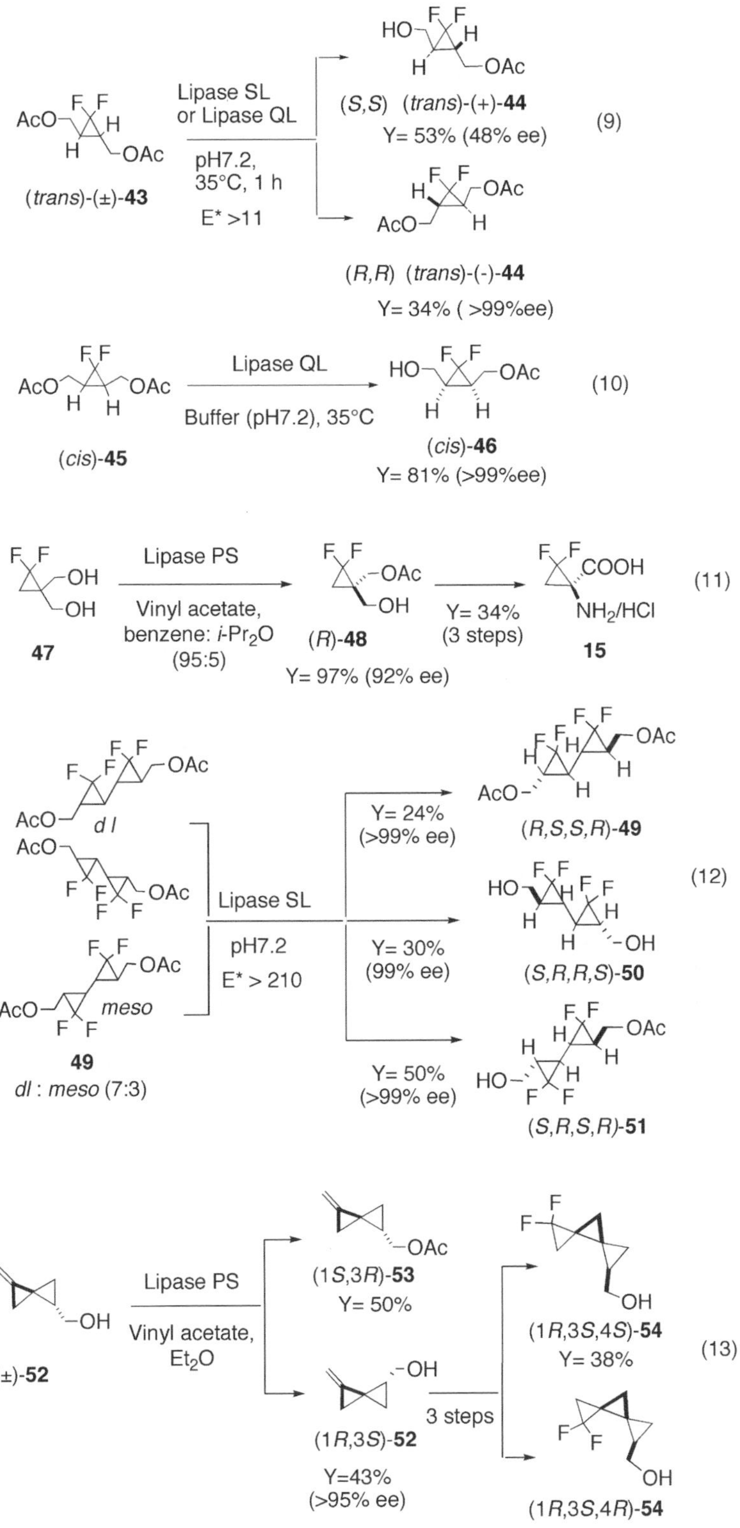

Scheme 12.4 *Preparation of optically active gem-difluorocyclopropanes using enzymatic resolution.*

12.5 Biologically Active *gem*-Difluorocyclopropanes: Anthracene–*gem*-Difluorocyclopropane Hybrids as DNA Cleavage Agents Switched on by Photoirradiation

As described above, *gem*-difluorocyclopropane analogues of structurally complex biologically active natural products reported to date have not shown enhanced activities [8, 14], while rather simple synthetic *gem*-difluorocyclopropanes have shown interesting biological activities [15, 16, 18–22]. We recently reported the synthesis of *gem*-difluorocyclopropanes that showed a strong DNA-cleaving property switched on by photoirradiation [22].

During the course of determining the stereochemistry of 1,3-bishydroxymethyl-2,2-difluorocyclopropane on the basis of CD (circular dischroism) spectroscopic analysis of the corresponding 9-anthracenecarboxylic acid diester **55**, we recognized the interesting fact that diester **55** was so unstable in dichloromethane solution under sunlight that it formed unidentified complex compounds by opening its *gem*-difluorocyclopropane ring [4g]. We also found that, similar to this mono-difluorocyclopropane **55**, the anthracenecarboxylic acid diester of bis-*gem*-difluorocyclopropane was rapidly decomposed by exposure to sunlight [4c]. It has been shown that anthracene or anthraquinone derivatives serve as potent DNA-cleavage agents [31–34]. For example, Schuster and co-workers reported that anthraquinonecarboxamide caused GG-selective cleavage of duplex DNA [32], while Kumar and Punzalan found that anthracene-substituted alkylamine derivatives possessed potent DNA-cleaving activity [33]. Toshima *et al.* reported that quinoxaline-carbohydrate hybrids acted as GG-selective DNA-cleaving or DNA-binding agents, depending on the structure of the sugar moiety [34]. From these results, we anticipated that our anthracene-*gem*-difluorocyclopropane hybrids would possess DNA-cleaving activity on photoirradiation (see Figure 12.9). It has been shown that decomposition of a *gem*-difluorocyclopropane ring proceeds via a radical species [35], which may cause DNA-damage [36]. We prepared optically active anthracenecarbonyl derivatives of *gem*-difluorocyclopropane, **19, 55**, and **56**, and evaluated their DNA-cleaving activities. As shown in Figure 12.9, (*S,S*)-**55** (lane II) and (*S,S*)-**56** (lane IV) showed only weak DNA-cleaving activities, while much stronger activity was exhibited by anthracenecarboxamidomethyl-*gem*-difluorocyclopropane **19**. In the presence of 100 μM of (*S,S*)-**19**, which corresponds to 1.3 molar equivalents to a DNA base pair, DNA was cleaved by photoirradiation and supercoiled ϕX174 DNA (Form I) was converted to the relaxed form (Form II) (see Figure 12.9, lane V). No significant difference in activity was observed between the enantiomers of **19**, i.e., (*S,S*)-**19** (lane V) and (*R,R*)-**19** (lane IV). Although nonfluorinated cyclopropane counterpart (*S,S*)-**57** also caused substantial DNA-cleavage (lane VI), the activity of *gem*-difluorocyclopropane (*S,S*)-**19** was obviously superior to that of (*S,S*)-**57** at the same concentration. However, no sequence specificity was observed in these DNA-cleavage experiments.

We also found an interesting difference in DNA-cleaving activity between the enantiomers of aminomethyl-*gem*-difluorocyclopropylmethyl anthracenecarboxylate, (*S,S*)-**18** and (*R,R*)-**18**. As Figure 12.10 shows, (*R,R*)-**18** exhibited ~10-fold stronger DNA-cleaving activity than (*S,S*)-**18** for supercoiled ϕX174 DNA. The result clearly indicates that the

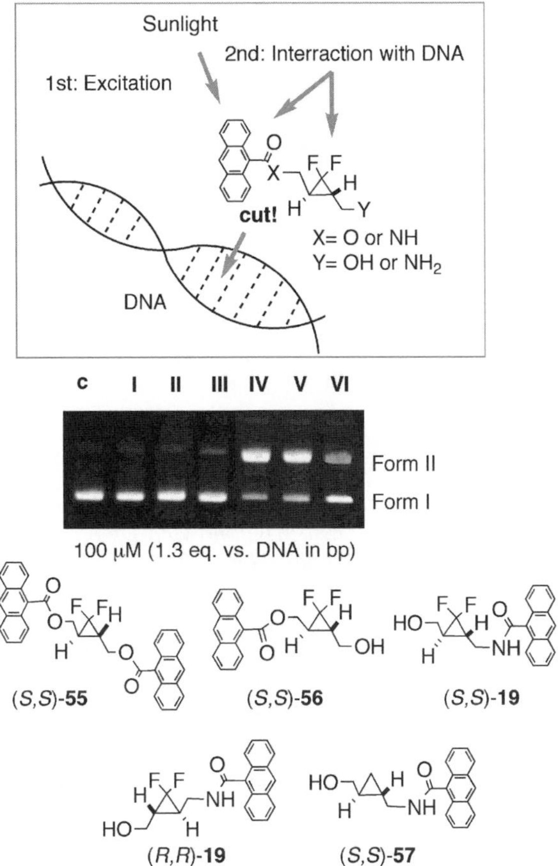

Figure 12.9 *Photocleavage of supercoiled ϕX174 plasmid DNA by cyclopropane derivatives. Lanes: c, control; I, ethyl 9-anthracenecarboxylate; II, (S,S)-***55***; III, (S,S)-***56***; IV, (R,R)-***19***; V, (S,S)-***19***; VI, (S,S)-***57***. Reaction buffer: 20 mM sodium phosphate pH 7.0 (20% DMSO) Samples were irradiated at 25 °C for 45 min at a distance of 6 cm using a xenon lamp with a polystyrene filter (365 nm). Electrophoresis was run using 0.8% agarose gel with TAE buffer at 100 V for 40 min. The gel was stained with EtBr and visualized by UV-B lamp (transilluminator).*

chirality of the *gem*-difluorocyclopropane moiety controls the DNA-cleaving activity of these anthracene–*gem*-difluorocyclopropane hybrids.

These results possibly suggest that the DNA-cleavage by these anthracene–*gem*-difluorocyclopropane hybrids is caused mainly by intercalation or minor-groove binding of the anthracenyl group to DNA and the *gem*-difluorocyclopropane component plays an important role in determining the DNA-binding mode. Although we found no differences in sequence specificity between (*S,S*)-**18** and (*R,R*)-**18**, we hypothesize that the terminal amino group may bind to the phosphate moiety of DNA, and then direct the intercalation

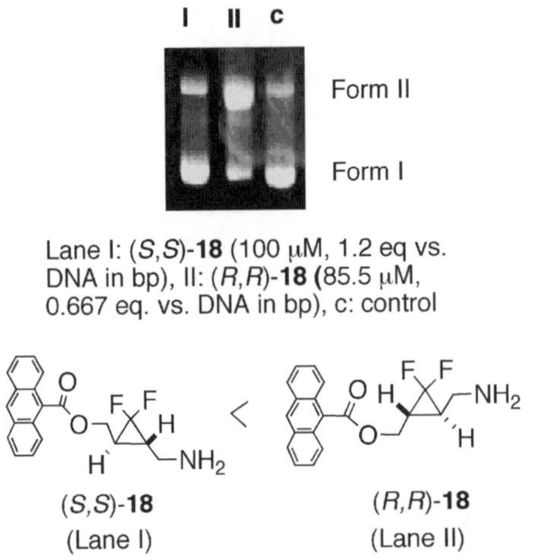

Figure 12.10 *Photocleavage of supercoiled ϕX174 plasmid DNA by cyclopropane derivatives (S,S)- and (R,R)-**18**. The (R,R)-isomer may bind with DNA more easily than the (S,S)-isomer via intercalative or groove binding. Reaction buffer: 25 mM sodium phosphate pH 7.0 (20% DMSO). Concentrations of photocleavers are 100 μM (1.2 equivalents of DNA in base pairs) for (S,S)-**18** and 85.5 μM (0.667 equivalents of DNA in base pairs) for (R,R)-**18**. Samples were irradiated at 25 °C for 40 min at a distance of 4 cm using a xenon lamp with a polystyrene filter (365 nm) Electrophoresis was run using 0.8% agarose gel with TAE buffer at 100 V for 40 min. The gel was stained with EtBr and visualized by UV-B lamp (transilluminator).*

or minor-groove binding of the anthracene moiety to certain DNA sequence (see Figure 12.11) [22b].

On the basis of these results, we set out to synthesize a molecular tool for designing novel DNA-markers. Thus, tetrakis(*gem*-difluorocyclopropane) **59** was prepared through homometathesis reaction of (*S,R,R,S*)-**58** using the Grubbs type II catalyst [37]. Olefin **59** was reacted with ethyl diazoacetate in the presence of $Cu(OAc)_2$ as catalyst in CH_2Cl_2 at 40 °C to give pentakis(cyclopropane) **60**. Hydrolysis and the subsequent esterification with pyrenylmethanol using EDC in the presence of DMAP afforded pyrenyl ester **61** as a mixture of diastereomers [22a] (see Scheme 12.5). Separation of diastereomers, derivatization, and evaluation of the activity of each isomer as sequence-specific DNA-marker is actively underway in our laboratory.

12.6 Conclusions and Outlook

Introduction of two fluorine atoms to a cyclopropane ring can alter its chemical reactivity and biological activity due to the strong electron-withdrawing nature of fluorine, and this can make it possible to create new molecules that exhibit unique biological activities or functions. As described above, numerous studies have been performed by focusing on preparing difluoro-mimics of biologically active cyclopropane-containing compounds. However, we found that rather simple 9-anthracenecarboxylic acid esters or amides of

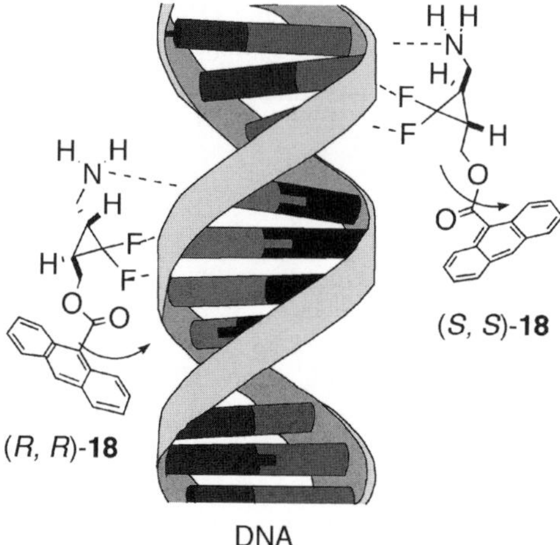

Figure 12.11 *Working hypothesis for the origin of difference in DNA-cleaving activities between (S,S)-**18** and (R,R)-**18**. The (R,R)-isomer may bind to DNA more tightly than the (S,S)-isomer via intercalation or minor-groove binding. See color Plate 12.11.*

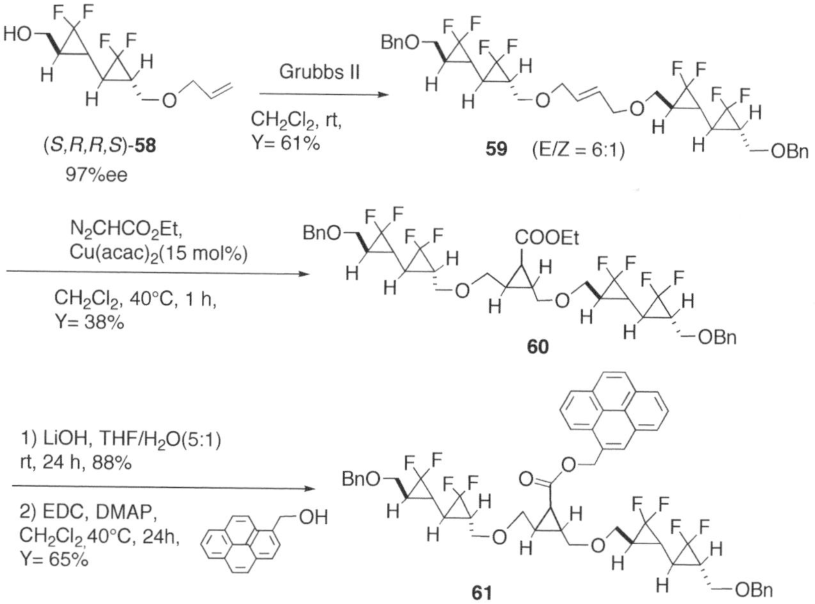

(S,R,R,S)-**58**
97%ee

Grubbs II
CH_2Cl_2, rt,
Y= 61%

59 (E/Z = 6:1)

N_2CHCO_2Et,
$Cu(acac)_2$(15 mol%)
CH_2Cl_2, 40°C, 1 h,
Y= 38%

60

1) LiOH, THF/H_2O(5:1)
rt, 24 h, 88%
2) EDC, DMAP,
CH_2Cl_2,40°C, 24h,
Y= 65%

61

Scheme 12.5 *Synthesis of a novel molecule 61 as a key building block for potential sequence-specific markers of DNA.*

gem-difluorocyclopropanes possess considerable DNA-cleaving activity that is switched on by photoirradiation. This finding suggests that *gem*-difluorocyclopropane-containing compounds may exhibit singular biological activities and that the *gem*-difluorocyclopropane structure may serve as a useful building block for a variety of unique functional molecules.

Acknowledgments

I am grateful to Dr. Keiko Ninomiya of Kyoto University for performing the DNA-cleavage assays for our anthracene–*gem*-difluorcyclopropane hybrids. I am also grateful to Professor Masahiko Sisido of Okayama University for helpful discussions throughout this work. I gratefully acknowledges my students, in particular, Dr. Koichi Ishida (Mitsukura) who is now an Assistant Professor at the Gifu University, Miss Nanane Ishida, Mr. Kuhihiko Tanimoto, and Mr. Manabu Kanbara, for their contributions to this work.

References

1. For reviews see: (a) De Meijere, A., Rademacher, P., Lebel, H., *et al.* (2003) Cyclopropanes and related rings. *Chem. Rev.*, **103**, 931–1448, 1603–1648. (b) Tozer, M. J. and Herpin, T. F. (1996) Methods for the synthesis of *gem*-difluoromethylene compounds. *Tetrahedron*, **52**, 8619–8683. (c) Wong, H. N. C., Hon, M-Y., Tse, C-H. and Yip, Y-C. (1989) Use of cyclopropanes and their derivatives in organic synthesis. *Chem. Rev.*, **89**, 165–198.
2. For reviews see. (a) Welch, J. T. (1987) Tetrahedron report number 221: Advances in the preparation of biologically active organofluorine compounds. *Tetrahedron*, **43**, 3123–3197. (b) Resnati, G. (1993) Synthesis of chiral and bioactive fluoroorganic compounds, *Tetrahedron*, **49**, 9385–9445. (c) Soloshonok, V. A. (ed.) (1999) *Enantiocontrolled Synthesis of Fluoroorganic Compounds: Stereochemical Challenges and Biomedicinal Targets*, John Wiley & Sons, Ltd, Chichester, UK. (d) Rosen, T. C., Yoshida, S., Kirk, K. L. and Haufe, G. (2004) Fluorinated phenylcyclopropylamines as inhibitors of monoamine oxidases. *ChemBioChem.*, **5**, 1033–1043.
3. (a) Dolbier, W. R., Jr. and Battiste, M. A. (2003) Structure, synthesis, and chemical reactions of fluorinated cyclopropanes and cyclopropenes. *Chem. Rev.*, **103**, 1071–1098. (b) Itoh, T. (2006) Optically active *gem*-difluorocyclopropanes; synthesis and application for novel molecular materials, current fluoroorganic chemistry, in *New Synthetic Directions, Technologies, Materials and Biological Applications*, ACS Symposium Series 949, Oxford University Press/ American Chemical Society, Washington D.C., Chapter 21, pp. 352–362 and references cited therein.
4. Examples for preparing optically active *gem*-difluorocyclopropanes. (a) Itoh, T., Mitsukura, K. and Furutani, M. (1998) Efficient synthesis of enantiopure 1,2-bis(hydroxymethyl)-3,3-difluorocyclopropane derivatives through lipase-catalyzed reaction. *Chem. Lett.*, 903–904. (b) Mitsukura, K., Korekiyo, S. and Itoh, T. (1999) Synthesis of optically active bisdifluorocyclopropanes through a chemo-enzymatic reaction strategy. *Tetrahedron Lett.*, **40**, 5739–5742. (c) Itoh, T., Mitsukura, K., Ishida, N. and Uneyama, K. (2000) Synthesis of bis- and oligo-*gem*-

difluorocyclopropanes using the olefin metathesis reaction, *Org. Lett.*, **2**, 1431–1434. (d) Itoh, T. (2000) Asymmetric synthesis of novel gem-difluorinated compounds using chemo-enzymatic methodology. *J. Synth. Org. Chem. Jpn.*, **58**, 316–326. (e) Itoh, T., Ishida, N., Mitsukura, K. and Uneyama, K. (2001) Synthesis of novel bis- and oligo-*gem*-difluorocyclopropanes. *J. Fluorine Chem.*, **112**, 63–69. (f) Itoh, T., Ishida, N., Ohashi, M., *et al.* (2003) Synthesis of novel liquid crystals which possess *gem*-difluorocyclopropane moieties, *Chem. Lett.*, **32**, 494–495. (g) Itoh, T., Ishida, N., Mitsukura, K., *et al.* (2004) Synthesis of optically active *gem*-difluorocyclopropanes through a chemo-enzymatic reaction strategy. *J. Fluorine Chem.*, **125**, 775–783.

5. (a) Holan, G., Johnson, W. M., Virgona, C. T. and Walser, R. A. (1986) Synthesis and biological activity of DDT-pyrethroid insecticides. *J. Agric. Food Chem.*, **34**, 520–524. (b) Wegner, P., Joppien, H., Hoemberger, G. and Koehn, A. (1989) European Patent Application, EP 318425 A1; CAN111:169387.
6. Pfister, J. R., Makra, F., Muehldorf, A. V., *et al.* (1995) Methanodibenzosuberylpiperazines as potent multidrug resistance reversal agents. *Bioorg. Med. Chem. Lett.*, **5**, 2473–2476.
7. Barnett, C. J., Huff, B., Kobierski, M. E., *et al.* (2004) Stereochemistry of C-6 nucleophilic displacements on 1,1-difluorocyclopropyldibenzosuberanyl substrates. An improved synthesis of multidrug resistance modulator LY335979 trihydrochloride. *J. Org. Chem.*, **69**, 7653–7660.
8. Boger, D. L. and Jenkins, T. J. (1996) Synthesis, X-ray structure, and properties of fluorocyclopropane analogs of the Duocarmycins incorporating the 9,9-difluoro-1,2,9,9a-tetrahydrocyclopropa[*c*]benzo[*e*]indol-4-one (F_2CBI) alkylation subunit. *J. Am. Chem. Soc.*, **118**, 8860–8870.
9. Taguchi, T., Kurishita, M., Shibuya, A. and Aso, K. (1997) Preparation of methylene-*gem*-difluorocyclopropanes and its reactivity as Michael acceptor. *Tetrahedron*, **53**, 9497–9508.
10. Haufe, G., Rosen, T. C., Meyer, O. G. J., *et al.* (2002) Synthesis, reactions and structural features of monofluorinated cyclopropanecarboxylates. *J. Fluorine Chem.*, **114**, 189–198.
11. For a review, see: Fedoryski, M. (2003) Syntheses of *gem*-dihalocyclopropanes and their use in organic synthesis. *Chem. Rev.*, **103**, 1099–1132.
12. Baker, G. B., Urichuk, L. J., McKenna, K. F. and Kennedy, S. H. (1999) Metabolism of monoamine oxidase inhibitors. *Cell. Mol. Neurobiol.*, **19**, 411–426.
13. Koizumi, N., Takegawa, S., Mieda, M. and Shibata, K. (1996) Antiandrogen. IV. C-17 Spiro 2-oxasteroids. *Chem. Pharm. Bull.*, **44**, 2162–2164.
14. Yang, Z-Q., Geng, X., Solit, D., *et al.* (2004) New efficient synthesis of resorcinylic macrolides via ynolides: Establishment of cycloproparadicicol as synthetically feasible preclinical anticancer agent based on Hsp90 as the target. *J. Am. Chem. Soc.*, **126**, 7881–7889.
15. (a) Taguchi, T., Shibuya, A., Sasaki, H., *et al.* (1994) Asymmetric synthesis of difluorocyclopropanes. *Tetrahedron Asymmetry*, **5**, 1423–1426. (b) Shibuya, A., Kurishita, M., Ago, C. and Taguchi, T. (1996) A highly diastereoselective synthesis of *trans*-3,4-(difluoromethano)glutamic acid. *Tetrahedron*, **52**, 271–278. (c) Shibuya, A., Sato, A. and Taguchi, T. (1998) Preparation of difluoro analogs of CCGs and their pharmacological evaluations. *Bioorg. Med. Chem. Lett.*, **8**, 1979–1984. (d) Saitoh, T., Ishida, M. and Shinozaki, A. (1998) Potentiation by DL-α-aminopimelate of the inhibitory action of a novel mGluR agonist (L-F_2CCG-I) on monosynaptic excitation in the rat spinal cord. *Br. J. Pharmacol.*, **123**, 771–779. (e) Ishida, M. and Shinozaki, H. (1999) Inhibition of uptake and release of a novel mGluR agonist (L-F_2CCG-I) by anion transport blockers in the rat spinal cord. *Neuropharmacology*, **38**, 1531–1541.
16. (a) Taguchi, T., Kurishita, M., Shibuya, A. and Aso, K. (1997) Preparation of methylene-*gem*-difluorocyclopropanes and its reactivity as Michael acceptor. *Tetrahedron*, **53**, 9497–9508.

(b) Taguchi, T., Kurishita, M. and Shibuya, A. (1999) Preparation of F_2MCPGs via selenoxide elimination. *J. Fluorine Chem.*, **97**, 157–159. (c) Shibuya, A., Okada, M., Nakamura, Y., *et al.* (1999) Preparation of methylenedifluorocyclopropanes *via* cyclopropyl anion promoted β-elimination. *Tetrahedron*, **55**, 10325–10340.

17. (a) Lai, M-T., Liu, L-D. and Liu, H-W. (1991) Mechanistic study on the inactivation of general acyl-CoA dehydrogenase by a metabolite of hypoglycin A. *J. Am. Chem. Soc.*, **113**, 7388–7397. (b) Li, D., Guo, Z. and Liu, H.-W. (1996) Mechanistic studies of the inactivation of crotonase by (methylenecyclopropyl)formyl-CoA. *J. Am. Chem. Soc.*, **118**, 275–276.
18. Pechacek, J. T., Bargar, T. M., Sabol, M. R. (1997) Inhibition of nematode induced root damage by derivatives of methylenecyclopropane acetic acid. *Bioorg. Med. Chem. Lett.*, **7**, 2665–2668.
19. (a) Kirihara, M., Takuwa, T., Kawasaki, M., *et al.* (1999) Synthesis of (+)-(*R*)-1-amino-2,2-difluorocyclopropane-1-carboxylic acid through lipase-catalyzed asymmetric acetylation. *Chem. Lett.*, 405–406. (b) Kirihara, M., Kawasaki, M., Takuwa, T., *et al.* (2003) Efficient synthesis of (*R*)- and (*S*)-1-amino-2,2-difluorocyclopropanecarboxylic acid via lipase-catalyzed desymmetrization of prochiral precursors. *Tetrahedron Asymmetry*, **14**, 1753–1761.
20. Song, Y., Yoshida, S., Fröhlich, R., *et al.* (2005) Fluorinated phenylcyclopropylamines. Part 4: Effects of aryl substituents and stereochemistry on the inhibition of monoamine oxidases by 1-aryl-2-fluoro-cyclopropylamines. *Bioorg. Med. Chem.*, **13**, 2489–2499.
21. Roe, R., Kallapur, V., Linderman, R. J. and Viviani, F. (2005) Organic synthesis and bioassay of novel inhibitors of JH III epoxide hydrolase activity from fifth stadium cabbage loopers, *Trichoplusia ni. Pestic. Biochem. Physiol.*, **83**, 140–154.
22. (a) Itoh, T. (2005) Optically active *gem*-difluorocyclopropanes; synthesis and application for novel molecular materials, in *Fluorine-Containing Synthons*, ACS Symposium Series 911, Oxford University Press/American Chemical Society, Washington D.C., Chapter 25, pp. 430–439. (b) Ninomiya, K., Tanimoto, K., Ishida, N., *et al.* (2006) Synthesis of novel gem-difluorinatedcyclopropane hybrids: applications for DNA cleavage agents switched by photo irradiation. *J. Fluorine Chem.*, **127**, 651–656.
23. (a) Csuk, R. and Eversmann, L. (1998) Synthesis of difluorocyclopropyl carbocyclic homo-nucleosides. *Tetrahedron*, **54**, 6445–6459. (b) Csuk, R. and Thiede, G. (1999) Preparation of novel difluorocyclopropane nucleosides. *Tetrahedron*, **55**, 739–750. (c) Csuk, R. and Eversmann, L. (2003) Synthesis of *trans*-configurated spacered nucleoside analogues comprising a difluorocyclopropane moiety. *Z. Naturforsch., B: Chem. Sci.*, **58**, 997–1004. (d) Csuk, R. and Thiede, G. (2003) Synthesis of spacered nucleoside analogues comprising a difluorocycloprop-nae moiety. *Z. Naturforsch., B: Chem. Sci.*, **58**, 853–860. (e) Csuk, R. and Eversmann, L. (2003) Synthese flexibler difluorieter cyclopropanoider nucleosidanaloga. *Z. Naturforsch., B: Chem. Sci.*, **58**, 1176–1186.
24. Wang, R., Ksebati, M. B., Corbett, T. H., *et al.* (2001) Methylene-*gem*-difluorocyclopropane analogues of nucleosides: Synthesis, cyclopropene-methylenecyclopropane rearrangement, and biological activity. *J. Med. Chem.*, **44**, 4019–4022.
25. (a) Nowak, I. and Robins, M. J. (2006) Addition of difluorocarbene to 4′,5′-unsaturated nucleosides: Synthesis and deoxygenation reactions of difluorospirocyclopropane nucleosides. *J. Org. Chem.*, **71**, 8876–8883. (b) Nowak, I., Cannon, J. F. and Robins, M. J. (2007) Addition of difluorocarbene to 3′,4′-unsaturated nucleosides: Synthesis of 2′-deoxy analogues with a 2-oxabicyclo[3.1.0]hexane framework. *J. Org. Chem.*, **72**, 532–537. (c) Nowak, I. and Robins, M. J. (2007) Synthesis of 3′-deoxynucleosides with 2-oxabicyclo[3.1.0]hexane sugar moieties: Addition of difluorocarbene to a 3′,4′-unsaturated uridine derivative and 1,2-dihydrofurans derived from D- and L-xylose. *J. Org. Chem.*, **72**, 3319–3325.

26. Kobayashi, Y., Taguchi, T., Morikawa, T., *et al.* (1980) Ring opening reaction of gem-difluorocyclopropyl ketones with nucleophiles. *Tetrahedron Lett.*, **21**, 1047–1050 and references cited therein.
27. Tian, F., Kruger, V., Bautista, O., *et al.* (2000) A novel and highly efficient synthesis of *gem*-difluorocyclopropanes. *Org. Lett.*, **2**, 563–564.
28. (a) Burton, D. J. and Naae, D. G. (1973) Bromodifluoromethylphosphonium salts. Convenient source of difluorocarbene. *J. Am. Chem. Soc.*, **95**, 8467–8468. (b) Bessard, Y., Müller, U. and Schlosser, M. (1990) gem-Difluorocyclopropanes: an improved method for their preparation. *Tetrahedron*, **46**, 5213–5221.
29. (a) Birmingham, J. M., Seyferth, D. and Wilkinson, G. (1954) A new preparation of bis-cyclopentadienyl-metal compounds. *J. Am. Chem. Soc.*, **76**, 4179–4179. (b) Seyferth, D. and Hopper, S. P. (1972) Halomethyl metal compounds. LX. Phenyl(trifluoromethyl)mercury. Useful difluorocarbene transfer agent. *J. Org. Chem.*, **37**, 4070–4075. (c) Seyferth, D., Hopper, S. P. and Murphy, G. J. (1972) Halomethyl-metal compounds : LVII. A convenient synthesis of aryl (trifluoromethyl)mercury compounds. *J. Organomet. Chem.*, **46**, 201–209. (d) Seyferth, D., Dertouzos, H., Suzuki, R. and Mui, J. Y. P. (1967) Halomethyl-metal compounds. XIII. Preparation of gem-difluorocyclopropanes by iodide ion-induced CF_2 transfer from trimethyl (trifluoromethyl)tin. *J. Org. Chem.*, **32**, 2980–2984.
30. Miyazawa, K., Yufit, D. S., Howard, J. A. K. and de Meijere, A. (2000) Cyclopropyl building blocks for organic synthesis, 60. Synthesis and properties of optically active dispiro[2.0.2.1]heptane derivatives as novel ferroelectric liquid crystalline compounds. *Eur. J. Org. Chem.*, **2000**, 4109–4117.
31. For reviews, see: (a) Da Ros, T., Spalluto G., Boutorine, A. S., *et al.* (2001) DNA-photocleavage agents. *Curr Pharm Des.*, **7**, 1781–1821. (b) Armitage, B. (1998) Photocleavage of nucleic acids. *Chem. Rev.*, **98**, 1171–1200.
32. (a) Ly, D., Kan, Y., Armitage, B. and Schuster, G. B. (1996) Cleavage of DNA by irradiation of substituted anthraquinones: Intercalation promotes electron transfer and efficient reaction at GG steps. *J. Am. Chem. Soc.*, **118**, 8747–8748. (b) Gasper, S. M., Armitage, B., Shi, X., *et al.* (1998) Three-dimensional structure and reactivity of a photochemical cleavage agent bound to DNA. *J. Am. Chem. Soc.*, **120**, 12402–12409.
33. Kumar, C. V., Punzalan, E. H. A. and Tan, W. B. (2000) Adenine–thymine base pair recognition by an anthryl probe from the DNA minor groove. *Tetrahedron*, **56**, 7027–7040.
34. (a) Toshima, K., Takano, R., Maeda, Y., *et al.* (1999) 2-Phenylquinoline–carbohydrate hybrids: Molecular design, chemical synthesis, and evaluation of a new family of light-activatable DNA-cleaving agents. *Angew. Chem. Int.* Ed., **38**, 3733–3735. (b) Toshima, K., Maeda, Y., Ouchi, H., *et al.* (2000) Carbohydrate-modulated DNA photocleavage: design, synthesis, and evaluation of novel glycosyl anthraquinones. *Bioorg. Med. Chem. Lett.*, **10**, 2163–2165. (c) Toshima, K., Takai, S., Maeda, Y., *et al.* (2000) Chemical synthesis and DNA photo-cleavage of the intercalator-carbohydrate hybrid moiety of the neocarzinostatin chromophore. *Angew. Chem. Int.* Ed. **39**, 3656–3658. (d) Toshima, K., Takano, R., Ozawa, T. and Matsumura, S. (2002) Molecular design and evaluation of quinoxaline-carbohydrate hybrids as novel and efficient photo-induced GG-selective DNA cleaving agents. *Chem. Commun.*, 212–213.
35. Tian, F., Bartberger, M. D. and Dolbier, W. R., Jr. (1999) Density functional theory calculations of the effect of fluorine substitution on the kinetics of cyclopropylcarbinyl radical ring openings. *J. Org. Chem.*, **64**, 540–545.
36. (a) Helissey, P., Bailly, C., Vishwakarma, J. N., *et al.* (1996) DNA minor groove cleaving agents: synthesis, binding and strand cleaving properties of anthraquinone-oligopyrrolecarboxamide hybrids. *Anticancer Drug Des.*, **11**, 527–551. (b) Amishiro, N., Nagamura, S., Murakata, C., *et al.* (2000) Synthesis and antitumor activity of duocarmycin derivatives: modification at

C-8 position of A-ring pyrrole compounds bearing the simplified DNA-binding groups. *Bioorg. Med. Chem.*, **8**, 381–391.

37. (a) Scholl, M., Ding, S., Lee, C. W. and Grubbs, R. H. (1999) Synthesis and activity of a new generation of ruthenium-based olefin metathesis catalysts coordinated with 1,3-dimesityl-4,5-dihydroimidazol-2-ylidene ligands. *Org. Lett.*, **1**, 953–956. (b) Ulman, M. and Grubbs, R. H. (1996) Ruthenium carbene-based olefin metathesis initiators: Catalyst decomposition and longevity. *J. Org. Chem.*, **64**, 7202–7207.

13

Fluorous Mixture Synthesis (FMS) of Drug-like Molecules and Enantiomers, Stereoisomers, and Analogues of Natural Products

Wei Zhang

13.1 Introduction

13.1.1 Natural Products and Synthetic Drug-like Compounds

Mother nature is a good resource for new molecules; over 10 000 natural products are isolated each year [1]. Historically, natural products have provided a good number of leads for the development of new drugs [2]. However, since natural products are commonly screened as an extraction mixture, deconvolution of an active component and structural characterization are difficult tasks. In addition, isolation of natural products has a long cycle time and is considered expensive [3]. These limitations have prompted efforts to synthesize natural product analogues and natural-product-like compounds for biological screening and quantitative structure–activity relationship (QSAR) studies.

Besides natural products, synthetic small molecules are another source of drug candidates. Drug-like synthetic molecules usually have privileged structures such as indole, benzopyran, and benzodiazepinone rings [4]. Compared with natural products, drug-like synthetic small compounds tend to have more aromatic rings, planar molecular structures, and few stereogenic centers. They usually have appropriate size and hydrogen-bonding capacity to bind in the active-site pockets of biological targets. Chemical stability and

Fluorine in Medicinal Chemistry and Chemical Biology Edited by Iwao Ojima

physical properties such as solubility and log*P* are always considered in the design of synthetic compounds. The drug-likeness of synthetic compounds can be estimated by the "Lipinski rule-of-five." [5].

Natural product analogues and drug-like compounds are commonly prepared by solution-phase synthesis. Solution-phase chemistry has favorable reaction kinetics and is easy for intermediate analysis. However, compounds are produced "one at a time," followed by a tedious purification process such as chromatography. This is not productive in the preparation of compound libraries for QSAR studies. The advance of solid-phase synthesis and solid-supported solution-phase synthesis has significantly increased the throughput of compound purification processes. Nevertheless, disadvantages such as unfavorable heterogeneous reaction kinetics and difficulty in monitoring the reaction process have limited the capacity of solid-phase synthetic technology to access many natural product analogues and complex drug-like molecules [6]. The recent development of fluorous technologies has provided a solution-phase approach for the synthesis of complicated molecules and their analogues.

13.1.2 The Concept of Fluorous Mixture Synthesis

Fluorous chemistry integrates the characteristics of solution-phase reactions and the phase tag strategy developed for solid-phase chemistry [7–15]. Perfluoroalkyl chains instead of polymer beads are used as the phase tags to facilitate the separation process. In 2001 the Curran group first reported the concept of fluorous mixture synthesis (FMS) for solution-phase library synthesis [16]. FMS is able to produce individual pure compounds without the effort of deconvolution. It adapts literature procedures to synthesize complex natural products, their enantiomers and diastereomers. FMS can also be used for the development of new synthetic protocols and to make novel drug-like molecules [17, 18].

Fluorous mixture synthesis consists of the following general steps (see Scheme 13.1).

1. Individually attach a set of substrates to a corresponding set of homologous fluorous tags with increasing fluorine content.
2. Mix the tagged substrates in one pot.
3. Conduct multistep mixture synthesis in one-pot or in split-parallel fashion.
4. Separate the mixtures (demixing) of tagged products by high-performance chromatography (HPLC) on a fluorous stationary phase (F-HPLC).
5. Detag to release final products.

The efficiency of FMS is directly proportional to the number of components mixed (step 2), the length of mixture synthesis (step 3), and the number of splits (step 3).

A fluorous stationary phase such as Fluoro*Flash*®, $Si(Me)_2CH_2CH_2C_8F_{17}$ is hydrophobic and lipophobic, but has strong affinity for fluorous compounds [19]. F-HPLC has the capability to separate mixtures of fluorous compounds according to their fluorine content [20–24]. A molecule with a long perfluoroalkyl (R^f) chain gives a longer retention time. A typical mobile phase for fluorous F-HPLC is a gradient of MeOH–H_2O with MeOH increasing up to 100%, which is similar to that in reversed-phase HPLC. Other solvents

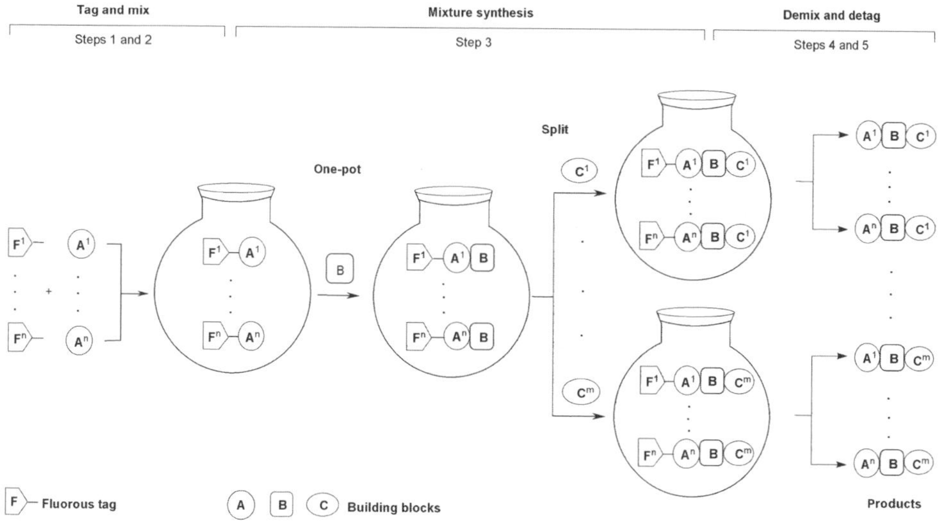

Scheme 13.1 *Schematic diagram of FMS.*

such as MeCN or THF can be used to replace MeOH for gradient elution. An F-HPLC trace shown in Figure 13.1 demonstrates a semipreparative-scale (~5 mg) separation of a mixture of seven fluorous-tagged mappicine analogues bearing different R^1 and R^f groups [25].

13.1.3 Quasi-racemic FMS

Enantiopure or enantioenriched compounds can be obtained by asymmetric synthesis or by separation of a racemic reaction mixture. Quasi-racemic FMS provides a new approach to enantiomeric compounds (see Scheme 13.2) [26, 27]. Quasi-racemic synthesis starts with two individual *R*- and *S*-enantiomers attached to two different fluorous tags. After steps of mixture synthesis followed by F-HPLC demixing and detagging, two individual products as enantiomers are obtained (see Sections 13.2.1 and 13.2.2). The separation and identification of the final quasi-enantiomers are ensured by the phase-tag-based F-HPLC. In a more complicated quasi-racemic FMS, additional enantiomerically pure building blocks and fluorous tags can be used to generate more chiral centers and more than two products as stereoisomers (see Sections 13.2.3 to 13.2.8).

13.1.4 Tags for FMS

Each fluorous tag (also called protecting group or linker) has two attachment points. One is permanently bound to a perfluoroalkyl (R^f) chain and the other is temporarily attached to a reaction substrate so that it can be cleaved to release the product from the support at

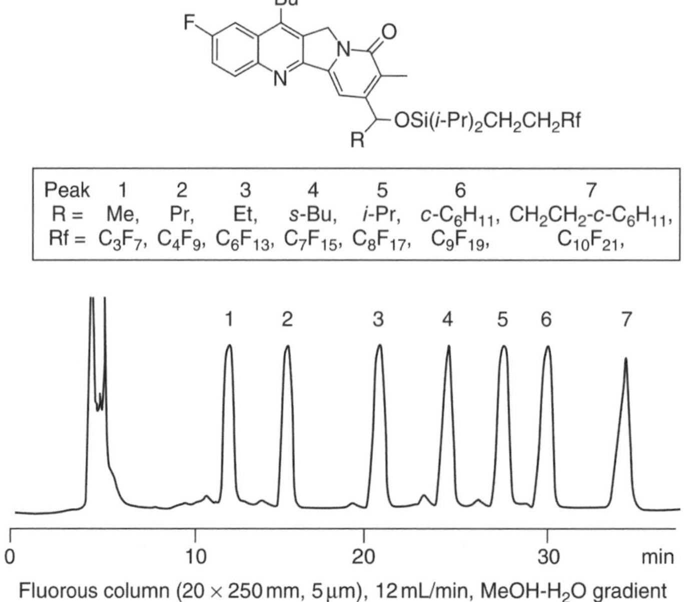

Figure 13.1 *Semi-preparative F-HPLC demixing of a seven-component mixture of mappicine analogues.*
(Source: *Reproduced with permission from Zhang, W., Luo, Z., Chen, C. H.-T. Curran, D. P., Solution-Phase Preparation of A 560-compound library of individually pure mappicine analogs by fluorous mixture synthesis, J. Am. Chem. Soc. (2002)* ***124****, 10443–10450. Copyright (2002) American Chemical Society.)*

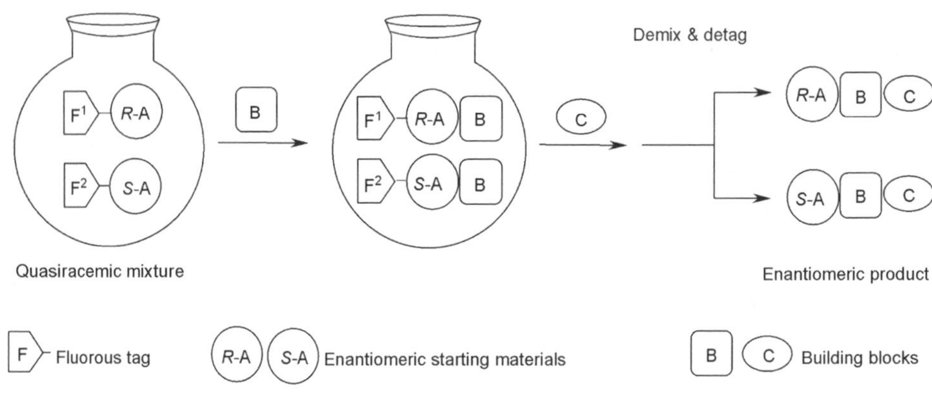

Scheme 13.2 *Schematic diagram of quasi-racemic FMS.*

the end of the synthesis. An ethylene or propylene spacer is used to minimize the electronic effect generated from the strong electron-withdrawing R^f group, which affects the reactivity of the functional group. Compared with solid-supported linkers, fluorous tags have the following unique features: (i) good solubility in many organic solvents at room or elevated

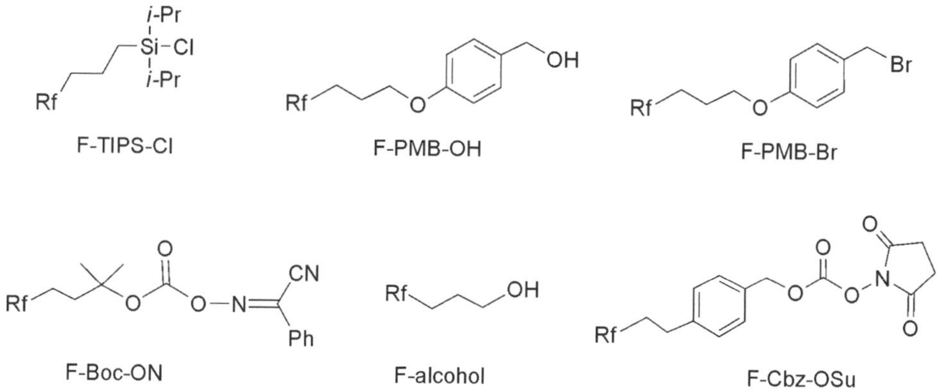

F-TIPS-Cl F-PMB-OH F-PMB-Br

F-Boc-ON F-alcohol F-Cbz-OSu

Scheme 13.3 *Tags for FMS.*

temperature; (ii) favorable homogeneous solution-phase reaction kinetics; (iii) easy intermediate analysis by conventional tools such as thin-layer chromatography (TLC), nuclear magnetic resonance (NMR), and liquid chromatography–mass spectrometry (LC-MS); and (iv) easy adoption of literature procedures for protecting group attachment and cleavage. Scheme 13.3 lists the fluorous protecting groups that have been developed for FMS [14]. They are the derivatives of common protecting groups such as TIPS, PMB, Boc, and Cbz, which are used to protect hydroxyl, amino, and carboxyl groups. These fluorous tags, with variation of the R^f groups, are commercially available from Fluorous Technologies, Inc. [19].

13.2 FMS of Natural Products and Drug-like Compounds and Libraries

13.2.1 Synthesis of Enantiomers of Pyridovericin

Pyridovericin was isolated from the entomopathogenic fungus *Beauveria bassiana* EPF-5 [28]. It is an inhibitor of the protein tyrosine kinase. The Curran group has demonstrated that enantiomers of pyridovericin can be prepared by quasi-racemic FMS (see Scheme 13.4) [29]. The (*S*)-**1a** and (*R*)-**1b** alcohols as the starting materials were attached to two fluorous silanes with C_6F_{13} and C_8F_{17} groups, respectively. They were then combined to form a quasi-enantiomeric mixture. The mixture was taken through a multistep synthesis through aldehydes M-**2** ("M" stands for mixture), β-ketoesters M-**3**, and then M-**4**. M-**4** was oxidized and then demixed by F-HPLC to give two quasi-enantiomers (*S*)-**5a** and (*R*)-**5b**. The fluorous tags were then removed to release the (*S*)-**6a** and (*R*)-**6b** enantiomers of pyridovericin. Quasi-racemic synthesis is the simplest version of FMS, with only two components in the mixture. Only one-pot, no split-parallel reactions are conducted in this quasi-racemic FMS.

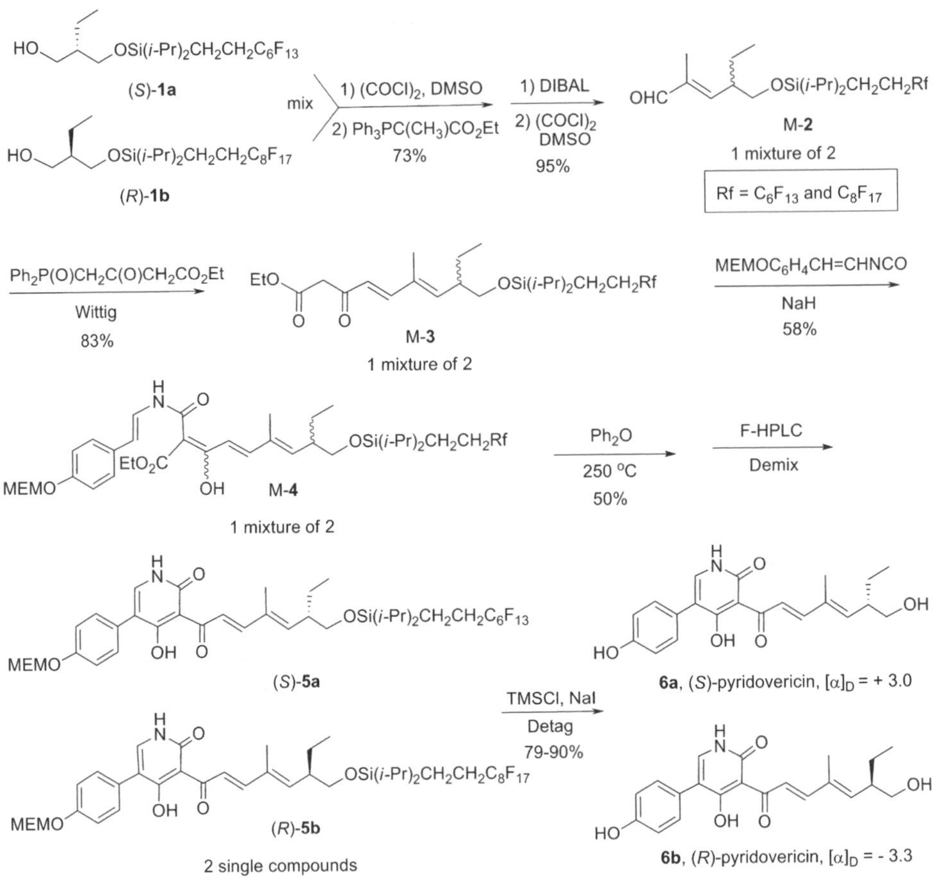

Scheme 13.4 *Quasi-racemic FMS of (S)- and (R)-pyridovericins.*

13.2.2 Synthesis of Enantiomers of Mappicine

The natural product (*S*)-mappicine was isolated from *Mappia foetida* [30]. Its analogue mappicine ketone, also known as nothapodytine B, was isolated from *nothapodytes foetida* and is active against herpes viruses (HSV) and human cytomegalovirus (HCMV) at a range of 3–13 μM [31]. One-pot total synthesis of both (*R*)- and (*S*)-mappicines has been accomplished by quasi-racemic FMS (see Scheme 13.5) [29]. Enantiomeric (*R*)- and (*S*)-alcohols tagged with silanes containing C_6F_{13} and C_8F_{17}, respectively, were converted to quasi-enantiomers (*R*)-**7a** and (*S*)-**7b**. The mixture of these two compounds in 1 : 1 molar ratio was subjected to TMS group exchange with ICl followed by demethylation with BBr_3 to form pyridine M-**8**. *N*-Propargylation and subsequent radical cyclization with phenyl isonitrile provided quasi-racemic mixture M-**9**. The separation of this mixture by F-HPLC yielded two quasi-enantiomers. Both (+)-**10a** and (−)-**10b** mappicine were obtained in enantiopure forms after deprotection with TBAF in THF. Similar chemistry for the synthesis of a substituted mappicine library containing 560 analogues is described in Section 13.2.9.

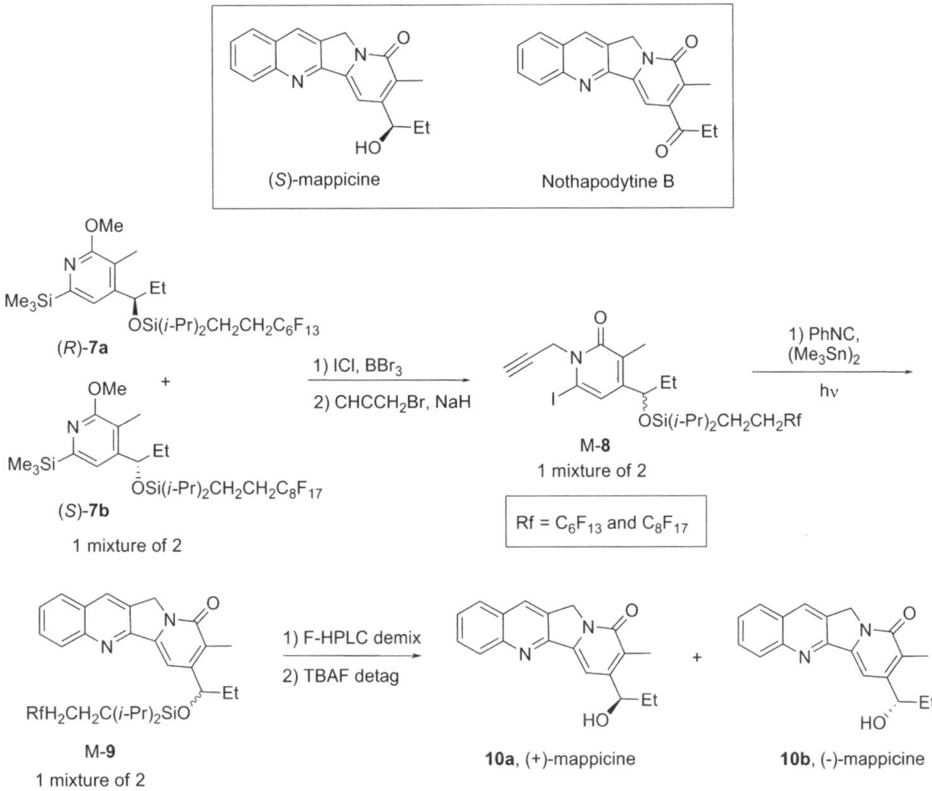

Scheme 13.5 *Quasiracemic FMS of (+)- and (−)-mappicines.*

13.2.3 Synthesis of Stereoisomers of Murisolin

The murisolin class of mono-tetrahydrofuran acetogenins has six stereocenters. Among the several known murisolin diastereomers, the most active one exhibits extremely high cytotoxicity (IC_{50}) at 1 femtomolar (fM) range, whereas the potency of other diastereomers may differ by factors of up to 1 billion [32].

The Curran group reported the FMS of 16 diastereomers of murisolin (see Scheme 13.6) [33]. Sixteen stereoisomers are derivatives of four stereocenters at C-15, C-16, C-19, and C-20 positions of dihydroxytetrahydrofuran fragment (shown in the small box of Scheme 13.6) with the 4(*R*) and 34(*S*) centers fixed. The FMS was started with M-**11,** a mixture of four enantiomerically pure compounds each tagged by a PMB with different R^f groups (C_2F_5, C_4F_9, C_6F_{13}, and C_8F_{17}) (see Scheme 13.3). M-**11** was then taken through a sequence of organic reactions to form M-**17**. Two split and parallel syntheses were conducted from M-**13** to M-**14**. Fluorous HPLC demixing of four mixtures of M-**17** followed by detagging provided 16 desired diastereomers of murisolin **18**.

The synthetic scope for making murisolin stereoisomers has been extended through the development of the first double tagging strategy [34, 35] A mixture of four stereoisomers of dihydroxytetrahydofuran encoded with four fluorous tags at C-19 and C-20 was

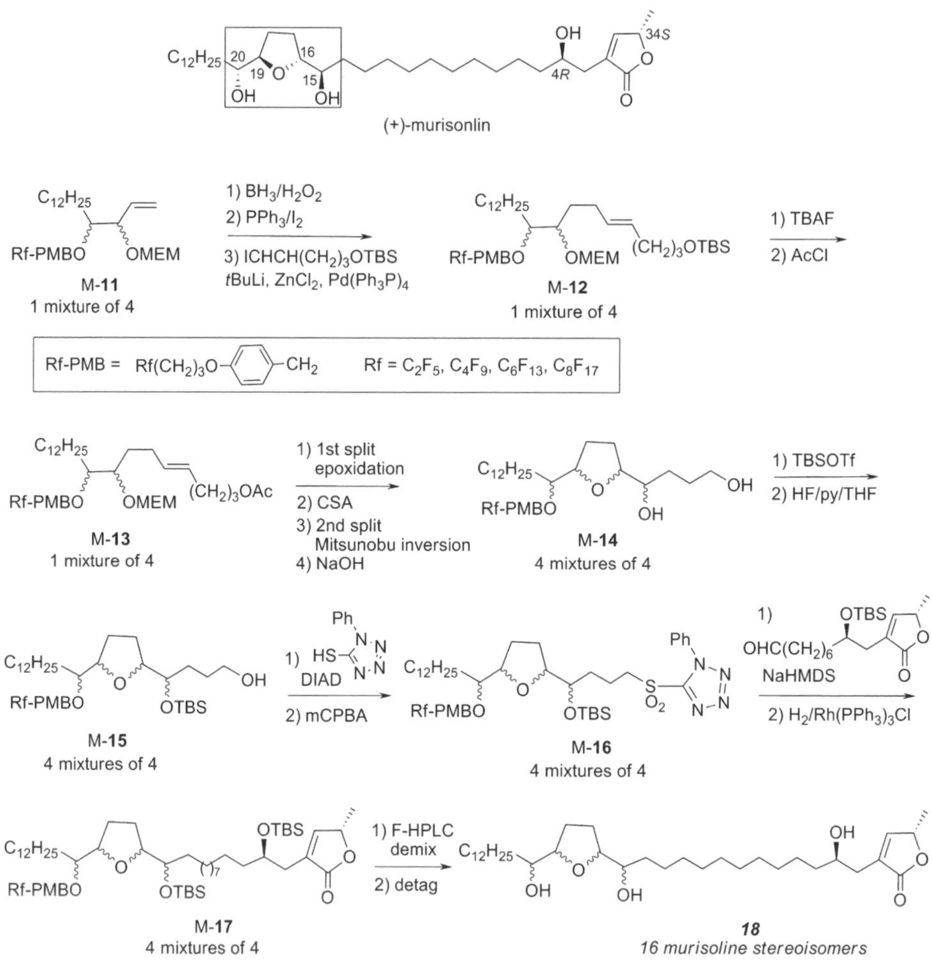

Scheme 13.6 *FMS of 16 stereoisomers of murisolin.*

coupled to a mixture of four stereoisomers of hydroxybutenolide fragments encoded with four oligoethylene glycol (OEG, $[OCH_2CH_2O]_n$) tags at C-4 and C-34 (see Scheme 13.7). The mixture containing 16 double-tagged murisolin diastereomers was first demixed by flash chromatography on normal silica gel to give four fractions according to the polarity (length) of the OEG tags. The second demixing with fluorous HPLC gave 16 individual double-tagged compounds. Both fluorous and OEG tags were cleaved by treatment with DDQ. Sixteen murisolin diastereomers were thus produced in a single solution-phase synthesis without splitting.

13.2.4 Synthesis of Stereoisomers of Pinesaw Fly Sex Pheromone

The propionate ester of 3,7,11-trimethyl-2-tridecanol is a female sex pheromone of the minor sawfly *Microdiprion pallipes* [36]. Pinesaw fly sex pheromones can be used as a

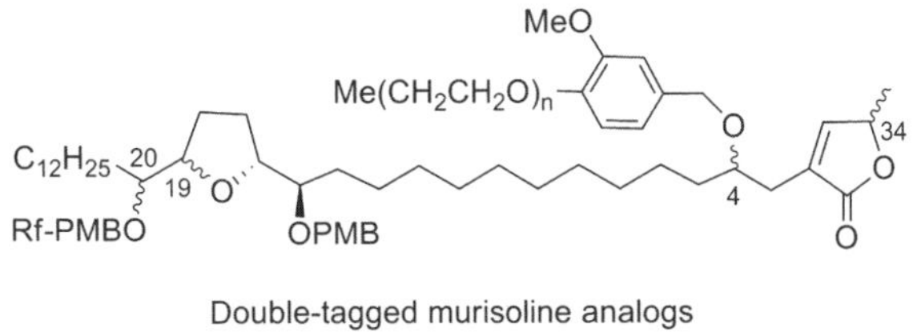

Double-tagged murisoline analogs

Scheme 13.7 *Double-tagged murisoline analogues.*

3,7,11-trimethyl-2-tridecanol ethyl ester
(pinesaw fly sex pheromone)

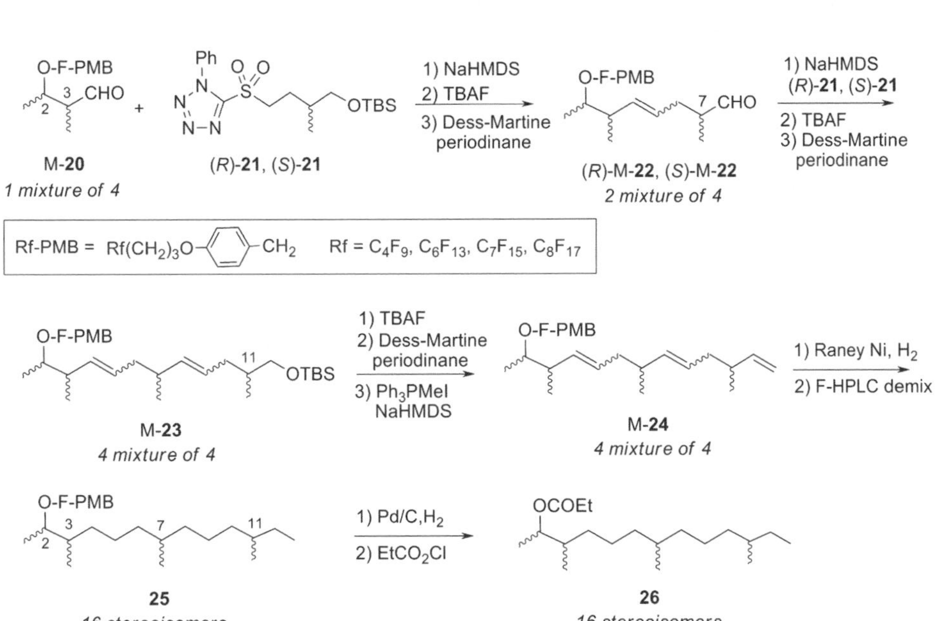

Scheme 13.8 *FMS of 16 stereoisomers of pinesaw fly sex pheromone.*

trap for pest control in infested pine tree areas. This molecule has four chiral centers and 16 possible stereoisomers. The Curran group accomplished the synthesis of all 16 stereoisomers of pinesaw fly sex pheromones by a four-component FMS (see Scheme 13.8) [37, 38]. A mixture (M-**20**) of four enantiomerically pure aldehydes attached to fluorous PMB with different R^f groups was split into two portions to react with sulfone (*R*)-**21** and (*S*)-**21**. This led to the formation of aldehyde (*R*)-M-**22** and (*S*)-M-**22**, respectively. Each of these two mixtures was subjected to reduction, coupling with (*R*)-**21** and (*S*)-**21**, TBS deprotection, oxidation, and Wittig reaction to give four mixtures of trienes M-**24**. The hydrogenation of alkene followed by F-HPLC demixing of M-**24** afforded 16 individual

F-PMB-attached tridecanols **25**. They were then converted to 16 pinesaw fly sex pheromone diastereomers **26** by parallel tag cleavage and acylation reactions.

13.2.5 Synthesis of Stereoisomers of (–)-Dictyostatin

(–)-Dictyostatin is a marine macrolactone that has potent anticancer activity [39]. This compound had been known for over a decade before its stereostructure was confirmed through total synthesis by the Paterson [40] and Curran [41] groups in 2004. Further biological testing on the synthetic sample showed that dictyostatin has equal or better activity against paclitaxel-resistant cell lines than its open-chain analogue discodermolide, radiolabeled paclitaxel, and epothilone B [42]. The Curran group modified their total synthesis route (over 20 steps) for FMS of (–)-dictyostatin and three C-6 and C-7 diastereomers (see Scheme 13.9) [43]. Instead of making these diastereomers in four parallel multistep syntheses, the FMS enables them to be produced in a single set. This is the longest multistep FMS conducted to date.

At the premix stage, a set of four enantiopure alcohols with chiral centers at C-6 and C-7 were individually tagged with a set of four fluorous TIPS-type silanes containing C_3F_7, C_4F_9, C_6F_{13}, and C_8F_{17} tags, respectively. The coded alcohols were then converted to fluorous esters **27a-d**. These four esters were blended in a ratio of 1.5 : 1 : 1 : 1.5 and then the resulting mixture M-**27a-d** was converted to M-**28** in three steps of FMS. M-**28** was coupled with an alkynyllithium and then reduced by (*S,S*)-Noyori's catalyst to give M-**29**. The alkyne group in M-**29** was reduced to the *cis*-alkene by Lindlar hydrogenation and the resulting secondary hydroxy group was protected with the TBS group. The cleavage of TES with dichloroacetic acid gave M-**30**. Dess–Martin oxidation of the primary alcohol followed by coupling with **31** gave the α,β-unsaturated ketone. The reduction of C-17–C-18 alkene with Stryker's reagent followed by reduction of the C-19 ketone with $LiAl(Ot\text{-}Bu)_3H$ gave β-alcohol M-**32** as the major product, which was isolated by silica gel chromatography. TBS protection of the C-19 hydroxy group, removal of the trityl group with $ZnBr_2$, oxidation of the allylic alcohol with the Dess–Martin reagent, then the Still–Gennari reaction provided (*E*),(*Z*)-diene M-**33**. The removal of PMB with DDQ, basic hydrolysis of the conjugated ester, followed by macrolactonization under Yamaguchi conditions gave a mixture of major (2*Z*),(4*E*) and minor (2*E*),(4*E*) macrolactones. F-HPLC demixing of the final mixture provided the four individual components. The desilylation with 3 N HCl in MeOH afforded dictyostatin (6*R*,7*S*)-**34a** and the other three C-6,C-7-*epi*-dictyostatin diastereomers after HPLC purification.

These four compounds were assayed against human ovarian carcinoma cells for their antiproliferative effects [43]. It was found that bis-*epi* diastereomer (6*S*,7*R*)-**34b** was less active than the other isomers, while monoepimer (6*R*,7*R*)-**34d** was equipotent to dictyostatin (6*R*,7*S*)-**34a**, and another monoepimer (6*S*,7*S*)-**34c** was four times more potent.

13.2.6 Synthesis of Stereoisomers of Passifloricins

In the total synthesis of an eight-membered stereoisomer library containing the enantiomer of passifloricin A and seven other stereoisomers at C-5, C-7, and C-9, an "en route" protocol was developed by the Curran group to introduce stereocenters and fluorous tags

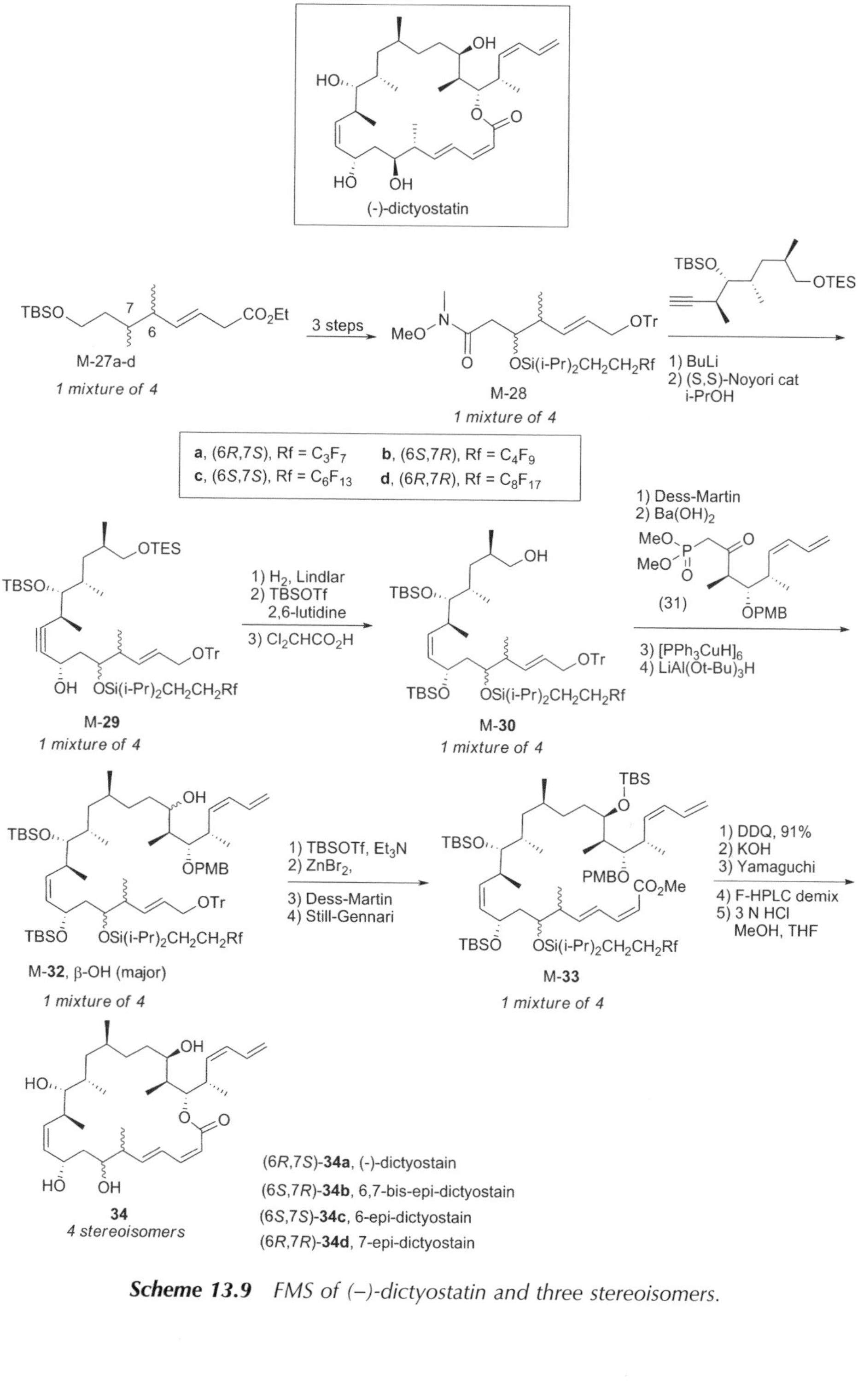

HO OH HO OH O O
(-)-dictyostatin

TBSO 7 6 CO_2Et
M-27a-d
1 mixture of 4

3 steps

MeO N O OTr
$OSi(i\text{-}Pr)_2CH_2CH_2Rf$
M-28
1 mixture of 4

TBSO OTES
1) BuLi
2) (S,S)-Noyori cat
i-PrOH

a, (6*R*,7*S*), Rf = C_3F_7	**b**, (6*S*,7*R*), Rf = C_4F_9
c, (6*S*,7*S*), Rf = C_6F_{13}	**d**, (6*R*,7*R*), Rf = C_8F_{17}

TBSO OTES OTr
OH $OSi(i\text{-}Pr)_2CH_2CH_2Rf$
M-**29**
1 mixture of 4

1) H_2, Lindlar
2) TBSOTf
2,6-lutidine
3) Cl_2CHCO_2H

TBSO OH OTr
TBSO $OSi(i\text{-}Pr)_2CH_2CH_2Rf$
M-**30**
1 mixture of 4

1) Dess-Martin
2) $Ba(OH)_2$
MeO P MeO O O
(31) OPMB
3) $[PPh_3CuH]_6$
4) $LiAl(Ot\text{-}Bu)_3H$

TBSO OH OPMB OTr
TBSO $OSi(i\text{-}Pr)_2CH_2CH_2Rf$
M-**32**, β-OH (major)
1 mixture of 4

1) TBSOTf, Et_3N
2) $ZnBr_2$,
3) Dess-Martin
4) Still-Gennari

TBS O
TBSO PMBO CO_2Me
TBSO $OSi(i\text{-}Pr)_2CH_2CH_2Rf$
M-**33**
1 mixture of 4

1) DDQ, 91%
2) KOH
3) Yamaguchi
4) F-HPLC demix
5) 3 N HCl
MeOH, THF

HO OH O O
HO OH
34
4 stereoisomers

(6*R*,7*S*)-**34a**, (-)-dictyostain
(6*S*,7*R*)-**34b**, 6,7-bis-epi-dictyostain
(6*S*,7*S*)-**34c**, 6-epi-dictyostain
(6*R*,7*R*)-**34d**, 7-epi-dictyostain

Scheme 13.9 *FMS of (–)-dictyostatin and three stereoisomers.*

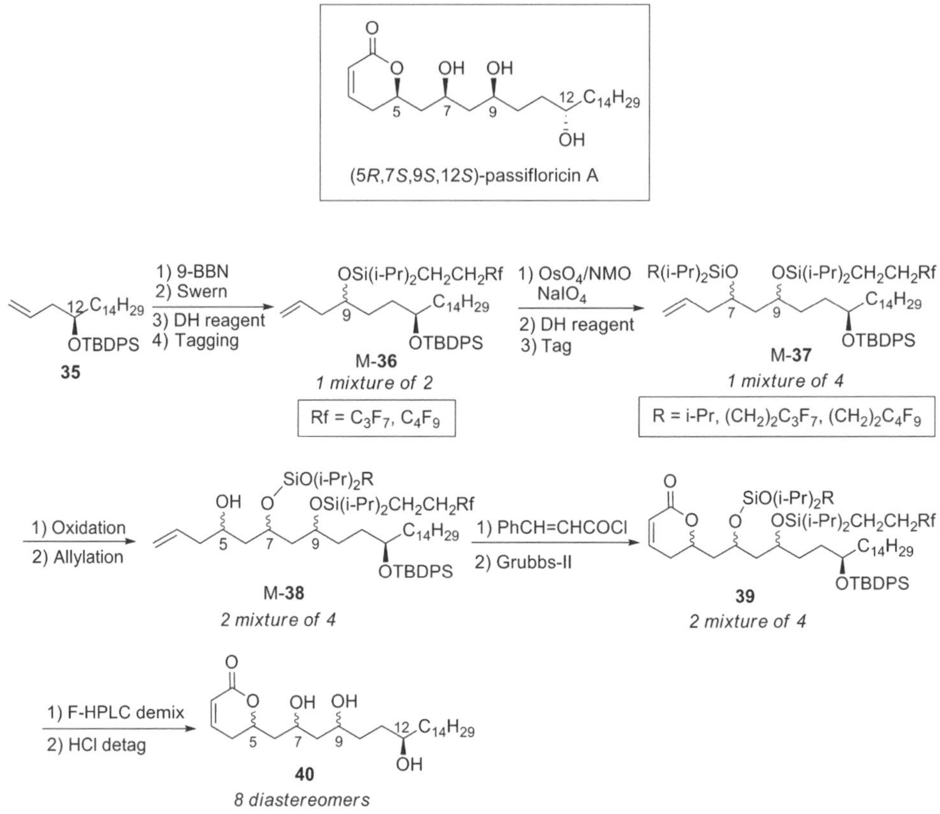

Scheme 13.10 *FMS of eight stereoisomers of passifloricin.*

during the synthesis [44]. This protocol is different from other FMS in which building blocks with coded stereocenters were premade and pretagged. In addition, the total fluorine content of each tagged molecule is defined by two tags, which allows using two fluorous tags for a four-component FMS. The number of fluorous tags is less than that of components in the mixture.

Enantiopure allyl silyl ether (*R*)-**35** was subjected to a sequence of hydroboration and oxidation reactions (see Scheme 13.10). Half of the resulting aldehyde was treated with the (*R,R*)-Duthaler–Hafner (DH) reagent, and the resulting alcohol was tagged with fluorous triisopropylsilyl trifluoromethanesulfonate (F-TIPSOTf) bearing a C_4F_9 group. The other half was treated with the (*S,S*)-DH reagent, and the alcohol was tagged with the F-TIPS group bearing C_3F_7. The mixed quasi-racemic M-**36** was split to two portions and subjected to oxidation, allylation, and then tagging reactions. The product from the (*R,R*)-DH reagent got a new nonfluorous TIPS tag, and the product from the (*S,S*)-DH reagent got the repeat C_3F_7 tag. A pair of two-compound mixtures was mixed to make a four-compound mixture M-**37**. Repeated split, oxidation, and allylation of M-**37** gave two four-component mixtures M-**38.** The mixtures were acylated, followed by ring-closing metathesis to provide the full stereoisomer library of protected passifloricins as two mixtures of four compounds M-**39**. Each of these two mixtures has four components with

C_3F_7, C_4F_9, two C_3F_7, and two C_4F_9 tags. Because the number of fluorine atoms on each component in the mixture is different, they were easily demixed by F-HPLC. All eight isomers of **39** were deprotected individually by exposure to 3 N HCl in EtOH to provide eight compounds **40** including the enantiomer of passifloricin A (5*S*,7*R*,9*R*,12*R*) and its seven diastereomers with R configurations at C-12 and all the possible configurations at C-5, C-7, and C-9.

13.2.7 Synthesis of Stereoisomers of Lagunapyrone B

Lagunapyrones A, B, and C were isolated during an investigation of the secondary metabolites of estuarine actinomycetes [45]. These natural products feature a 24-carbon chain consisting of an α-pyrone ring with two adjacent stereocenters (C-6,7) separated by 11 carbon atoms from a second group of three stereocenters (C-19–21). All seven of the double bonds in the backbone of the lagunapyrones are trisubstituted, and lagunapyrones A, B, and C differ only in the nature of the group attached to C-2 (R = Me, Pr, Bu). Lagunapyrone B exhibits moderate activity (ED_{50} = 3.5 μg/mL) against a human colon cancer cell line [45].

Despite the novel skeleton and interesting biological activity, no synthetic efforts toward the lagunapyrones had been published until the Curran group's work on FMS (see Scheme 13.11) [46]. Quasi-racemic FMS of M-**41** was conducted to construct the left fragment, which has two stereocenters at C-6 and C-7 and an α-pyrone ring. The quasi-racemic product mixture of the left fragment was demixed by F-HPLC and detagged with HF to give (*R,S*)-**42** and (*S,R*)-**42**. The right fragment M-**44** containing C-19–21 stereocenters was also prepared by quasi-racemic FMS. The quasi-racemic mixture of the right fragment M-**44** was reacted with enantiomers (*R,S*)-**42** and (*S,R*)-**42** through the Stille coupling reaction to give two quasi-racemic mixtures of tagged products. These two mixtures were separated by F-HPLC and detagged to afford four lagunapyrone B stereoisomers **45**.

13.2.8 Synthesis of Truncated Analogues of (+)-Discodermolide

FMS of truncated analogues of the natural product (+)-discodermolide at the C-22 position has been accomplished by the Curran group [47]. Four starting materials with different R group (H, CH=CH_2, Et, Ph) were protected with the corresponding fluorous PMB ($R^f = C_4F_9$, C_6F_{13}, C_8F_{17}, $C_{10}F_{21}$) (see Scheme 13.12) and mixed to form M-**46**. 4-Component FMS converted M-**46** sequentially to phosphonium salt M-**47**, the Wittig reaction product M-**48**, carbamate M-**49,** and then tagged product M-**50**. Four truncated discodermolide analogues **51** were produced after demixing and detagging of M-**50**.

13.2.9 Synthesis of a Mappicine Library

The power of FMS has been further demonstrated in the preparation of a 560-member library of mappicine analogues (see Scheme 13.13) [25]. In this case, the fluorous tags encoded substrates with different substitutions rather than substrates with different

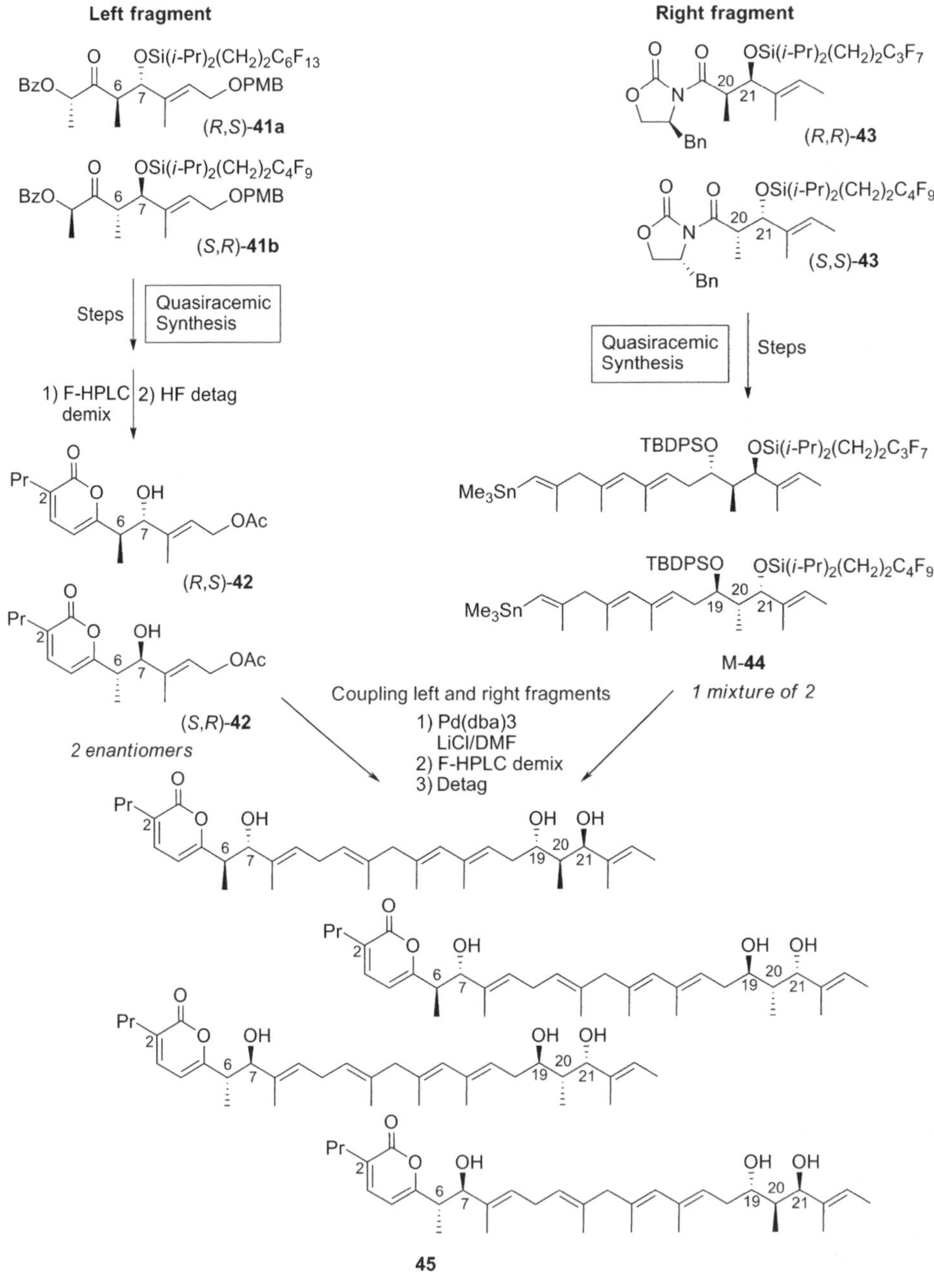

Scheme 13.11 *FMS of four stereoisomers of lagunapyrone B.*

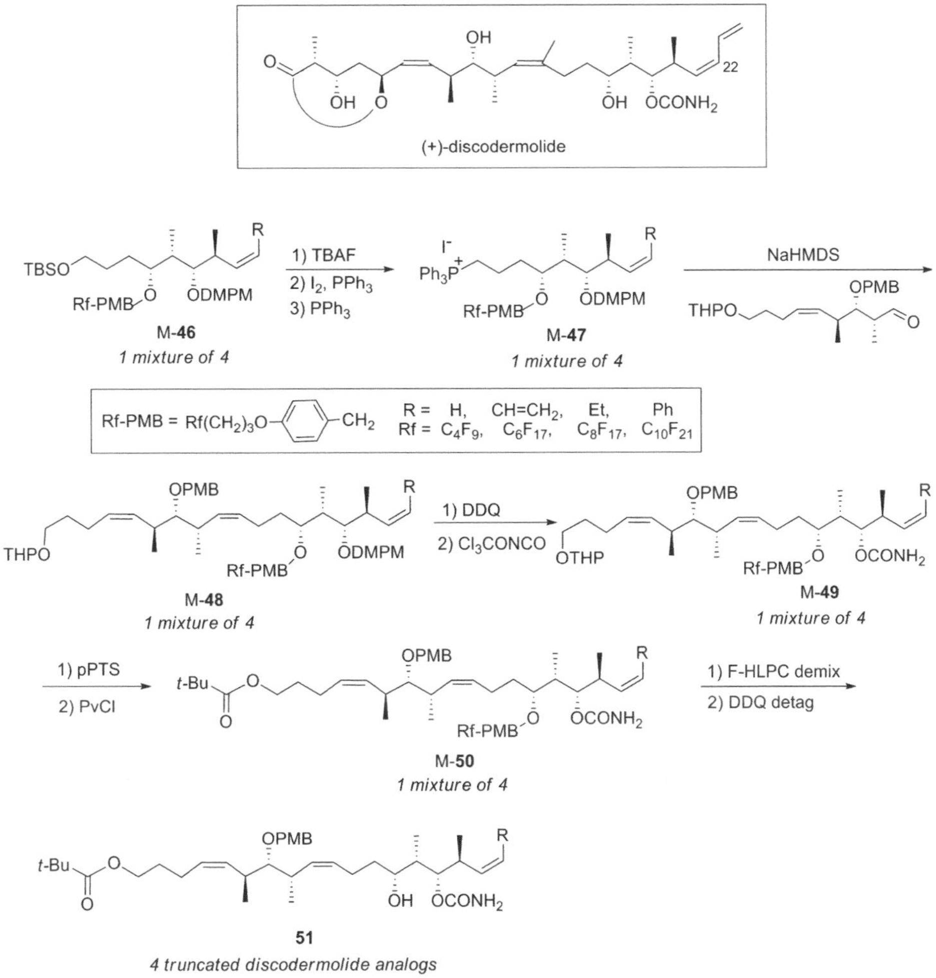

Scheme 13.12 *FMS of four truncated discodermolide analogues.*

stereochemistry as was described in Section 13.2.2 [29]. A seven-component mixture M-**52** was prepared by iodo-exchange followed by demethylation using the same procedures described in Scheme 13.5. M-**52** was split into 8 portions and subjected to *N*-propargylations with 8 different bromides to give 8 mixtures of M-**53**. Each of the 8 mixtures of M-**53** was further split into 10 portions for radical annulation reaction with isonitriles. The resulting 80 mixtures of M-**54** were demixed by F-HPLC and then detagged by HF-pyridine to give a 560-member mappicine library **55** (see Figure 13.1). After each step, the reaction mixture could be analyzed by fluorous HPLC and the by-products or unreacted starting materials were removed by flash column chromatography with normal-phase silica gel. The synthesis of this 560-membered library is accomplished in 90 reactions (not including the parallel detagging reactions), while 630 steps are needed for a corresponding parallel synthesis. The overall separations required only 90 chromatography steps compared with 630 chromatography steps needed for a parallel synthesis.

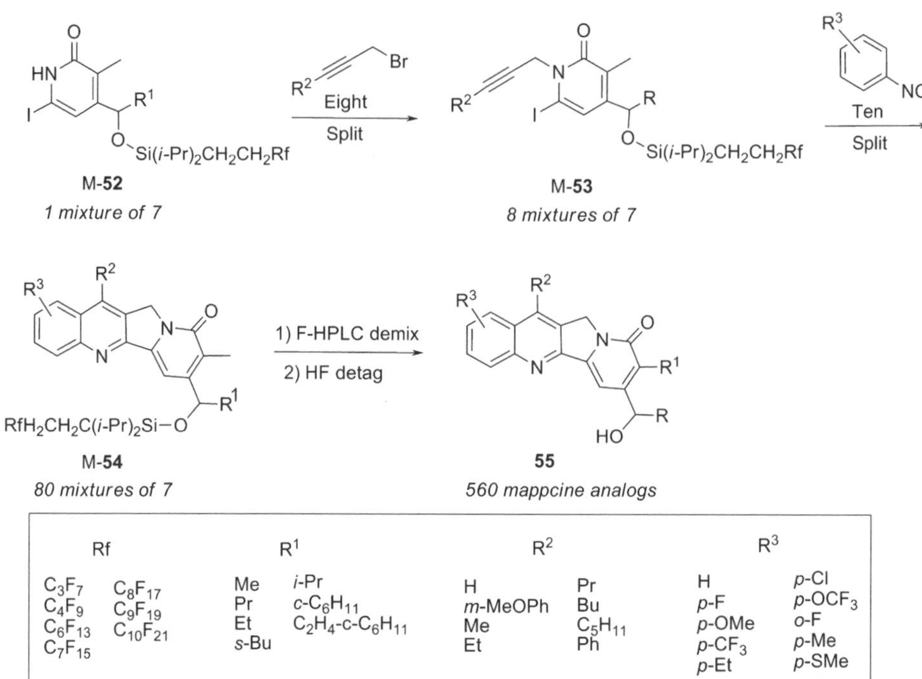

Scheme 13.13 *Seven-component FMS of a 560-membered mappicine library.*

13.2.10 Synthesis of Tecomanine-like Compounds

Tecomanine belongs to a family of natural alkaloids. This compound has shown powerful hypoglycemic activity [48]. The 4-alkylidine cyclopentenone scaffold **62** has a ring skeleton similar to tecomanine. These types of cyclopentenones can be considered as novel bicyclic α-amino acid derivatives that can potentially be useful in the synthesis of peptidomimetics. The Brummond [49] and Curran [50] groups developed a four-component FMS protocol to produce a 16-compound library of the 4-alkylidine cyclopentenone via a rhodium-catalyzed [2 + 2 + 1] allenic Pauson–Khand cycloaddition (see Scheme 13.14). Four F-CBzs containing C_4F_7, C_6F_{13}, C_8F_{17}, and C_9F_{19} groups were individually attached to a set of amino acids **56** with four different R^1. The resulting F-CBz-protected amino acids were mixed in equimolar amounts and then reacted in single pot with a propargyl alcohol to form a four-component mixture of esters M-**57**. The allenic amino acid mixture M-**59** was obtained through the Claisen rearrangement of propargyl esters M-**58**. The allene mixture M-**59** was split into four portions and reacted with one of four different propargyl bromides to afford four mixtures of alkynyl allenes M-**60**. The [2 + 2 + 1] cycloaddition of allenes M-**60** followed by F-HPLC demixing afforded 16 individual 4-alkylidene cyclopentenones **61**. The fluorous Cbz tag was removed by treatment with dimethyl sulfide in the presence of BF_3-Et_2O to afford 16 final products **62**.

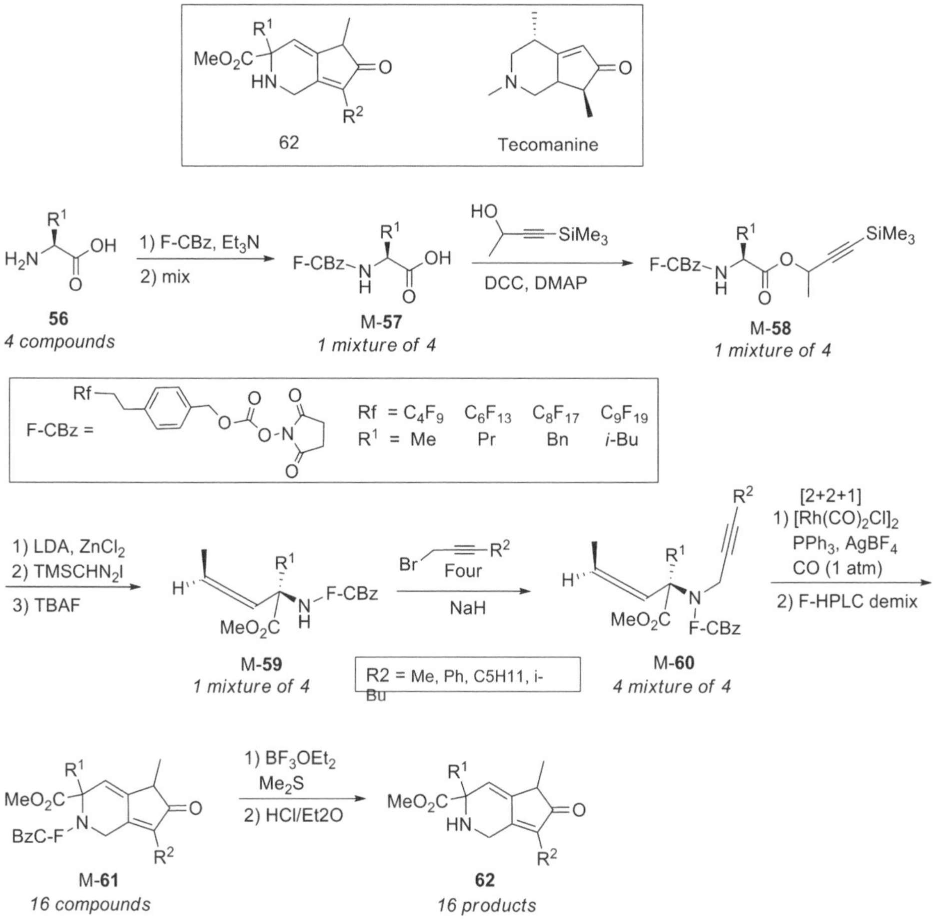

62

Tecomanine

56
4 compounds

1) F-CBz, Et_3N
2) mix

M-57
1 mixture of 4

DCC, DMAP

M-58
1 mixture of 4

F-CBz =

Rf = C_4F_9 C_6F_{13} C_8F_{17} C_9F_{19}
R^1 = Me Pr Bn *i*-Bu

1) LDA, $ZnCl_2$
2) $TMSCHN_2I$
3) TBAF

M-59
1 mixture of 4

Four
NaH

R2 = Me, Ph, C5H11, i-Bu

M-60
4 mixture of 4

[2+2+1]
1) $[Rh(CO)_2Cl]_2$
PPh_3, $AgBF_4$
CO (1 atm)
2) F-HPLC demix

M-61
16 compounds

1) BF_3OEt_2
Me_2S
2) HCl/Et2O

62
16 products

Scheme 13.14 *FMS of 16 tecomanine-like compounds.*

13.2.11 Synthesis of Fused-Tricyclic Hydantoins

In addition to the "en route" protocol described in Section 13.2.6 for the synthesis of passifloricins, the use of "redundant tags" for the synthesis of fused-tricyclic hydantoins is another example of FMS in which the number of fluorous tags is less than that of components in the mixture. Since molecules for FMS have fluorine atoms both on the parent structure and on the tag, each component in the mixture has different total fluorine content – even the tag may be redundant.

The six amino acids **63** were prepared individually, and mixed to give acids M-**64** (see Scheme 13.15) [51]. Esterification of M-**64**, zinc-chelated ester enolate Claisen rearrangement of M-**65**, *tert*-butyl esterification, and removal of the Me_3Si group yielded M-**66**. The alkynyl allenes M-**67** were obtained by *N*-propargylation. The allenic Pauson–Khand reaction of M-**67** afforded three products; (*R*)-alkylidenecyclopentenone

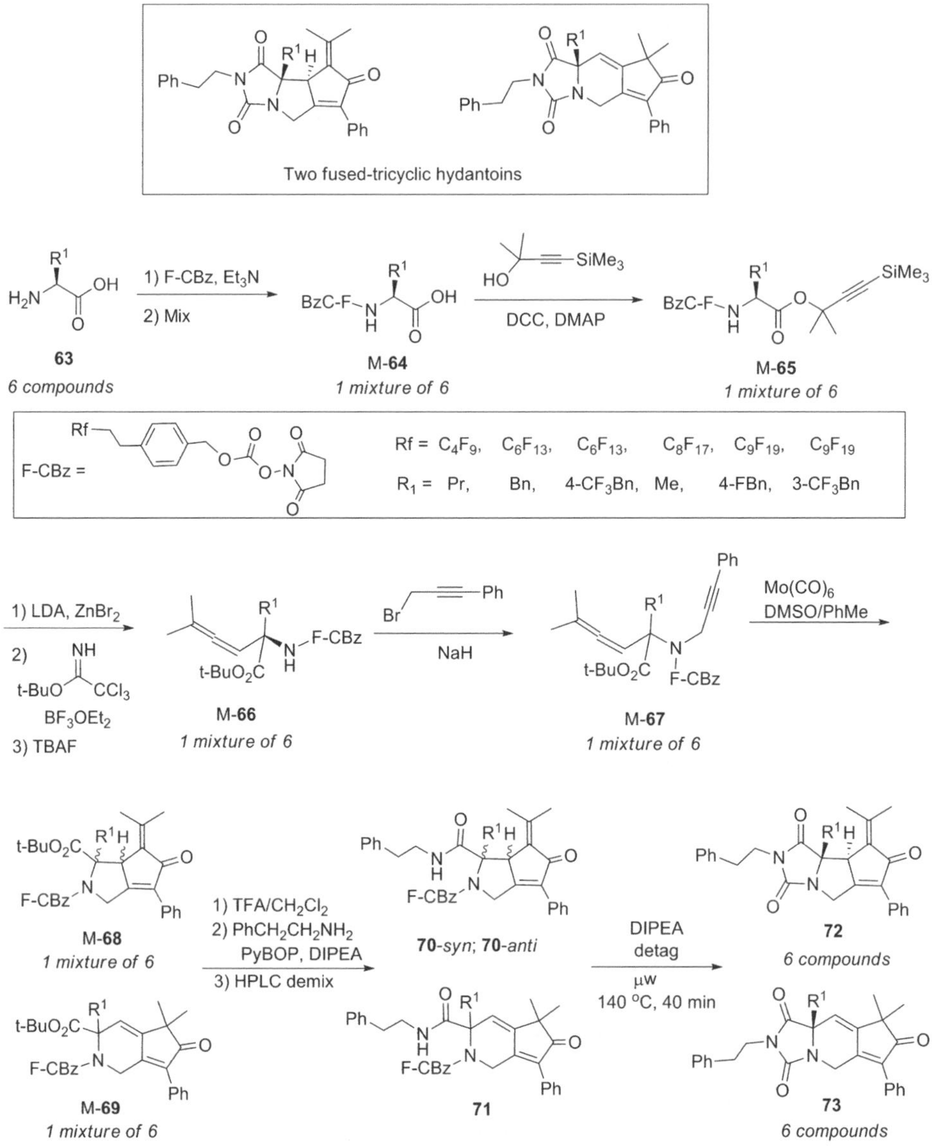

Two fused-tricyclic hydantoins

F-CBz = Rf = C_4F_9, C_6F_{13}, C_6F_{13}, C_8F_{17}, C_9F_{19}, C_9F_{19}

R_1 = Pr, Bn, 4-CF_3Bn, Me, 4-FBn, 3-CF_3Bn

Scheme 13.15 *FMS of two fused-tricyclic hydantoin scaffolds.*

stereoisomers M-**68**-*syn* and M-**68**-*anti* along with 4-alkylidenecyclopentanone regioisomer M-**69**. This complex mixture showed only three main spots in a standard silica TLC analysis, and it was purified by flash chromatography over regular silica gel. The single mixture of M-**68**-*syn*, M-**68**-*anti*, and M-**69** was treated with TFA to cleave the *tert*-butyl group, and the resulting acid mixture was coupled with phenethylamine. The mixture was separated by normal silica gel flash chromatography to give three submixtures containing predominantly M-**70**-*syn*, M-**70**-*anti*, and M-**71**. These mixtures were then demixed by

F-HPLC to give 17 crude individual products **70**-*syn/anti* and **71**. These crude products were not isomerically pure. Removal of the fluorous tag and hydantoin formation was achieved by treatment of the individual amides **70**-*syn/anti* and **71** with diisopropylethylamine (DIPEA) under microwave conditions. The cyclative cleavage reactions of **70**-*syn* and **70**-*anti* provided the same products **72**. Normal-phase HPLC purification gave 11 of 12 possible final products **72** and **73**.

13.2.12 Synthesis of Novel Heterocyclic Compound Libraries

Zhang and co-workers recently developed a [3 + 2] cycloaddition/detag/cyclization protocol for diversity-oriented synthesis (DOS) of hydantoin-, piperazinedione-, and benzodiazepine-fused heterocyclic scaffolds **76, 77**, and **78** (see Scheme 13.16) [52]. Each of these three scaffolds has four stereocenters on the central pyrrolidine ring and up to four points of diversity (R^1 to R^4). The ring skeleton of these compounds resembles the structures of some known biologically active compounds; the structure of compound **76** is similar to tricyclic thrombin inhibitors [53], the structure of compound **77** is similar to diketopiperazine-based inhibitors of human hormone-sensitive lipase [54], and compound **78** contains a privileged benzodiazepine moiety which exists in numerous pharmaceutically interesting compounds [55]. The key intermediate **75** for DOS was made by one-pot [3 + 2] cycloaddition of fluorous amino ester **74** with slight excess of benzaldehydes and maleimides. The cycloaddition was highly stereoselective. The stereostructure of **75** was confirmed by an x-ray crystal analysis.

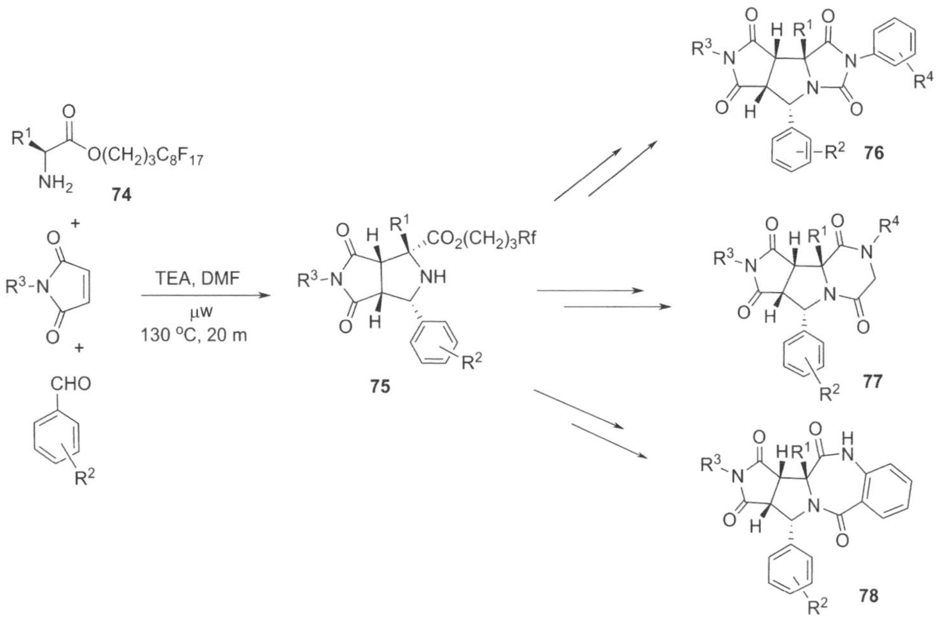

Scheme 13.16 *DOS of three novel heterocyclic scaffolds.*

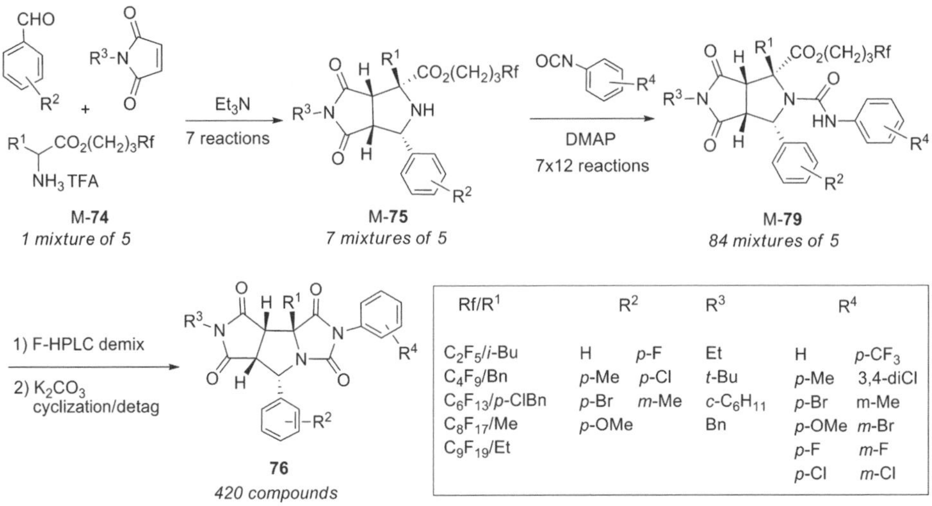

Rf/R^1	R^2		R^3	R^4	
C_2F_5/*i*-Bu	H	*p*-F	Et	H	*p*-CF_3
C_4F_9/Bn	*p*-Me	*p*-Cl	*t*-Bu	*p*-Me	3,4-diCl
C_6F_{13}/*p*-ClBn	*p*-Br	*m*-Me	*c*-C_6H_{11}	*p*-Br	m-Me
C_8F_{17}/Me	*p*-OMe		Bn	*p*-OMe	*m*-Br
C_9F_{19}/Et				*p*-F	*m*-F
				p-Cl	*m*-Cl

Scheme 13.17 *Five-component FMS of a 420-member fused-hydantoin library.*

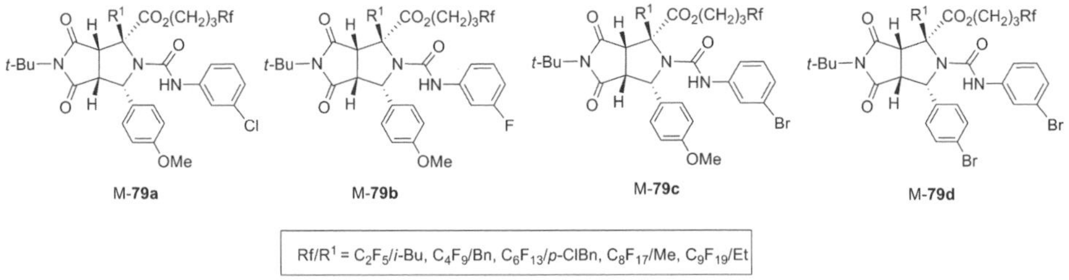

Rf/R^1 = C_2F_5/*i*-Bu, C_4F_9/Bn, C_6F_{13}/*p*-ClBn, C_8F_{17}/Me, C_9F_{19}/Et

Scheme 13.18 *Four FMS samples for parallel demixing.*

A 420-membered hydantoin-fused tricyclic compound library was synthesized by a five-component FMS (see Scheme 13.17) [56]. Five α-amino acids bearing different R^1 groups were paired with five perfluoroalkyl alcohols in such combinations as C_2F_5/*i*-Bu, C_4F_9/Bn, C_6F_{13}/*p*-ClBn, C_8F_{17}/Me, C_9F_{19}/Et. An equimolar mixture of five fluorous amino esters M-**74** was split into seven portions for 1,3-dipolar cycloaddition reaction with one of the seven benzaldehydes and one of the four maleimides. The resulting seven mixtures of M-**75** were each split into 12 portions and reacted with one of the 12 phenylisocyanates to form 84 mixtures of M-**79**. F-HPLC demixing followed by parallel detagging produced a 420-member library of **76**.

The incorporation of four-column parallel F-HPLC coupled with a multichannel MS interface increased the speed both for sample analysis and sample demixing [56]. A 5 min analysis method for baseline separation of five components of M-**79a–d** was applied to four-channel parallel LC-MS analysis (see Scheme 13.18 and Figure 13.2). Four mixture samples containing a total of 20 compounds could be separated in 5 min, which dramatically improves the efficiency of FMS.

The synthesis of a 60-member benzodiazepine-fused tetracyclic compound library **78** has also been accomplished by the Zhang group [56]. A mixture of two fluorous amino

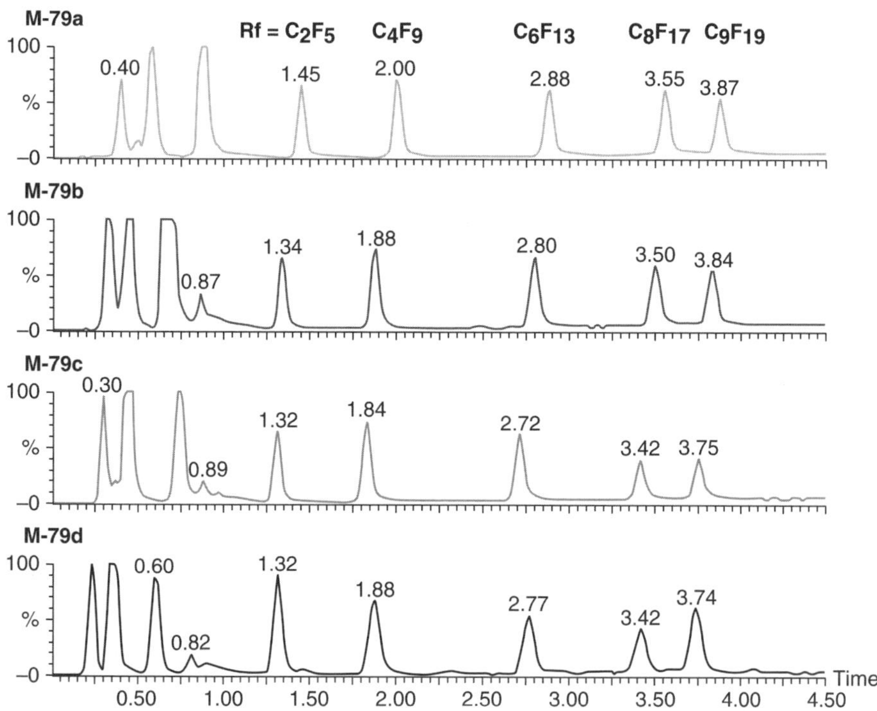

Figure 13.2 *Four-column parallel LC-MS of four FMS reaction mixtures (M-**79a–d**).* (Source: *Reproduced with permission from Zhang, W., Lu, Y., Chen, C. H.-T., Zeng, L., Kassel, D. B., Fluorous mixture synthesis of two libraries with hydantoin- and benzodiazepinedione-fused heterocyclic scaffolds,* J. Comb. Chem. *(2006)* **8**, *687–695. Copyright (2006) American Chemical Society.)*

acids (R^1 = H and Me) attached to C_6F_{13} and C_8F_{17} was reacted with 10 aldehydes and three maleimides to give 30 mixtures of M-**75** (see Scheme 13.19). *N*-Acylation followed by nitro group reduction with zinc dust in acetic acid under sonication gave 30 mixtures of M-**81**. F-HPLC demixing of M-**81** gave 60 individual compounds. These compounds then underwent cyclative tag cleavage with 1,8-diazabicyclo[4.3.0]non-5-ene (DBU) to form the corresponding benzodiazepine-fused tetracyclic compounds **78**.

13.3 Conclusions

FMS is a new and highly efficient solution-phase synthetic technology for making individually pure products. Fluorous tag-based HPLC ensures the identification and separation of reaction mixtures. FMS has advantages of homogeneous reaction environment, easy analysis and purification of reaction intermediates, easy adoption of traditional solution-phase reaction conditions, no requirement to use large excess of reagents for reaction completion, and good chemical and thermal stability of fluorous tags. It has been

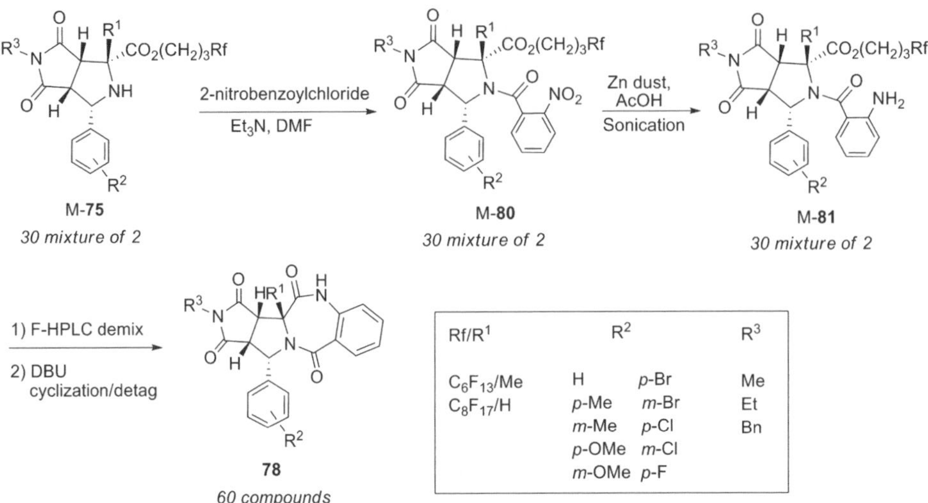

Scheme 13.19 *Two-component FMS of a 60-member fused-benzodiazapindione library.*

demonstrated in the synthesis of drug-like molecules, complex natural products and their enantiomers, diastereomers, and analogues. As a promising new technology, FMS will play increasingly important roles in organic chemistry, medicinal chemistry, and compound library synthesis.

Acknowledgments

I thank Professor Dennis P. Curran and his group for their pioneering and continuous work on FMS, my former co-workers Dr. Zhiyong Luo, Ms. Christine Chen, and Mr. Yimin Lu for their contributions to the FMS projects conducted at FTI, and National Institutes of General Medical Sciences for Phase I and Phase II SBIR grants (1R43GM062717-01 and 2R44GM062717-02).

References

1. Wessjohann, L. A. (2000) Synthesis of natural-product-based compound libraries. *Curr. Opin. Chem. Biol.*, **4**, 303–309.
2. McGee, P. (2006) Natural products re-emerge. Drug Discov. *Develop.*, **5**, 18–26.
3. Bindseil, K. U., Jakupovic, J., Wolf, D., *et al.* (2001) Pure compound libraries; a new perspective for natural product based drug discovery. *Drug Discov. Today* **6**, 840–847.
4. Horton, D. A., Bourne, G. T. and Smythe, M. L. (2003) The combinatorial synthesis of bicyclic privileged structures or privileged substructures. *Chem. Rev.*, **103**, 893–930.

5. Lipinski, C. A., Lombardo, F., Dominy, B. W. and Feeney, P. J. (1997) Experimental and computational approaches to estimate solubility and permeability in drug discovery and development settings. *Adv. Drug Deliv. Rev.*, **23**, 3–25.
6. Geysen, H. M., Schoenen, F., Wagner, D. and Wagner, R. (2003) Combinatorial compound libraries for drug discovery: An ongoing challenge. *Nat. Rev. Drug Discov.* **2**, 222–230.
7. Gladysz, J. A., Hovath, I., Curran. D. (eds) (2004) *The Handbook of Fluorous Chemistry*, Wiley-VCH Verlag GmbH, Weinheim.
8. Studer, A., Hadida, S., Ferritto, S.Y., *et al.* (1997) Fluorous synthesis: A fluorous-phase strategy for improving separation efficiency in organic synthesis. *Science* **275**, 823–826.
9. Curran, D. P. (1998) Strategy-level separations in organic synthesis: from planning to practice. *Angew. Chem. Int. Ed. Eng.*, **37**, 1174–1196.
10. Tzschucke, C. C., Markert, C., Bannwarth, W., *et al.* (2002) Modern separation techniques for the efficient workup in organic synthesis. *Angew. Chem. Int.* Ed., **41**, 3964–4000.
11. Two special issues on fluorous chemistry. For the Editorial articles see: (a) Gladysz, J. A. and Curran, D. P. (2002) Fluorous chemistry: from biphasic catalysis to a parallel chemical universe and beyond. *Tetrahedron*, **58**, 3823–3825. (b) Zhang, W. (2006) Fluorous chemistry: from biphasic catalysis to high-throughout synthesis of small molecules and separation/immobilization of biomolecules. *QSAR Comb. Sci.*, **25**, 679.
12. Zhang, W. (2003) Fluorous technologies for solution-phase high-throughput organic synthesis. *Tetrahedron* **59**, 4475–4489.
13. Zhang, W. (2004) Fluorous synthesis of heterocyclic systems. *Chem. Rev.*, **104**, 2531–2556.
14. Zhang, W. (2004) Fluorous tagging strategy for solution-phase synthesis of small molecules, peptides and oligosaccharides. *Curr. Opin. Drug Discov. Develop.*, **7**, 784–797.
15. Curran, D. P. (2006) Organic synthesis with light-fluorous reagents, reactants, catalysts, and scavengers. *Aldrichimica Acta*, **39**, 1–9.
16. Luo, Z. Y., Zhang, Q. S., Oderaotoshi, Y. and Curran, D. P. (2001) Fluorous mixture synthesis: A fluorous-tagging strategy for the synthesis and separation of mixtures of organic compounds. *Science* **291**, 1766–1769.
17. Curran, D. P. (2004) Light fluorous chemistry – A user's guide, in *The Handbook of Fluorous Chemistry* (eds J. A. Gladysz, D. P. Curran and I. T. Horvath), Wiley-VCH Verlag GmbH, Weinheim, pp. 128–156.
18. A short review article on FMS: Zhang, W. (2004) Fluorous mixture synthesis (FMS) of enantiomers, diastereomers, and compound libraries. *Arkivoc*, (**i**) 101–109.
19. Fluorous Technologies, Inc.: www.fluorous.com.
20. Curran, D. P. (2004) Separations with fluorous silica gel and related materials, in *Handbook of Fluorous Chemistry* (eds J. A. Gladysz, D. P. Curran and I. T. Horvath), Wiley-VCH Verlag GmbH, Weinheim, pp. 101–127.
21. Curran, D. P. (2001) Fluorous reverse phase silica gel. A new tool for preparative separations in synthetic organic and organofluorine chemistry. *Synlett*, 1488–1496.
22. Curran, D. P. and Oderaotoshi, Y. (2001) Thiol additions to acrylates by fluorous mixture synthesis: Relative control of elution order in demixing by the fluorous tag and the thiol substituent. *Tetrahedron*, **57**, 5243–5253.
23. Glatz, H., Blay, C., Engelhardt, H. and Bannwarth, W. (2004) New fluorous reversed phase silica gel for HPLC separations of perfluorinated compounds. *Chromatographia*, **59**, 567–570.
24. Zhang, W. and Curran, D. P. (2006) Synthetic applications of fluorous solid-phase extraction (F-SPE). *Tetrahedron*, **62**, 11837–11865.
25. Zhang, W., Luo, Z., Chen, C. H.-T. and Curran, D. P. (2002) Solution-phase preparation of a 560-compound library of individually pure mappicine analogs by fluorous mixture synthesis. *J. Am. Chem. Soc.*, **124**, 10443–10450.

26. Zhang, Q. S. and Curran, D. P. (2005) Quasienantiomers and quasiracemates: new tools for identification, analysis, separation, and synthesis of enantiomers. *Chem. Eur. J.*, **11**, 4866–4880.
27. Curran, D. P., Amatore, M., Guthrie, D., *et al.* (2003) Synthesis and reactions of fluorous carbobenzyloxy ((f)cbz) derivatives of alpha-amino acids. *J. Org. Chem.*, **68**, 4643–4647.
28. (a) Takahashi, S., Uchida, K., Kakinuma, N., *et al.* (1998) The structures of pyridovericin and pyridomacrolidin, new metabolites from the entomopathogenic fungus, *Beauveria bassiana. J. Antibiot.*, **51**, 1051–1054. (b) Takahashi, S., Kakinuma, N., Uchida, K., *et al.* (1998) Pyridovericin and pyridomacrolidin: Novel metabolites from entomopathogenic fungi, Beauveria bassiana. *J. Antibiot.*, **51**, 596–598.
29. Zhang, Q. S., Rivkin, A. and Curran, D. P. (2002) Quasiracemic synthesis: concepts and implementation with a fluorous tagging strategy to make both enantiomers of pyridovericin and mappicine. *J. Am. Chem. Soc.*, **124**, 5774–5781.
30. Govindachari, T, R., Ravindranath, K. R. and Viswanathan, N. (1974) Mappicine, a minor alkaloid from *Mappia foetida* Miers. *J. Chem. Soc., Perkin Trans.* **1**, 1215–1217.
31. (a) Pendrak, I., Barney, S., Wittrock, R., *et al.* (1994) Synthesis and anti-HSV activity of A-ring-deleted mappicine ketone analog. *J. Org. Chem.*, **59**, 2623–2625. (b) Pendrak, I., Wittrock, R. and Kingsbury, W. D. (1995) Synthesis and anti-HSV activity of methylenedioxy mappicine ketone analogs. *J. Org. Chem.*, **60**, 2912–2915.
32. Woo, M. H., Zeng, L., Ye, Q., *et al.* (1995) 16,19-*cis*-Murisolin and murisolin a, two novel bioactive monotetrahydrofuran annonaceous acetogenins from *Asimina triloba* seeds. *Bioorg. Med. Chem. Lett.*, **5**, 1135–1140.
33. Zhang, Q. S., Lu, H. J., Richard, C. and Curran, D. P. (2004) Fluorous mixture synthesis of stereoisomer libraries: Total syntheses of (+)-murisolin and fifteen diastereoisomers. *J. Am. Chem. Soc.*, **126**, 36–37.
34. Wilcox, C. S., Gudipati, V., Lu, L., *et al.* (2005) Solution-phase mixture synthesis with double-separation tagging: double demixing of a single mixture provides a stereoisomer library of 16 individual murisolins. *Angew. Chem. Int. Ed.*, **44**, 6938–6940.
35. Gudipati, V., Curran, D. P. and Wilcox, C. S. (2006) Solution-phase parallel synthesis with oligoethylene glycol sorting tags. Preparation of all four stereoisomers of the hydroxybutenolide fragment of murisolin and related acetogenins. *J. Org. Chem.*, **71**, 3599–3607.
36. Larsson, M., Nguyen, B.-V., Hogberg, H.-E. and Hedenstrom, E. (2001) Syntheses of the sixteen stereoisomers of 3,7,11-trimethyl-2-tridecanol, including the (2*S*,3*S*,7*S*,11*R*) and (2*S*,3*S*,7*S*,11*S*) stereoisomers identified as pheromone precursors in females of the pine sawfly *Microdiprion pallipes* (Hymenoptera: Diprionidae). *Eur. J. Org. Chem.*, 353–363.
37. Dandapani, S., Jeske, M. and Curran, D. P. (2004) Stereoisomer libraries: Total synthesis of all 16 stereoisomers of the pine sawfly sex pheromone by a fluorous mixture-synthesis approach. *Proc. Natl. Acad. Sci. U. S. A.*, **101**, 12008–12012.
38. Dandapani, S., Jeske, M. and Curran, D. P. (2005) Synthesis of all 16 stereoisomers of pinesaw fly sex pheromones tools and tactics for solving problems in fluorous mixture synthesis. *J. Org. Chem.*, **70**, 9447–9462.
39. Pettit, G. R., Cichacz, Z. A., Gao, F., *et al.* (1994) Isolation and structure of the cancer cell growth inhibitor dictyosatin 1. *J. Chem. Soc., Chem. Commun.*, 1111–1112.
40. Paterson, I., Britton, R., Delgado, O., *et al.* (2004) Total synthesis and configurational assignment of (−)-dictyostatin, a microtubule-stabilizing macrolide of marine sponge origin. *Angew. Chem. Int. Ed.*, **43**, 4629–4633.
41. Shin, Y., Fournier, J. H., Fukui, Y., *et al.* (2004) Total synthesis of (−)-dictyostatin: confirmation of relative and absolute configurations. *Angew. Chem., Int. Ed.*, **43**, 4634–4637.
42. (a) Isbrucker, R. A., Cummins, J., Pomponi, S. A., *et al.* (2003) Tubulin polymerizing activity of dictyostatin-1, a polyketide of marine sponge origin. *Biochem. Pharm.*, **66**, 75–82.

(b) Madiraju, C., Edler, M. C., Hamel, E., *et al.* (2005) Tubulin assembly, taxoid site binding, and cellular effects of the microtubule-stabilizing agent dictyostatin. *Biochemistry*, **44**, 15053–15063.

43. Fukui, Y., Brückner, A. M., Shin, Y., *et al.* (2006) Fluorous mixture synthesis of (–)-dictyostatin and three stereoisomers. *Org. Lett.*, **8**, 301–304.
44. Curran, D. P., Moura-Letts, G. and Pohlman, M. (2006) Solution-phase mixture synthesis with fluorous tagging en route: total synthesis of an eight-member stereoisomer library of passifloricins. *Angew. Chem. Int. Ed.*, **45**, 2423–2426.
45. Lindell, T., Jenson, P. R. and Fenical, W. (1996) Lagunapyrones A–C: Cytotoxic acetogenins of a new skeletal class from a marine sediment bacterium. *Tetrahedron Lett.*, **37**, 1327–1330.
46. Yang, F. and Curran, D. P. (2006) Structure assignment of lagunapyrone b by fluorous mixture synthesis of four candidate stereoisomers. *J. Am. Chem. Soc.*, **128**, 14200–14205.
47. Curran, D. P. and Furukawa, T. (2002) Simultaneous preparation of four truncated analogues of discodermolide by fluorous mixture synthesis. *Org. Lett.*, **4**, 2233–2235.
48. Lins, A. P. and Felicio, J. D. (1993) Monoterpene alkaloids from *Tecoma stans*. *Phytochemistry*, **34**, 876–878.
49. Brummond, K. M. and Mitasev, B. (2004) Allenes and transition metals: A diverging approach to heterocycles. *Org. Lett.*, **6**, 2245–2248.
50. Manku, S. and Curran, D. P. (2005) Fluorous mixture synthesis of 4-alkylidene cyclopentenones via a rhodium-catalyzed [2 + 2 + 1] cycloaddition of alkynyl allenes. *J. Comb. Chem.*, **7**, 63–68.
51. Manku, S. and Curran, D. P. (2005) Fluorous mixture synthesis of fused-tricyclic hydantoins. Use of a redundant tagging strategy on fluorinated substrates. *J. Org. Chem.*, **70**, 4470–4473.
52. Zhang, W., Lu, Y., Chen, C. H.-T., *et al.* (2006) Fluorous synthesis of hydantoin-, piperidinedione-, and benzodiazepinedione-fused tricyclic and tetracyclic ring systems. *Eur. J. Org. Chem.*, 2055–2059.
53. Olsen, J., Seiler, P., Wagner, B., *et al.* (2004) A fluorine scan of the phenylamidinium needle of tricyclic thrombin inhibitors: effects of fluorine substitution on pK_a and binding affinity and evidence for intermolecular C-F … CN interactions. *Org. Biomol. Chem.*, **2**, 1339–1352.
54. Slee, D. H., Bhat, A. S., Nguyen, T. N., *et al.* (2003) Pyrrolopyrazinedione-based inhibitors of human hormone-sensitive lipase. *J. Med. Chem.*, **46**, 1120–1122.
55. Kamal, A., Reddy, K. L., Devaiah, V., *et al.* (2006) Recent advances in the solid-phase combinatorial synthetic strategies for the benzodiazepine based privileged structures. Mini-Rev. *Med. Chem.*, **6**, 53–68.
56. Zhang, W., Lu, Y., Chen, C. H.-T., *et al.* (2006) Fluorous mixture synthesis of two libraries with hydantoin- and benzodiazepinedione-fused heterocyclic scaffolds. *J. Comb. Chem.*, **8**, 687–695.

14

Fluorine-18 Radiopharmaceuticals

Michael R. Kilbourn and Xia Shao

14.1 Introduction

PET (positron emission tomography) is a powerful *in vivo* imaging technique capable of providing dynamic information on biochemical processes in the living human subject. Applications of PET in oncology, neurology, psychiatry, cardiology, and other medical specialties continue to grow each year, and this modern imaging technique has become an important part of medical care. PET has also emerged as an important tool in drug research and development, where it is increasingly used to measure pharmacokinetics and pharmacodynamics. The use of PET is driven by the characteristics and availability of appropriately labeled radiopharmaceuticals that are specifically designed for measurement of targeted biochemical processes. Of the handful of positron-emitting radionuclides that have been utilized for PET imaging, fluorine-18 has particularly appealing properties. Addition of a fluorine atom to a large molecule can often (but not always) be accomplished without significant changes to the physiochemical or biological properties of the compound. Fluorine-18 is a positron-emitting radionuclide (97% positron emission, 0.635 MeV maximum energy, mean range 2.39 mm in water) with a relatively long half-life (109.6 min) compared to the other often-used positron-emitting radionuclides (e.g., carbon-11, 20.4 min; nitrogen-13, 9.96 min; oxygen-15, 2 min). The half-life permits both longer synthesis times (typically 1–2 half-lives, or nearly 4 hours), thus opening up possibilities for multistep radiolabeling procedures, as well as the option of preparation of the radiopharmaceutical at a location remote from the site of use.

A large number of ^{18}F-labeled organic molecules have now been synthesized, ranging from the simplest ([^{18}F]fluoromethane) to quite large (radiolabeled proteins). A thorough review of compounds that have now been synthesized in fluorine-18 form is impossible.

Fluorine in Medicinal Chemistry and Chemical Biology Edited by Iwao Ojima

In 1982, a comprehensive review by Fowler and Wolf listed 88 fluorine-18 radiochemicals [1]; by 1990, a subsequent review [2] had listed over 280 compounds in a similar table, and although there are no more recent attempts to compile a comprehensive listing of such radiochemicals, at the most recent International Symposium on Radiopharmaceutical Chemistry, there were over 140 individual abstracts involving fluorine-18 alone [3], indicative of the continued growth in the field of fluorine-18 radiochemistry. A few recent reviews are available for fluorine-18 radiopharmaceuticals [4–9], and the reader is directed to modern literature-searching methods to obtain a more complete listing of currently available radiochemicals.

The labeling methods for ^{18}F-fluorination reactions, described in the following sections, have been continuously expanded and improved upon through the years, leading to the proliferation of useful new ^{18}F-labeled compounds for *in vivo* studies of biochemistry in animals and humans. A few ^{18}F-labeled radiochemicals will be presented as examples of the issues involved in their design and synthesis. Due to the growing importance of PET, and the unrelenting need for new and better radiopharmaceuticals, continued development of new and improved methods for ^{18}F-labeling has been pursued and some of the most recent developments in that area will also be reviewed.

14.2 Fluorine-18 Chemistry

14.2.1 General Considerations

Fluorine-18 is a short-lived (109.6 min) radionuclide that has to be generated anew for each synthesis. Production of the radionuclide is these days almost exclusively by the use of a small cyclotron (typical energies of 11–30 MeV for protons), a particle accelerator which can often be accommodated within a medical center or other research facility, although the concept of producing and shipping the radionuclide to distant sites for use in radiochemical syntheses has been accomplished. The majority of fluorine-18 is produced via the proton irradiation of an oxygen-18-enriched water target (^{18}O(p,n) ^{18}F) with the radionuclide product obtained as [^{18}F]fluoride ion in aqueous solution. A variety of rapid and sometimes ingenious means have been developed to convert the aqueous [^{18}F]fluoride ion to both reactive nucleophilic fluoride (usually in an organic solvent; but see recent developments described below) or electrophilic fluorination reagents. Fluorine-18 can be generated by a number of other nuclear reactions (Table 14.1), and radiolabeled fluorine gas ([^{18}F]F_2) is generated directly in gas cyclotron targets, but these methods are of less utility in the field of fluorine-18 radiochemistry.

For the synthesis of high-specific-activity radiopharmaceuticals (specific activity is defined as radioactivity per unit mass) the use of [^{18}F]fluoride ion is preferred if not required. The theoretical specific activity for fluorine-18 is 1.71×10^9 Ci/mol, but that value has never been achieved and in practice the specific activity of ^{18}F-labeled radiochemicals has been significantly lower. Typically, finished radiochemicals are produced with specific activities in the 1000–10 000 Ci/mmol. Some of this difference has been attributed to physical dilution of the fluorine-18 by the presence of fluorine-19 in

Table 14.1 *Nuclear reactions used to produce fluorine-18*

$^{20}Ne(d,\alpha)^{18}F$
$^{20}Ne(p,2n)^{18}F$
$^{16}O(^{3}He,p)^{18}F$
$^{16}O(\alpha,pn)^{18}F$
$^{18}O(p,n)^{18}F$
$^{20}Ne(^{3}He,n)^{18}Ne,^{18}Ne\text{-}^{18}F$

reagents and equipment (e.g., glassware); in other instances the lower specific activities may better represent the ability of analytical methods such as high-pressure liquid chromatography (HPLC) to identify and quantify cold mass associated with a radiochemical product peak. Thus, the production of true "carrier-free" fluorine-18 radiopharmaceuticals has not been achieved, but in current practice specific activities of 1000–10 000 Ci/mol (37–370 GBq/mmol) are routinely achieved, and such specific activities are suitable for even the most demanding *in vivo* applications of the radiolabeled compounds.

The availability of fluorine-18 from high yield oxygen-18-enriched cyclotron water targets, the development of automated or remotely controlled synthesis apparatus capable of yielding hundreds of milliCuries to Curies of products, and the desire to produce new radiopharmaceuticals with the potential for widespread (and even commercial) distribution from the point of synthesis has resulted in a problem not often seen in earlier days of fluorine-18 chemistry: radiolytic decomposition. Syntheses can now be done using multi-Curie amounts of the radionuclide, often in small reaction volumes (commonly several hundred microliters of solvent, but sometimes even much lower than that), resulting in very high rates of irradiation of chemical components by both the positrons and the 511 keV gamma rays produced by their annihilation. Yields and radiochemical purities of reactions runs using low amounts of radionuclide, as is commonly done in development of new radiotracers, may not necessarily translate into high-level production runs. Radiolytic decomposition can also occur when high amounts (levels of several hundreds of milliCuries) of final products are formulated for injection, leading to the need in many cases to add an antioxidant or free-radical scavenger, such as ethanol or ascorbic acid, to the formulation to reduce decomposition of the radiochemical. This represents a unique problem not encountered in the usual use of fluorine in medicinal or bioorganic chemistry.

In the following sections are examples of typical methods for ^{18}F-fluorination using nucleophilic and electrophilic reaction mechanisms, with selected examples of their use in preparation of PET radiopharmaceuticals.

14.2.2 Radiochemistry: Yields and Specific Activities

In the field of radiochemistry, and particularly with the use of high-specific-activity radionuclides, syntheses are almost uniformly done with an excess of unlabeled precursor and reagents with respect to the radionuclide, in this review fluorine-18. For that reason the efficiencies of reactions are generally given in terms of radiochemical yields instead of chemical yields. However, radiochemical yields in the literature are defined in many different ways: as final product (amount of radioactive product) divided by the amount of radioactivity produced in the cyclotron target; as that perhaps rendered reactive by an intermediate preparation in the solvent of choice (e.g., preparing [^{18}F]fluoride ion in organic solvent); or even relative to a small-molecule intermediate prepared for the radiolabeling reaction (e.g., yields based on electrophilic fluorination reagents prepared from the product from the cyclotron target, [^{18}F]fluorine gas or [^{18}F]fluoride ion). Furthermore, radiochemical yields can be expressed either decay-corrected or without decay correction, depending on whether the investigator chooses to correct the yield for the time of synthesis. Because radiochemical yields can be expressed in so many different ways, and certainly are in the literature, in this review yields will not be routinely discussed unless they are significantly higher (or lower) than usually found, or where a single publication has reported multiple yields all calculated the same way such that comparisons can be properly made.

In the synthesis of fluorine-18 radiochemicals an important attribute of the final product is also the measured specific activity. Unfortunately, the literature is also replete with different means of estimating or measuring specific activity. The majority of fluorine-18 radiochemicals are today analyzed by some means of chromatography, such as gas or high-pressure liquid chromatography, leading to a specific activity measurement relating the observed mass in the chromatographic spectrum (often UV absorbance) relative to injection of standard solutions of known mass. Specific activity measurements can thus vary between investigators, for reasons relating to the quality of the radionuclide used in the synthesis but also depending on the precision and accuracy of the chromatographic analyses. In this review there are general discussions of the importance of no-carrier-added versus carrier-added fluorine-18 radiochemical syntheses: with a few exceptions, most radiopharmaceuticals are best if prepared in no-carrier-added form, which does not mean they are free of the presence of any fluorine-19 but simply that no cold carrier fluorine has been deliberately added either during the generation of the radionuclide or during the radiochemical synthesis.

14.3 Nucleophilic ^{18}F-fluorination: General Aspects

By far the majority of fluorine-18 radiotracers are synthesized using some type of nucleophilic fluorination reaction. As fluorine-18 is most widely produced by cyclotrons as an aqueous solution of [^{18}F]fluoride ion, considerable effort was expended in the development of rapid methods to convert this to a source of reactive [^{18}F]fluoride in organic solvents appropriate for subsequent radiochemical syntheses. This was accomplished by removal of water, most often by azeotropic evaporation in the presence of a cationic counterion; to enhance solubility of metallic cation–fluoride ion salts, often a cryptand is included, with the most popular pairing being the aminopolyether 8-crown-6 (Kryptofix) and potassium ions [10]. Other metal salts such as cesium and rubidium salts, or organic cations such as tetraalkylammonium salts, have been employed as alternatives. Finally, [^{18}F]fluoride ion can be efficiently trapped on anion exchange resins, which can be directly used for on-column nucleophilic reactions [11]. In general the reactions using [^{18}F]fluoride ion are conducted in aprotic dipolar organic solvents, such as dimethyl sulfoxide or acetonitrile, although recently a novel solvent system has been applied (see Section 14.5).

14.3.1 2-[^{18}F]Fluoro-2-deoxy-D-glucose: Two-step Aliphatic Nucleophilic Substitution

For both basic science animal research, clinical medical research, and clinical medical care, 2-[^{18}F]fluoro-2-deoxy-D-glucose ([^{18}F]FDG) is the single most used PET radiopharmaceutical in the world. As an analogue of glucose, the major energy source of biological systems, it has found applications in nearly every area of nuclear medicine imaging, with exceptionally widespread application in imaging of the location and characterization of tumors.

The synthesis of [^{18}F]FDG has been very well developed and refined since its introduction by Ido and co-workers in 1978 [12]. Whereas the original synthesis of [^{18}F]FDG was done using electrophilic fluorination of an alkene precursor, yields were limited by the maximum 50% possible from an electrophilic reagent (see later discussion on electrophilic ^{18}F-fluorinations), and much higher-yield nucleophilic fluorination methods were subsequently developed [13] and have been adopted by virtually every investigator at present.

The commonly used synthesis of [^{18}F]FDG (Figure 14.1) involves two steps: reaction of [^{18}F]fluoride ion with an appropriately substituted mannose derivative, followed by removal of the hydroxyl-protecting groups. Use of the tetraacetate esters of the hydroxyl groups, present in the sugar structure but also found in the hydroxyl or amino groups of many compounds of interest for radiolabeling with fluorine-18, is sometimes necessary as otherwise the reactive [^{18}F]fluoride ion acts as a strong base to extract the acidic hydrogen to form H[^{18}F]F, and prevents reaction of the [^{18}F]fluoride ion in a nucleophilic reaction. As nucleophilic fluorination to displace the 2-triflate leaving group proceeds

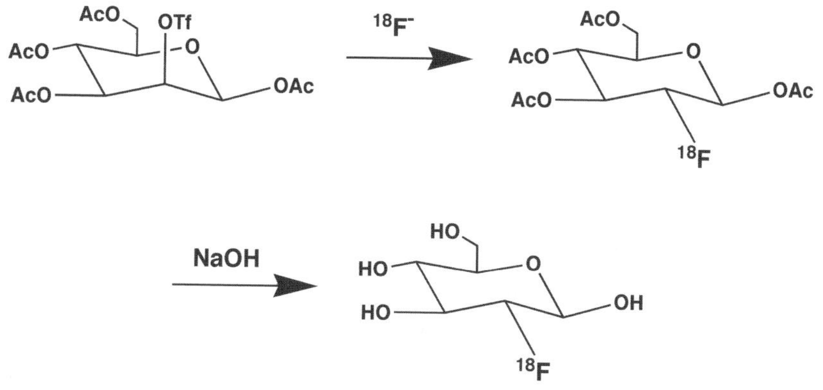

Figure 14.1 *Synthesis of 2-[^{18}F]fluoro-2-deoxy-D-glucose ([^{18}F]FDG) using two-step procedure of nucleophilic substitution of [^{18}F]fluoride ion for a triflate ester leaving group followed by base-catalyzed hydrolysis of acetyl protecting groups.*

by inversion of stereochemistry at the C-2 carbon center, a suitably protected mannose precursor rather than one derived from glucose itself becomes the substrate. Although syntheses using different cationic counterions and even solid-phase supported ^{18}F-fluorinations have been applied to [^{18}F]FDG syntheses, the state of the art utilizes the potassium ion/Kryptofix in acetonitrile system for ^{18}F-fluorination followed by acid or base hydrolysis of the protecting groups (base hydrolysis has been found to be faster and milder). The synthesis of [^{18}F]FDG is so well worked out, and so reliable, that relatively simple entirely computer-controlled automatic apparatus is marketed for the production of this important radiopharmaceutical. Purification of the final product is accomplished using a series of disposable bonded-phase chromatographic columns, neutralization of the base with an appropriate buffer solution, and sterilization; automated equipment to make [^{18}F]FDG is now available which produces the radiopharmaceutical in high yields (>60% corrected for decay) and overall synthesis times of less than 26 minutes (starting from end of cyclotron target irradiation). [^{18}F]FDG is clearly the most refined fluorine-18 synthesis yet developed.

14.3.2 [^{18}F]Fluoropropyldihydrotetrabenazine (FP-DTBZ): One-step Aliphatic Nucleophilic Substitution

For the *in vivo* imaging of the vesicular monoamine transporter type 2 (VMAT2), a monoamine-neuron specific protein in the brain and other organs, the clinically used drug tetrabenazine and several derivatives were labeled with carbon-11 leading to a final widely used radiopharmaceutical, (+)-α-[^{11}C]dihydrotetrabenazine (DTBZ) [14], which is in fact a single enantiomer (tetrabenazine is used clinically as a racemic mixture) of the predominant metabolite of tetrabenazine formed in the human body. Numerous studies have utilized this radiopharmaceutical in a variety of neurological and psychiatric diseases [15]. Although useful, the 20-minute half-life of the radionuclide carbon-11 significantly limited the application of this radiopharmaceutical to institutions possessing an on-site cyclotron and radiochemistry expertise in carbon-11 syntheses. The development of [^{18}F]fluoropropyldihydrotetrabenazine ([^{18}F]FP-DTBZ) represents an excellent example of how a ^{18}F-labeled alternative can be readily designed and implemented by building on the carbon-11 experience. [^{18}F]FP-DTBZ also represents an example of a radiopharmaceutical that can be prepared in a single step but which requires much more effort in purification and quality control.

As the *in vivo* imaging application of [^{18}F]FP-DTBZ requires a high-specific-activity radiopharmaceutical, due to the limited capacity of the high-affinity VMAT2 binding sites in the brain, a synthesis starting with [^{18}F]fluoride ion was deemed necessary. The stereochemical requirements of the ligand had been evaluated in the syntheses of the carbon-11 radiopharmaceuticals, and thus a straightforward synthesis of 9-*O*-fluoroalkyl derivatives of (+)-α-9-desmethyldihydrotetrabenazine provided high-affinity radioligands with excellent *in vitro* properties and with binding affinities equal to or better than the parent tetrabenazine [16, 17]. Radiolabeling of the fluoroalkyl groups could then be readily achieved by a one-step aliphatic nucleophilic substitution, with [^{18}F]fluoride ion substitution of a mesylate ester leaving group (Figure 14.2). Purification of the desired [^{18}F]fluoroalkyl-DTBZ requires application of high-pressure liquid chromatography (HPLC) to effectively

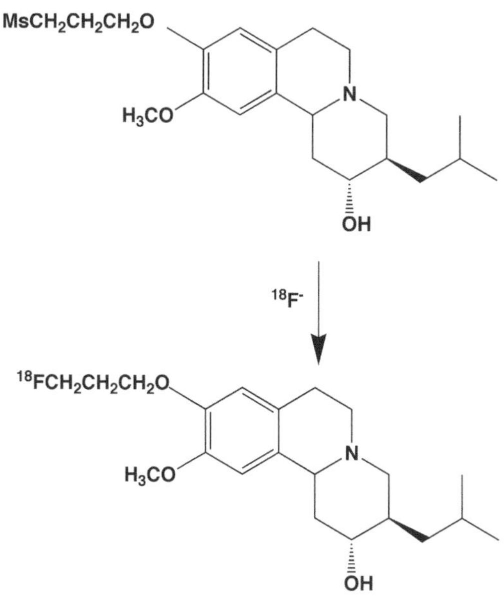

Figure 14.2 *Synthesis of 9-(3-[^{18}F]fluoropropyl)-9-desmethyl-(+)-(α)-dihydrotetrabenazine ([^{18}F]FP-DTBZ) by a single-step nucleophilic aliphatic substitution of a mesylate ester leaving group.*

remove unreacted mesylate precursor as well as any hydroxyl derivative formed by hydrolysis of the sulfonate ester. There are actually few radiochemical impurities in many simple nucleophilic ^{18}F-fluorination reactions: the unreacted [^{18}F]fluoride ion is often rapidly removed using relatively simple column chromatography (bonded-phase chromatography, such as Sep-Paks) and the HPLC purification is mostly necessary to obtain the required chemical purity. The radiolabeling strategy for [^{18}F]FP-DTBZ is an example of that used for many fluorine-18 radiopharmaceuticals and utilizes the displacement of a sulfate ester (mesylate, tosylate) group by [^{18}F]fluoride ion. Interestingly, this is an example where the free hydroxyl group in the molecule did not have to be protected as an ester or ether to allow reactivity of the [^{18}F]fluoride ion with the mesylate leaving group. Syntheses of radiopharmaceuticals incorporating [^{18}F]fluoroalkyl groups have mostly been limited to [^{18}F]fluoromethyl, [^{18}F]fluoroethyl and [^{18}F]fluoropropyl groups, likely due to the increasing lipophilicity of longer alkyl chains, which can lead to undesired *in vivo* pharmacokinetics.

[^{18}F]FP-DTBZ is an example of a ^{18}F-labeled radiopharmaceutical where high specific activity is a requirement because of the limited capacity of the *in vivo* binding site, and the desire for any *in vivo* radiotracer to occupy a negligible number of such sites. For other radiopharmaceuticals, the need for high specific activity is not only rooted in this desire for minimal occupancy but is also due to potential unwanted pharmacological or toxicological effects of the higher masses of radioligand if administered at a low specific activity. This is exemplified by such radiotracers as [^{18}F]fluoroethoxybenzovesamicol [18] and ^{18}F-labeled nicotinic acetylcholinergic receptor agonists [19–22], two classes of compounds with significant peripheral pharmacological effects that severely limit the allowed administered mass of compound to prevent such physiological effects. Fortunately, sufficiently high specific activities can now be attained routinely with fluorine-18 radiotracers such that human studies of these compounds are quite feasible.

14.3.3 Aromatic Nucleophilic Substitution: Activated Aryl Rings

A large number of compounds containing aromatic rings have now been radiolabeled with fluorine-18. Most often, nucleophilic substitutions of [^{18}F]fluoride ion is done using a suitable leaving group (e.g., halogen, nitro, or trimethylammonium group) on a benzene ring bearing an electron-withdrawing substituent (e.g., nitro, ketone, aldehyde, ester, nitrile), through a classical aromatic nucleophilic substitution reaction. Syntheses of [^{18}F]fluoroheterocyclic rings proceeds in a similar fashion. Nucleophilic substitution by [^{18}F]fluoride also proceeds successfully with aryl rings bearing both electron-withdrawing and electron-donating substituents [23].

In using simple ^{18}F-fluorination of an activated benzene ring, radiotracers are usually designed such that the activating group is easily incorporated into the final radiotracer design, converted rapidly into a functional group needed in the final product, or even removed entirely. An example of incorporating the activating group into the final desired radiopharmaceutical is 4-[^{18}F]fluorobenzoyl hexadecanoate ([^{18}F]HFB), a long-chain ester developed for radiolabeling of stem cells as it is rapidly and simply absorbed and retained in the lipid bilayer of a cell [24]. Synthesis of [^{18}F]HFB was accomplished by reaction of [^{18}F]fluoride ion, using the potassium/Kryptofix system in dimethyl sulfoxide, with a 4-trialkylammonium-substituted benzoyl ester precursor (Figure 14.3). The synthesis is thus one-step: the desired ^{18}F-labeled ester is easily separated from the cationic precursor as well as unreacted [^{18}F]fluoride ion by simple column filtration. As an alternative, a synthesis of a long-chain fatty benzoic acid ester with the fluorine-18 on the alkyl chain could also have been employed, by utilizing an appropriately placed leaving group as in the synthesis of fluorine-18 fatty acids such as 16-[^{18}F]fluorohexadecanoic and 14-[^{18}F]fluoro-6-thiaheptadecanoic acid [25]. This exemplifies how fluorine-18 chemistry often offers several options for incorporation into different positions of desired target molecules.

Nucleophilic aromatic ^{18}F-fluorination has also been used in the synthesis of small prosthetic groups which are then incorporated into larger molecules, or attached to very large molecules such as peptides or proteins. This approach is often taken if the larger organic compound is unsuitable for direct ^{18}F-fluorination reactions (for example, due to

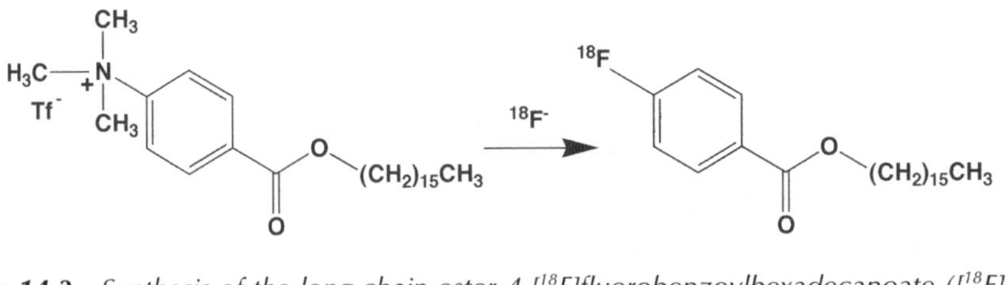

Figure 14.3 Synthesis of the long-chain ester 4-[^{18}F]fluorobenzoylhexadecanoate ([^{18}F]HFB) using [^{18}F]fluoride ion in a nucleophilic aromatic substitution of a trimethylammonium leaving group, with the ring activated by the ester function.

instability under usual conditions of nucleophilic or electrophilic ^{18}F-fluorination, or insolubility in solvent systems typically used), or if the needed synthetic precursor for direct substitution proves too hard to prepare. Examples of small aromatic molecules used as prosthetic groups include 4-[^{18}F]fluorophenacyl bromide [17], *N*-succimidyl-4-[^{18}F]fluorobenzoate [27], *N*-succimidyl-4-([^{18}F]fluoromethyl)benzoate [28], and 4-[^{18}F]fluorobenzaldehyde [29], all prepared using aromatic nucleophilic ^{18}F-fluorination as the first synthetic step. The single drawback to this approach is that all of these small molecules are themselves synthesized using multiple steps and some purification is needed before they are usable in the final desired reaction to label the target molecule.

Nucleophilic aromatic substitution of aryl rings followed by removal of the ring-activating electron-withdrawing group offers a more lengthy but feasible route to the ^{18}F-labeling of aromatic rings bearing no activating groups or, even more challenging, bearing electron-donating groups that normally inhibit aromatic nucleophilic substitution reactions [23, 30]. Although useful fluorine-18 radiopharmaceuticals have been prepared using these methods [31, 32], a certain loss of overall radiochemical yield is expected due to radionuclide decay during these required extra steps after ^{18}F-fluorination. As an alternative there are more direct methods for radiofluorination of electron-rich aryl rings, as discussed in the next section.

14.3.4 ^{18}F-Labeling of Electron-Rich Aryl Rings

As not all biological molecules or drug candidates present a suitable activated aryl ring for simple nucleophilic aromatic substitutions, the direct ^{18}F-labeling of electron-rich aromatic rings represent a significant challenge in the synthesis of radiolabeled pharmaceuticals. Electrophilic reagents such as: [^{18}F]F_2 and acetyl [^{18}F]hypofluorite (AcOF), have been used to label electron-rich aromatics [28]. These electrophilic reagents are generally produced via carrier-added methods and thus provide final products with low to at best moderate specific activity. This has limited the use of electrophilic ^{18}F-fluorination reactions to the production of radiopharmaceuticals for which there is no need for high specific activity and where the chemical species are not toxic [5]. In addition, these reagents are very reactive and non-regioselective, often leading to a mixture of ^{18}F-labeled products and a requirement for careful separation and purification. This reduces the radiochemical yield of any single desired product.

Recently, a few improved methods of direct ^{18}F-fluorination on electron-rich aromatics have been reported, using nucleophilic substitution [33–36] or electrophilic substitution [37–39]. These improved methods are encouraging solutions to the complex problem of ^{18}F-fluorination of electron-rich aryl rings.

14.3.4.1 Nucleophilic ^{18}F-Fluorinations Using Iodonium Salts

For years, diaryliodonium salts have been used for the direct ^{18}F-fluorinations in nucleophilic substitution reaction on aromatic rings [34, 35, 40, 41]. The introduction of no-carrier-added [^{18}F]fluoride into an aryl substituent of diaryliodonium salt results in a [^{18}F]fluorinated arene and a corresponding iodoarene (Figure 14.4).

The nucleophilic attack by [^{18}F]fluoride ion on the diaryliodonium salt occurs preferably at the more electron-deficient ring [42]. Therefore, a more electron-rich ring such as

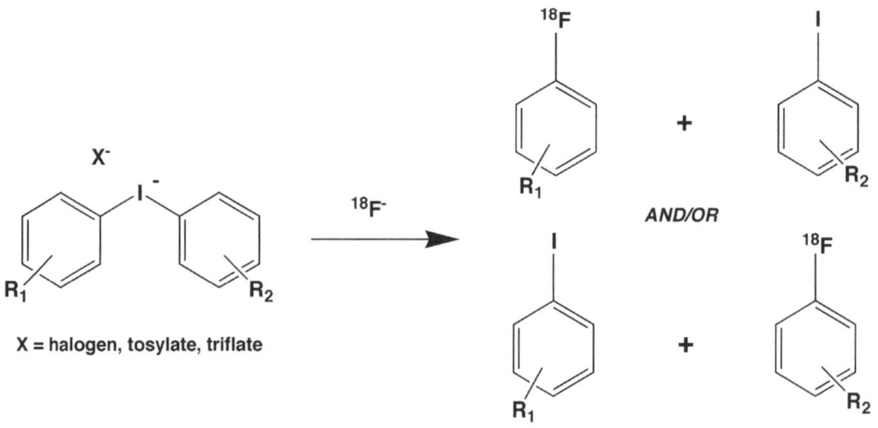

X = halogen, tosylate, triflate

Figure 14.4 Reaction of [^{18}F]fluoride ion with diaryliodonium salts. Proportions of the four possible products ([^{18}F]fluoroarenes vs. iodoarenes) is dependent on the ring substituents.

Table 14.2 Radiochemical yields (RCY) for nucleophilic ^{18}F-fluorination of aryl(2-thienyl)iodonium bromides

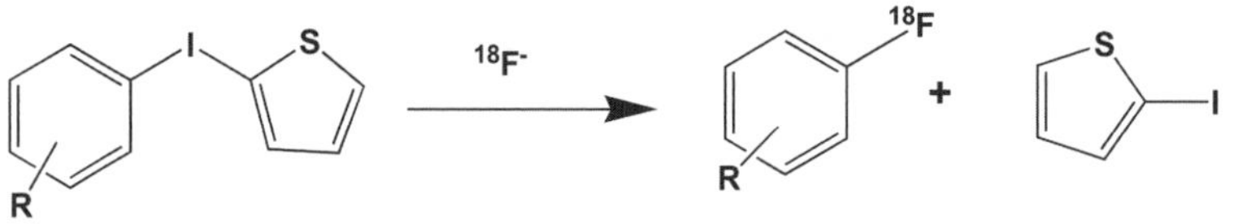

R	RCY (%)
4-OBn	36 ± 3
4-OMe	29 ± 3
4-Me	32 ± 2
H	64 ± 4
3-OMe	20 ± 3
4-Cl	62 ± 4
4-Br	70 ± 5
4-I	60 ± 8
2-OMe	61 ± 5

a heteroaromatic moiety in the iodonium salt should allow the direct nucleophilic ^{18}F-labeling of a second aryl group that is comparatively less but still electron-rich [36]. It has been reported by Ross *et al.* using the 2-thienyl group as a highly electron-rich heteroaromatic ring to direct the nucleophilic ^{18}F-fluorination to the second, less electron-rich aryl ring [34]. This heteroaromatic system provides a regiospecific single radioactive product and yields of the corresponding aryl substituents via this method are shown in Table 14.2. As expected, the radiochemical yields were generally high with a decrease of the electron density, except for the *ortho*-methoxy derivative (61%). The huge difference between *ortho*- and *para*-substituted rings is not explainable by the regular character of the aromatic nucleophilic substitution reaction mechanism, and the *meta*-derivative shows a slightly higher initial reaction rate than the *ortho*- and *para*-derivatives but ends in a lower radiochemical yield. These might be due to a so-called Meisenheimer complex formed as an intermediate in this reaction, which is stabilized by resonance structures [34].

14.3.5 Prosthetic Groups: Fluoroalkyl and Fluoroacyl Substituents

In the design of radiopharmaceuticals labeled with fluorine-18, it has become fashionable to simply alter the structure of a known parent drug or biochemical substrate by the substitution of a fluoroalkyl group for a methyl group. This often provides minimal disruption to the physiochemical properties of a molecule (lipophilicity, molecular weight, pK_a) and has been found in many cases to provide molecules with equivalent or even enhanced biological activity *in vivo*. Although in many cases appropriate precursors can be prepared with leaving groups pre-positioned at the end of short alkyl chains (see [^{18}F]fluoropropyl-DTBZ, above), the development of simple ^{18}F-fluoroalkylating [43] and ^{18}F-fluoroacylating agents [44, 45] also allows the application of this approach to families of compounds without the need to synthesize for each one an appropriate precursor for a single-step nucleophilic fluorination. Most useful radiotracers have incorporated [^{18}F]fluoromethyl, [^{18}F]fluoroethyl or [^{18}F]fluoropropyl alkyl groups, or [^{18}F]fluoroacetyl and [^{18}F]fluoropropionyl acyl groups: larger groups begin to introduce considerable size and lipophilicity to the original molecule.

14.4 Electrophilic ^{18}F-fluorination: General Aspects

Although the majority of radiopharmaceuticals labeled with fluorine-18 have been prepared using nucleophilic fluorination reactions, in a few instances the application of electrophilic fluorination reactions has proved quite suitable. The most significant limitation of electrophilic ^{18}F-fluorinations is the relatively low specific activities (less than 10 Ci/mmol) commonly obtained for final products. This is a result of the fact that electrophilic ^{18}F-fluorination reagents (perchloryl fluoride, acetyl hypofluorite, xenon difluoride, *N*-fluoro-*N*-alkylsulfonamides, diethylaminosulfur trifluoride) are prepared in low specific activity from ^{18}F-labeled fluorine gas, which in itself produced in a carrier-added fashion. A second drawback of electrophilic fluorination is that the maximum radiochemical yield obtainable is 50%, as only one of the two fluorine atoms in fluorine gas can end up in the product (or, for preparation of electrophilic reagents such as acetyl [^{18}F]hypofluorite, the maximum yield of preparing the reagent from [^{18}F]F_2 is 50%).

Production of [^{18}F]F_2, from which numerous subsequent electrophilic fluorination reagents have been prepared, can be accomplished in two fashions. The first is the irradiation of cyclotron targets containing small amounts of carrier fluorine gas in neon [46]; fluorine-18 is produced by the nuclear reaction ^{20}Ne(d,α)^{18}F, with essentially exchange of the fluorine-18 and fluorine-19 atoms to produce carrier-added [^{18}F]F_2. As an alternative to direct production of fluorine-18 gas in a neon target, it is also possible to form [^{18}F]F_2 using a two-step procedure of first irradiation of oxygen-18 (the high-yield method for fluorine-18 production), followed by a second irradiation of a gas mixture containing carrier fluorine gas to initiate exchange of the fluorine atoms, producing carrier-added fluorine gas [47, 48].

Despite the low specific activity obtained with electrophilic ^{18}F-fluorination reagents, important radiopharmaceuticals have been prepared in this fashion.

14.4.1 [^{18}F]FluoroDOPA: One-step Electrophilic Fluorination

L-3,4-dihydroxy-6-[^{18}F]fluorophenylalanine ([^{18}F]FDOPA) is a amino acid derivative primarily utilized to follow the biosynthesis of dopamine in the brain, as it is taken up by dopaminergic neurons and decarboxylated to 6-[^{18}F]fluorodopamine and stored in vesicles alongside endogenous dopamine. It was one of the first fluorine-18 radiopharmaceuticals developed for clinical investigations of movement disorders (e.g., Parkinson's disease) and its synthesis has drawn much attention and been discussed numerous times [4, 49]. The synthesis of [^{18}F]FDOPA can be achieved using both electrophilic and nucleophilic routes, and provides a good example of how these different methods each have advantages and drawbacks. The synthesis of FDOPA is also a general example of methods for preparation of ^{18}F-labeled aromatic amino acids and derivatives, and thus has been extended to desired molecules such as fluorinated phenylalanines and fluorotyrosines.

The initial syntheses of [^{18}F]FDOPA were done using direct electrophilic fluorination of DOPA [49], which was soon replaced with methods for regioselective ^{18}F-fluorination of DOPA using organomercuric [50] or organostannane precursors [51] (Figure 14.5).

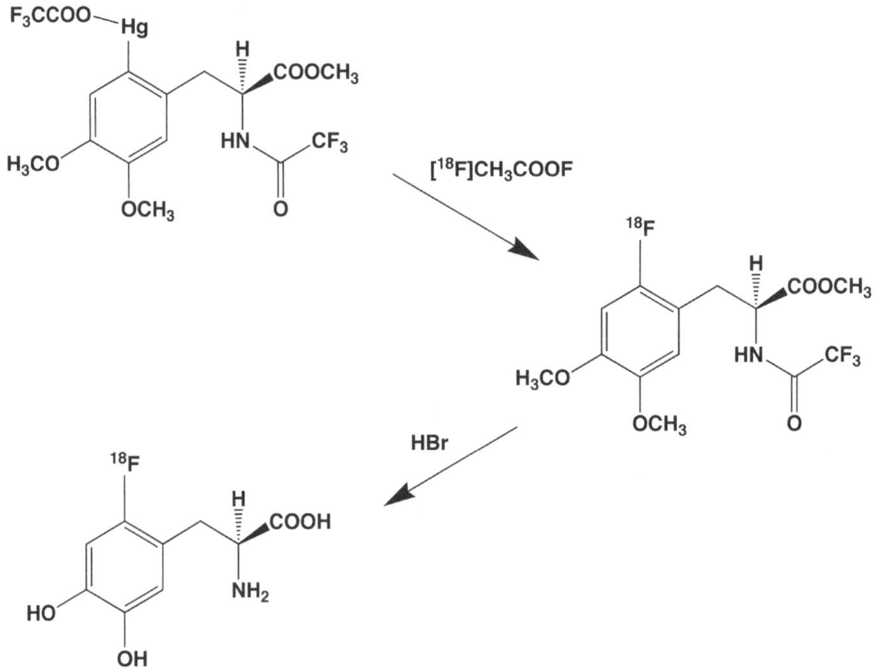

Figure 14.5 Electrophilic ^{18}F-fluorination of an organomercurial derivative of dihydroxyphenylalanine as a route to synthesis of 6-[^{18}F]fluoro-3,4-dihydroxyphenylalanine (6-[^{18}F]fluoroDOPA, FDOPA). Reaction is stereospecific and regioselective but requires a step for deprotection of functional groups.

The methods using organometallics are relatively simple and can provide enantiomerically pure [^{18}F]FDOPA in yields as high as 25% (decay-corrected). These methods have even been automated [51]. The use of the organometallic approach exemplifies a simplification of the radiochemistry, which, however, complicates the quality control analysis, as in addition to the normal requirements for a radiopharmaceutical (radiochemical purity, radionuclidic purity, chemical purity), there is also the need to develop and implement sensitive analytical techniques to ensure that there are no residual amounts of the metals (mercury or tin) in the final product. The advantage of the electrophilic approach to [^{18}F]FDOPA is that the product is obtained in two steps in enantiomerically pure form; the disadvantages are the limited radiochemical yield, the low specific activity inherent in electrophilic reactions, and the need for extensive quality control analysis. Fortunately, [^{18}F]FDOPA is one of the radiochemicals for which high specific activity is not required, as the transporters and enzymes involved in [^{18}F]FDOPA uptake and retention in neurons are not saturated by the chemical amounts of FDOPA associated with the radiochemical product, and it has also been demonstrated there is no toxicity associated with the chemical. Efforts continue for improvement and simplification of methods for electrophilic fluorination to yield FDOPA [52].

In contrast to the electrophilic approach, the nucleophilic methods [53–55] for synthesis of [^{18}F]FDOPA involve four synthetic steps (Figure 14.6) starting with [^{18}F]fluoride ion, require a significantly longer total synthesis time, and in the end probably do not result

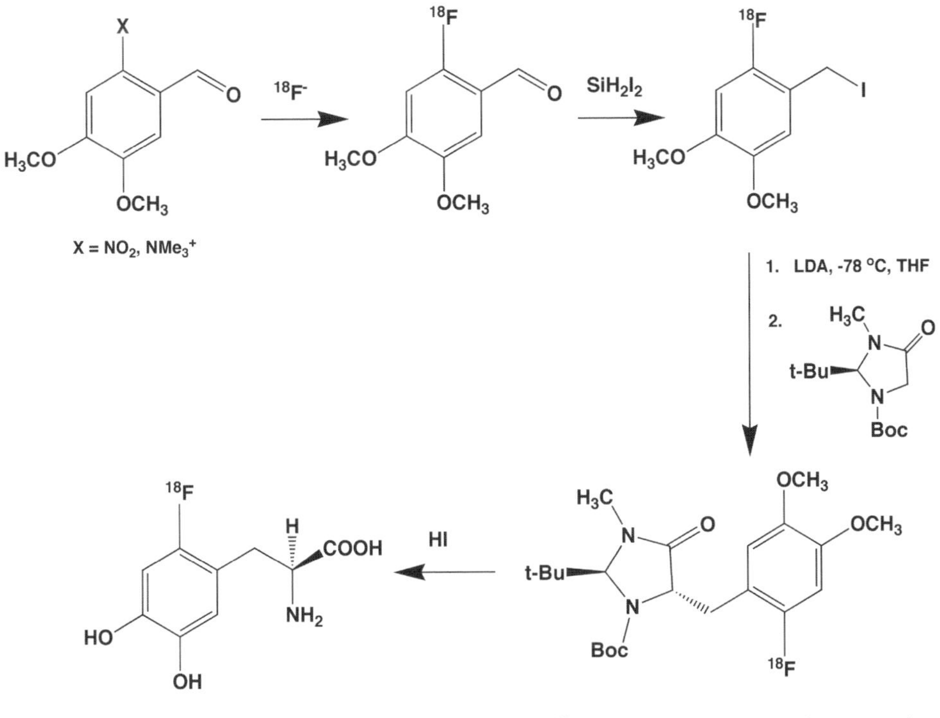

Figure 14.6 *Nucleophilic approach to synthesis of 6-[^{18}F]fluoroDOPA. The multistep synthesis introduces fluorine-18 in an initial nucleophilic aromatic substitution reaction.*

in a much higher yield than that obtained by electrophilic routes. The syntheses require the use of chiral auxiliaries or chiral phase-transfer reagents to produce [^{18}F]FDOPA in appropriate enantiomeric purity [54]; otherwise a chiral column chromatography step is necessary to separate the two isomers formed in the alkylation of a simple glycine synthon, which of course results in loss of some product as the wrong isomer. Use of the nucleophilic route does produce higher specific activities, but this is not a clear advantage for a radiopharmaceutical that does not require such. In contrast, [^{18}F]fluorodopamine serves as an excellent example where the choice of a synthetic strategy is very important; the synthesis via nucleophilic aromatic ^{18}F-fluorination [56] provides a high specific activity product without *in vivo* hemodynamic effects, whereas the same compound made by an electrophilic route [57] yields a carrier-added radiopharmaceutical that is not suitable for radiotracer studies of *in vivo* biochemistry.

14.4.2 Fluoro-destannylation

Trimethyltin substituents are known to react with [^{18}F]fluorine in a regioselective manner [58]. Several important radiopharmaceuticals such as [^{18}F]fluoro-L-DOPA [39] [^{18}F]fluoro-L-tyrosine [37], and [^{18}F]fluorometaraminol [59] have been synthesized using this method. The limitation of this method is again the low specific activity from [^{18}F]fluorine gas. Hopefully, with development of advanced technology, a better electrophilic [^{18}F]fluorine source might be available. Another concern of this method is the residue of potentially toxic metals, but tin has less toxicity than mercury. Recently, resin-bonded aryltin precursors have been reported for ^{125}I-labeling by destannylation [60, 61]. The desired radioactive product will only be released from resin via destannylation, while unreacted precursor and tin-containing side-product remains bonded to the resin which is easily removed by filtration. This method may have great potential for use in ^{18}F-fluorination in a similar fashion.

14.5 New Directions in ^{18}F-labeling

Labeling of biologically active molecules with fluorine-18 has taken great strides in the last two decades, largely due to the successful application of well-known reaction methods – nucleophilic and electrophilic fluorinations – to an increasingly diverse assortment of chemical structures. Two recent developments, the application of "click" chemistry and the use of protic solvents in nucleophilic fluorinations, deserve mention here as new directions in ^{18}F-labeling that have the potential to significantly impact both the types of compounds labeled and the yields obtained.

14.5.1 Application of "Click" Chemistry to ^{18}F-labeling

"Click" chemistry is a generic term that describes chemical reactions that are easy to perform and work up, are high yielding, and are insensitive to oxygen or water [62, 63]. However, there has arisen one particular reaction that has become the leader of the field

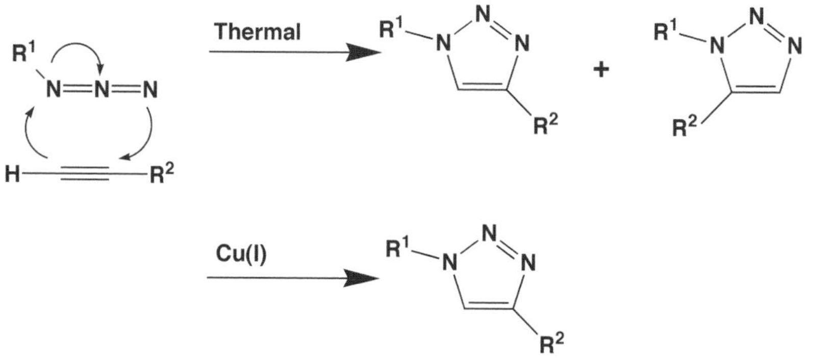

Thermal
Cu(I)
1,4-regioisomer 1,5-regioisomer

Figure 14.7 *The 1,3-cycloaddition of an azide and a terminal alkene to form a 1,2,3-triazole. The thermal reaction produces both the possible regioisomers, whereas the copper-catalyzed reaction yields the single 1,4-regioisomer.*

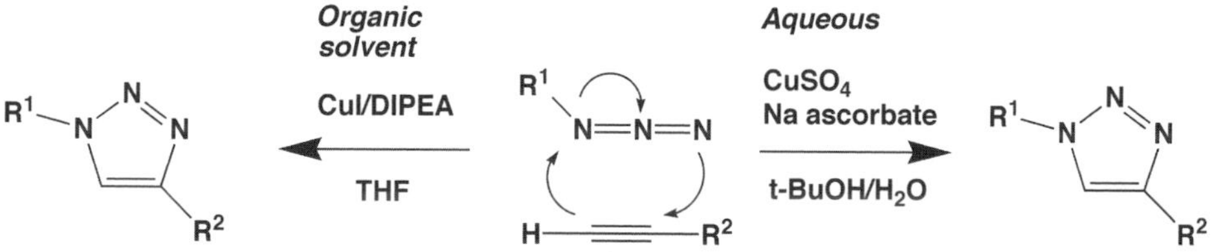

Organic solvent
CuI/DIPEA
THF
Aqueous
$CuSO_4$
Na ascorbate
t-BuOH/H_2O

Figure 14.8 *Use of the copper(I)-mediated alkene-azide 1,3-cycloaddition ("click" chemistry) in both organic solvent and aqueous solvent systems. The aqueous system, with in situ generation of copper(I) from copper(II), is less sensitive to oxygen.*

and that is the Huisgen 1,3-dipolar cycloaddition of terminal alkynes with azides to give 1,2,3-triazoles (Figure 14.7) [64].

Prior to the use of copper(I), the cycloaddition required elevated temperatures for prolonged periods. Typically, reactions required refluxing in toluene or carbon tetrachloride for 10–48 h. Under these conditions, the cycloaddition was non-regiospecific with two possible regioisomers (1,4 and 1,5) formed. Some control of regiospecificity was obtainable when electron-withdrawing groups were substituted on the alkyne, favoring production of 1,4-regioisomer. Meanwhile, the addition of electron-withdrawing groups to the azide favors production of the 1,5-regioisomer. In practice, the cycloaddition yielded mixtures of product and exclusive production of one regioisomer remained elusive [65]. Thus, the cycloaddition under these conditions failed to fulfill the requirement of a "click" chemistry [64].

The utility of the Huisgen 1,3-dipolar cycloaddition was discovered when it was realized that copper(I) not only catalyzes the reaction but also promotes the regiospecificity, with exclusive production of the 1,4-triazole regioisomer (Figure 14.8) [64]. The first publication by Meldal *et al.*, in 2002, outlined the use of copper(I) in the cycloaddition reaction for triazole synthesis on a solid phase [66]. This was an organic-solvent-based procedure that used copper(I) iodide with *N,N*-diisopropylethylamine (DIPEA) in various solvent such as acetonitrile, dichloromethane, tetrahydrofuran, toluene, or *N,N*-dimethylformamide, with the terminal alkyne immobilized on a swollen solid support. This was closely followed by the report of Folkin and Sharpless and colleagues showing that the reaction could be carried out in water using copper sulfate and sodium ascorbate [67]. Both methods have recently become very popular and have made the Huisgen 1,3-dipolar cycloaddition the essential "click" chemistry.

The facility of this water-based method earned it many applications. The azide and terminal alkyne are mixed in a mixture of *tert*-butanol and water (1:1 or 2:1). Then sodium ascorbate (5–10% mol) is added followed by a copper(II) sulfate solution (1–5% mol), and the flask is sealed and stirred vigorously at ambient temperature [56]. Copper(I) is generated *in situ* by reduction of the copper(II) with an excess of sodium ascorbate and under these condition the normally oxygen-sensitive copper(I) survives. These reactions are typically run overnight, but a mild thermal or microwave-assisted heating shortens reaction time to 10–15 minutes [68]. A number of modified reaction conditions have been reported, using copper(I) species directly, as CuI, $CuOTf{\cdot}C_6H_6$ or $[Cu(CH_3CN)_4PF_6]$, with a nitrogen base such as triethylamine or pyridine [69–71]. The reaction mechanism is still under investigation and appears quite complex [53, 64 72]. A recent analysis suggests that both azide and alkyne are activated by copper, possibly within a multinuclear copper-acetylide species, supporting earlier reports of two copper centers participating in the catalysis [72, 73].

Organic solvent-based procedures have been used in situations when the reactants are not soluble in aqueous media. Copper(I) is supplied directly to the reaction in form of CuI with DIPEA, and co-solvents such as acetonitrile, dichloromethane, tetrahydrofuran, toluene, or *N,N*-dimethylformamide are used. Some alternative protocols, using THF with $Cu(PPh_3)Br$ and DIPEA, or CuBr in DMF with bipyridine, have also been reported [74, 76].

The simplicity and efficiency of the "click" chemistry is attractive to fluorine-18 chemistry, where time plays an important role in synthesis due to the relative short half-life of fluorine-18. This one-pot reaction provides a versatile tool for coupling drug-like fragments in high yield and under mild conditions. The product 1,2,3-triazole formed from cycloaddition is biologically stable with polarity and size similar to an amide group that is a common functional group in many radiopharmaceuticals [77].

^{18}F-labeled peptides are a rapidly emerging field for targeted PET imaging probes. Although a variety of ^{18}F-labeled prosthetic groups have been developed in the past decade, only a limited number of chemical reactions have been utilized to incorporate the prosthetic groups into peptides, including acylation [78–83], alkylation [84], and oxime formation [85, 86]. Acylation is the most commonly used approach and requires multistep protection and deprotection of other functional groups within the peptide sequence, which otherwise would be acylated. For both the alkylation and oxime formation reactions, the reagents used have potential to react with other functional groups within the peptides as well. The products formed by acylation, alkylation and oxidation are often species that are susceptible to hydrolysis or oxidation. Therefore the "click" chemistry of azides and terminal alkynes is expected to be a superior method for the preparation of ^{18}F-labeled peptides because of the following advantages: (i) the reaction can be performed in an aqueous media using readily accessible reagents and without exclusion of atmospheric

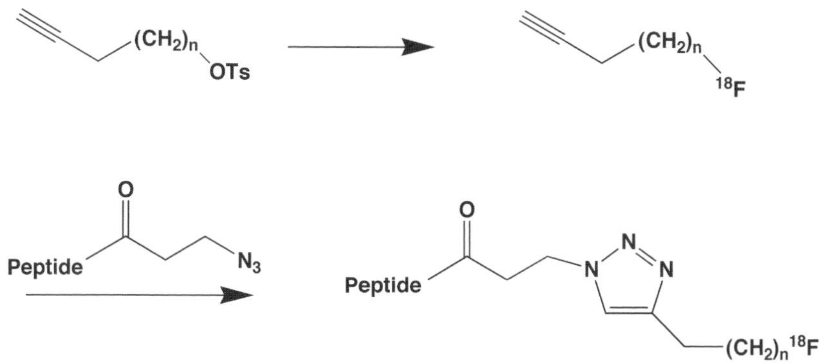

Figure 14.9 *Use of "click" chemistry of 1,3-dipolar cycloadditions of [¹⁸F]fluoroalkynes to radiolabel* N-*(3-azidopropionyl)peptides.*

oxygen; (ii) the relatively mild reaction conditions are tolerant of peptide bonds, and neutralization of the reaction media is not required before or after reaction; (iii) since the reaction between alkyne and azide is orthogonal to any functional groups [87–91] it is not necessary to protect other functional groups within the peptide sequence; (iv) the reaction is highly regioselective, leading to 1,4-disubstituted 1,2,3-triazole in high yield with simple work-up and purification; and (v) the product is relatively stable. The large dipole moment and the nitrogen atoms in positions 2 and 3 of the triazole serve as weak hydrogen bond acceptors and improve the solubility of the product in water [63].

The first application of copper(I)-catalyzed 1,3-dipolar cycloaddition in preparation of [^{18}F]fluoropeptides was reported by Marik and Sutcliffe in 2006 (Figure 14.9) [92]. Three [^{18}F]fluoroalkynes (n = 1, 2, and 3) were prepared in yields ranging from 36% to 80% by nucleophilic substitution of a *p*-toluenesulfonyl moiety with [^{18}F]fluoride ion. Reaction of these [^{18}F]fluoroalkynes with various peptides (previously derivatized with 3-azidopropionic acid) via the Cu(I)-mediated 1,3-dipolar cycloaddition provided the desired ^{18}F-labeled peptides in 10 minutes at room temperature with yields of 54–99% and great radiochemical purity (81–99%) [82].

The kinetically driven copper(I)-catalyzed cycloaddition of azides and alkynes requires hours of reaction time to obtain quantitative yields [63]. However, in the case of no-carrier added radiochemical synthesis the ratio of reactants and catalysts differs considerably from that in traditional chemistry. In particular, the azide component and catalyst are in huge excess compared with the [^{18}F]fluoroalkyne. The quantitative incorporation of [^{18}F]fluoroalkyne could be achieved in 10 minutes when an optimized catalytic system was used [81]. Reversed-phase HPLC analysis of all ^{18}F-labeled peptides showed only a single product, indicating that the reaction proceeded regioselectively to yield 1,4-disubstituted 1,2,3-triazoles as previously reported [64].

In another study, coupling of [^{18}F]fluoroethylazide with various alkyne substrates that included a peptide to form the corresponding 2-[^{18}F]fluoroethyl-1, 2, 3-triazoles has been recently reported by Glaser and Arstad (Figure 14.10) [93].

After nucleophilic fluorination of 2-azidoethyl-4-tolunenesulfonate, 2-[^{18}F]fluoroethylazide was reacted with a small library of terminal alkynes in the presence of excess Cu(II) or copper powder (method A or B). The radiochemical yields were

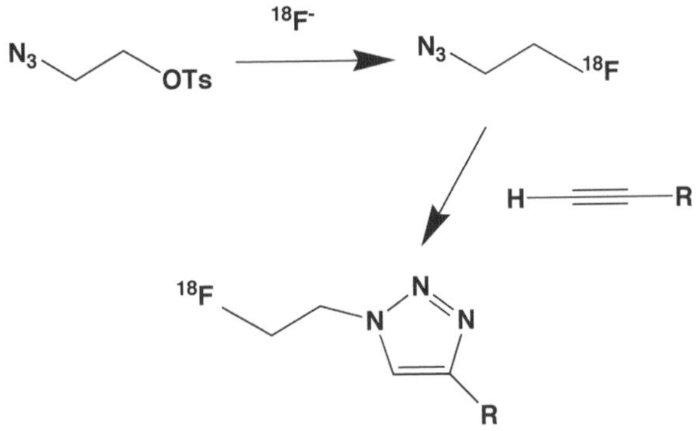

Figure 14.10 *Application of "click" chemistry of 1,3-dipolar cycloaddition of [^{18}F]fluroethylazide to a terminal alkene as a route to one-step radiolabeling of larger molecules.*

measured by analytical HPLC 15 minutes post reaction at ambient temperature and also after subsequent heating to 80 °C for 15 minutes (Table 14.3) [93, 94]. At ambient temperature, the degree of incorporation of 2-[^{18}F]fluoroethylazide varied from no product to greater than 98% yield of product, depending on the alkyne substrate and the catalytic system. Following heating to 80 °C for 15 minutes, nearly complete conversions of 2-[^{18}F]fluoroethylazide to corresponding 1,2,3-triazoles were observed for a majority of the substrates. It should be noted that the "click" reaction has a high tolerance for other functional groups in the terminal alkynes, including N- and C-terminal amides, which is very attractive for labeling of peptides. With a suitable match of catalytic system and alkyne substrate, high product yields can be obtained within a short reaction time under mild conditions, which opens up the possibility of labeling fragile biomolecules that otherwise cannot be labeled with fluorine-18 [95].

The "click" chemistry has been rapidly adopted by radiochemists in ^{18}F-labeling. There is a wide selection of ^{18}F-labeled prosthetic alkynes and azides, as well as commercially available alkynes and azides, providing numerous possible combinations of this azide–alkyne cycloaddition to form various ^{18}F-labeled biomolecules with numerous functional groups. In addition to the continuously growing ^{18}F-labeled peptide synthesis [24, 81–84], ^{18}F-labeled folic acid, folates [96, 97], glucose analogues [98], oligonucleotides [99], and lipids [99] have been prepared using "click" chemistry. In all cases, 1,2,3-triazole formation was completed in less than 30 minutes at room temperature in aqueous media in good yields. The products could be purified using simple HPLC or solid-phase extraction with no multistep purification method needed.

In conclusion, the Huisgen 1,3-dipolar cycloaddition of terminal alkynes with azides to form 1,2,3-triazoles, referred to as "click" chemistry, provides a simple, flexible, and highly efficient method for ^{18}F-labeling. This powerful linking reaction opens numerous possibilities for combining alkynes with azides, forming all varieties of ^{18}F-labeled, highly functionalized biomolecules. The improvement of peptide and lipid labeling using "click" chemistry will further benefit the ^{18}F-labeling of a variety of possible drug carries such as dendrimers, micelles, microbubbles, liposomes, and cells. Nuclear medicine imaging will be able to take advantages of the novel ^{18}F-labeling methodologies, where fluorine-18 could be simply clicked on these multifunctional vehicles without damaging other

Table 14.3 *Alkyne substrates and radiochemical yields used in a 1,3-dipolar cycloaddition reaction with [¹⁸F]fluoroethylazide*

Substrate	Radiochemical yield (%)[a]	
	Method A	Method B
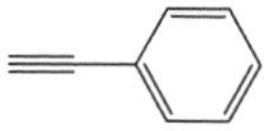	84 (7)	56 (10)
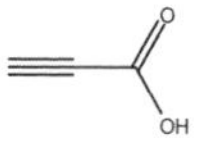 OH, O	61 (15)	29 (0)
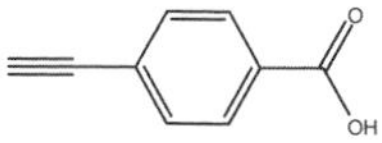 OH, O	93 (40)	15 (6)
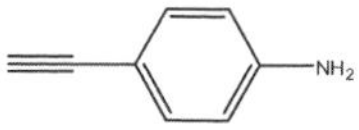 NH_2	>98 (12)	96 (33)
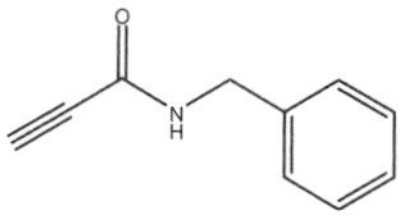 O, N H	>98 (>98)	>98 (70)
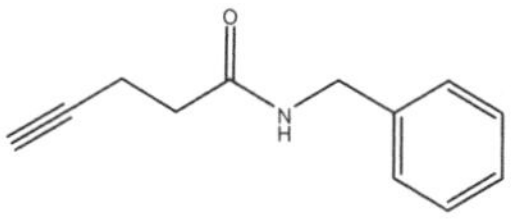 O, N H	98 (9)	76 (26)
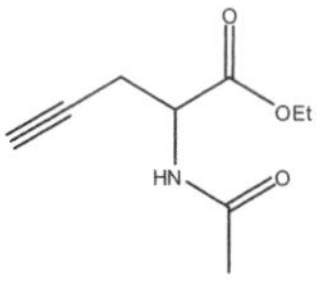 O, OEt, HN, O	>98 (31)	97 (90)
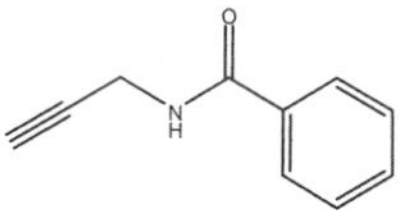 O, N H	>98 (34)	92 (51)
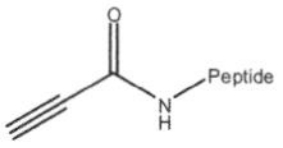 O, N H, Peptide	(92)	

[a]Numbers in parentheses are the yields at ambient temperature.

functional groups. This would extend the power of PET imaging to many new applications in pharmaceutical research and clinical application.

14.5.2 Nucleophilic ^{18}F-fluorination in Protic Solvents

Nucleophilic fluorination using [^{18}F]fluoride ion is a typical displacement reaction that is generally performed in a polar aprotic solvent such as acetonitrile, dimethyl sulfoxide, or dimethylformamide [100]. In polar aprotic solvents, the nucleophilicity of anions including [^{18}F]fluoride is enhanced by the selective solvation of the cations. Conversely, solvation in protic solvents likely reduces an anion's nucleophilicity by interaction with the partial

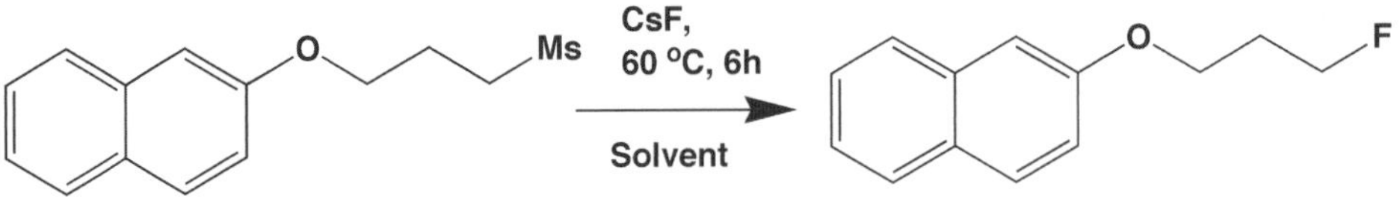

Solvent	Yield
t-BuOH	92%
t-Amyl alcohol	93%
Dimethylformamide	48%
Acetonitrile	7%

Figure 14.11 *Nucleophilic aliphatic substitution of a mesylate ester leaving group using fluoride ion as cesium salt in protic and aprotic organic solvents.*

positive charge of the protic solvent. Conventional theory suggests that polar aprotic solvents are much better than protic solvents for nucleophilic substitution [100].

Recently, a new investigation into nucleophilic displacement reactions using fluoride ion in protic solvents has been reported by Kim *et al.* [101]. In that study, they showed that sterically hindered protic solvents such as tertiary alcohols, in the absence of any kind of catalyst, actually enhanced the nucleophilicity of an alkali metal fluoride (CsF). This enhanced nucleophilicity dramatically increased the rate of nucleophilic fluorination compared with conventional methods and reduced formation of typical reaction byproducts (Figure 14.11).

This finding is very interesting because it is in direct contrast with the predictions made from the conventional reaction mechanism. The detailed reaction mechanism of nucleophilic fluorination in a protic solvent system is presently not clear. However, a series of compounds have been investigated (Table 14.4) and the corresponding fluorine-substituted compounds are produced using this method in comparable or greater yields than previously reported by other methods [101].

The fluorination reaction of mesylates or tosylates has been reported to be less than twice as fast as that of iodides [101]. However, in the protic solvent system, the fluorination rate of a tosylate is approximately 12 times faster than that of the corresponding iodide (Table 14.4). This result suggests that the reaction rate is determined not only by the nature of the leaving group but also by other types of interactions, such as those between the solvent (*tert*-alcohol) and the leaving group. For example, hydrogen bonding between the alcohol solvent and the oxygen atoms in the alkanesulfonate leaving group may enhance its nucleophilic (leaving group) character. Remarkably, using this fluorination procedure, a fluoroproline derivative was prepared in good yield after only 1.5 hours at ambient temperature from the corresponding triflate precursor (Table 14.4). It is notable that a triflate group has six available sites for H-bonding with the solvent alcohol (three oxygens and three fluorines). In contrast, fluorination of reactants with halide groups in the *tert*-amyl alcohol media required more than 12 hours as well as vigorous conditions, although these reactions did eventually produce the fluorine-substituted product in high yields (72% and 88%).

The fluorination of haloethyl or alkanesulfonyloxyethyl aromatic compounds in conventional methods usually uses a "naked" fluoride ion generated from phase-transfer catalyst (i.e., potassium ion complexed by a cryptand, or a tetraalkylammonium salt) to

Table 14.4 *Reactions of alkyl halide mesylates with [^{18}F]fluoride ion in protic solvent systems*

	Temperature (°C)	Time (h)	Yield (%)	Comment
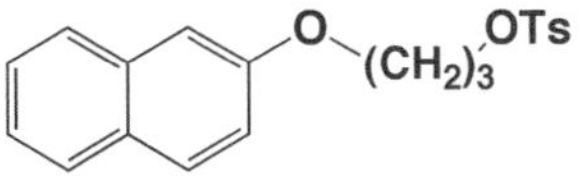 O $(CH_2)_3$ OTs	90	2	93	trace tosylate
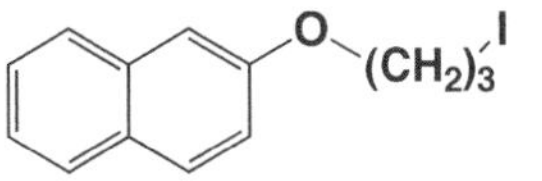 O $(CH_2)_3$ I	90	24	73	18% alkene trace iodide
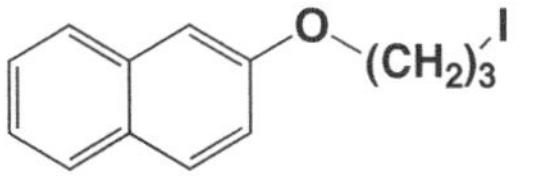 O $(CH_2)_3$ I	reflux	12	72	20% alkene
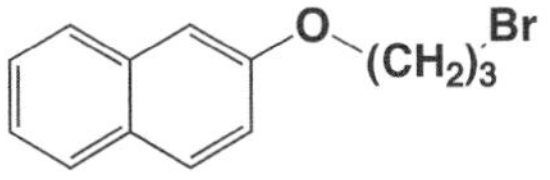 O $(CH_2)_3$ Br	reflux	18	88	6% alkene
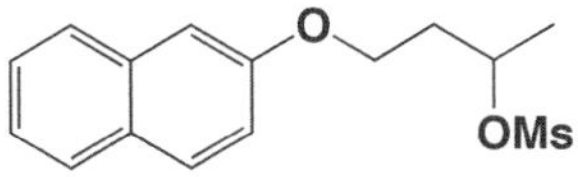 O OMs	90	3.5	81	12% alkene
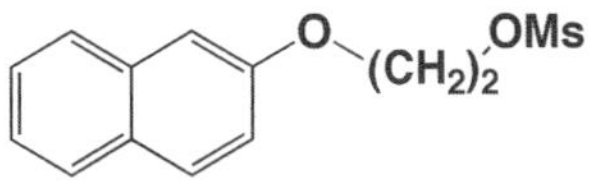 O $(CH_2)_2$ OMs	90	2.5	92	trace alkene
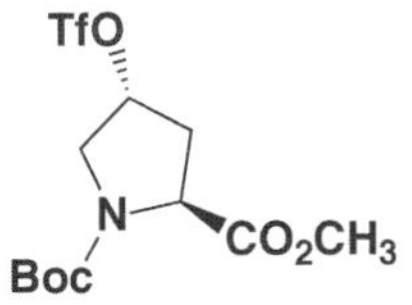 TfO N Boc CO_2CH_3	25	1.5	69	15% alkenes

enhance the nucleophilicity and solubility of fluoride ions. The "naked" fluoride ion, as a strong base, results in the competing elimination reaction to give the vinylarene by-product. In Table 14.4, the fluorination of 2-(2-mesyloxyethyl)naphthalene to 2-(2-fluoroethyl)naphthalene in a protic solvent proceeds almost to completion, producing the corresponding fluoroalkane in 92% yield with only trace quantities of alkene byproduct. The protic solvents may suppress the formation of side-products by weakening the basicity of the fluoride through the formation of hydrogen bonds between fluoride ion and alcohol.

Further studies indicated that product yields from protic solvent systems are also highly dependent on the cations, where cesium generated higher reaction yields than tetrabutylammonium (TBA) and potassium cations produced <1% product [101]. This suggests the protic solvent used may increase the nucleophilicity of the fluoride ion in the cesium or tetrabutylammonium complex by weakening ionic bonds between cation and anion through the formation of hydrogen bonds between fluoride ion and alcohol. Thus, the problem of low reactivity due to the strong ionic bond of fluoride salt in conventional methods may be overcome. Surrounded by a large size cation (Cs^+ and TBA^+), fluoride ion could form only weak hydrogen bonds with bulk alcohol solvent. This prevents the reduction of nucleophilicity caused by solvation in protic solvents. Fuller understanding

Table 14.5 *Yields of example radiopharmaceuticals labeled with [18F]fluoride ion using nucleophilic substitution reactions performed in protic solvent systems compared with literature methods utilizing aprotic organic solvents*

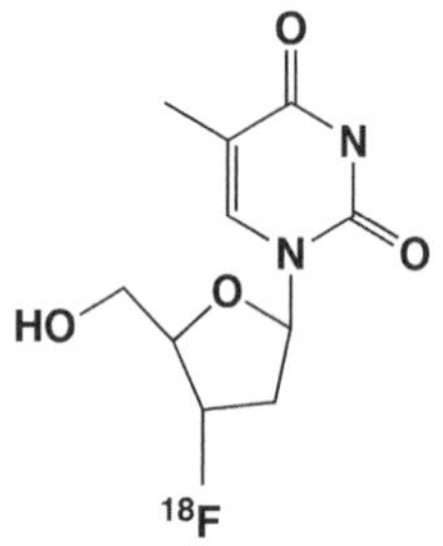

[18F]FLT [18F]FP-CIT [18F]FMISO

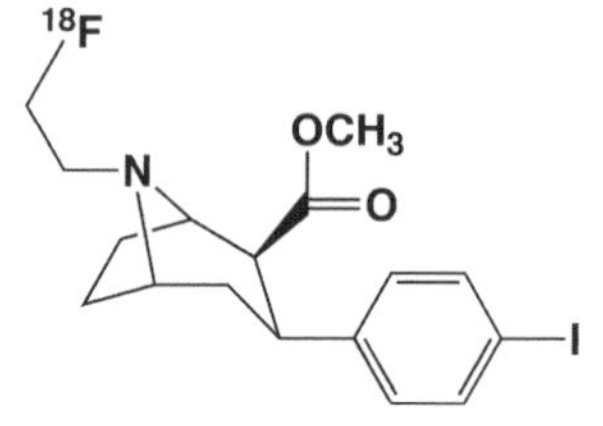

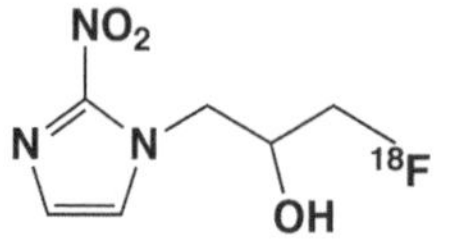

	Method	Temperature (°C)	Time (min)	Yield (%)
	Method	Temp (°C)	time (min)	yield (%)
2-[18F]FDG	Cs[18F]/alcohol	100	10	82 +/− 7.8 (n = 10)
	K[18F]/acetonitrile	100	5	60–70%
[18F]FLT	Cs[18F]/alcohol	120	10	65 +/− 5.4 (n = 10)
	literature	110	7.5	15 +/− 5.4 (n = 3)
[18F]FP-CIT	Cs[18F]/alcohol	100	20	35 +/− 5.2 (n = 14)
	literature	90	10	1%
[18F]FMISO	Cs[18F]/alcohol	100	10	69.6 +/− 1.8 (n = 10)
	literature	105	6	15.0 +/− 5.4 (n = 3)

of the reaction mechanism of protic solvent systems and testing of various other solvents would be of interest.

Applying this novel fluorination method to ^{18}F-fluorination provides great improvement in radiochemical yields (Table 14.5) [101]. The use of a protic solvent in nucleophilic ^{18}F-fluorination not only improved the radiochemical yield of widely used radiopharmaceuticals such as [^{18}F]FLT (3′-deoxy-3′-[^{18}F]fluorothymidine) [101, 102], [^{18}F]FP-CIT (3-[^{18}F]fluoropropyl-2-β-carboxymethoxy-3-β-(4-iodophenyl) nortropane) [101, 103], and [^{18}F]FMISO ([^{18}F]fluoromisonidazole) [101, 104], which were difficult to prepare in conventional methods under mild conditions. This method has advanced the availability of these important radiopharmaceuticals, and should help the development of new pharmaceuticals for research and clinical application in the future.

14.6 The Future of Fuorine-18

The increasing availability of positron emission (PET) imaging equipment as part of the normal clinical care in nuclear medicine, and incorporation of PET into the routine tools of the pharmaceutical industry, portends a growing demand for new ^{18}F-labeled radiopharmaceuticals. It is thus expected that the methods for incorporation of the radionuclide into simple and complex molecules of interest in what has recently been termed "molecular imaging" will continue to be a very active and fruitful area of chemical research.

References

1. Fowler, J. S. and Wolf, A. P. (1982) *The Synthesis of Carbon-**11**, Fluorine-18, and Nitrogen-13 Labeled Radiotracers for Biomedical Applications*, Technical Information Center, U. S. Department of Energy, New York.
2. Kilbourn, M. R. (1990) *Fluorine-***18** *Labeling of Radiopharmaceuticals*, National Academy Press, Washington, D.C., pp. 149.
3. International Symposium on Radiopharmaceutical Sciences. (2007) *J Labeled Comp. Radiopharm.*, **50**, S1–S526.
4. Stocklin, G. (1995) Fluorine-18 compounds, in *Principles of Nuclear Medicine* (eds H. N. Wagner, Z. Szabo and J. W. Buchanan), W.B. Saunders Company, Philadelphia, pp. 178.
5. Snyder, S. E. and Kilbourn, M. R. (2003) Chemistry of fluorine-18 radiopharmaceuticals, in *Handbook of Radiopharmaceuticals* (eds M. J. Welch, M. J. and C. S. Redvanly), John Wiley & Sons, Ltd, Chichester.
6. Stocklin, G. (1992) Tracers for metabolic imaging of brain and heart. Radiochemistry and radiopharmacology. *Eur. J. Nuclear Med.*, **19**, 527–551.
7. Fowler, J. S. (1993) The synthesis and application of F-18 compounds in positron emission tomography, in *Organofluorine Compounds in Medicinal Chemistry* (ed. R. Filler), Elsevier Science Publishers, Amsterdam, pp. 309–338.
8. Ding, Y.-S. and Fowler J.S. (1996) Fluorine-18 labeled tracers for PET studies in the neurosciences, in *Biomedical Frontiers of Fluorine Chemistry* (eds I. Ojima, J. McCarthy and J. Welch), American Chemical Society Publishers, Washington, D.C., pp. 328–343.
9. Iwata, R., (2004) *Reference Book for PET Radiopharmaceuticals*, Tohoku University, Sendai, 224 pp. http://kakuyaku.cyric.tohoku.ac.jp/public/Reference%20Book%202004c%20for%20 web.pdf.
10. Coenen, H. H., Klatte, B., Knoechel, A., *et al.* (1986) Preparation of n.c.a. [17-^{18}F]-fluoroheptadecanoic acid in high yields via aminopolyether-supported, nucleophilic fluorination. *J Labeled Comp. Radiopharm.*, **23**, 455–466.
11. Toorongian, S. A., Mulholland, G. K., Jewett, D. M., *et al.* (1990) Routine production of 2-deoxy-2-[^{18}F]fluoro-D-glucose by direct nucleophilic exchange on a quaternary 4-aminopyridinium resin. *Nucl. Med. Biol.*, **17**, 273–279.
12. Ido, T., Wan, C. N., Casella, V., *et al.* (1978) Labeled 2-deoxy-D-glucose analogs. Fluorine-18-labeled 2-deoxy-2-fluoro-D-glucose, 2-deoxy-2-fluoro-D-mannose and ^{14}C-2-deoxy-2-fluoro-D-glucose. *J Labeled Com. Radiopharm.* **14**, 175–183.
13. Hamacher, K., Coenen, H. H. and Stocklin, G. (1986) Efficient stereospecific synthesis of no-carrier-added 2-[^{18}F]-fluoro-2-deoxy-D-glucose using aminopolyether supported nucleophilic substitution. *J Nucl. Med.* **27**, 235–238.
14. Kilbourn, M. R. (1997) In vivo radiotracers for vesicular neurotransmitter transporters. *Nucl. Med. Biol.*, **24**, 615–619.
15. Frey, K. A., Wieland, D. M. and Kilbourn, M. R. (1997) Imaging of monoaminergic and cholinergic vesicular transporters, in *Catecholamines: Bridging Basic Science with Clinical Medicine* (eds D. Goldstein, G. Eisenhofer and R. McCarty), Academic Press, San Diego, pp. 269–272.

16. Goswami, R., Kung, M.-P., Ponde, D. E., *et al.* (2006) Fluoro alkyl derivatives of dihydrotetrabenazine as positron emission tomography imaging agents targeting vesicular monoamine transporters. *Nucl. Med. Biol.*, **33**, 685–694.
17. Hostetler, E. D., Patel, S., Guenther, I., *et al.* (2007) Characterization of a novel F-18 labelled radioligand for VMAT2. *J Labeled Comp. Radiopharm.*, **50**, S330.
18. Mulholland, G. K., Jung, Y. W., Wieland, D. M., *et al.* (1993) Synthesis of [^{18}F]-fluoroethoxybenzovesamicol, a radiotracer for cholinergic neurons. *J Labeled Comp. Radiopharm.*, **33**, 583–591.
19. Horti, A. G., Koren, A. O., Ravert, H. T., *et al.* (1998) Synthesis of a radiotracer for studying nicotinic acetylcholine receptors: 2-[^{18}F]fluoro-3-(2(*S*)-azetidinylmethoxy)pyridine (2-[^{18}F]A-85380). *J Labeled Comp. Radiopharm.*, **41**, 309–318.
20. Dolle, F., Dolci, L., Valette, H., *et al.* (1999) Synthesis and nicotinic acetylcholine receptor in vivo binding properties of 2-fluoro-3-[2(S)-2-azetidinylmethoxy]pyridine: A new positron emission tomography ligand for nicotinic receptors. *J Med. Chem.* **42**, 2251–2259.
21. Molina PE, Ding Y-S, Carroll FI, *et al.* (1997) Fluoronorchloroepibatidine-Preclinical assessment of acute toxicity. *Nucl. Med. Biol.*, **24**, 743–747.
22. Ding, Y-S., Liu, N., Wang, T., *et al.* (2000) Synthesis of 6-[^{18}F]fluoro-3-(2(*S*)-azetidinylmethoxy)pyridine for PET studies of nicotinic acetylcholine receptors. *Nucl. Med. Biol.*, **27**, 381–389.
23. Ding, Y.-S., Shiue, C.-Y., Fowler, J.S., *et al.* (1990) No-carrier-added (NCA) aryl[^{18}F]fluorides via the nucleophilic aromatic substitution of electron rich aromatic rings. *J Labeled Comp. Radiopharm.*, **48**, 189–205.
24. Ma, B., Hankenson, K. D., Dennis, J. E., *et al.* (2005) A simple method for stem cell labeling with fluorine-18. *Nucl. Med. Biol.*, **32**, 701–705.
25. Knust, E. J., Kupfernagel, C. and Stoecklin, G. (1979) Long-chain F-18 fatty acids for the study of regional metabolism in heart and liver; odd-even effects of metabolism in mice. *J. Nucl. Med.*, **20**, 1170–1175.
26. Kilbourn, M. R., Dence, C. S., Welch, M. J. and Mathias, C. J. (1987) Fluorine-18 labeling of proteins. *J. Nucl. Med.* **28**, 462–470.
27. Vaidyanathan, G. and Zalutsky, M. R. (1992) Labeling proteins with fluorine-18 using *N*-succinimidyl 4-[^{18}F]fluorobenzoate. *Nucl. Med. Biol.*, **19**, 275–281.
28. Lang, L. and Eckelman, W. C. (1994) One-step synthesis of ^{18}F labeled [^{18}F]-*N*-succinimidyl 4-(fluoromethyl)benzoate for protein labeling. *Appl. Rad. Isot.*, **45**, 1155–1163.
29. Lee, Y.-S., Jeong, J. M., Kim, H. W., *et al.* (2006) An improved method of ^{18}F peptide labeling: hydrazone formation with HYNIC-conjugated c(RGDyK). *Nucl. Med. Biol.*, **33**, 677–683.
30. Chakraborty, P. K. and Kilbourn, M. R. (1991) Fluorine-18 fluorination/decarbonylation: A new route to aryl [^{18}F]fluorides. *Appl. Rad. Isot.*, **42**, 1209–1213.
31. Chakraborty, P. K. and Kilbourn, M. R. (1991) Oxidation of substituted 4-fluorobenzaldehydes: application to the no-carrier-added syntheses of fluorine-18 labeled 4-[^{18}F]fluoroguaiacol and 4-[^{18}F]fluorocatechol. *Appl. Rad. Isot.*, **42**, 673–681.
32. Ludwig, T., Ermert, J. and Coenen, H. H. (2002) 4- [^{18}F]fluoroarylalkylethers via an improved synthesis of n.c.a. 4- [^{18}F]fluorophenol. *Nucl. Med. Biol.*, **29**, 255–262.
33. Coenen, H. H., Franken, K., Kling, P. and Stoecklin, G. (1988) Direct electrophilic radiofluorination of phenylalanine, tyrosine and dopa. *Appl. Rad. Isot.*, **39**, 1243–1250.
34. Ross, T. L., Ermert, J., Hocke, C. and Coenen, H. H. (2006) Nucleophilic ^{18}F-fluorination of heteroaromatic iodonium salts with no-carrier added [^{18}F]fluoride. *J. Amer. Chem. Soc.*, **129**, 8018–8025.
35. Ermert, J., Hocke, C., Ludwig, T., *et al.* (2004) Comparison of pathways to the versatile synthon of no-carrier-added 1-bromo-4-[^{18}F]fluorobenzene. *J Labeled Comp. Radiopharm.*, **47**, 429–441.

36. Zhang, M.-R., Kumata, K. and Suzuki, K. (2007)A Practical Route for Synthesizing a PET ligand containing [18F]fluorobenzene using reaction of diphenyliodonium salt with [18F]F-. *Tetrahedron Lett.* **48**, 8632–8635.
37. Hess, E., Sichler, S., Kluge, A. and Coenen, H. H. (2002) Synthesis of 2-[18F]fluoro-L-tyrosine via regiospecific fluoro-de-stannylation. *Appl. Rad. Isot.*, **57**, 185–191.
38. Teare, H., Robins, E. G., Arstrad, E., *et al.* (2007) Synthesis and reactivity of [18F]-*N*-fluorobenzenesulfonimide. *Chem. Commun.* 2330–2332.
39. Fuchtner, F. and Steinbach, J. (2003) Efficient synthesis of the 18F-labelled 3-*O*-methyl-6-[18F]fluoro-L-DOPA. *Appl. Rad. Isot.*, **58**, 575–578.
40. Shah, A., Pike, V. W. and Widdowson, D. A. (1998) Synthesis of [18F]fluoroarenes from the reaction of cyclotron-produced [18F]fluoride ion with diaryliodonium salts. *J. Chem. Soc., Perkin Trans. 1.*, **13**, 2043–2046.
41. Pike, V. W. and Aigbirhio, F. I. (1995) Reactions of cyclotron-produced [18F]fluoride with diaryliodonium salts - a novel single-step route to no-carrier-added [18F]fluoroarenes. *J. Chem. Soc. Chem. Commun.*, **21**, 2215–2216.
42. Martin-Santamaria, S., Carroll, M. A., Carroll, C. M., *et al.* (2000) Fluorination of heteroaromatic iodonium salts: experimental evidence supporting theoretical prediction of the selectivity of the process. *Chem. Commun.*, **8**, 649–650.
43. Block, D., Coenen, H. H., Laufer, P. and Stocklin, G. (1986) Nca (18F)-fluoroalkylation via nucleophilic fluorination of disubstituted alkanes and application to the preparation of *N*-(18F)-fluoroethylspiperone. *J Labeled Comp. Radiopharm.*, **23**, 1042.
44. Block, D., Coenen, H. H. and Stoecklin, G. (1987) The N.C.A. nucleophilic 18F-fluorination of 1,*N*-disubstituted alkanes as fluoroalkylation agents. *J Labeled Comp. Radiopharm.*, **24**, 1029–1042.
45. Kilbourn, M. R., Dence, C. S. and Welch, M. J. (1986) Reagents for fluorine-18 labeling of proteins. *J Labeled Comp. Radiopharm.*, **23**, 1059–1061.
46. Casella V., Ido T., Wolf A.P., *et al.* (1980) Anhydrous 18F labeled elemental fluorine for radiopharmaceutical production, *J. Nucl. Med.*, **21**,750–757.
47. Nickels, R.J., Daube, M.E. and Ruth, T.J. (1984). An O_2 target for the production of [18F]F_2, *Int. J. Appl. Rad. Isot.*, **35**,117–122.
48. Solin, O. and Bergman, J. (1986) Production of 18F-F2 from 18O_2. *J Labeled Comp. Radiopharm.*, **23**,1202–1204.
49. Luxen, A., Guillaume, M., Melega, W. P., *et al.* (1992) Production of 6-[18F]fluoro-L-DOPA and its metabolism in vivo: a critical review. *Nucl. Med. Biol.*, **19**, 149–158.
50. Adam, M. J. and Jivan, S. (1988) Synthesis and purification of L-6-[18F]fluorodopa. *Appl. Rad. Isot.*, **39**, 1203–1210.
51. Luxen, A., Perlmutter, M., Bida, G. T., *et al.* (1990) Remote, semiautomated production of 6-[fluorine-18]fluoro-L-dopa for human studies with PET. *Appl. Rad. Isot.*, **41**, 275–281.
52. Azad, B. B., Chirakal, R. and Schrobilgen, G. J. (2007) Trifluoromethanesulfonic acid, an alternative solvent medium for the direct electrophilic fluorination of DOPA: new syntheses of 6-[18F]fluoro-L-DOPA and 6-[18F]fluoro-D-DOPA. *J Labeled Comp. Radiopharm.*, **50**, 1454.
53. Lemaire, C., Guillouet, S., Plenevaux, A., *et al.* (1999) The synthesis of 6-[18F]fluoro-L-dopa by chiral catalytic phase-transfer alkylation. *J Labeled Comp. Radiopharm.*, **42**, S113–S115.
54. Lemaire, C., Guillaume, M., Cantineau, R., *et al.* (1991) An approach to the asymmetric synthesis of L-6-[18F]fluorodopa via NCA nucleophilic fluorination. *Appl. Rad. Isot.*, **42**, 629–635.
55. Lemaire, C., Damhaut, P., Plenevaux, A. and Comar, D. (1994) Enantioselective synthesis of 6-[fluorine-18]-fluoro-L-dopa from no-carrier-added fluorine-18-fluoride. *J. Nucl. Med.*, **35**, 1996–2002.

56. Ding, Y.-S., Fowler, J.S., Dewey, S.L., *et al.* (1993) Comparison of high specific activity (–) and (+)-6-[^{18}F]fluoronorepinephrine and 6-[^{18}F]fluorodopamine in baboons: Heart uptake, metabolism and the effect of desipramine. *J. Nucl. Med.*, **34**, 619–629.
57. Goldstein, D.S., Chang, P.C., Eisenhofer, G., *et al.* (1990) Positron emission tomographic imaging of cardiac sympathetic innervation and function. *Circulation*, **81**,1606–1621.
58. Coenen, H. H. and Moerlein, S. M. (1987) Regiospecific aromatic fluorodemetalation of Group IVA metalloarenes using elemental fluorine or acetyl hypofluorite. *J. Fluorine Chem.*, **36**, 63–75.
59. Eskola, O., Gronroos, T., Bergman, J., *et al.* (2004) A novel electrophilic synthesis and evaluation of medium specific radioactivity (1*R*,2*S*)-4-[^{18}F]fluorometaraminol, a tracer for the assessment of cardiac sympathetic never integrity with PET. *Nucl. Med. Biol.*, **31**, 103–110.
60. Donovan, A., Forbes, J., Dorff, P., *et al.* (2006) A new strategy for preparing molecular imaging and therapy agents using fluorine-rich (fluorous) soluble supports. *J. Amer. Chem. Soc.*, **128**, 3536–3537.
61. Mcintee, J. W. and Valliant, J. F. (2007) Expanding the utility of the fluorous labeling method. Trialkylstannylation of functionalized aryl and vinyl halides with a fluorous distannane. *J Labeled Comp. Radiopharm.*, **50**, S198.
62. Kolb, H. C., Finn, M. G. and Sharpless, K. B. (2001) Click chemistry: diverse chemical function from a few good reactions. *Angew. Chem. Int. Ed.*, **40**, 2004–2021.
63. Kolb, H. C. and Sharpless, K. B. (2003) The growing impact of click chemistry on drug discovery. *Drug Discov. Today.*, **8**, 1128–1137.
64. Evans, R. A. (2007) The rise of azide-alkyne 1,3-dipolar "click" cycloaddition and its application to polymer science and surface modification. *Aust. J. Chem.*, **60**, 384–395.
65. Lwowski, W. (1984) *Azides and nitrous oxide, in 1,3-Dipolar Cycloaddition Chemistry* (ed. A. Padwa), John Wiley & Sons, Inc., New York, pp. 621–634.
66. Tornoe, C. W., Christensen, C. and Meldal, M. (2002) Peptidotriazoles on solid phase: [1,2,3]-triazoles by regiospecific copper(I)-catalyzed 1,3-dipolar cycloadditions of terminal alkynes to azides. *J. Org. Chem.*, **67**, 3057–3064.
67. Rostovtsev, V. V., Green, L. G., Fokin, V. V. and Sharpless, K. B. (2002) A stepwise Huisgen cycloaddition process: copper(I)-catalyzed regioselective "ligation" of azides and terminal alkynes. *Angew. Chem. Int. Ed.*, **41**, 2596–2599.
68. Sharpless, W. D., Wu, P., Hansen, T. V. and Lindberg, J. G. (2005) Just click it: undergraduate procedures for the copper(I)-catalyzed formation of 1,2,3-triazoles from azides and terminal acetylenes. *J Chem Educ.*, **82**, 1833–1836.
69. Chan, T. R., Hilgraf, R., Sharpless, B. K. and Fokin, V. V. (2004) Polytriazoles as copper(I)-stabilizing ligands catalysis. *Org. Lett.*, **6**, 2853–2855.
70. Wu, P., Feldman, A. K., Nugent, A. K., *et al.* (2004) Efficiency and fidelity in a click-chemistry route to triazole dendrimers by the copper(I)-catalyzed ligation of azides and alkynes. *Angew. Chem. Int. Ed.*, **40**, 3928–3932.
71. van Steenis, D., Jan V.C, David, O. R. P., *et al.* (2005) Click-chemistry as an efficient synthetic tool for the preparation of novel conjugated polymers. *Chem. Commun.*, **34**, 4333–4335.
72. Rodionov, V. O., Fokin, V. V. and Finn, M. G. (2005) Mechanism of the ligand-free CuI-catalyzed azide-alkyne cycloaddition reaction. *Angew. Chem. Int. Ed.*, **44**, 2210–2215.
73. Lewis, W. G., Magallon, F. G., Fokin, V. V. and Finn, M. G. (2004) Discovery and characterization of catalysts for azide-alkyne cycloaddition by fluorescence quenching. *J. Amer. Chem. Soc.*, **126**, 9152–9153.
74. Malkoch, M., Schleicher, K., Drockenmuller, E., *et al.* (2005) Structurally diverse dendritic libraries: A highly efficient functionalizatin approach using click chemistry. *Macromolecules*, **38**, 3663–3678.

75. Gujadhur, R., Venkataraman, D. and Kintigh, J. T. (2001) Formation of aryl-nitrogen bonds using a soluble copper(I) catalyst. *Tetrahedron Lett.*, **42**, 4791–4793.
76. Laurent, B. A. and Grayson, S. M. (2006) An efficient route to well-defined Macrocyclic polymers via "Click" cyclization. *J. Amer. Chem. Soc.*, **128**, 4238–4239.
77. Bock, V. D., Hiemstra, H. and van Maarseveen, J. H. (2006) Cu(I)-catalyzed alkyne-azide "click" cycloadditons from a mechanistic and synthetic perspective. *Eur. J. Org. Chem.*, 51–68.
78. Jagoda, E. M., Aloj, L., Seidel, J., *et al.* (2002) Comparison of an ^{18}F labeled derivative of vasoactive intestinal peptide and 2-deoxy-2-[^{18}F]fluoro-D-glucose in nude mice bearing breast cancer xenografts. *Mol. Imag. Biol.*, **4**, 369–379.
79. Fredriksson, A., Johnstrom, P., Stone-Elander, S., *et al.* (2001) Labeling of human C-peptide by conjugation with *N*-succinimidyl-4-[^{18}F]fluorobenzoate. *J Labeled Comp. Radiopharm.*, **44**, 509–519.
80. Chen, X., Park, R., Hou, Y., *et al.* (2004) MicroPET imaging of brain tumor angiogenesis with ^{18}F-labeled PEGylated RGD peptide. *Eur. J. Nucl. Med. Mol. Imag.*, **31**, 1081–1089.
81. Chen, X., Park, R., Shahinian, A. H., *et al.* (2004) ^{18}F-labeled RGD peptide: initial evaluation for imaging brain tumor angiogenesis. *Nucl. Med. Biol.*, **31**, 179–189.
82. Toretsky, J., Levenson, A., Weinberg, I. N., *et al.* (2004) Preparation of F-18 labeled annexin V: a potential PET radiopharmaceutical for imaging cell death. *Nucl. Med. Biol.*, **31**, 747–752.
83. Wust, F., Hultsch, C., Bergmann, R., *et al.* (2003) Radiolabeling of isopeptide Ne-(γ-glutamyl)-L-lysine by conjugation with *N*-succinimidyl 4-[^{18}F]fluorobenzoate. *Appl. Rad. Isot.*, **59**, 43–48.
84. Glaser, M., Karlsen, H., Solbakken, M., *et al.* (2004) ^{18}F-Fluorothiols: A new approach to label peptides chemoselectively as potential tracers for positron emission tomography. *Bioconj. Chem.*, **15**, 1447–1453.
85. Poethko, T., Schottelius, M., Thumshirn, G., *et al.* (2004) Two-step methodology for high-yield routine radiohalogenation of peptides: ^{18}F-labeled RGD and octreotide analogs. *J. Nucl. Med.*, **45**, 892–902.
86. Poethko, T., Schottelius, M., Thumshirn, G., *et al.* (2004) Chemoselective pre-conjugate radiohalogenation of unprotected mono- and multimeric peptides via oxime formation. *Radiochim. Acta.*, **92**, 317–327.
87. Mocharla, V. P., Colasson, B., Lee, L. V., *et al.* (2004) In situ click chemistry: enzyme-generated inhibitors of carbonic anhydrase II. *Angew. Chem. Int. Ed.*, **44**, 116–120.
88. Wang, Q., Chan, T. R., Hilgraf, R., *et al.* (2003) Bioconjugation by copper(I)-catalyzed azide-alkyne [3 + 2] cycloaddition. *J. Amer. Chem. Soc.*, **125**, 3192–3193.
89. Lee, L. V., Mitchell, M. L., Huang, S.-J., *et al.* (2003) A potent and highly selective inhibitor of human α-1,3-fucosyltransferase via click chemistry. *J. Amer. Chem. Soc.*, **125**, 9588–9589.
90. Link, A. J. and Tirrell, D. A. (2003) Cell surface labeling of *Escherichia coli* via copper(I)-catalyzed [3 + 2] cycloaddition. *J. Amer. Chem. Soc.*, **125**, 11164–11165.
91. Speers, A. E., Adam, G. C. and Cravatt, B. F. (2003) Activity-based protein profiling in vivo using a copper(I)-catalyzed azide-alkyne [3 + 2] cycloaddition. *J. Amer. Chem. Soc.*, **125**, 4686–4687.
92. Marik, J. and Sutcliffe, J. L. (2006) Click for PET: rapid preparation of [^{18}F]fluoropeptides using Cu^I catalyzed 1,3-dipolar cycloaddition. *Tetrahedron Lett.*, **47**, 6681–6684.
93. Glaser, M. and Arstad, E. (2007) "Click labeling" with 2-[^{18}F]fluoroethylazide for positron emission tomography. *Bioconj. Chem.*, **18**, 989–993.
94. Glaser, M. and Arstad, E. (2007) Click labelling with 2-(^{18}F)fluoroethylazide for PET. *J Labeled Comp. Radiopharm.*, **50**, S36.

95. Wust, F., Ramenda, T. and Bergmann, R. (2007) Synthesis of ^{18}F-labeled neurotensin(8–13) via copper-mediated 1,3-dipolar (3 + 2)cycloaddition reaction. *J Labeled Comp. Radiopharm.*, **50**, S38.
96. Ross, T. L., Mindt, T. L., Baumann, S., *et al.* (2007) Radiosynthesis of ^{18}F-labelled folic acid derivatives using a direct method and "click chemistry." *J Labeled Comp. Radiopharm.*, **50**, S35.
97. Mindt, T. L., Ross, T. L. and Schibli, R. (2007) Efficient strategy for the parallel development of multiple modality imaging probes using "click chemistry." *J Labeled Comp. Radiopharm.*, **50**, S34.
98. Kim, D. H., Choe, Y. S., Lee, I., *et al.* (2007) A ^{18}F-labeled glucose analog: synthesis via click chemistry and evaluation *J Labeled Comp. Radiopharm.*, **50**, S40.
99. Sirion, U., Kim, H. J., Lee, B. S., *et al.* (2007) An efficient ^{18}F-labeling method for PET study: Huisgen 1,3-dipolar cycloaddition of bioactive substances and ^{18}F-labeled compounds. *J Labeled Comp. Radiopharm.*, **50**, S37.
100. Smith, M. D. and March, J. (2001) *Advanced Organic Chemistry*, 5th edn., Wiley-Interscience, New York.
101. Kim, D. W., Ahn, D.-S., Oh, Y.-H., *et al.* (2006) A new class of S_N2 reactions catalyzed by protic solvents: facile fluorination for isotopic labeling of diagnostic molecules. *J. Amer. Chem. Soc.*, **128**, 16394–16397.
102. Oh, S. J., Lee, S. J., Chi, D. Y., *et al.* (2007) Radiochemical yield dependence of 3′-deoxy-3′-[^{18}F]fluorothymidine in various protic solvent systems. *J Labeled Comp. Radiopharm.*, **50**, S1.
103. Lee, S. J., Oh, S. J., Chi, D. Y., *et al.* (2007) One-step high-radiochemical-yield synthesis of [^{18}F]FP-CIT using a protic solvent system. *Nucl. Med. Biol.*, **34**, 345–351.
104. Chi, D. Y., Kim, D. W., Oh, S. J., *et al.* (2007) New nucleophilic ^{18}F-fluorination in protic solvents: *tert*-alcohols. *J Labeled Comp. Radiopharm.*, **50**, S2.

Applications of Fluorinated Amino Acids and Peptides to Chemical Biology and Pharmacology

15

Application of Artificial Model Systems to Study the Interactions of Fluorinated Amino Acids within the Native Environment of Coiled Coil Proteins

Mario Salwiczek, Toni Vagt, and Beate Koksch

15.1 Introduction

Because of the unique physicochemical properties of carbon-bound fluorine and due to the fact that it does not appear within the pool of ribosomally encoded amino acids, fluorinated analogues can be used as powerful analytical labels to investigate protein structure and protein–ligand interactions [1]. Furthermore, fluorine's impact on the structure, biological activity, and stability of polypeptides makes it an interesting substituent for protein modification [2–5] as well as for the *de novo* design of artificial proteins [6]. In this context, fluorinated analogues of hydrophobic amino acids show great promise as modulators for the stability and self-organization of protein folding motifs whose interactions are largely based on complementary hydrophobic side-chain packing. Global substitution of apolar residues by fluorinated analogues within the hydrophobic core of α-helical coiled coils usually results in a strong thermodynamic stabilization and the specific formation of fluorous interfaces that strongly direct the self-sorting of these peptides [7]. In addition to such "*teflon*" proteins [8] per se, an interesting task is also to investigate how fluorinated amino

Fluorine in Medicinal Chemistry and Chemical Biology Edited by Iwao Ojima

acids interact with their naturally occurring counterparts. However systematic approaches toward this goal have been published only very recently [9]. Numerous attempts at investigating the interactions of comparably small nonpeptidic organic molecules with enzymes indicate that, in addition to the hydrophobicity of fluoroalkyl groups, their polarity also plays an important role [10–14]. Polar interactions of carbon bound fluorine were shown to induce a favorable binding of a potential inhibitor to the thrombin active site [13]. Moreover, fluorine-induced polarity may result in disadvantageous, inverse fluorous effects such as a decrease in lipophilicity [10, 15]. Regarding polar interactions of fluorine, it is also important to note that fluorine scientists have not yet fully agreed on whether carbon-bound fluorine may accept hydrogen bonds, especially from the functional groups of proteins [16]. Accordingly, to this day, there is no consistent opinion on fluorine's behavior as a nonnative "functional group" in amino acid side-chains. Such a specification, however, would be the most important precondition for enabling a directed application of fluorine's unique properties in the engineering of peptides and proteins and their interactions with one another.

Scientific approaches that attempt to rationalize fluorine's effects on the interaction of polypeptides with native proteins usually rely on model systems with a precisely defined interaction pattern that mimics a native environment. With the objective of unraveling the effects of even single fluorine substitutions, the structural homogeneity and stability of such models are indispensable prerequisites. It is also very important that the chosen model system and its analogues are easy to synthesize. Although successful attempts at incorporating fluorinated amino acids by diverse protein expression methods have been reported [17, 18], most synthetic strategies for peptides bearing nonnatural substitutions rely on solid-phase peptide synthesis (SPPS). While linear SPPS is restricted by the achievability of long sequences [19], convergent synthetic routes applying various peptide ligation methods [20] as well as expressed protein ligation [21] pave the way to large modified proteins. Nevertheless, fast synthetic approaches are desirable for the synthesis of a broad variety of different modified peptides. In addition, comparably small model systems allow for a more comprehensive interpretation of experimental data. Consequently, model systems are often significantly smaller than natural proteins and, thus, have to be very carefully designed to efficiently mimic a natural protein environment. In this respect, α-helical coiled coil peptides have greatly gained in importance in recent years [7]. Coiled coils are ubiquitous small proteins that show broad biological activities [22]. As the structural components of many DNA-binding proteins, they play an important role in gene transcription, cell growth, and proliferation. Larger coiled coil assemblies provide molecular scaffolds and networks for the cytoskeleton as well as important structural components of so-called "motor proteins" [23]. Such naturally occurring coiled coils are usually composed of two to five monomeric α-helices whose primary structure is characterized by a repetitive alignment of seven amino acids $(abcdefg)_n$ called a "heptad repeat". Positions *a* and *d* are mostly hydrophobic and harbor leucine, valine, and in some cases isoleucine and methionine. In the folded state, these positions point to one side of the helix, whereas the predominantly hydrophilic positions *b*, *c*, and *f* point to the other side. This spatial separation of hydrophobic and hydrophilic residues imparts significant amphiphilicity to the molecule. Due to segregation of hydrophobic surface area from the aqueous solvent, the helices usually fold into left-handed superhelical oligomers that bury the hydrophobic residues within the so-called "hydrophobic core" (Figure 15.1). In dimeric coiled coils,

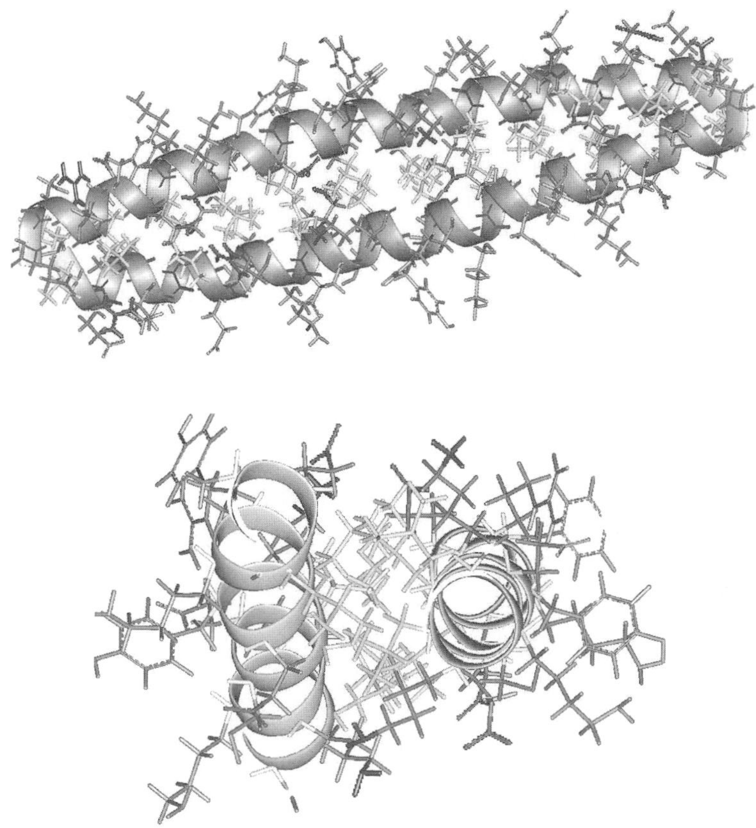

Figure 15.1 Schematic representation of a modelled antiparallel coiled coil homodimer in both a side view (top) and a view along the superhelical axis (bottom). Hydrophobic side-chains (Leu) are represented in yellow, complementary salt bridges in red (Glu), and blue (Lys or Arg). See color plate 15.1.

positions *e* and *g* are usually occupied by charged residues such as glutamic acid, lysine, or arginine. These residues additionally stabilize the folded structure by attractive interhelical coulomb interactions. Besides its stabilizing role, this charged interaction domain is an important determinant of folding specificity [24]. This detailed knowledge about their structure enables the *de novo* design of coiled coils that predictably fold into a specific oligomeric structure that is best suited for a certain investigation. In the subsequent sections, we will summarize how the coiled coil can be used to study fluorine as a side-chain substituent in different native-like polypeptide environments. The incorporation of fluorinated amino acids at different sites within the heptad repeat allows one to probe the impact of fluorination in both hydrophobic and polar environments. Furthermore, the position of fluorinated residues as well as the folding specificity of the coiled coil determines the orientation of the side-chains. Accordingly, the interactions of fluorinated amino acids depend not only on the nature of the environment but also on the way they "look" at it.

15.2 Hydrophobicity, Spatial Demand and Polarity – Fluorine in a Hydrophobic and a Polar Polypeptide Environment

A rationally designed antiparallel, homodimeric α-helical coiled coil peptide served as a model system for the first systematic approach to investigating the interaction characteristics of fluorinated amino acids with native residues (Figure 15.2). The hydrophobic core of the model is exclusively composed of leucine, whereas complementary coulomb interactions between positions *e* and *e′* as well as *g* and *g′* control the antiparallel alignment of the helices within the dimer [25]. In both strands, position *a9* in the hydrophobic core

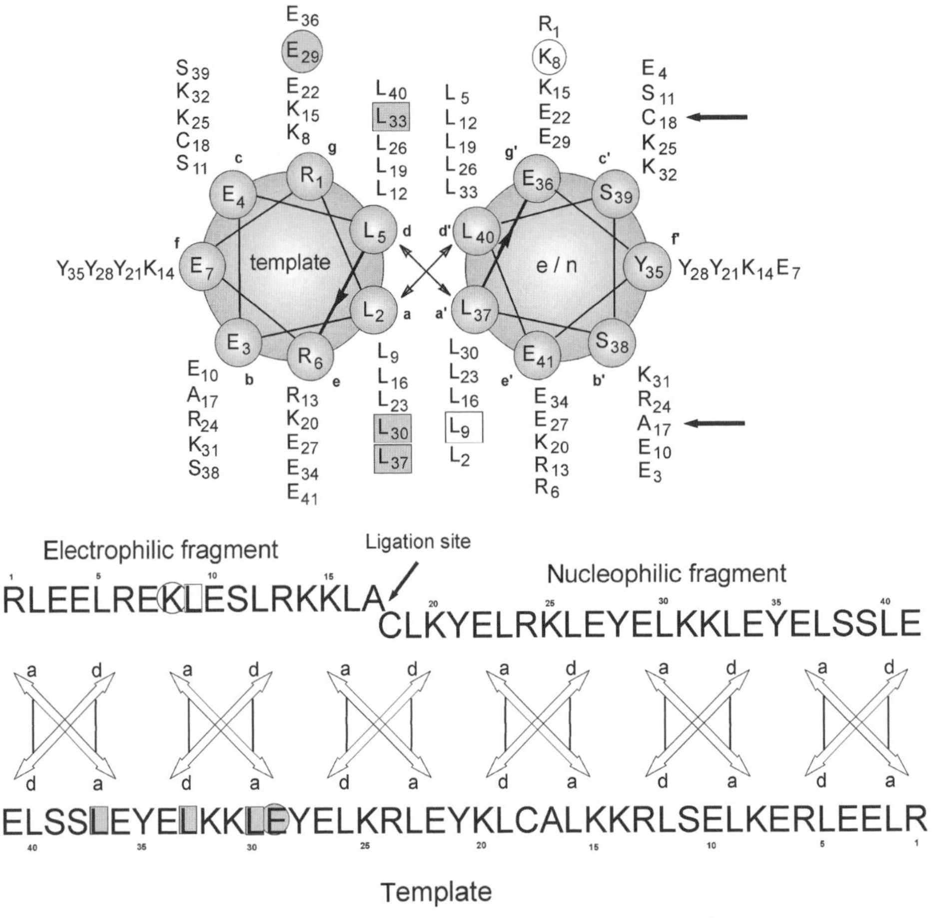

Figure 15.2 Helical wheel and sequence representation of the antiparallel, dimeric model system. The substitution positions are highlighted in one strand with an open square for the hydrophobic core and an open circle for the charged domain. Their direct interaction partners in the opposite strand are shaded in gray squares or circles, respectively. The ligation site is marked with an arrow.

and position *g8* in the charged interface served as substitution positions for fluorinated amino acids. This substitution pattern enabled an investigation of the effects of fluorination in both a hydrophobic and a hydrophilic coiled coil environment. The antiparallel orientation of the two 41-amino-acid peptide chains ensures that fluorinated residues interact exclusively with native residues in the opposite strand. As the peptides were to be used in an assay of the self-replication rate of these peptides (see next section), a convergent synthesis strategy based on native chemical ligation [26] was applied. Three *f*-positions contain tyrosine as an analytical label.

The strength of hydrophobic interactions largely correlates with the hydrophobic surface area but also depends on packing effects of the interacting residues. Regarding the hydrophobic core of the coiled coil model, the first important step was to choose the amino acids that would be appropriate for a systematic study of the impact of fluorination. Single fluorine substitutions for hydrogen often behave bio-isosterically to hydrogen [27], that is, they alter neither the conformation nor the activity of the molecule. Nevertheless, there is no linear correlation between the degree of fluorination and the spatial demand of fluorinated alkyl groups. A comparison of the van der Waals volumes shows a trifluoromethyl group (42.8 Å^3) to be approximately twice as large as a methyl group (24.5 Å^3). These volumes are calculated on a per-molecule basis according to published procedures [28]. Furthermore, the steric effects of a trifluoromethyl group, which are determined by both spatial demand *and* conformational flexibility, show similarities to an isopropyl group [29] rather than to a methyl group.

Accordingly, the investigations were based on (*S*)-aminobutyric acid (Abu) derivatives. Stepwise fluorination of the Abu side-chain yields (*S*)-difluoroethylglycine (DfeGly) and (*S*)-trifluoroethylglycine (TfeGly). The spatial demand of the side-chain can be further increased by elongation by one methyl group, yielding (*S*)-difluoropropylglycine (DfpGly), which may be expected to have a spatial demand that is comparable to that of leucine. Alanine, with the smallest side-chain in this series, served as a control substitution. Figure 15.3 summarizes the discussion of steric size/effects and the resulting substitution pattern.

The analysis of temperature-induced unfolding monitored by circular dichroism (CD) spectroscopy yields the midpoint of thermal denaturation. The melting point (T_m) is defined as the temperature at which 50% of the oligomer is unfolded. Given that the coiled coil maintains its folding specificity when single substitutions are performed, this melting point serves as a qualitative, yet not absolute, measure of stability. Small differences in T_m do not necessarily imply a difference in thermodynamic stability. Significant changes, however, may be interpreted in terms of stabilization or destabilization when comparing structurally equivalent coiled coils since, in many cases, an increase in T_m is associated with an increase in thermodynamic stability [30]. Figure 15.4 shows the melting curves of all variants that carry substitutes for leucine at position *a9* within the hydrophobic core. The respective melting points are given in Table 15.1. With the exception of the DfpGly-variant, the thermal stability in this series of dimers increases along with the increasing spatial demand of the side-chain in position *a9* in the order Ala > Abu > DfeGly > TfeGly > Leu.

Several factors have to be taken into account for a conclusive interpretation of these results. Recent investigations have shown that some fluorinated amino acids exhibit less favorable helix-forming propensities [31]. Thus, structural perturbations of the helical backbone may contribute to some extent to the general destabilization upon incorporation

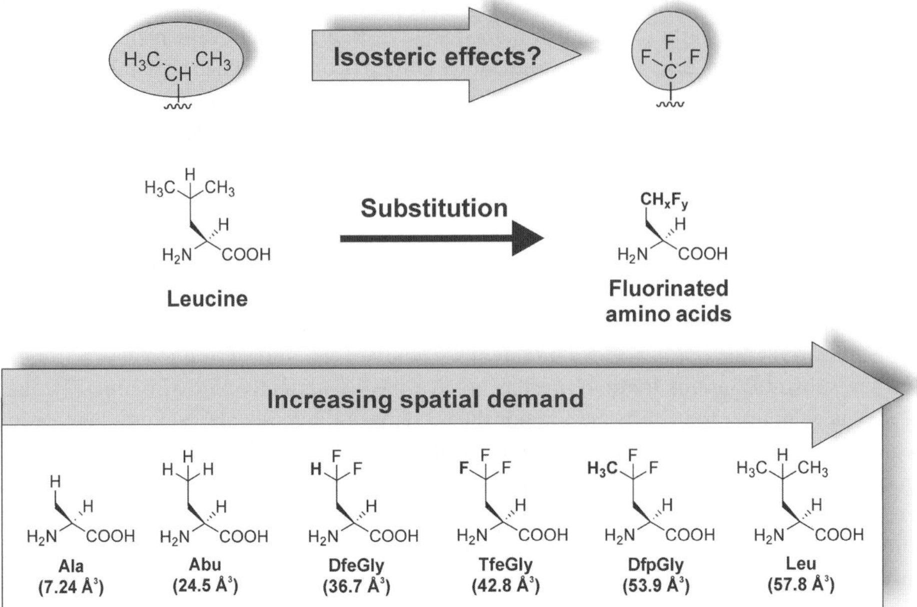

Figure 15.3 *Based on the proposed comparability of isopropyl and trifluoromethyl, one native leucine residue within the hydrophobic core is substituted by amino acids bearing fluorinated ethyl side-chains. Starting from Abu the spatial demand was increased by stepwise fluorination and elongation of the side-chain from the left to the right. Alanine served as a control substitution. The van der Vaals volumes in parentheses were calculated according to reference [28] and refer to the group attached to the β-carbon of the side-chain.*

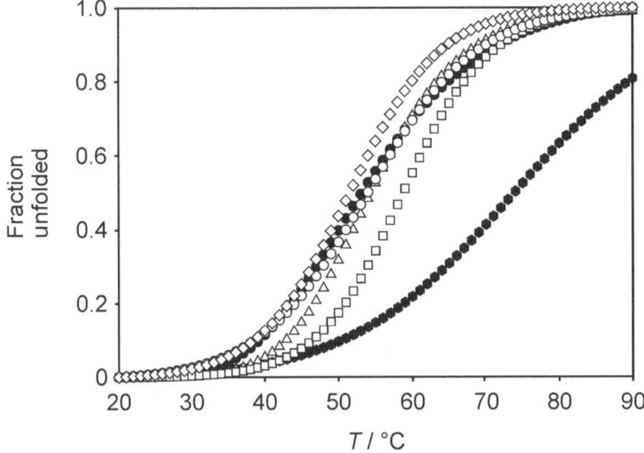

Figure 15.4 *Melting curves of the coiled -coil dimers substituted in position a9. Parent peptide (Leu) (⬢), Ala (●), Abu (○), DfeGly (△),TfeGly (□), and DfpGly (◇).Recorded at peptide concentration of 20 μM at a pH of 7.4 (phosphate buffer containing 5 M GdnHCl).*

Table 15.1 *Melting points of the Leu9 variants*

Amino acid at position *a9*	T_m (°C)
Leu	73.9
Ala	53.2
Abu	54.0
DfeGly	54.4
TfeGly	59.9
DfpGly	51.6

of fluorinated residues. However, with the exception of TfeGly these effects have not been studied for the amino acids used here. Moreover, helix-forming propensity is not the sole determinant of the stability of the coiled coil folding motif whose structure formation is based on intermolecular interactions. The thermodynamic driving force for the oligomerization of coiled coils largely originates from hydrophobic interactions and side-chain packing effects within the hydrophobic core. Due to such stabilizing intermolecular interactions, the effects of helical propensity are often smaller within the interface of coiled coils then they are in single helices [32]. In addition, many investigations have shown that helix-forming propensity can only partly account for conformational preferences of amino acids and that other interactions such as hydrophobic side-chain packing may also play a key role in this regard [33]. For example: the helix propensity for Abu is 1.22 while the value for TfeGly is only 0.05 [31]. Nevertheless, the thermal stability of the folding motif presented here increases upon replacing Abu by TfeGly. The increase in surface area and hydrophobicity upon stepwise fluorination of the Abu side-chain has a favorable effect on hydrophobic interactions and thus appears to stabilize the folding motif. The DfpGly variant is excluded from this trend as, surprisingly, it represents the most destabilizing substitution. Although DfpGly bears a side-chain closest to leucine regarding steric size, its presence in the hydrophobic interior results in an even stronger destabilization than that caused by alanine in the same position.

This effect can be ascribed to the highly polarized methyl group at the γ-carbon (Figure 15.3). Even though the side-chain should be bulky enough to gain sufficient packing within the hydrophobic core, its polarity appears to prevent a favorable interaction in this nonpolar environment. The interpretation of this result gains support from two other findings. The formal γ-difluorination of the Abu side-chain, although it increases the spatial demand, shows only a marginal effect on T_m ($\Delta T_m = +0.4\,K$). One could argue that the stabilizing effect of increased steric bulk is offset by the destabilization that is introduced by a more polarized γ-hydrogen atom. Furthermore, the TfeGly substitution, although expected to show steric effects comparable to leucine, results in a dimer that is roughly 15 K less stable than the parent peptide. As the strong inductive effect of fluorine affects the β-methylene groups, this finding can also be explained by the polarization of hydrogen atoms within the side-chain.

Nevertheless, another important aspect regarding the stability of the folded state must be taken into consideration. In comparison to leucine, TfeGly may exhibit a comparable spatial demand but it lacks two carbon atoms within the side-chain. Therefore, the conformational flexibility and, consequently, the entropy and stability of the folded dimer are reduced. However, as the thermal stability of the model system decreases along with the increasing number of polarized hydrogen atoms buried within the hydrophobic core, the interpretation in terms of polarity is conclusive. The stability follows the order TfeGly > -DfeGly > DfpGly. DfpGly bears the highest number of polarized hydrogen atoms and, accordingly, represents the most destabilizing substitution.

The next important step was to investigate how the same amino acids would behave as substitutes for lysine in position *g8* of the charged domain. This substitution replaces the salt-bridge to glutamic acid in the opposite helix by introducing fluorinated amino acids as noncharged interaction partners. As shown in Figure 15.5, the resulting impact on the thermal stability of the dimer is much lower than that observed for the hydrophobic core (and see Table 15.2).

The decrease in melting temperature compared to the Lys-variant ranges from 2 K for Abu to roughly 6 K for DfeGly. There is no clear trend pointing to a correlation between solvent-exposed hydrophobic surface area and thermal stability. However, there are very interesting effects arising from fluorination of the Abu side-chain. Fluorine substitutions in position *g8* generally result in a somewhat stronger destabilization than the incorporation of Ala and Abu. As shown for substitutions within the hydrophobic core, fluorinated

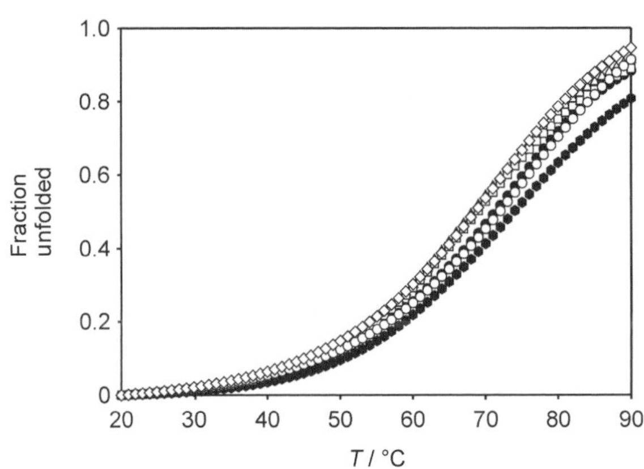

Figure 15.5 *Melting curves of the coiled coil dimers substituted in position g8. Parent peptide (Lys) (⬢), Ala (●), Abu (○), DfeGly (△), TfeGly (□), and DfpGly (◇). Recorded at peptide concentrations of 20 µM at pH 7.4 (phosphate buffer in 5 M GdnHCl).*

Table 15.2 *Melting points of the Lys8 variants*

Amino acid at position *g8*	T_m (°C)
Lys	73.9
Ala	71.3
Abu	71.9
DfeGly	68.3
TfeGly	68.9
DfpGly	68.6

alkyl groups exhibit the unique property of being polar *and* hydrophobic. This "polar hydrophobicity" [34] of fluorinated compounds is one reason for the preferential formation of fluorous phases. Following this argument, the fluorine-substituted peptides may unfold more readily than their nonfluorinated analogues to enable fluorous interactions between the unfolded peptide chains. This assumption gains support from investigations on the self-replication properties of the model system [9] presented in the next section.

15.3 Do Fluorine–Fluorine Interactions Interfere with Coiled Coil Association and Dissociation?

The analysis of temperature-induced unfolding as described in the preceding section explains how single substitutions of proteinogenic residues by fluorinated amino acids affect the dimer's overall stability. The study of the impact of fluorine substitutions on the replicase activity of coiled coils provides some valuable additional insight. Primary structures based on the heptad repeat motif exhibit the ability to promote the condensation of two monomeric fragments whose amino acid sequence follows a complementary heptad repeat. Thus, depending on the primary structure of the fragments, coiled coil peptides can act as either ligases or replicases [35]. The investigations presented in this section are based on the replicase activity, which explains the convergent route for the synthesis of the model peptide (see Section 15.2).

For the establishment of such a replicase cycle, two peptide fragments are required, one of which carries a cysteine at its N-terminus (nucleophilic fragment). The electrophilic fragment usually carries a C-terminal benzyl thioester [36]. A noncatalyzed peptide bond formation between the nucleophilic and the electrophilic fragment that yields a full-length monomer initiates the process of self-replication. The first step in this reaction is a thioester-exchange reaction between the C-terminal benzylthioester moiety of the electrophilic fragment and the N-terminal cysteine of the nucleophilic fragment followed by an irreversible S $\rightarrow$ N acyl migration to yield a native peptide bond (native chemical ligation) [26]. Based on the complementary coiled coil interactions, the monomer catalyzes the first catalytic cycle, acting as a template for the annealing of the nucleophilic and the electrophilic fragments. The specific association of the fragments with the template brings their reactive functional groups into close proximity and thereby promotes the fragment condensation as described above. The cycle is finalized by the dissociation of the oligomer, releasing a newly formed template for further reaction cycles. For the investigation of replicase activity, the proteinogenic residues in both substitution positions (*g8* and *a9*) of the model peptide were replaced by three different amino acids that bear a side-chain of identical length and vary only in fluorine content (Abu, DfeGly, and TfeGly). Figure 15.6 shows the turnover for peptides that contain these residues within the hydrophobic as well as within the charged interaction domain.

The substitution of Abu for leucine in the hydrophobic core position *a9* shows that a significant decrease in spatial demand and hydrophobicity that thermally destabilizes the folding motif also reduces the rate of product formation. The γ-di- and trifluorination of Abu further decelerates rather than accelerates the reaction. This trend correlates with the number of fluorine atoms within the side-chain. Interestingly, the same kind of

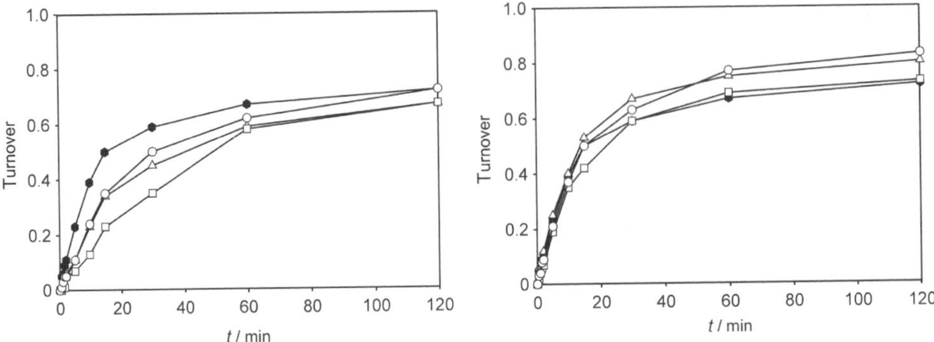

Figure 15.6 *Time dependent turnover of the fragments substituted in position a9 (left panel) and g8 (right panel). The turnover vs. time plot is shown for the parent peptide (⬢) and its Abu (○), DfeGly (△), and TfeGly (□) substituted analogues.*

substitutions in the charged domain yields comparable results. In general, the removal of one salt-bridge of the charged domain slightly accelerates the reaction, as shown for the substitution of Abu for Lys in position *g8*. The association of coiled coils is thermodynamically driven by the formation of the hydrophobic core. Since for substitutions in the charged interface the hydrophobic core remains unchanged, the association is presumably not affected. Therefore, the acceleration upon the incorporation of Abu into position *g8* may be interpreted as an increased dissociation rate in the last step of the cycle. The subsequent fluorination, however, appears to decrease the rate of product formation. Again, this trend correlates with the fluorine content.

According to the unfolding experiments (see preceding section), the stability of hydrophobic interactions follows the order Abu < DfeGly < TfeGly. However, in the charged domain of the dimer the fluorinated residues generally represent the more destabilizing substitutions compared with the non-fluorinated residues. Most importantly, the decrease in replication rate follows the same order as for substitutions in the hydrophobic core (Abu > DfeGly > TfeGly). Being hydrophobic *and* lipophilic, the fluoroalkyl groups in the unfolded peptides segregate from both the water and hydrophobic coiled coil residues. This fluorous effect is comparable to the well-known hydrophobic effect that is accompanied by a lipophobic effect for partly fluorinated alkyl groups [10]. A plausible explanation would be that fluorinated amino acids exhibit fluorine–fluorine 'interactions' in the unfolded state of the peptides and therefore interfere with folding. A similar explanation has also been proposed as the reason for the destabilization of chloramphenicol acetyltransferase upon global fluorination [37]. Fluorous 'interactions' may indeed be rather specific. The association of transmembrane helices in lipid bilayers, for example, is much more efficient when native leucine residues are replaced by hydrophobic *and* lipophobic fluorinated analogues [38]. Accordingly, the same effect may be responsible for inhibiting the coiled coil association in the first step of the replicase cycle. It may be argued that, as a contrary effect, the dissociation within the last step of the cycle should consequently be enhanced. Nevertheless, the fluorous effect of a single amino acid is very likely to be more efficient within the short fragments so that the effect on the association rate is much more pronounced. Figure 15.7 provides an image of how, following the argu-

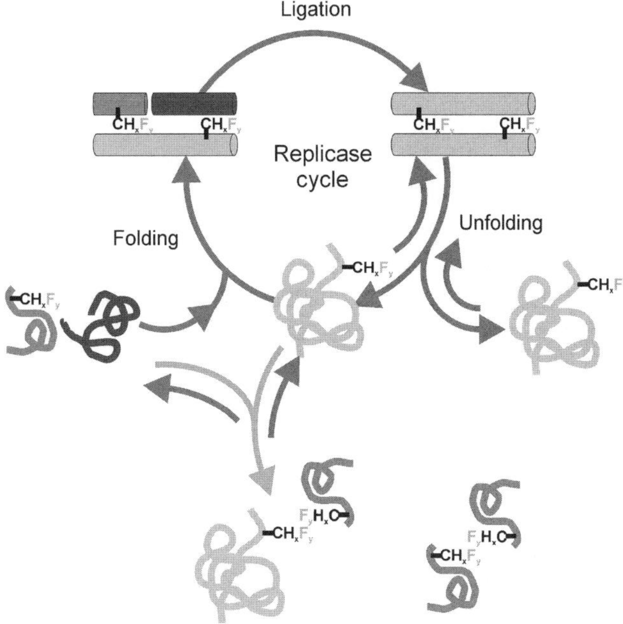

Figure 15.7 *The rate of replication depends on the association–dissociation equilibrium. The data for time-dependent turnover raise the question whether fluorous interactions interfere with the association of the electrophilic fragment with the template thereby inhibiting catalysis [9]. See color plate 15.7.*

ments above, we assume the fluorous effect to compete with coiled coil formation and, thus, to alter replicase activity.

The findings gained with the model system show that two contrary effects characterize the interactions of fluorinated amino acids within the hydrophobic core: spatial demand and hydrophobicity on one side, and fluorine-induced polarity on the other. While the increase in hydrophobic surface area upon fluorination may be favorable for hydrophobic interactions, fluorine's inductive effect appears to interfere with the formation of an intact hydrophobic core. In addition, the investigations of fluoroalkyl side-chains in the charged domain as well as the analysis of fluorine's effect on replicase activity indicate that contacts between fluorinated residues may also have an impact on peptide and protein folding.

The antiparallel homodimeric model system described above, however, represents a rather specific example. Although the fluorinated amino acids here interact exclusively with native residues, *two* fluorinated residues are present in the folded state due to *homodimer* formation. The next goal was to study how subtle changes in the environment of the fluorine substitution – that is, its position within the hydrophobic core – affect its impact on folding and stability. The investigations described in the next section represent a further important step.

15.4 Effects of Fluorination Are Difficult to Predict Because the Impact of Spatial Demand, Hydrophobicity, and Polarity on Structural Stability Depends on the Environment

A new coiled coil model system was designed so that the fluorinated peptide interacted with an exclusively native interaction partner that, itself, was not intrinsically affected by fluorine. Therefore, the folding specificity of the coiled coil system had to be reversed by design. The question was whether the findings for this new model system would be comparable to those for the antiparallel model system described in the preceding sections.

As mentioned before, coiled coil folding specificity can be directed by design. Peptide chains based on the heptad repeat can fold into homo- or heterooligomers either in a parallel or antiparallel fashion. For the new model system, homooligomerization had to be prevented to make fluorine–fluorine contacts impossible. This condition was achieved by designing two complementary peptides [39]. Peptide VPE presents negative charges (glutamic acid) at its *e*- and *g*-positions while peptide VPK presents positive charges (lysine) at the same positions. Under physiological conditions (pH 7.4), heterodimerization is favored because the Coulomb interactions between these positions are repellant in a homodimeric arrangement of VPE and VPK, respectively. Since in this case the charged positions were needed to control heterodimerization, they could not be used to control the orientation of the dimer as well. Thus, a redesign of the hydrophobic core was necessary. The comparison of naturally occurring heptad repeats shows that β-branched amino acids such as valine are highly conserved in *a*-positions of parallel coiled coil dimers but that they are barely found in their antiparallel dimeric counterparts [40]. Placing valine at *a*- and leucine at *d*-positions, therefore, favors a specific, parallel orientation of the dimer. Peptide VPE was used as a template to screen various fluorinated variants of VPK. In this study, the fluorinated amino acids and Abu (see Figure 15.3) were incorporated into the hydrophobic core either at position *a16* or at position *d19* of VPK (Figure 15.8). Two different hydrophobic core positions were chosen to study the impact of the immediate environment of the substitution on the interactions of the fluorinated side-chains. Two peptides that contain leucine instead of fluorinated amino acids at the respective substitution position served as the reference peptides.

The stability of all dimers was assessed by temperature-induced unfolding experiments monitored by circular dichroism spectroscopy in the absence of guanidinium hydrochloride. It was therefore possible to derive the standard free energy of unfolding from data fitting using the Gibbs–Helmholtz equation (15.1) adapted to a two state monomer–dimer equilibrium.

$$\Delta G^\circ = \Delta H^{T_m}\left(1-\frac{T}{T_m}\right)+\Delta c_p\left[T-T_m-T\ln\left(\frac{T}{T_m}\right)\right]-RT\ln K^{T_m} \tag{15.1}$$

K is defined in terms of fraction unfolded f_u by equation 15.2:

$$K=\frac{4f_u^2[\mathrm{D}]_0}{1-f_u} \tag{15.2}$$

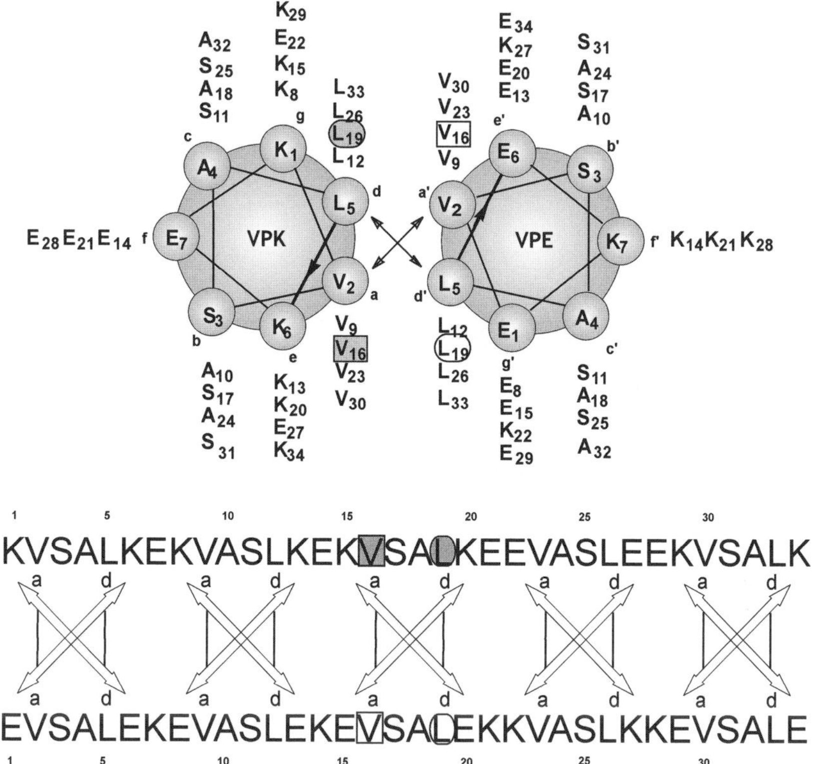

Figure 15.8 *Helical wheel and sequence representation of the parallel, dimeric model system. The substitution positions are highlighted in gray. Their main interaction partners are encircled in black.*

with $[D]_0$ being the concentration of the dimer at the starting temperature where the peptides were assumed to be fully folded. The change in heat capacity ΔC_p was calculated to be 0.94 ± 0.1 kcal/mol [39].

Figure 15.9 shows the fitted melting curves as a plot of fraction unfolded against temperature for the peptides substituted at position *a16*. The respective melting points and standard free energies of unfolding are given in Table 15.3.

While the di- and trifluorination of the Abu-side-chain have only marginal effects on the stability of folding, the bulky DfpGly side-chain stabilizes the structure by roughly 0.8 kcal/mol. Thus, a significant increase in spatial demand of the fluorinated side-chain is able to stabilize the interactions at position *a16*. Nevertheless, all peptides are less stable than the control peptide carrying leucine at this position. Except for DfpGly, these results are qualitatively in good agreement with those gained for the antiparallel model system. The significant difference is that DfpGly represents the most stabilizing substitution at position *a16* within this series of fluorinated peptides. The packing of hydrophobic side-chains has a much stronger impact on the stability of antiparallel coiled coils than it has for their parallel analogues because they are more tightly buried within the hydrophobic

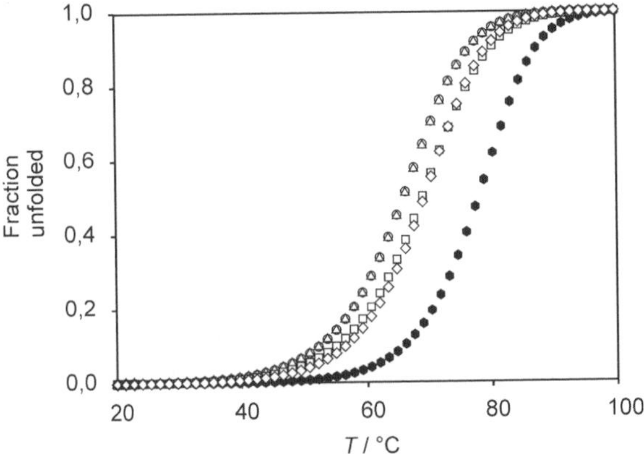

Figure 15.9 *Melting curves of the coiled coil dimers substituted in position a16. Leu (●), Abu (○), DfeGly (△),TfeGly (□), and DfpGly (◇). Recorded at overall peptide concentrations of 20 μM at pH 7.4 (phosphate buffer).*

Table 15.3 *Standard[a] free energy of unfolding and melting points of the a16 substituted dimers*

Amino acid at position *a16*	T_m (°C)	$\Delta G°$ [kcal/mol)
Leu	77.9	13.83
Abu	65.9	11.48
DfeGly	66.0	11.46
TfeGly	69.0	11.51
DfpGly	69.3	12.27

[a] Standard state = 1 M, 101 325 Pa, 25 °C.

core of antiparallel coiled coils [41]. Obviously, the DfpGly side-chain in position *a16* of the parallel model is more flexible than it is in position *a9* of its antiparallel counterpart. Consequently, the polarized γ-methyl group of DfpGly is less tightly buried within the core, which prevents it from disturbing the hydrophobic interactions. The side-chains of DfeGly and TfeGly are shorter and, thus, less flexible. Thus, unlike for the γ-methyl group of DfpGly, the proximity of the polarized β-hydrogen atoms to the rigid backbone does not allow for a more favorable positioning toward the hydrophobic interaction partners and, therefore, disturb the formation of an intact hydrophobic core.

The results for substitutions in position *d19* additionally support the finding that the orientation of fluorinated side-chains and the way they are packed against their interaction partners determines their impact on hydrophobic polypeptide environments. The substitution of Leu at position *d19* by Abu and its fluorinated analogues shows effects that are comparable to those in position *a16* with respect to loss in unfolding free energy (Figure 15.10 and Table 15.4). The formal difluorination of Abu slightly increases the spatial demand and, accordingly, the stability of the coiled coil dimer.

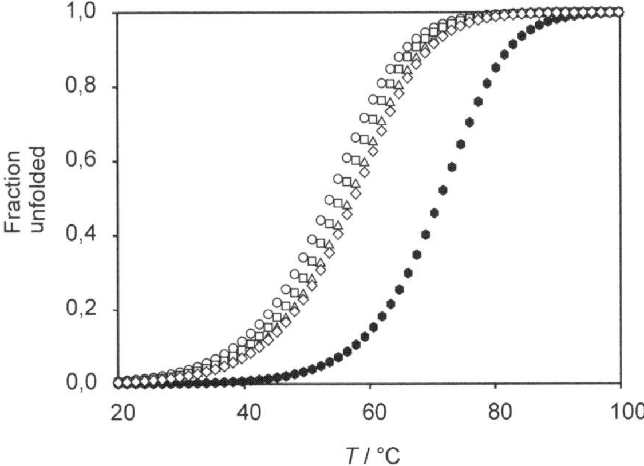

Figure 15.10 *Melting curves of the coiled coil dimers substituted in position d19. Leu (●), Abu (○), DfeGly (△),TfeGly (□), and DfpGly (◇). Recorded at overall peptide concentrations of 20 μM at a pH of 7.4 (phosphate buffer).*

Table 15.4 *Standard[a] free energy of unfolding and melting points of the d19 substituted dimers*

Amino acid at position *d19*	T_m (°C)	$\Delta G°$ (kcal/mol)
Leu	71.3	11.66
Abu	53.7	9.64
DfeGly	56.9	9.99
TfeGly	55.3	9.87
DfpGly	57.5	10.00

[a] Standard state = 1 M, 101 325 Pa, 25 °C.

The incorporation of an additional fluorine substituent (TfeGly) or a methyl group (DfpGly), however, shows only marginal effects. The finding for DfpGly, again, is in sharp contrast not only with the finding for the antiparallel coiled coil but also with results for position *a16* within the same peptide. At position *a16* the additional methyl group was able to stabilize the structure, while this effect is much less pronounced at position *d19*.

Unfortunately, no high-resolution structures of the fluorine containing coiled coil dimers are available at present. Nevertheless, a feasible explanation for the different results for the two substitution positions *a16* and *d19* can still be found by analyzing the high-resolution coiled coil structure of a sequence-equivalent interaction motif, the GCN4 leucine zipper [42]. A comparison of the *a*- and *d*-positions within the crystal structure of GCN4 reveals important differences in terms of side-chain packing. The side-chains in position *a* pack parallel against the side-chains in position *a′*, while those in position *d* pack against position *d′* of the opposite strand. The differences between positions *a* and *d* arise from the C_α–C_β vectors of the side-chains pointing in different directions. In the case of

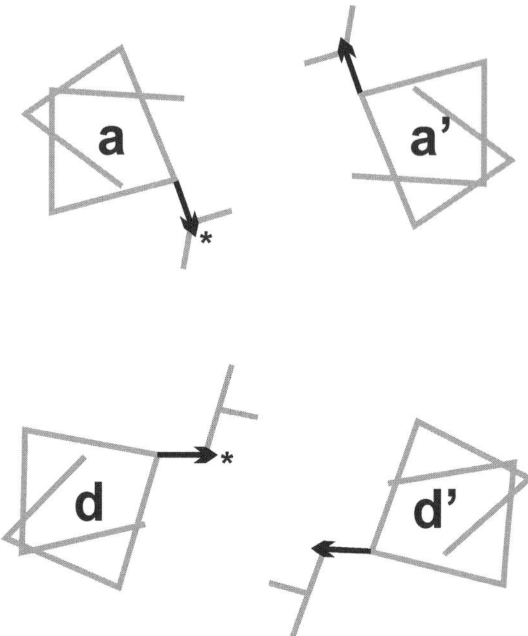

Figure 15.11 Schematic representation of the side-chains at a- and d-positions within the hydrophobic core of the parallel GCN4 coiled coil dimer (according to reference [42]). The asterisk marks the β-carbon atoms that would be polarized in the case of the fluorinated amino acids used for these investigations.

position *a*, this vector points away from the hydrophobic core as well as from the interaction partner in the opposite strand. In contrast, the vectors point into the core and toward their interaction partners within the opposite strand for *d*-positions (Figure 15.11).

This difference in side-chain orientation highly affects the packing as well as the interactions of the fluorinated side-chains with their proteinogenic counterparts. The amino acids studied here all share one similarity: they all carry fluorine atoms at the γ-carbon atom. Consequently, they strongly polarize the β-methylene groups that point in different directions – away from the interaction partner for the *a*-position and toward it for the *d*-position. It is plausible that, accordingly, fluorine-induced polarity at the β-methylene groups has a stronger impact at the *d*-position of the parallel model system. These arguments may explain why a significant increase in spatial demand from TfeGly to DfpGly has the potential to stabilize the interactions within the *a*-position while it seems impossible at the *d*-position.

These findings indicate that although fluorine may impart unique and general properties to hydrocarbons, these properties can have different effects at different positions within a hydrophobic protein environment. Coiled coils share a unique and very specific interaction pattern. Nevertheless, their stability is susceptible to subtle structural modifications, which makes them valuable candidates for systematically investigating the effects of fluorine on peptide–protein interactions.

15.5 Summary and Prospects

The studies reviewed in this chapter describe the impact of single fluorinated amino acid substitutions on protein stability in terms of hydrophobic as well as polar interactions within coiled coil-based model systems. Aiming at a general description of possible molecular interactions of fluorinated amino acids within native protein environments, we can summarize our results as follows.

In contrast to their canonical counterparts, fluorinated amino acids uniquely combine two contrary properties within their side-chain: hydrophobicity *and* polarity. Thus, fluorinated analogues of hydrophobic amino acids affect peptide and protein stability in a way that is not easily predictable, because it is difficult to evaluate the impact of each effect separately. Nevertheless, in contrast to global fluorination of protein interaction domains, single amino acid substitutions generally have a destabilizing effect on the hydrophobic as well as the charged interface of coiled coil peptides. Most importantly, the environment of the substitution itself plays a key role in this context. In these studies, the environment is defined by either a parallel or an antiparallel orientation of coiled-coil dimers as well as by the substitution position, either in the hydrophobic core or in the charged interaction domain. The specificity of coiled coil folding determines the orientation of amino acid side-chains, which results in divergent impacts of the dipoles that are induced by fluorine substituents (*a*- versus *d*-position). Accordingly, a general prediction of the influence of fluorinated amino acids on the nature and stability of protein folding seems to be difficult at present.

The model systems described here deliver useful information about the effect of single fluorinated amino acids on coiled coil folding but, due to the versatility of natural folding motifs, individual proof of structural and thermodynamic effects caused by incorporation of fluorinated amino acids is needed in order to utilize their full potential.

A future goal would be to screen for fluorophilic interaction partners within the pool of the proteinogenic amino acids to guide an optimal application of fluorinated analogues in peptide and protein engineering.

References

1. Jäckel, C. and Koksch, B. (2005) Fluorine in peptide design and protein engineering. *European Journal of Organic Chemistry*, **2005**(21), 4483–4503.
2. Horng, J. C. and Raleigh, D. P. (2003) ϕ-Values beyond the ribosomally encoded amino acids: kinetic and thermodynamic consequences of incorporating trifluoromethyl amino acids in a globular protein. *Journal of the American Chemical Society*, **125**(31), 9286–9287.
3. Cornilescu, G., Hadley, E. B., Woll, M. G., *et al.* (2007) Solution structure of a small protein containing a fluorinated side-chain in the core. *Protein Science*, **16**(1), 14–19.
4. Meng, H. and Kumar, K. (2007) Antimicrobial activity and protease stability of peptides containing fluorinated amino acids. *Journal of the American Chemical Society*, **129**(50), 15615–15622.
5. Smits, R. and Koksch, B. (2006) How C^{α}-fluoroalkyl amino acids and peptides interact with enzymes: studies concerning the influence on proteolytic stability, enzymatic resolution and peptide coupling. *Current Topics in Medicinal Chemistry*, **6**(14), 1483–1498.

6. Montclare, J. K. and Tirrell, D. A. (2006) Evolving proteins of novel composition. *Angewandte Chemie International Edition*, **45**(27), 4518–4521.
7. Hakelberg, M. and Koksch, B. (2007) Coiled coil model systems as tools to evaluate the influence of fluorinated amino acids on structure and stability of peptides. *Chemistry Today*, **25**(4), 48–53.
8. Budisa, N., Pipitone, O., Siwanowicz, I., *et al.* (2004) Efforts towards the design of "teflon" proteins: in vivo translation with trifluorinated leucine and methionine analogues. *Chemistry & Biodiversity*, **1**(10), 1465–1475.
9. Jäckel, C., Salwiczek, M. and Koksch, B. (2006) Fluorine in a native protein environment – How the spatial demand and polarity of fluoroalkyl groups affect protein folding. *Angewandte Chemie International Edition*, **45**(25), 4198–4203.
10. Smart, B. E. (2001) Fluorine substituent effects (on bioactivity). *Journal of Fluorine Chemistry*, **109** (1), 3–11.
11. Fischer, F. R., Schweizer, W. B. and Diederich, F. (2007) Molecular torsion balances: evidence for favorable orthogonal dipolar interactions between organic fluorine and amide groups. *Angewandte Chemie International Edition*, **46**(43), 8270–8273.
12. Olsen, J. A., Banner, D. W., Seiler, P., *et al.* (2003) A fluorine scan of thrombin inhibitors to map the fluorophilicity/fluorophobicity of an enzyme active site: evidence for C–F···C=O interactions. *Angewandte Chemie International Edition*, **42**(22), 2507–2511.
13. Olsen, J. A., Banner, D. W., Seiler, P., *et al.* (2004) Fluorine interactions at the thrombin active site: protein backbone fragments H–C_α–C=O comprise a favorable C-F environment and interactions of C-F with electrophiles. *ChemBioChem*, **5**(5), 666–675.
14. Parlow, J. J., Stevens, A. M., Stegeman, R. A., *et al.* (2003) Synthesis and crystal structures of substituted benzenes and benzoquinones as tissue factor VIIa inhibitors. *Journal of Medicinal Chemistry*, **46**(20), 4297–4312.
15. Hoffmann-Röder, A., Schweizer, E., Egger, J., *et al.* (2006) Mapping the fluorophilicity of a hydrophobic pocket: synthesis and biological evaluation of tricyclic thrombin inhibitors directing fluorinated alkyl groups into the P pocket. *ChemMedChem*, **1**(11), 1205–1215.
16. Dunitz, J. D. and Taylor, R. (1997) Organic fluorine hardly ever accepts hydrogen bonds. *Chemistry – A European Journal*, **3**(1), 89–98.
17. Budisa, N. (2004) Prolegomena to future experimental efforts on genetic code engineering by expanding its amino acid repertoire. *Angewandte Chemie International Edition*, **43**(47), 6426–6463.
18. Wang, L. and Schultz, P. G. (2005) Expanding the genetic code. *Angewandte Chemie International Edition*, **44**(1), 34–66.
19. Tam, J. P., Yu, Q. and Miao, Z. (1999) Orthogonal ligation strategies for peptide and protein. *Peptide Science*, **51**(5), 311–332.
20. Tam, J. P., Xu, J. and Eom, K. D. (2001) Methods and strategies of peptide ligation. *Peptide Science*, **60**(3), 194–205.
21. David, R., Richter, M. P. and Beck-Sickinger, A. G. (2004) Expressed protein ligation. Method and applications. *European Journal of Biochemistry*, **271**(4), 663–677.
22. Mason, J. M. and Arndt, K. M. (2004) Coiled coil domains: stability, specificity, and biological implications. *ChemBioChem*, **5**(2), 170–176.
23. Rose, A. and Meier, I. (2004) Scaffolds, levers, rods and springs: diverse cellular functions of long coiled coil proteins. *Cellular and Molecular Life Sciences*, **61**(16), 1996–2009.
24. Woolfson, D. N. (2005) The design of coiled coil structures and assemblies. *Advances in Protein Chemistry*, **70**, 79–112.
25. Pagel, K., Seeger, K., Seiwert, B., *et al.* (2005) Advanced approaches for the characterization of a de novo designed antiparallel coiled coil peptide. *Organic & Biomolecular Chemistry*, **3**(7), 1189–1194.

26. Dawson, P. E., Muir, T. W., Clark-Lewis, I. and Kent, S. B. (1994) Synthesis of proteins by native chemical ligation. *Science*, **266**(5186), 776–779.
27. O'Hagan, D. and Rzepa, H. S. (1997) Some influences of fluorine in bioorganic chemistry. *Chemical Communications*, **1997**(7), 645–652.
28. Zhao, Y. H., Abraham, M. H. and Zissimos, A. M. (2003) Fast calculation of van der Waals volume as a sum of atomic and bond contributions and its application to drug compounds. Journal of Organic *Chemistry*, **68**(19), 7368–7373.
29. Leroux, F. (2004) Atropisomerism, biphenyls, and fluorine: a comparison of rotational barriers and twist angles. *ChemBioChem*, **5**(5), 644–649.
30. Acharya, A., Rishi, V. and Vinson, C. (2006) Stability of 100 homo and heterotypic coiled coil a-a′ pairs for ten amino acids (A, L, I, V, N, K, S, T, E, and R). *Biochemistry*, **45**(38), 11324–11332.
31. Chiu, H. P., Suzuki, Y., Gullickson, D., *et al.* (2006) Helix propensity of highly fluorinated amino acids. *Journal of the American Chemical Society*, **128**(49), 15556–15557.
32. Kwok, S. C., Mant, C. D. and Hodges, R. S. (1999) Effects of α-helical and β-sheet propensities of amino acids on protein stability. In *Peptides 98,* 25th European Peptide Symposium, Budapest, 1998, (eds. S. Bajusz and F. Hudecz), Akadémiai Kiadó, Budapest, pp 34–35.
33. Chakrabartty, A., Kortemme, T. and Baldwin, R. L. (1994) Helix propensities of the amino acids measured in alanine-based peptides without helix-stabilizing side-chain interactions. *Protein Science*, **3**(5), 843–852.
34. Biffinger, J. C., Kim, H. W. and DiMagno, S. G. (2004) The polar hydrophobicity of fluorinated compounds. *ChemBioChem*, **5**(5), 622–627.
35. Severin, K., Lee, D. H., Martinez, J. A. and Ghadiri, M. R. (1997) Peptide self-replication via template-directed ligation. *Chemistry – A European Journal*, **3**(7), 1017–1024.
36. Ingenito, R., Bianchi, E., Fattori, D. and Pessi, A. (1999) Solid phase synthesis of peptide C-terminal thioesters by Fmoc/t-Bu chemistry. *Journal of the American Chemical Society*, **121**(49), 11369–11374.
37. Panchenko, T., Zhu, W. W. and Montclare, J. K. (2006) Influence of global fluorination on chloramphenicol acetyltransferase activity and stability. *Biotechnology and Bioengineering*, **94**(5), 921–930.
38. Naarmann, N., Bilgicer, B., Meng, H., *et al.* (2006) Fluorinated interfaces drive self-association of transmembrane alpha helices in lipid bilayers. *Angewandte Chemie International Edition*, **45**(16), 2588–2591.
39. Salwiczek M., Samsonov S., Vagt T., *et al.* (2009) Position dependent effects of fluorinated amino acids on hydrophobic core formation of a coiled coil heterodimer. *Chemistry – A European Journal* (2009) in press.
40. McKnight, S. L. (1991) Molecular zippers in gene regulation. *Scientific American*, **264**(4), 54–64.
41. Monera, O. D., Zhou, N. E., Kay, C. M. and Hodges, R. S. (1993) Comparison of antiparallel and parallel two-stranded α-helical coiled coils. Design, synthesis, and characterization. *Journal of Biological Chemistry*, **268**(26), 19218–19227.
42. O'Shea, E. K., Klemm, J. D., Kim, P. S. and Alber, T. (1991) X-ray structure of the GCN4 leucine zipper, a two-stranded, parallel coiled coil. *Science*, **254**(5031), 539–544.

16

Fluorinated Amino Acids and Biomolecules in Protein Design and Chemical Biology

He Meng, Ginevra A. Clark, and Krishna Kumar

16.1 Introduction

While materials scientists have long appreciated the usefulness of fluorinated compounds, chemical biologists and protein scientists have just begun to uncover their value. Fluorine has unique and fascinating properties. The judicious replacement of hydrogen with fluorine can provide tools for unraveling the workings of biological systems. Rational design using fluorinated amino acids has been successful in deciphering collagen stability, in mapping ligand–receptor interactions, in unveiling protein folding and dynamics, and in creating extra-biological structures. In addition, the third-phase properties of perfluorocarbons have been utilized in biomolecule enrichment, artificial membranes, and small molecule micro-arrays. Unusual properties of fluorinated molecules will continue to proffer novel ways to perturb and observe biological systems.

16.1.1 Fluorine in the Context of Biological Systems

Noncovalent interactions are central to many key biological functions. The self-assembly of the plasma and organelle membranes, association between receptors and ligands, and folding of RNA and protein molecules are all controlled by noncovalent interactions. Noncovalent forces in the design of proteins have proved useful in gaining insight into

Fluorine in Medicinal Chemistry and Chemical Biology Edited by Iwao Ojima

sequence/structure and function relationships. We and others have employed fluorocarbons as building blocks to construct biomolecule analogues and exploited their unique characteristics in biological systems. Organic fluorine exhibits unusual properties. At the level of substitution of hydrogen by a single fluorine atom or a few fluorine atoms (e.g., H to F, CH_3 to CF_3), the effects are mainly manifested in electronic, steric, and hydrophobic properties of the resulting compounds. On the other hand, for perfluorocarbon chains, third-phase properties, as distinct from water and hydrocarbons, become more relevant. We first review here the enrichment and self-assembly of biomolecules based on fluorous phase separation. We then highlight some recent uses of fluorinated amino acids in protein design and chemical biology.

16.1.2 Unique Properties of Fluorine

Fluorine is the most electronegative element. While fluorine forms the strongest hydrogen bonds in the ionic state, it does not readily participate in hydrogen bonding once covalently bound to carbon [1]. This observation was at first surprising, but careful scrutiny of crystal structures has revealed few organic molecules displaying F···H intermolecular distances required for hydrogen bonding. These interactions only occur in the absence of better hydrogen-bond acceptors. This property illustrates an essential feature of fluorine: it holds its unshared electrons tightly, evident in its low polarizability ($\alpha = 0.557 \times 10^{-24}\,cm^{-3}$, $\sigma_\alpha = +0.13$ kcal/mol) [2, 3]. The C–F bond (485 kJ/mol) [4] is the strongest formed by carbon bonded singly to any element. Perfluorocarbons are chemically inert and thermally stable. The C–F bond has a reversed and large dipole moment ($\mu = 1.85$ D) [5] compared with that of a C–H bond ($\mu = 0.4$ D) [2]. Despite electron localization on fluorine, perfluorocarbons are minimally polarizable, resulting in weak intermolecular interactions and high vapor pressures.

16.1.3 Phase Separation Properties of Fluorocarbons

The most striking feature of perfluorocarbons is their insolubility. They phase separate from nonpolar organic solvents and water at room temperature. This property was exploited by Horváth in 1994 with the introduction of fluorous biphasic catalysis (FBC) [6]. This new paradigm for separation and purification of catalysts and organic molecules has been used widely [7–9]. Gladysz and Curran [10] have formalized the definition of the term "fluorous" in analogy to "aqueous" as "of, relating to, or having characteristics of highly fluorinated saturated organic materials, molecules or molecular fragments. Or, more simply (but less precisely), 'highly fluorinated' or 'rich in fluorines' and based upon sp^3-hybridized carbon."

The Hilderbrand–Scatchard solubility parameter δ (Equation 16.1) can be used to estimate the miscibility of fluorocarbons with organic solvents [11–13].

$$\delta = \left(\Delta E^{v}/v\right)^{1/2} \qquad (16.1)$$

ΔE^v is the vaporization energy of the pure component (cal/mol) and v is the molar volume (cm^3/mol) at temperature T. Perfluorocarbons have relatively small δ values due to large molar volumes and exceedingly low propensities for intermolecular interactions [11]. In contrast, water has a δ value [14] of 23.5 $cal^{1/2}\,cm^{-3/2}$. The mutual solubility of two nonpolar liquids (1 and 2) can be estimated primarily on the difference between δ values $|\delta_1 - \delta_2|$. Solubility also depends on the temperature (T) and molar volumes (v_1 and v_2) (Equation 16.2).

$$(v_1 + v_2)\cdot(\delta_1 - \delta_2)^2 < 4RT \qquad (16.2)$$

In general, when $\delta_1 = \delta_2$, there is no heat of mixing and two liquids are miscible in all proportions. When $|\delta_1 - \delta_2|$ is less than 3.5 $cal^{1/2}\,cm^{-3/2}$ for an average molar volume of 100 mL, the liquids are still completely miscible at room temperature [12]. When the differences in δ become substantial, phase separation occurs.

16.2 Fluorous-based Methods in Chemical Biology

16.2.1 Fluorous Enrichment of Peptides

Fluorous affinity separation was originally used to remove catalysts from complex reaction mixtures [6]. A perfluoroalkyl moiety (generally no shorter than $-C_6F_{13}$) is appended to a compound of interest. Tagged molecules are then rapidly separated from other components in the mixture by either liquid–liquid extraction or liquid–solid-phase extraction. Fluorous affinity-based separation has recently been used in biomolecule purification, proteomics, and microarray experiments [15–20].

Solid-phase peptide synthesis (SPPS) [21] has greatly facilitated the preparation of these important biomolecules. The final products often reside in a mixture containing similar compounds, due to incomplete coupling of certain residues. Purification of the desired product from the crude mixture is tedious, costly, and environmentally unfriendly. Our laboratory has developed a new fluorous capping reagent, a trivalent iodonium salt, to simplify the purification of polypeptides synthesized on solid support (Figure 16.1a) [15, 16]. In analogy to routine capping steps using acetic anhydride (Ac_2O)/diisopropyl ethylamine (DIEA), this reagent reacts aggressively with free amines of α-amino acids to deliver secondary amines with fluoroalkyl chains. The newly formed bonds between fluoroalkyl tags and peptides are stable enough to survive subsequent coupling, deprotection, and cleavage conditions in both *t*-Boc and Fmoc chemistry. The capped peptides can be removed either by centrifugation or by fluorous solid-phase extraction (FSPE). Removal of deletion products greatly simplifies the purification of desired peptides on reversed-phase high performance liquid chromatography (RP-HPLC).

Complementary to our capping approach, Boom and Overkleeft have pursued a different strategy using the tagging method (Figure 16.1b) [22, 23]. This method is suitable for SPPS using Fmoc chemistry. Peptides are assembled with routine capping steps (Ac_2O/DIEA). Upon the completion of the peptide chain, benzyloxycarbonyl- (or

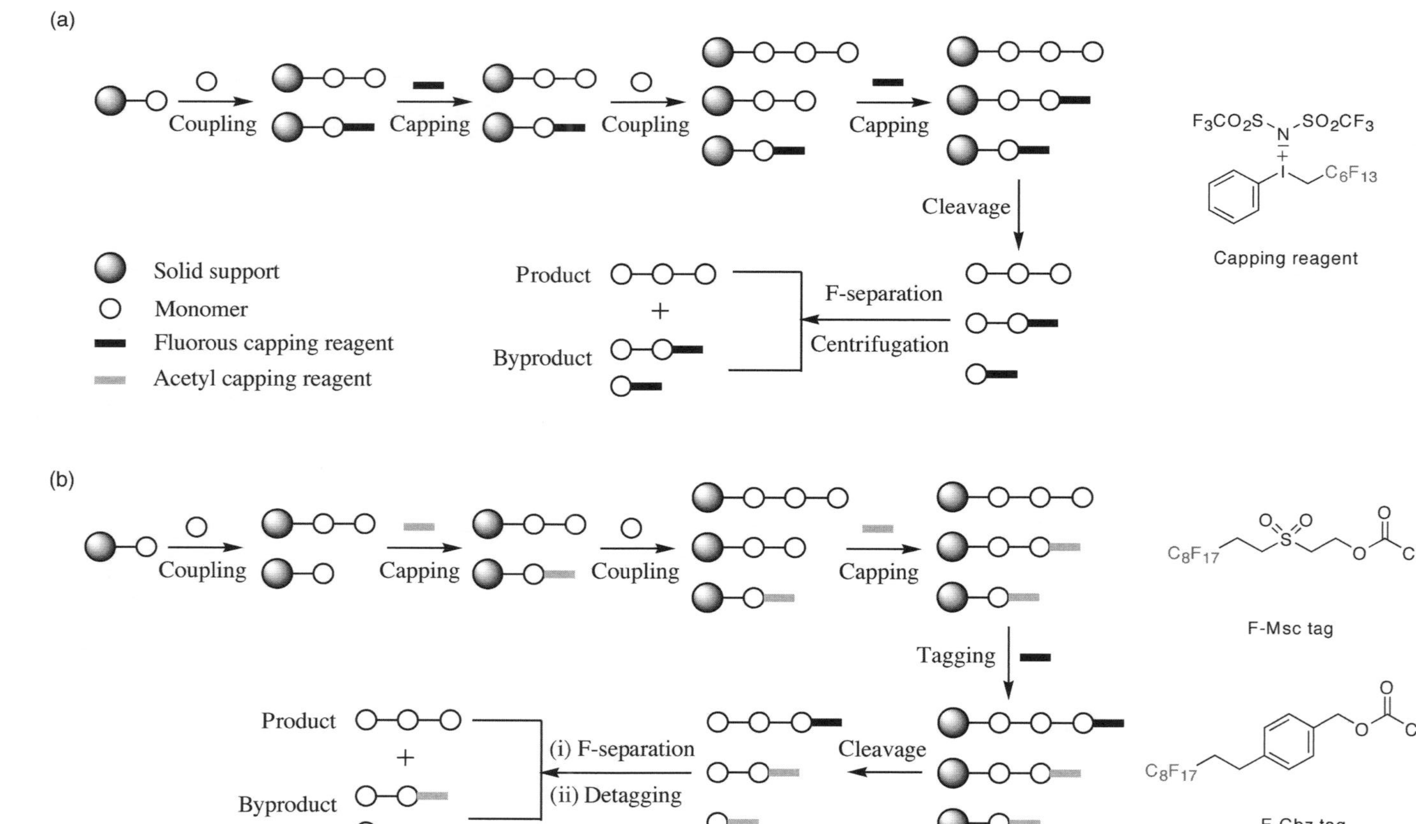

Figure 16.1 *Fluorous purification of peptides synthesized on solid phase. (a) Capping method and trivalent iodonium salt reagent used in both t-Boc and Fmoc chemistry. (b) Tagging method and reagents for Fmoc chemistry.*

methylsulfonylethoxycarbonyl-) based fluorous ponytails are appended to the full-length products. The cleavage step frees all peptides from the resin. Tagged full-length peptides are easily purified on Fluophase™ columns or by FSPE. An additional required detagging step then delivers the final products.

In addition, fluorous tags have been developed for purification of oligonucleotides [18] and oilgosaccahrides [19] synthesized on solid support.

Peters and colleagues have implemented a similar strategy in proteomics [24]. Proteins of interest were first subjected to enzymatic digestion. Peptide fragments were then selectively labeled with fluorous tags via cysteine residues or by β-elimination/Michael addition to phosphorylated peptides. Tagged peptides were significantly enriched and isolated from complex mixtures by FSPE. These samples can be directly analyzed by MALDI-MS or ESI-MS. Fluoroalkyl tags do not undergo fragmentation, providing a distinct advantage over biotin-based affinity reagents. This fluorous derivatization and enrichment strategy has been extended to analyze small molecules and peptides using desorption/ionization on silicon mass spectrometry (DIOS-MS) [25].

16.2.2 Fluorous Small-Molecule Microarrays

Microarrays have tremendously accelerated gene sequencing and biological sample screening [26, 27]. The distinct advantages are the high-throughput nature and the minuscule amounts of sample required. In contrast, small molecules of interest in pharmacology and biology are discovered by tedious and expensive procedures. Small-molecule microarrays (SMMs) offer an attractive alternative for this discovery process. One challenge confronting SMMs is the immobilization of small molecules onto glass slides. This step conventionally relies on covalent modifications of small molecules, often requiring multiple chemical steps. Fluorous SMMs appear to be promising for alleviating this difficulty.

Pohl and co-workers have developed fluorous-based carbohydrate microarrays [28]. The noncovalent interactions between fluorous tagged carbohydrates and fluorous functionalized slides are the focus here in the array fabrication. Sugars (mannose, galactose, *N*-acetylglucosamine, and fucose) were selected as initial targets. The C_8F_{17} tails were appended to protected trichloroacetimides of these monosaccharides to produce fluorous tagged sugars. Next, fluorous sugars were spotted onto a commercially available glass slide coated with a Teflon/epoxy mixture. The fluorous-based microarrays were then interrogated using fluorescein isothiocyanate-labeled jack bean lectin concanavalin A (FITC-ConA). FITC-ConA bound exclusively to mannose as expected. This experiment demonstrated that the C_8F_{17} groups anchored fluorous sugars onto the glass slide, and that this type of microarray could be used for investigating carbohydrate–protein interactions.

Spring and co-workers have applied the same principle for fabricating fluorous-based SMMs to illustrate the recognition of small-molecule ligands by proteins [29]. The prototype binding pair of biotin–avidin was employed, in which biotin was conjugated to fluorous tags and “printed” onto fluoroalkyl-coated glass slides. Avidin interacted with biotin on the glass surface as judged by the fluorescence emanating from dyes linked to avidin.

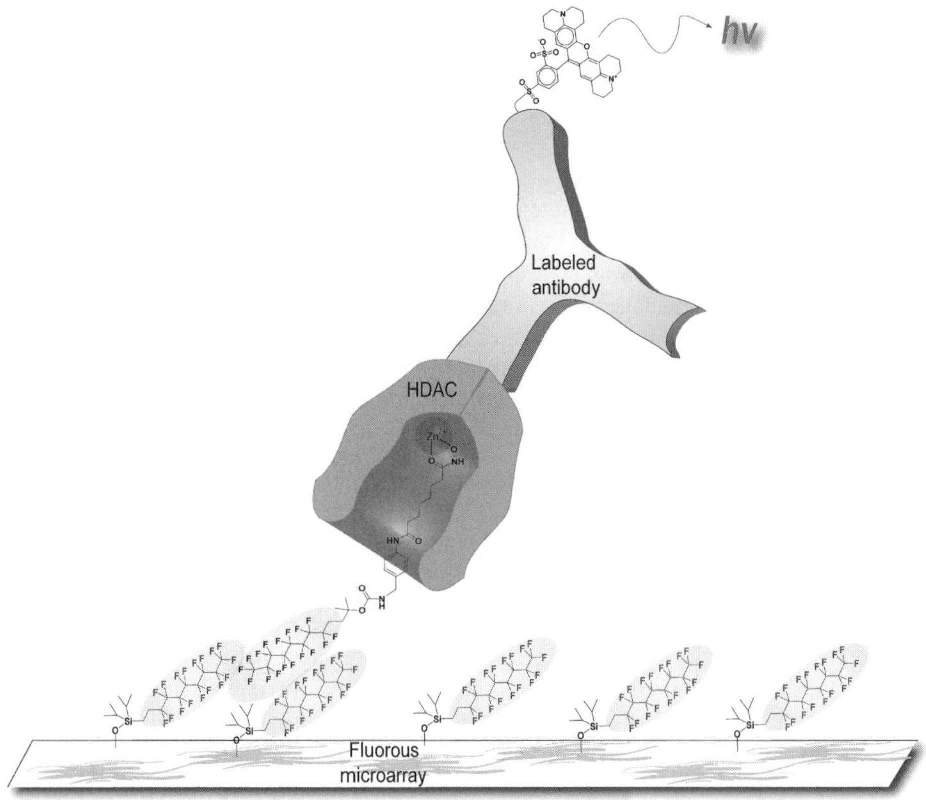

Figure 16.2 *Fluorous small-molecule microarrays. Small-molecule histone deacetylase (HDAC) binders are noncovalently immobilized onto a glass slide coated with fluorocarbon compounds. An antibody labeled with a fluorescent dye recognizes HDAC proteins. See color plate 16.2.*
(Source: *Vegas, A. J., Brander, J. E., Tang, W. et al., Fluorous-based small-molecule microarrays for the discovery of histone deacetylase inhibitors, Angew. Chem. Int. Ed. (2007),* **46**, *7960–7964. Copyright Wiley-VCH Verlag GmbH & Co. KGaA. Reproduced with permission.)*

Schreiber and co-workers have also developed fluorous-based SMMs (Figure 16.2) for the discovery of histone deacetylase (HDAC) inhibitors [30]. Twenty small molecules comprised of putative active and inactive HDAC inhibitors were appended to C_8F_{17}-based anchors. Among them was included suberoylanilide hydroxamic acid (SAHA), a known inhibitor of multiple members of the HDAC family. Once the array was produced as above, purified His-tag fused HDAC2, HDAC3/NCoR2, and HDAC8 were allowed to interact with the array. Detection by Alexa-647-labeled anti-His antibody revealed several binders for each enzyme. The nonfluorous tagged compounds were examined to check their inhibitory capacity against their targets in a fluorescence-based biochemical activity assay. Further, these compounds were also studied by surface plasmon resonance (SPR) for binding to HDAC3/NCoR2. It is important to note that the results obtained from these three different techniques are in good agreement, validating the fluorous-based SMM method.

16.2.3 Fluorinated Lipids

Phospholipids are ubiquitous components of biological membranes and play key roles in confining cellular components, regulating cell–cell interactions, and supporting membrane protein and lipid functions. They are therefore commonly used to fabricate model membranes such as vesicles and supported lipid bilayers for studies in cell biology, drug delivery, and materials science. Clustered display of ligands and receptors has been postulated to be mediated by phase-separated microdomains (or "lipid rafts"). Using fluorocarbons to control lipid self-assembly and present clustered biological ligands would be very useful for probing such mechanisms.

Bendas, Schmidt, and Tanaka have designed fluorinated glycolipids with a glycerol diether core. The glycolipid sLexFL88 (sialyl Lewis^x hexasaccharide with a *n*-perfluorooctanyl tail) self-associates into atypical clusters in 1,2-distearoyl-*sn*-glycero-3-phosphocholine (DSPC) lipids [31]. With 1 mol% of sLexFL88 in DSPC, the domains average ~300 nm in diameter. The size of these clusters increases with increasing concentrations of sLexFL88. CHO (Chinese hamster ovary) cells expressing E-selectins were used to explore leukocyte rolling on these model membranes. The size of sLexFL88 clusters and the distance between clusters dramatically influenced cell binding and rolling [32]. Clustering of sLexFL88 enhanced the cell-binding behavior; however, cells migrated more slowly on membranes composed of sLexFL88 and LacFL88 than on those made from sLexEO6 and DSPC [32].

Our group has fabricated supported lipid bilayers using 2-dipalmitoyl-*sn*-glycero-3-phosphocholine (DPPC) and its fluorinated derivative [33]. The fluorinated derivative was synthesized by replacing the terminal *n*-hexyl groups of acyl chains with *n*-perfluorohexyl groups. Such membranes contain intricate composition-dependent structures, which are stripes of ~50–100 nm interspersed between ~1 μm sized domains. Furthermore, variable-temperature atomic force microscopy (AFM) revealed that domains and stripes are intrinsic features of two gel phases of DPPC and its fluorinated counterpart at low temperatures. These results suggest that fluorinated lipids self-assemble into segregated microdomains in biologically relevant lipid environments. Clustering derived from fluorination could be used for investigating polyvalent interactions and producing biomaterials with specific surface functionalities.

16.3 Fluorinated Amino Acids in Protein Design

16.3.1 Unique Properties of Fluorinated Amino Acids

There are myriad ways by which fluorine can be exploited to perturb the properties of biomolecules. Fluorine is scarce in living systems and the use of ^{19}F NMR for studying protein structure and dynamics is extremely useful. Fluorine is massively electron withdrawing and significantly perturbs acidity constants of amino acids [34]. For example, the pK_a values for pentafluorophenylalanine (2.2 and 8.3) [35] and hexafluorovaline (3.17 and 6.30) [36] are greatly shifted from their hydrocarbon counterparts (2.16 and 9.1 for Phe; 2.61 and 9.71 for Val) [35]. It can also influence hydrogen-bonding properties of the backbone. The inductive properties reverse the quadrupole of aromatic rings, altering magnitudes of π–π interactions and π–cation interactions [37]. Fluorine substitution also influences the conformational preferences of proline, resulting in a change in the *trans/cis* ratio of the peptide bond [38]. Finally, substitution of hydrogen by fluorine influences the α-helical propensities of amino acids [39].

While a C–F substitution dramatically changes the electronic properties of a C–H bond, it exerts only a minor steric influence [40]. The van der Waals radius of fluorine (1.47 Å) is only slightly larger than that of hydrogen (1.2 Å) [41]. In general, a C–F group is nearly isosteric with a C–H group. A CH_3 to CF_3 substitution, on the other hand, significantly increases the bulk. The volume of the van der Waals hemisphere changes from 16.8 $Å^3$ for CH_3 to 42.6 $Å^3$ for CF_3 [4]. The van der Waals volume of CF_3 is comparable to an ethyl group and is only slightly smaller than that of an isopropyl group [42].

Another important consideration is how substitution of CH_3 with CF_3 influences hydrophobicity. The driving force for folding of globular proteins, and many ligand binding events, is hydrophobic interactions [43, 44]. The strength of the hydrophobic effect is related to the size and the molecular nature of the hydrophobic group. Whitesides and co-workers have demonstrated that the interaction of a hydrophobic ligand with carbonic anhydrase is directly proportional to the surface area of the hydrophobic group [45]. Fluorocarbons were more hydrophobic than hydrocarbons as the surface area is larger for fluorocarbons. (Hansch parameters for CF_3 ($\Pi = 1.07$) and CH_3 ($\Pi = 0.5$) groups) [46]. The Hansch parameters for side-chains of hexafluoroleucine and Leu are 1.87 and 1.58 [47], pointing to the superior hydrophobicity of the fluorinated amino acid. As such, the substitution of CH_3 with CF_3 should result in an increase in the conformational stability of a protein if the CF_3 group is shielded from exposure to solvent in the folded structure.

16.3.2 Incorporation of Fluorinated Amino Acids into Proteins

In order to use fluorinated amino acids to study biological systems, they need to be synthesized and incorporated. Despite the challenges in both steps, there are several methods available. For instance, enantiomerically pure fluorinated amino acids may be prepared by asymmetric synthesis or by stereochemical resolution using enzymatic methods [48]. Fluorinated amino acids can be introduced into proteins biosynthetically, or chemically by SPPS. Several reviews that detail the synthesis of enantiomerically pure fluorinated amino acids and incorporation methods into proteins are available [48–51].

16.3.3 Fluorine NMR for Structure Determination

Nuclear magnetic resonance (NMR) methods have emerged as an important complement to X-ray crystallography for determining protein structure. Structural details can be obtained for proteins that are not readily crystallized, such as membrane proteins or molten globules. However, even with NMR methods, it is inherently difficult to obtain structural information on dynamic protein states. One major problem is that ^{1}H NMR resonances tend to overlap, making interpretation of spectra difficult. Because ^{19}F has a large magnetogyric ratio, is present in 100% natural abundance, and displays a broad dispersion in chemical shifts, it represents an invaluable probe for exploring protein dynamics and protein–protein interactions [52].

Rat intestinal fatty acid binding protein (IFABP) binds and transfers fatty acids to their metabolic destination. Structures of apo- and holo-IFABP have been solved by NMR and X-ray crystallography. IFABP consists of two short α-helices and 10 antiparallel β-sheets arranged to form a ligand-binding cavity (Figure 16.3). The side-chains of several aromatic amino acids in IFABP play an important role in both structure and ligand binding. Knowledge of the details of the conformation and dynamics of these side-chains would allow a deeper understanding of the protein–ligand interaction and protein folding. To extract such details, Frieden and co-workers replaced eight phenylalanine (Phe) residues with 4-F-Phe [53]. One-dimensional (1D) ^{19}F resonance assignments were based on the

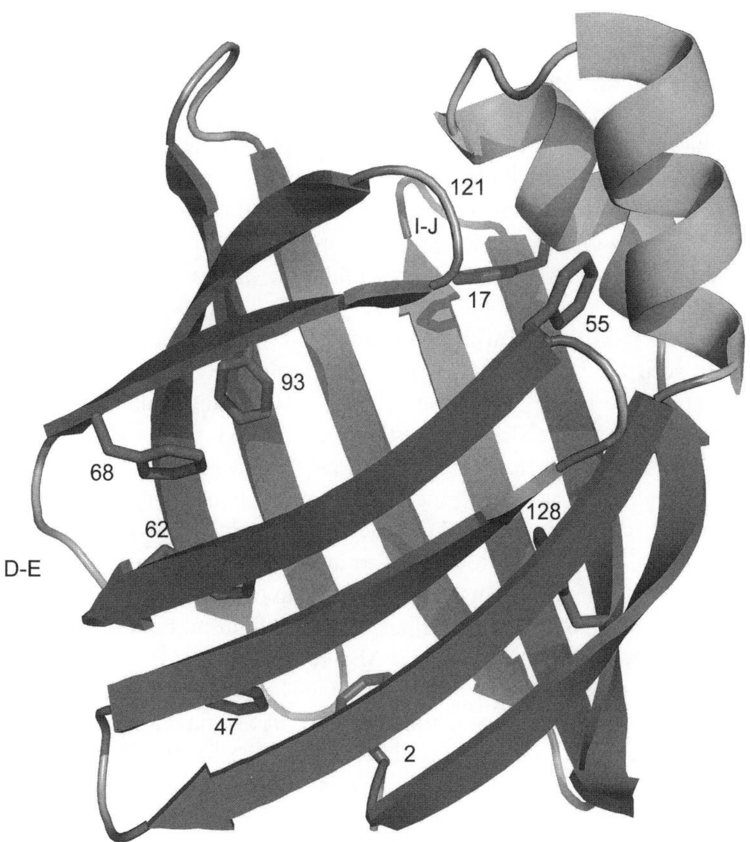

Figure 16.3 *The structure of apo-IFABP (PDB code: 1IFB). Eight Phe residues (shown in stick representation), the D–E and I–J regions, and location of G121 are indicated by labels. The structure was generated using MacPyMOL (DeLano Scientific LLC, Palo Alto, CA, U.S.A.). See color plate 16.3.*

chemical shift of individual proteins containing one or two 4-F-Phe residues. The ligand-bound (oleic acid) and ligand-free forms were then analyzed by 2D ^{19}F–^{19}F NOE (nuclear Overhauser effect) spectra and linewidth measurements. Upon ligand binding, the NOE spectrum revealed more exchange cross-peaks, and the 1D ^{19}F spectrum indicated that most of the peaks were broadened. These results suggested that aromatic side-chains in the binding cavity surprisingly become more flexible upon ligand binding [54], even though the backbone is more rigid in the ligand-bound form [55].

Fluorine labels have also been used to explore the folding process of IFABP. Protein folding involves interactions among side-chains and backbone hydrogen bonds [56]. One challenge in elucidating this process is the characterization of "intermediates" between folded and unfolded states. The "acid state" of IFABP may well resemble such an intermediate – also referred to as a "molten globule." Pulsed-field gradient (PFG) NMR indicated that apparent hydrodynamic radii of the acid state IFABPs were larger than that of the native state, but smaller than that of the fully denatured state. One-dimensional ^{19}F spectra showed that IFABP was structured at pH 2.8, whereas it was mostly unfolded at pH 2.3. It is of note that even at pH 2.3, there was still a small portion of the protein in its native-like structure. More importantly, at pHs lower than 4.8, significant changes in chemical shift and linewidths were observed for Phe128, 17, 68, and 93, suggesting that the D–E turn and I–J regions are involved in the early stages of unfolding. The overall structure of the hydrophobic core remained intact at pH 2.8, as indicated by ^{19}F–^{19}F NOE between Phe68 and Phe93 as well as circular dichroism (CD) measurements. Collectively, changes in side-chain orientations occur before structural changes in the backbone upon lowering the pH, even though these side-chains are buried in the hydrophobic core. Upon oleic acid binding, the spectrum at pH 2.3 resembles the spectrum at higher pH, suggesting that the ligand binding may shift the unfolded protein to a native-like structure [57]. The folding kinetics of IFABP were also monitored by ^{19}F NMR on a slow-folding mutant in which Val replaced Gly121 [58]. This single-site replacement is postulated to disrupt a normal nucleation site in the I–J region. The G121V mutant folds more slowly, is less stable, and exhibits the same structure and dynamics as wild type, allowing for a closer examination of its folding behavior by stop-flow ^{19}F NMR and CD. In the process of refolding, the secondary structures formed twice as quickly as the stabilized conformation of side-chains. A local, nonnative-like structure involving Phe62, Phe68, and Phe93 appeared within milliseconds, followed by arrangement of Phe2 and Phe17, and finally Phe47 into their conformations. This is followed by an overall rearrangement. Without ^{19}F NMR, it would have been a serious challenge to shed light on both side-chain dynamics and intermediate states during folding of IFABP [58].

Mehl and co-workers have incorporated 4-CF_3-Phe into proteins to elucidate structural changes in nitroreductase upon addition of the cofactor flavin mononucleotide. When 4-CF_3-Phe was introduced in the active site (Phe124), the ^{19}F signal shifted upon ligand addition. If it was introduced at a distant site (Phe36), the change in chemical shift was quite subtle. In addition, substitution of 4-CF_3-Phe did not alter the activity of the enzyme [59, 60]. These results suggest that 4-CF_3-Phe could be a sensitive probe for probing ligand binding to proteins.

Membrane proteins are good pharmaceutical targets, since they are displayed on the cell surface where drugs can bind without permeating lipid bilayers. Furthermore, they are involved in cell-signaling interactions, viral entry, and a host of other processes important for therapeutics. However, it is challenging to obtain structural information on these proteins, as they are difficult to crystallize or analyze by solution NMR methods. ^{19}F NMR studies in lipid bilayers have advanced our understanding of both membrane proteins and antimicrobial peptides. In one approach, residues in the TM1 domain (residues 32–48) of diacylglycerol kinase (DAGK) were mutated to Cys, which were then thioalkylated with 3-bromo-1,1,1-trifluoropropanone [61]. Oxygen is distributed nonhomogeneously in the membrane, and exerts an influence on the chemical shift of fluorine, hence allowing for determination of the depth of the residue within the bilayer. The DAGK constructs were assembled in micelles and the 1D solution-state ^{19}F NMR spectrum was obtained in the presence of ambient and increased O_2 partial pressures. The change in chemical shift induced by oxygen ($\Delta\sigma^P$) was plotted for each residue. The resultant plot showed an oscillation of $\Delta\sigma^P$ with a period of 3.6 residues, suggesting that this domain forms an α-helix. Further, the magnitude of changes of oxygen-induced chemical shifts indicated that only one side of the helix interfaces with the lipids.

Antimicrobial peptides such as magainin form cationic and hydrophobic domains upon α-helix formation. These peptides are attracted to anionic cell surfaces and insert their hydrophobic domains into the lipid bilayer, eventually compromising the integrity of the membrane [62]. There are several proposed mechanisms of action for this class of antimicrobial peptides. In the "carpet mechanism," the α-helical axis is parallel to the membrane surface (S-state) and the peptide is monomeric. In the "barrel-stave" and "toroidal pore" mechanisms, the axes of α-helices axis are perpendicular to the membrane surface (I-state). Ulrich and co-workers have used ^{19}F NMR to determine the orientation of helices in the membrane [63]. The strength of ^{19}F dipolar coupling is large, and distances upto 17.5 Å can be measured assuming a resolvable 30 Hz homonuclear coupling [64]. A series of peptide variants of PGLa, peptidyl-glycylleucine-carboxylamide, a member of the magainin family with 4-CF_3-phenylglycine substitutions were synthesized. Residual dipolar coupling between ^{19}F nuclei was used to determine helix orientation in a macroscopically aligned lipid environment. At low concentrations (peptide/lipid molar ratio ~1 : 200), PGLa sits parallel to the membrane surface (90 °, S-state). However, at peptide concentrations (peptide/lipid molar ratio >1 : 50) required for activity, PGLa dimerizes in a tilted orientation (120 °, T-state) (Figure 16.4). This newly discovered T-state may represent an intermediate between the S-state and I-state and is also supported by ^{15}N NMR.

Hong and co-workers have used ^{19}F NMR to investigate the antimicrobial peptide protegrin-1 (PG-1) in lipid bilayers. They first determined that PG-1 forms a dimer in 1-palmitoyl-2-oleoyl-*sn*-glycero-3-phosphocholine (POPC) bilayers using a ^{19}F CODEX (centerband-only detection of exchange) method, evident by two 4-F-Phe12 residues that were within a 15 Å distance [65]. However, the dimer interface could not be described in sufficient detail. Modeling of the PG-1 dimer was based on its solution structure (PDB code: 1PG1) and indicated that the F–F distance in both the parallel and antiparallel dimers is within 11–14 Å. Hong then employed resonance-echo double-resonance (REDOR) solid-state NMR [66] in an attempt to resolve this orientational ambiguity. Phe12 in PG-1 was labeled with ^{19}F, Val16 with ^{13}C, and Cys15 with ^{15}N and ^{13}C. Several intermolecular and intramolecular ^{19}F–^{13}C, or ^{1}H–^{13}C, ^{15}N–^{13}C distances were obtained from both experimental REDOR data fitting and modeling of PG-1. It was determined that PG-1 adopts a parallel dimer in POPC (Figure 16.5), where the C-terminal regions form the dimer

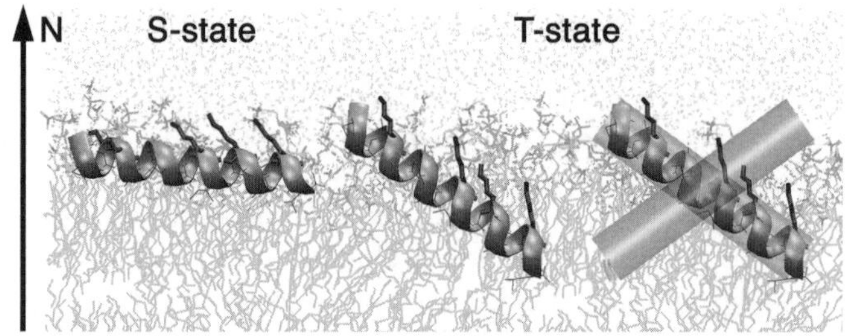

Figure 16.4 *Illustration of the S-state and T-state of the antimicrobial peptide PGLa in a DMPC membrane. At low concentrations, PGLa adopts an S-state. At high concentrations, PGLa assumes a tilted T-state. PGLa forms a dimer in the T-state (shown in purple). See color plate 16.4.*
(Source: *Glaser, R. W., Sachse, C., Durr, U. H. N. et al., Concentration-dependent realignment of the antimicrobial peptide PGLa in lipid membranes observed by solid-state F-19-NMR, Biophys. J. (2005)* **88**, *3392–3397. Reproduced with permission from the Biophysical Society.)*

interface. The design of membrane-disrupting antimicrobial peptides could benefit from the identification and characterization of the dimer interfacial region.

16.3.4 ^{18}F Amino Acids as PET Tracers

Reliable methods for early identification of tumors are crucial for successful treatment regimens. Positron emission tomography (PET) is used to detect radiolabeled compounds that have specificity for tumors [67, 68]. The half-life of ^{18}F (~110 min) is longer than those of other PET isotopes such as ^{11}C, ^{13}N, and ^{15}O. This makes it a popular choice, since radioactive samples may be synthesized and tranferred to a PET facility, allowing greater access to this technology. By far the most common tracer in oncology is 2-[^{18}F]fluoro-2-deoxy-D-glucose; however, ^{18}F-labeled amino acids have also shown promising results in recent studies [69]. The uptake of amino acids by cancer cells is faster than that by normal cells [70, 71], as active transport systems are upregulated in tumor cells [71]. Studies with (*O*-(2-[^{18}F]fluoroethyl-L-tyrosine) showed higher specificity for uptake by tumors than for other conditions such as inflammation [72, 73]. The amino acid L-DOPA is important for brain function and [^{18}F]fluoro-L-DOPA has been used to study dopamine synthesis – an important factor in diseases such as Parkinson's [74].

16.3.5 Fluorinated Aromatic Amino Acids in Receptors and Enzymes

Fluorinated aromatic amino acids have been used to identify and characterize cation–π molecular recognition events in biology. This type of interaction often occurs in protein–cation ligand binding pairs. Addition of fluorine(s) to aromatic rings decreases the negative electrostatic potential on the ring and weakens cation–π interactions. Systematic mutation of aromatic (Phe, Tyr, Trp) residues in proteins with fluorinated analogues and subsequent

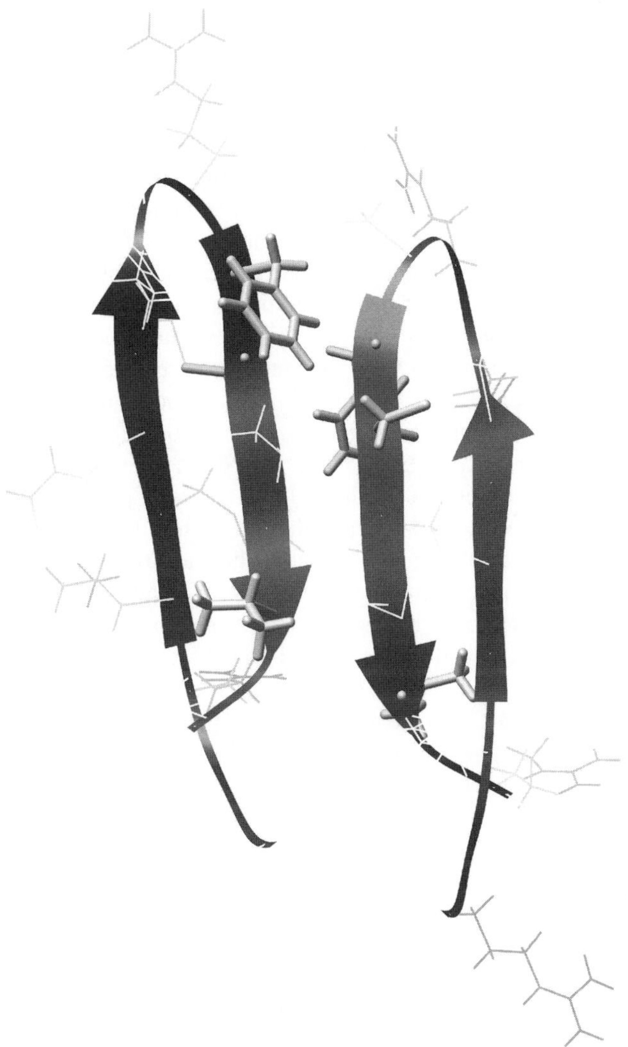

Figure 16.5 *The dimer structure of PG-1 (PDB code: 1ZY6) in POPC bilayers as determined by solid-state NMR. Isotopically labeled amino acids are shown in stick format; Phe12 was labeled with ^{19}F, Cys12 with ^{15}N and ^{13}C, and Val16 with ^{13}C. Inter- and intramolecular distances were used to determine the relative position of the two monomers. See color plate 16.5.*

functional assays have provided insights in identifying cation–π interactions and mapping the binding sites in ligand–receptor pairs that utilize such interactions [75].

The nicotinic acetylcholine receptor (nAChR) belongs to a superfamily of ligand-gated ion channels that includes glycine, 5-hydroxytryptamine-3A ($5HT_{3A}$), and γ-amino-butyric acid (GABA) receptors. These proteins are pentameric, with five homologous subunits arranged around a central pore. Ligand binding sites are located on the extracellular N-terminal domains. Of particular interest are a number of aromatic residues near the ligand-binding site. To assess whether these aromatic side-chains are involved in the recognition of the quaternary ammonium group of acetylcholine (ACh), Dougherty and co-workers incorporated a series of fluorinated tryptophan derivatives into four Trp sites (α86, α149, α184, and γ55/δ57) in nAChR [76]. These fluorinated aromatic residues are introduced into receptors through the site-directed nonsense suppression method [51]. The successful heterologous expression of mutated receptors onto the cell surface allowed for subsequent electrophysiological investigations. Tetra-fluorinated Trp at α86, α184, and γ55/δ57 gave receptor activation (EC_{50}) values that did not differ significantly from wild type (wt) (<2-fold). This observation indicated that the steric perturbation on the receptor due to fluorine substitution is tolerated, and that these sites do not direct strong cation–π interactions. However, tetra-fluorinated Trp at α149 dramatically shifted the EC_{50} value from 50 μM (wt) to 2700 μM. More importantly, the EC_{50}values were strongly correlated with the level of fluorination on the indole ring at position α149. A linear relationship between $\log[EC_{50}/EC_{50}(wt)]$ and the calculated gas-phase cation–π binding affinity was established (Figure 16.6) [76]. These results suggest that the ammonium group of ACh makes van der Waals contact with the indole ring of Trp149. In fact, a constitutively active receptor resulting from incorporation of Tyr–O–$(CH_2)_3$–$N(CH_3)_3^+$ at α149 validated this finding. Similar experiments have led to Trp183 of the 5-hydroxytryptamine-3A receptor ($5HT_{3A}R$) being implicated in a cation–π interaction with the primary ammonium ion of serotonin (Figure 16.6) [77]. Along these lines, fluorinated phenylalanine analogues have been introduced into the $GABA_A$ receptor. Functional measurements revealed a novel cation–π interaction between GABA and Tyr97 in the β_2 subunit of the receptor [78].

Stubbe, Nocera, and co-workers have employed fluorinated tyrosines to probe the mechanism of proton-coupled electron transfer (PCET) in ribonucleotide reductases (RNRs) [79]. The *E. coli* RNR, composed of two subunits (R1 and R2), catalyzes the conversion of nucleotides to deoxynucleotides. Substrate reduction requires a Cys439 radical in R1, propagated from a Tyr122 radical in R2. Three residues from each subunit participate in this long-distance (>35 Å) radical propagation [Tyr122 → Trp48 → Tyr356 within R2, then → Try731 → Tyr730 → Cys439 within R1]. Residue Tyr356 is invisible in the crystal structure of R2 [80] and the docking model of R1 and R2 [81], but is postulated to be on the radical propagation pathway. To examine its role, Tyr356 in R2 was replaced by di-, tri-, and tetra-fluorinated Tyr analogues by the intein-mediated peptide ligation. The fluorinated variants retained their reductive activity. In addition, no significant difference in reduction rates was observed at a pH at which the variants (pK_a ~5.6–7.8 for Ac-FnTyr-NH_2) are deprotonated but the wild type is not (pK_a 9.9 for Ac-Tyr-NH_2). Furthermore, the reduced enzymatic activity was related to the elevated reduction potential of fluorinated tyrosines (E_p (Y/Y^-) ~755–968 mV for analogues vs. 642 mV for Ac-Tyr-NH_2) [82]. These measurements point to a redox-active role for Tyr356 and further suggest that the phenolic proton is not essential in the radical propagation pathway [79].

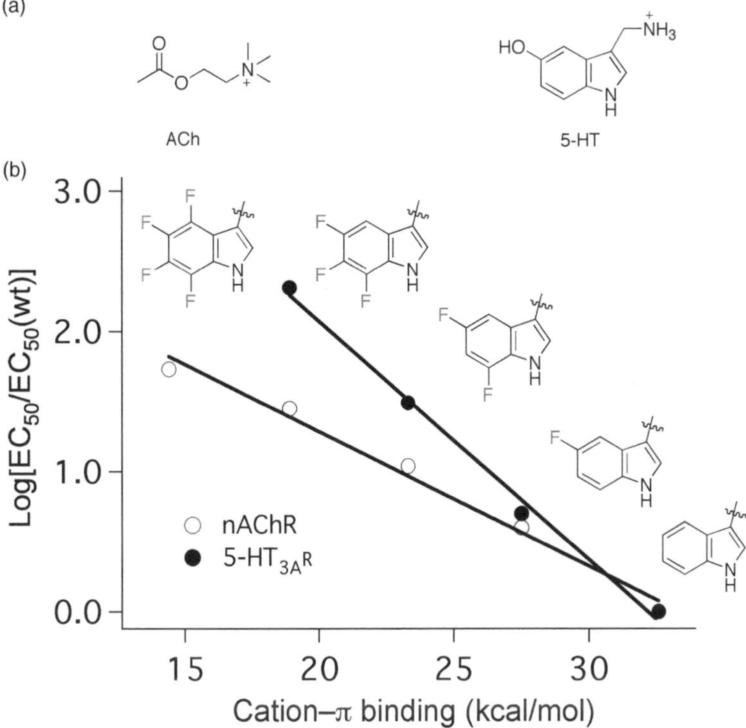

Figure 16.6 *Modulation of cation–π interactions using fluorinated amino acids. (a) Structures of acetylcholine (ACh) and 5-hydroxytryptamine (5-HT). (b) Receptor activation (log[EC_{50}/EC_{50}(wt)]) vs. calculated gas phase cation–π binding ability. Data from references [76] and [77].*

16.3.6 Fluorinated Aromatic Amino Acids in Protein Design

The inductive effect of fluorine on aromatic amino acids can effect not only π–cation interactions, but also π–π interactions. In addition to enhanced hydrophobicity, perfluorinated aromatics exhibit other interesting properties. The solubility parameter values (δ) for benzene and perfluorobenzene are similar, 9.2 and 8.1 $cal^{1/2}\,cm^{-3/2}$ respectively [11]. This suggests that unlike in the case of hexane and perfluorohexane, these two liquids should be miscible at room temperature. The ΔH of mixing is −1.98 kJ/mol at 25 °C and the melting point for crystals of the mixture is about 15 °C higher than for either pure component [1]. The powder diffraction map for the mixture indicates that benzene and perfluorobenzene stack with the H and F atoms aligned. This alignment is quite different from that proposed in hydrocarbon π–π interactions, where a hydrogen interacts with the negative quadrupole of the benzene ring. In order to achieve this interaction, the rings can be aligned perpendicular to one another (an edge–face geometry), or parallel but staggered (face-to-face). The quadrupole moment for perfluorobenzene ($32 \times 10^{-40}\,C\,m^2$) is nearly equal in magnitude and opposite to that of benzene ($-29 \times 10^{-40}\,C\,m^2$). It has been suggested that a quadrupole interaction is responsible for the observed ΔH of mixing, though others have argued that it is a result of coulombic interactions. Gas-phase calculations suggest that the stacking energy between benzene and hexafluorobenzene is –3.7 kcal/mol at a stacking distance of 3.6 Å [37]. Regardless of the origin of the interaction, its magnitude suggests that it can be harnessed to stabilize protein folding.

Pentafluorophenylalanine (f_5-Phe) has been incorporated into several constructs, but clear evidence for π–π stacking has not yet been obtained. Waters and co-workers studied the effect of f_5-Phe on the α-helical stability of a peptide [83]. They designed a series of peptides with aromatic groups at the i and $(i + 4)$ positions, so that upon α-helix formation, the side-chains could interact. Phe–Phe constructs and Phe–f_5-Phe constructs showed similar interaction energies when the aromatic residues were introduced in the center of the sequence ($\Delta G = -0.27$ kcal/mol). When the aromatic residues were introduced near the C-terminus, Phe–Phe constructs displayed higher interaction energies ($\Delta G = -0.8$ kcal/mol) than Phe–f_5-Phe constructs ($\Delta G = -0.55$ kcal/mol). Nevertheless, peptides containing either a Phe–Phe or Phe–f_5-Phe interacting pair displayed higher helicities than their respective control peptides that placed aromatic amino acids at noninteracting positions [i and $(i + 5)$]. Waters has argued that the Phe–f_5-Phe pair cannot achieve ideal geometry for interaction in this construct, thereby reducing the overall increase in stability of the folded form.

Gellman and co-workers investigated a 35-residue peptide, the chicken villin headpiece subdomain (cVHP), that contains three buried Phe residues. They made mutants in which one, two, or all of the Phe residues are changed to f_5-Phe [84]. They also made other protein modifications to facilitate measurement of the folding free energy. One mutant, Phe 10 $\rightarrow$ f_5-Phe was more stable than the unmodified sequence by 0.6 kcal/mol. The NMR solution structure of the Phe 10 $\rightarrow$ f_5–Phe mutant revealed little perturbation in the geometry of these residues [85]. Fluorinated phenyl rings can be used to stabilize protein folds, though the stabilizing magnitude might be different from the level expected from ideal quadrupolar interactions.

It is important to note that the electron-withdrawing effects of fluorine can influence not only side-chains but also the backbone. Side-chain electron-withdrawing substituents can perturb the hydrogen-bonding potential of the peptide bond, and alter backbone conformation. Using a peptoid construct, Blackwell and co-workers discovered that the placement of (*S*)-*N*-(1-(pentafluorophenyl)ethyl)glycine reduces the stability of a threaded loop structure that depends upon specific hydrogen-bonding interactions [86], and that the placement of this residue at the N-terminus results in increased stability. They reasoned that, by withdrawing electrons, fluorine increases the acidity of the ammonium ion, hence increasing its hydrogen-bonding potential. Organic fluorine is unique in its ability to perform this task, since other electron-withdrawing atoms (oxygen and nitrogen) could potentially compete with the hydrogen-bond acceptors and alter structure.

16.3.7 Collagen Stability Revealed Using Fluorine Substitution

Raines and co-workers have used fluorinated proline to explore stereoelectronic influences on the stability of collagen. Collagen is the most abundant protein in mammals and consists of three polypeptide chains that form an extended triple helix. Each polypeptide is composed of ~300 (Xaa-Yaa-Gly) repetitive motifs, where Xaa is often proline (Pro) and Yaa is often 4(*R*)-hydroxy-L-proline (Hyp). It has long been known that the 4-hydroxy group in Hyp is critical to the stability of collagen and about 10% of residues are Hyp in common collagen proteins. A crystal structure of collagen [87] revealed that these residues were involved in interchain hydrogen bonding through structured water molecules. Initially it was thought that these hydrogen bonds imparted stability to collagen [88]. Raines challenged this notion, arguing that the entropic cost of sequestering water molecules would likely be prohibitive in noncrystalline conditions [89, 90], and further proposed that the hydroxyl groups might impart stability by imposing conformational constraints on proline.

To demonstrate that water-mediated hydrogen bonds are not responsible for the structural stability of collagen, Raines and co-workers used 4(*R*)-fluoro-L-proline (Flp) at Yaa positions to construct collagen mimics. Carbon-bound fluorine does not form hydrogen bonds and is very electronegative – and therefore a suitable probe for investigating the proposed stereoelectronic effects [90]. Early studies revealed that the Cγ-*exo* ring pucker is predominant in Hyp residues at Yaa positions. By placing fluorine at the *pro-R* position, a Cγ-*exo* ring conformation should be favored through the "gauche effect." The Cγ-*exo* ring pucker predetermines the main-chain torsion angles (φ, ψ, ω) of Flp residues and they are close to the angles found in collagen (Figure 16.7). For example, the ψ angle in crystalline AcFlpOMe is 141°, very close to the main-chain ψ angle of ~150° in collagen. Moreover, in such a preorganized ring, an $n \rightarrow \pi^*$ interaction between O_0 and $C_1{=}O_1$ along the peptide chain occurs [38, 91, 92]. Again, in crystalline AcFlpOMe, the angle and distance of $O_0 \cdots C_1 = O_1$ are 98° and 2.76 Å, reminiscent of the Bürgi–Dunitz trajectory [93, 94] of nucleophile attack on carbonyl groups (109° ± 10° and in the range ~1.5–3.0 Å) [95]. This $n \rightarrow \pi^*$ interaction not only stabilizes the ideal ψ angle for Flp in the triple helix but also contributes to the required *trans* amide bond (ω = 180°) configuration. Based on the ability to preorganize the proline ring conformation in order of electronegativity F > OH > H, the trend of thermal stability of collagen and collagen mimics could be predicted. Indeed, Flp stabilized collagen triple helices when incorporated in the Yaa position, where the T_m of $(\text{Pro-Flp-Gly})_{10}$ is 22 °C higher than that of $(\text{Pro-Hyp-Gly})_{10}$ ($T_m = 69$ °C) [90] in 50 mM acetic acid ($\Delta T_m = 20$ °C in PBS). The T_m of $(\text{Pro-Pro-Gly})_{10}$ is 41 °C. Additional work by Raines' laboratory has demonstrated that both 4(*S*)-fluoro-L-proline (flp) and 4(*S*)-hydroxy-L-Proline (Hyp) at Yaa positions destabilized collagen [38], essentially by imposing a Cγ-*endo* ring pucker. These results demonstrate that stereoelectronic effects are crucial for the extra stability of collagen, and not the bridging water molecules.

Molecular models indicate that residues at the Xaa position that promote the Cγ-*endo* conformation would stabilize collagen because of better packing in the triple helical form [96, 97]. However, substituting Pro at Xaa positions with either Cγ-*exo* constraining Hyp or Cγ-*endo* constraining Hyp in the $(\text{XaaProGly})_7$ construct diminishes their stabilities [96, 97]. One plausible reason could be that hydroxyl groups experience nonbonded steric interactions in the collagen triple helix. Raines and co-workers again utilized fluorine to probe whether there was a stereoelectronic effect in operation at the Xaa position. The rationale was that –F is smaller than –OH and therefore can be used to avert a potential steric clash. Residue flp favors the Cγ-*endo* ring conformation, whereas Flp favors the Cγ-*exo* ring conformation. As expected, flp, not Flp, at Xaa positions was able to stabilize

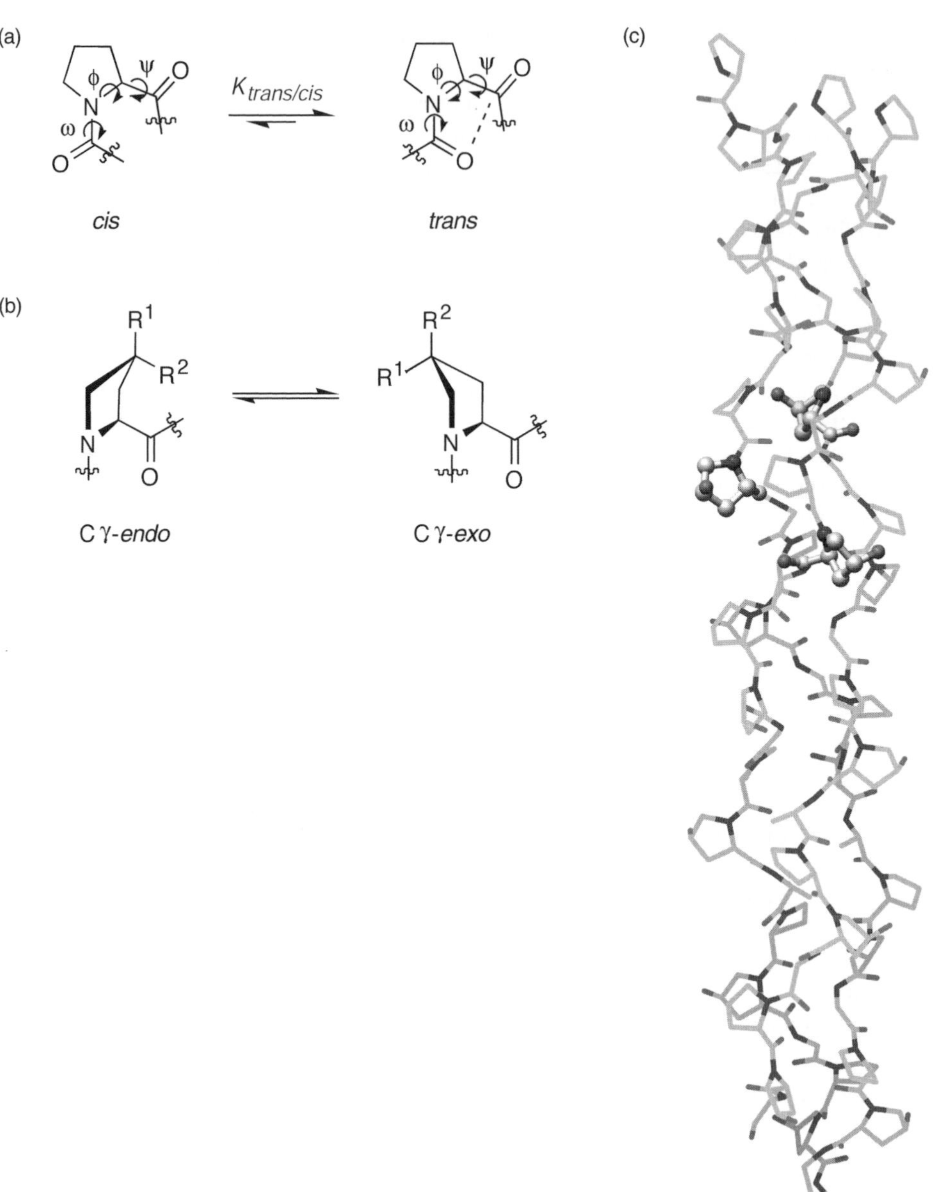

Figure 16.7 Fluorinated prolines in collagen. (a) The trans/cis isomerization of amide bonds and main-chain angles of proline residues. The $n \rightarrow \pi^*$ interaction, depicted by a dashed line, helps stabilize backbone dihedral angles. (b) Electron-withdrawing groups at Cγ influence the ring conformation of proline residues through the "gauche effect." The Cγ-endo pucker is favored when $R^1 = H$ and $R^2 = F$, OH, or H. The Cγ-exo pucker is favored when $R^1 = OH$ or F, and $R^2 = H$. Preorganization of the pyrrolidine ring contributes to the thermal stability of collagen and mimics. Flp: $R^1 = H$ and $R^2 = F$; flp: $R^1 = F$ and $R^2 = H$; Hyp: $R^1 = H$ and $R^2 = OH$; hyp: $R^1 = OH$ and $R^2 = H$. (c) Model structure of collagen (PDB code: 1CAG). Three hydroxyprolines at Yaa positions from each peptide chain are shown in ball-and-stick representation. Carbon = gray; Nitrogen = blue; Oxygen = red. See color plate 16.7.

the triple helix. The peptide (flp-Pro-Gly)$_7$ (T_m = 33 °C) forms a triple helix and is more stable than (Pro-Pro-Gly)$_7$ (T_m = ~6–7 °C) [96]. The peptide (Flp-Pro-Gly)$_7$ does not self-associate into triple helices. These findings suggest that stereoelectronic perturbations at the Xaa position can be used to stabilize artificial collagen.

Next, both flp and Flp were introduced into a triplet repeat sequence. The peptide (flp-Flp-Gly)$_7$ does not form a triple helix [98]. Models suggest that the two fluorine atoms abut one another in the folded structure. These clashes can be averted if two strands of (flp-Flp-Gly)$_7$ are mixed with one strand of (Pro-Pro-Gly)$_7$. With this information in hand, one can begin to design complex structures that oligomerize in 2 : 1 or 1 : 1 : 1 assemblies. Natural collagen is diverse in structure. For example, basement-membrane collagen is rich in 3(*R*)-hydroxy-L-proline, which destabilizes its structure and may influence complex interactions that occur in this matrix. As a result of these studies, we are better equipped to design synthetic collagen for biomedical applications.

16.3.8 Trifluoromethyl-containing Amino Acids in Protein Design

Koksch and co-workers have investigated the effects of replacing a single residue with several fluorinated derivatives in the context of a coiled-coil system, where the amino acid sequence can be described as a heptad repeat. The first and fourth residues (*a* and *d* residues) of the heptad consist of hydrophobic amino acids. The coiled coil is formed when two or more strands oligomerize, where the individual strands are α-helices and the strands wind around each other with a left-handed superhelical twist. Residues at the *e* and *g* positions form interhelical salt-bridges, which can impart parallel or antiparallel specificity to the coiled coil. Koksch and co-workers designed a coiled coil to fold in an antiparallel manner [99]. A single core residue (L9) was replaced by a series of fluorinated amino acids and changes in stability were monitored. Experiments were designed to probe the size of a trifluoromethyl group. The authors posit that if the trifluoromethyl group is as large as an isopropyl group, trifluoroethyl glycine (TEG) should be as large as leucine. If this is so, one expects that TEG substitution would not change the stability of the peptide. In fact, this substitution drastically reduced the stability of coiled coils [99, 100]. However, a series of additional experiments may be needed in order to ascribe these observations entirely to size effects as distinct from other parameters, such as packing, dihedral angle preferences, hydrophobicity, helical propensities, and stereoelectronics.

Substitution of (2*S*, 4*S*)-5-fluoroleucine at two positions in ubiquitin decreases its thermal stability by about 8 °C [101]. Differential scanning calorimetry revealed a similar curve for the fluorinated and wild-type derivatives, where ΔC_p is similar but ΔH_{unf} was smaller for the fluorinated derivative. However, numerous studies have been performed where trifluoromethyl substitutions improve thermal stabilities. In one example, Raleigh and Horng introduced a single trifluorovaline into the N-terminal domain of ribosomal protein L9 (NTL9) to investigate its influence on the kinetics and thermodynamics of protein folding [102]. Val3 and Val21 are mostly buried in wild type and occupy positions in adjacent β-sheets. These residues were separately replaced with trifluorovaline to deliver two variants tfV3 and tfV21. CD and NMR experiments indicated that both variants folded into their native states with limited structural perturbation. The folding free energy ($\Delta G°$) determined by guanidinium hydrochloride (Gdn·HCl) denaturation was 4.17 kcal/mol for

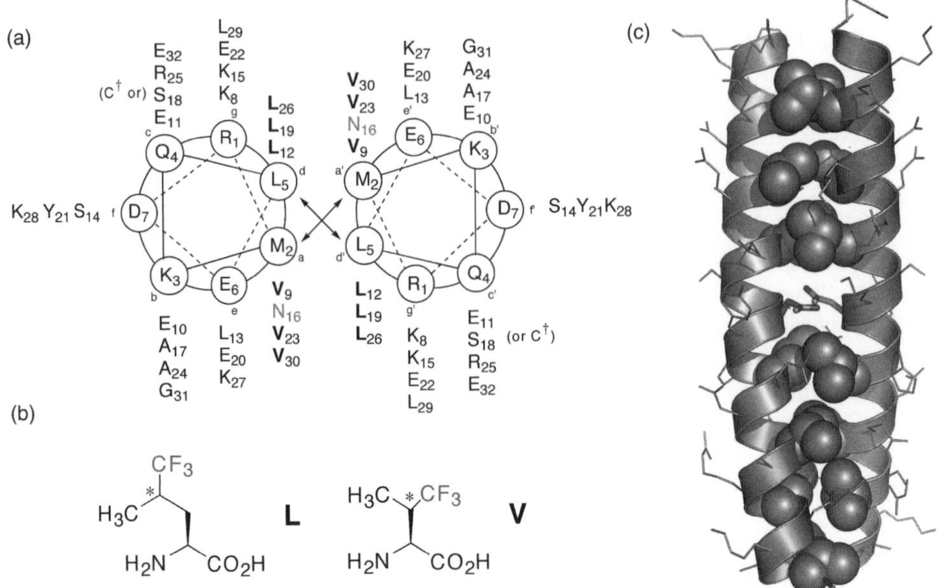

Figure 16.8 Model GCN4-p1 peptides. (a) Helical wheel diagram and sequence of GCN4-p1 analogue. C† = acetamidocysteine. (b) Structures of trifluoroleucine (L) and trifluorovaline (V) used to stabilize peptide ensembles. The asterisk indicates unresolved stereochemistry. (c) Model structure of GCN4-p1 (PDB code: 2ZTA). Side-chains of V and L residues at a and d positions are shown as spheres. Side-chains of Asn residues are shown in stick representation. The structure was generated using MacPyMOL (DeLano Scientific LLC, Palo Alto, CA, U.S.A.).

wild type, 4.96 and 5.61 kcal/mol for tfV3 and tfV21, respectively. The elevated stability by a single CF_3 group incorporation is quite remarkable and significantly larger than what has been observed on coiled-coil systems on a per-residue basis.

Our laboratory [103, 104] and that of David Tirrell [105–107] have independently designed protein folds with super thermal and chemical stability. GCN4-p1, the dimerization domain of the yeast transcriptional activator protein bZip, served as a starting point for the engineering efforts. GCN4-p1 peptides [108] pack against each other to form a homodimeric coiled coil with valine in *a* positions and leucine in every *d* position. We envisioned that replacing the core residues with fluorinated counterparts would increase the driving force for self-association (Figure 16.8). Indeed, incorporation of 4,4,4-trifluorovaline and 5,5,5-trifluoroleucine at *a* and *d* positions resulted in a coiled coil with higher stability. The melting temperature (T_m) for the fluorocarbon peptide was 62 °C, compared with 47 °C for the hydrocarbon peptide. Chaotropic denaturation with Gdn·HCl showed that the apparent free energy of unfolding of fluorinated peptides was ~1.0 kcal/mol higher than the control. The increased thermal and chemical stability could be directly attributed to the higher hydrophobicity of the CF_3 over the CH_3 group [103].

Tirrell and co-workers substituted all four leucines with trifluoroleucines at the *d* position in GCN4-p1 and observed a similar improvement in thermal stability ($\Delta T_m = 13$ °C at a peptide concentration of 30 μM) [106]. Next, they set out to fluorinate the 56-residue bZip miniprotein derived from C-terminal region of GCN4. Gel-retardation assays indicated that the binding affinity and specificity of fluorinated bZip proteins to target DNA were essentially identical to wild type and thermal stability was improved by 8 °C [106]. A higher increase of thermal stability (ΔT_m = 22 °C) was achieved when eight hexafluoroleucines were incorporated into a 74-residue protein A1 [105]. Recently, bZip peptides have been expressed with 5,5,5-trifluoroisoleucine (TFI) or (2*S*, 3*R*)-4,4,4-trifluorovaline (2S,3R-TFV) at the *a* positions. Both mutants were stabilized to different extents compared to the respective hydrocarbon peptides (ΔT_m = 27 °C for the TFI mutant; ΔT_m = 4 °C for the TFV mutant) [109]. The larger increase in T_m imparted by TFI over TFV is quite remarkable, considering that four CF_3 groups are buried in each mutant. The authors reasoned that the side-chains packing in the core might contribute substantially to this difference. According to the GCN4-p1 X–ray structure, the γ–methyl groups of valine interact with the δ–methyl groups from two consecutive leucines in the neighboring helix [108]. The increased size of CF_3 groups in valine may place additional constraints on core packing, thus offsetting the stabilization derived from enhanced hydrophobicity of the CF_3 groups [109]. An alternative explanation is that homotypic *a–a′* interactions between isoleucines are more favorable in this dimeric interface [110]. Packing of the TFI side-chain might allow a more complete burial of the CF_3 groups in TFI than in the TFV variants [109].

Marsh and co-workers have redesigned an antiparallel 4-helix bundle protein by incorporation of hexafluoroleucines into two, four, or six layers in the core. The free energy of unfolding increased by 0.3 kcal/mol per hexafluoroleucine for repacking of the central two layers and by an additional 0.12 kcal/mol for other layers [47, 111]. NMR studies suggested a more structured backbone and a less fluid hydrophobic core in the fluorinated proteins, relative to the hydrocarbon control [111].

The reports described in the last section attribute the increased stability of the protein folds to the enhanced hydrophobicity of trifluoromethyl-containing amino acids, assuming similar α-helical propensities for fluorinated amino acids compared with natural ones. Cheng and co-workers have reported that some fluorinated amino acids have a lower helical propensity when incorporated into monomeric α-helical peptides [39]. For this study, a monomeric construct was designed in which the fluorinated side-chain is solvent-exposed in the folded state. It was found that the helical propensity of ethylglycine was 20-fold higher than that of trifluoroethylglycine. Helical propensities of hexafluoroleucine (8.3-fold lower than leucine) and pentafluorophenylalanine (4.1-fold lower than phenylalanine) are likewise lower. These results suggest that the stabilizing effects engendered by fluorinated amino acids may be larger but are in fact offset by their lower helical propensities. Nonetheless, several groups have reported increased stability by incorporation of CF_3 groups in the side-chain provided the group is not solvent exposed in the folded form.

16.3.9 Self-sorting Fluorinated Peptides

Inspired by the phase behavior of perfluorocarbons, we explored the potential of utilizing fluorinated amino acids to direct specific protein–protein interactions [112]. A parallel coiled coil (**H**) with seven leucines and a single asparagine on each strand in the core was prepared. Glutamic acids and lysines were included at the *e* and *g* positions to ensure a parallel topology. A fluorinated version (**F**) was assembled with replacements of all seven leucines with hexafluoroleucines. A disulfide exchange assay was used to evaluate the self-sorting behavior of fluorocarbon and hydrocarbon peptides. Due to the limited affinity of fluorinated surfaces for hydrocarbon surfaces, we anticipated that the fluorinated surfaces would preferentially pack against each other. Indeed, the disulfide-linked heterodimer (HF) disproportionated nearly completely into homo-oligomers (HH and FF) under equilibrium conditions. The free energy of specificity (ΔG_{spec}) was determined to be –2.1 kcal/mol. Such a high selectivity of homo-oligomerization is derived from both the hyperstability of the fluorinated oligomers and the relative instability of HF.

16.3.10 Fluorinated Membrane Peptides

The *de novo* design of membrane protein architectures presents a formidable challenge. Soluble proteins are folded with hydrophobic residues segregated inside and polar/charged residues outside. However, for the majority of integral membrane proteins, this type of asymmetry in distribution of side-chains is not seen. Energetic requirements of transferring backbone amides from water to the lipid bilayer necessitate hydrogen bond formation between carbonyl and amide N–H groups. Therefore, membrane proteins mostly consist of β-barrels and bundles of α-helices [113]. Our design efforts have been focused on the latter case based on the self-sorting behavior of fluorinated α-helices. Further, by virtue of the simultaneously hydrophobic and lipophobic nature of fluorocarbons, we expected to construct higher-order protein assemblies within lipid environments.

We envisioned a two-step process [114, 115] for the insertion and association of fluorinated transmembrane (TM) helices. First, the hydrophobic TM peptides would partition into micelles or vesicles and form α-helices by main-chain hydrogen bonding. Upon secondary structure formation, one face of the helices would present a highly fluorinated surface. Second, phase separation of fluorinated surfaces within hydrophobic environments would mediate helix–helix interactions, driving bundle formation.

Degrado and Lear [116] and Engelman [117] have successfully converted GCN4-p1 dimers into membrane-soluble forms by switching hydrophilic residues to hydrophobic ones. They found that a single asparagine residue in the sequence could trigger aggregation of helices. In the absence of this critical Asn, the control sequences were monomeric [116, 117]. To explore the potential of fluorinated interfaces to regulate self-association of membrane-embedded helices, we designed four peptides with the presumed helix–helix interface resembling soluble coiled coils (Figure 16.9). All peptides had a membrane-spanning region composed of 20 residues and a polylysine tail appended to the C-terminus to facilitate handling and purification. Sequences TH1 and TH2 are identical except for a single asparagine at position 14 in TH1. TF1 and TF2 were designed by replacing all leucines at *a* and *d* positions with hexafluoroleucines in TH1 and TH2.

Equilibrium centrifugation sedimentation experiments were used to characterize the oligomeric states of designed peptides. TF1 and TF2 assembled to give tetramers and dimers in octaetlylene glycol monododecyl ether (C12E8) micelles, while TH1 and TH2 were found to be dimeric and monomeric, respectively. The presence of the higher-order oligomers was also confirmed using Förster resonance energy transfer (FRET) measure-

(a)

	1234567.... d e f g abcd ef g	14 abcd efg a b c d	
TH1	BQLWIA LLLLIAV	NLILLIA LARLKKK	$KK\cdot CONH_2$
TH2	BQLWIA LLLLIAV	LLILLIA LARLKKK	$KK\cdot CONH_2$
TF1	BQLWIA LLLLIAV	NLILLIA LARLKKK	$KKK\cdot CONH_2$
TF2	BQLWIA LLLLIAV	LLILLIA LARLKKK	$KKK\cdot CONH_2$

Membrane spanning region (3-22)

(b)

H_2N CO_2H B

NO_2 NBD

$(H_3C)_2N$ O ⊕ Cl⊖ $N(CH_3)_2$ COOH HOOC TAMRA

Figure 16.9 *(a) Sequences of transmembrane peptides TH1, TF1, TH2, and TF2. Asn (N) residues are assumed to form interhelical hydrogen bonds in the core. L (green): hexafluoroleucine. (b) Structures of β-alanine, NBD (donor fluorophore), and TAMRA (acceptor fluorophore). See color plate 16.9.*

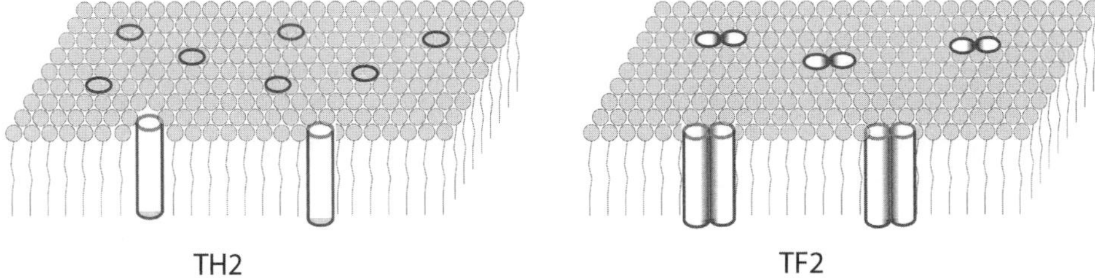

Figure 16.10 *Oligomerization states of TH2 (left) and TF2 (right) in micelles. Without the aid of hydrogen bonding, only fluorinated TF2 peptides self-associate into dimers, while TH2 remains monomeric.*

ments on peptides equipped with either a donor (NBD; 7-nitrobenz-2-oxa-1,3-diazole) or an acceptor (TAMRA, 5-(and 6-) carboxytetramethylrhodamine) on the N-terminus. Dissociation constants were extracted by data fitting for a monomer–trimer equilibrium for TH1 ($K_d = 3.42 \times 10^{-6}\,MF^2$) and TF1 ($K_d = 1.38 \times 10^{-6}\,MF^2$), and a monomer–dimer equilibrium for TH2 ($K_d = 5.75 \times 10^{-3}\,MF$) and TF2 ($K_d = 1.12 \times 10^{-3}\,MF$) (MF = mole fraction of peptides in detergents) [118]. These observations demonstrated that the fluorinated interface in TF2 is able to mediate helix–helix association without the help of hydrogen bonding in micelles (Figure 16.10). When the hydrogen-bonding functionality is added, the fluorinated interfaces result in a stronger interaction between helices in membranes. These results may be explained by studies performed on fluorinated amphiphiles. The incremental changes in the free energy of transfer ($\Delta\Delta G°$) from water to an air–water interface are −620 cal/mol for adding a CH_2 group to a long-chain hydrocarbon molecule and −1220 cal/mol for adding a CF_2 group into a fluorocarbon molecule. From water to a hexane–water interface, $\Delta\Delta G°$ values are −820 cal/mol for CH_2 and −1210 cal/mol for CF_2, while $\Delta\Delta G°$ values from water to a perfluorohexane–water interface are −690 and −1280 cal/mol for CH_2 and CF_2, respectively [119, 120]. These data illustrate that replacement of H by F significantly increases hydrophobicity; more importantly, the interaction between CH_2 and hexane is much stronger than that between CF_2 and perfluorohexane. Thus, the affinity of hydrocarbon lipids for transmembrane peptides and the relative disaffinity of fluorinated surfaces for acyl chains of lipids seem to be responsible for the self-assembly of fluorinated transmembrane peptides in micelles.

FRET experiments further showed that TF1 has a greater tendency to self-associate into trimers in vesicles formed by the zwitterionic lipid 1-palmitoyl-2-oleoyl-*sn*-glycero-3-phosphocholine (POPC) [121] than its hydrocarbon control TH1. Dissociation constants for a monomer–trimer equilibrium were extracted from global fitting of FRET results, with $K_d = (1.8 \pm 0.2) \times 10^{-5}\,\mathrm{MF}^2$ for TH1 and $K_d = (0.56 \pm 0.05) \times 10^{-5}\,\mathrm{MF}^2$ for TF1. Taking the findings together, we have demonstrated that fluorinated interfaces are able to drive the self-association of transmembrane helices with or without the aid of hydrogen bonding in membrane-like environments. Such supramolecularly presented fluorinated interfaces are orthogonal to hydrocarbon interfaces employed by nature and could be very useful for *de novo* design or re-design of membrane-embedded proteins.

16.3.11 Bioactive Peptides

Substitution of fluorine into bioactive peptides has a significant impact on their function. In addition, fluorination results in peptides with improved protease stabilities, from which many therapeutic peptides could benefit. Peptide hormone–receptor interactions play a key role in cell signaling and in the treatment of many human diseases. Thus, a variety of hormonal peptides containing fluorine have been synthesized and evaluated for their biological activities. Angiotensin II (AII: Asp-Arg-Val-Tyr-Val-His-Pro-Phe), a vasoconstrictor, has been subjected to fluorination. Fluorination of AII highlights the diversity of functional perturbations that fluorinated amino acids confer on hormonal peptides. Replacement of Tyr4 and Phe8 with 4-F-Phe resulted in two AII analogues with minimal chemical modifications (–OH to –F, or –H to –F) [122]. Remarkably, the two analogues had distinct profiles in functioning as an oxytocic and modulating blood pressure in rats, and in prostaglandin release assays. The former is a competitive inhibitor, while the latter is an equipotent agonist. The dramatic difference from such subtle modifications uncovers the importance of the hydroxyl group of Tyr4 and the tolerance to fluorination of the phenyl ring of Phe8. However, pentafluoroPhe8-AII, generated by substituting Phe8 with pentafluorophenylalanine, appeared to have a reduced contractile activity in rat uterus and rabbit aorta [123]. The inverted quadrupole moment of the phenyl ring has been implicated in being responsible for the decreased activity [123]. Substitution of Val5 by hexafluorovaline resulted in a more potent vasoconstrictor that showed 133% activity with respect to contraction in rat uterus [124]. However, incorporation of hexafluorovaline at position 8 (in Ac-Asn1-AII) produced a long-acting angiotensin II inhibitor (HFV8-AII) that was resistant to proteolytic degradation by carboxypeptidase A and α–chymotrypsin [125]. Other than AII, fluorinated amino acids have also been substituted into enkephalin, thyrotropin-releasing hormone (TRH), *N*-formylated tripeptide fMFL, a chemoattractant [126], and others [48, 127]. Fluorination of peptide hormones has proved useful for modulating activities and delineating binding events.

Another group of bioactive peptides of interest are the cationic antimicrobial peptides (AMPs). AMPs display great potential as antibiotics in combating microorganisms that have developed resistance to conventional drugs. Gramicidin S (GS) is a cyclic decapeptide [cyclo(-Val-Orn-Leu-D-Phe-Pro-)$_2$] and is effective against a broad spectrum of Gram-positive and Gram-negative bacteria, as well as several pathogenic fungi [128]. The antiparallel β-sheets are joined together by two β-turns and four intramolecular hydrogen

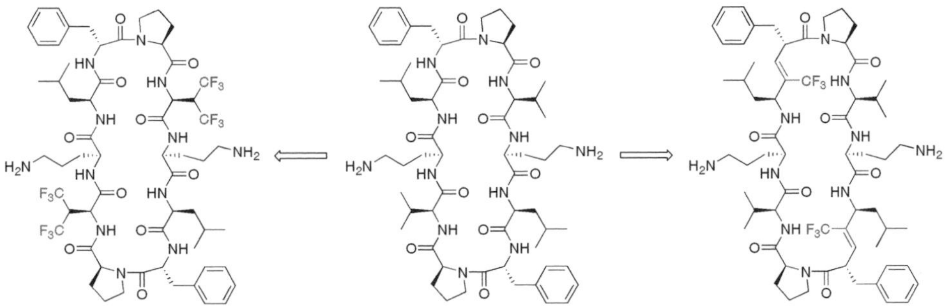

Figure 16.11 *Gramicidin S (GS, center), analogue containing hexafluorovalines (left), and analogue containing CF_3-(E)-alkene isosteres (right).*

bonds between valines and leucines. Segregation of side-chains of Val and Leu on one face and those of Orn on another face of the compact framework results in an amphipathic structure – essential for perturbing bacterial membranes. The more hydrophobic hexafluorovaline (Figure 16.11) was introduced into GS to explore peptide–lipid interactions and the effect of fluorination on antimicrobial activity. CD and ^{1}H NMR studies established that the conformation of [hexafluorovalyl$^{1,1'}$]-GS was similar to that of GS. Fluorinated GS was able to inhibit bacterial growth, though with lower potency relative to the hydrocarbon one [129].

Wipf and co-workers synthesized GS analogues by replacing two amides with CF_3-substituted (*E*)-alkene peptide isosteres and CH_3-substituted (*E*)-alkenes as controls (Figure 16.11) [130]. CF_3-(*E*)-alkenes provide a reasonable fit for the $C_{i(\alpha)}$–$C_{i+1(\alpha)}$ distance (3.8 Å). Moreover, CF_3-(*E*)-alkenes ($\mu = 2.3$ D) match the dipole moment of the amide bond ($\mu = 3.6$ D) better than most other alkene isosteres [131]. The replacements were made at critical hydrogen-bond acceptor positions for β-turns to probe the mimicry of amides by such isosteres. Both solution and solid-state conformational analysis indicated that the bistrifluoromethylated GS retained the overall structure of the parent GS. However, the plane of the double bond was tilted out 70° relative to the parent amide. The analogue was equipotent to GS against *B. subtilis*. This retention of activity has been attributed to the structural similarity [130]. In another system, replacement of leucines with trifluoroleucines in cell lytic melittin led to an augmentation in binding to 1,2-dioleoyl-*sn*-glycero-3-phosphocholine (DOPC) lipid bilayers [132]. Partitioning of fluorinated melittins from water into membranes was increased up to ~2-fold compared to wild type, again highlighting the elevated hydrophobicity of the fluorinated congener.

Our laboratory has modified potential therapeutic peptides using hexafluoroleucines in order to simultaneously improve potency and increase proteolytic stability. We chose to fluorinate two model antimicrobial peptides, magainin 2 amide (M2) [133, 134] and buforin II (BII) [135], peptides that have distinct modes of action. M2 lyses bacteria by forming toroidal pores in lipid bilayers [62], whereas BII kills bacteria by entering cells and binding intracellular DNA and RNA [136]. Four fluorinated analogues in the buforin series and two in the magainin series were synthesized by introducing hexafluoroleucines onto the nonpolar face of model peptides. Five out of six fluorinated analogues had significantly enhanced (up to 25-fold) or similar antimicrobial activity [137]. Hemolytic activity of fluorinated buforins was essentially null, while fluorinated magainins displayed an increase in hemolysis compared to M2. The increase in hydrophobicity of the peptides due to fluorination was evaluated by RP-HPLC. CD measurements using trifluoroethanol (TFE) titrations indicated that fluorination promoted helical structure in both BII and M2. The increased hydrophobicity and elevated secondary structure content are likely responsible for the increase in antimicrobial activity of the BII analogues. In contrast, these two factors correlate with the increased hemolytic activity of M2 analogues. Fluorinated peptides also showed higher protease stability against trypsin, relative to the parent peptides. The increase in side-chain volume of hexafluoroleucine and greater structural stability may be responsible for increased proteolytic stability. These results suggest that fluorination is an effective strategy to increase potency of known antimicrobials and to improve stability of bioactive peptides when proteolytic degradation limits therapeutic value [138, 139].

16.3.12 Functional Proteins

Tirrell and colleagues have fluorinated functional proteins using trifluoromethyl amino acids. They first incorporated 5,5,5-trifluoroisoleucines (TFI) into murine interleukin-2 (mIL-2) [140]. Cytokine mIL-2, resembling human IL-2, contains five isoleucines in the core. Refolding of fluorinated mIL-2 led to a fully functional protein. The potency of mutants was slightly diminished as evaluated by H2-T cell proliferative responses (EC_{50} 2.70 vs. 3.87 ng/mL for the wt and mutant). But the maximal responses were essentially identical for the fluorinated and wt mIL-2. These results indicated that the fluorinated variants were able to retain native folding and remain functional [140]. Tirrell and Montclare then embarked on evolving thermally stable enzymes by introduction of trifluoroleucines (TFL) [141, 142]. Thermostable enzymes are useful for applications at high temperatures [143]. The enzyme of choice was chloramphenicol acetyltransferase (CAT) [141, 142]. CAT catalyzes the transfer of acetyl groups from acetyl-coenzyme A to hydroxyl groups of chloramphenicol and in doing so makes the bacterium resistant against the antibiotic. CAT functions as a homotrimer and each monomer contains 13 leucines. Of these well-dispersed leucines, six are buried in the hydrophobic core. In light of hydrophobic forces stabilizing protein folds, incorporation of the more hydrophobic TFL would be expected to lead to a more thermally stable CAT. Primary global replacement of 13 leucines with trifluoroleucines was achieved using a Leu auxotrophic *E. coli* strain. The mutated protein was an active enzyme. However, it exhibited a 20-fold decrease in the half-life of thermal deactivation [143] at 60 °C, $t_{½,60}$ (101 min and 5 min for wt and variant), and a 9 °C decrease in T_{50}, the temperature at which 50% of the enzyme activity is lost during an incubation period of 30 min. This loss in stability was recovered after two rounds of random mutagenesis. The best mutant containing TFL at 13 sites (L2-A1) exhibited a $t_{½,60}$ of 133 min, 32 min longer than that of the wild type, and an identical T_{50} to the wt CAT (66 °C). Moreover, the catalytic efficiency of L2-A1 was comparable to that of wt CAT. Three different mutations in L2-A1 were K46M, S87N, and M142I (Figure 16.12), as well as a nonsense mutation that deleted one residue from the evolved proteins. Intriguingly, none of mutations was proximal (<15 Å) to the chloramphenicol binding site and none of them made direct contact with any of the 13 Leu/TFL residues. Although incorporation of fluorinated amino acids at multiple sites in natural proteins might compromise

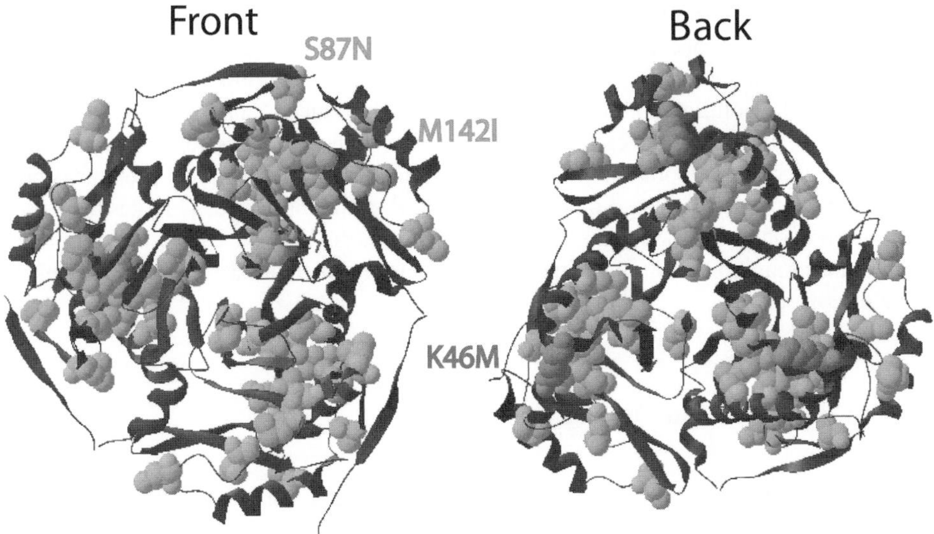

Figure 16.12 *Front and back views of the CAT trimer. The three stabilizing mutations in L2-A1 are S87N, M142I in orange, and K46M in pink. Trifluorolecuines/leucines are colored blue and chloramphenicol red. See color plate 16.12.*
(Source: Montclare, J. K., Tirrell, D. A., Evolving proteins of novel composition. Angew. Chem. Int. Ed. (2006) **45***, 4518–4521. Copyright Wiley-VCH Verlag GmbH & Co. KGaA. Reproduced with permission.)*

their activities, this report certainly indicates a new direction for engineering functional proteins with novel compositions [141].

Another template chosen for evolving a functional protein with fluorinated side-chains was the 238-residue green fluorescent protein (GFP) with 19 leucines [144]. The starting protein (GFPm) originated from the "cycle 3" variant [145], in which two additional mutations (S65G and S72A) [146] were introduced for appropriate spectroscopic signatures. The initial replacements of all 19 leucines by TFL resulted in a mutant (GFPm-T) that was mostly insoluble and did not fluoresce. Misfolding and aggregation were likely responsible for the observed insolubility. To recover the folding and function of GFPm-T, mutagenesis by error-prone PCR (polymerase chain reaction) and screening by FACS (fluorescence-activated cell sorting) were conducted. After 11 rounds, a fluorinated GFP mutant (11.3.3-T) containing 20 amino acid substitutions was obtained. The median fluorescence of cells expressing this mutant (11.3.3-T) exhibited a ~650-fold increase over that of cells expressing GFPm-T. This increase could in principle be due to two reasons: (1) the enhancement of spectral characteristics, (2) the elevated level of expressed, properly folded, functional GFP. SDS-PAGE experiments showed that both the expression level and soluble fraction of variants were increased during the course of laboratory evolution. In addition, the spectroscopic properties were also improved. The brightness of the purified protein 11.3.3-T was 1.6-fold higher than that of GFPm-L (GFPm expressed in Leu media). Sedimentation velocity experiments revealed that GFPm-L exists in a monomer–dimer equilibrium, while both 11.3.3-T and its hydrocarbon version largely exist as dimers. These results suggested that directed evolution was able to fine-tune the folding behavior of the fluorinated variant. Indeed, the refolding rates of fluorinated variants were increased over the evolution and that of 11.3.3-T was comparable to that of GFPm-L. It is worth noting that the overall evolution eliminated 6 leucines and introduced 2 new ones. The 11.3.3-T containing 15 trifluoroleucines folds and fluoresces intensely. This report again underscores that global incorporation of fluorinated amino acids into functional proteins is possible and is of practical utility. So far, only one fluorinated amino acid, TFL, has been used in directed evolution experiments. With incorporation of different and/or multiple types of fluorinated amino acids, other unexpected properties of fluorinated proteins are likely to be discovered.

16.4 Conclusion and Outlook

The bioorthogonal properties of fluorinated biomolecules have made them an attractive weapon in diverse applications due to their fundamentally distinct physical and chemical properties. Third-phase properties of fluorocarbons allow for patterning lipid bilayers and simplifying biomolecule enrichment. Fluorinated amino acids have been utilized for designing peptides/proteins with extra-biological properties, probing ligand–receptor interactions, modulating bioactivities of peptides, structural characterization, and probing mechanisms of enzymatic catalysis. Recently, these building blocks have also been incorporated into functional proteins. With our growing ability to use biosynthetic pathways to construct complex structures, many new biomacromolecular design avenues using fluorinated amino acids and other fluorinated precursors (e.g., fluorinated lipids, nucleotides, and carbohydrates) are certain to emerge.

Acknowledgments

Work in our laboratory is partially supported by the NIH, NSF and DuPont de Nemours & Co. We thank Professor Anne S. Ulrich (Universität Karlsruhe), Professor Jin K. Montclare (Polytechnic University) and Arturo J. Vegas (Harvard University) for providing high-resolution figures.

References

1. Dunitz, J. D. (2004) Organic fluorine: Odd man out. *ChemBioChem*, **5**(5), 614–621.
2. Kirsch, P. (2004) *Modern Fluoroorganic Chemistry: Synthesis, Reactivity, Applications*. Wiley-VCH Verlag GmbH, Weinheim.
3. Hehre, W. J., Pau, C. F., Headley, A. D., *et al.* (1986) A scale of directional substituent polarizability parameters from abinitio calculations of polarizability potentials. *Journal of the American Chemical Society*, **108**(7), 1711–1712.

4. Seebach, D. (1990) Organic synthesis – Where now. *Angewandte Chemie International Edition in English*, **29**(11), 1320–1367.
5. Biffinger, J. C., Kim, H. W. and DiMagno, S. G. (2004) The polar hydrophobicity of fluorinated compounds. *ChemBioChem*, **5**(5), 622–627.
6. Horváth, I. T. and Rabai, J. (1994) Facile catalyst separation without water – Fluorous biphase hydroformylation of olefins. *Science*, **266**(5182), 72–75.
7. de Wolf, E., van Koten, G. and Deelman, B. J. (1999) Fluorous phase separation techniques in catalysis. *Chemical Society Reviews*, **28**(1), 37–41.
8. Fish, R. H. (1999) Fluorous biphasic catalysis: A new paradigm for the separation of homogeneous catalysts from their reaction substrates and products. *Chemistry – A European Journal*, **5**(6), 1677–1680.
9. Studer, A., Hadida, S., Ferritto, R., *et al.* (1997) Fluorous synthesis: A fluorous-phase strategy for improving separation efficiency in organic synthesis. *Science*, **275**(5301), 823–826.
10. Gladysz, J. A. and Curran, D. P. (2002) Introduction – Fluorous chemistry: from biphasic catalysis to a parallel chemical universe and beyond. *Tetrahedron*, **58**(20), 3823–3825.
11. Scott, R. L. (1948) The solubility of fluorocarbons. *Journal of the American Chemical Society*, **70**(12), 4090–4093.
12. Scott, R. L. (1958) The anomalous behavior of fluorocarbon solutions. *Journal of Physical Chemistry*, **62**(2), 136–145.
13. Hildebrand, J. H. and Cochran, D. R. F. (1949) Liquid–liquid solubility of perfluoromethylcyclohexane with benzene, carbon tetrachloride, chlorobenzene, chloroform and toluene. *Journal of the American Chemical Society*, **71**(1), 22–25.
14. Myers, K. E. and Kumar, K. (2000) Fluorophobic acceleration of Diels–Alder reactions. *Journal of the American Chemical Society*, **122**(48), 12025–12026.
15. Montanari, V. and Kumar, K. (2004) Just add water: A new fluorous capping reagent for facile purification of peptides synthesized on the solid phase. *Journal of the American Chemical Society*, **126**(31), 9528–9529.
16. Montanari, V. and Kumar, K. (2006) A fluorous capping strategy for Fmoc-based automated and manual solid-phase peptide synthesis. *European Journal of Organic Chemistry*, (**4**), 874–877.
17. Filippov, D. V., van Zoelen, D. J., Oldfield, S. P., *et al.* (2003) Fluorophilic tagging reagents for solid phase peptide synthesis. *Biopolymers*, **71**(3), 346–346.
18. Pearson, W. H., Berry, D. A., Stoy, P., *et al.* (2005) Fluorous affinity purification of oligonucleotides. *Journal of Organic Chemistry*, **70**(18), 7114–7122.
19. Palmacci, E. R., Hewitt, M. C. and Seeberger, P. H. (2001) "Cap-Tag"-novel methods for the rapid purification of oligosaccharides prepared by automated solid-phase synthesis. *Angewandte Chemie International Edition*, **40**(23), 4433–4437.
20. Miura, T., Goto, K. T., Hosaka, D. and Inazu, T. (2003) Oligosaccharide synthesis on a fluorous support. *Angewandte Chemie International Edition*, **42**(18), 2047–2051.
21. Merrifield, R. B. (1963) Solid phase peptide synthesis .1. Synthesis of a tetrapeptide. *Journal of the American Chemical Society*, **85**(14), 2149–2154.
22. Filippov, D. V., van Zoelen, D. J., Oldfield, S. P., *et al.* (2002) Use of benzyloxycarbonyl (Z)-based fluorophilic tagging reagents in the purification of synthetic peptides. *Tetrahedron Letters*, **43**(43), 7809–7812.
23. de Visser, P. C., van Helden, M., Filippov, D. V., *et al.* (2003) A novel, base-labile fluorous amine protecting group: synthesis and use as a tag in the purification of synthetic peptides. *Tetrahedron Letters*, **44**(50), 9013–9016.
24. Brittain, S. M., Ficarro, S. B., Brock, A. and Peters, E. C. (2005) Enrichment and analysis of peptide subsets using fluorous affinity tags and mass spectrometry. *Nature Biotechnology*, **23**(4), 463–468.

25. Go, E. P., Uritboonthai, W., Apon, J. V., *et al.* (2007) Selective metabolite and peptide capture/mass detection using fluorous affinity tags. *Journal of Proteome Research*, **6**(4), 1492–1499.
26. Hoheisel, J. D. (2006) Microarray technology: beyond transcript profiling and genotype analysis. *Nature Reviews Genetics*, **7**(3), 200–210.
27. Kingsmore, S. F. (2006) Multiplexed protein measurement: technologies and applications of protein and antibody arrays. *Nature Reviews Drug Discovery*, **5**(4), 310–320.
28. Ko, K. S., Jaipuri, F. A. and Pohl, N. L. (2005) Fluorous-based carbohydrate microarrays. *Journal of the American Chemical Society*, **127**(38), 13162–13163.
29. Nicholson, R. L., Ladlow, M. L. and Spring, D. R. (2007) Fluorous tagged small molecule microarrays. *Chemical Communications*, (**38**), 3906–3908.
30. Vegas, A. J., Brander, J. E., Tang, W., *et al.* (2007) Fluorous-based small-molecule microarrays for the discovery of histone deacetylase inhibitors. *Angewandte Chemie International Edition*, **46**(42), 7960–7964.
31. Gege, C., Schneider, M. F., Schumacher, G., *et al.* (2004) Functional microdomains of glycolipids with partially fluorinated membrane anchors: Impact on cell adhesion. *ChemPhysChem*, **5**(2), 216–224.
32. Schumacher, G., Bakowsky, U., Gege, C., *et al.* (2006) Lessons learned from clustering of fluorinated glycolipids on selectin ligand function in cell rolling. *Biochemistry*, **45**(9), 2894–2903.
33. Yoder, N. C., Kalsani, V., Schuy, S., *et al.* (2007) Nanoscale patterning in mixed fluorocarbon-hydrocarbon phospholipid bilayers. *Journal of the American Chemical Society*, **129**(29), 9037–9043.
34. Walborsky, H. M. and Lang, J. H. (1956) Effects of the trifluoromethyl group .4. The pKs of ω-trifluoromethyl amino acids. *Journal of the American Chemical Society*, **78**(17), 4314–4316.
35. Filler, R., Ayyangar, N. R., Gustowsk, W. and Kang, H. H. (1969) New reactions of polyfluoroaromatic compounds . Pentafluorophenylalanine and tetrafluorotyrosine. *Journal of Organic Chemistry*, **34**(3), 534–538.
36. Eberle, M. K., Keese, R. and Stoeckli-Evans, H. (1998) New synthesis and chirality of (−)-4,4,4,4′,4′,4′-hexafluorovaline. *Helvetica Chimica Acta*, **81**(1), 182–186.
37. West, A. P., Mecozzi, S. and Dougherty, D. A. (1997) Theoretical studies of the supramolecular synthon benzene ··· hexafluorobenzene. *Journal of Physical Organic Chemistry*, **10**(5), 347–350.
38. Bretscher, L. E., Jenkins, C. L., Taylor, K. M., *et al.* (2001) Conformational stability of collagen relies on a stereoelectronic effect. *Journal of the American Chemical Society*, **123**(4), 777–778.
39. Chiu, H. P., Suzuki, Y., Gullickson, D., *et al.* (2006) Helix propensity of highly fluorinated amino acids. *Journal of the American Chemical Society*, **128**(49), 15556–15557.
40. Mikami, K., Itoh, Y. and Yamanaka, M. (2004) Fluorinated carbonyl and olefinic compounds: Basic character and asymmetric catalytic reactions. *Chemical Reviews*, **104**(1), 1–16.
41. Bondi, A. (1964) Van Der Waals volumes + radii. *Journal of Physical Chemistry*, **68**(3), 441–451.
42. Leroux, F. (2004) Atropisomerism, biphenyls, and fluorine: A comparison of rotational barriers and twist angles. *ChemBioChem*, **5**(5), 644–649.
43. Rose, G. D., Geselowitz, A. R., Lesser, G. J., *et al.* (1985) Hydrophobicity of amino-acid residues in globular-proteins. *Science*, **229**(4716), 834–838.
44. Southall, N. T., Dill, K. A. and Haymet, A. D. J. (2002) A view of the hydrophobic effect. *Journal of Physical Chemistry B*, **106**(3), 521–533.

45. Gao, J. M., Qiao, S. and Whitesides, G. M. (1995) Increasing binding constants of ligands to carbonic-anhydrase by using greasy tails. *Journal of Medicinal Chemistry*, **38**(13), 2292–2301.
46. Fujita, T., Hansch, C. and Iwasa, J. (1964) New substituent constant pi derived from partition coefficients. *Journal of the American Chemical Society*, **86**(23), 5175–&.
47. Lee, K. H., Lee, H. Y., Slutsky, M. M., *et al.* (2004) Fluorous effect in proteins: De novo design and characterization of a four-alpha-helix bundle protein containing hexafluoroleucine. *Biochemistry*, **43**(51), 16277–16284.
48. Kukhar', V. P. and Soloshonok, V. A. (eds.) (1995) *Fluorine-containing Amino Acids: Synthesis and Properties*, John Wiley & Sons Ltd., Chichester.
49. Qiu, X. L., Meng, W. D. and Qing, F. L. (2004) Synthesis of fluorinated amino acids. *Tetrahedron*, **60**(32), 6711–6745.
50. Link, A. J., Mock, M. L. and Tirrell, D. A. (2003) Non-canonical amino acids in protein engineering. *Current Opinion in Biotechnology*, **14**(6), 603–609.
51. Wang, L. and Schultz, P. G. (2005) Expanding the genetic code. *Angewandte Chemie International Edition*, **44**(1), 34–66.
52. Danielson, M. A. and Falke, J. J. (1996) Use of F-19 NMR to probe protein structure and conformational changes. *Annual Review of Biophysics and Biomolecular Structure*, **25** 163–195.
53. Li, H. and Frieden, C. (2005) Phenylalanine side chain behavior of the intestinal fatty acid-binding protein – The effect of urea on backbone and side chain stability. *Journal of Biological Chemistry*, **280**(46), 38556–38561.
54. Li, H. and Frieden, C. (2005) NMR studies of 4-F-19-phenylalanine-labeled intestinal fatty acid binding protein: Evidence for conformational heterogeneity in the native state. *Biochemistry*, **44**(7), 2369–2377.
55. Hodsdon, M. E. and Cistola, D. P. (1997) Ligand binding alters the backbone mobility of intestinal fatty acid-binding protein as monitored by N-15 NMR relaxation and H-1 exchange. *Biochemistry*, **36**(8), 2278–2290.
56. Rose, G. D., Fleming, P. J., Banavar and J. R., Maritan, A. (2006) A backbone-based theory of protein folding. *Proceedings of the National Academy of Sciences of the U. S. A.*, **103**(45), 16623–16633.
57. Li, H. and Frieden, C. (2006) Fluorine-19 NMR studies on the acid state of the intestinal fatty acid binding protein. *Biochemistry*, **45**(20), 6272–6278.
58. Li, H. L. and Frieden, C. (2007) Observation of sequential steps in the folding of intestinal fatty acid binding protein using a slow folding mutant and F-19 NMR. Proceedings of the National Academy of Sciences of the U. S. A., 104(29), 11993–11998.
59. Jackson, J. C., Hammill, J. T. and Mehl, R. A. (2007) Site-specific incorporation of a F-19-amino acid into proteins as an NMR probe for characterizing protein structure and reactivity. *Journal of the American Chemical Society*, **129**(5), 1160–1166.
60. Jackson, J. C., Duffy, S. P., Hess, K. R. and Mehl, R. A. (2006) Improving nature's enzyme active site with genetically encoded unnatural amino acids. *Journal of the American Chemical Society*, **128**(34), 11124–11127.
61. Luchette, P. A., Prosser, R. S. and Sanders, C. R. (2002) Oxygen as a paramagnetic probe of membrane protein structure by cysteine mutagenesis and F-19 NMR spectroscopy. *Journal of the American Chemical Society*, **124**(8), 1778–1781.
62. Zasloff, M. (2002) Antimicrobial peptides of multicellular organisms. *Nature*, **415**(6870), 389–395.
63. Glaser, R. W., Sachse, C., Durr, U. H. N., *et al.* (2005) Concentration-dependent realignment of the antimicrobial peptide PGLa in lipid membranes observed by solid-state F-19-NMR. *Biophysical Journal*, **88**(5), 3392–3397.

64. Ulrich, A. S. (2005) Solid state F-19 NMR methods for studying biomembranes. *Progress in Nuclear Magnetic Resonance Spectroscopy*, **46**(1), 1–21.
65. Buffy, J. J., Waring, A. J. and Hong, M. (2005) Determination of peptide oligomerization in lipid bilayers using F-19 spin diffusion NMR. *Journal of the American Chemical Society*, **127**(12), 4477–4483.
66. Mani, R., Tang, M., Wu, X., *et al.* (2006) Membrane-bound dimer structure of a beta-hairpin antimicrobial peptide from rotational-echo double-resonance solid-state NMR. *Biochemistry*, **45**(27), 8341–8349.
67. Couturier, O., Luxen, A., Chatal, J. F., *et al.* (2004) Fluorinated tracers for imaging cancer with positron emission tomography. *European Journal of Nuclear Medicine and Molecular Imaging*, **31**(8), 1182–1206.
68. Le Bars, D. (2006) Fluorine-18 and medical imaging: Radiopharmaceuticals for positron emission tomography. *Journal of Fluorine Chemistry*, **127**(11), 1488–1493.
69. Laverman, P., Boerman, O. C., Corstens, F. H. M. and Oyen, W. J. G. (2002) Fluorinated amino acids for tumour imaging with positron emission tomography. *European Journal of Nuclear Medicine and Molecular Imaging*, **29**(5), 681–690.
70. Isselbac, K.J., Deykin, D., Kaminska, E., *et al.* (1972) Sugar and amino-acid transport by cells in culture – Differences between normal and malignant cells. *New England Journal of Medicine*, **286**(17), 929–933.
71. Jager, P. L., Vaalburg, W., Pruim, J., *et al.* (2001) Radiolabeled amino acids: Basic aspects and clinical applications in oncology. *Journal of Nuclear Medicine*, **42**(3), 432–445.
72. Rau, F. C., Weber, W. A., Wester, H. J., *et al.* (2002) O-(2- F-18 fluoroethyl)-L-tyrosine (FET): a tracer for differentiation of tumour from inflammation in murine lymph nodes. *European Journal of Nuclear Medicine and Molecular Imaging*, **29**(8), 1039–1046.
73. Wester, H. J., Herz, M., Weber, W., *et al.* (1999) Synthesis and radiopharmacology of O-(2-F-18 fluoroethyl)-L-tyrosine for tumor imaging. *Journal of Nuclear Medicine*, **40**(1), 205–212.
74. Remy, P., Samson, Y., Hantraye, P., *et al.* (1995) Clinical correlates of F-18 fluorodopa uptake in 5 grafted Parkinsonian-patients. *Annals of Neurology*, **38**(4), 580–588.
75. Dougherty, D. A. (1996) Cation-pi interactions in chemistry and biology: A new view of benzene, Phe, Tyr, and Trp. *Science*, **271**(5246), 163–168.
76. Zhong, W. G., Gallivan, J. P., Zhang, Y. O., *et al.* (1998) From ab initio quantum mechanics to molecular neurobiology: A cation-pi binding site in the nicotinic receptor. *Proceedings of the National Academy of Sciences of the U. S. A.*, 95(21), 12088–12093.
77. Beene, D. L., Brandt, G. S., Zhong, W. G., *et al.* (2002) Cation-pi interactions in ligand recognition by serotonergic (5-HT3A) and nicotinic acetylcholine receptors: The anomalous binding properties of nicotine. *Biochemistry*, **41**(32), 10262–10269.
78. Padgett, C. L., Hanek, A. P., Lester, H. A., *et al.* (2007) Unnatural amino acid mutagenesis of the GABA(A) receptor binding site residues reveals a novel cation-pi interaction between GABA and beta(2)Tyr97. *Journal of Neuroscience*, **27**(4), 886–892.
79. Seyedsayamdost, M. R., Yee, C. S., Reece, S. Y., *et al.* (2006) pH rate profiles of FnY356-R2s (n = 2, 3, 4) in *Escherichia coli* ribonucleotide reductase: Evidence that Y-356 is a redox-active amino acid along the radical propagation pathway. *Journal of the American Chemical Society*, **128**(5), 1562–1568.
80. Nordlund, P., Sjoberg, B. M. and Eklund, H. (1990) 3-Dimensional structure of the free-radical protein of ribonucleotide reductase. *Nature*, **345**(6276), 593–598.
81. Stubbe, J., Nocera, D. G., Yee, C. S. and Chang, M. C. Y. (2003) Radical initiation in the class I ribonucleotide reductase: Long-range proton-coupled electron transfer? *Chemical Reviews*, **103**(6), 2167–2201.

82. Seyedsayamdost, M. R., Reece, S. Y., Nocera, D. G. and Stubbe, J. (2006) Mono-, di-, tri-, and tetra-substituted fluorotyrosines: New probes for enzymes that use tyrosyl radicals in catalysis. *Journal of the American Chemical Society*, **128**(5), 1569–1579.
83. Butterfield, S. M., Patel, P. R. and Waters, M. L. (2002) Contribution of aromatic interactions to alpha-helix stability. *Journal of the American Chemical Society*, **124**(33), 9751–9755.
84. Woll, M. G., Hadley, E. B., Mecozzi, S. and Gellman, S. H. (2006) Stabilizing and destabilizing effects of phenylalanine → F-5-phenylalanine mutations on the folding of a small protein. *Journal of the American Chemical Society*, **128**(50), 15932–15933.
85. Cornilescu, G., Hadley, E. B., Woll, M. G., *et al.* (2007) Solution structure of a small protein containing a fluorinated side chain in the core. *Protein Science*, **16**(1), 14–19.
86. Gorske, B. C. and Blackwell, H. E. (2006) Tuning peptoid secondary structure with pentafluoroaromatic functionality: A new design paradigm for the construction of discretely folded peptoid structures. *Journal of the American Chemical Society*, **128**(44), 14378–14387.
87. Bella, J., Eaton, M., Brodsky, B. and Berman, H. M. (1994) Crystal-structure and molecular-structure of a collagen-like peptide at 1.9-angstrom resolution. *Science*, **266**(5182), 75–81.
88. Bella, J., Brodsky, B. and Berman, H. M. (1995) Hydration structure of a collagen peptide. *Structure*, **3**(9), 893–906.
89. Holmgren, S. K., Taylor, K. M., Bretscher, L. E. and Raines, R. T. (1998) Code for collagen's stability deciphered. *Nature*, **392**(6677), 666–667.
90. Holmgren, S. K., Bretscher, L. E., Taylor, K. M. and Raines, R. T. (1999) A hyperstable collagen mimic. *Chemistry & Biology*, **6**(2), 63–70.
91. Hodges, J. A. and Raines, R. T. (2006) Energetics of an n – > pi* interaction that impacts protein structure. *Organic Letters*, **8**(21), 4695–4697.
92. DeRider, M. L., Wilkens, S. J., Waddell, M. J., *et al.* (2002) Collagen stability: Insights from NMR spectroscopic and hybrid density functional computational investigations of the effect of electronegative substituents on prolyl ring conformations. *Journal of the American Chemical Society*, **124**(11), 2497–2505.
93. Burgi, H. B., Dunitz, J. D. and Shefter, E. (1973) Geometrical reaction coordinates .2. Nucleophilic addition to a carbonyl group. *Journal of the American Chemical Society*, **95**(15), 5065–5067.
94. Burgi, H. B., Dunitz, J. D., Lehn, J. M. and Wipff, G. (1974) Stereochemistry of reaction paths at carbonyl centers. *Tetrahedron*, **30**(12), 1563–1572.
95. Hinderaker, M. P., Raines, R. T. (2003) An electronic effect on protein structure. *Protein Science*, **12**(6), 1188–1194.
96. Hodges, J. A. and Raines, R. T. (2003) The effect of fluoroproline in the X-position on the stability of the collagen triple helix. *Biopolymers*, **71**(3), 312–312.
97. Jenkins, C. L., Bretscher, L. E., Guzei, I. A. and Raines, R. T. (2003) Effect of 3-hydroxyproline residues on collagen stability. *Journal of the American Chemical Society*, **125**(21), 6422–6427.
98. Hodges, J. A. and Raines, R. T. (2005) Stereoelectronic and steric effects in the collagen triple helix: Toward a code for strand association. *Journal of the American Chemical Society*, **127**(45), 15923–15932.
99. Jackel, C., Seufert, W., Thust, S. and Koksch, B. (2004) Evaluation of the molecular interactions of fluorinated amino acids with native polypepticles. *ChemBioChem*, **5**(5), 717–720.
100. Jackel, C., Salwiczek, M. and Koksch, B. (2006) Fluorine in a native protein environment – How the spatial demand and polarity of fluoroalkyl groups affect protein folding. *Angewandte Chemie International Edition*, **45**(25), 4198–4203.

101. Alexeev, D., Barlow, P. N., Bury, S. M., *et al.* (2003) Synthesis, structural and biological studies of ubiquitin mutants containing (2S, 4S)-5-fluoroleucine residues strategically placed in the hydrophobic core. *ChemBioChem*, **4**(9), 894–896.
102. Horng, J. C. and Raleigh, D. P. (2003) Phi-values beyond the ribosomally encoded amino acids: Kinetic and thermodynamic consequences of incorporating trifluoromethyl amino acids in a globular protein. *Journal of the American Chemical Society*, **125**(31), 9286–9287.
103. Bilgicer, B., Fichera, A. and Kumar, K. (2001) A coiled coil with a fluorous core. *Journal of the American Chemical Society*, **123**(19), 4393–4399.
104. Bilgicer, B., Xing, X. and Kumar, K. (2001) Programmed self-sorting of coiled coils with leucine and hexafluoroleucine cores. *Journal of the American Chemical Society*, **123**(47), 11815–11816.
105. Tang, Y. and Tirrell, D. A. (2001) Biosynthesis of a highly stable coiled-coil protein containing hexafluoroleucine in an engineered bacterial host. *Journal of the American Chemical Society*, **123**(44), 11089–11090.
106. Tang, Y., Ghirlanda, G., Vaidehi, N., *et al.* (2001) Stabilization of coiled-coil peptide domains by introduction of trifluoroleucine. *Biochemistry*, **40**(9), 2790–2796.
107. Tang, Y., Ghirlanda, G., Petka, W. A., *et al.* (2001) Fluorinated coiled-coil proteins prepared in vivo display enhanced thermal and chemical stability. *Angewandte Chemie International Edition*, **40**(8), 1494–1496.
108. Oshea, E. K., Klemm, J. D., Kim, P. S. and Alber, T. (1991) X-ray structure of the GCN4 leucine zipper, a 2-stranded, parallel coiled coil. *Science*, **254**(5031), 539–544.
109. Son, S., Tanrikulu, I. C. and Tirrell, D. A. (2006) Stabilization of bzip peptides through incorporation of fluorinated aliphatic residues. *ChemBioChem*, **7**(8), 1251–1257.
110. Acharya, A., Ruvinov, S. B., Gal, J., *et al.* (2002) A heterodimerizing leucine zipper coiled coil system for examining the specificity of a position interactions: Amino acids I, V, L, N, A, and K. *Biochemistry*, **41**(48), 14122–14131.
111. Lee, H. Y., Lee, K. H., Al-Hashimi, H. M. and Marsh, E. N. G. (2006) Modulating protein structure with fluorous amino acids: Increased stability and native-like structure conferred on a 4-helix bundle protein by hexafluoroleucine. *Journal of the American Chemical Society*, **128**(1), 337–343.
112. Bilgicer, B. and Kumar, K. (2002) Synthesis and thermodynamic characterization of self-sorting coiled coils. *Tetrahedron*, **58**(20), 4105–4112.
113. Popot, J. L. and Engelman, D. M. (2000) Helical membrane protein folding, stability, and evolution. *Annual Review of Biochemistry*, **69** 881–922.
114. Popot, J. L. and Engelman, D. M. (1990) Membrane-protein folding and oligomerization – the 2-stage model. *Biochemistry*, **29**(17), 4031–4037.
115. Engelman, D. M., Chen, Y., Chin, C. N., *et al.* (2003) Membrane protein folding: beyond the two stage model. *Febs Letters*, **555**(1), 122–125.
116. Choma, C., Gratkowski, H., Lear, J. D. and DeGrado, W. F. (2000) Asparagine-mediated self-association of a model transmembrane helix. *Nature Structural Biology*, **7**(2), 161–166.
117. Zhou, F. X., Cocco, M. J., Russ, W. P., *et al.* (2000) Interhelical hydrogen bonding drives strong interactions in membrane proteins. *Nature Structural Biology*, **7**(2), 154–160.
118. Bilgicer, B. and Kumar, K. (2004) De novo design of defined helical bundles in membrane environments. *Proceedings of the National Academy of Sciences of the U. S. A.*, 101(43), 15324–15329.
119. Mukerjee, P. and Handa, T. (1981) Adsorption of fluorocarbon and hydrocarbon surfactants to air-water, hexane-water, and perfluorohexane-water interfaces – Relative affinities and fluorocarbon-hydrocarbon nonideality effects. *Journal of Physical Chemistry*, **85**(15), 2298–2303.
120. Mukerjee, P. (1994) Fluorocarbon hydrocarbon interactions in micelles and other lipid assemblies, at interfaces, and in solutions. *Colloids and Surfaces A – Physicochemical and Engineering Aspects*, **84**(1), 1–10.

121. Naarmann, N., Bilgicer, B., Meng, H., *et al.* (2006) Fluorinated interfaces drive self-association of transmembrane alpha helices in lipid bilayers. *Angewandte Chemie International Edition*, **45**(16), 2588–2591.
122. Vine, W. H., Brueckner, D. A., Needleman, P. and Marshall, G. R. (1973) Synthesis, biological-activity, and F-19 nuclear magnetic resonance spectral of Angiotensin-II analogs containing fluorine. *Biochemistry*, **12**(8), 1630–1637.
123. Bovy, P. R., Getman, D. P., Matsoukas, J. M. and Moore, G. J. (1991) Influence of polyfluorination of the phenylalanine ring of Angiotensin-II on conformation and biological-activity. *Biochimica et Biophysica Acta*, **1079**(1), 23–28.
124. Vine, W. H., Hsieh, K. H. and Marshall, G. R. (1981) Synthesis of fluorine-containing peptides – analogs of Angiotensin-II containing hexafluorovaline. *Journal of Medicinal Chemistry*, **24**(9), 1043–1047.
125. Hsieh, K. H., Needleman, P. and Marshall, G. R. (1987) Long-acting Angiotensin-II inhibitors containing hexafluorovaline in position-8. *Journal of Medicinal Chemistry*, **30**(6), 1097–1100.
126. Houston, M. E., Harvath, L. and Honek, J. F. (1997) Synthesis of and chemotactic responses elicited by fMET-Leu-Phe analogs containing difluoro- and trifluoromethionine. *Bioorganic & Medicinal Chemistry Letters*, **7**(23), 3007–3012.
127. Jäckel, C. and Koksch, B. (2005) Fluorine in peptide design and protein engineering. *European Journal of Organic Chemistry*, (**21**), 4483–4503.
128. Kondejewski, L. H., Farmer, S. W., Wishart, D. S., *et al.* (1996) Modulation of structure and antibacterial and hemolytic activity by ring size in cyclic gramicidin S analogs. *Journal of Biological Chemistry*, **271**(41), 25261–25268.
129. Arai, T., Imachi, T., Kato, T., *et al.* (1996) Synthesis of hexafluorovalyl(1,1′) gramicidin S. *Bulletin of the Chemical Society of Japan*, **69**(5), 1383–1389.
130. Xiao, J. B., Weisblum, B. and Wipf, P. (2005) Electrostatic versus steric effects in peptidomimicry: Synthesis and secondary structure analysis of gramicidin S analogues with (E)-alkene peptide isosteres. *Journal of the American Chemical Society*, **127**(16), 5742–5743.
131. Wipf, P., Henninger, T. C. and Geib, S. J. (1998) Methyl- and (trifluoromethyl)alkene peptide isosteres: Synthesis and evaluation of their potential as beta-turn promoters and peptide mimetics. *Journal of Organic Chemistry*, **63**(18), 6088–6089.
132. Niemz, A. and Tirrell, D. A. (2001) Self-association and membrane-binding behavior of melittins containing trifluoroleucine. *Journal of the American Chemical Society*, **123**(30), 7407–7413.
133. Zasloff, M. (1987) Magainins, a class of antimicrobial peptides from Xenopus skin – isolation, characterization of 2 active forms, and partial cDNA sequence of a precursor. *Proceedings of the National Academy of Sciences of the U. S. A.*, 84(15), 5449–5453.
134. Chen, H. C., Brown, J. H., Morell, J. L. and Huang, C. M. (1988) Synthetic magainin analogs with improved antimicrobial activity. *Febs Letters*, **236**(2), 462–466.
135. Park, C. B., Yi, K. S., Matsuzaki, K., *et al.* (2000) Structure-activity analysis of buforin II, a histone H2A-derived antimicrobial peptide: The proline hinge is responsible for the cell-penetrating ability of buforin II. *Proceedings of the National Academy of Sciences of the U. S. A.*, 97(15), 8245–8250.
136. Park, C. B., Kim, H. S. and Kim, S. C. (1998) Mechanism of action of the antimicrobial peptide buforin II: Buforin II kills microorganisms by penetrating the cell membrane and inhibiting cellular functions. *Biochemical and Biophysical Research Communications*, **244**(1), 253–257.
137. Meng, H. and Kumar, K. (2007) Antimicrobial activity and protease stability of peptides containing fluorinated amino acids. *Journal of the American Chemical Society*, **129**(50), 15615–15622.

138. Giménez, D., Andreu, C., del Olmo, M. L., *et al.* (2006) The introduction of fluorine atoms or trifluoromethyl groups in short cationic peptides enhances their antimicrobial activity. *Bioorganic & Medicinal Chemistry*, **14**(20), 6971–6978.
139. Sato, A. K., Viswanathan, M., Kent, R. B. and Wood, C. R. (2006) Therapeutic peptides: technological advances driving peptides into development. *Current Opinion in Biotechnology*, **17**(6), 638–642.
140. Wang, P., Tang, Y. and Tirrell, D. A. (2003) Incorporation of trifluoroisoleucine into proteins in vivo. *Journal of the American Chemical Society*, **125**(23), 6900–6906.
141. Montclare, J. K. and Tirrell, D. A. (2006) Evolving proteins of novel composition. *Angewandte Chemie International Edition*, **45**(27), 4518–4521.
142. Panchenko, T., Zhu, W. W. and Montclare, J. K. (2006) Influence of global fluorination on chloramphenicol acetyltransferase activity and stability. *Biotechnology and Bioengineering*, **94**(5), 921–930.
143. Zhao, H. M. and Arnold, F. H. (1999) Directed evolution converts subtilisin E into a functional equivalent of thermitase. *Protein Engineering*, **12**(1), 47–53.
144. Yoo, T. H., Link, A. J. and Tirrell, D. A. (2007) Evolution of a fluorinated green fluorescent protein. *Proceedings of the National Academy of Sciences of the U. S. A.*, **104**(35), 13887–13890.
145. Crameri, A., Whitehorn, E. A., Tate, E. and Stemmer, W. P. C. (1996) Improved green fluorescent protein by molecular evolution using DNA shuffling. *Nature Biotechnology*, **14**(3), 315–319.
146. Cormack, B. P., Valdivia, R. H. and Falkow, S. (1996) FACS-optimized mutants of the green fluorescent protein (GFP). *Gene*, **173**(1), 33–38.

17

Effects of Fluorination on the Bioorganic Properties of Methionine

John F. Honek

17.1 Introduction

Amino acids, peptides and proteins play essential roles in living systems. As chemists, we have the ability to fabricate new structures of great complexity. Yet frequently the synthesis and application of even the simplest chemical structures can contribute substantial information on living systems. This has certainly been true for the application of fluorine chemistry to the synthesis of fluorinated amino acids [1, 2]. These molecules are not only interesting in their own right but may act as potent mechanism-based enzyme inhibitors that may have application in medicine or diagnostics, or they can be valuable probes that, incorporated into peptides or proteins, elucidate fundamental biological chemistry or uncover new aspects of biochemical structure and function [3–5].

Why fluorine? Its van der Waals radius is 1.47 Å compared with 1.2 Å for hydrogen and 1.57 Å for oxygen and with small molecules such as amino acids, the introduction of a large substituent would likely be completely disruptive to its biological function [6]. However, a relatively small atom such as fluorine introduces a minor alteration in sterics to the molecule. In addition, the presence of a fluorine atom introduces a powerful nuclear magnetic resonance (NMR) probe (^{19}F) into the molecule. Fluorinated analogues of tyrosine, phenylalanine, and tryptophan have been and continue to be utilized as biological probes in biological chemistry and as important reporter groups in protein folding and ligand binding studies [7–10]. In addition, the electronic perturbations that can occur upon addition of the fluorine nucleus into a structure such as an amino acid can also be used to

Fluorine in Medicinal Chemistry and Chemical Biology Edited by Iwao Ojima

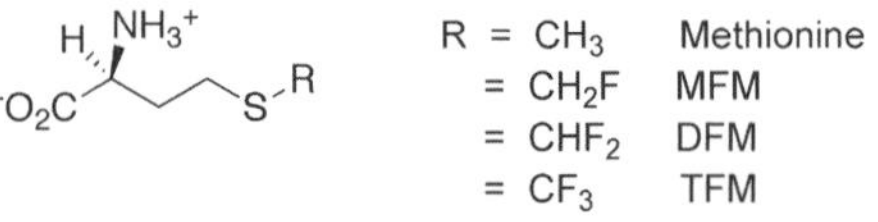

R = CH_3	Methionine
= CH_2F	MFM
= CHF_2	DFM
= CF_3	TFM

Figure 17.1 *The chemical structures of L-methionine, and monofluoro- (MFM), difluoro- (DFM) and trifluoromethionine (TFM).*

evaluate the electronic aspects involved in molecular recognition. The synthetic preparation of fluorinated amino acids and their application in medicinal chemistry and biochemistry has contributed much to our fundamental understanding of molecular recognition as well. Although it is easy to replace a hydrogen atom by a fluorine atom in a chemical structure when drawing this structure, the chemistry involved in actually making the fluorinated target molecule can usually be quite challenging. Frequently new synthetic approaches may be necessary to accommodate either available fluorinated starting materials or fluorinating reagents. Regardless of the synthetic difficulties that might be encountered in the preparation of various fluorinated amino acids, the large impact that fluorine can play by its presence continues to be a major reason for chemists and biochemists to include the use of fluorinated amino acids in their research programs.

This chapter focuses on how fluorinated methionine analogues interact with biological systems. Specifically the compounds L-monofluoromethionine (L-(*S*)-(monofluoromethyl) homocysteine; MFM), L-difluoromethionine (L-(*S*)-(difluoromethyl)homocysteine; DFM), and L-trifluoromethionine (L-(*S*)-(trifluoromethyl(homocysteine); TFM) will be the subject of this article (Figure 17.1). A discussion of their syntheses and chemical and conformational properties will be presented. This will be followed by a discussion of their incorporation into peptides and proteins and the properties of the resulting fluorinated biomolecules. The ^{19}F NMR spectroscopic characteristics of these biomolecules will be covered along with how these resonances have revealed information on the properties of the difluoro- and trifluoromethyl moiety (and on the proteins as well!). Further elaboration of these amino acids with metalloenzymes completes the survey.

17.2 Syntheses and Characteristics of Fluorinated Methionines

A number of methods for preparing protected and unprotected MFM, DFM, and TFM have been reported. In the case of MFM, protected versions of MFM have been reported utilizing either xenon difluoride on the sulfide or diethylaminosulfur trifluoride (DAST) on the protected methionine sulfoxide [11, 12]. In the DAST approach, the *N*-acetylated methyl ester of methionine sulfoxide was reacted with DAST and ZnI_2 (60 °C, 12 h) to produce the targeted protected monofluoromethyl sulfide in 67% yield. Unfortunately, the lack of stability of the monofluoromethyl sulfide group to chromatography and to the aqueous conditions required for deprotection prevented the isolation of deprotected MFM. This is further experimental evidence for the lack of stability of an α-monofluoroalkyl sulfide group as aqueous instability has been observed for a nucleoside containing a similar functional group [13].

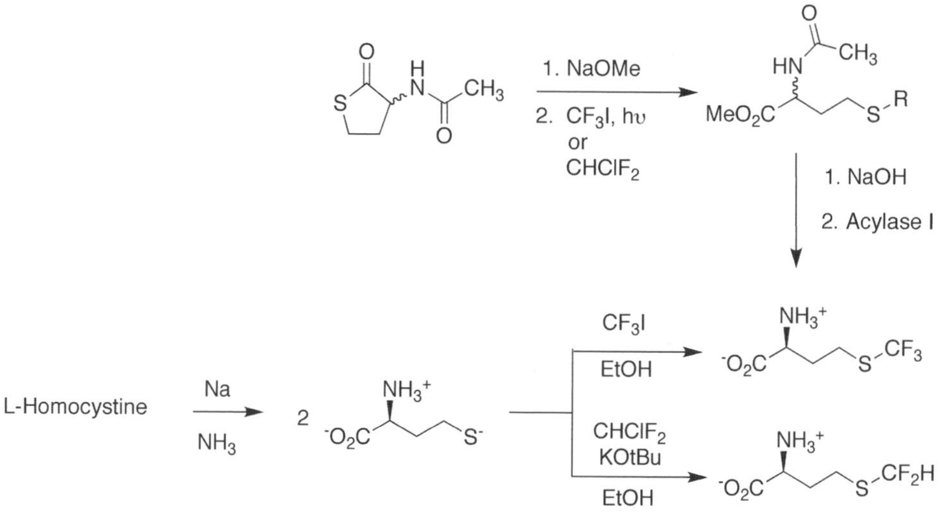

Figure 17.2 *Summary of some of the synthetic routes to DFM and TFM.*

Dannley and Taborsky first reported the synthesis of racemic D,L-TFM utilizing $(CF_3S)_2Hg$ as the source of the trifluoromethylthio group [14]. Unfortunately, low yields were reported with this approach (11% in five chemical steps). Improved synthetic routes have been reported and are dependent upon the photochemical reaction of various CF_3-containing reagents with homocysteine or *N*-acetylthiolactone [12, 15]. Reagents such as trifluoromethyl iodide or *S*-trifluoromethyl diaryl sulfonium salts are able to transfer a trifluoromethyl group to thiols, and these reagents have been useful in this regard [12, 15–17]. Starting with inexpensive racemic *N*-acetylhomocysteine thiolactone, the racemic *N*-acetyltrifluoromethionine methyl ester was deprotected by base followed by enzymatic resolution using acylase I. Similar approaches have been reported for the successful synthesis of DFM, that is, by reaction of L-homocysteine or racemic *N*-acetylhomocysteine thiolactone under basic conditions with Freon-22 (chlorodifluoromethane) [12, 18]. A summary of several of the approaches to preparation of DFM and TFM is shown in Figure 17.2.

The introduction of fluorine into the methyl group of methionine alters a number of properties. For example, calculations have indicated that the overall hydrophobicity would be altered upon fluorination of the thiomethyl group ($\log P_{CH_3SCH_3} = 0.46$; $\log P_{CF_2HSCH_3} = 1.77$; $\log P_{CF_3SCH_3} = 2.39$) [19]. Computational modeling of the side-chain of TFM indicates a shortening of the carbon–sulfur bond of CF_3–S and a lengthening of the S–CH_2 bond [20]. This results in a lower energy barrier for dihedral angle rotation in the TFM side-chain, with dihedral angles of 80°, 180°, and 280° being lower-energy conformations. The energy barrier between these minima are on the order of 0.8 kJ/mol compared with the methionine side-chain, which has a calculated energy barrier on the order of 1.5 kJ/mol on the basis of relaxed potential energy scans at the B3LYP/6–31+G(d,p) level [20].

One of the goals for the study of methionine biochemistry utilizing these fluorinated methionines was to control, in a stepwise fashion, the electron density and the nucleophilicity of the thioether moiety by increasing fluorination adjacent to the sulfur atom. With respect to the effect that fluorination makes on the steric properties of the methyl group in methionine, it has been suggested that a trifluoromethyl moiety may be similar in size to the isopropyl group [6]. Computational calculations at the *ab initio* level (RHF/6–31G*) have resulted in the calculated volume enclosed by the electron density surface computed at the 0.002 electrons/(atomic unit)^3 level to be 71.2 Å^3, 80.9 Å^3, and 84.4 Å^3 for the simple molecules CH_3SCH_3, CH_3SCHF_2, and CH_3SCF_3, respectively [19]. Electronic properties of the thiomethyl function in methionine are also seen to change upon fluorination. Calculations using density functional theory at B3LYP/6–31+G(d,p)// B3LYP/6–31+G(d,p) levels have determined the Mulliken charges on the sulfur atoms of CH_3SCH_3, CH_3SCHF_2, and CH_3SCF_3 to be 0.079e, 0.139e, and 0.212e respectively and the CHelpG charges [21] fitted to the electrostatic potential according to the CHelpG scheme are –0.286e, –0.221e, and –0.149e for the sulfur atoms in CH_3SCH_3, CH_3SCHF_2, and CH_3SCF_3 respectively (unpublished results). Hence, the introduction of fluorine into the methionine structure will perturb the sterics as well as the electronic properties of the side-chain. However, these alterations are made in a stepwise direction DFM → TFM when both analogues are studied in parallel in a particular biological system.

17.3 Replacement of Methionine Residues in Peptides and Proteins by DFM and TFM

To our knowledge, the only report of the application of DFM and TFM to peptides has been on the effect of these replacements on the chemotactic activity of the tripeptide formyl-Met-Leu-Phe [19]. The methionine residue in this peptide has been shown to be critical to the bioactivity of the peptide, as replacement by other methionine analogues such as ethionine, norleucine, *S*-methylcysteine, *S*-ethylcysteine or 2-aminoheptanoic acid results in ~4-fold to 264-fold decrease in chemotactic activity in human and rabbit neutrophil migration assays [22–24]. Interestingly the DFM- and TFM-containing peptides were found to enhance by ~10-fold the directed migration responses of human blood neutrophils. Studies of the effects of DFM and TFM on the bioactivity of other peptides would certainly be worthy of further investigation.

In contrast to the lack of application of DFM and TFM to peptide biochemistry, bioincorporation of these analogues into several proteins has been reported. DFM and TFM have been successfully incorporated into recombinant bacteriophage lambda lysozyme (DFM and TFM) [15, 25], *Pseudomonas aeruginosa* azurin (DFM and TFM) [26], *P. aeruginosa* alkaline protease (DFM) [27], calmodulin (DFM) [28], *Escherichia coli* leucine-isoleucine-valine (LIV) binding protein (DFM) [29], and a mutant of the *Aequorea victoria* green fluorescent protein (TFM) [30]. The presence of these fluorinated amino acids in those proteins does not negatively affect either the enzymatic activity or the bioactivity of the proteins, an important ingredient in the characteristics that an amino acid analogue must have if it is to serve as a useful biophysical probe. The bioincorporation of DFM or TFM into the above proteins (except for the *P. aeruginosa* azurin) has been readily accomplished by induction of the target gene's expression (induced by isopropyl-β-D-thiogalactopyranoside, IPTG) in the presence of either DFM or TFM in minimal growth media. This approach, which is frequently utilized for the bioincorporation of other fluorinated amino acids such as those of tryptophan, phenylalanine, and tyrosine, is applicable to the introduction of DFM and TFM into almost any protein whose gene has been isolated and whose expression can be controlled [9]. In the case of the DFM and TFM analogues, placing the gene in an *E. coli* methionine auxotroph such as B834(λDE3) reduces competition from the presence of intracellular biosynthesized methionine [15]. The incorporation levels of DFM into proteins are higher than those for TFM even though the analogues differ by only a single fluorine atom [25]. For example, for bacteriophage lambda lysozyme, which contains three methionine residues (Met1, Met14, and Met107), incorporation levels for DFM approach wild-type levels (>20 mg/L of cell growth) with close to 100% incorporation (all three methionine residues replaced by DFM). TFM incorporation, however, ranges from 31% TFM-labeled protein (~15 mg/L) to 70% incorporation (2 mg/L) with the requirement that small amounts of methionine be added to the growth media to reduce TFM toxicity to *E. coli* cells (DFM was not found to be toxic to *E. coli* under these bioincorporation experiments) [15]. The toxicity of TFM compared with DFM, an analogue having only one less fluorine, is quite interesting but as yet the underlying molecular details for the toxicity have not been elucidated, although their differential processing by the enzyme methionine aminopeptidase (MAP) could be a contributing factor (see below).

17.4 ^{19}F NMR Spectroscopy of Proteins Incorporating DFM and TFM

^{19}F NMR spectroscopy is an extremely useful technique for the study of protein dynamics and structure. The presence of the fluorine nucleus in these proteins due to the presence of DFM or TFM enables the use of ^{19}F NMR to study their behavior in the presence of ligands and inhibitors. Since there is no natural background of fluorine in proteins and as the sensitivity of the ^{19}F nucleus to NMR detection approaches that of ^{1}H (83%), DFM/TFM-labeled proteins can readily be investigated through this NMR method and several interesting observations have already been made [7, 8]. TFM-labeled bacteriophage lysozyme, because of partial TFM incorporation, is composed of an ensemble of partially labeled proteins that are not readily separable by chromatographic techniques. Four distinct ^{19}F resonances (two of which integrate together to one TFM residue) are detected for this ensemble, although only three methionines are present in the protein. Based on a series of site-directed mutagenesis studies (M14L and M107L mutants labeled with TFM) as well as use of the paramagnetic NMR line-broadening agent gadolinium(III) ethylenediaminetetraacetic acid (Gd(III)EDTA), which affects surface exposed TFM residues, the four ^{19}F NMR resonances were assigned to specific residues in the protein [31]. Two ^{19}F resonances occurring at −39.32 ppm and −39.82 ppm were assigned to Met1 and Met14, respectively. However, Met107 was found to occur at either −39.99 ppm or −40.11 ppm depending upon the extent of TFM incorporation. It was concluded that the position of the ^{19}F resonance for TFM107 was dependent upon whether TFM was present at position 14 (producing the −40.11 ppm resonance) or not (resulting in the −39.99 ppm resonance for TFM107). This implied that the presence of the larger TFM residue at position 14 compared with the smaller methionine residue somehow affects the environment

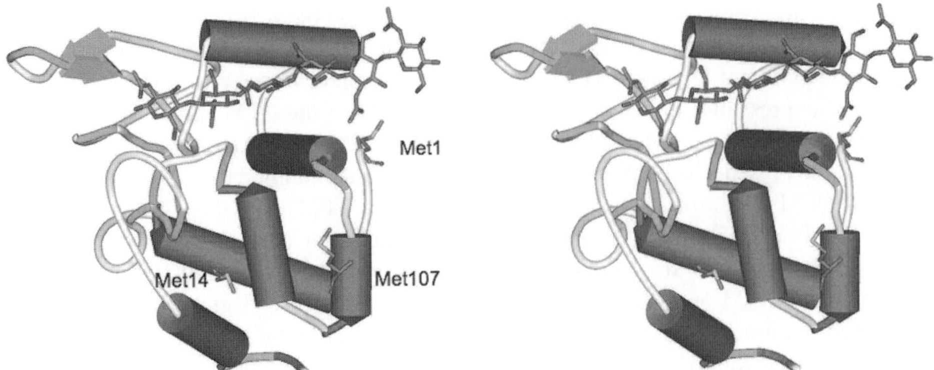

Figure 17.3 Cartoon stereo diagram of the x-ray structure of wild-type bacteriophage lambda lysozyme as solved for the $(GlcNAc)_6$ complex and showing the position of the three methionine residues in the enzyme (PDB 1d9u) [32]. See color plate 17.3.

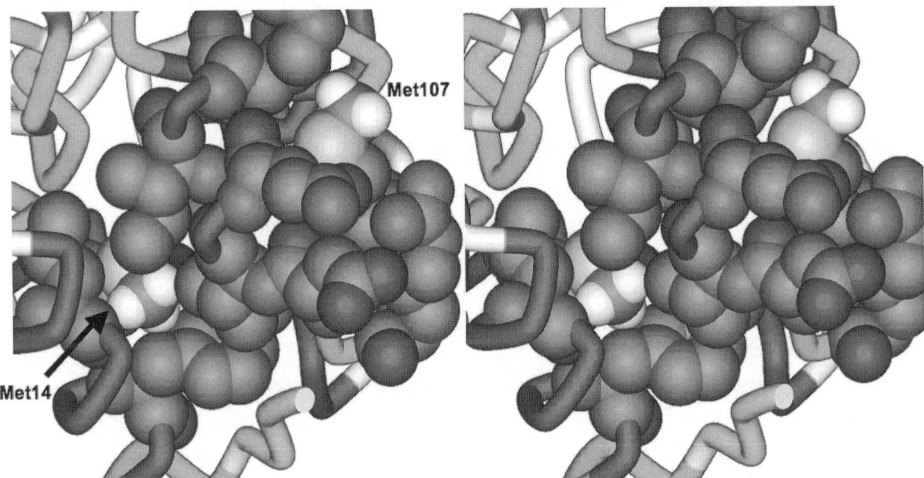

Figure 17.4 Close-up of the detailed interactions between Met14 and adjacent residues and the location of Met107 (PDB 1d9u) [32]. See color plate 17.4.

surrounding position 107. When the x-ray structure of the bacteriophage lysozyme was determined in collaboration with the group of Professor Albert Berghuis, it became clear how this subtle change at position 14 might be transmitted to position 107 [32]. As shown in Figure 17.3, the protein was crystallized in complex with the hexasaccharide $(GlcNAc)_6$, an oligosaccharide that is not hydrolyzed by this bacteriophage lysozyme. The positions of the three methionines are also shown in Figure 17.3. Figure 17.4 presents a view of the intervening region between Met14 and Met107. As can be seen from the structure, Met14 is in the hydrophobic core of the protein and tightly surrounded by several side-chains (Ile113, Ile117, Leu142, and Phe146) that appear to contact the thiomethyl group of Met14.

As the size of the thiomethyl group is increased when TFM is introduced at this position, the proximal residues would need to accommodate the increased steric demand produced by the three fluorine atoms. Ile113 and Ile117 are positioned on a helix that contains Arg110, Asp112, Gln115, and Arg119, and interestingly these residues surround Met107. It is possible that the alteration of the ^{19}F resonance of TFM107 occurs due to the steric interaction between TFM14 and proximal residues, and that these interactions are transmitted through the helix to eventually result in a slightly altered environment for TFM107. These observations indicate that TFM can be a very subtle probe of protein environment and that the trifluoromethyl moiety does slightly perturb protein structure, although not sufficiently to alter enzymatic activity. In this case, bacteriophage lambda lysozyme enzymatic activity was measured by turbidimetric assay using chloroform-treated *Escherichia coli* cells that lyse upon reaction with the lysozyme. Unfortunately, standard chromophore-labeled oligosaccharides such as *p*-nitrophenol-linked saccharides, used extensively in the study of hen egg white lysozyme, are not substrates for the phage enzyme.

Another interesting observation resulted when DFM was incorporated into the bacteriophage lysozyme. As mentioned previously, almost 100% incorporation of DFM into the lysozyme occurs, which should simplify the ^{19}F spectrum as no protein ensemble exists to complicate this spectrum. However, the ^{19}F NMR spectrum of the DFM-labeled lysozyme exhibited the expected resonances for DFM1 (–91.2 ppm; doublet) and DFM107 (–92.5 ppm; broader doublet) but a complex set of resonances (two quartets centered at –94.2 ppm and –95.5 ppm) for the ^{19}F resonances associated with DFM14 (assignments were made by site-directed mutagenesis, paramagnetic line broadening experiments, and two-dimensional ^{19}F–^{19}F COSY [correlation spectroscopy] NMR experiments) [25]. The explanation for the NMR complexity of the ^{19}F signals for DFM14 is due to the fact that the two fluorine atoms in DFM are diasterotopic due to the chirality of the amino acid, and hence these two fluorine atoms are chemical shift inequivalent. As indicated above, the tightly packed environment surrounding the DFM14 position will likely slow the side-chain rotation of DFM14 compared with DFM at positions 1 and 107. This situation would increase the observed chemical shift inequivalence of the two fluorines on DFM14, resulting in the two sets of quartets in the proton-coupled ^{19}F NMR spectrum (in addition to the chemical shift difference, there would also be present both ^{19}F–^{19}F and ^{1}H–^{19}F coupling). This effect is also observed in the study of the temperature effects on chemical shifts for DFM itself, with low temperatures increasing the complexity of the ^{19}F resonances of the difluoromethyl group. The DFM-labeled bacteriophage lambda lysozyme ^{19}F spectrum is a stunning example of NMR chemical shift inequivalence.

17.5 DFM and TFM: Interactions with Platinum and as Probes of Metalloenzymes

Although DFM and TFM are important contributions to the range of spectroscopic probes available for application to biochemical studies, other properties of these fluorinated amino acid analogues can also be of use. For example, the subtle change in size of the thiomethyl group and the alteration in the electronic properties of the sulfur atom in these methionine analogues could be useful properties to explore biological systems. In order to obtain additional fundamental information on these analogues, their interaction with metal complexes was undertaken. For example, the reaction of methionine with K_2PtCl_4 to produce a methionine–platinum complex is well known (Figure 17.5) [33]. In addition, the rate of sulfur inversion when complexed to platinum has been measured previously (Figure 17.6) [34, 35]. It was therefore of interest to determine the effect, if any, of fluorination on the rate of complex formation as well as on the rate and energy barriers to sulfur inversion. A preliminary comparison of the rate at which a series of methionine analogues react with K_2PtCl_4 resulted in the following values (in $mM\,s^{-1}$): selenomethionine 129 ± 4; methionine 32 ± 2; ethionine 26.1 ± 0.9; DFM 0.50 ± 0.02; TFM 0.18 ± 0.04; norleucine 0.053 ± 0.001; control 0.021 ± 0.001 [36]. These observations signify that increasing fluorination reduces the rate of interaction with the platinum metal center and that TFM is especially slow in this regard. Nevertheless, reaction of DFM and TFM with K_2PtCl_4 did result in the formation of the corresponding platinum complexes, which were then studied by NMR spectroscopy to determine the effect of fluorination on the rate of inversion of the sulfur center [36]. The x-ray structure of the methionine–platinum complex is shown in Figure 17.7 for reference [37]. Detailed dynamic NMR studies indicated that, within experimental error, fluorination does not affect the rate of the sulfur inversion process in these complexes as the inversion barriers for the DFM– and TFM–platinum complexes were determined to be 16.4 ± 0.2 and 18 ± 1 kcal/mol, respectively, compared with a value of 17 ± 1 kcal/mol for the methionine complex. The similarity of the inversion barriers was also predicted from detailed computational studies at the B3LYP/SDD level.

Another aspect of the application of DFM and TFM analogues in protein engineering is their relatively subtle steric changes compared to methionine. This alteration would be

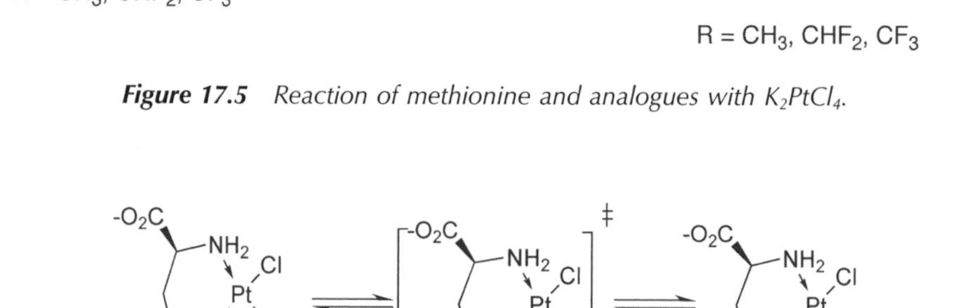

Figure 17.5 Reaction of methionine and analogues with K_2PtCl_4.

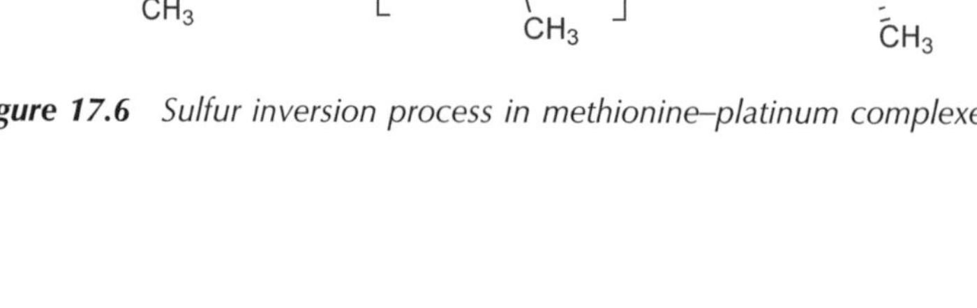

Figure 17.6 Sulfur inversion process in methionine–platinum complexes.

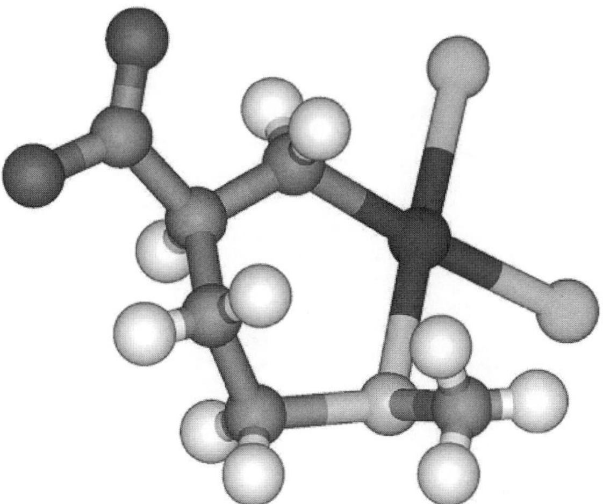

Figure 17.7 *Crystal structure of the methionine–$PtCl_2$ complex (coordinates from Wilson et al. [37]). See color plate 17.7.*

difficult to accomplish by site-directed mutagenesis using only the standard 20 amino acids. An example of the use of this characteristic is in the study of the key active site methionine (Met214) in the *P. aeruginosa* alkaline protease (Figure 17.8) [38]. This enzyme is a member of the metzincin family of zinc metalloproteases, which contain an invariant methionine residue in their active site [39–41]. There are over 700 protein members in this class. The role that the invariant methionine plays in this class of protease is still uncertain, although a contribution to catalytic activity and/or protein stability for some of the metzincins may occur [42, 43]. It was of interest to explore the capability of DFM to substitute for this invariant methionine in the active site of the alkaline protease. Successful incorporation of DFM could then be applied to the use of ^{19}F NMR spectroscopy to investigate this class of protease. The invariant methionine residue is in close proximity to the metal ligands of the protein (His176 and His186) (Figure 17.9), and replacement of this residue by fluorinated methionines should also be interesting from a protein structural perspective. Bioincorporation of DFM into the Met1 and Met214 positions of the protein was successfully accomplished after protein expression was optimized in *E. coli* [27]. The presence of DFM in the invariant 214 position was found to be well tolerated by the enzyme. Enzyme activity, as measured by hydrolysis of the artificial substrate Z-ArgArg-*p*-nitroanalide, was found to be comparable to that of the wild-type enzyme [44]. Protein stability, as measured by differential scanning calorimetry, was also found to be relatively unaffected by the presence of DFM. An interesting observation was that the cellular removal of the N-terminal methionine from the alkaline protease by the enzyme methionine aminopeptidase (MAP) was reduced when DFM replaced methionine. It has been reported that the presence of TFM at the N-terminal of the green fluorescent protein also affects normal N-terminal methionine processing [30].

Recently DFM, TFM and other methionine analogues were utilized to explore the contribution that methionine makes to the reduction-oxidation potential of the *P. aeruginosa* metalloprotein azurin [26, 45]. Azurin is a copper metalloprotein that is involved in

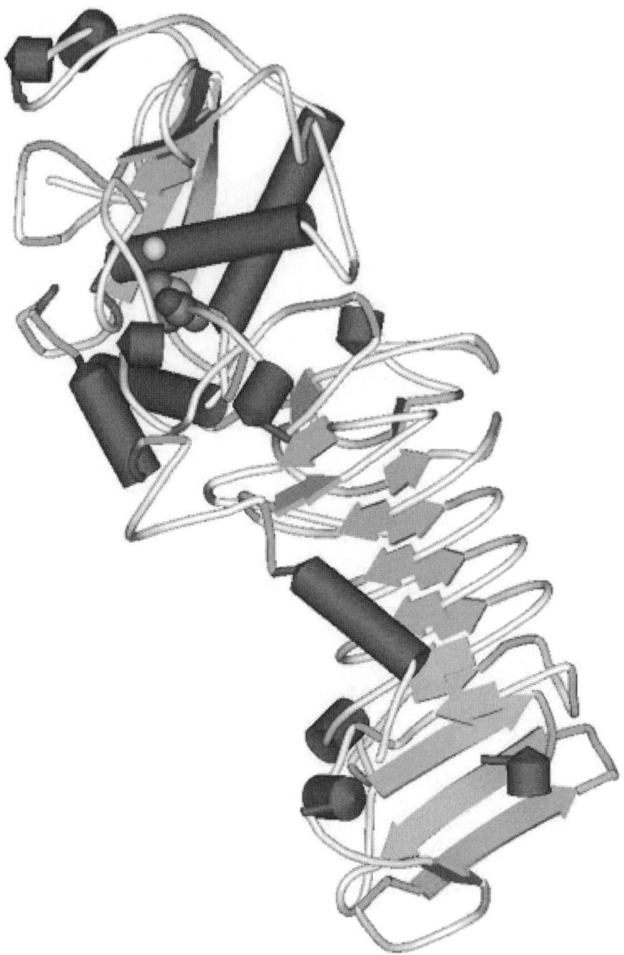

Figure 17.8 *Ribbon diagram of the* P. aeruginosa *alkaline protease showing the active site Zn^{2+} and the invariant methionine residue (PDB 1kap. [38]). See color plate 17.8.*

cellular redox reactions. In *P. aeruginosa* azurin, the active-site metal is ligated by His46, His117, Cys112, the peptide amide of Gly45 and a more distal interaction with Met121 (Figure 17.10) [46]. There has been much speculation as to the contribution that Met makes to the redox potential of this protein. The replacement of Met121 with a variety of non-natural methionine analogues was undertaken to probe this interaction. In order to incorporate these analogues into the recombinant protein, expressed protein ligation was employed [45]. Peptides (17-mer) corresponding to residues 112–128 (H_2N-CysThrPhePro GlyHisSerAlaLeuMet*LysGlyThrLeuThrLeuLys-COOH) and containing the methionine analogues were synthetically prepared by standard peptide synthesis and were linked to recombinant C-terminal thioester activated recombinant azurin corresponding to residues 1–111 [26]. This methodology allowed for the semisynthetic preparation of azurin mutants and is a convenient method to prepare site-selected modifications containing unnatural

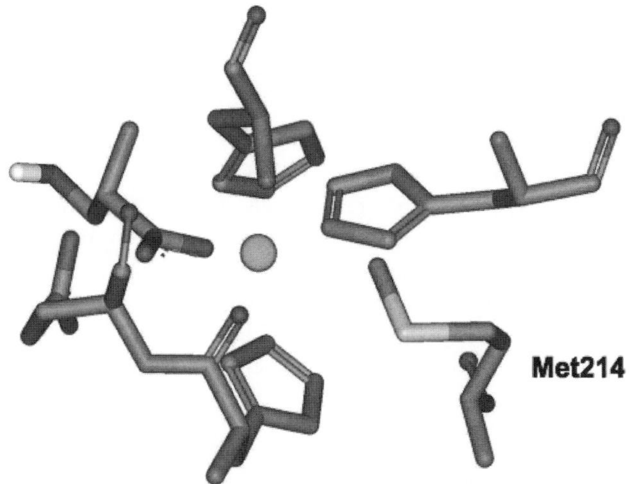

Figure 17.9 *Close-up of the active-site region of* P. aeruginosa *alkaline protease showing* Zn^{2+} *and protein ligands and the position of the thiomethyl group in close proximity to these side-chains (PDB 1kap. [38]). See color plate 17.9.*

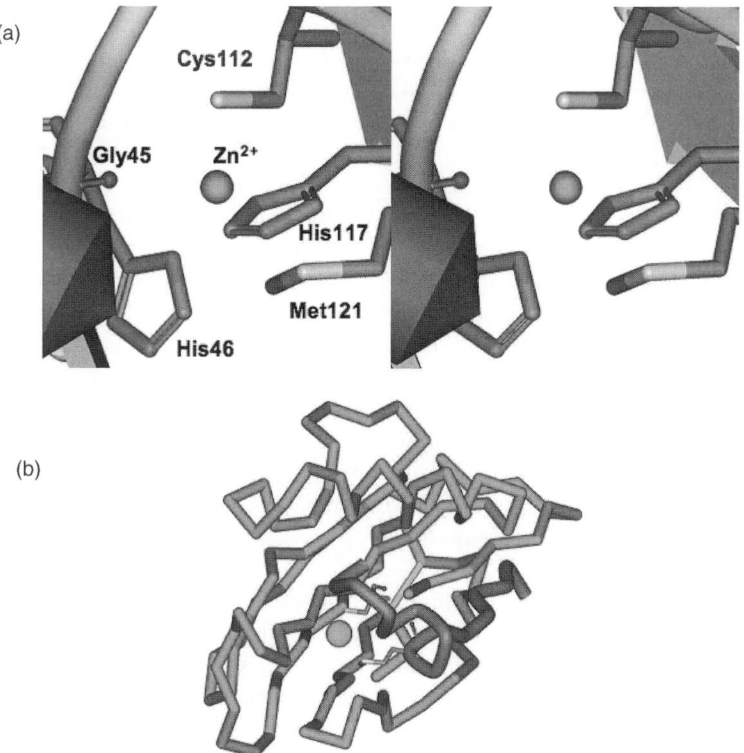

Figure 17.10 *(a) Stereo view of the active-site structure of* P. aeruginosa *azurin (PDB 4azu) [46]. (b) Tube diagram of the* P. aeruginosa *azurin showing the section of protein that was contributed by intein ligation. See color plate 17.10.*

amino acids, which standard bioincorporation methods do not achieve. There was very little structural alteration due to the presence of the methionine analogues as determined by analysis of the S(Cys)-Cu charge-transfer bands in the electronic spectra as well as the copper hyperfine coupling constants in the X-band electron paramagnetic resonance spectra. Azurin containing methionine at position 121 had a measured redox potential of 341 ± 3 mV (vs. NHE), whereas DFM- and TFM-substituted azurin had redox potential values of 329 ± 5 mV and 379 ± 4 mV, respectively. It was found through this approach that there appears to be a linear relationship between the redox potential of the mutant azurins and hydrophobicity of the amino acid analogues, a property difficult to identify without the use of unnatural methionine analogues with varying properties. The application of DFM and TFM to other metalloenzymes would be of interest both from a fundamental perspective as well as in the engineering of altered electronic characteristics for other redox proteins.

17.6 Structural Studies of DFM/TFM–Enzyme Complexes

Unfortunately, there is no x-ray structure of a DFM- or TFM-substituted protein available to allow detailed study of fluorine–protein interactions in spite of efforts to accomplish this (unpublished results). However, a few molecular structures have been determined for DFM and TFM bound as substrates/inhibitors to two key enzymes involved in methionine biochemistry, *E. coli* Met-tRNA synthetase and *E. coli* methionine aminopeptidase. Met-tRNA synthetase is responsible for the ATP-dependent coupling of methionine on to its cognate tRNA [47]. It would be the enzyme quintessential for the bioincorporation of DFM/TFM into proteins. Insight into the preference of Met-tRNA synthetase for DFM compared to TFM was obtained when the x-ray structures of the *E. coli* Met-tRNA synthetase co-crystallized with DFM and with TFM were compared to that found with methionine [48]. DFM was observed to bind in the same fashion as methionine, with several key residues interacting in a similar fashion in the complexed active site (Figure 17.11). Both Met and DFM binding results in the movement of Tyr15, Trp253, and Phe300 toward the amino acid. This results in additional movement of other residues such as Trp229 and Phe304 and stacking of Phe300 and Trp253, which is shown in Figure 17.11a. However the structure of the TFM-tRNA synthetase complex exhibited substantial differences (Figure 17.11b) and no stacking interaction between Phe300 and Trp253 is present in this complex. This difference alters the binding pocket shape and size for Met/DFM compared to TFM. This may explain the lower bioincorporation levels of TFM compared to DFM and, in addition, these x-ray structures provide critical insight into the enzymatic process itself.

A report has also appeared detailing the molecular interactions between TFM and the enzyme *E. coli* methionine aminopeptidase (MAP) [49]. The structure of MAP with bound TFM shows some of the detailed interactions that exist between this fluorinated amino acid and the enzyme and its interaction with the hydrophobic pocket composed of Cys59, Tyr62, Tyr65, Cys70, His79, Phe177, and Trp221 (Figure 17.12). Overall the structure of the TFM complex appears similar to that of the Met complex. Further analyses are required to understand the factors that contribute to the apparently lower processing rates

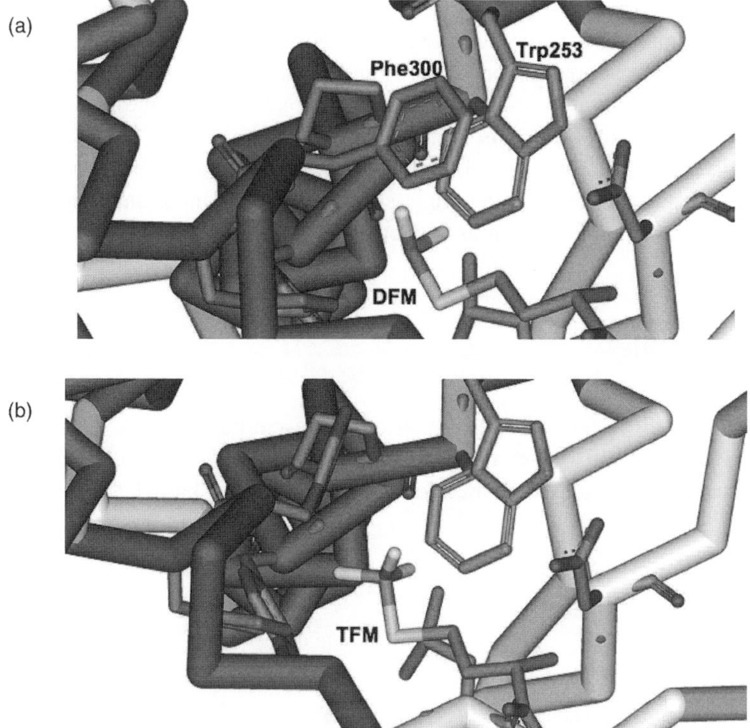

Figure 17.11 *X-ray structures of (a) DFM (PDB 1pfv) [48] and (b) TFM (PDB 1pfw) [48] complexes with Met-tRNA synthetase from* E. coli. *See color plate 17.11.*

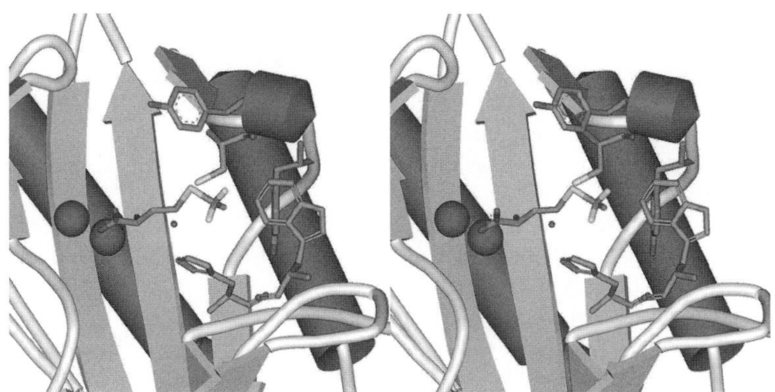

Figure 17.12 *Stereo view of the active-site interactions of TFM with* E. coli *methionine aminopeptidase (PDB 1c22) [49]. See color plate 17.12.*

of fluorinated N-terminal residues. Nevertheless these structures contribute additional insight into the molecular recognition between fluorinated amino acids and proteins.

In conclusion, the application of DFM and TFM to bioorganic chemistry has led to new information on how fluorinated compounds interact with enzymes, has provided novel ^{19}F NMR biophysical probes for use in biochemistry, and has led to subtle structurally and electronically modified analogues of methionine with which to probe enzymes. The addition of several fluorine atoms to a simple structure such as methionine has certainly led to an interesting series of adventures in bioorganic chemistry.

Acknowledgments

The author gratefully acknowledges the past and present contributions of his undergraduate and graduate students, research associates, and collaborators. Funding from NSERC (Canada) and the University of Waterloo is also gratefully acknowledged.

References

1. Budisa, N. (2006) *Engineering the Genetic Code*, Weinheim: Wiley-VCH Verlag GmbH.
2. Kukhar, V. P. and, Soloshonok, V. A. (eds.) (1995) *Fluorine-containing Amino Acids*, John Wiley & Sons, Ltd, Chichester.
3. Welch, J. T. and Eswarakrishnan, S. (1991) *Fluorine in Bioorganic Chemistry*, John Wiley & Sons, Inc., New York.
4. Welch, J. T. (ed.) (1991) *Selective Fluorination in Organic and Bioorganic Chemistry*, American Chemical Society, Washington D.C.
5. Soloshonok, V. A., Mikami, K., Yamazaki, T., *et al.* (eds.) (2007) *Current Fluoroorganic Chemistry*, American Chemical Society, Washington D.C.
6. O'Hagan, D. and Rzepa, H. S. (1997) Some influences of fluorine in bioorganic chemistry. *J. Chem. Soc. Chem. Commun.*, 645–652.
7. Gerig, J. T. (1994) Fluorine NMR of proteins. *Prog. NMR Spectrosc.*, **26**, 93–370.
8. Danielson, M. A. and Falke, J. J. (1996) Use of 19F NMR to probe protein structure and conformational changes. *Annu. Rev. Biophys. Biomol. Struct.*, **25**, 63–95.
9. Frieden, C., Hoeltzli, S. D. and Bann, J. G. (2004) The preparation of ^{19}F-labeled proteins for NMR studies. *Methods Enzymol.*, **380**, 400–415.
10. Shu, Q. and Frieden, C. (2005) Relation of enzyme activity to local/global stability of murine adenosine deaminase: ^{19}F NMR studies. *J Mol Biol.*, **345**(3), 599–610.
11. Janzen, J. F., Wang, P. M. C. and Lemire, A. E. (1983) Fluorination of methionine and methionylglycine derivatives with xenon difluoride. *J. Fluorine Chem.*, **22**, 557–559.
12. Houston, M. E. and Honek, J. F. (1989) Facile synthesis of fluorinated methionines. *J. Chem. Soc. Chem. Commun.*, 761–762.
13. Robins, M. J., Wnuk, S. F., Mullah, K. B. and Dalley, N. K. (1994) Nucleic Acid related compounds. 80. Synthesis of 5′-*S*-(alkyl and aryl)-5′-fluoro-5′-thioadenosines with xenon difluoride or (diethylamido)sulfur trifluoride, hydrolysis in aqueous buffer, and inhibition of *S*-adenosyl-L-homocysteine hydrolase by derived "adenosine 5′-aldehyde species." *J. Org. Chem.*, **59**, 544–555.

14. Dannley, R. L. and Taborsky, R. G. (1957) Synthesis of DL-*S*-trifluoromethylhomocysteine (trifluoromethylmethionine). *J. Org. Chem.*, **22**, 1275–1276.
15. Duewel, H., Daub, E., Robinson and V., Honek, J. F. (1997) Incorporation of trifluoromethionine into a phage lysozyme: implications and a new marker for use in protein 19F NMR. *Biochemistry*, **36**(11), 3404–3416.
16. Kieltsch, I., Eisenberger, P. and Togni, A. (2007) Mild electrophilic trifluoromethylation of carbon- and sulfur-centered nucleophiles by a hypervalent iodine(III)-CF_3 reagent. *Angew. Chem. Int. Ed.*, **46**, 754–757.
17. Soloshonok, V. A., Kukhar, V., Pustovit, Y. and Nazaretian, V. (1992) A new and convenient synthesis of *S*-trifluoromethyl-containing amino acids. *SynLett.*, 657–658.
18. Tsushima, T., Ishihara, S. and Fujita, Y. (1990) Fluorine-containing amino acids and their derivatives. synthesis and biological activities of difluoromethylhomocysteine. *Tetrahedron Lett.*, **31**, 3017–3018.
19. Houston, M. E., Harvath, L. and Honek, J. F. (1997) Syntheses of and chemotactic responses elicited by fMet-Leu-Phe analogs containing difluoro- and trifluoromethionine. *Bioorg. Med. Chem. Lett.*, **7**, 3007–3012.
20. Vaughan, M. D., Sampson, P. B., Daub, E. and Honek, J. F. (2005) Investigation of bioisosteric effects on the interaction of substrates inhibitors with the methionyl-tRNA synthetase from Escherichia coli. *Med Chem.*, **1**, 227–237.
21. Breneman, C. M. and Wiberg, K. B. (1990) Determining atom-centered monopoles from molecular electrostatic potentials. The need for high sampling density in formamide conformational analysis. *J. Comp. Chem.*, **11**(3), 361–373.
22. Freer, R. J., Day, A. R., Radding, J. A., *et al.* (1980) Further studies on the structural requirements for synthetic peptide chemoattractants. *Biochemistry*, **19**(11), 2404–2410.
23. Fruchtmann, R., Kreisfeld, K., Marowski, C. and Opitz, W. (1981) Synthetic tripeptides as chemotaxins or chemotaxin antagonists. *Hoppe-Seyler's Z. Physiol. Chem.*, **362**(2), 163–174.
24. Freer, R. J., Day, A. R., Muthukumaraswamy, N., *et al.* (1982) Formyl peptide chemoattractants: a model of the receptor on rabbit neutrophils. *Biochemistry*, **21**(2), 257–263.
25. Vaughan, M. D., Cleve, P., Robinson, V., *et al.* (1999) Difluoromethionine as a novel ^{19}F NMR structural probe for internal amino acid packing in proteins. *J. Am. Chem. Soc.*, **121**, 8475–8478.
26. Garner, D. K., Vaughan, M. D., Hwang, H. J., *et al.* (2006) Reduction potential tuning of the blue copper center in Pseudomonas aeruginosa azurin by the axial methionine as probed by unnatural amino acids. *J. Am. Chem. Soc.*, **128**(49), 15608–15617.
27. Walasek, P. and Honek, J. F. (2005) Nonnatural amino acid incorporation into the methionine 214 position of the metzincin *Pseudomonas aeruginosa* alkaline protease. *BMC Biochem.*, **6**, 21.
28. McIntyre, D. D., Yuan, T. and Vogel, H. J. (1997) Fluorine-19 and proton NMR studies of difluoromethionine calmodulin. *Prog. Biophys. Mol. Biol.*, **65**, P-A1–27.
29. Salopek-Sondi, B., Vaughan, M. D., Skeels, M. C., *et al.* (2003) ^{19}F NMR studies of the leucine-isoleucine-valine binding protein: evidence that a closed conformation exists in solution. *J. Biomol. Struct. Dyn.*, **21**(2), 235–246.
30. Budisa, N., Pipitone, O., Siwanowicz, I., *et al.* (2004) Efforts toward the design of "Teflon" proteins: in vivo translation with trifluorinated leucine and methionine analogues. *Chem. Biodivers.*, **1**, 1465–1475.
31. Duewel, H. S., Daub, E., Robinson, V. and Honek, J. F. (2001) Elucidation of solvent exposure, side-chain reactivity, and steric demands of the trifluoromethionine residue in a recombinant protein. *Biochemistry*, **40**(44), 13167–13176.
32. Leung, A. K., Duewel, H. S., Honek, J. F. and Berghuis, A. M. (2001) Crystal structure of the lytic transglycosylase from bacteriophage lambda in complex with hexa-*N*-acetylchitohexaose. *Biochemistry*, **40**(19), 5665–5673.

33. Appleton, T. G., Connor, J. W. and Hall, J. R. (1988) *S,O*- versus *S,N*-chelation in the reactions of the cis-diamminediaquaplatinum(II) cation with methionine and *S*-methylcysteine. *Inorg. Chem.*, **27**, 130–137.
34. Gummin, D. D., Ratilla, E. M. A. and Kostic, N. M. (1986) Variable-temperature platinum-195 NMR spectroscopy, a new technique for the study of stereodynamics. sulfur inversion in a platinum(II) complex with methionine. *Inorg. Chem.*, **25**, 2429–2433.
35. Galbraith, J. A., Menzel, K. A., Ratilla, E. M. A. and Kostic, N. M. (1987) Study of stereodynamics by variable-temperature platinum-195 NMR spectroscopy. Diastereoisomerism in platinum(II) thioether complexes and solvent effects. *Inorg. Chem.*, **26**, 2073–2078.
36. Vaughan, M. D., Robertson, V. J. and Honek, J. F. (2007) Experimental and theoretical studies on inversion dynamics of dichloro(L-difluoromethionine-*N,S*)platinum(II) and dichloro(L-trifluoromethionine-*N,S*)platinum(II) complexes. *J. Fluorine Chem.*, **128**, 65–70.
37. Wilson, C., Scudder, M. L., Hambley, T. W. and Freeman, H. C. (1992) Structures of dichloro[(*S*)-methionine-*N,S*]platinum(II) and chloro[glycyl-(*S*)-methioninato-*N,N′,S*]platinum(II) monohydrate. *Acta Crystallogr., Sec. C: Cryst. Struct. Commun.*, **48**, 1012–1015.
38. Baumann, U., Wu, S., Flaherty, K. M. and McKay, D. B. (1993) Three-dimensional structure of the alkaline protease of *Pseudomonas aeruginosa:* a two-domain protein with a calcium binding parallel beta roll motif. *EMBO J.*, **12**(9), 3357–3364.
39. Stocker, W. and Bode, W. (1995) Structural features of a superfamily of zinc-endopeptidases: the metzincins. *Curr. Opin. Struct. Biol.*, **5**(3), 383–390.
40. Stocker, W., Grams, F., Baumann, U., *et al.* (1995) The metzincins – topological and sequential relations between the astacins, adamalysins, serralysins, and matrixins (collagenases) define a superfamily of zinc-peptidases. *Protein Sci.*, **4**(5), 823–840.
41. Gomis-Ruth, F. X. (2003) Structural aspects of the metzincin clan of metalloendopeptidases. *Mol. Biotechnol.*, **24**(2), 157–202.
42. Butler, G. S., Tam, E. M. and Overall, C. M. (2004) The canonical methionine 392 of matrix metalloproteinase 2 (gelatinase A) is not required for catalytic efficiency or structural integrity: probing the role of the methionine-turn in the metzincin metalloprotease superfamily. *J. Biol. Chem.*, **279**(15), 15615–15620.
43. Hege, T. and Baumann, U. (2001) The conserved methionine residue of the metzincins: a site-directed mutagenesis study. *J. Mol. Biol.*, **314**(2), 181–186.
44. Louis, D., Kohlmann, M. and Wallach, J. (1997) Spectrophotometric assay for amidolytic activity of alkaline protease from *Pseudomonas aeruginosa*. *Anal. Chim. Acta*, **345**, 219–225.
45. Berry, S. M., Ralle, M., Low, D. W., *et al.* (2003) Probing the role of axial methionine in the blue copper center of azurin with unnatural amino acids. *J. Am. Chem. Soc.*, **125**, 8760–8768.
46. Nar, H., Messerschmidt, A., Huber, R., *et al.* (1991) Crystal structure analysis of oxidized *Pseudomonas aeruginosa* azurin at pH 5.5 and pH 9.0. A pH-induced conformational transition involves a peptide bond flip. *J. Mol. Biol.*, **221**(3), 765–772.
47. Ibba, M., Francklyn, C. and Cusack, S. (eds.) (2005) *The Aminoacyl-tRNA Synthetases*, Georegetown: Landes Bioscience.
48. Crepin, T., Schmitt, E., Mechulam, Y., *et al.* (2003) Use of analogues of methionine and methionyl adenylate to sample conformational changes during catalysis in *Escherichia coli* methionyl-tRNA synthetase. *J. Mol. Biol.*, **332**(1), 59–72.
49. Lowther, W. T., Zhang, Y., Sampson, P. B., *et al.* (1999) Insights into the mechanism of *Escherichia coli* methionine aminopeptidase from the structural analysis of reaction products and phosphorus-based transition-state analogues. *Biochemistry*, **38**(45), 14810–14819.

18

Structure Analysis of Membrane-Active Peptides Using ^{19}F-labeled Amino Acids and Solid-State NMR

Parvesh Wadhwani and Erik Strandberg

18.1 Introduction

Fluorine, one of the most abundant elements on earth, is rarely found to participate in biological processes. With its small size and high electronegativity, the physiochemical and pharmacological properties of fluorinated molecules can be widely tuned to obtain substances that range from metabolically stable drugs to very poisonous compounds. During the last two decades, there has been an increasing use of fluorine in bioorganic and medicinal chemistry [1, 2]. Incorporation of fluorine or a CF_3-group as a hydrogen atom replacement has led to a substantial increase in the bioactivity and bioavailability of a variety of pharmaceuticals. Similarly, there has been an increase in the use of fluorine in other areas of chemistry, such as agricultural and materials sciences. Likewise, the high sensitivity of fluorine as an NMR-active nucleus and the large ^{19}F–^{1}H or ^{19}F–^{19}F dipolar coupling strengths have also found tremendous application. Consistent development of new synthetic methods to produce ^{19}F-labeled amino acids, complemented by simultaneous development in the field of ^{19}F NMR hardware, has led to a situation in which ^{19}F NMR can be routinely used to study ^{19}F-labeled peptides and proteins and address a variety of open questions [3–7]. By performing ^{19}F NMR on a suitably labeled peptide or protein, one can easily obtain information about conformational changes, the structure and orientation of peptides, or the kinetics of ligand binding, properties that are intimately related to the function of the peptide. The pre-requirement of such experiments is the

Fluorine in Medicinal Chemistry and Chemical Biology Edited by Iwao Ojima

inclusion of a specific ^{19}F-reporter group. This chapter highlights the use of ^{19}F-labeled amino acids and their incorporation into peptides and proteins using biosynthetic and chemical approaches. A description is given of solid-state ^{19}F NMR experimental procedures and how experimental data may be used to determine the structure and orientation of peptides in membranes, giving information that is fundamental to understanding and improving their function. Finally, some results are presented where several membrane-active peptides prepared with various ^{19}F-labeled amino acids have been investigated using solid-state ^{19}F NMR. The chapter focuses on the use of orientational NMR constraints and will not go into details about NMR distance constraints.

18.1.1 Fluorine-containing Peptides

Incorporation of fluorine into the peptides can be straightforwardly achieved using ^{19}F-labeled amino acids and synthesizing the peptide in question by following standard solid-phase peptide synthesis (SPPS) protocols. There have been numerous reports on the synthesis of ^{19}F-labeled amino acids that are beyond the scope of this chapter and have been described in detail elsewhere [1, 2]. Generally, a ^{19}F-labeled amino acid is an analogue of a naturally occurring amino acid. Fluorine is the element with the second smallest known atomic radius, and the replacement of a hydrogen atom by a fluorine atom is expected to cause minimum steric perturbations; therefore, a variety of amino acids can be synthesized in which a single hydrogen atom is replaced with a fluorine atom in the side chain. Alternatively, ^{19}F-labeled peptides may also be obtained by direct fluorination of peptides, using a variety of fluorinating agents, but most of these efforts have been aimed to fluorinate cyclic or aromatic amino acids (see reference 8 and references therein).

For example, synthesis of α-fluoroglycine has been attempted by various researchers, but this amino acid is unstable and spontaneous elimination of HF results in an acyl-iminium intermediate that further decomposes; however, syntheses of protected α-fluoroglycines have been reported [9]. Most of the efforts so far have concentrated on the synthesis of either β-fluoro-amino acids, such as β-fluoroalanine and their di- and trifluoro derivatives, or of γ-fluorinated amino acids, as discussed in detail in the literature [9, 10]. Replacement of a methyl group (CH_3-, with a volume of 54 Å^3) by a trifluoromethyl group (CF_3-, with a volume of 94 Å^3) leads to some structural changes due to the increased spatial requirement, but often such substitutions are justified by comprehensive structural and functional studies on the peptide in question [11, 12]. Substitution of a CH_3-group by a CF_3-group in alanine, isoleucine, leucine and valine is well known. These ^{19}F-labeled analogues have been found to change the solubility properties of peptides or, for example, to enhance the stability of coiled coil motifs (see references 12–16 and references therein).

Several aromatic amino acids have also been converted to their ^{19}F-labeled analogues by replacement of a single H atom on the aromatic ring. These include fluorotryptophan, fluorophenylalanine, fluorophenylglycine, fluorotyrosine, and their various isomers. Certain solid-state ^{19}F NMR structural approaches require specific ^{19}F-labeled amino acids bearing a fluorine atom or CF_3-group that is rigidly attached to the peptide backbone (collinear with the C_α–C_β bond). These include 4-F-phenylglycine (4F-Phg), 4-CF_3-

phenylglycine (CF_3-Phg), 3-CF_3-bicyclopentylglycine (CF_3-Bpg), and 2-CF_3-alanine analogues. For other analytical methods of investigation (solution NMR, circular dichroism [CD], analytical centrifugation, and thermodynamic methods) several other ^{19}F-labeled amino acids have been employed [17–22]. These ^{19}F-labeled amino acids may contain a single fluorine atom next to a hydroxyl group (i.e., serine and threonine), a thiol group (cysteine and methionine), a carboxyl group (glutamic acid and aspartic acid), or a nitrogen functionality (glutamine, histidine, lysine, proline, and tryptophan). Likewise, there are reports of trifluoro- and hexafluoro-derivatives of isoleucine, leucine, and valine having been employed in protein design and engineering [12–16].

18.1.2 Membrane-Active Peptides

The cellular membrane is a hydrophobic barrier that surrounds the cytoplasm of every cell and is involved in complex cellular processes, such as signaling and transport, which are essential to maintain the normal life cycle of a cell; major components of this cellular membrane are lipids, proteins, peptides, and carbohydrates. Peptides that interact with cellular membranes are referred to as membrane-active peptides and can be broadly divided into three major classes: antimicrobial, cell-penetrating, and fusogenic peptides. Any of these may have variable lengths, hydrophobicities, and secondary structures, but they often exhibit similar effects on membranes. For example, some antimicrobial peptides have cell-penetrating properties, and vice versa [23, 24]; these peptides usually cause some degree of membrane destabilization.

Antimicrobial peptides (AMPs), or host defense peptides, are short (10–40 amino acids) and positively charged and they usually form an amphipathic structure (α-helix, β-sheet, cyclic or flexible). Upon microbial invasion, these peptides move into action and kill the invading microbes by rupturing their membrane or by making pores in the membrane [25–27]. In higher organisms, they can also activate further host defense pathways. Several studies suggest that these peptides are initially unstructured in solution. Upon membrane binding, which is governed by initial electrostatic attraction between the positively charged peptide and the negatively charged bacterial membrane, these peptides acquire their secondary amphipathic structure and lie on the membrane surface in an orientation that compensates for charge and hydrophobicity (see Figure 18.1). As the concentration of the peptides is increased, they either break the membrane barrier (“carpet mechanism” [28]) or insert into the membrane and form organized peptide assemblies which permeabilize the membrane by forming pores. In the “barrel-stave pore” model, peptides tilt into a transmembrane orientation and form a bundle, resulting in a hole in the membrane and causing cytoplasmic leakage [29]. In the “toroidal wormhole pore” model, some negatively charged lipid molecules also participate in the formation of a hole, thereby compensating the charges of the peptides [30]. Figure 18.1 illustrates some of the models that are often invoked to explain the mechanism of action of AMPs. The peptides often also show hemolytic activity against human erythrocytes, which has so far limited their therapeutic use.

Cell-penetrating peptides (CPPs) can internalize into cells without disturbing the membrane potential and, therefore, are an attractive research target for drug delivery. These peptides are also short, cationic, and usually amphiphilic, and are able to bring a “functional cargo” into the cell [24, 31–33]. There is no mechanism unanimously agreed upon,

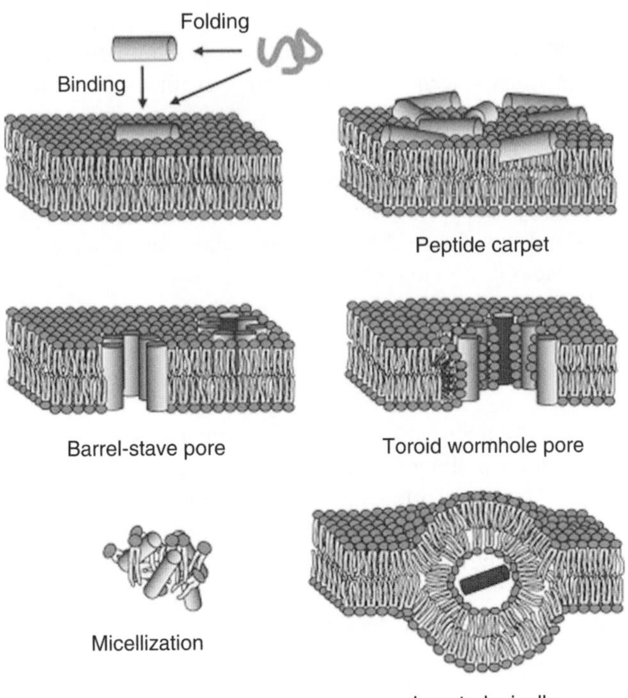

Figure 18.1 *Models for different modes of peptide–lipid interaction of membrane-active peptides. The peptide remains unstructured in solution and acquires an amphipathic structure in the presence of a membrane. The hydrophobic face of the amphipathic peptide binds to the membrane, as represented by the grayscale. At low concentration, the peptide lies on the surface. At higher peptide concentrations the membrane becomes disrupted, either by the formation of transmembrane pores or by destabilization via the "carpet mechanism." In the "barrel-stave pore" the pore consists of peptides alone, whereas in the "toroidal wormhole pore" negatively charged lipids also line the pore, counteracting the electrostatic repulsion between the positively charged peptides. The peptide may also act as a detergent and break up the membrane to form small aggregates. Peptides can also induce inverted micelle structures in the membrane.*

and various cellular processes related to endocytosis may be involved. For direct peptide uptake across the plasma membrane, or just as well for escape from an endosome, an inverted micellar mechanism has been proposed to explain the release of cargo into the cytosol (see Figure 18.1).

Fusogenic peptides, on the other hand, are instrumental at the membrane surface and mediate the fusion of two lipid bilayers [34], which is vital for fertilization and is an essential step in various viral infections. Figure 18.2 illustrates how the two opposing membranes are perturbed by the presence of a fusogenic peptide to form an intermediate state, which then results in complete intermixing of the cytoplasmic material. In nature these peptides are usually smaller fragments of larger proteins, but *in vitro* the smaller fragments are self-sufficient in executing membrane fusion.

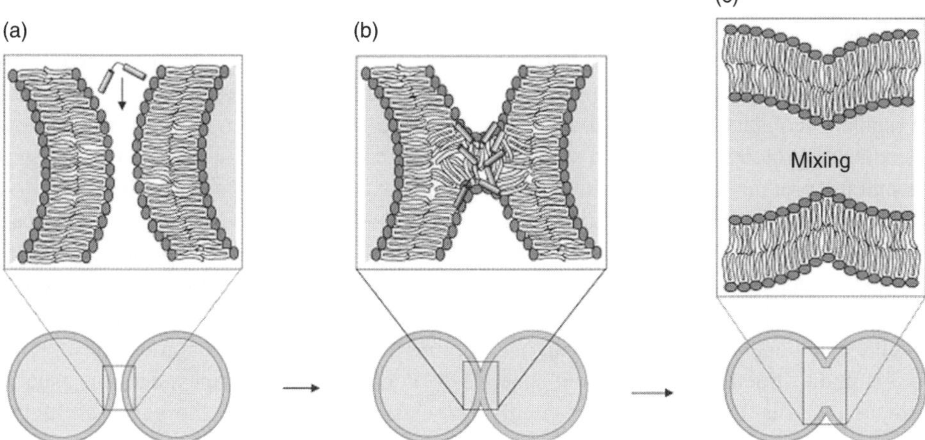

Figure 18.2 (a) Two membranes and a fusogenic peptide before fusion. (b) Fusion of the two lipid bilayers, mediated by fusogenic peptides. (c) Merged compartments after fusion.

18.1.3 Structure and Orientation of Peptides in Membranes

To understand the function of membrane-active peptides, it is important to know the structure and orientation of the peptide in the membrane. As is evident from Figure 18.1, it is possible to distinguish between, for example, carpet and pore mechanisms of action by determining the peptide's orientation in the membrane. Various techniques, such as electron spin resonance (ESR) [35], infrared (IR) spectroscopy [36–38], circular dichroism (CD) [35, 39, 40], and solid-state NMR (SSNMR) [4–7] are used to investigate membrane-active peptides in a quasi-native lipid bilayer environment. In the following sections, methods to determine peptide structure and orientation are presented.

18.1.3.1 Peptide Structure

The first step in these investigations is usually to determine or to confirm the secondary structure of the peptide. The conventional methods used to determine the three-dimensional structure of biomolecules are x-ray crystallography and solution-state NMR, but neither method is suitable for small peptides bound to membranes. While it is relatively easy to crystallize peptides by themselves, it is very difficult to crystallize them in a membrane, due to the high intrinsic mobility of the system. It is easier and more meaningful to crystallize large membrane proteins, which are less mobile and whose structure is more determined by internal constraints. Solution-state NMR works well with peptides in solution or in small detergent micelles that tumble sufficiently rapidly, but this conformation can be very different from the structure in a membrane and no orientational information is gained. SSNMR does not have these limitations and is one of the best methods for study of the structure of peptides in membranes [41–43]. SSNMR usually requires isotopic peptide labeling, and, in order to determine a detailed structure, a large number of constraints are needed.

The structure of a peptide inside a hydrophobic membrane is highly constrained. There is a high cost of free energy for exposing polar groups of the peptide backbone, such as CO and NH, to the hydrophobic environment; therefore, these groups need to form hydrogen bonds. In the membrane, such hydrogen bonds are most readily accommodated in regular secondary structures, namely α-helices or β-sheets. In fact, most transmembrane domains of proteins with known structure are α-helical, and there are also a few integral membrane proteins forming β-sheet structures, for example the β-barrel porins. As a first approximation of the peptide structure, an α-helical or a β-stranded type of structure can readily be identified by CD in lipid vesicles. It is also possible to use solution-state NMR methods to determine the peptide structure in micelles to serve as a model for the structure analysis in lipid membranes. For some peptides with strong internal constraints, such as small cyclic peptides or peptides with one or more disulfide bridges, it is most meaningful to obtain the structure from solution-state NMR or x-ray crystallography, and it is usually safe to assume that the resultant highly stable structure will be the same as in the membrane.

18.1.3.2 Peptide Orientation

Once the structure of a membrane-active peptide is known, it is then possible to determine the orientation of the peptide in the membrane. This orientation can be defined in terms of a tilt angle (τ) and an azimuthal angle (ρ) (see Figure 18.3). The tilt angle is the angle between the peptide's long axis and the bilayer normal; for helical peptides, the helix axis is conveniently defined as the long axis. The azimuthal angle describes the rotation around the long axis, defined from some individually defined point of reference. The two most

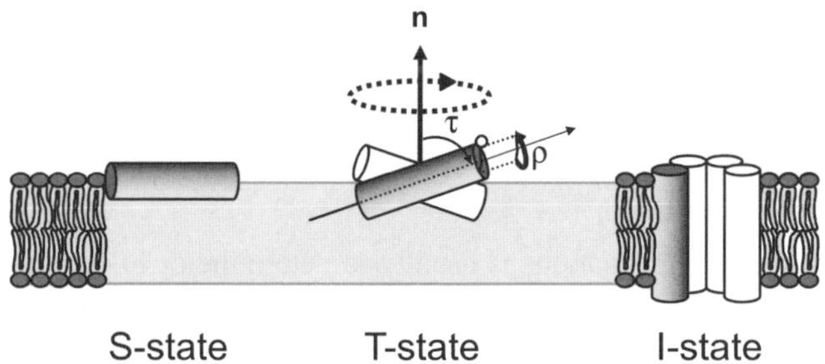

Figure 18.3 *The orientation of a peptide in the membrane can be described by the tilt angle τ and the azimuthal angle ρ. τ is the angle between the bilayer normal (**n**) and the peptide long axis. ρ describes a rotation around the peptide long axis and must be defined with respect to a reference group as indicated by the white circle. In liquid-crystalline bilayers, peptides can usually also rotate around the membrane normal (shown by the dashed arrow). Three characteristic peptide orientations are shown: in the S-state the peptide lies flat on the membrane surface with charged amino acids facing the water; in the T-state the peptide is inserted with an oblique tilt into the membrane, possibly in a dimeric state (shown as a second peptide in white); and in the inserted I-state the peptide has a transmembrane orientation. In this state, the peptide may self-assemble into pores (shown here as a barrel-stave pore together with additional white peptides).*

extreme orientations of membrane-active peptides distinguish between peptides lying flat on the membrane surface, and inserted peptides with a transmembrane orientation. The first case is referred to as the "surface" state or S-state, and the second case is called the "inserted" state or I-state. In the S-state, the tilt angle is approximately 90° and in the I-state it is approximately 0°. From SSNMR on membrane-active peptides, an obliquely "tilted" state has also been detected in several cases; this is called the T-state, in which peptides are tilted by approximately 125° [44–47]. These three states are shown in Figure 18.3.

Several methods are available for measuring the orientation of a membrane-active peptide in a lipid bilayer and each of these differs in accuracy. One such method is oriented circular dichroism (OCD), which gives an approximate tilt angle of α-helical peptides, and which can monitor the reorientation of β-sheet peptides [39]. This method does not require any labeling, but on the other hand it provides no information about the azimuthal angle. More accurate values of the tilt, as well as the azimuthal angle, can be obtained using SSNMR [4]. Many NMR parameters depend on the orientation of certain NMR tensors with respect to the static magnetic field, which provides a natural reference frame. Examples of such tensors are the chemical shift anisotropy, dipolar and quadrupolar tensors [48]. These tensors are fixed in the peptide molecule and, by determining the orientation of several tensors with respect to the magnetic field, it is possible to calculate the orientation of the molecules with respect to the magnetic field and also to the membrane. This process is described in more detail in the next section. In this chapter we will focus on the use of solid-state ^{19}F NMR to characterize ^{19}F-labeled peptides in model membranes.

18.2 Solid-State ^{19}F NMR of Membrane-Active Peptides

This section briefly describes the ^{19}F NMR experiments performed on ^{19}F-labeled peptides, the fundamental NMR parameters that are used to obtain orientational constraints, and the means by which peptide structure is determined from these constraints.

18.2.1 Advantages of ^{19}F NMR

SSNMR methods usually rely on isotopic labeling of the peptide, in contrast to solution-state NMR which often makes use of naturally occurring ^{1}H nuclei. The dipolar couplings between these protons are motionally averaged in solution, but this is not the case in the solid state. These strong interactions give broad lines in SSNMR spectra, which makes it impossible to resolve the signals of individual ^{1}H nuclei. Accordingly, other NMR-active nuclei are introduced into the molecule and observed by SSNMR. The most common isotopes used for labeling are ^{15}N, ^{13}C and ^{2}H. However, the first has low sensitivity, and the other two have considerable natural abundance, which can give problems with background signals, especially at low concentrations of the peptide of interest. Instead, for studies of membrane-active peptides our group has developed methods using ^{19}F as an NMR label (see references 6, 7 and references therein) with high sensitivity and no background

Table 18.1 *Comparison of selected NMR labels used for peptide structure analysis*

Information	Nucleus		
	^{15}N	^{2}H	^{19}F
Label	^{15}N in backbone	Ala-d_3	CF_3-group
Sensitivity	Low (<1%)	Medium (5%)	**High (100%)**
Minimum *P/L[a]	1:200	1:300	**1:3000**
Structure analysis	**Single label gives helix tilt**, but azimuthal rotation is not generally accessible	Ambiguous sign of splitting can give multiple solutions, hence many labels are required	**Full information on tilt and azimuthal rotation,** but several labels are required
Accuracy	Low (±30°)	**High (±5°)**	Medium (±15°)
Structural perturbation	**Entirely unperturbing**	**Unperturbing at alanine positions,** risky at other positions	Risk of perturbation

* P/L[a] = peptide-to-lipid molar ratio

signals. Table 18.1 summarizes advantages and disadvantages of some selected NMR labels. It is evident from Table 18.1 that using highly sensitive ^{19}F-labels it is possible to investigate concentration regimes that are not easily accessible using conventional isotope labels. It is also possible to use smaller amounts of material or shorter measurement times. ^{19}F NMR offers complete information about peptide orientation and dynamics.

Hardware requirements are slightly different from those for traditional NMR measurements; NMR probe heads must be absolutely free from fluorine, but such probes are now commercially available. It is usually necessary to have two high-frequency channels on the NMR spectrometer to allow fluorine measurements with proton decoupling, and good RF-filters must be used to separate the ^{19}F and ^{1}H channels.

18.2.2 Solid-State NMR Constraints

In order to determine the structure and alignment of a peptide, it is necessary to determine the orientation of several NMR interaction tensors in the magnetic field; however, first the orientation of the tensors should be known in the molecular frame. The peptide structure and its alignment in the membrane can then be determined, given that the membrane orientation in the magnetic field is known. Thus, it is often advantageous to use macroscopically oriented membrane samples in which the lipid membrane normal is aligned parallel to the external magnetic field.

In the case of ^{19}F-labels, the relevant tensors are the ^{19}F chemical shift anisotropy (CSA), and for CF_3-groups the ^{19}F–^{19}F homonuclear dipolar couplings. Both interactions have a $\frac{1}{2}(3\cos^2\theta - 1)$ dependence, where θ is the angle between the interaction tensor and the magnetic field direction. Both tensors can be used to determine peptide orientation; however, the orientation of the CSA tensor is not well characterized in all ^{19}F-labeled amino acids (see e.g. 4F-Phg). In practice, it is also difficult to measure the chemical shift accurately, since referencing of ^{19}F NMR signals is not straightforward [49]; it is therefore often easier to use the dipolar couplings of CF_3-groups. The orientation of the dipolar

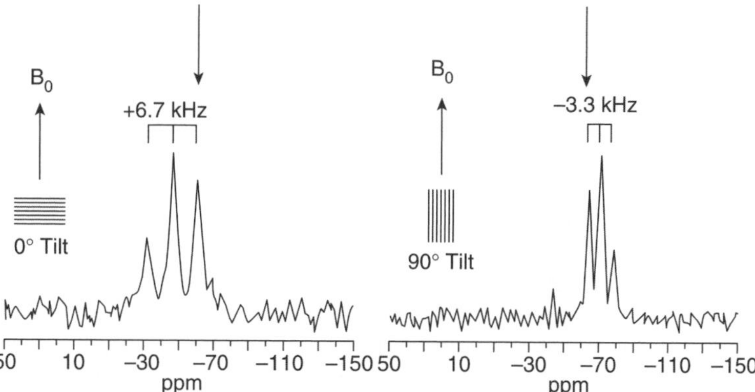

Figure 18.4 *^{19}F NMR spectra of MSI-103, where Ile-13 is replaced with CF_3-Phg in DMPC at P/L = 1 : 400. The left spectrum was measured with the bilayer normal parallel to the magnetic field and the right spectrum was measured with the bilayer normal perpendicular to the magnetic field. Because of fast rotation of the peptide around the bilayer normal, the splittings at are scaled by a factor of −1/2. The sign of the dipolar coupling can be deduced from the position of the triplet relative to the isotropic chemical shift (close to −60 ppm, marked by an arrow). When the triplet is downfield of the isotropic chemical shift, the coupling is positive, and if it is shifted upfield, the coupling is negative.*

tensor is well known and it is uniaxial, making calculations easier. The dipolar coupling can be accurately determined from the splitting in the NMR spectrum, without need for referencing. In our recent peptide studies we have therefore used various CF_3-labeled amino acids, which will be described below in Section 18.3.2.

The CF_3-group of these amino acids is rigidly attached to the peptide backbone and reflects the overall peptide orientation. Due to fast rotation of the CF_3-group, the three pairwise dipolar interactions among the three fluorine nuclei are all averaged in the same way and give rise to an average dipolar tensor, which is oriented along the C_α–C_β-bond vector. From the measured dipolar coupling, the orientation of the tensor can be determined, which fixes the orientation of the bond in the magnetic field [49]. Normally, only the absolute value of the dipolar coupling can be determined from the spectrum; however, for a CF_3-group the chemical shift change with orientation is of a magnitude similar to that of the change in dipolar coupling. The sign of the splitting can thus be determined directly from the position of the chemical shift [49], as described in Figure 18.4.

18.2.3 Structure and Orientation Calculations

Using a number of orientational constraints from NMR measurements, the orientation of the peptide in the membrane can be determined, assuming that the structure of the peptide is known. The peptides we have studied mostly have well defined structures, such as α-helices or rigid cyclic conformations. The validity of the proposed peptide structure as it interacts with the membrane can also be tested using the fit of the data [45, 49]. For example, the data can be fitted to different helical models, such as α- and 3_{10}-helix: a good fit can only occur for one structure and will thus give an indication that this is the actual structure inside the membrane.

In principle, the peptide orientation is fully defined by the tilt and azimuthal angles described in Figure 18.3. However, peptides interacting with a membrane are mobile, and to account for this motion a simplified order parameter, S_{mol}, is introduced, which has the effect of scaling all the calculated splittings by a factor between 0 and 1 [4, 23, 44–47, 49–51]. $S_{mol} = 0$ would correspond to complete isotropic averaging, where all the orientational information would be lost, and $S_{mol} = 1$ corresponds to a completely immobile peptide. The peptides in our studies usually have S_{mol} values between 0.6 and 0.8. This gives information about the mobility of the peptides and can also be used to estimate the size of aggregates. More elaborate motional models have recently been investigated (E. Strandberg *et al.*, submitted), but have shown that the calculated values of tilt and azimuthal angles for the systems investigated were virtually identical to those obtained using the more simple approach described here. For our peptides, we thus determine the tilt angle, the azimuthal angle, and S_{mol}. In principle, three constraints are needed to determine these three parameters, each constraint being obtained from a ^{19}F NMR measurement on a singly ^{19}F-labeled peptide. In practice, at least four ^{19}F-labeled peptides are used in order to get a more reliable result, to make sure all the data are consistent, and to rule out any possible structural perturbations due to introduction of ^{19}F-labeled amino acids.

After measuring several local dipolar splittings, these data are used to calculate the global orientation of the full peptide. From the known structure of the peptide, theoretical curves show which dipolar couplings are expected for different labeled positions, depending on tilt angle τ, azimuthal angle ρ, and S_{mol}. The calculated values of splittings are then compared with the experimentally obtained values, and the root mean square deviation (RMSD) is calculated.

An example of how the procedure works is shown in Figure 18.5. The peptide in this example is MSI-103, which has been labeled with ^{19}F at the four positions indicated in Figure 18.5a. Four different peptides were synthesized where a single amino acid (Ala-7, Ile-9, Ala-10, or Ile-13) was replaced with CF_3-Phg. For each label a ^{19}F NMR spectrum is recorded (see Figure 18.5a); in this example, spectra are from MSI-103 in DMPC (1,2-dimyristoyl-*sn*-glycero-3-phosphocholine) at a peptide-to-lipid molar ratio (P/L) = 1 : 400,

→

Figure 18.5 *(a) The peptide MSI-103 was labeled with CF_3-Phg at four positions, marked in lighter gray. For each label a ^{19}F NMR spectrum was recorded in DMPC at P/L = 1 : 400. From these spectra the dipolar couplings were measured, giving the values shown next to each spectrum. (b) The measured dipolar couplings are compared with theoretical curves for different orientations of the peptide. The best-fit curve is here shown together with experimental data for the different labeled positions (filled squares). (c) The RMSD plot shows the root mean square deviation between experimental and calculated splittings, for all possible combinations of tilt and azimuthal angles. In this case, the best-fit tilt angle (τ) is 101° and the azimuthal angle (ρ) is 130°. (d) Side view of the peptide in the membrane, which is represented by a gray box. The tilt angle defines the angle between the peptide long axis and the membrane normal. (e) View of the peptide along the helical axis. The azimuthal angle defines how much the peptide is rotated around its axis, with the starting point defined as the vector from the helical axis to C_α of residue Lys-12 being parallel with the bilayer surface.*

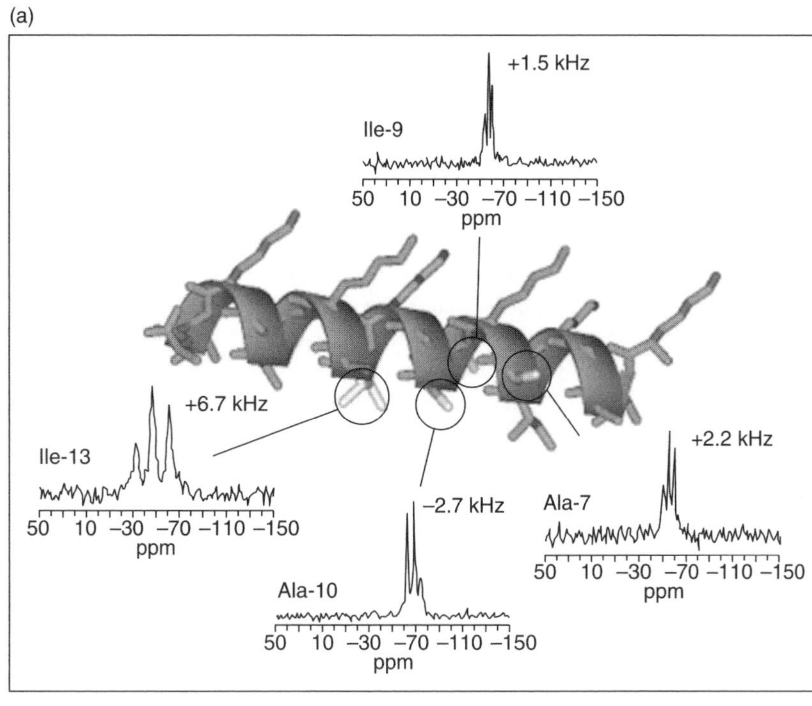
(a)
+1.5 kHz
Ile-9
50 10 −30 −70 −110 −150
ppm
+6.7 kHz
Ile-13
50 10 −30 −70 −110 −150
ppm
−2.7 kHz
Ala-10
50 10 −30 −70 −110 −150
ppm
+2.2 kHz
Ala-7
50 10 −30 −70 −110 −150
ppm

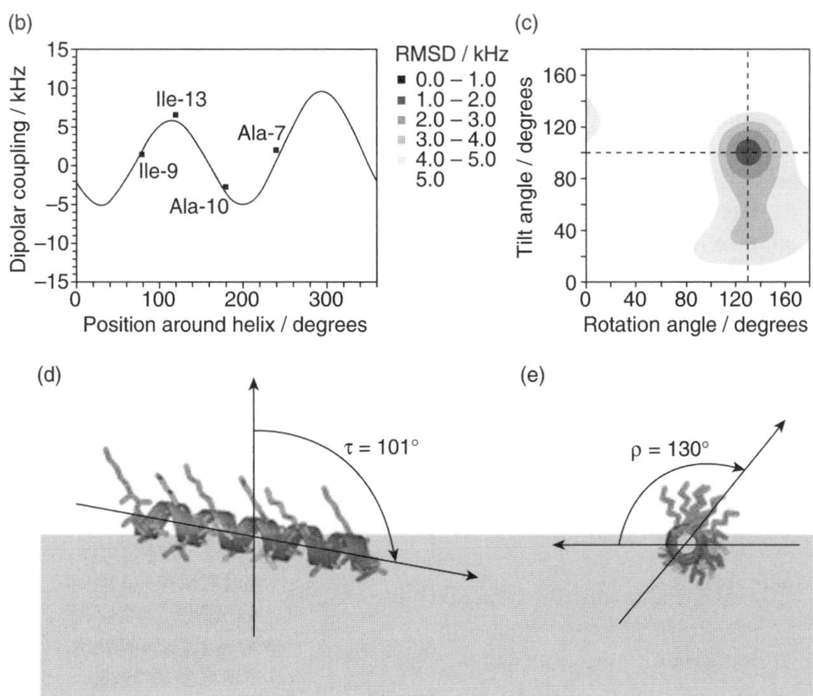
(b)
Dipolar coupling / kHz
15
10
5
0
−5
−10
−15
Ile-13
Ala-7
Ile-9
Ala-10
0 100 200 300
Position around helix / degrees
RMSD / kHz
0.0 – 1.0
1.0 – 2.0
2.0 – 3.0
3.0 – 4.0
4.0 – 5.0
5.0
(c)
Tilt angle / degrees
160
120
80
40
0
0 40 80 120 160
Rotation angle / degrees
(d)
τ = 101°
(e)
ρ = 130°

and dipolar couplings are measured. Each coupling gives information about the orientation of the respective C_α–C_β-bond vector of the labeled residue with respect to the magnetic field. The measured couplings are then used to determine the best-fit orientation of the whole peptide. The experimentally observed splittings (filled squares in Figure 18.5b) are then compared with theoretical splittings, calculated for different peptides orientations. The *x*-axis in Figure 18.5b shows the angle of the label as in a helical wheel projection, with the first amino acid at 0°, the second amino acid at 100° and so on; angles larger than 360° are projected back into the 0–360° region. A change in azimuthal angle will shift the dipolar curve along the *x*-axis of the plots, keeping the same shape, while a change in tilt angle will change the amplitude and shape of the curve. Figure 18.5b shows the best-fit curve of calculated splittings, which corresponds to the darkest area in the RMSD plot in Figure 18.5c. In this plot, for each possible combination of τ and ρ, the RMSD between experimental and calculated splittings is shown in a gray-scale code. In this case, the best-fit tilt angle (τ) is 101°, the azimuthal angle (ρ) is 130°, and the corresponding orientation of the peptide in the membrane is illustrated in Figures 18.5d and 18.5e.

For a given secondary structure of the peptide, the shape of the dipolar curve is characteristic of the peptide tilt. Figure 18.6 shows the theoretical dipolar curves for a surface-bound S-state ($\tau \approx 90°$), an inserted I-state ($\tau \approx 0°$), and a tilted T-state ($0° \leq \tau \leq 90°$) orientation of a helical peptide. The positions in the τ/ρ-plot corresponding to the different states are indicated. (For exact values of τ and ρ used in the calculations, see the figure legend.) It can be noted that the curve in Figure 18.6a is very similar to the curve of Figure 18.5b, both of which correspond to an S-state orientation of the peptide.

Additional dynamic information can be obtained from oriented samples. The lipid membranes can be oriented with the bilayer normal either parallel to the magnetic field (0° tilt) or perpendicular to the magnetic field (90° tilt). If the peptides rotate quickly around the bilayer normal, then, according to simple theory of motional averaging, the measured splittings at 90° tilt should be −1/2 times the splitting at 0° tilt. Figure 18.4 shows such ^{19}F NMR-spectra of a peptide with fast rotation. On the other hand, if the peptide does not rotate, then the splitting at 0° tilt will be the same, but at 90° tilt a superposition of different orientations around the membrane normal will lead to a more complex broad lineshape looking like a powder pattern.

If the peptide is not oriented in the membrane but forms large immobilized aggregates, this is easily seen in the NMR spectrum. In this case, all different orientations of the peptide are present in the sample and a very broad lineshape is seen, as for a peptide powder. When this is the case, no orientation of the peptide can be determined. Looking at the NMR spectral lineshape is a useful method for investigating the aggregation behavior of peptides interacting with membranes, which may be intimately related to the function or malfunction of the peptide.

18.2.4 ^{19}F NMR Experimental Considerations

To obtain orientational constraints, peptide/lipid samples for ^{19}F NMR studies are usually prepared with lipid bilayers that are macroscopically oriented on glass plates [52–54]. Lipids and peptides in appropriate amounts are co-dissolved in organic solvent and spread onto thin glass plates. After removal of the solvent, the plates are stacked and placed in a

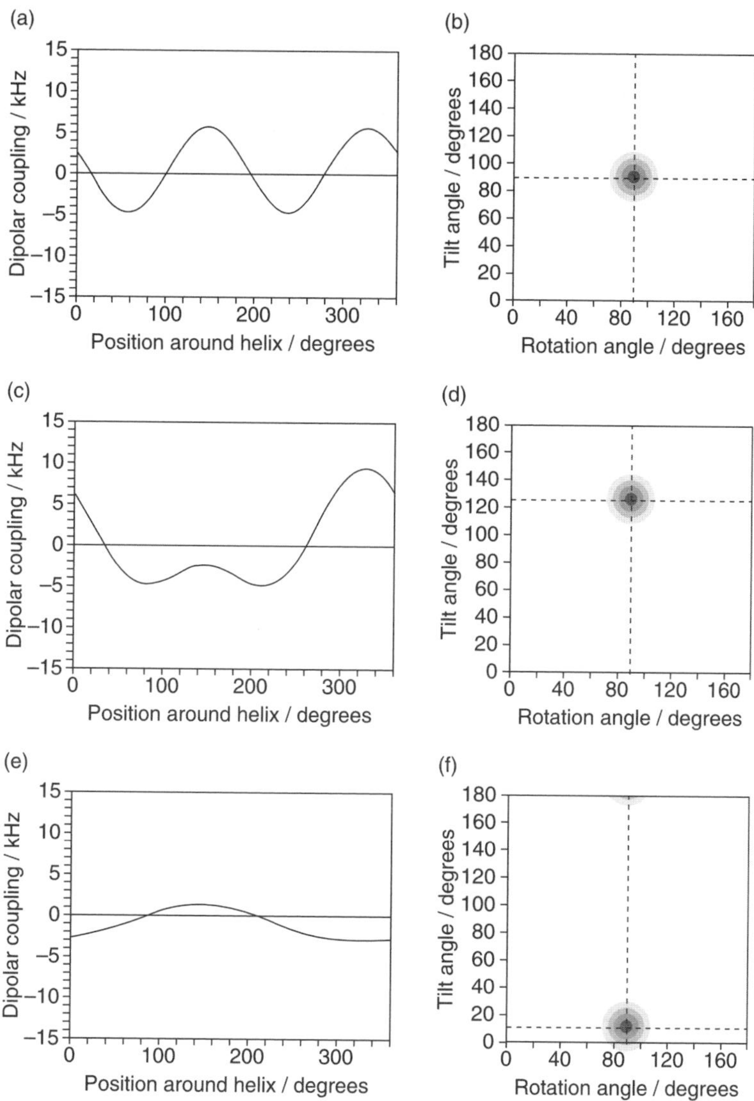

Figure 18.6 *Dipolar wave curves calculated for different orientational states of α-helical peptides (a, c, e), showing the theoretical dipolar splittings for different positions around the helical axis. (b, d, f) The orientation of the peptide is determined from a plot of the tilt angle τ versus the azimuthal angle ρ, where each point in the τ/ρ-plot corresponds to a specific orientation. (a) Helical curve corresponding to an S-state orientation with τ = 90°, ρ = 90°; (b) the position of this S-state orientation in the τ/ρ-plot is marked by the center of the concentric circles. (c) Helical curve for a tilted T-state orientation with τ = 125°, ρ = 90°; (d) the position of this T-state orientation in the τ/ρ-plot. (e) Helical curve for an I-state orientation with τ = 10°, ρ = 90°; (f) the position of this I-state orientation in the τ/ρ-plot.*

hydration chamber to take up water and equilibrate. The hydrated plates are then stacked, wrapped and stored at −20 °C until use. The quality of the peptide/lipid sample is checked with ^{31}P NMR, which gives a signal from the phospholipid head groups from which the degree of orientation can be estimated, as illustrated in Figure 18.7. ^{31}P NMR also gives information about the lipid phase behavior and can be used to confirm that the sample is in a membrane-like lamellar phase. ^{19}F NMR measurements are then conducted with a ^{1}H-decoupled single-pulse ^{19}F NMR experiment in an NMR probe in which the orientation of the glass plates can be varied. Usually samples are measured with the bilayer normal oriented parallel to the magnetic field, as shown in Figure 18.4 for the peptide MSI-103 labeled with CF_3-Phg. A second spectrum can then be acquired with a perpendicular sample alignment in order to detect the occurrence of long-axis rotation around the membrane normal. Samples can also be prepared as unoriented, multilamellar vesicles (MLV) by co-dissolving peptides and lipids, and hydrating them after removing the solvent. If peptides are averaged by long-axial rotation about the bilayer normal, the same orientational information can be obtained from MLV samples as from oriented samples [46, 55].

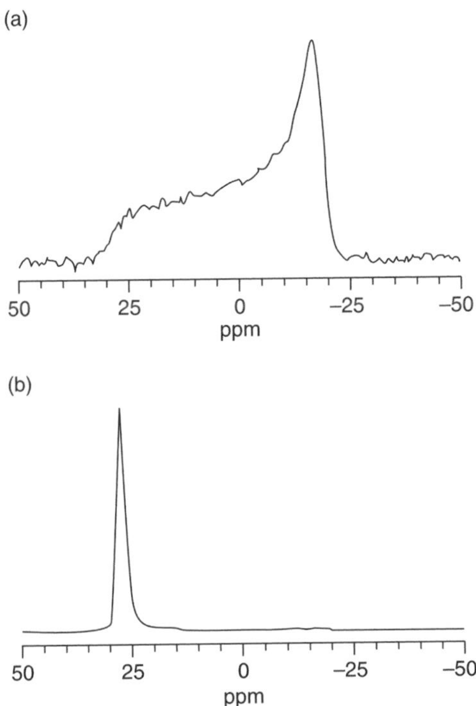

Figure 18.7 *^{31}P NMR spectra of MSI-103 in DMPC. (a) P/L = 1 : 200, nonoriented multilamellar vesicles sample. The broad signal shows the typical powder spectrum of a liquid crystalline lamellar lipid phase. (b) P/L = 1 : 20, macroscopically oriented sample. The peak around 28 ppm shows that the sample is well oriented with the lipid bilayer normal oriented parallel to the magnetic field.*

18.3 ^{19}F Labeling of Peptides

^{19}F-labeled peptides can be used to study ligand binding, unfolding, mobility, aggregation and orientation by ^{19}F NMR in solution or in their native membrane-bound environment, and they could also be used to study the effect of fluorination on thermal stability, solubility, selectivity, and activity of such biological systems. Highly fluorinated amino acids have been employed by various research groups to stabilize proteins for different applications (see reference 56 and references therein). Since this chapter focuses on investigating membrane-active ^{19}F-labeled peptides by SSNMR, we will restrict ourselves to specifically ^{19}F-labeled amino acids (see Figure 18.8), and their incorporation into peptides at a desired position. We will then discuss the suitability of various labels for measuring peptide orientation and conformation, mobility, and assembly, and intra- and intermolecular distances between two fluorine labels in lipid bilayers.

18.3.1 Labeling Strategies

Incorporation of fluorine into peptides and proteins is usually achieved through biosynthetic routes, chemical synthesis, or a combination of these two methods. Each method has its own advantages and shortcomings, but each requires a ^{19}F-labeled amino acid. The most commonly used ones are commercially available analogues of aromatic amino acids, such as tryptophan, phenylalanine, tyrosine, and phenylglycine. Aliphatic ^{19}F-labeled amino acids are not commonly available and usually have to be synthesized. The synthesis of most fluorinated amino acids is described in detail in the literature [1, 9, 10]. For structure analysis of peptides and proteins, it is important that (i) the fluorine label is rigidly attached to the peptide backbone, (ii) the label does not alter the structure or function of the peptide, (iii) the extent of fluorination is restricted to avoid multiple signals, and (iv)

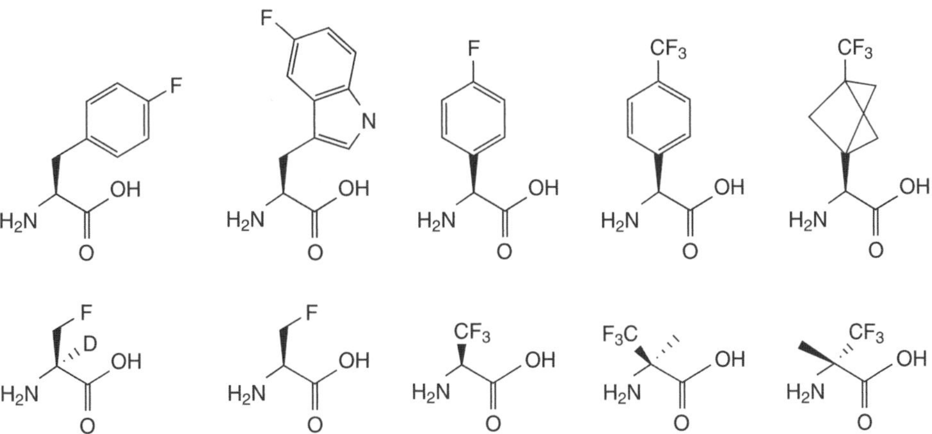

Figure 18.8 *^{19}F-labeled amino acids. Top row: 4F-Phe, 5F-Trp, 4F-Phg, CF_3-Phg, CF_3-Bpg. Bottom row: 2D-3F-Ala, 3F-Ala, F_3-Ala, CF_3-Ala (R-form), CF_3-Ala (S-form).*

the label is stable and not prone to chemical side-reactions such as HF-elimination. Figure 18.8 shows various commonly used ^{19}F-labeled amino acids.

18.3.1.1 Biosynthetic Labeling

Biosynthetic incorporation of fluorinated amino acids is a convenient way to obtain ^{19}F-labeled peptides or proteins. With the advancement in biotechnological protocols using various bacterial and yeast machineries, it has been possible to produce proteins of different origin quickly and economically. This process requires an auxotrophic strain of bacteria for the desired ^{19}F-amino acid, which is added to the growth medium. Alternatively, glyphosate can be used to inhibit the metabolic synthesis of all aromatic amino acids in bacterial culture so that the desired ^{19}F-labeled amino acid along with the other aromatic ones have to be externally supplemented. Using such approaches, it has been possible to produce desired sequences with fluorotryptophan [17, 18], fluorotyrosines [57], fluorophenylalanine [58, 59], trifluoroisoleucine [60], and trifluoroleucine [15].

Biosynthetic incorporation has several limitations: (i) only fluorinated analogues of natural amino acids may be used, provided they are nontoxic to the culture; (ii) site-selective labeling is not possible; (iii) relatively low uptake of ^{19}F-labeled amino acids results in lower yield of the expressed peptide or protein; and (iv) some proteins might still contain the natural amino acid that is devoid of fluorine. All these limitations can be avoided by using chemical peptide synthesis, as described below.

18.3.1.2 Solid-Phase Peptide Synthesis

Using SPPS protocols, ^{19}F-labeled amino acids can be readily incorporated into membrane-active peptides at any desired site. Standard Fmoc SPPS protocols are detailed in the literature [61]. Chemical synthesis has an intrinsic limitation in peptide chain length, depending on the sequence; this limits the synthesis of the fluorine-containing peptides to a maximum of 60 amino acids. Most membrane-active peptides have typically 10–40 amino acids and are well suited for chemical synthesis. Using synthetic ^{19}F-labeled peptides, we have investigated many different systems in model membranes by ^{19}F NMR; these are listed in Table 18.2, in Section 18.4. In practice, special care has to be taken at different stages of the synthetic protocol to avoid problems of fluorine contamination, elimination, or racemization, all of which are discussed in the following paragraphs.

Fluorine contamination. Trifluoroacetic acid (TFA), which is used to cleave the full-length peptide from the solid support and remove the side-chain protections in the final step of peptide synthesis, is often found to be tightly bound to the peptide. Usually up to a 5-fold molar excess of TFA is found to be associated with the peptide, and cannot be removed by repeated lyophilization [62]. Using an ion-exchange column, it is easy to replace the triflate ion with another suitable counteranion, but such purification steps are accompanied by extensive peptide loss. TFA is also universally used as an ion-pairing agent in HPLC purification of peptides. For hydrophobic peptides, many researchers use fluorinated alcohols, such as trifluoroethanol and hexafluoroisopropanol, to dissolve the crude peptide or even as a constituent of the mobile phase during HPLC purification. These fluorinated solvents often contaminate the "purified" peptide with substantial amounts of undesired fluorine, which interferes in the ^{19}F NMR spectra. For example, TFA is seen as an intense signal around −75 ppm, and this signal overlaps with signals of aromatic fluorine

Table 18.2 *Peptide sequences labeled with* ^{19}F*-labeled amino acids*

Peptide	Sequence[a]	Label	Reference
Antimicrobial peptides			
Gramicidin S	Cyclo-(PVOLf)$_2$	4F-Phg	62, 66, 80
		CF_3-Phg	79
PGLa	GMASKAGAIAGKIAKVALKAL-NH_2	4F-Phg	44, 62
		CF_3-Phg	44, 49, 84, 85
		CF_3-Bpg	68
MSI-103	(KIAGKIA)$_3$-NH_2	3F-Ala	71
		CF_3-Phg	45, 89
Cell-penetrating peptide			
MAP [110]	KLALKLALKALKAALKLA-NH_2	CF_3-Phg	Submitted
Fusogenic peptides			
B18	LGLLLRHLRHHSNLLANI	4F-Phg	50, 62
FP23	AVGIGALFLGFLGAAGSTMGARS-NH_2	CF_3-Phg	Submitted

[a] One-letter code. O = ornithine; f = D-isomer of Phe.

and CF_3-groups. It is thus essential to exclude any traces of fluorinated solvents and ion-pairing agents from the NMR sample. We have shown that TFA from the cleavage step of peptide synthesis can be efficiently removed by employing HCl as an ion-pairing agent in HLPC purification protocols [62]; therefore, in a single step, it is possible to obtain ^{19}F-labeled peptide that is free of any undesired fluorine background.

Fluorine elimination. The high electronegativity of fluorine makes it prone to certain side reactions that result in HF-elimination. These side reactions are very uncommon for a fluorine atom attached to an aromatic ring; however, the presence of labile hydrogen in the vicinity of an aliphatic fluorine substituent often leads to spontaneous HF-elimination, which results either in an olefinic bond or a cyclized structure. Furthermore, elimination reactions are accelerated in the presence of base which is typically used in amino acid coupling procedures. 3-Fluoroalanine (3F-Ala), or 3,3,3-trifluoroalanine (F_3-Ala) in particular eliminate HF almost invariably when incorporated via the usual coupling protocols employing basic conditions. It has been possible to minimize the loss of fluorine by using 2-deutero-3-fluoroalanine (2D-3F-Ala) and by coupling at low temperatures using base-free coupling strategies [63].

Racemization. The high electronegativity of fluorine also makes many ^{19}F-labeled amino acids susceptible to base-induced racemization. In the case of aromatic side-chains, this effect is even more pronounced, due to the electron-withdrawing effect of the ring. Under base-catalyzed coupling procedures, fluorophenylglycine undergoes complete racemization, leading to epimeric peptides (see Figure 18.9). It has nevertheless been possible to couple this enantiomerically pure amino acid using base-free protocols employing diisopropylcarbodiimide/hydroxybenzotriazol to avoid racemization [110]. Most ^{19}F-labeled amino acids are in any case commercially available only as racemic mixtures; hence the use of standard synthetic protocols involving basic conditions is acceptable and pragmatic. In this case, the resulting epimeric peptides have to be separated on an HPLC column and isolated in high purity. In our investigations, we often study both epimeric

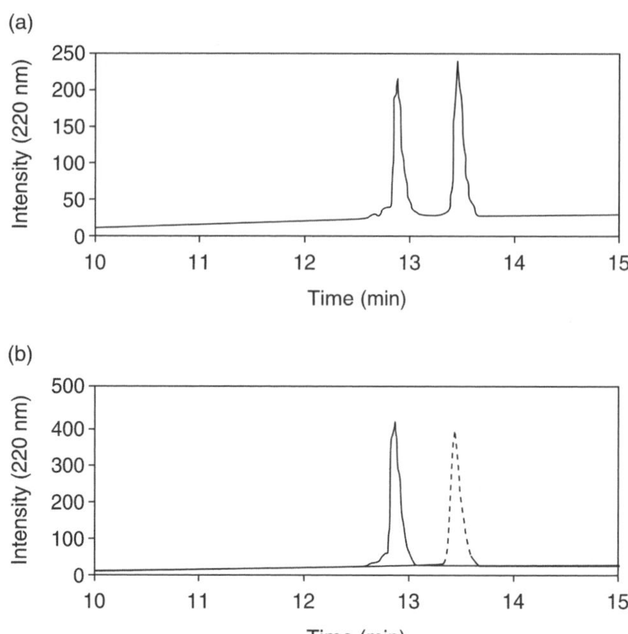

Figure 18.9 *(a) HPLC chromatogram showing the presence of two epimeric peptides, resulting from incorporation of a racemic mixture of CF_3-Phg, and (b) analytical HPLC chromatograms (solid and dashed lines) after preparative separation of the epimers.*

peptides separately after identifying the chirality of the ^{19}F-labeled amino acid, in order to obtain additional structural information [44]. To identify which epimer contains the D- and L-forms of the ^{19}F-labeled amino acid, an aliquot of these separated peptides should be hydrolyzed to their constituent amino acids and derivatized using Marfey's reagent to identify the D- and L-forms [62]. Alternatively, the racemic mixture may be acetylated to obtain the *N*-acetyl amino acids, which can be subsequently enzymatically deacetylated. Porcine kidney acylase 1 selectively deacetylates the L-amino acid, leaving the D-enantiomer unchanged [64, 65].

Several different ^{19}F-labels with a rigid connection to the peptide backbone, which qualify for ^{19}F NMR according to the criteria described above, were used in our investigations and will now be described in more detail. These amino acids were incorporated into various membrane-active peptides as NMR labels, thereby providing structural information that will be described in Section 18.4.

18.3.2 ^{19}F-labeled Amino Acids

Various ^{19}F-labeled amino acids have been used for SSNMR structural studies of membrane-active peptides. Their chemical structures are shown in Figure 18.8, including the aromatic 4-fluorophenylalanine, 5-fluorotryptophan, 4-fluorophenylglycine, and 4-trifluoromethylphenylglycine. Apart from the cyclic side chain of 3-trifluoromethylbicyclopentylglycine, the other aliphatic ^{19}F-labeled amino acids can be considered as alanine derivatives, namely 2-deutero-3-fluoroalanine, 3-fluoroalanine, 3,3,3-trifluoroalanine, (*R*)-3-trifluoromethylalanine, and (*S*)-3-trifluoromethylalanine. We will now describe these ^{19}F-labeled amino acids in more detail.

18.3.2.1 4-Fluorophenylglycine

4-Fluorophenylglycine (4F-Phg) has a fluorine atom attached at the *para*-position of the phenyl ring, which in turn is directly attached to a glycine skeleton. This results in an amino acid with fluorine rigidly attached to the peptide backbone. In our earlier studies, 4F-Phg was extensively used as an NMR label, since it provides information about the orientation and dynamics of peptides in the membrane [50, 62, 66]. Some advantages of this label are that both enantiomers are commercially available, there is no fluorine elimination, and the peptide concentration can easily determined by UV-Vis spectroscopy. Using standard synthetic protocols, extensive racemization of 4F-Phg is observed, but the epimeric peptides can be isolated and characterized, and both peptides bearing L- and D-forms of 4F-Phg, respectively, can then be used to obtain constraints for NMR structure analysis [62]. 4F-Phg is rather useful to determine ^{19}F-^{19}F distances from their homonuclear dipolar coupling [66]. For orientational analysis, on the other hand, this label is less useful because the side-chain C_α–C_β torsion angle of the phenyl ring is usually not known. This means that the relative orientation of the nonsymmetric CSA tensor of the ^{19}F-substituent relative to the peptide backbone is not known. When analyzing the anisotropic chemical shift of 4F-Phg as an orientational constraint, it is therefore important to determine the torsion angle of the phenyl group, which has been shown to have different conformational preferences for α-helical or β-sheet peptides [50]. Finally, it is also important to carefully reference the spectral scale to obtain an accurate reading of the single ^{19}F-resonance in the spectrum.

18.3.2.2 4-Trifluoromethylphenylglycine

4-Trifluoromethylphenylglycine (CF_3-Phg) has a similar skeleton to 4F-Phg except that it contains a CF_3-group at the *para*-position of the phenyl ring. The presence of a rotationally averaged CF_3-group removes the ambiguity resulting from the torsion angle of the phenyl ring. Therefore, the effective CSA tensor is axially symmetric along the C_α–C_β direction, which allows us to use this label to obtain orientational constraints. As an NMR label, CF_3-Phg offers many advantages over 4F-Phg. It yields not only the chemical shift but also the intra-CF_3 dipolar coupling, which gives the same information about the orientation of a peptide, free of referencing errors. For the CF_3-label it is also easy to measure the sign of the dipolar coupling by observing the simultaneous chemical shift relative to the isotropic position. This removes the ambiguity due to the unknown sign of the splitting, and results in rather precise and accurate structural constraints compared to other dipolar coupling analyses. This amino acid is commercially available, but only as a racemic mixture. Since CF_3-Phg is sensitive to racemization, we have almost always used the racemic mixture in peptide synthesis, and purified the resulting epimers using HPLC (see Figure 18.9) followed by identification with Marfey's derivatization [49, 62]. Some of the peptides containing CF_3-D-Phg have been found to show interesting effects, such as suppression of aggregation via β-sheet formation [110].

18.3.2.3 3-Trifluoromethylbicyclopentylglycine

Once the advantages of using a CF_3-group over a monofluoro-group, and the disadvantage of an aromatic system in allowing racemization became clear, 3-trifluoromethylbicyclopentylglycine (CF_3-Bpg) was specifically designed as an NMR label for peptides [67, 68]. CF_3-Bpg contains a bicyclo[1.1.1]pentane skeleton, bearing a CF_3-group that is collinear with C_α–C_β and is devoid of any aromatic ring, which substantially reduces the inductive effect that makes the proton at C_α acidic. CF_3-Bpg is easily incorporated into peptides without racemization, and the absence of a labile hydrogen atom adjacent to a fluorine atom also ensures that there is no HF-elimination. As described for CF_3-Phg, the dipolar coupling (and its sign) of the CF_3-group is easily measured, and gives well-defined orientational constraints. These qualities make CF_3-Bpg an ideal label for investigating peptides with solid-state ^{19}F NMR, and it is found to be essentially nonperturbing when replacing aliphatic hydrophobic amino acids such as leucine, valine, alanine, or isoleucine [67].

18.3.2.4 Fluoroalanines

Fluoroalanine derivatives (see Figure 18.8, bottom row, 2D-3F-Ala, 3F-Ala, F_3-Ala) are well suited to replace alanine and other aliphatic hydrophobic amino acids. The synthesis of most of the fluoroalanine derivatives is well known [69, 70], and F_3-Ala is commercially available as a racemic mixture. Considering their size and steric perturbations, fluoroalanines would be the ideal choice as NMR labels, and they have previously been used for ^{19}F REDOR (rotational echo double resonance) measurements [71]. One of the main disadvantages is that fluoroalanines are extremely prone to HF-elimination, resulting in partial or complete loss of fluorine, and introducing a dehydroalanine unit in the peptide. Using 2D-3F-Ala and performing the coupling at low temperature, it is possible to minimize "DF-elimination" due to the kinetic isotope effect. Yet the presence of an acidic H in the close vicinity, even in the form of a hydroxyl group, again leads to HF-elimination and formation of cyclic structures in the peptide backbone. In peptides labeled with F_3-Ala, there is an additional problem resulting from the reduced nucleophilicity of the free amine. Incorporation of F_3-Ala into a growing peptide chain is relatively easy, but it severely reduces the nucleophilicity of the free amine, making the coupling of the subsequent amino acid extremely slow and inefficient. In biosynthetic applications, certain analogues of fluoroalanine are found to be toxic to bacteria or are poorly taken up by the bacterial machinery.

18.3.2.5 2-Trifluoromethylalanine

Synthesis of 2-trifluoromethylalanine (CF_3-Ala), or trifluoroaminoisobutyric acid, has been reported previously [72]. In contrast to fluoroalanine, an advantage of using CF_3-Ala analogues is that it is devoid of the hydrogen atom at the α-position, which means that HF-elimination is no longer a problem. Since the members of a large family of membrane-active peptides, the so-called "peptaibols," contain aminoisobutyric acid units, CF_3-Ala serves as an ideal label for investigating these peptides using ^{19}F NMR. As in the case with F_3-Ala, the incorporation of CF_3-Ala into a growing peptide chain also makes coupling of the subsequent amino acid slow and inefficient. To avoid this problem, tripeptides were first synthesized in which CF_3-Ala is flanked by two other amino acid residues. The resulting epimeric peptides could then be separated and their stereochemistry identified, and the individual tripeptides could subsequently be used as a building block in the synthesis of the full peptide sequence. This approach was used to elucidate the structure of the peptaibol alamethicin [73].

18.3.2.6 Fluorotryptophan

Fluorotryptophan (F-Trp) was one of the first labels used for investigation of peptides and proteins by ^{19}F NMR. Both 5- and 6-fluorotryptophan are commercially available. In this case, orientational constraints and distance measurements are very useful for obtaining information about the side-chain conformation [74, 75]. The tryptophan side-chain is often found to anchor transmembrane helices in the lipid head-group region of the membrane and it often plays a functional role in peptides that form channels in the membrane, such as gramicidin A [74, 76] or the influenza virus peptide M2 [75]. Fluorotryptophans are easily incorporated into peptides and proteins using biosynthetic and SPPS protocols.

18.4 Applications

In this section we provide an overview of structural results from our previous ^{19}F NMR analysis of a number of different membrane-active peptides in lipid bilayers. The results are grouped according to the different biological functions attributed to these peptides as discussed in Section 18.1.2. All the peptides are listed in Table 18.2, together with the ^{19}F-labeled amino acids used and with the original references. For all ^{19}F-labeled peptide analogues the secondary structure was ascertained using solution CD and the effect of fluorine substitution was validated by performing appropriate functional tests on the labeled peptide. Only those ^{19}F-labeled analogues that were found to retain their secondary structure and biological activity were used for structure elucidation.

18.4.1 Antimicrobial Peptides

Membrane-active antimicrobial peptides (AMPs), or host-defense peptides, kill microorganisms by permeabilizing their membrane. They often form amphipathic structures upon binding to lipid membranes. At low peptide concentrations they are normally in a monomeric surface-bound S-state, and at higher concentrations they may self-assemble and insert into the bilayer in a functionally active T- or I-state (see Figures 18.1 and 18.3). In our previous ^{19}F NMR investigations we have compared three such AMPs, which are described below.

18.4.1.1 Gramicidin S

Gramicidin S (GS) is an amphiphilic, cyclic β-sheeted decapeptide obtained from *Bacillus brevis*. Its polar face consists of two ornithines and the hydrophobic face is comprised of two valines and two leucines (see Table 18.2) [77, 78]. The symmetric peptide was labeled by simultaneous replacement of both valines or both leucines with 4F-Phg [62, 66] and later with CF_3-Phg (79). These ^{19}F-labels had little or no effect on the secondary structure of the peptides, which were found to be biologically as active as the native peptide [62]. Since the analogues were symmetrically labeled with two substituents, it was possible to measure their internuclear ^{19}F–^{19}F distances, which were later used for elucidation of the GS structure in lipid membranes [66, 79]. In DMPC membranes at a low peptide-to-lipid molar ratio (P/L) of 1 : 80 (mol/mol), GS showed a single resonance signal, which corresponded to the peptide lying on the membrane surface, representative of an S-state, as is biophysically expected [66]. The hydrophobic groups of valine (or 4F-Phg) or leucines (or 4F-Phg) were embedded in the nonpolar lipid bilayer, and the polar residues (ornithines) pointed to the polar head group region of the membrane. The high mobility of the peptide ($S_{mol} = 0.3$) suggests that GS is monomeric and highly mobile within the lipid bilayer; the presence of a single resonance signal is suggestive of fast rotation of the peptide along its symmetry axis, which is parallel to the membrane normal [66]. As the concentration of GS increased (P/L = 1 : 20), a second signal appeared in the oriented ^{19}F NMR spectrum, indicating a realignment of the peptide. Data analysis showed that the peptide had flipped nearly 90° into the membrane and had virtually no mobility ($S_{mol} = 1.0$). This suggests that several peptides have assembled to form a transmembrane pore [80]. Indeed, the possibility of forming hydrogen bonds along the edge of the β-stranded peptide supports the notion that it may have oligomerized as a β-barrel structure.

18.4.1.2 PGLa

PGLa is a 23-residue peptide, belonging to the magainin family of AMPs found in the skin of the African frog *Xenopus laevis* (see Table 18.2) [81, 82]. It is unstructured in aqueous solution, but forms an amphipathic α-helix when bound to lipid vesicles [49, 83]. The peptide was labeled by 4F-Phg or CF_3-Phg at single alanine or isoleucine positions in the sequence [49, 62]. Using four selective CF_3-Phg labels, the helix orientation could be determined in DMPC. At low peptide concentration (P/L = 1 : 200) an S-state was found [49], as previous ^{15}N NMR results have also shown [83]. A concentration-dependent reorientation was observed, toward a tilted state at P/L = 1 : 50 [44]. Using ^{19}F NMR it was possible to characterize this T-state for the first time. The same T-state was also observed under a wide variety of conditions, such as different hydration levels, different peptide concentrations, and in the presence of negatively charged DMPG lipids (47). This T-state was confirmed by nonperturbing ^2H-labels and is attributed to peptide dimerization [46].

Remarkably, an extended analysis of the helix alignment as a function of temperature showed that it is highly dependent on the lipid phase state. Namely, the S-state was found at very high temperatures, the T-state at lower temperatures above the lipid chain melting transition, and the transmembrane I-state at temperatures where the lipids were in the gel phase. Finally, some disordered and presumably aggregated peptides were found at very low temperatures in crystalline lipids [84].

More recently PGLa was also labeled with CF_3-Bpg at the same positions as was previously labeled with CF_3-Phg. This study demonstrated that CF_3-splittings from CF_3-Bpg are just as readily analyzed, and that this designer-made label did not perturb the conformation of the peptide [68]. CF_3-Bpg has many advantages regarding peptide synthesis and thus promises to be a suitable choice for future peptide studies [67].

One of the main advantages of ^{19}F as an NMR label is the high sensitivity. Signals are readily detected at low peptide concentrations down to P/L = 1 : 3000 [54] or when only small amounts of material are used. This has made it possible to observe ^{19}F-labeled PGLa in real bacterial or erythrocyte membranes and, as a consequence, to obtain for the first time orientational information about membrane-active peptides in biologically relevant membrane systems [85].

18.4.1.3 MSI-103

MSI-103 is a designer-made AMP based on the PGLa sequence [86, 87]. It has 21 amino acids in heptameric repeats (see Table 18.2). Its antimicrobial activity effect is higher than that of PGLa, while the hemolytic side-effects are similarly low [23, 86, 87]. MSI-103 was labeled with 3F-Ala in a single position, and spectral changes showed a change in alignment between P/L = 1 : 200 and 1 : 20 [71]. An extensive set of ^{19}F- and ^{13}C-labeled peptides were studied using REDOR to obtain intra- and intermolecular ^{13}C–^{19}F distances, which were combined with ^{13}C–^{31}P and ^{15}N–^{31}P distances between peptides and lipids. From these distance constraints, the peptide was proposed to form parallel dimers, and contacts between the peptides and the lipid head groups and acyl chains were indicated [71, 88]. These results are compatible with peptides forming toroidal pores, but the orientation of the peptide in the membrane was not determined.

More recently, MSI-103 was selectively labeled with CF_3-Phg at five different positions, and the helix orientation in DMPC bilayers was monitored using ^{19}F NMR over a series of peptide-to-lipid ratios from 1 : 800 to 1 : 20 [45, 89]. The ^{19}F NMR analysis showed the peptide to be in the S-state at concentrations up to P/L = 1 : 200. At 1 : 50 the dipolar splittings changed, indicating a change of orientation, and at 1 : 20 the peptide aggregated without any preferential orientation [45]. ^{2}H NMR on ^{2}H-labeled peptides in the same study showed that the peptide assumed a T-state between P/L = 1 : 50 and 1 : 20, with almost identical tilt and rotation angles as PGLa at the same concentration [45].

18.4.2 Cell-Penetrating Peptides

In the last decade, a variety of small cationic and often amphipathic peptides of different origins, including synthetic designer-made peptides, have been used as delivery vectors to transport different types of cargo into a cell without causing any leakage of the cellular membrane [24, 31–33]. Such peptides, commonly known as cell-penetrating peptides, cross the lipid bilayer via a mechanism that is still poorly understood. SSNMR offers an excellent tool for investigating the uptake mechanism of these peptides by observing not only the peptide but also its effect on the phospholipids membrane using ^{31}P NMR.

18.4.2.1 Model Amphipathic Peptide

The so-called "model amphipathic peptide" (MAP) was originally designed to form an amphipathic α-helix and was later found to be cell-penetrating [90, 91]. Four leucines were selectively replaced with CF_3-Phg, and ^{19}F NMR on these different analogues showed the same kind of concentration-dependent realignment of the helix from S- to T-state as was observed for the other helical peptides discussed above. Additionally, we could address the aggregation tendency by monitoring the immobilization and loss of alignment as a function of peptide concentration [110].

18.4.2.2 HIV-TAT

HIV-TAT is a 13-residue peptide derived from the HIV protein TAT. It is rich in arginine and therefore highly charged [92]. Although this peptide does not possess any hydrophobic side-chains that could be ^{19}F-labeled, it was chosen as an example to highlight the use of solid state ^{31}P NMR on phospholipids, which is suitably combined with ^{19}F NMR on ^{19}F-labeled peptides. Here we investigated the effect of unlabeled HIV-TAT on the phospholipid bilayer. It was observed from the isotropic ^{31}P NMR signals that HIV-TAT probably induces inverted micelles, which are short and assemble into bundles where the guanidinium group of arginines complexes with the phosphate groups [93].

18.4.3 Fusogenic Peptides

Fusogenic peptides work at the interface of two cells or vesicular compartments and are responsible for merging the two membranes, which leads to the mixing of their cytoplasmic contents (see Figure 18.2). Such a process is involved in fertilization and it constitutes an important step in viral infection and multiplication. We have investigated two fusogenic peptides with ^{19}F NMR, which are described below.

18.4.3.1 B18

A short 18-amino-acid sequence within the sea urchin fertilization protein bindin, the B18 peptide [94, 95], is involved in promoting fusion of oocyte and sperm. The histidine-rich B18 has also been shown *in vitro* to promote the fusion of DMPC vesicles in the presence of Zn^{2+} [96]. For most fusogenic peptides, conformational flexibility allows them to adopt different structures, depending on pH and membrane environment [97–99]. It has been shown that B18 adopts a helix–turn–helix structure. Various analogues were synthesized in which each hydrophobic residue was selectively replaced with 4F-Phg [50, 62, 100]. ^{19}F NMR of B18 in DMPC/DMPG bilayers at $P/L = 1:150$ showed that the peptide is monomeric and well folded. The C-terminal helix lies on the membrane surface with a tilt angle of about 90°, and the N-terminal helix has a tilt angle of about 50°. Such a model results in a peptide alignment that satisfies the expected orientation of the hydrophobic and hydrophilic residues. Under prolonged storage or at high peptide concentrations, B18 has been shown to form fibrils. Kinetic analysis of vesicle fusion by lipid-mixing assays has shown that the active state of the peptide is highly flexible and, therefore, independent of a single mutation involving a D- or L-enantiomer of 4F-Phg.

18.4.3.2 FP23

The HIV protein gp41 contains a relatively small 23-residue fusion peptide, FP23, which mediates the fusion of the virus membrane with the target cell [101, 102]. Different studies have shown that FP23 has a pronounced conformational plasticity and can change between α-helix and β-sheet, depending upon membrane composition and/or peptide concentration. Previously an α-helical, T-state-like structure [103–105], but also a β-sheeted assembly of FP23 have been described [106–108]. Five analogues of FP23, specifically labeled with CF_3-Phg, were synthesized and all analogues were found to retain their fusogenic activity [109]. At low peptide concentration (P/L = 1 : 300), the peptide was rather mobile, but at high concentrations the ^{19}F NMR spectrum became typical of a powder pattern, indicating that either the peptide was aggregated or it had formed higher-order oligomers that were disordered and immobile in the membrane. The ^{19}F NMR data did not fit the usual, known secondary structures, which indicates that an unusual peptide conformation with many β-turns might be the active species that is involved in the fusion process (D. Grasnick *et al.*, unpublished results).

18.5 Conclusions

Fluorine NMR is an extremely useful method for study of membrane-active peptides, giving information about conformation, orientation, mobility, self-assembly, and aggregation of membrane-active peptides when bound to a lipid membrane. For the purpose of ^{19}F NMR, ^{19}F-labeled amino acids are easy to obtain either commercially or synthetically. In most cases, the substitution of a single hydrophobic residue by a ^{19}F-labeled amino acid did not perturb the structure or function of the peptide. Problems associated with incorporation of ^{19}F-labeled amino acids, like HF-elimination, racemization, and slow coupling, can be easily overcome. In structural studies, labeling is most preferable with CF_3-groups that are rigidly attached to the peptide backbone, using CF_3-Phg or CF_3-Bpg, for example. ^{19}F has a higher sensitivity than traditional NMR labels, which reduces the amount of material and/or the measurement time needed. It is the only technique that can be used to investigate a wide concentration regime, ranging from P/L = 1 : 3000 to 1 : 10. Since concentration-dependent realignment of peptides is related to their biological function, this approach can contribute to better understanding of peptide function and thus make it possible to design peptides with improved activity.

Acknowledgments

We thank collaborators and former and present members of the group of Professor Anne S. Ulrich who have contributed to the work presented in this review. Special thanks are due to Professor Anne S. Ulrich and Rebecca Klady for critically reading the manuscript. This work was supported by the Deutsche Forschungsgesellschaft (HBFG, and CFN E1.2) and by Forschungszentrum Karlsruhe.

References

1. Kukhar, V. P. and Soloshonok, V. A. (1995) *Fluorine-Containing Amino Acids: Synthesis and Properties*, John Wiley & Sons. Ltd, Chichester, New York.

2. Soloshonok, V. A., Mikami, K., Yamazaki, T., *et al.* (eds.) (2007) *Current Fluoroorganic Chemistry: New Synthetic Directions, Technologies, Materials, and Biological Applications.* American Chemical Society, Washington, DC.
3. Grage, S. L., Salgado, J. B., Dürr, U. H. N., *et al.* (2001) Solid state ^{19}F-NMR of biomembranes, in *Perspectives on Solid State NMR in Biology* (eds. S. R. Kiihne and H. J. M. de Groot), Kluwer Academic Publishers, Dordrecht/Boston/London, pp. 83–91.
4. Strandberg, E. and Ulrich, A. S. (2004) NMR methods for studying membrane-active antimicrobial peptides. *Concepts in Magnetic Resonance Series A*, **23A**, 89–120.
5. Ulrich, A. S., Wadhwani, P., Dürr, U. H. N., *et al.* (2006) Solid-state ^{19}F-nuclear magnetic resonance analysis of membrane-active peptides, in *NMR Spectroscopy of Biological Solids* (ed. A. Ramamoorthy), CRC Press, Boca Raton, FL, pp. 215–236.
6. Ulrich, A. S. (2005) Solid state ^{19}F-NMR methods for studying biomembranes. *Progress in Nuclear Magnetic Resonance Spectroscopy*, **46**, 1–21.
7. Wadhwani, P., Tremouilhac, P., Strandberg, E., *et al.* (2007) Using fluorinated amino acids for structure analysis of membrane-active peptides by solid-state ^{19}F-NMR, in *Current Fluoroorganic Chemistry: New Synthetic Directions, Technologies, Materials, and Biological Applications* (eds. V. A. Soloshonok, K. Mikami, T. Yamazaki, *et al.*), American Chemical Society, Washington, D.C., pp. 431–446.
8. Levine-Pinto, H., Bouabdallah, B., Morgat, J. L., *et al.* (1981) Specific and direct fluorination of an histidine-containing peptide: thyroliberin. *Biochemical and Biophysical Research Communications*, **103**, 1121–1130.
9. Sutherland, A. and Willis, C. L. (2000) Synthesis of fluorinated amino acids. *Natural Product Reports*, **17**, 621–631.
10. Qiu, X. L., Meng, W. D. and Qing, F. L. (2004) Synthesis of fluorinated amino acids. *Tetrahedron*, **60**, 6711–6745.
11. Israelachvili, J. N., Mitchell, D. J. and Ninham, B. W. (1977) Theory of self-assembly of lipid bilayers and vesicles. *Biochimica et Biophysica Acta*, **470**, 185–201.
12. Jaeckel, C. and Koksch, B. (2005) Fluorine in peptide design and protein engineering. *European Journal of Organic Chemistry*, **2005**, 4483–4503.
13. Tang, Y., Ghirlanda, G., Petka, W. A., *et al.* (2001) Fluorinated coiled-coil proteins prepared in vivo display enhanced thermal and chemical stability. *Angewandte Chemie International Edition*, **40**, 1494–1496.
14. Bilgicer, B., Fichera, A. and Kumar, K. (2001) A coiled coil with a fluorous core. *Journal of the American Chemical Society*, **123**, 4393–4399.
15. Tang, Y., Ghirlanda, G., Vaidehi, N., *et al.* (2001) Stabilization of coiled-coil peptide domains by introduction of trifluoroleucine. *Biochemistry*, **40**, 2790–2796.
16. Tang, Y. and Tirrell, D. A. (2001) Biosynthesis of a highly stable coiled-coil protein containing hexafluoroleucine in an engineered bacterial host. *Journal of the American Chemical Society*, **123**, 11089–11090.
17. Danielson, M. A. and Falke, J. J. (1996) Use of ^{19}F NMR to probe protein structure and conformational changes. *Annual Review of Biophysics and Biomolecular Structures*, **25**, 163–195.
18. Gerig, J. T. (1994) Fluorine NMR of proteins. *Progress in Nuclear Magnetic Resonance Spectroscopy*, **26**, 293–370.
19. Sun, S.-Y., Pratt, E. A. and Ho, C. (1996) ^{19}F-labeled amino acids as structural and dynamic probes in membrane-associated proteins, in *Biomedical Frontiers of Fluorine Chemistry* (eds. I. Ojima, J. R. McCarthy and J. T. Welch), American Chemical Society, Washington, D.C., pp. 296–310.
20. Salopek-Sondi, B. and Luck, L. A. (2002) ^{19}F NMR study of the leucine-specific binding protein of *Escherichia coli*: mutagenesis and assignment of the 5-fluorotryptophan-labeled residues. *Protein Engineering*, **15**, 855–859.

21. Luck, L. A. and Johnson, C. (2000) Fluorescence and ^{19}F NMR evidence that phenylalanine, 3-L-fluorophenylalanine and 4-L-fluorophenylalanine bind to the L-leucine specific receptor of *Escherichia coli*. *Protein Science*, **9**, 2573–2576.
22. Luck, L. A., Vance, J. E., O'Connell, T. M. and London, R. E. (1996) ^{19}F NMR relaxation studies on 5-fluorotryptophan- and tetradeutero-5-fluorotryptophan-labeled *E. coli* glucose/ galactose receptor. *Journal of Biomolecular NMR*, **7**, 261–272.
23. Strandberg, E., Tiltak, D., Ieronimo, M., *et al.* (2007) Influence of C-terminal amidation on the antimicrobial and hemolytic activities of cationic α-helical peptides. *Pure and Applied Chemistry*, **79**, 717–728.
24. Langel, U. (2002) *Cell-Penetrating Peptides: Processes and Applications*, CRC Press, Boca Raton, FL.
25. Boman, H. G. (2003) Antibacterial peptides: basic facts and emerging concepts. *Journal of Internal Medicine*, **254**, 197–215.
26. Reddy, K. V., Yedery, R. D. and Aranha, C. (2004) Antimicrobial peptides: premises and promises. *International Journal of Antimicrobial Agents*, **24**, 536–547.
27. Brogden, K. A. (2005) Antimicrobial peptides: pore formers or metabolic inhibitors in bacteria? Nature Reviews on *Microbiology*, **3**, 238–250.
28. Oren, Z. and Shai, Y. (1998) Mode of action of linear amphipathic α-helical antimicrobial peptides. *Biopolymers*, **47**, 451–463.
29. Huang, H. W. and Wu, Y. (1991) Lipid-alamethicin interactions influence alamethicin orientation. *Biophysical Journal*, **60**, 1079–1087.
30. Huang, H. W. (2000) Action of antimicrobial peptides: two-state model. *Biochemistry*, **39**, 8347–8352.
31. Duchardt, F., Fotin-Mleczek, M., Schwarz, H., *et al.* (2007) A comprehensive model for the cellular uptake of cationic cell-penetrating peptides. *Traffic*, **8**, 848–866.
32. Fischer, R., Fotin-Mleczek, M., Hufnagel, H. and Brock, R. (2005) Break on through to the other side: biophysics and cell biology shed light on cell-penetrating peptides. *Chembiochem*, **6**, 2126–2142.
33. Fischer, R., Kohler, K., Fotin-Mleczek, M. and Brock, R. (2004) A stepwise dissection of the intracellular fate of cationic cell-penetrating peptides. Journal of *Biological Chemistry*, **279**, 12625–12635.
34. Shai, Y. (2000) Functional domains within fusion proteins: prospectives for development of peptide inhibitors of viral cell fusion. *Bioscience Reports*, **20**, 535–555.
35. Ragona, L., Molinari, H., Zetta, L., *et al.* (1996) CD and NMR structural characterization of ceratotoxins, natural peptides with antimicrobial activity. *Biopolymers*, **39**, 653–664.
36. Andrushchenko, V. V., Vogel, H. J. and Prenner, E. J. (2006) Solvent-dependent structure of two tryptophan-rich antimicrobial peptides and their analogs studied by FTIR and CD spectroscopy. *Biochimica et Biophysica Acta*, **1758**, 1596–1608.
37. Mangoni, M. L., Papo, N., Mignogna, G., *et al.* (2003) Ranacyclins, a new family of short cyclic antimicrobial peptides: biological function, mode of action, and parameters involved in target specificity. *Biochemistry*, **42**, 14023–14035.
38. Seto, G. W., Marwaha, S., Kobewka, D. M., *et al.* (2007) Interactions of the Australian tree frog antimicrobial peptides aurein 1.2, citropin 1.1 and maculatin 1.1 with lipid model membranes: Differential scanning calorimetric and Fourier transform infrared spectroscopic studies. *Biochimica et Biophysica Acta*, **1768**, 2787–2800.
39. Huang, H. W. (2006) Molecular mechanism of antimicrobial peptides: the origin of cooperativity. *Biochimica et Biophysica Acta*, **1758**, 1292–1302.
40. Ladokhin, A. S., Selsted, M. E. and White, S. H. (1999) CD spectra of indolicidin antimicrobial peptides suggest turns, not polyproline helix. *Biochemistry*, **38**, 12313–12319.
41. Ketchem, R. R., Hu, W. and Cross, T. A. (1993) High-resolution conformation of gramicidin A in a lipid bilayer by solid-state NMR. *Science*, **261**, 1457–1460.

42. Kovacs, F., Quine, J. and Cross, T. A. (1999) Validation of the single-stranded channel conformation of gramicidin A by solid-state NMR. *Proceedings of the National Academy of Sciences of the U. S. A.*, **96**, 7910–7915.
43. Ketchem, R. R., Hu, W., Tian, F. and Cross, T. A. (1994) Structure and dynamics from solid state NMR spectroscopy. *Structure*, **2**, 699–701.
44. Glaser, R. W., Sachse, C., Dürr, U. H. N., *et al.* (2005) Concentration-dependent realignment of the antimicrobial peptide PGLa in lipid membranes observed by solid-state ^{19}F-NMR. *Biophysical Journal*, **88**, 3392–3397.
45. Strandberg, E., Kanithasen, N., Tiltak, D., *et al.* (2008) Solid-State NMR analysis comparing the designer-made antibiotic MSI-103 with its parent peptide PGLa in lipid bilayers. *Biochemistry*, **47**, 2601–2116.
46. Strandberg, E., Wadhwani, P., Tremouilhac, P., *et al.* (2006) Solid-state NMR analysis of the PGLa peptide orientation in DMPC bilayers: structural fidelity of ^{2}H-labels versus high sensitivity of ^{19}F-NMR. *Biophysical Journal*, **90**, 1676–1686.
47. Tremouilhac, P., Strandberg, E., Wadhwani, P. and Ulrich, A. S. (2006) Conditions affecting the re-alignment of the antimicrobial peptide PGLa in membranes as monitored by solid state ^{2}H-NMR. *Biochimica et Biophysica Acta*, **1758**, 1330–1342.
48. Levitt, M. H. (2001) *Spin Dynamics: Basics of Nuclear Magnetic Resonance*. John Wiley & Sons Ltd, Chichester, New York.
49. Glaser, R. W., Sachse, C., Dürr, U. H. N., *et al.* (2004) Orientation of the antimicrobial peptide PGLa in lipid membranes determined from ^{19}F-NMR dipolar couplings of 4-CF_3-phenylglycine labels. *Journal of Magnetic Resonance*, **168**, 153–163.
50. Afonin, S., Dürr, U. H. N., Glaser, R. W. and Ulrich, A. S. (2004) 'Boomerang'-like insertion of a fusogenic peptide in a lipid membrane revealed by solid-state ^{19}F NMR. *Magnetic Resonance in Chemistry*, **42**, 195–203.
51. Tremouilhac, P., Strandberg, E., Wadhwani, P. and Ulrich, A. S. (2006) Synergistic transmembrane alignment of the antimicrobial heterodimer PGLa/magainin. *Journal of Biological Chemistry*, **281**, 32089–32094.
52. Moll, F., III and Cross, T. A. (1990) Optimizing and characterizing alignment of oriented lipid bilayers containing gramicidin D. *Biophysical Journal*, **57**, 351–362.
53. Hallock, K. J., Wildman, K. H., Lee, D. K. and Ramamoorthy, A. (2002) An innovative procedure using a sublimable solid to align lipid bilayers for solid-state NMR studies. *Biophysical Journal*, **82**, 2499–2503.
54. Glaser, R. W. and Ulrich, A. S. (2003) Susceptibility corrections in solid-state NMR experiments with oriented membrane samples. Part I: applications. *Journal of Magnetic Resonance*, **164**, 104–114.
55. Strandberg, E., Özdirekcan, S., Rijkers, D. T. S., *et al.* (2004) Tilt angles of transmembrane model peptides in oriented and non-oriented lipid bilayers as determined by ^{2}H solid state NMR. *Biophysical Journal*, **86**, 3709–3721.
56. Chiu, H. P., Suzuki, Y., Gullickson, D., *et al.* (2006) Helix propensity of highly fluorinated amino acids. *Journal of the American Chemical Society*, **128**, 15556–15557.
57. Minks, C., Huber, R., Moroder, L. and Budisa, N. (2000) Noninvasive tracing of recombinant proteins with "fluorophenylalanine-fingers." *Analytical Biochemistry*, **284**, 29–34.
58. Dunn, T. F. and Leach, F. R. (1967) Incorporation of p-fluorophenylalanine into protein by a cell-free system. *Journal of Biological Chemistry*, **242**, 2693–2699.
59. Lian, C., Le, H., Montez, B., Patterson, J., *et al.* (1994) ^{19}F nuclear magnetic resonance spectroscopic study of fluorophenylalanine- and fluorotryptophan-labeled avian egg white lysozymes. *Biochemistry*, **33**, 5238–5245.
60. Wang, P., Tang, Y. and Tirrell, D. A. (2003) Incorporation of trifluoroisoleucine into proteins *in vivo*. *Journal of the American Chemical Society*, **125**, 6900–6906.

61. Fields, G. B. and Noble, R. L. (1990) Solid-phase peptide synthesis utilizing 9-fluorenylmethoxycarbonyl amino acids. *International Journal of Peptide and Protein Research*, **35**, 161–214.
62. Afonin, S., Glaser, R. W., Berditchevskaia, M., *et al.* (2003) 4-Fluorophenylglycine as a label for ^{19}F-NMR structure analysis of membrane-associated peptides. *Chembiochem*, **4**, 1151–1163.
63. Eisenhuth, R. (2003) *Verbesserung der Kopplingseffizienz in der Peptidsynthese durch FEP, und Darstellung von ^{19}F-markierten Peptiden*, Diploma Thesis, University of Karlsruhe.
64. Morgan, J., Pinhey, J. T. and Sherry, C. J. (1997) Reaction of organolead triacetates with 4-ethoxycarbonyl-2-methyl-4,5-dihydro-1,3-oxazol-5-one. The synthesis of α-aryl- and α-vinyl-*N*-acetylglycines and their ethyl esters and their enzymic resolution. *Journal of the Chemical Society Perkin Transactions* **1**, 613–619.
65. Chenault, H. K., Dahmer, J. and Whitesides, G. M. (1989) Kinetic resolution of unnatural and rarely occurring amino acids: enantioselective hydrolysis of N-acyl amino acids catalyzed by acylase I. *Journal of the American Chemical Society*, **111**, 6354–6364.
66. Salgado, J., Grage, S. L., Kondejewski, L. H., *et al.* (2001) Membrane-bound structure and alignment of the antimicrobial β-sheet peptide gramicidin S derived from angular and distance constraints by solid state ^{19}F-NMR. *Journal of Biomolecular NMR*, **21**, 191–208.
67. Mikhailiuk, P. K., Afonin, S., Chernega, A. N., *et al.* (2006) Conformationally rigid trifluoromethyl-substituted α-amino acid designed for peptide structure analysis by solid-state ^{19}F NMR spectroscopy. *Angewandte Chemie International Edition*, **45**, 5659–5661.
68. Afonin, S., Mikhailiuk, P. K., Komarov, I. V. and Ulrich, A. S. (2007) Evaluating the amino acid CF_3-bicyclopentylglycine as a new label for solid-state ^{19}F-NMR structure analysis of membrane-bound peptides. *Journal of Peptide Science*, **13**, 614–623.
69. Fustero, S., Navarro, A., Pina, B., *et al.* (2001) Enantioselective synthesis of fluorinated α-amino acids and derivatives in combination with ring-closing metathesis: intramolecular π-stacking interactions as a source of stereocontrol. *Organic Letters*, **3**, 2621–2624.
70. Crucianelli, M., Battista, N., Bravo, P., *et al.* (2002) Facile and stereoselective synthesis of non-racemic 3,3,3-trifluoroalanine. *Molecules*, **5**, 1251–1258.
71. Toke, O., O'Connor, R. D., Weldeghiorghis, T. K., *et al.* (2004) Structure of $(KIAGKIA)_3$ aggregates in phospholipid bilayers by solid-state NMR. *Biophysical Journal*, **87**, 675–687.
72. Koksch, B., Sewald, N., Hofmann, H. J., *et al.* (1997) Proteolytically stable peptides by incorporation of alpha-Tfm amino acids. *Journal of Peptide Science*, **3**, 157–167.
73. Maisch, D. (2008) *Synthese und Strukturuntersuchungen des membranaktiven Peptaibols Alamethicin mittels ^{19}F-Festkörper-NMR*, Ph.D. thesis, University of Karlsruhe.
74. Grage, S. L., Wang, J., Cross, T. A. and Ulrich, A. S. (2002) Structure analysis of fluorine-labelled tryptophan side-chains in gramicidin A by solid state ^{19}F-NMR. *Biophysical Journal*, **83**, 3336–3350.
75. Witter, R., Nozirov, F., Sternberg, U., *et al.* (2008) Solid-state ^{19}F NMR spectroscopy reveals that Trp41 participates in the gating mechanism of the M2 proton channel of influenza A virus. *Journal of the American Chemical Society*, **130**, 918–924.
76. Cotten, M., Tian, C., Busath, D. D., *et al.* (1999) Modulating dipoles for structure-function correlations in the gramicidin A channel. *Biochemistry*, **38**, 9185–9197.
77. Gause, G. F. and Brazhnikova, M. G. (1944) Gramicidin S. and its use in the treatment of infected wounds. *Nature*, **154**, 703.
78. Gause, G. F. and Brazhnikova, M. G. (1944) Gramicidin S. Origin and mode of action. *Lancet*, **247**, 715–716.
79. Grage, S. L., Suleymanova, A. V., Afonin, S., *et al.* (2006) Solid state NMR analysis of the dipolar couplings within and between distant CF_3-groups in a membrane-bound peptide. *Journal of Magnetic Resonance*, **183**, 77–86.

80. Afonin, S., Dürr, U. H. N., Wadhwani, P., *et al.* (2008) Solid state NMR structure analysis of the antimicrobial peptide gramicidin S in lipid membranes: concentration-dependent re-alignment and self-assembly as a β-barrel, in *Topics in Current Chemistry*, **273**, Bioactive Conformation (ed. T. Peteas) Springer, Berlin, pp. 139–154.
81. Soravia, E., Martini, G. and Zasloff, M. (1988) Antimicrobial properties of peptides from Xenopus granular gland secretions. *FEBS Letters*, **228**, 337–340.
82. Zasloff, M. (1987) Magainins, a class of antimicrobial peptides from *Xenopus* skin: isolation, characterization of two active forms, and partial cDNA sequence of a precursor. *Proceedings of the National Academy of Sciences of the U. S. A.*, **84**, 5449–5453.
83. Bechinger, B., Zasloff, M. and Opella, S. J. (1998) Structure and dynamics of the antibiotic peptide PGLa in membranes by solution and solid-state nuclear magnetic resonance spectroscopy. *Biophysical Journal*, **74**, 981–987.
84. Afonin, S., Grage, S. L., Ieronimo, M., *et al.* (2008) Temperature-dependent transmembrane insertion of the amphiphilic PGLa in lipid bilayers observed by solid state ^{19}F-NMR. *Journal of the American Chemical Society*, **130**, 16512–16514.
85. Ieronimo, M. (2008) *Towards the activity of the antimicrobial peptide PGLa in cell membranes. Solid-state NMR studies of PGLa and magainin2*, Ph.D. thesis, University of Karlsruhe.
86. Blazyk, J., Wiegand, R., Klein, J., *et al.* (2001) A novel linear amphipathic β-sheet cationic antimicrobial peptide with enhanced selectivity for bacterial lipids. *Journal of Biological Chemistry*, **276**, 27899–27906.
87. Maloy, W. L. and Kari, U. P. (1995) Structure-activity studies on magainins and other host-defense peptides. *Biopolymers*, **37**, 105–122.
88. Toke, O., Maloy, W. L., Kim, S. J., *et al.* (2004) Secondary structure and lipid contact of a peptide antibiotic in phospholipid bilayers by REDOR. *Biophysical Journal*, **87**, 662–674.
89. Kanithasen, N. (2005) *^{2}H- and ^{19}F-solid-state NMR studies of the antimicrobial peptide (KIAGKIA)$_3$ in phospholipid bilayers*, Diploma thesis, University of Karlsruhe.
90. Steiner, V., Schar, M., Bornsen, K. O. and Mutter, M. (1991) Retention behaviour of a template-assembled synthetic protein and its amphiphilic building blocks on reversed-phase columns. *Journal of Chromatography*, **586**, 43–50.
91. Oehlke, J., Scheller, A., Wiesner, B., *et al.* (1998) Cellular uptake of an α-helical amphipathic model peptide with the potential to deliver polar compounds into the cell interior non-endocytically. *Biochimica et Biophysica Acta*, **1414**, 127–139.
92. Vives, E., Brodin, P. and Lebleu, B. (1997) A truncated HIV-1 Tat protein basic domain rapidly translocates through the plasma membrane and accumulates in the cell nucleus. Journal of *Biological Chemistry*, **272**, 16010–16017.
93. Afonin, S., Frey, A., Bayerl, S., *et al.* (2006) The cell-penetrating peptide TAT(48–60) induces a non-lamellar phase in DMPC membranes. *Chemphyschem*, **7**, 2134–2142.
94. Hofmann, A. and Glabe, C. (1994) Bindin, a multifunctional sperm ligand and the evolution of a new species. *Seminars in Developmental Biology*, **5**, 233–242.
95. Vacquier, V. D., Swanson, W. J. and Hellberg, M. E. (1995) What have we learned about sea-urchin sperm bindin. *Development Growth & Differentiation*, **37**, 1–10.
96. Ulrich, A. S., Tichelaar, W., Förster, G., *et al.* (1999) Ultrastructural characterization of peptide-induced membrane fusion and peptide self-assembly in the bilayer. *Biophysical Journal*, **77**, 829–841.
97. Binder, H., Arnold, K., Ulrich, A. S. and Zschörnig, O. (2000) The effect of Zn^{2+} on the secondary structure of a histidine-rich fusogenic peptide and its interaction with lipid membranes. *Biochimica et Biophysica Acta*, **1468**, 345–358.
98. Binder, H., Arnold, K., Ulrich, A. S. and Zschörnig, O. (2001) Interaction of Zn^{2+} with phospholipid membranes. *Biophysical Chemistry*, **90**, 57–74.

99. Barre, P., Zschörnig, O., Arnold, K. and Huster, D. (2003) Structural and dynamical changes of the bindin B18 peptide upon binding to lipid membranes. A solid-state NMR study. *Biochemistry*, **42**, 8377–8386.
100. Glaser, R. W., Grüne, M., Wandelt, C. and Ulrich, A. S. (1999) NMR and CD structural analysis of the fusogenic peptide sequence B18 from the fertilization protein bindin. *Biochemistry*, **38**, 2560–2569.
101. Weissenhorn, W., Dessen, A., Harrison, S. C., *et al.* (1997) Atomic structure of the ectodomain from HIV-1 gp41. *Nature*, **387**, 426–430.
102. Nieva, J. L., Nir, S., Muga, A., *et al.* (1994) Interaction of the HIV-1 fusion peptide with phospholipid-vesicles – different structural requirements for fusion and leakage. *Biochemistry*, **33**, 3201–3209.
103. Kliger, Y., Aharoni, A., Rapaport, D., *et al.* (1997) Fusion peptides derived from the HIV type 1 glycoprotein 41 associate within phospholipid membranes and inhibit cell-cell fusion. Structure-function study. *Journal of Biological Chemistry*, **272**, 13496–13505.
104. Martin, I., Defrise-Quertain, F., Decroly, E., *et al.* (1993) Orientation and structure of the NH_2-terminal HIV-1 gp41 peptide in fused and aggregated liposomes. *Biochimica et Biophysica Acta*, **1145**, 124–133.
105. Martin, I., Schaal, H., Scheid, A. and Ruysschaert, J. M. (1996) Lipid membrane fusion induced by the human immunodeficiency virus type 1 gp41 N-terminal extremity is determined by its orientation in the lipid bilayer. *Journal of Virology*, **70**, 298–304.
106. Yang, J., Gabrys, C. M. and Weliky, D. P. (2001) Solid-state nuclear magnetic resonance evidence for an extended beta strand conformation of the membrane-bound HIV-1 fusion peptide. *Biochemistry*, **40**, 8126–8137.
107. Yang, J., Prorok, M., Castellino, F. J. and Weliky, D. P. (2004) Oligomeric β-structure of the membrane-bound HIV-1 fusion peptide formed from soluble monomers. *Biophysical Journal*, **87**, 1951–1963.
108. Yang, J. and Weliky, D. P. (2003) Solid-state nuclear magnetic resonance evidence for parallel and antiparallel strand arrangements in the membrane-associated HIV-1 fusion peptide. *Biochemistry*, **42**, 11879–11890.
109. Reichert, J., Grasnick, D., Afonin, S., *et al.* (2007) A critical evaluation of the conformational requirements of fusogenic peptides in membranes. *European Biophysics Journal with Biophysics Letters*, **36**, 405–413.
110. Wadhwani, P., Bürck, J., Strandberg, E., *et al.* (2008) Using a sterically restrictive amino acid as a ^{19}F NMR label to monitor and to control peptide aggregation in membranes. *Journal of the American Chemical Society*, **130**, 16515–16517.

19

Study of Metabolism of Fluorine-containing Drugs Using *In Vivo* Magnetic Resonance Spectroscopy

Erika Schneider and Roger Lin

19.1 Introduction

Fluorinated compounds are widely used in medicine and treat a range of conditions including those in psychiatry (e.g., mood disorders, anxiety disorders, psychotic disorders) [1, 2], cancer (e.g., antimetabolite agents) [3–10], infection and inflammation (e.g., anti-inflammatory agents, antimalarial drugs, antiviral agents, antihistamines, steroids) [3, 4, 13], and anesthetics. Inclusion of a fluorine atom can alter drug disposition as well as modify its interaction with the pharmacological target [18], making fluorination of compounds a common drug development technique.

In drug development, it is important to know that the compound has reached its intended target as well as to understand the absorption, excretion, bioavailability, metabolism, and distribution (e.g., the pharmacodynamics and pharmacokinetics) of the compound. Direct monitoring of human *in vivo* drug and metabolite concentrations in the target organs or in adverse event-related organs, such as the brain, liver, or heart, is often needed to understand the therapeutic impact, potential adverse events, or other effects.

Magnetic resonance (MR) is routinely used in clinical practice as a powerful, noninvasive diagnostic tool. While human *in vivo* MR is mainly an anatomical and morphological imaging technique, it can also provide functional information using both imaging and spectroscopic methods. MR spectroscopy has proved to be an important technique in

Fluorine in Medicinal Chemistry and Chemical Biology Edited by Iwao Ojima

drug design [13, 60, 61] and preclinical studies [14–16]. Extension of these laboratory techniques into the clinic can be challenging, however ^{19}F MR spectroscopy has proved to be highly specific for identifying fluorinated drugs and their metabolites *in vivo* [1, 3–10, 17]. MR methods *in vivo* are very useful despite a relatively low detection sensitivity, because they are noninvasive, because the effects of one or more doses can be measured, and because they do not expose patients to ionizing radiation, thereby allowing repeat measurements to be performed as often as needed and outcomes to be measured in longitudinal studies. Competing methods of measuring the *in vivo* biochemistry of fluorine compounds use expensive radioactive labeling with detection by either positron emission tomography (PET) or single-photon emission computerized tomography (SPECT). The radioisotopes decay quickly, which limits the evaluation to only the initial kinetics, metabolism, and biodistribution of a single dose. However, fluorodeoxyglucose (FDG) PET is a factor of 10^6–10^9 more sensitive than ^{19}F MR methods, but has no ability to differentiate metabolites from the parent compound [17]. Thus, when it is essential to differentiate metabolic products that are not present in blood or urine, or to measure biodistribution after administration of multiple doses, only biopsy or MR techniques are of use.

Since 1987 when Wolf and colleagues [6] published the first use of ^{19}F MR to monitor the human *in vivo* metabolism of a drug, a range of fluorine-containing pharmaceuticals and metabolites have been evaluated in patients. The most frequently examined organs have been the brain and the liver, although heart, liver and extremity muscle, as well as bone marrow, have also been evaluated. Fluorine MR spectroscopy is able to measure both the relative and absolute concentration of administered fluorinated compounds due in part to the lack of naturally occurring MR-visible fluorine metabolites. However, the accuracy of ^{19}F MR assessments is critically dependent on the MR visibility of the compound, which in turn is determined by the spin–lattice (T_1) and spin–spin (T_2) relaxation properties as well as *in vivo* processes and interactions [19, 20].

A number of excellent review articles exist, including several that explain MR spectroscopy and detail the impact of MR spectroscopy in drug design and in preclinical studies, and others that provide overviews of human *in vivo* metabolism and pharmacokinetics measured by ^{19}F MR in fluoropyrimidine compounds as well as in psychotropic compounds [1, 3–5, 7–10, 59, 60–63]. In keeping with other facets of drug development, the number of ^{19}F MR clinical investigations is much smaller than the number of animal model, specimen, cell line, or solution experiments.

In this chapter, assessment of fluorine-containing drugs using human *in vivo* fluorine (^{19}F) MR spectroscopy will be reviewed along with the technical challenges.

19.2 MR Properties of ^{19}F *In Vivo*

The fluorine MR signal *in vivo* will depend upon both the fluorine concentration and the compound's MR visibility [1, 3, 5, 7–9, 19, 20, 71]. Like protons, ^{19}F nuclei have spin ½ with 100% natural abundance; however, ^{19}F has sensitivity of 83% relative to protons (see Table 19.1) [71]. Unlike protons, which are ubiquitous, there are no natural occurrences of MR-visible fluorine at sufficiently high concentration to be observed by MR in biological systems and hence there is no background fluorine signal. Physiological fluorine

Table 19.1 *Nuclear spin properties of fluorine compared with protons [21, 22, 71]*

Nucleus	Spin	Gyromagnetic ratio ($rad\,s^{-1}\,T^{-1}$)	Larmor frequency, (MHz) at 3.0 T	Natural abundance (%)	Sensitivity relative to ^{1}H	Chemical shift range (ppm)
^{1}H	½	26.7519	127.7	99.98	1.0	16
^{19}F	½	25.181	120.1	100	0.83	400

is immobilized in bones and teeth, and is thus invisible to clinical or research MR imaging systems. The technology for obtaining ^{19}F MR spectra is identical to that for obtaining proton spectra, but the MR system must be equipped with fluorine Larmor frequency-specific excitation and detection packages.

The frequency separation between resonances of different chemical species (chemical shift) varies linearly with the magnetic field strength [71] and is commonly indicated relative to a reference compound in parts per million (ppm). Since chemical shift is a normalized quantity, it has no dependence upon magnetic field strength. For example, water protons resonate at 4.7 ppm relative to tetramethylsilane (TMS) and fat protons in CH_2 and CH_3 groups resonate at about 1.4 ppm at every magnetic field strength. However, at 1.5 Tesla (T) the water and fat proton resonance frequencies are separated by about 220 Hertz (Hz), and at 3 T they are separated by ~440 Hz.

The chemical shift range of fluorine nuclei is over an order of magnitude larger than that of protons (see Table 19.1) because the fluorine nucleus is surrounded by nine electrons. Due to this large range, the fluorine chemical shift is much more sensitive to the local environment than proton chemical shift. This means that fluorine chemical shift changes are often observed when fluorinated small molecules bind to proteins or receptors [14–16, 20] and it is important to investigate the compound under the experimental conditions expected *in vivo*. In particular, the potentially large shift difference between the free and bound species means that it is necessary to acquire ^{19}F MR spectra of the compound in plasma or a cellular suspension to evaluate the potential impact of associative complexes, protein binding, or lipophilic interactions [11, 14, 15, 23].

Fluorine MR reference compounds are used for frequency and/or concentration standards and can also be used for monitoring chest motion due to breathing and thereby reducing motion artifacts (respiratory compensation) [11]. A reference compound should have only one spectral resonance line and it is usually selected on the basis of the chemical shift (or chemical) similarity to the compound being evaluated (see Table 19.2). Since fluorine chemical shifts span a wide range, there is no one compound that will be suitable for universal use as a reference compound. Many suitable reference compounds exist; the reader is referred to the *CRC Handbook of Basic Tables for Chemical Analysis* [22]. Often it is necessary to integrate the reference compound into the radiofrequency (RF) coil [5, 9, 11, 23–26, 28]; in other cases, the reference can be the organ water content [13, 25, 68–70] or intrinsically contained in the ^{19}F MR spectrum. If the reference compound is external to the patient, it is important to ensure that it is housed in an unbreakable vial that will not react with either the solvent or the reference compound, and that will not create a background signal. For respiration compensation the reference standard

Table 19.2 *Examples of potential ^{19}F reference compounds (chemical shift measured relative to trifluoroacetic acid)*

Compound	Chemical shift (ppm)
Trifluoroacetic acid (CF_3CO_2H)	0.0
Fluorobenzene (C_6H_5F)	−34.0
2,5-Difluoro benzophenone ($C_{13}H_8F_2O$)	−41.0
Potassium fluoride (KF)	−44.0
Tricholorfluoromethane ($CFCl_3$)	−78.5
Hexafluorobenzene (C_6F_6)	−87.3

that performs best has a high concentration, whereas for quantitative analysis the best concentration would have similar peak amplitude to the administered compound *in vivo*. The reference compound, its concentration, solvent, and temperature need to be reported in publications to enable the results to be reproduced.

19.2.1 Signal

For all MR imaging and spectroscopy applications, the signal-to-noise ratio (SNR) is directly proportional to the Larmor frequency and hence to the magnetic field strength [71]. This linear dependence motivates the use of high magnetic fields, because the greater SNR improves sensitivity and measurement precision. MR SNR also has a square-root gain achieved by increasing the number of acquisitions (or number of excitations, NEX). To increase the SNR by a factor of 2, the NEX and hence the acquisition time must be quadrupled.

SNR is particularly important in ^{19}F MR spectroscopy, because the maximum *in vivo* concentration of fluorinated drugs administered to humans is often only a few micromolar [1, 11, 27–33]. In addition, patients can only tolerate about 60–90 min MR examinations, inclusive of patient and coil positioning, proton imaging localization, and magnet shimming as well as the fluorine spectroscopic acquisition. Today, whole-body MR systems for human examinations range from 0.2 T to 9.4 T, with most spectroscopy examinations performed between 1.5 T and 4 T. Based on trends over the past two decades, the magnetic field strength will continue to increase in the future. This trend is part of the continuous drive to obtain higher spatial resolution motivated in part by pharmaceutical development [16], but also to detect lower *in vivo* concentrations with MR spectroscopy [1, 5, 11, 38].

Development of high-field MR is limited by safety concerns including exposure to magnetic fields [34], exposure to time-varying gradient fields [58], and RF power deposition [35, 36]. Power deposition, also called specific absorption rate (SAR), is related to the potential increase in tissue temperature due to the RF exposure during acquisition. At low magnetic field strengths, power deposition increases with the square of the magnetic field [35, 36, 37] while at higher magnetic field strengths the dependence on magnetic field strength is lower [36].

There are other costs associated with utilizing high magnetic fields. In general, the price of the MR system increases linearly with the magnetic field strength (e.g., a 7 T whole-body MR system currently costs about twice that of a comparably configured 3 T MR system). In addition, the T_1 relaxation time is lengthened, the sensitivity to magnetic susceptibility is increased (e.g., the T_2* can be shorter), and the impact of chemical shift artifacts with spatial encoding is also increased.

The SNR and sensitivity gain at high magnetic fields (3 T and 4 T) has recently been shown to outweigh the disadvantages [11, 38, 39]. However, optimization of the ^{19}F detection hardware and spectroscopic acquisition is important at any magnetic field strength, since SNR improvement by up to a factor of 2 can result [24, 71].

19.2.2 Relaxation

Fluorine spins exchange energy through interactions with the surrounding proton and fluorine spins. Interactions through chemical bonds (spin–spin coupling) and over short distances through space (dipole–dipole coupling) are powerful relaxation mechanisms [1, 13–16, 71] that are tissue specific. The T_2 relaxation time depends upon the rate of energy exchange between neighboring nuclei (spin–spin interaction). The T_1 relaxation time depends upon the time required for the spins to align with the magnetic field, or equivalently, the time required for the spins to transfer their energy to the surrounding nuclei. A critical aspect of both the spin–spin or transverse (T_2) and spin–lattice or longitudinal (T_1) relaxation processes is the local mobility of the structure containing the fluorine nucleus. The T_2 relaxation times of small, rapidly tumbling molecules is generally long compared with T_2 relaxation times for large, slowly reorienting molecules such as proteins. As a result, the resonance lines in the MR spectra of small molecules, or of gas-phase compounds, are much narrower than those from macromolecules or from small molecules that are bound to a receptor or proteins [13–16, 20, 21, 23, 40, 71]. Conversely, the T_1 relaxation times of small, rapidly tumbling molecules should be proportional to the T_2 value and moderately long, whereas the T_1 relaxation times for large, slowly reorienting molecules such as proteins tend to be longer.

A second important mechanism for fluorine spin–lattice and spin–spin relaxation is produced by the chemical shielding anisotropy (CSA) [13, 14, 21, 71]. The magnetic field experienced by a nuclear spin depends on both the electronic structure of the molecule and how easily the electrons can move in the molecular orientations. In addition, the CSA depends on how the molecule is oriented in the magnetic field. Like spin–spin and dipole–dipole interactions, the CSA of small, rapidly tumbling molecules will be an averaged value (the chemical shift). However, these tumbling motions cause fluctuations of the local magnetic field that lead to relaxation. Also slower reorientation, or an environment that restricts the molecular motion, will result in broader lines due to CSA.

Broad resonance lines, accompanied by T_1 relaxation times much longer than the T_2 relaxation times, are indications that the fluorine nucleus is in a restricted-mobility environment [14, 20, 21, 71]. The specific MR characteristics of a compound in its *in vivo* state must be evaluated and the acquisition and analysis adjusted to account for the values of these characteristics prior to performing human ^{19}F spectroscopy studies. Effective T_2 values (T_2*) that are shorter than 2 ms have spectra that are difficult to measure using whole-body MR systems. In general, acquisition of spectra that contain short T_2*-value species require the use of short (<0.5 ms) excitation RF pulses and careful attention to minimizing the lost time between the end of the excitation pulse and the first useful detection time point. Even when signal decay is sampled appropriately, short T_2* values can limit the ability to perform quantification due to broader lines and reduced signal-to-noise levels.

The differences in local fluorine environment are reflected in the T_1 [20] and T_2* values reported in Table 19.3. In theory, the trifluoromethyl groups found in dexfenfluramine, fluoxetine, fluvoxamine, and niflumic acid undergo less restrictive internal rotation and should also be less sensitive to their local environment than monofluorinated aromatic rings such as in tecastemizole or 5-FU. However, these compounds all have quite short *in vivo* T_2* values except for niflumic acid [12]. Because these T_2* values are significantly shorter than those found in solution, it is fair to anticipate that all the compounds, except potentially niflumic acid, interact with their environment. However, as demonstrated with fluoxetine and FBAL (Table 19.3), reported T_1 values may vary considerably. This illustrates the difficulty of making a robust *in vivo* T_1 measurement under conditions of low SNR and broad resonance lines [8, 11, 12, 20, 24, 28, 31, 32, 35, 41–43, 50, 56, 68]. In addition, the variable bioavailability, due to the small number of patients generally evaluated, compounds the measurement difficulty.

Spin–spin coupling, dipole–dipole coupling, and CSA also have the potential to split and broaden the ^{19}F spectral lines, both which reduce the fluorine signal amplitude and sensitivity. These mechanisms can also probe the chemical structure of complex molecules such as proteins [13, 14, 15, 72]. If proton excitation and polarization transfer and/or decoupling sequences can be implemented [1, 5, 13–16, 56], increases in the ^{19}F SNR can be realized by removal of the two couplings. Spin–spin polarization transfer techniques such as those used for nuclear Overhauser effect (NOE) spectroscopy should be used judiciously and the acquisition parameters should be tuned *in vivo*, because either signal decreases or increase can result [2, 5, 13, 16, 56]. These dual-frequency (fluorine–proton) methods can increase SNR and may provide additional information about structure and physiology as they have in liquid and solid-state samples in the laboratory setting [13–15, 71, 72], but are RF energy intensive [35, 36] and SAR constraints may be encountered even at 1.5 T [24].

19.2.3 Spatial Localization

Proton MR spectroscopy has limited spatial resolution compared to morphological imaging due to low *in vivo* metabolite concentrations. In the clinic, morphological MR imaging has moderate spatial resolution, with typical two-dimensional (2D) acquisitions having 3–5 mm slice thickness with in-plane resolution on the order of 0.3–0.5 mm × 0.3–0.5 mm. In clinical research, MR imaging is often a higher spatial resolution technique, with three-dimensional (3D) acquisitions having 0.5–1.5 mm slice thickness with in-plane resolution on the order of 0.25–0.5 mm × 0.25–0.5 mm. This relatively high spatial resolution can be achieved because the tissue proton concentration is approximately 55 M [13]. However, since proton metabolite concentrations are 1–10 mM, voxel volumes of 1–10 cm^3 are required for proton MR spectroscopy at 1.5 T [1]. Even at these high concentrations, the MR spectroscopic spatial resolution is poor compared with PET which can detect much lower concentrations ~10^{-12} M of the radiolabeled compound [17]. In general, PET has

Table 19.3 *Examples of human* in vivo *^{19}F relaxation times. While T_2 relaxation times are relatively consistent,* in vivo *T_1 values (e.g., for fluoxetine and FBAL) vary considerably, indicating the difficulty of implementing a robust T_1 measurement method with low signal levels and broad spectral linewidths*

Compound	Organ	Field strength (T)	Group	T_1 (ms)	T_2* (ms)	Reference
Dexfenfluramine (Redux)	Brain	1.5	CF_3	–	2.89 ± 0.55	31
Dexfenfluramine (Redux)	Brain	1.5	CF_3	–	2.7–3.5	32
Fluoxetine (Prozac®)	Brain	1.5	CF_3	190	–	41
Fluoxetine (Prozac®)	Brain	1.5	CF_3	650	–	28
Fluoxetine (Prozac®)	Brain	1.5	CF_3	–	2.6 (2.2–3.3)	42
Fluvoxamine (Luvox®, Faverin®, Fevarin®, Dumyrox®)	Brain	1.5	CF_3	140–230	2.7–3.5	30
Niflumic acid	Liver	1.59	CF_3	–	33	12
Tecastemizole (Soltara™)	Liver/heart	4.0	4-Fluorophenyl	300	3.1	11
Haloperidol decanoate	Muscle	1.5	3-Fluorophenyl	365	–	25
Fluorouracil (5-FU)	Liver	1.5	5-Fluoropyrimidine	1600 ± 200	5–11	50, 56
α-Fluoro-β-alanine (FBAL)	Liver	1.5	CHF	1730 ± 130	4–7	50, 56
FBAL	Liver	1.5	CHF	380 ± 80	–	24
FBAL	Liver	3.0	CHF	400 ± 140	–	68
Capecitabine	Liver	3.0	5-Fluoropyrimidine	250 ± 70	–	68

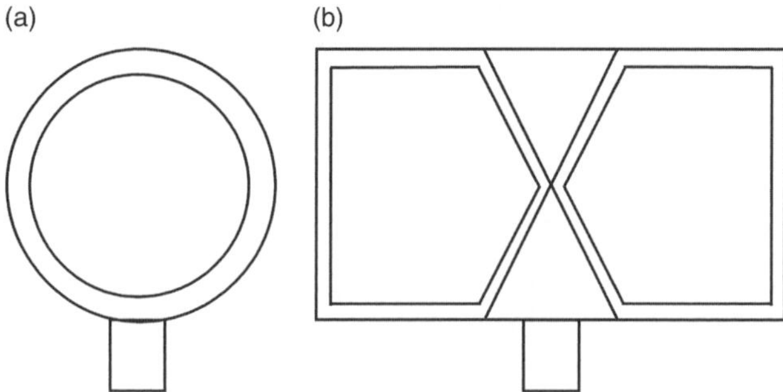

Figure 19.1 *Surface coils can be as simple as a loop of copper wire or tubing. They are placed directly over the organ of interest and have a sensitive depth of detection which is approximately half the coil diameter. Examples of surface coils that may be used for either transmit/receive or receive-only applications: (a) simple loop coil; (b) figure-eight coil.*

both higher spatial resolution (~3–5 mm) and higher temporal resolution, which results from this higher sensitivity. Because of the potential for MR spectroscopic techniques to resolve metabolite resonances and assess the impact of multiple doses, this is a complementary modality to the PET kinetic information, even for first-pass or single-dose assessments.

Since the human *in vivo* concentration of exogenous fluorinated drugs is often in the micromolar range, spatial localization is often restricted to either the sensitive volume of the detection coil [5, 9, 11, 27–33, 42, 45], a single voxel in the organ of interest [5], or slabs of differing depth [5, 23, 56]. However, should the *in vivo* fluorine concentration be in the millimolar range, spatially localization using 2D or 3D chemical shift imaging (CSI) techniques become feasible [5, 56]. If spatial localization is feasible, SNR can be traded off either for increased spatial resolution or for decreased acquisition time to achieve better temporally resolved spectra.

Surface RF coils (see Figure 19.1) are receiver coils (often made only of a loop of copper wire or tubing) that are placed directly over the organ of interest and have a sensitive depth of detection that is approximately half the coil diameter [5, 11, 15, 23]. If the region of interest is superficial, this approach works well. However, if the organ is deep, the subcutaneous fat and any other tissue within the coil's sensitive volume may contribute to the detected signal. All surface coils suffer from inhomogeneous receive fields. If used to excite the nuclear spins, surface coils will also have spatially nonuniform excitation RF fields. This means that the same flip angle cannot be obtained throughout the coil's sensitive volume. The impact of the inhomogeneities is magnified if the same coil is used for both excitation and detection. A larger (larger than the receive coil), more homogeneous excitation surface coil is preferred to minimize the impact of the nonuniform limited volume excitation RF fields [5, 56].

Transmit–receive surface coils are generally used for MR spectroscopic studies in organs other than the brain. If dual-frequency techniques such as NOE, polarization transfer, or decoupling are to be utilized with surface coil excitation, or if relaxation time measurements are to be performed, RF pulses that are insensitive to spatial and amplitude inhomogeneity must be employed. Such RF pulses can have high power requirements and may encounter specific absorption rate (SAR) limitations. Brain ^{19}F MR spectroscopy generally utilizes a transmit–receive volume coil, which does not suffer nearly as much from RF field spatial inhomogeneities; however, SAR is particularly of concern with dual-frequency sequences.

The inhomogeneity of the excitation RF fields from transmit–receive surface coils can be used to advantage for selecting a sensitive volume slab based on approximate distance from the coil. Limited spatial selectivity can be achieved by adjusting the flip angle to best select the depth of interest.

19.2.4 Metabolism, Binding, and Association

The signal intensity of an MR-visible fluorinated compound is directly proportional to the number of fluorine spins present, similarly to other spin ½ nuclei. However, a number of processes can restrict compound MR visibility and thereby impede quantification from spectra.

The appearance of fluorine MR spectra can change if exchange, binding, or metabolism is present [1, 9, 11, 14, 15, 19, 20, 23, 30–33, 42, 44, 59, 63, 68]. If the process is slow compared with the Larmor resonance frequency difference between the two states or species, then two distinct sets of resonances will be observed. However, if the process is fast compared with the difference in Larmor frequency, the spectrum will have a single resonance located at the weighted average of the chemical shifts of the two states or species [14].

Another common impact of *in vivo* processes is that the compound and its metabolic products or states will undergo line broadening [1, 9, 11, 14, 15, 19, 20, 23, 30–33, 42, 44, 62, 63]. When one or more compounds have broad spectral lines, measurement of drug metabolism *in vivo* may also be limited. Overlapping fluorine resonances of the parent drug and those of active and inactive metabolites [1, 11, 23, 41] at magnetic field strengths used for human MR spectroscopy may also occur, further confounding the results.

19.3 Applications of *In Vivo* Fluorine MR in Medicine and Drug Development

Prior to using ^{19}F MR spectroscopy on a new compound in the clinic, the potential and limitations should first be explored *in vitro* and then in animals [19, 20]. If quantitative fluorine MR measures are desired, laboratory research may be needed prior to going into the clinic with a new compound to reveal any potential difficulties that might be encountered *in vivo*. Even with extensive pre-work, cross-species variation as well as selective organ uptake may lead to different amounts of MR-visible signal [1, 31, 32, 41, 42, 44]. Without the proper preliminary *in vitro* studies, spectroscopic signal changes measured *in vivo* may be misinterpreted [12, 41, 44].

Examples of spectral changes obtained from laboratory solutions, *ex vivo* specimens and *in vivo* clinical measurements are shown in Figures 19.2–19.6 for tecastemizole

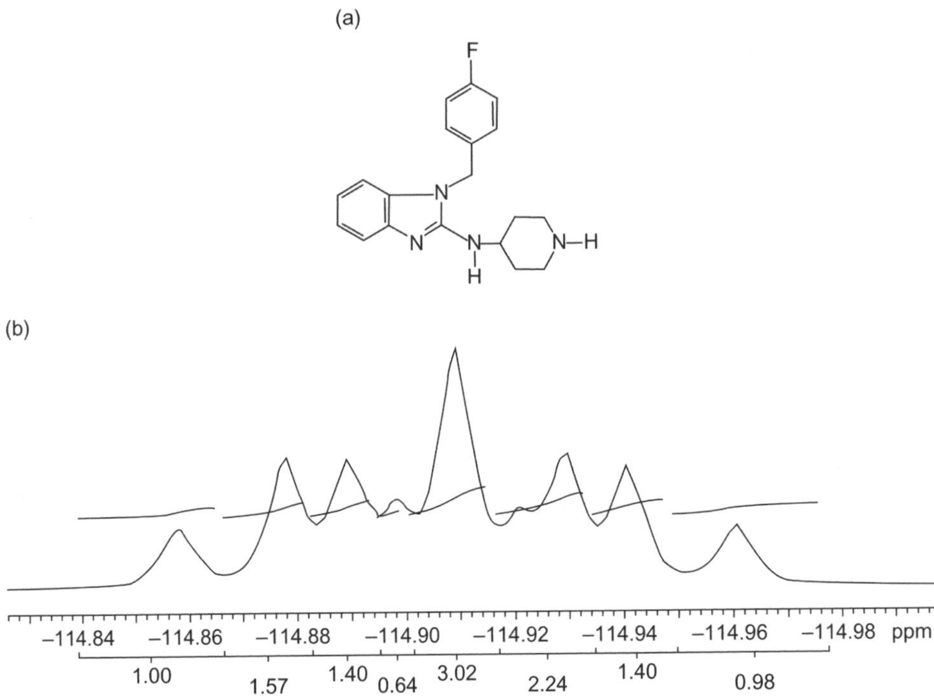

Figure 19.2 *(a) Chemical structure of tecastemizole. All known metabolites also have the fluorine atom in the para-position of the phenyl ring. (b) High-resolution, solution-state ^{19}F spectrum without proton decoupling at 7 T of tecastemizole in DMSO.*
(Source: Schneider E, Bolo NR, Frederick B et al., Magnetic resonance spectroscopy for measuring the biodistribution and in situ in vivo pharmacokinetics of fluorinated compounds: validation using an investigation of liver and heart disposition of tecastemizole, J. Clin. Pharm. Ther. *(2006)* **31**, *261–273. Copyright (2006) John Wiley & Sons. Reprinted with permission.)*

(Soltara™), a monofluorinated metabolite of the antihistamine astemizole (Hismanal) [11]. The spectrum of tecastemizole (see Figure 19.2) and its primary metabolite, 2-hydroxynorestemizole, are composed of a complex multiplet with several nonequivalent ^{1}H-^{19}F couplings. At 7 T and in dimethyl sulfoxide (DMSO), the parent compound can be distinguished from the metabolites; however, the four known metabolites are co-resonant and cannot be differentiated. Since tecastemizole and its metabolites are all biologically active, the lack of spectral resolution does not adversely impact biodistribution measurements. Ideally, separable ^{19}F MR resonances would be observed for the parent compound as well as its active and inactive metabolites. If they are not spectrally resolved, *in vivo* measurement of fluorinated drug metabolism by ^{19}F MR spectroscopy might have limited utility.

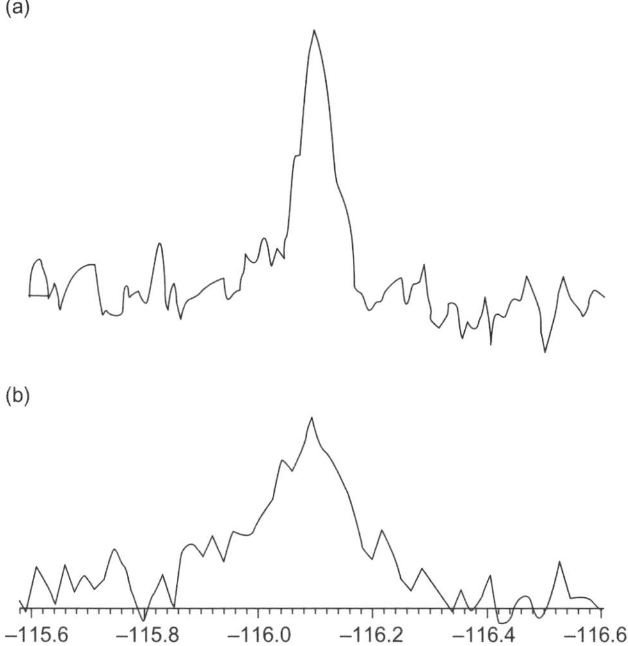

Figure 19.3 *Solution-state ^{19}F spectra at 7 T of 2.7 mM tecastemizole in a 50% whole blood solution. (a) The integrated fluorine signal is only 45% of that measured in DMSO because of protein binding. (b) After 20 min, 70% of the integrated signal was lost.*

Interactions between compounds and tissue constituents may also interfere with detection by MR spectroscopy. In particular, compounds that interact with proteins and/or other cellular components will have altered T_1 [20] and decreased T_2 [11] relaxation times. In a bovine serum albumin (BSA) solution, protein interactions caused the tecastemizole ^{19}F linewidth to increase (i.e., T_2^* and possibly T_2 decreased) and a 35% loss of total signal (integrated area) was found in addition to a small frequency shift. After introduction to a whole-blood solution (see Figure 19.3), protein binding also occurred, with 45% of the integrated signal being lost immediately and 70% of the integrated fluorine signal gone in 20 minutes. In a mixture of intact and disrupted human liver cells, an 85% loss of the ^{19}F integrated signal area from tecastemizole occurred within one hour due to interactions with the cellular components. Interactions between this compound with protein and cellular constituents were strong enough to increase the correlation time, thus causing line broadening and an overall loss of integrated signal intensity compared with that observed in solution.

At 4 T, the fluorine resonance frequency difference between tecastemizole and 2-hydroxynorestemizole in isotonic saline solution was found to be 6.7 Hz and a linear response of the measured ^{19}F integrated signal (see Figure 19.4) was calibrated against analytically validated concentrations (see Figure 19.4b). When incorporated into either a dog heart (see Figure 19.5a) or liver, tecastemizole and its fluorinated metabolites were co-resonant in a single broad resonance ($T_2^* = 1.4$ ms; $T_1 = 300$ ms). A linear response of

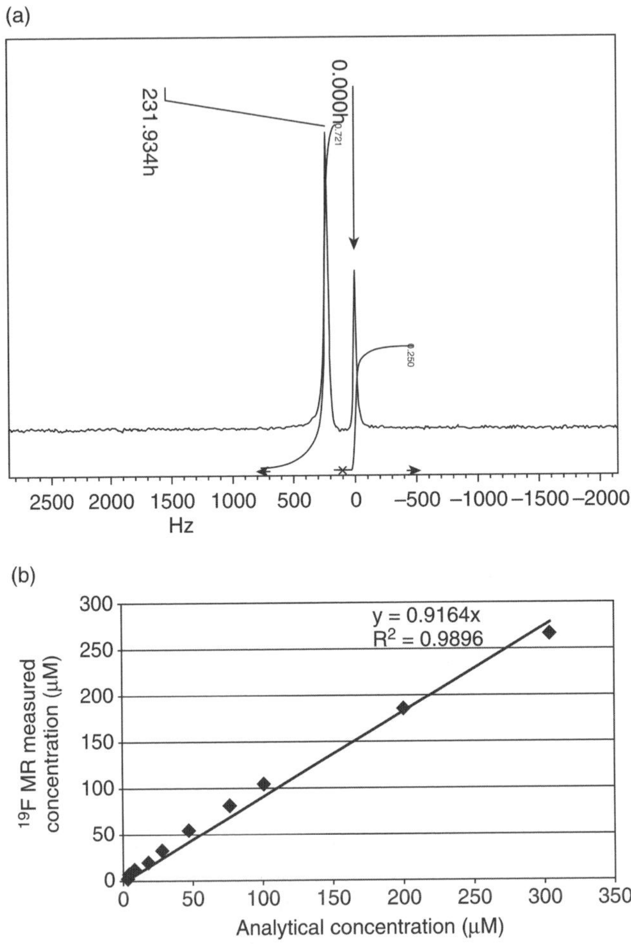

Figure 19.4 *Typical phantom calibration curve measurement. (a) ^{19}F MR spectrum of compound and reference standard: (left) tecastemizole; (right) potassium fluoride. (b) Calibration curve of tecastemizole concentration measured analytically by LC-MS versus that measured using ^{19}F MR.*

integrated signal (area under the curve or AUC, see Figure 19.5a-c) with tecastemizole total fluorine measured using liquid chromatography–mass spectrometry (LC-MS) was found (see Figure 19.5d) in the *ex vivo* specimen [11]. The calibration curve presented in Figure 19.5d, based on *ex vivo* spectroscopic measurements, provides a good indication that the *in vivo* compound visibility should be linear with concentration and that sequestration of the compound is not expected over the dose ranges investigated. *In vivo* ^{19}F MR spectra were also similar to those measured in the *ex vivo* specimens: the parent compound resonance was unresolvable from that of its metabolites (see Figure 19.6), and the broad single resonance had similar relaxation times to that found *ex vivo* (see Table 19.3).

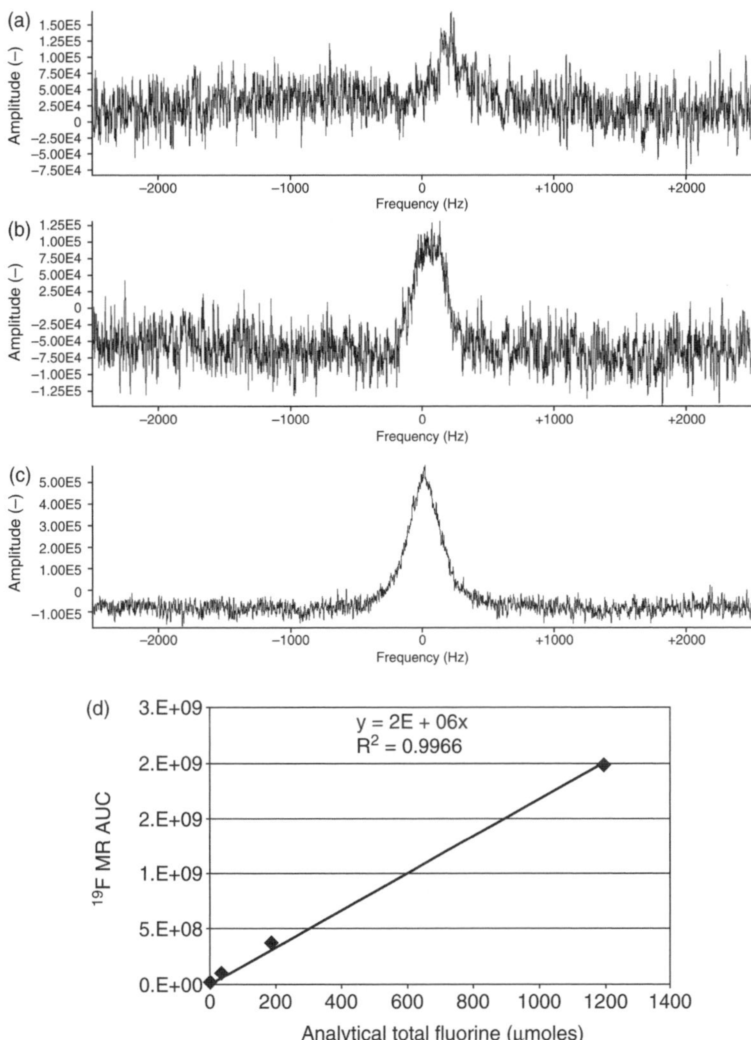

Figure 19.5 *Typical* ex vivo *specimen calibration curve measurements.* ^{19}*F MR spectra of* ex vivo *dog hearts at 4 T as a function of analytically measured tecastemizole concentration: (a) 1.1 μM; (b) 359 μM; (c) 2149 μM. (D) Calibration curve of the ex vivo specimen total fluorine content measured analytically by LC-MS* versus *integrated AUC measured using* ^{19}*F MR.*

19.4 Fluorine MR Spectroscopy in Cancer: Fluoropyrimidines

The impact of fluorine MR spectroscopy as an *in vivo* analytical technique was demonstrated by observations made with fluorouracil (5-FU) and has been the subject of excellent review articles [3–5, 7–10, 46, 47, 59]. 5-FU is an antimetabolite that has been used for over 45 years in the treatment of several common cancers; however, significant side-effects

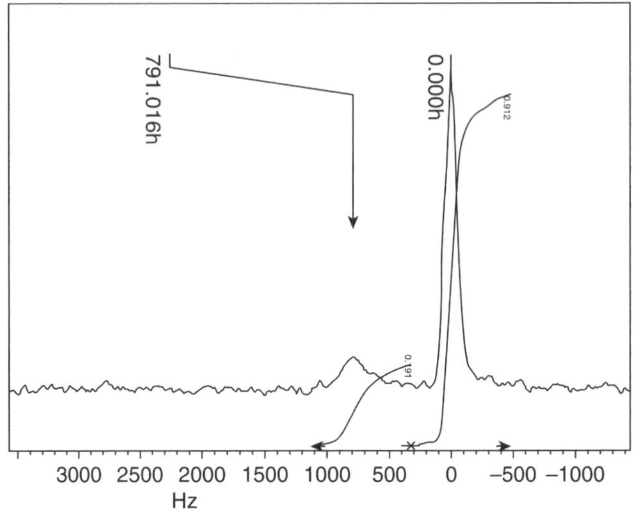

Figure 19.6 *^{19}F MR spectrum of tecastemizole* in vivo *at 4 T from the liver of a normal human volunteer.*

can occur, including cardiotoxicity and neurotoxicity. On the basis partly of evidence obtained by *in vivo* ^{19}F MR spectroscopy, it was eventually hypothesized that the side-effects resulted from transformation of FBAL, the primary catabolic metabolite, into fluoroacetate, which has known cardiotoxicity and neurotoxicity [3–5, 7–10, 46]. To our knowledge there have been no new clinical ^{19}F MR studies published on fluoropyrimidine drugs since the most recent review (2006) [4]. For an extensive discussion of the clinical utilization of fluoropyrimidine drugs (including 5-FU), the metabolic pathway and modulation by other medications, the reader is referred to these articles.

The metabolic pathways of 5-FU and its prodrugs have been well characterized in the liver as well as in liver tumors and metastases by *in vivo* ^{19}F MR spectroscopy. The liver and extrahepatic spaces catabolize 5-FU, which is subsequently excreted in the urine [3–5, 7–10, 46, 59]. Active metabolites (fluoronucleotides) are created by anabolism in tumors. Even though clinical studies showed significant individual subject variations [3, 5, 7–10, 26, 46, 47], spectral characteristics such as resonance frequency, linewidth, relaxation time, and amplitude were not related to the therapeutic response. However, dynamic processes, specifically the accumulation and retention of 5-FU in the tumor, were indicative of response. Patients showing tumor half-lives of free 5-FU of 20 minutes or longer observed in patients were characterized by Wolff *et al.* [47, 53] as "trappers." While over 50% of the evaluated population were nontrappers, approximately 60% of the patients who responded to therapy were trappers [3–5, 7–10].

Use of *in vivo* ^{19}F MR spectroscopy to directly measure the pharmocokinetics of 5-FU in liver tumors and metastases enabled identification of drug mechanism of action and investigation of tumor pathophysiology. Similar studies were undertaken to evaluate the ability of a range of medications to modulate the *in vivo* effectiveness of fluorouracil and its prodrugs [3–5, 7–10]. These studies have served as a model for other *in vivo* drug MR spectroscopic studies. In many respects, the analytic characterization of a potential responder has evolved into the concept of personalized medicine.

In ^{19}F MR examinations of patients with liver tumors or metastases, the receive coil sensitive volume typically encompassed the tumor, a significant portion of normal tissue, and possibly surrounding organs (such as the gallbladder, spleen, kidney, etc.). Initial findings of tumor trapping were based predominantly on large superficial liver tumors or metastases that were evaluated by a surface coil placed in close proximity to the tumor. However, since the sensitive region of the coil encompasses a considerable amount of normal liver tissue, unequivocal proof of drug retention by the tumor was not possible. This information required spatial localization and motivated technological improvements to increase SNR and resolve the problems associated with chemical shift artifact [5, 24, 38, 56, 59].

Because of the relatively low concentration of 5-FU metabolites *in vivo*, it was challenging to perform volume-selective ^{19}F MR spectroscopy prior to technological advances. An average 2-fluoro-β-alanine (FBAL) concentration of 0.92 ± 0.26 mmol/kg liver for the first 50 min post infusion was observed for doses varying from 750 mg to 2000 mg [43], with a mean maximum concentration of 1.31 ± 0.33 mmol/kg liver, or, in a separate study, 1.0 ± 0.2 mM FBAL in the liver at 60 ± 10 min post infusion [56].

To detect 5-FU metabolism in the different liver regions, single-voxel acquisitions, and 1D, 2D, or 3D chemical shift imaging (CSI) have been employed. CSI utilizes gradient magnetic fields for spatial localization, identical to imaging techniques. Because the first gradient encoding (slab selection or 1D CSI) is based on frequency selectivity, the large Larmor frequency differences between the 5-FU and its metabolites results in chemical shift artifacts along the slice select direction. This artifact means that the FBAL signal arises from a spatially different slice from where the 5-FU signal originates. At 1.5 T, the spatial shift between 5-FU and FBAL was 2.3 cm [5]. Methods to circumvent chemical shift artifacts include frequency-selective excitation RF pulses or frequency selective presaturation RF pulses to isolate ^{19}F MR signal detection to only one chemical species [48–50, 56].

In spite of the challenge presented by low *in vivo* concentrations and the chemical shift artifact, spatially localized ^{19}F spectroscopy studies in the liver of patients receiving 5-FU chemotherapy enabled advancement of the understanding of 5-FU metabolism and trapping. In addition, a catabolic resonance originating from the gallbladder was identified [57], confirming extrahepatic catabolism of 5-FU. At 1.5 T, 2D CSI techniques with 8×8 localization voxel volumes of $6\,cm \times 6\,cm \times 4\,cm$ ($144\,cm^3$) were acquired in 12.8 min using a pulse repetition time (TR) of 60 ms and 12 800 excitations [50, 56]. 3D CSI localization with voxel volumes of $4\,cm \times 4\,cm \times 4\,cm$ ($64\,cm^3$) were acquired in 8.5 min using a TR of 1 s and 512 excitations [50] or in 45 min using a TR of 260 ms and 10 240 excitations [35]. Even smaller voxel volumes of $3\,cm \times 3\,cm \times 3\,cm$ ($27\,cm^3$) were achieved in 45 min with double resonance (^{1}H–^{19}F) spectroscopy and an $8 \times 8 \times 8$ resolution 3D CSI acquisition [50]. However, the long acquisition times (45 min) for the $27\,cm^3$ 3D CSI acquisition precluded assessment of 5-FU pharmacodynamics *in vivo*.

In vivo application of double resonance (^{1}H–^{19}F) spectroscopic techniques [50, 73, 74], including a combination of NOE and proton decoupling, have proved to increase SNR, which can be traded off for improved visibility or improved spatial or temporal resolution of fluorinated compounds. The double resonance decoupling, NOE, and polarization transfer techniques used *in vivo* are similar to those used in solid state

[71, 72] and animal spectroscopic evaluations. *In vivo* challenges include transmit coil uniformity as well as high-power RF pulses that may limit duty cycle (TR) due to potential heating (SAR) concerns [35, 36]. In addition, these techniques can only be used *in vivo* when sufficient signal (or sufficient information) is present to correctly adjust the parameters.

Klomp *et al.* [24] were able to double the SNR of their ^{19}F MR spectra at 1.5 T by systematically reducing the noise factor of each component in the signal detection of the MR system. With this increase, a $4\,cm \times 4\,cm \times 4\,cm$ ($64\,cm^3$) 3D CSI acquisitions were acquired in 4 min without polarization transfer or decoupling and these acquisitions were subsequently used to evaluate five patients receiving 5-FU chemotherapy for treatment of superficial and central liver metastases. The *in vivo* T_1 for 5-FU in the liver following a bolus injection was 380 ± 80ms, which was considerably lower than in earlier publications (see Table 19.3).

A dramatic increase of *in vivo* ^{19}F MR SNR (a factor of 1.3–3) was then again achieved by evaluating capecitabine metabolism in the liver at 3 T magnetic field strength compared with 1.5 T [38, 67]. At 3 T, the spectral resolution was also increased and allowed differentiation of the FBAL–bile acid conjugate [57] from the primary FBAL resonance. Due to increased SNR, which resulted both from use of optimized hardware and use of 3 T technology, Klomp *et al.* [68] were able to quantify the spatial distribution of capecitabine and its prodrugs (including FBAL) in the liver of patients with advanced colorectal cancer using a 3D CSI acquisition with $10 \times 10 \times 10$ resolution in a 9 min acquisition (see Figure 19.7). This study found a nonhomogeneous spatial distribution of capecitabine and FBAL, in agreement with Li *et al.* [56]. The order of magnitude of the FBAL concentration [68] was also in agreement [56]; however, the other capecitabine metabolite concentrations differed significantly. The finding of spatial heterogeneity of compound distribution and metabolism must be accounted for when *in vivo* concentrations are determined [56, 68].

Kamm *et al.* [26] used the higher SNR *in vivo* ^{19}F MR configuration at 1.5 T [24] to examined the uptake and metabolism of 5-FU modulated by trimetrexate in liver metastases of colorectal cancer. All patients had one or more liver metastases of at least 2 cm diameter within 8 cm of the skin surface. These criteria enabled inclusion of patients with large superficial tumors, as well as patients with smaller metastases. Treatment response was assessed during week 6 of the first chemotherapy cycle using unlocalized ^{19}F MR spectroscopy as well as before and after each 8-week treatment cycle with either computed tomography or ultrasound. The patients were divided into two groups, those with a larger contribution of the ^{19}F signal from normal liver tissue (smaller tumors) and those with a smaller contribution (larger tumors). In patients with larger tumors, a correlation was found between increase in tumor size and the catabolite FBAL concentration, and the poorer response to treatment was hypothesized to be correlated with higher degradation of 5-FU into catabolites. However, this relationship was not found when smaller tumors were studied [26]. It is unclear whether these discrepant results were found because of the large contribution of anabolic signal from the liver (e.g. partial volume effects due to lack of spatial localization) or because of true differences in tumor drug metabolism. In addition to examining patients with smaller liver metastases, the ^{19}F measurements were performed after 5 weeks of chemotherapy and were timed to potentially detect only tumor cells refractory to 5-FU [26].

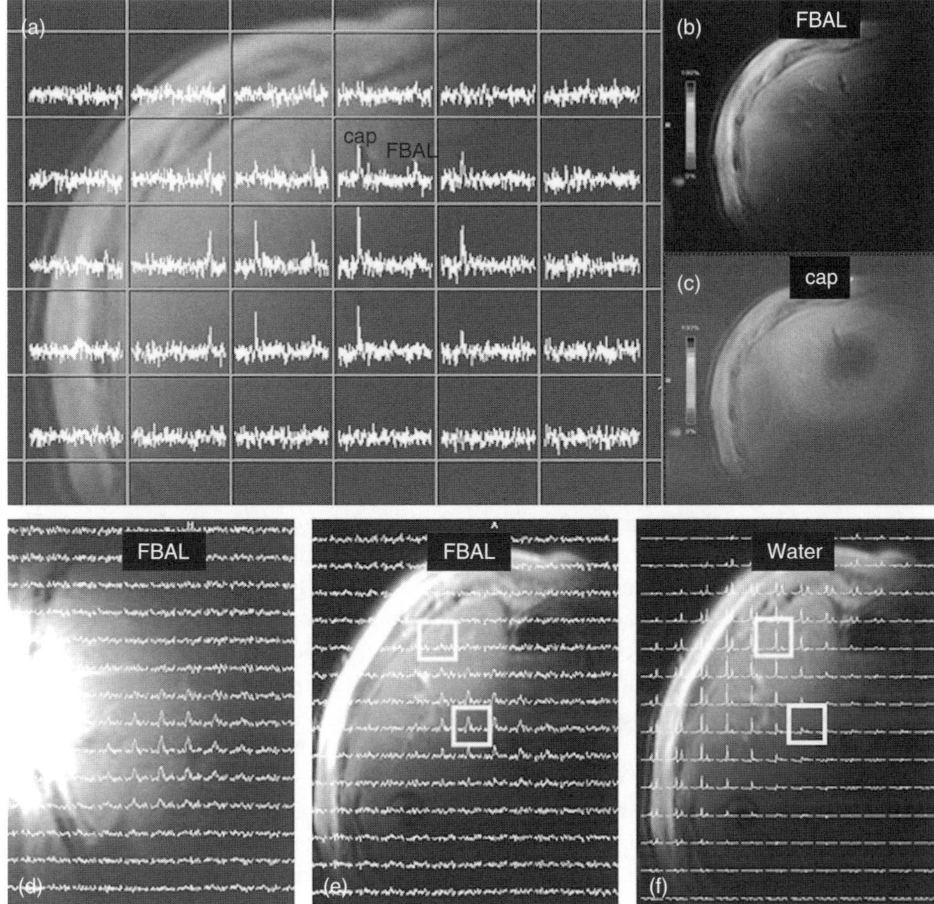

Figure 19.7 *Distribution of capecitabine and its metabolite FBAL in the liver of a patient treated with oral capecitabine at 3 T. (a) Spatially localized* ^{19}*F MR CSI spectra overlaid on the axial proton image of the liver acquired using the same surface coil. (b) and (c) Color depiction of distribution of FBAL in the axial plane and capecitabine in the coronal plane, respectively. (d) and (e) Distribution of FBAL in the coronal and axial planes, respectively, depicted by CSI spectra. (f) Distribution of water signal in the axial plane. See color plate 19.7.*
(Source: *Klomp D., van Laarhoven H., Scheenen T., et al. Quantitative* ^{19}*F MR spectroscopy at 3 T to detect heterogeneous capecitabine metabolism in human liver,* NMR in Biomedicine *(2007)* **20**, *485–492. Copyright (2007) John Wiley & Sons. Reprinted with permission.)*

In addition to 5-FU, its fluorinated prodrugs such as gemzar, floxuridine, capecitabine, tegafur uracil, etc. have also been evaluated using ^{19}F MR spectroscopy, at least in the laboratory or in animal models [3, 7–10]. Improved efficacy of 5-FU has been achieved by using it in combination with other medications that either modulate its uptake or/and increase its metabolism. ^{19}F MR has been used to measure the modulation of 5-FU metabolism and tumor half lives by a range of compounds including metholtrexate, α-interferon and Leucovorin [3, 7–10].

19.5 Fluorine MR Spectroscopy in Neuropharmacology

MR-visible fluorinated compounds that can be measured in the human brain comprise a large number of psychiatric medications including most of the serotonin-specific reuptake inhibitors (SSRIs) such as fluoxetine (Prozac®) [1, 28, 41], as well as pharmaceuticals with mechanisms of action outside the central nervous system such as dexfenfluramine (fen-phen, a serotoninergic anorectic drug) [1, 31]. A recent review [1] covers the impact of MR spectroscopy in psychiatry.

In some compounds with long elimination half-lives, such as fluoxetine, the plasma concentration reaches steady-state levels after 4–5 weeks of dosing. As a result of 30 days of dosing at 40 mg/day, plasma concentrations of fluoxetine and its active metabolite, norfluoxetine, are in the range 91–302 ng/mL and 72–258 ng/mL, respectively [51]. The total fluorine serum concentration is thus 1.5–5.4 μM. However, tissue accumulation occurs with many fluorinated compounds, in part due to their lipophilicity and pH trapping in acidic vesicles in the brain [1], and allows the fluorine concentration to be in a range detectable using MR spectroscopy [1, 33]. Even with tissue accumulation, the vast majority of brain ^{19}F MR spectroscopy assessments are unlocalized (i.e., the entire brain is contained within a transmit/receive volume coil) [27–33, 41, 42, 45] due to low *in vivo* fluorine concentrations.

Since MR spectroscopy detects signals from unbound forms of fluorinated drugs, tissue accumulation in which the drug is bound may pose compound visibility problems. For example, fluoxetine is known to undergo extensive protein binding (94.5%) [1, 51], whereas dexfenfluramine/dexnorfenfluramine have more moderate protein binding (36%) [52]. Despite moderate protein binding in solution, *in vivo* binding was not found for dexfenfluramine/dexnorfenfluramine. The *in vivo* ^{19}F MR concentration in the brain was compared with that determined post-mortem by LC-MS [31, 44]. In these primate experiments, dexfenfluramine/dexnorfenfluramine was found to be fully detectable *in vivo*. In contrast, significant fluoxetine binding and MR invisibility in the brain were found from an autopsy of a patient who received 40 mg/day [32]. However, in the latter study [32] the tissue was not preserved cryogenically but was fixed in formalin for 2 weeks prior to performing the ^{19}F MR measurement. Potential changes in the cellular environment caused by use of formalin [32] raise questions about the accuracy of "bound fraction" of fluoxetine measured post-mortem.

Strauss and Dager [42] also measured the relative contribution of unbound versus bound fluoxetine and metabolites to the ^{19}F MR visible signal *in vivo* in the human brain by applying magnetization transfer methods to acquire ^{19}F MR spectra. Signals from the bound form of fluoxetine/norfluoxetine were found to be approximately 14.2% of those detected from the unbound form [42]. A similar binding or interaction between fleroxacin in plasma *in vitro* and in tissue *in vivo* was found [23]. Understanding the nature and characteristics of the *in vivo* spectrum was a prerequisite for quantitation of the resultant fleroxacin ^{19}F spectra.

While Christensen *et al.* [31, 44] clearly demonstrated that *in vivo* ^{19}F MR spectroscopy can provide reliable drug concentration estimates in the brain, the brain to serum drug ratios may vary widely between species [1, 31]. For irrefutable determination of MR visibility of a compound, an *in vivo* ^{19}F MR spectroscopy study must be performed either with organ biopsy or with sacrifice and necropsy, for drug concentration correlation. Such validation studies need to be planned and performed carefully since the method of tissue handling can dramatically influence results. Karson *et al.* [41] reported that levels of drug detected in formalin-fixed, postmortem brain tissue from a single subject were much lower than the level of drug detected immediately following extraction. For neuropharmaceuticals, ^{19}F MR spectroscopy may provide one of the only direct *in vivo* measurement of drugs and their metabolites in the target organ.

Sassa *et al.* [2] used *in vivo* ^{19}F MR to evaluate the haloperidol decanoate intramuscular injection site in four well-controlled schizophrenia patients. An initial MR assessment of the deltoid muscle with a 15 cm diameter surface coil was performed immediately after injection with subsequent measurements at 1 and 2 weeks later. For comparison, haloperidol plasma concentrations were analyzed with high-performance liquid chromatography at the same time (half-life 3 weeks). A broad fluorine signal was detectable up to 16 days post injection. The fluorine signal decreased during the first week and then appeared to stabilize; this decline was much faster than expected based on the increase in plasma concentration. The local distribution of haloperidol was assessed with 2D CSI on the day post injection, but the signal was too low for CSI after 8 days. On day 1, fluorine CSI could detect the drug's oil carrier, but the oil was not visible on proton imaging.

19.6 Fluorine MR Spectroscopy for Measuring Biodistribution

Long-term sequestration of fluorinated psychotropic compounds has been found by ^{19}F MRS at 3 T in peripheral tissue several months after cessation of treatment [39]. The high-field-strength (3 T) MR system significantly enhanced the ability to detect the compound-specific resonance in the lower limbs of patients with major depressive disorder treated with fluoxetine or fluvoxamine. ^{19}F MR CSI several months after complete disappearance of compound signal from the plasma and brain [27] suggested that the post-drug withdrawal signal originated from bone marrow (see Figure 19.8). Mean *in vivo* concentrations of the trifluorinated compounds in the lower-extremity bone marrow, estimated by an external reference standard, were $38 \pm 17\,\mu M$ for fluoxetine and $40 \pm 17\,\mu M$ for fluvoxamine [39].

Conversely, long-term accumulation of tecastemizole in the liver was not found. After 8 days of tecastemizole treatment at a daily dose of 270mg, ^{19}F MR signal was found in the livers, but not the hearts, of healthy volunteers with a concentration range of 2–30 μM [11]. In contrast to the psychotropic compounds, no fluorine signal was found in the liver after 28 days of washout, suggesting that tecastemizole and its fluorinated metabolites were eliminated from liver.

^{19}F MR spectroscopy has also been used for evaluation of the short-term accumulation and elimination of niflumic acid, a nonsteroidal anti-inflammatory drug (NSAID), in the

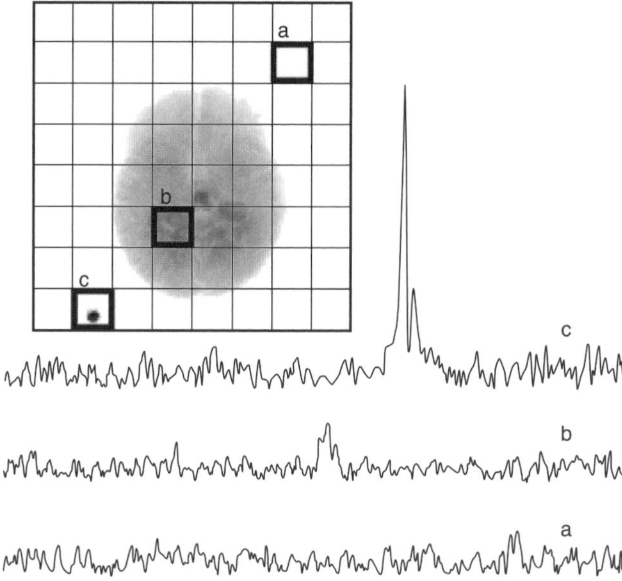

Figure 19.8 *Distribution of fluvoxamine at steady-state concentration in the brain of a volunteer measured by* ^{19}*F MR CSI at 3 T. (a) Spatially localized* ^{19}*F MR CSI spectra overlaid on the axial proton image of the brain and including the external reference sample.* ^{19}*F spectra of a voxel containing (b) the background (e.g., no structure and no signal), (c) posterior brain tissue, and (d) external reference compound (trifluoroethanol). The density of the compounds is reflected by peak amplitude and area under the curve (AUC).*
*(*Source: *Reprinted by permission from Macmillan Publishers Ltd from Bolo, N. R., Hode, Y., Nedelec, J. F.,* et al. *Brain pharmacokinetics and tissue distribution in vivo of fluvoxamine and fluoxetine by fluorine magnetic resonance spectroscopy.* Neuropsychopharmacology *(2000)* **23**, *428–438.)*

liver [12]. At about 55 min following oral administration, the parent compound was visible in the liver. After about 127 min, at least one MR-visible metabolite compound was also visible in the liver. The trifluorinated parent compound and its MR-visible metabolite both had linewidths of 30 Hz (T_2* ~33 ms). An external standard was used to estimate *in vivo* concentrations of the two compounds, with the minimum detectable concentration for niflumic acid of 1.4 μM (4.2 μM in fluorine). The standard deviation for the concentration measurements was estimated to be ±40–50% across the detected range (4.2–376 μM in fluorine). The rapid compound visibility, as well as the narrow *in vivo* fluorine linewidth suggest that the niflumic acid and metabolites may have been located in plasma rather than in tissue.

Payne *et al.* [13] used ^{19}F MR to measure the hepatic concentration of sitafloxacin, a fluoroquinolone antimicrobial agent. Elevated liver enzymes have been reported after use of sitafloxacin, even though it is eliminated predominantly by renal excretion (>99%). This evaluation was to determine whether drug accumulation in the liver occurs. Sitafloxacin has two nonequivalent fluorines (*cis*-oriented (1*R*,2*S*)-2-fluorocyclopropylamine moiety) with resonances separated by 92 ppm. The experiments were focused only on one resonance with smaller ^1H–^1F splitting and longer T_2* and T_1 values. The calculated liver half-life from the patients with evaluable fluorine spectroscopic data was 5.85 ± 5.3 h compared with 5.9 ± 0.7 h in plasma. Because of the large liver standard deviation, it is possible that there was a difference between liver and plasma half-lives; however, these study subjects were not found to have any significant changes in liver enzymes. It was concluded that there was no evidence of hepatic accumulation of sitifloxacin [13].

Seddon *et al.* [25] examined the pharmacodynamics of a tissue hypoxia probe (SR-4554) in a phase 1 dose escalation study. Three patients, with tumors of at least 3 cm diameter within 4 cm depth from the skin surface, underwent unlocalized *in vivo* ^{19}F MR spectroscopy at 1.5 T to determine whether SR-4554 could be detected. Fluorine signal was detected in all patients immediately after the intravenous infusion, in the highest two doses within 8 h (~2.5 times the mean plasma half-life), and no signal was detected even at the highest dose at 27.5 h (~9.5 plasma half-lives). The tumor concentration of SR-4554 in the tumor was of the same order as the plasma concentration; hence it is unclear whether the fluorine signal represented tumor blood flow or uptake by the tumor [25].

Long term *in vivo* biodistribution and metabolism cannot be measured using radio-pharmaceuticals. MR spectroscopy and tissue biopsy remain the only choices for investigating the long-term disposition of pharmaceuticals. While challenging, deep tissue *in vivo* biodistribution and metabolism can be measured over time using ^{19}F MR after multiple doses, after the initial washout phase, and noninvasively.

19.7 Analysis of *In Vivo* ^{19}F MR Spectra

Considerable processing must occur following collection of the MR signal decay (free induction decay or FID) and careful attention must be paid throughout the steps. Each step has the potential to alter the ability to quantify the signal based on the assumptions made and the rigor with which the parameters are chosen and applied.

First, either a Lorentzian or Gaussian filter is applied to the FID to reduce the amount of noise. The choice of lineshape will depend on the shape of the frequency domain spectrum, the lineshape is related to how the fluorine spins interact with their environment. The filter linewidth is generally similar to or slightly less than the T_2* value (T_2* can be estimated from the spectral linewidth). After application of the time domain filter, a fast Fourier transform (FFT) is performed. The resultant frequency domain spectrum will then need to undergo phase adjustment to obtain a pure absorption spectrum. The amount of receiver dead time (time lost between the end of the excitation pulse and the first useful detection time point) will determine the presence and extent of baseline artifact present as well as how difficult phase adjustment will be to accomplish.

Each resonance in the *in vivo* frequency domain spectrum is then fitted to either a Gaussian or Lorentzian lineshape. The fits of the individual resonances are separately integrated to achieve an area under the curve (AUC). By using a reference for calibration and accounting for different relaxation rates on the integrated AUC, it is possible to obtain either relative or absolute concentration from *in vivo* ^{19}F MR spectroscopy [1, 5–7, 11, 12, 23, 25, 27, 28, 39, 68–70].

Accounting for different T_2* and T_1 relaxation times of the different resonances corrects for signal loss due to incomplete T_1 relaxation of the fluorine spins. This step is required if the T_1 of the compound and reference differ and if the TR is less than 5 multiples of the longest T_1 value. In addition, signal from short-T_2* species require correction for signal loss owing to the time between the midpoint of the excitation pulse and the start of the first useful detection time point.

To obtain normalized MR signal, the compound/metabolite resonance AUC is divided by the AUC of a reference compound. A linear response of ^{19}F MR integrated signal intensity as a function of concentration (see Figure 19.5d) has been shown [5, 11, 43, 46]. Alternatively, it is also possible to normalize the ^{19}F MR spectral resonances to the proton water resonance because the water content of liver is estimated to be ~75% at 55 M [13, 25, 68–70].

To derive absolute organ concentrations from the raw spectral data, it is necessary to segment the organ of interest and to account for the effect of the ^{19}F coil sensitivity profile [5, 11, 23, 28, 43, 56, 69, 70]. Several additional experiments are required to perform this calculation. If the coil is tuned to both proton and fluorine resonance frequencies [13, 25, 68–70], direct measurement of coil sensitivity for both ^{1}H and ^{19}F is needed. This method is preferable at high magnetic field strength due to tissue-specific dielectric resonance [35] and RF power absorption [36]. The relaxation time corrected AUCs of the proton CSI spectra are then used to normalize the ^{19}F spectra. With either reference method, for quality assurance and for accuracy estimates, the transmit–receive sensitivity of the ^{19}F RF coil is mapped using a low-flip-angle proton 3D acquisition on a uniform phantom.

For external reference sample methods, a further step is required whereby the AUC of the ^{19}F MR signal using the same coil on a series of different concentration phantoms (see Figure 4B) is used to help determine the absolute fluorine calibration concentration. Then, for each subject these maps are aligned with the segmented images to calculate the effective volumes of the target organ. The normalized signal multiplied by the slope of the calibration curve and divided by the effective organ volume will provide the absolute organ concentration. The key to successful phantom calibration is to ensure that coil tuning and matching on the patients is identical to that on the phantoms.

Within the previously described limitations of co-resonant metabolites and parent compounds [11, 41], few studies have measured the accuracy of *in vivo* ^{19}F MR concentration determination because of the need for a tissue sample. However accuracy measurements have been made for other spin-½ nuclei in other species (for example, ^{31}P [69]) and found no significant difference. Even with preclinical preparation and clinical quality control, SSRIs have been documented to be underestimated by *in vivo* ^{19}F MR spectroscopy due to their protein interactions. Strauss and Drager [42] estimated the MR-invisible component of fluoxetine/norfluoxetine to be 14.2%. In primates, with the external reference sample method, Christensen *et al.* [44] measured the accuracy of brain ^{19}F MR concentration determination of dexfenfluoramine/norfenfluoramine concentrations. In contrast to SSRIs, they [44] determined that the MR concentrations of dexfenfluoramine/norfenfluoramine were greater than or equivalent to, but in all cases within one standard deviation of, those measured using gas chromatography. In addition to carefully designed and executed ^{19}F MR examinations, quality assurance throughout the experiment, preliminary *in vitro* and preclinical animal experiments are needed to help interpret *in vivo* spectroscopic signal changes in multiple investigations [23, 31, 32, 41, 42, 44]. The accuracy of quantification will always be best at the highest SNR levels.

19.8 Conclusions and Perspectives

In vivo ^{19}F MR spectroscopy has been shown to be a highly specific, noninvasive tool for identifying and monitoring fluorinated compounds and metabolites. This technology is complementary to other structural and functional imaging tools, and often a combination of approaches is required to address specific questions. Although almost all ^{19}F MR examinations are limited by the relatively low SNR of the spectra, fluorine MR spectroscopy has a unique role in clinical research applications and can provide a direct measurement of drug uptake in tissue.

While *in vivo* ^{19}F MR has provided unique insight into fluoropyrimidine metabolism, most spectroscopic studies have limited numbers of patients and thus are subject to potential bias due to the variability in individual biometabolism. The small number of subjects can further limit generalization of the results to a larger population. In addition, studies of deep tissues or small lesions are challenging using ^{19}F MR spectroscopy because they cannot be adequately evaluated using surface coil techniques and the low signal levels preclude use of a volume coil.

The number of facilities capable of performing *in vivo* ^{19}F MR spectroscopy is limited, in part due to the substantial hardware and software requirements associated with data acquisition and analysis. Because there is no reimbursement for MR spectroscopic examinations in the United States [54, 55], the number of clinical applications will continue to lag behind development of structural MR imaging.

Instrumentation improvements for ^{19}F and other multinuclear MR spectroscopy examinations have focused on increasing the SNR using similar technologies that proved valuable in clinical MR imaging. SNR gains are sought both to obtain higher spatial resolution [16] and to detect lower *in vivo* concentrations with MR spectroscopy [1, 5, 11, 38]. Increases in SNR have been achieved by performing spectroscopic examinations at higher magnetic field strengths, usually 3 T or 4 T [11, 38, 39], and by optimizing the detection hardware [24]. Detection systems for multinuclear MR spectroscopy now use digital receivers and frequency-specific components including tuned preamplifiers and switches [11, 24].

On the basis of trends over the past two decades, the availability and utilization of higher strength magnets should continue to increase, and thus increases in SNR will continue to be realized in the future. As lower-noise components become available and are incorporated into the receiver chain, incremental SNR gains will also be achieved. Use of cooled receivers and high-temperature superconducting coils [64] may also be of interest to gain SNR for low-gamma nuclei that resonate at low frequencies. In addition, use of phased-array detection coils [65] or *in situ* detection coils such as endovascular coils [66] may further improve SNR, depending upon the anatomical target. Interestingly, the ^{19}F MR spectroscopic signature has been recently used in the laboratory setting as a valuable tool for identifying potentially counterfeit or otherwise illegally manufactured pharmaceuticals such as those with lower than advertised concentrations of active ingredient [67].

In spite of the challenges, the *in vivo* biodistribution as well as absolute concentrations of fluorinated compounds can be measured noninvasively using ^{19}F MR. Human studies require careful planning and knowledge of the interactions of the molecules with the environment *in vivo*. All compounds should be assessed for the extent of protein binding, compound sequestration, or compartmentalization prior to going into the clinic [1, 11, 20, 31, 41, 42]. With a modest amount of preparation and planning, assessment of individual drug and metabolite levels is feasible using ^{19}F MR, as is measurement of the pharmacokinetics and pharmacodynamics. Since long-term *in vivo* biodistribution cannot be measured using radiopharmaceuticals, MR spectroscopy and tissue biopsy remain the only choices for investigating the disposition of pharmaceuticals. Due to the widespread use of fluorinated compounds in medicine, fluorine MR spectroscopy is a useful noninvasive method for assessing absorption, biodistribution, metabolism, and excretion during the drug development process because ^{19}F MR allows repeat measurement and accurate *in vivo* evaluation of metabolism of fluorine-containing compounds.

Acknowledgments

We thank Scott Wilkinson, MS, Steve Jones, PhD, Tom Rollins, and other members of the Sepracor Inc. Clinical and Pre-Clinical Development teams, and Blaise Frederick, PhD, Nick Bolo, PhD and Perry Renshaw, MD, PhD at McLean Hospital Brain Imaging Center in Boston, MA for their scientific contributions to the tecastemizole work. In addition, we are grateful to Sepracor, Inc. for permission to publish the tecastemizole spectra and data.

In addition, we thank Dennis Klomp, PhD (University Medical Center Utrecht, Utrecht, the Netherlands) and Nick Bolo, PhD (McLean Hospital, Boston, MA, U.S.A.) for their generous contribution of the materials for Figures 19.7 and 19.8, respectively.

Abbreviations

1D	one-dimensional
2D	two-dimensional
3D	three-dimensional
5-FU	5-fluorouracil
AUC	area under the curve
CSA	chemical shielding anisotropy
CSI	chemical shift imaging
DMSO	dimethyl sulfoxide
fen-phen	dexfenfluramine
FBAL	2-fluoro-β-alanine
FDG	fluorodeoxyglucose
FID	free induction decay
FFT	fast Fourier transform
LC-MS	liquid chromatography-mass spectrometry

MR	magnetic resonance
NEX	number of acquisitions
NOE	nuclear Overhauser effect
NSAID	nonsteroidal anti-inflammatory drug
PET	positron emission tomography
RF	radiofrequency
SAR	specific absorption rate
SNR	signal-to-noise ratio
SPECT	single-photon emission computerized tomography
SR-4554	SR-4554 (compound number; sensitive to tissue hypoxia)
SSRI	serotonin-specific reuptake inhibitor
T_1	spin–lattice relaxation time
T_2	spin–spin relaxation time
T_2*	effective T_2 relaxation time
TR	pulse repetition time

References

1. Lyoo, I. K. and Renshaw, P. F. (2002) Magnetic resonance spectroscopy: current and future applications in psychiatric research. *Biological Psychiatry*, **51**, 195–207.
2. Sassa, T., Suhara, T., Ikehira, H., *et al.* (2002) ^{19}F-magnetic resonance spectroscopy and chemical shift imaging for schizophrenic patients using haloperidol decanoate. *Psychiatry and Clinical Nuerosciences*, **56**, 637–642.
3. Port, R. E. and Wolf, W. (2003) Noninvasive methods to study drug distribution. *Investigational New Drugs*, **21**, 157–168.
4. Martino, R., Gilard, V., Desmoulin, F. and Malet-Martino, M. (2006) Interest of fluorine-19 nuclear magnetic resonance spectroscopy in the detection, identification and quantification of metabolites of anticancer and antifungal fluoropyrimidine drugs in human biofluids. *Chemotherapy*, **52**(5), 215–219.
5. Bachert, P. (1998). Pharmacokinetics using fluorine NMR *in vivo*. *Progress in NMR Spectroscopy*, **33**, 1–56.
6. Wolf, W., Albright, M. J., Silver, M. S., *et al.* (1987) Fluorine-19 NMR spectroscopic studies of the metabolism of 5-fluorouracil in the liver of patients undergoing chemotherapy. *Magnetic Resonance Imaging*, **5**, 165–169.
7. Wolf, W., Presant, C. A. and Waluch, V. (2000) 19F-MRS studies of fluorinated drugs in humans. *Advanced Drug Delivery Reviews*, **41**, 55–74.
8. Malet-Martino, M., Gilard, V., Desmoulin, F. and Martino, R. (2006) Fluorine nuclear magnetic resonance spectroscopy of human biofluids in the field of metabolic studies of anticancer and antifungal fluoropyrimidine drugs. *Clinica Chimica Acta*, **366**, 61–73.
9. van Laarhoven, H. W. M., Punt, C. J. A., Kamm, Y. J. L. and Heerschap, A. (2005) Monitoring fluoropyrimidine metabolism in solid tumors with *in vivo* ^{19}F magnetic resonance spectroscopy. *Critical Reviews in Oncology/Hematology*, **56**, 321–343.
10. Martino, R. G., Filard, V., Desmoulin, F. and Malet-Martino, M. (2005) Fluorine-19 or phosphorus-31 NMR spectroscopy: a suitable analytical technique for quantitative *in vitro* metabolic studies of fluorinated or phosphorylated drugs. *Journal of Pharmaceutical and Biomedical Analysis*, **38**(5), 871–891.

11. Schneider, E., Bolo, N. R., Frederick, B., *et al.* (2006) Magnetic resonance spectroscopy for measuring the biodistribution and *in situ in vivo* pharmacokinetics of fluorinated compounds: validation using an investigation of liver and heart disposition of tecastemizole. *Journal of Clinical Pharmacy and Therapeutics*, **31**, 261–273.
12. Bilecen, D., Sculte, A. C., Kaspar, A., *et al.* (2003) Detection of the non-steroidal anti-inflammatory drug niflumic acid in humans: a combined ^{19}F MRS *in vivo* and *in vitro* study. *NMR in Biomedicine*, **16**, 144–151.
13. Pellecchia, M., Sem, D. S. and Wüthrich, K. (2002) NMR in drug discovery. *Nature Reviews Drug Discovery*, **1**, 211–219.
14. Marzola, P., Osculati, F. and Sbarbati, A. (2003) High field MRI in preclinical research. *European Journal of Radiology*, **48**, 165–170.
15. Beckmann, N., Mueggler, T., Allegrini, P. R., *et al.* (2001) From anatomy to the target: contributions of magnetic resonance imaging to preclinical pharmaceutical research. *Anatomical Record*, **265**(2), 85–100.
16. Payne, G. S., Collins, D. J. and Loynds, P., *et al.* (2004) Quantitative assessment of the hepatic pharmacokinetics of the antimicrobial sitafloxacin in humans using *in vivo* ^{19}F magnetic resonance spectroscopy. *British Journal of Clinical Pharmacology*, **59**(2), 244–248.
17. Brix, G., Bellemann, M. E., Haberkorn, U., *et al.* (1996) Assessment of the biodistribution and metabolism of 5-fluorouracil as monitored by ^{18}F PET and ^{19}F MRI: a comparative animal study. *Nuclear Medicine and Biology*, **23**, 897–906.
18. Park, B. K., Kitteringham, N. R. and O'Neill, P. M. (2001) Metabolism of Fluorine-containing Drugs. *Annual Reviews of Pharmacology and Toxicology*, **41**, 443–470.
19. Danielson, M. A. and Falke, J. J. (1996) Use of ^{19}F NMR to probe protein structure and conformational change. *Annual Reviews of Biophysics and Biomolecular Structure*, **25**, 163–195.
20. Dzik-Jurasz, A. S. K., Leach, M. O. and Rowland, I. J. (2004) Investigation of microenvironmental factors influencing the longitudinal relaxation times of drugs and other compounds. *Magnetic Resonance Imaging*, **22**(7), 973–982.
21. Harris, R. K. (1983) *NMR Spectroscopy: A Physicochemical View*, Pitman, London.
22. Bruno, T. J. and Svoronos, P. D. N. (2003) *CRC Handbook of Basic Tables for Chemical Analysis*, 2nd edn., CRC Press, Taylor and Francis Group, Boca Raton, FL.
23. Jynge, P., Skjetne, T., Gribbestad, I., *et al.* (1990) *In vivo* tissue pharmacokinetics by fluorine magnetic resonance spectroscopy: A study of liver and muscle disposition of fleroxacin in humans. *Journal of Clinical Pharmacy and Therapeutics*, **48**, 481–489.
24. Klomp, D. W. J., van Laarhoven, H. W. M., Kentgens, A. P. M. and Heerschap, A. (2003) Optimization of localized ^{19}F magnetic resonance spectroscopy for the detection of fluorinated drugs in the human liver. *Magnetic Resonance in Medicine*, **50**, 303–308.
25. Seddon, B. M., Payne, G. S., Simmons, L., *et al.* (2003) A Phase I study of SR-4554 via intravenous administration for noninvasive investigation of tumor hypoxia by magnetic resonance spectroscopy in patients with malignancy. *Clinical Cancer Research*, **9**, 5101–5112.
26. Kamm, Y. J. L., Heerschap, A., van den Bergh, E. J. and Wagener, D. J. T. (2004) ^{19}F-magnetic resonance spectroscopy in patients with liver metastases of colorectal cancer treated with 5-fluorouracil. *Anti-Cancer Drugs*, **15**, 229–233.
27. Bolo, N. R., Hode, Y., Nedelec, J. F., *et al.* (2000) Brain pharmacokinetics and tissue distribution *in vivo* of fluvoxamine and fluoxetine by fluorine magnetic resonance spectroscopy. *Neuropsychopharmacology*, **23**, 428–438.
28. Miner, C. M., Davidson, J. R. T., Potts, N. L. S., *et al.* (1995) Brain fluoxetine measurements using ^{19}F MRS in social phobia. *Biological Psychiatry*, **38**, 696–698.
29. Strauss, W. L., Layton, M. E., Hayes, C. E. and Dager, S. R. (1997) ^{19}F MRS of acute and steady state brain fluvoxamine levels in obsessive compulsive disorder. *American Journal of Psychiatry*, **151**, 516–522.

30. Strauss, W. L., Layton, M. E. and Dager, S. R. (1998) Brain elimination half life of fluvoxamine ^{19}F MRS. *American Journal of Psychiatry*, **155**, 380–384.
31. Christensen, J. D., Yurgelun-Todd, D. A., Babb, S. M., *et al.* (1999) Measurement of human brain dexfenfluramine by ^{19}F MRS. *Brain Research*, **831**, 1–5.
32. Komoroski, R., Newton, J. E., Cardwell, D., *et al.* (1994) *In vivo* ^{19}F spin relaxation and localized spectroscopy of fluoxetine in human brain. *Magnetic Resonance in Medicine*, **31**, 204–211.
33. Renshaw, P., Guimaraes, A. R., Fava, M., *et al.* (1992) Accumulation of fluoxetine and norfluoxetine in human brain during therapeutic administration. *American Journal of Psychiatry*, **149**, 1592–1594.
34. Schenck, J. F. (2000) Safety of strong, static magnetic fields. *Journal of Magnetic Resonance Imaging*, **12**, 3–19.
35. Bottomley, P. A. and Andrew, E. R. (1978) RF magnetic field penetration, phase shift and power dissipation in biological tissue: implication for NMR imaging. *Physics in Medicine and Biology*, **23**, 630–643.
36. Hoult, D. I. (2000) Sensitivity and power deposition in a high-field imaging experiment. *Journal of Magnetic Resonance Imaging*, **12**, 46–67.
37. Hoult, D. I. and Lauterbur, P. C. (1979) The sensitivity of the zeugmatographic experiment involving human samples. *Journal of Magnetic Resonance*, **34**, 425–433.
38. van Laarhoven, H. W. M., Klomp, D. W., Kamm, Y. J. L., *et al.* (2003) *In vivo* monitoring of capecitabine metabolism in human liver by ^{19}Fluorine magnetic resonance spectroscopy at 1.5 and 3T field strength. *Cancer Research*, **63**, 7609–7612.
39. Bolo, N. R., Hode, Y. and Macher, J. P. (2004) Long-term sequestration of fluorinated compounds in tissues after fluvoxamine or fluoxetine treatment: a fluorine magnetic resonance spectroscopy study *in vivo*. *Magnetic Resonance in Materials Physics, Biology and Medicine*, **16**, 268–276.
40. Purcell, E. M., Torrey, E. C. and Pound, R. V. (1946) Resonance absorption by nuclear magnetic moments in a solid. *Physics Review*, **69**, 37–38.
41. Karson, C. N., Newton, J. E., Livingston, R., *et al.* (1993) Human brain fluoxetine concentrations. *Journal of Neuropsychiatry and Clinical Neurosciences*, **5**, 322–329.
42. Strauss, W. L. and Dager, S. R. (2001) Magnetization transfer of fluoxetine in the brain with ^{19}F MRS. *Biological Psychiatry*, **49**, 798–802.
43. Schlemmer, H. P., Bachert, P., Semmler, W., *et al.* (1994) Drug monitoring of 5-fluorouracil: *in vivo* ^{19}F NMR study during 5-FU chemotherapy in patients with metastases of colorectal adenocarcinoma. *Magnetic Resonance Imaging*. **12**, 497–511.
44. Christensen J. D., Babb, S. M., Cohen, B. M. and Renshaw, P. F. (1998) Quantitation of dexfenfluramine/d-norfenfluramine concentration in primate brain using ^{19}F NMR spectroscopy. *Magnetic Resonance in Medicine*, **39**, 149–154.
45. Henry, M. E., Moore, C. M., Kaufman, M. J., *et al.* (2000) Brain kinetics of paraxetine and fluoxetine. *American Journal of Psychiatry*, **157**, 1506–1508.
46. Martino, R., Malet-Martino, M. and Gilard, V. (2000) Fluorine nuclear magnetic resonance, a privileged tool for metabolic studies of fluoropyrimidine drugs. *Current Drug Metabolism* **1**(3), 271–303.
47. Wolf, W., Waluch, V. and Presant, C. A. (1998) Non-invasive ^{19}F NMRS of 5-fluorouracil in pharmacokinetics and pharmacodynamic studies. *NMR in Biomedicine*, **11**, 380–387.
48. Haase, A., Frahm, J., Hänicke, W. and Matthaei, D. (1985) ^{1}H NMR chemical shift selective (CHESS) imaging. *Physics in Medicine and Biology*, **30**, 341–344.
49. Frahm, J., Haase, A., Hänicke, W., *et al.* (1985) Chemical shift selective MR imaging using a whole-body magnet. *Radiology*, **156**, 441–444.
50. Gonen, O., Murphy-Boesch, J., Li, C. W., *et al.* (1997) Simultaneous 3D NMR spectroscopy of proton-decoupled fluorine and phosphorus in human liver during 5-fluorouracil chemotherapy. *Magnetic Resonance in Medicine*, **37**, 164–169.

51. Prozac, package insert. Eli Lilly and Company, Indianapolis, IN, U.S.A. www.lilly.com.
52. Redux, package insert. Wyeth-Ayerst Laboratories, Philadelphia, PA, U.S.A.
53. Wolf, W., Presant, C. A., Servis, K. L., *et al.* (1990) Tumor trapping of 5-fluorouracil: *in vivo* ^{19}F NMR spectroscopic pharmacokinetics in tumor-bearing humans and rabbits. *Proceedings of the National Academy of Sciences U.S.A.* **87**(1), 492–496.
54. Centers for Medicare & Medicaid Services. Decision Memo for Magnetic Resonance Spectroscopy for Brain Tumors (CAG-00141N). January, **29**, 2004. http://www.cms.hhs.gov/mcd/viewdecisionmemo.asp?id=52
55. Centers for Medicare & Medicaid Services. National Coverage Determination for Magnetic Resonance Spectroscopy (220.2.1). September 2004. http://www.cms.hhs.gov
56. Li, C. W., Negendank, W. G., Padavic-Shaller, K. A., *et al.* (1996) Quantitation of 5-fluorouracil catabolism in human liver *in vivo* by three-dimensional localized ^{19}F magnetic resonance spectroscopy. *Clinical Cancer Research*, **2**, 339–345.
57. Dzik-Jurasz, A. S. K., Collins, D. J., Leach, M. O. and Rowland, I. J. (2000) Gallbladder localization of ^{19}F MRS catabolite signals in patients receiving bolus and protracted venous infusional 5-fluorouracil. *Magnetic Resonance in Medicine*, **44**, 516–520.
58. Schaefer, D. J., Bourland, J. D. and Nyenhuis, J. A. (2000) Review of patient safety in time-varying gradients fields. *Journal of Magnetic Resonance Imaging*, **12**, 20–29.
59. Griffiths, J. R. and Glickson, J. D. (2000) Monitoring pharmacokinetics of anticancer drugs: non-invasive investigation using magnetic resonance spectroscopy. *Advances in Drug Delivery Review*, **41**, 75–89.
60. Beckmann, N. (ed.) (2006) *In vivo MR Techniques in Drug Discovery and Development*, Taylor and Francis Group, New York.
61. Beckmann, N., Laurent D, Tigani B, *et al.* (2004) Magnetic resonance imaging in drug discovery: lessons from disease areas. *Drug Discovery Today*, **9**, 35–42.
62. Mason, G. F. and Krystal, J. H. (2006) MR spectroscopy: its potential role for drug development for the treatment of psychiatric diseases. *NMR in Biomedicine*, **19**, 690–701.
63. Workman, P., *et al.* (2006) Minimally invasive pharmacokinetic and pharmacodynamic technologies in hypothesis-testing clinical trials of innovative therapies. *Journal of the National Cancer Institute*, **98**, 580–598.
64. Miller, J. R., Zhang K., Ma, Q. Y., *et al.* (1996) Superconducting receiver coils for sodium magnetic resonance imaging. *IEEE Transactions on Biomedical Engineering*, **43**(12),1197–1199.
65. Xu, D., Chen, A. P., Cunningham, C., *et al.* (2006) Spectroscopic imaging of the brain with phased-array coils at 3.0T. *Magnetic Resonance Imaging*, **24**(1), 69–74.
66. Wacker, F. K., Hillenbrand, C. M., Duerk, J. L. and Lewin, J. S. (2005) MR-guided endovascular interventions: device visualization, tracking, navigation, clinical applications, and safety aspects. *Magnetic Resonance Imaging Clinics of North America*, **13**(3), 431–439.
67. Trefi, S., Gilard, V., Balayssac, S., *et al.* (2008) Quality assessment of fluoxetine and fluvoxamine pharmaceutical formulations purchased in different countries or via the Internet by ^{19}F and 2D DOSY ^{1}H NMR. *Journal of Pharmaceutical and Biomedical Analysis*, **46**(4), 707–722.
68. Klomp D., van Laarhoven H., Scheenen T., *et al.* (2007) Quantitative ^{19}F MR spectroscopy at 3T to detect heterogeneous capecitabine metabolism in human liver. *NMR in Biomedicine*, **20**(5), 485–492.
69. Song, S., Hotchkiss, R. S. and Ackerman, J. J. H. (1992) Concurrent quantification of tissue metabolism and blood flow via ^{2}H/^{31}P NMR *in vivo*. I. Assessment of absolute metabolite quantification. *Magnetic Resonance in Medicine*, **25**(1), 45–55.
70. Thulborn, K. R. and Ackerman, J. J. H. (1983) Absolute molar concentrations by NMR in inhomogeneous B1 – a scheme for analyses of *in vivo* metabolites. *Journal of Magnetic Resonance*, **55**(3), 357–371.

71. Abragam, A. (1961) *Principles of Nuclear Magnetism*, Oxford University Press, New York.
72. Nagayama, K., Bachmann, P., Ernst, R. R. and Wüthrich, K. (1979) Selective spin decoupling in the *J*-resolved two-dimensional ^{1}H NMR spectra of proteins. *Biochemical and Biophysical Research Communications*, **86**(1), 218–225.
73. Krems, B., Bachert, P., Zabel, H. J. and Lorenz, W. J. (1995) ^{19}F-^{1}H nuclear Overhauser effect and proton decoupling of 5-fluoruracil and alph-fluoro-beta-alanine. *Journal of Magnetic Resonance B* **108**(2), 155–164.
74. Li, B. S., Payne, G. S., Collins, D. J. and Leach, M. O. (2000) ^{1}H decoupling for *in vivo* ^{19}F MRS studies using the time-share modulation method on a clinical 1.5T NMR system. *Magnetic Resonance in Medicine*, **44**(1), 5–9.

Appendix

Approved Active Pharmaceutical Ingredients Containing Fluorine

Elizabeth Pollina Cormier, Manisha Das, and Iwao Ojima*

The following tables list drugs containing fluorine that have been approved by the U.S. Food and Drug Administration (FDA) for use in either humans or animals. For human products, we used the FDA Orange Book, 27th edition (http://www.fda.gov/cder/orange/), which covers drug products approved under Section 505 of the Federal Food, Drug, and Cosmetic Act up to 2007. Thus, the updated version may include more fluorine-containing drugs. Specifically, sections 3.0, 4.0, and 6.0, covering prescription, OTC, and discontinued drug products, respectively, were considered. Information was verified for completeness using the Electronic Orange Book (http://www.fda.gov/cder/ob/), which is updated on a monthly basis, as well as Drugs@FDA (http://www.accessdata.fda.gov/scripts/cder/drugsatfda/).

For veterinary products, we used the FDA Green Book (http://www.fda.gov/cvm/Green_Book/elecgbook.html), which covers animal drug products approved under section 512 of the Federal Food, Drug and Cosmetic Act up to 2007. Thus, the updated version may include more fluorine-containing drugs. Sections 2 and 6 containing active ingredient and voluntary withdrawals were reviewed to compile the relevant compounds. Information was elaborated with the Database of Approved Animal Drugs (http://dil.vetmed.vt.edu/) as well as the Freedom of Information (FOI) Summaries (http://www.fda.gov/cvm/FOI/foidocs.htm).

Other chemical information was obtained from:

*Dr. Cormier's contribution was performed on her own time, and does not represent official policy or views of the U.S. Food and Drug Administration. No official support or endorsement by the FDA is intended or should be inferred.

Fluorine in Medicinal Chemistry and Chemical Biology Edited by Iwao Ojima

- National Library of Medicine's ChemIDplus (http://chem.sis.nlm.nih.gov/chemidplus/chemidlite.jsp)
- National Center for Biotechnology Information's PubChem (http://pubchem.ncbi.nlm.nih.gov/)
- The Merck Index (http://themerckindex.cambridgesoft.com)

Section 1 Fluorine-containing Drugs for Human Use Approved by FDA in the United States

Section 2 Fluorine-containing Drugs for Veterinary Use Approved by FDA in the United States

Section 1. Fluorine-containing Drugs for Human Use Approved by FDA in the United States

Fluorine in Medicinal Chemistry and Chemical Biology Edited by Iwao Ojima

Active ingredient/ trade name	Chemical name	Structure
Alatrofloxacin Mesylate/ Trovan	7-[(1*R*,5*S*)-6-[[(2*S*)-2-[[(2*S*)-2-amino-propanoyl]amino] propanoyl]amino]-3-azabicyclo[3.1.0]hexan-3-yl]-1-(2,4-difluorophenyl)-6-fluoro-4-oxo-1,8-naphthyridine-3-carboxylic acid; methanesulfonic acid	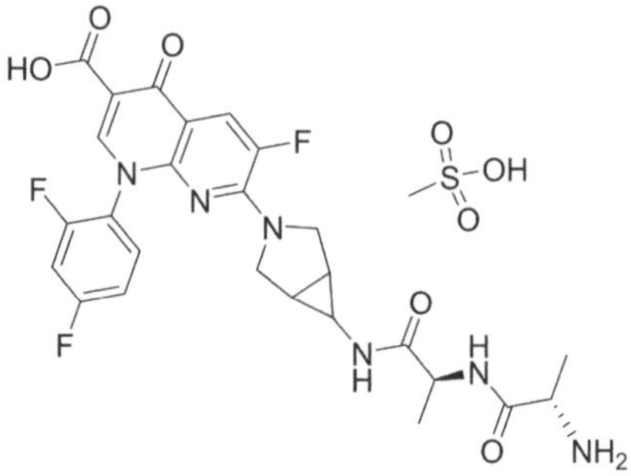
Amcinonide	pregna-1,4-diene-3,20-dione, 21-(acetyloxy)-16,17-[cyclopentylidenebis (oxy)] -9-fluoro-11-hydroxy-,(11β, 16α)	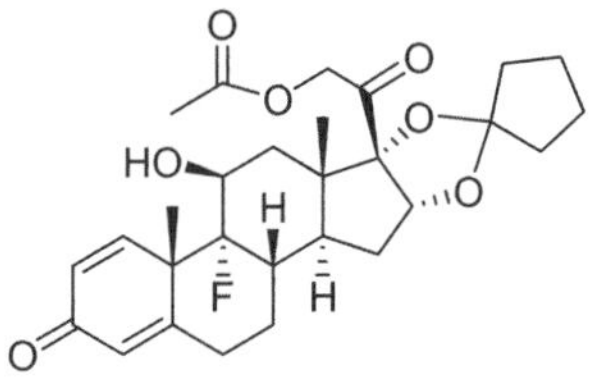
Aprepitant/ Emend	5-[[(2*R*,3*S*)-2-[(1*R*)-1-[3,5-bis-(trifluoromethyl)phenyl] ethoxy]-3-(4-fluorophenyl)-4-morpholinyl]methyl]-1,2-dihydro-3H-1,2,4-triazol-3-one	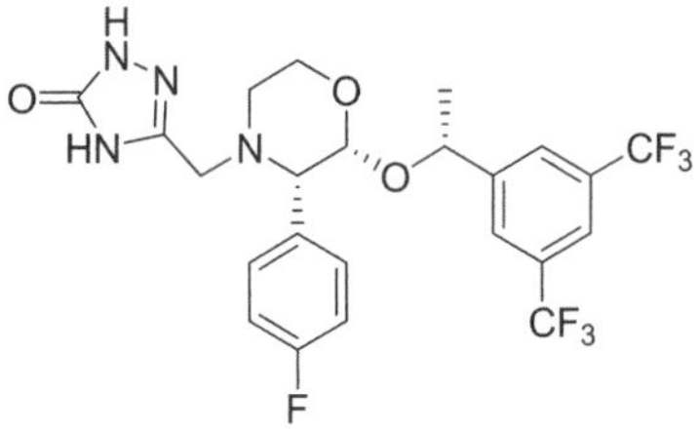
Atorvastatin Calcium/ Lipitor Caduet	[*R*-(*R**,*R**)]-2-(4-fluorophenyl)-β, δ-dihydroxy-5-(1-methylethyl)-3-phenyl-4-[(phenyl-amino)carbonyl]-1*H*-pyrrole-1-heptanoic acid,calcium salt (2 : 1) trihydrate	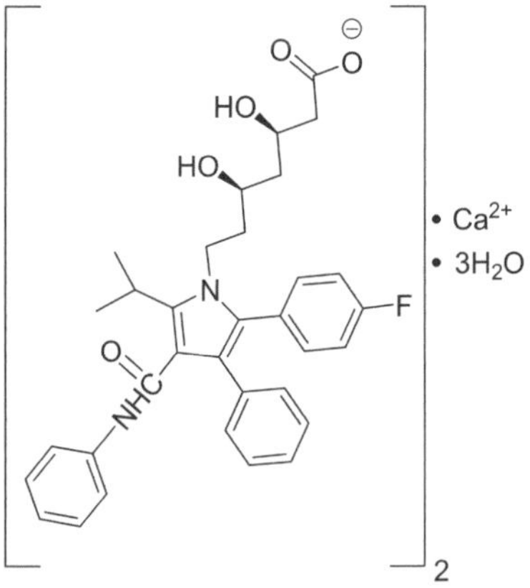 • Ca^{2+} • $3H_2O$
Bendroflumethiazide/ Corzide	3-benzyl-5,5-dioxo-9-(trifluoromethyl)-5λ^6-thia-2,4-diaza-bicyclo[4.4.0] deca-7,9,11-triene-8-sulfonamide	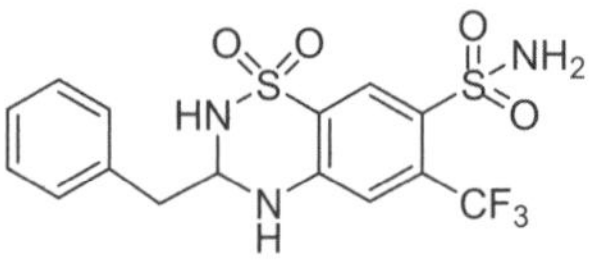

Formula	Pharmaceutical action	Dosage form	CAS No.	Approval date/ drug sponsor
$C_{27}H_{29}F_3N_6O_8S$	Anti-infective agent	Injectable	157605-25-9	Dec. 1997/ Pfizer
$C_{28}H_{35}FO_7$	Anti-inflammatory agent Antipruritic agents Glucocorticoid	Topical cream, lotion and ointment	51022-69-6	Oct. 1979/ Altana Taro Pharm Inds
$C_{23}H_{21}F_7N_4O_3$	Antiemetic	Oral capsule	170729-80-3	March 2003/ Merck
$(C_{33}H_{34}FN_2O_5)_2Ca \cdot 3H_2O$	Anticholesteremic agent	Oral tablet	344423-98-9	Dec. 1996/ Pfizer
$C_{15}H_{14}F_3N_3O_4S_2$	Diuretic Anihyertensive agent	Oral tablet	73-48-3	May 1983/ King Pharms Impax Labs

Active ingredient/ trade name	Chemical name	Structure
Betamethasone Benzoate/ Uticort[Δ]	[(8*S*,9*R*,10*S*,11*S*,13*S*,14*S*, 16*S*,17*R*)-9-fluoro-11-hydroxy-17-(2-hydroxy-acetyl)-10,13,16-trimethyl-3-oxo-6,7,8,11,12,14,15,16-octahydro-cyclo-penta[a] phenanthren-17-yl] benzoate	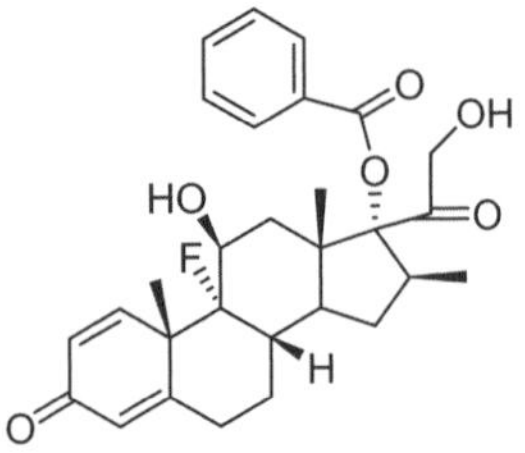
Betamethasone Dipropionate[V]/ Diprolene Lotrisone Taclonex	[(8*S*,9*R*,10*S*,11*S*,13*S*,14*S*, 6*S*,17*R*)-9-fluoro-11-hydroxy-10,13,16-trimethyl-3-oxo-17-(2-propanoyloxyacetyl)-6,7,8, 11,12,14,15, 16-octahydrocyclo-penta[a]phenanthren-17-yl] propanoate	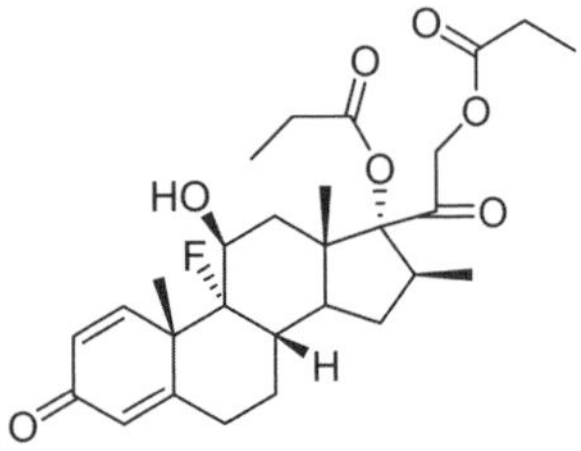
Betamethasone Sodium Phosphate[V]/ Celestone Soluspan	[2-[(8*S*,9*R*,10*S*,11*S*,13*S*, 14*S*,16*S*,17*R*)-9-fluoro-11,17-dihydroxy-10,13,16-trimethyl-3-oxo-6,7,8,11, 12,14,15,16-octahydrocy-clopenta[*a*]phenanthren-17-yl]-2-oxo-ethyl] phosphate	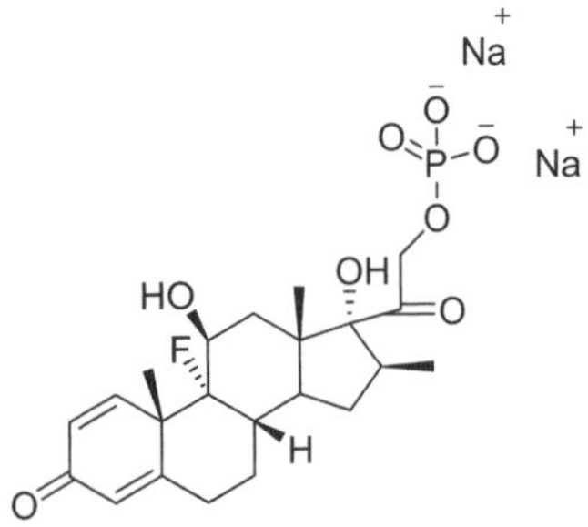
Betamethasone Valerate[V]/ Luxiq Valnac Beta-Val Dermabet	(1) pregna-1,4-diene-3,20-dione, 9-fluoro-11,21-dihydroxy-16-methyl-17-[(1-oxopentyl)oxy]-, (11β, 16β)-; (2) 9-fluoro-11 β,17,21-trihydroxy-16 β-methylpregna-1,4-diene-3,20-dione 17 valerate	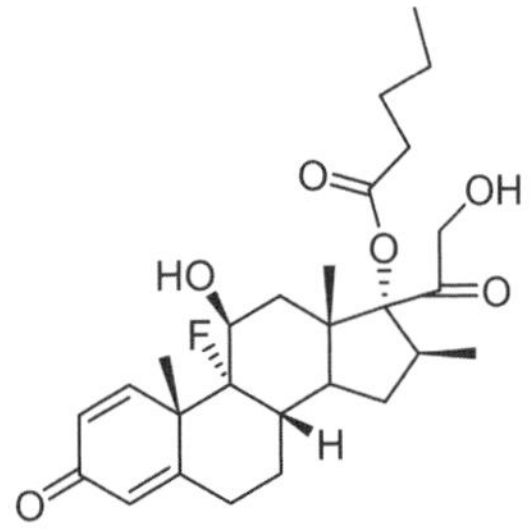
Betamethasone/ Celestone	9-fluoro-11,17-dihydroxy-17-(2-hydroxyacetyl)-10,13,16-trimethyl-6,7,8,9,10, 11,12,13,14,15,16,17-dodecahydrocyclopenta [*a*]phenanthren-3-one	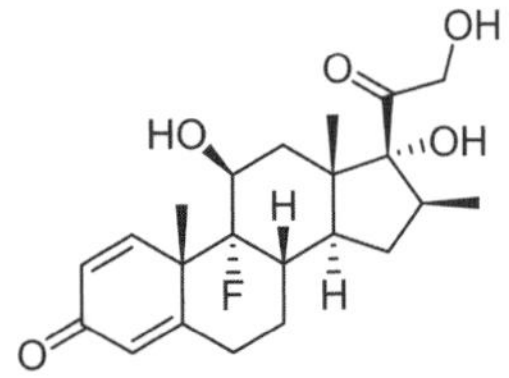

Formula	Pharmaceutical action	Dosage form	CAS No.	Approval date/ drug sponsor
$C_{29}H_{33}FO_6$	Glucocorticoid	Topical cream, gel, lotion, and ointment	22298-29-9	Prior to 1982/ Parke-Davis
$C_{28}H_{37}FO_7$	Anti-inflammatory agent Glucocorticoid	Topical cream, gel, lotion, and ointment	5593-20-4	Prior to 1982/ Actavis Mid Atlantic Fougera Taro Teva Perrigo New York QLT USA Schering Schering Plough Res
$C_{22}H_{28}FNa_2O_8P$		Injectable	151-73-5	Prior to 1982/ Schering
$C_{27}H_{37}FO_6$	Anti-inflammatory agent	Topical cream, lotion, ointment	2152-44-5	Prior to 1982/ Connetics Actavis Mid Atlantic Fougera Taro Teva
$C_{22}H_{29}FO_5$	Anti-inflammatory agent Glucocorticoid Anti-asthmatic agent	Oral syrup	378-44-9	April 1961/ Schering

Active ingredient/ trade name	Chemical name	Structure
Bicalutamide/ Casodex	*N*-[4-cyano-3-(trifluoromethyl)phenyl]-3-[(4-fluorophenyl) sulfonyl]-2-hydroxy-2-methyl-, (+/–)	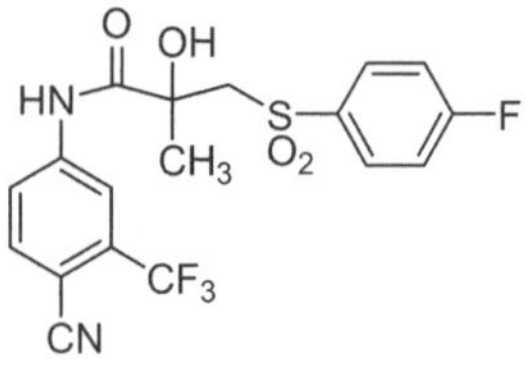
Bromperidol/ Impromen Tesoprel Azuren	4-[4-(4-bromo-phenyl)-4-hydroxy-1-piperidyl]-1-(4-fluorophenyl)butan-1-one	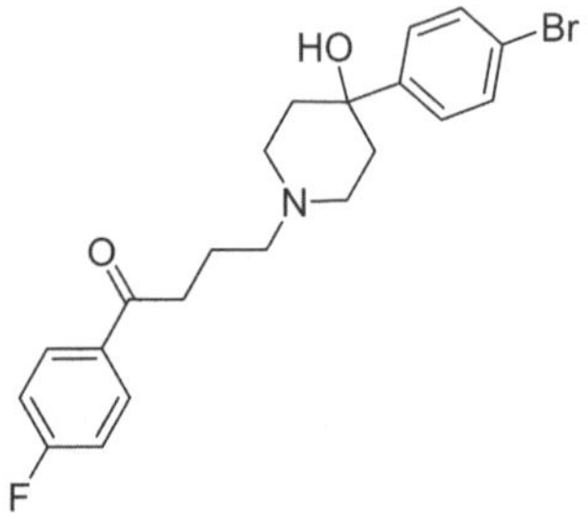
Capecitabine/ Xeloda	5′-deoxy-5-fluoro-*N*-[(pentyloxy) carbonyl]cytidine	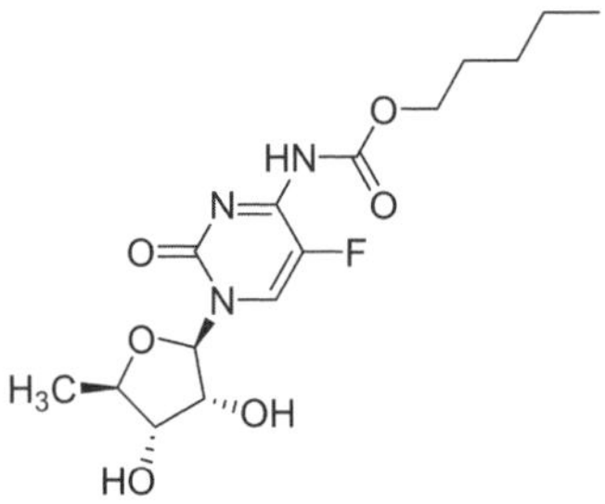
Celecoxib/ Celebrex	4-[5-(4-methylphenyl)-3-(trifluoromethyl)-1*H*-pyrazol-1-yl] benzenesulfonamide	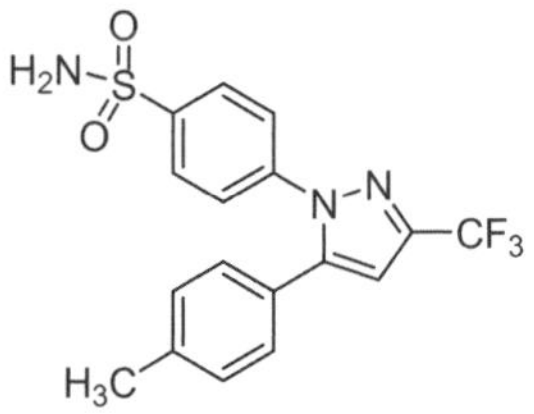
Cerivastatin Sodium/ Baycol$^{\Delta}$	sodium [*S*-[*R**,*S**-(*E*)]]-7-[4-(4-fluorophenyl)-5-methoxymethyl)-2,6-bis (1-methylethyl)-3-pyridinyl]-3,5-dihydroxy-6-heptenoate	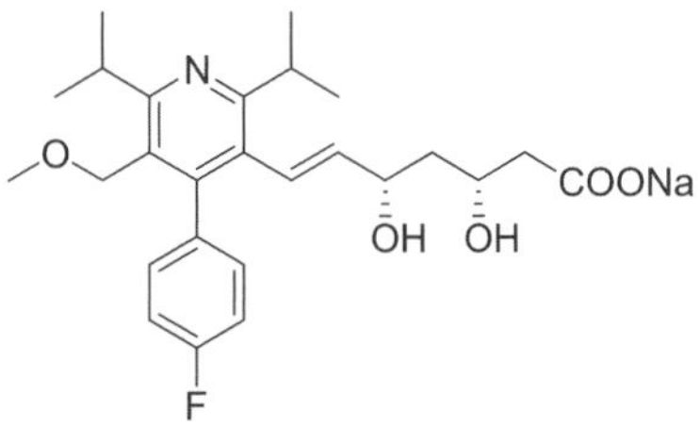
Cinacalcet Hydrochloride/ Sensipar	*N*-[1-(*R*)-(-)-(1-naphthyl)ethyl]-3-[3-(trifluoromethyl)phenyl]-1-aminopropane hydrochloride	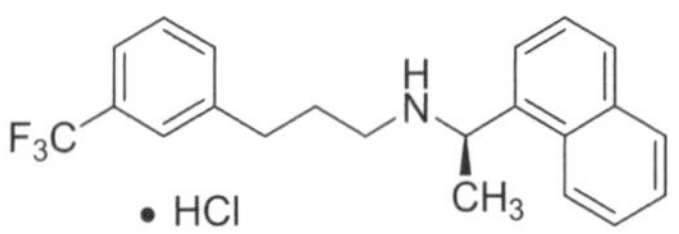

Formula	Pharmaceutical action	Dosage form	CAS No.	Approval date/ drug sponsor
$C_{18}H_{14}N_2O_4F_4S$	Androgen antagonist Antineoplastic agent	Oral tablet	90357-06-5	Oct. 1995/ Astrazeneca
$C_{21}H_{23}BrFNO_2$	Antipsychotic agent		10457-90-6	Janssen Organon Cilag-Chemi
$C_{15}H_{22}FN_3O_6$	Antineoplastic agent	Oral tablet	154361-50-9	April 1998/ HLR
$C_{17}H_{14}F_3N_3O_2S$	Anti-inflammatory Nonsteroidal cyclooxygenase inhibitor	Oral capsule	169590-42-5	Dec. 1998/ GD Searle
$C_{26}H_{33}FNO_5Na$	Hydroxymethylglutaryl-CoA reductase inhibitor	Oral tablet	143201-11-0	June 1997/ Bayer Pharms
$C_{22}H_{22}F_3N \cdot HCl$	Calcimimetic agent	Oral tablet	226256-56-0	March 2004/ Amgen

Active ingredient/ trade name	Chemical name	Structure
Ciprofloxacin Hydrochloride/ Ciloxan Cipro Proquin XR	monohydrochloride monohydrate salt of 1-cyclopropyl-6-fluoro-1, 4-dihydro-4-oxo-7-(1-piperazinyl)-3-quinolinecarboxylic acid	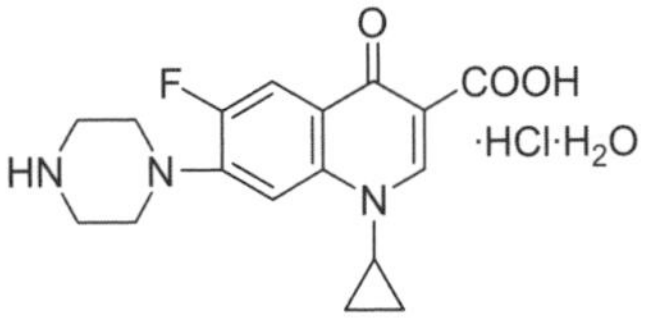 F, COOH, $\cdot HCl \cdot H_2O$, HN, N, N, O
Ciprofloxacin/ Cipro	1-cyclopropyl-6-fluoro-1,4-dihydro-4-oxo-7-(1-piperazinyl)-3-quinolinecarboxylic acid	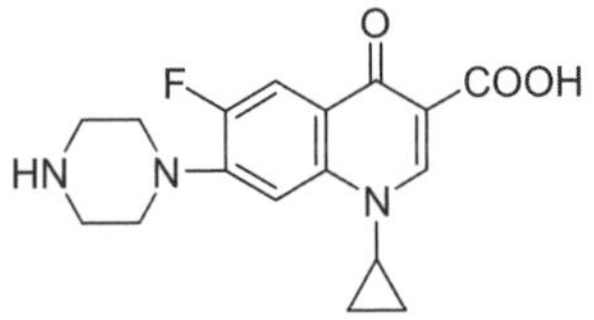 F, COOH, HN, N, N, O
Cisapride Monohydrate[Δ]/ Propulsid	4-amino-5-chloro-*N*-[1-[3-(4-fluorophenoxy)propyl]-3-methoxypiperidin-4-yl]-2-methoxybenzamide hydrate	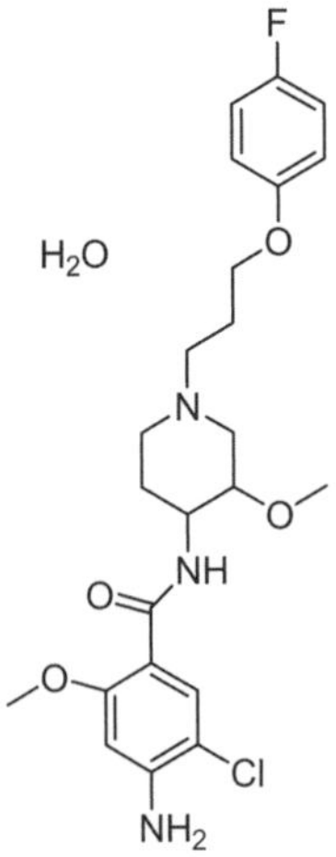 F, H_2O, O, N, O, NH, O, O, Cl, NH_2

Formula	Pharmaceutical action	Dosage form	CAS No.	Approval date/ drug sponsor
$C_{17}H_{18}FN_3O_3 \cdot HCl \cdot H_2O$	Antibacterial agent	Ophthalmic ointment and drops Oral tablet	86393-32-0	Oct. 1987/ Alcon, Bausch and Lomb Hitech Pharma Novex Bayer Pharms Barr Carlsbad Cobalt Dr Reddys Labs Ltd Genpharm Hikma IVAX Pharms Martec USA LLC Mylan Pliva Ranbaxy Sandoz Taro Teva Torpharm Unique Pharm Labs Esprit Pharma
$C_{17}H_{18}FN_3O_3$	Antimicrobial agent	Oral suspension Injectable Ophthalmic drops	85721-33-1	Oct. 1987/ Bayer Pharms Abraxis Pharm Bedford Labs Hospira Sicor Pharms Nexus Pharms
$C_{23}H_{31}ClFN_3O_5$	Anti-ulcer agent Gastrointestinal agent Serotonin agonist	Oral suspension and tablet	81098-60-4	July 1993/ Janssen Pharma

Active ingredient/ trade name	Chemical name	Structure
Citalopram Hydrobromide/ Celexa	1-[3-(dimethylamino)-propyl]-)-1-(p-fluorophenyl)- 5-phthalancarbonitrile monohydrobromide	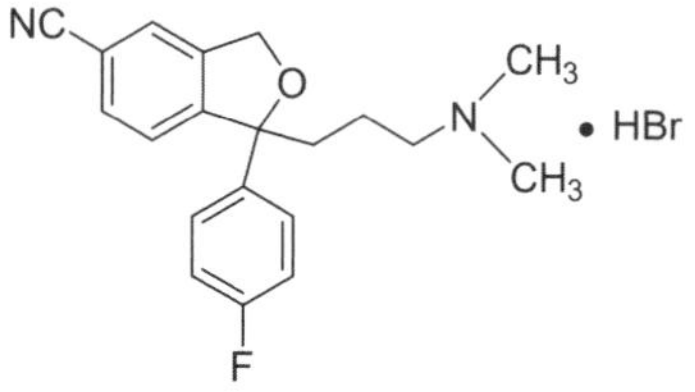 NC, O, N, CH_3, CH_3, • HBr, F
Clobetasol Propionate/ Olux Temovate Embeline Cormax Clobex	(11β, 16β)-21-chloro-9-fluoro-11-hydroxy-16-methyl-17-(1-oxopropoxy)-pregna-1,4-diene-3,20-dione	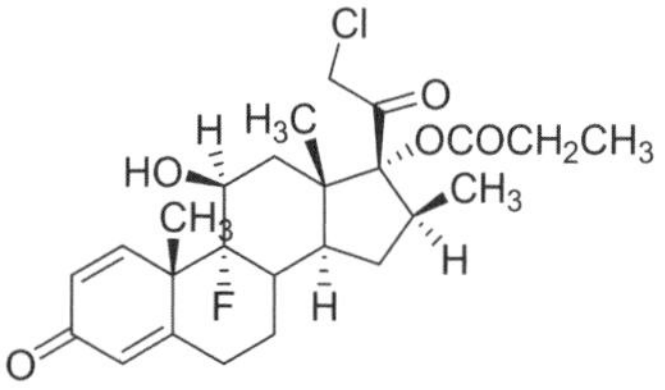 Cl, O, H, H_3C, HO, $OCOCH_2CH_3$, CH_3, CH_3, F, H, H, O
Clocortolone Pivalate/ Cloderm	9-chloro-6α-fluoro-11β, 21-dihydroxy-16α-methylpregna-1,4-diene-3,20-dione-21-pivalate	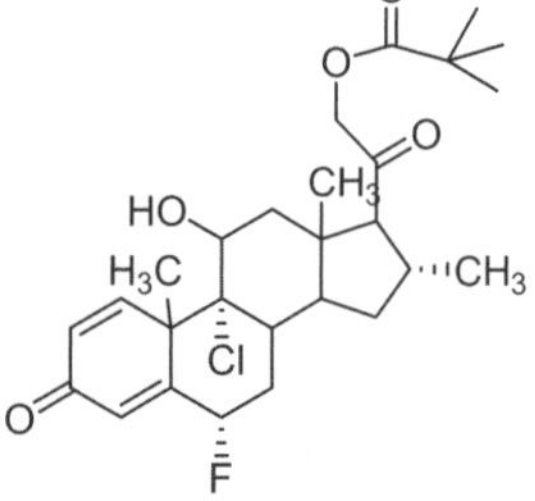 O, O, O, CH_3, HO, H_3C, CH_3, Cl, O, F

Formula	Pharmaceutical action	Dosage form	CAS No.	Approval date/ drug sponsor
$C_{20}H_{21}FN_2O \cdot HBr$	Antidepressant Serotonin uptake inhibitor	Oral tablet and solution	59729-32-7	July 1998/ Alphapharm Apotex Inc Aurobindo Pharma Ltd Forest Labs Actavis Elizabeth Akyma Pharms Apotex Inc Caraco Cobalt Corepharma Dr Reddys Labs Ltd Interpharm Invagen IVAS Pharms Kali Labs Mylan Pliva Sandoz Taro Teva Pharms Torrent Pharms Watson Labs Biovail Labs Intl
$C_{25}H_{32}ClFO_5$	Anti-inflammatory Glucocorticoid	Topical foam, cream, gel, ointment, spray, and solution	25122-46-7	Dec. 1985/ Connetics Actavis Mid Atlantic Altana Fougera Healthpoint Stiefel Taro Teva Pharms Perrigo Galderma Labs LP DPT Tolmar Morton Grove
$C_{27}H_{36}ClFO_5$	Glucocorticoid	Topical	34097-16-0	Aug. 1977/ Healthpoint

Active ingredient/ trade name	Chemical name	Structure
Clofarabine	2-chloro-9-(2-deoxy-2-fluoro-β-D-arabinofuranosyl)-9 *H*-purin-6-amine	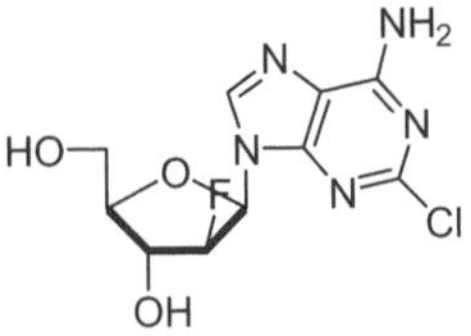
Desflurane	(±)1,2,2,2-tetrafluoroethyl difluoromethyl ether	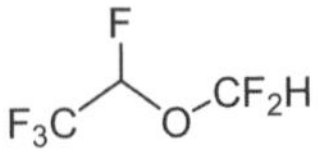
Desoximetasone/ Topicort	pregna-1, 4-diene-3, 20-dione, 9-fluoro-11, 21-dihydroxy-16-methyl-, (11β, 16α)-	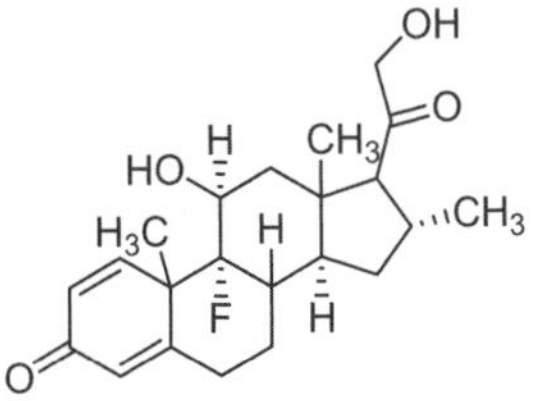
Dexamethasone Acetate[ΔV]/ Decadron-LA	[(8*S*,9*R*,10*S*,11*S*,13*S*,14*S*,16*R*,17*R*)-9-fluoro-11-hydroxy-17-(2-hydroxyacetyl)-10,13,16-trimethyl-3-oxo-6,7,8,11,12,14,15,16-octahydrocyclopenta[a]phenanthren-17-yl] acetate hydrate	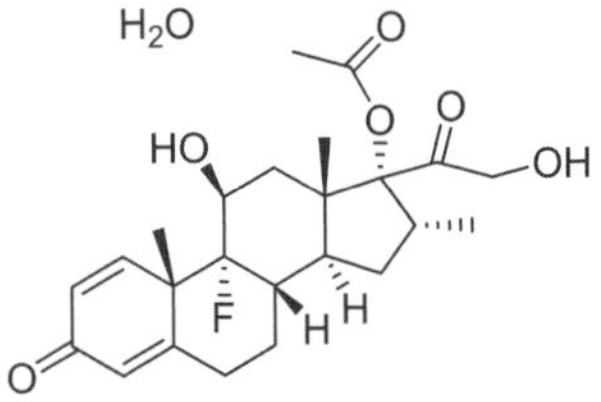
Dexamethasone Sodium Phosphate[V]	sodium [2-[(8*S*,9*R*,10*S*,11*S*,13*S*,14*S*,16*R*,17*R*)-9-fluoro-11,17-dihydroxy-10,13,16-trimethyl-3-oxo-6,7,8,11,12,14,15,16-octahydrocyclopenta[*a*]phenanthren-17-yl]-2-oxoethyl] phosphate	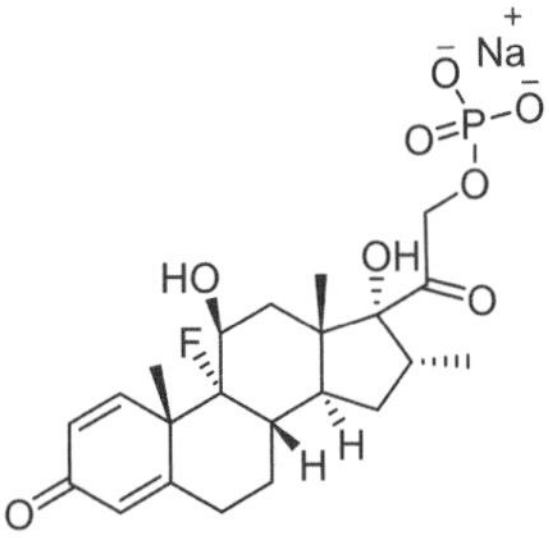
Dexamethasone[V] Mymethasone	9-fluoro-11β,17,21-trihydroxy-16α-methylpregna-1,4-diene-3,20-dione	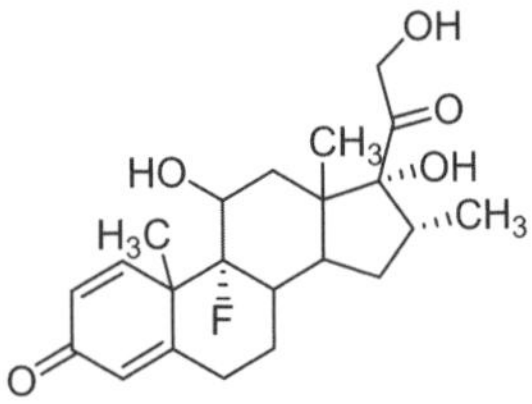

Formula	Pharmaceutical action	Dosage form	CAS No.	Approval date/ drug sponsor
$C_{10}H_{11}ClFN_5O_3$	Anti-metabolite	Injectable	123318-82-1	Dec. 2004/ Genzyme
$C_3H_2F_6O$	Anesthetic	Inhalation	57041-67-5	Sept. 1992/ Baxter Hlthcare Corp
$C_{22}H_{29}FO_4$	Anti-inflammatory Anti-pruritic agent	Topical cream, gel and ointment	382-67-2	March 1982/ Taro Pharms North Perrigo New York Taro
$C_{24}H_{33}FO_7$	Synthetic adrenocortical steroid	Injectable	55812-90-3	Sept. 1973/ Watson Labs Merck
$C_{22}H_{28}FNaO_8P$	Antiasthmatic Glucocorticoid	Ophthalmic drops Injectable	50-02-2	Sept. 1959/ Merck$^{\Delta}$ Abraxis Pharn Baxter Hlthcare Luitpold Teva Parenteral Alcon Universal Baush and Lomb
$C_{22}H_{29}FO_5$	Synthetic adrenocortical steroid	Oral tablet, elixir, and solution Ophthalmic drops	50-02-2	Dec.1960/ Organon USA Inc$^{\Delta}$ Morton Grove Roxane Par Pharm

Active ingredient/ trade name	Chemical name	Structure
Diflorasone Diacetate/ Psorcon Florone$^{\Delta}$	6α,9-difluoro-11β,17,21-trihydroxy-16β-methylpregna-1,4-diene-3,20-dione 17,21-diacetate	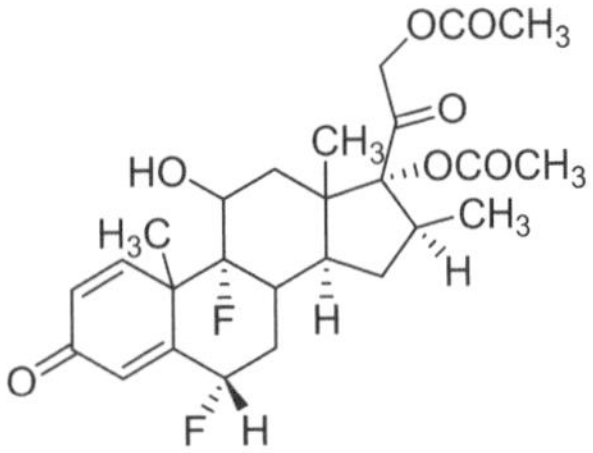
Diflunisal/ Dolobid$^{\Delta}$	5-(2,4-difluorophenyl)-2-hydroxy-benzoic acid	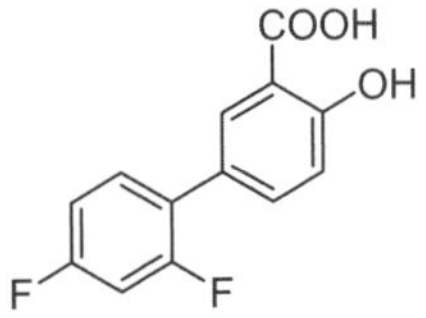
Droperidol^{V}/ Inaspine	3-[1-[4-(4-fluorophenyl)-4-oxobutyl]-3,6-dihydro-2*H*-pyridin-4-yl]-1*H*-benzimidazol-2-one	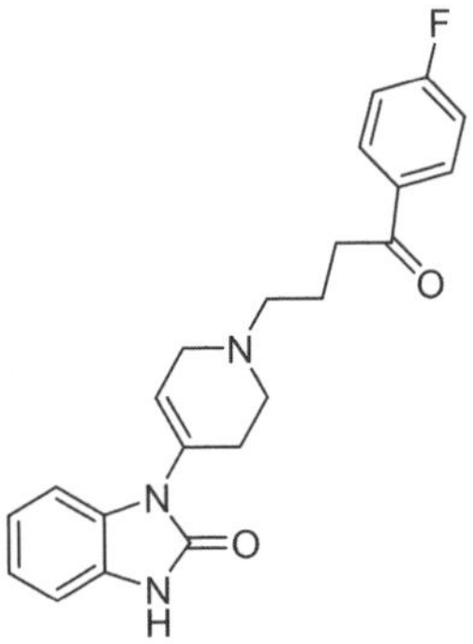
Dutasteride/ Avodart	(5α,17β)-*N*-{2,5 bis (trifluoromethyl)phenyl}-3-oxo-4-azaandrost-1-ene-17-carboxamide	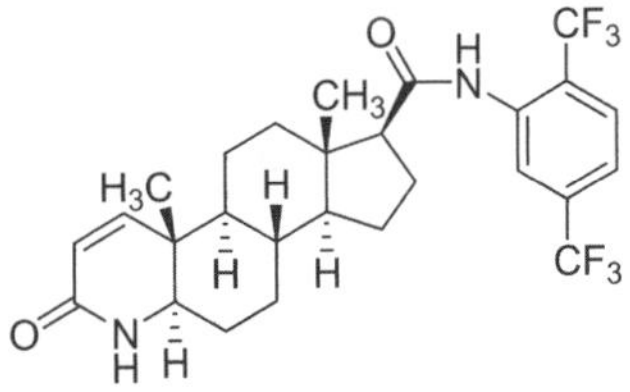
Efavirenz/ Sustiva	(*S*)-6-chloro-4-(cyclopropylethynyl)-1,4-dihydro-4-(trifluoromethyl)-2*H*-3,1-benzoxazin-2-one	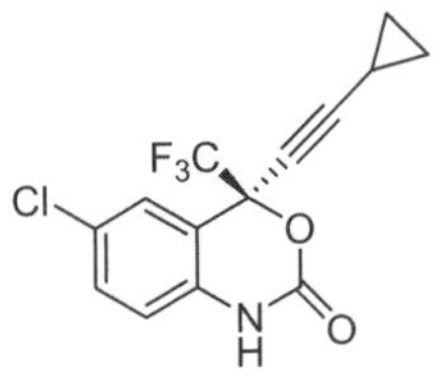
Eflornithine Hydrochloride/ Vaniqa	(+)-2-(difluoromethyl) ornithine monohydrochloride monohydrate	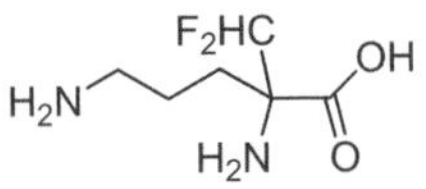

Formula	Pharmaceutical action	Dosage form	CAS No.	Approval date/ drug sponsor
$C_{26}H_{32}F_2O_7$	Anti-inflammatory Glucocorticoid	Topical ointment and cream	1869-92-7	March 1978/ Sanofi Aventis US Altana Taro Pharmacia and Upjohn$^{\Delta}$
$C_{13}H_8F_2O_3$	Analgesic Anti-Inflammatory Nonsteroidal cyclooxygenase inhibitor	Oral tablet	22494-42-4	April 1982/ TEVA Watson Labs Sandoz Merck$^{\Delta}$
$C_{22}H_{22}FN_3O_2$	Anesthetic Antiemetic Antipsychotic agent Dopamine antagonist	Injectable	548-73-2	June 1970/ Akorn Hospira Luitpold
$C_{27}H_{30}F_6N_2O_2$	Enzyme inhibitor Anti-prostatic hypertrophy	Oral capsule	164656-23-9	Nov. 2001/ GlaxoSmithKline
$C_{14}H_9ClF_3NO_2$	Reverse transcriptase inhibitor Anti-HIV agent	Oral capsule and tablet	154598-52-4	Sept. 1998/ Bristol Myers Squibb
$C_6H_{12}F_2N_2O_2 \cdot HCl \cdot H_2O$	Antineoplastic agent Enzyme inhibitors Trypanocidal agent	Topical cream Injectable$^{\Delta}$	96020-91-6	Nov. 1990/ Skinmedica Sanofi Aventis US$^{\Delta}$

Active ingredient/ trade name	Chemical name	Structure
Emtricitabine/ Atripla Emtriva Truvada	5-fluoro-1-(2*R*,5*S*)-[2-(hydroxymethyl)-1,3-oxathiolan-5-yl]cytosine ((–) enantiomer)	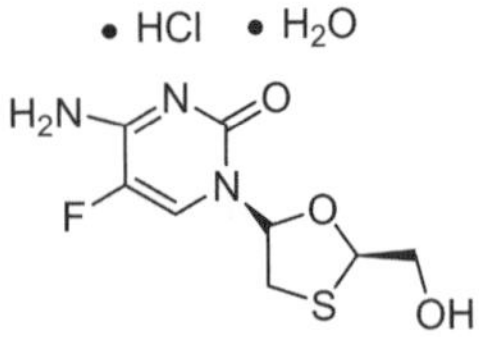 • HCl • H_2O
Enflurane/ Entrane	2-chloro-1,1,2-trifluoroethyl difluoromethyl ether	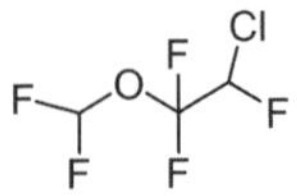
Enoxacin$^{\Delta}$/ Penetrex	1-ethyl-6-fluoro-4-oxo-7-piperazin-1-yl-1,8-naphthyridine-3-carboxylic acid	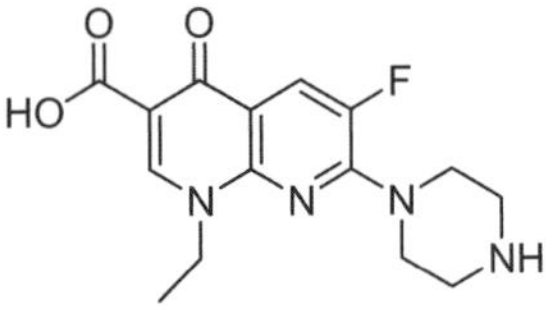
Escitalopram Oxalate/ Lexapro	*S*-(+)-1-[3-(dimethylamino)propyl]-1-(*p*-fluorophenyl)-5-phthalancarbonitrile oxalate	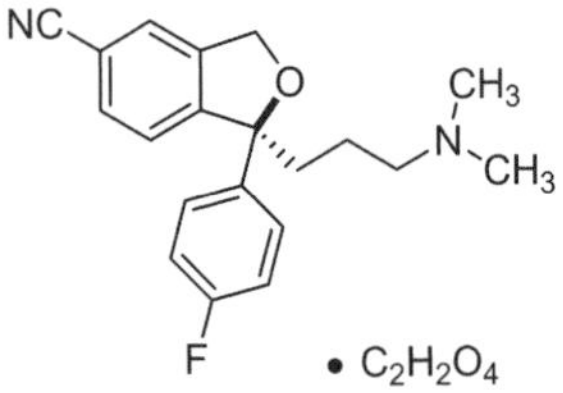 • $C_2H_2O_4$
Ezetimibe/ Zetia Vytorin	1-(4-fluorophenyl)-3(*R*)-[3-(4-fluorophenyl)-3(*S*)-hydroxypropyl]-4(*S*)-(4-hydroxyphenyl)-2-azetidinone	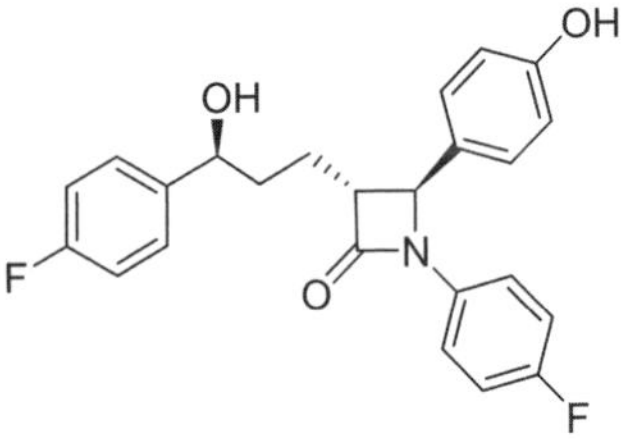
Flecainide Acetate/ Tambocor	*N*-(2-piperidylmethyl)-2,5-bis(2,2,2-trifluoroethoxy) benzamide	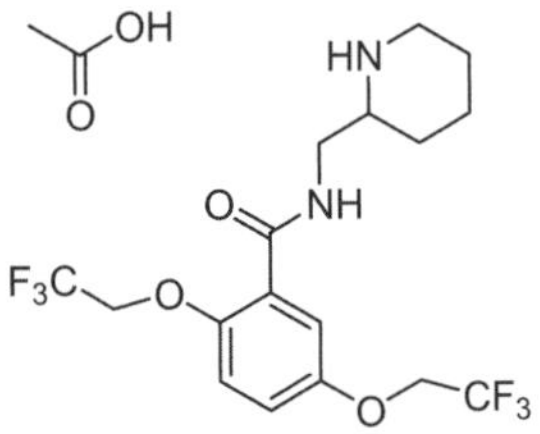
Floxuridine/ Fudr	5-fluoro-1-[4-hydroxy-5-(hydroxymethyl) tetrahydrofuran-2-yl]-1*H*-pyrimidine-2,4-dione	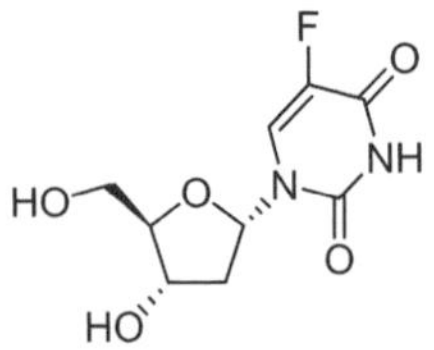

Formula	Pharmaceutical action	Dosage form	CAS No.	Approval date/ drug sponsor
$C_8H_{10}FN_3O_3S$	Antiviral agent Anti-HIV Reverse transcriptase inhibitor	Oral capsule, tablet, solution	143491-57-0	July 2003/ Gilead
$C_3H_2ClF_5O$	Anesthetic	Inhalation	13838-16-9	Aug. 1972/ Baxter Hlthcare Corp. Abbott$^{\Delta}$ Minrad
$C_{15}H_{17}FN_4O_3$	Anti-infective agent	Oral tablet	74011-58-8	Dec. 1991/ Sanofi Aventis US
$C_{20}H_{21}FN_2O \cdot C_2H_2O_4$	Serotonin uptake inhibitor Antidepressant	Oral tablet, capsule, and solution	219861-08-2	Aug. 2002/ Forest Labs IVAX Pharms Alphapharm
$C_{24}H_{21}F_2NO_3$	Antilipemic	Oral tablet	163222-33-1	Oct. 2002/ MSP Singapore
$C_{17}H_{20}F_6N_2O_3.C_2H_4O_2$	Anti-arrhythmia agent	Oral tablet	54143-56-5	Oct. 1985/ Graceway Alphapharm Barr Ranbaxy Roxane Sandoz
$C_9H_{11}FN_2O_5$	Antineoplastic Antiviral	Injectable	50-91-9	Dec. 1970/ Hospira Bedford Abraxis Pharm

Active ingredient/ trade name	Chemical name	Structure
Fluconazole/ Diflucan	2,4-difluoro-α,α[1]-bis(*1H*-1,2,4-triazol-1-ylmethyl)	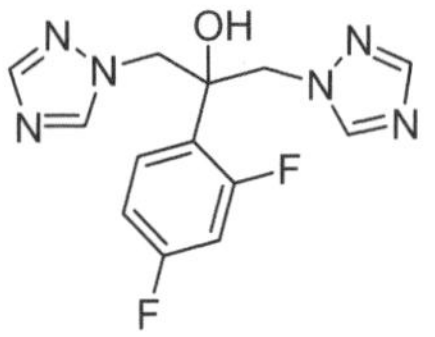
Flucytosine/ Ancobon	5-fluorocytosine	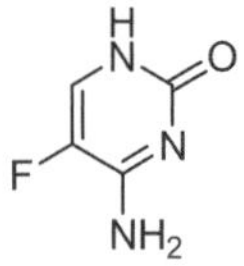
Fludarabine Phosphate/ Fludara	9*H*-purin-6-amine, 2-fluoro-9-(5-0-phosphono-β-D-arabinofuranosyl) (2-fluoro-ara-AMP)	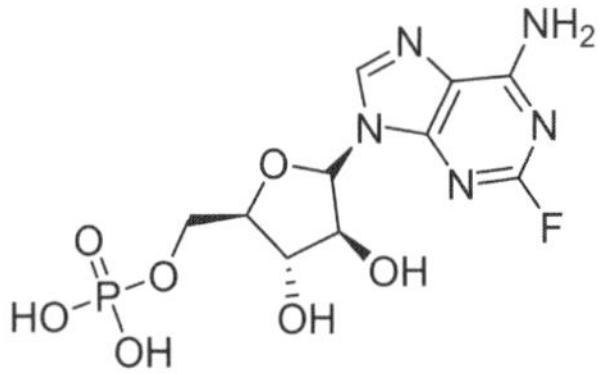
Fludeoxyglucose F-18	(2*S*,3*R*,4*S*,5*S*,6*R*)-3-fluoro-6-(hydroxymethyl)oxane-2,4,5-triol	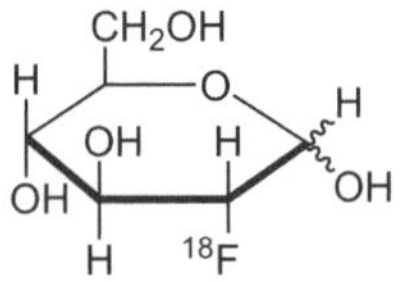
Fludrocortisone Acetate/ Flornef	2-[(8*S*,9*S*,10*S*,11*S*,13*S*,14*S*,17*R*)-9-fluoro-11,17-dihydroxy-10,13-dimethyl-3-oxo-1,2,6,7,8,11,12, 14,15,16-decahydrocyclopenta[a] phenanthren-17-yl]-2-oxo-ethyl] acetate	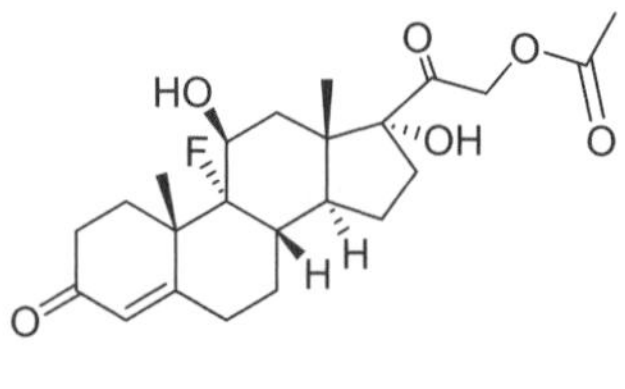

Formula	Pharmaceutical action	Dosage form	CAS No.	Approval date/ drug sponsor
$C_{13}H_{12}F_2N_6O$	Antifungal agents	Oral tablet and suspension Injectable	86386-73-4	Jan. 1990/Pfizer IVAX Pharms Ranbaxy Roxane Taro Pharm Inds Abraxis Pharm Apotex Baxter Hlthcare Bedford Hikma Farmaceutica Hospira Teva Parenteral Dr Reddys Lab Inc. Genpharm Glenmark Pharma Mylan Pliva Sandoz Torpharm Unique Pharm Labs Teva Taro
$C_4H_4FN_3O$	Antifungal agent	Oral capsule	2022-85-7	Nov. 1971/ Valeant
$C_{10}H_{13}FN_5O_7P$	Antineoplastic Antimetabolite Immunosuppressive agent	Injectable	75607-67-9	April 1991/ Bayer Hlthc Hospira Teva Parenteral
$C_6H_{11}{}^{18}FO_5$	Diagnostic aid	Injectable	105851-17-0	Aug. 2004/ Weill Medcl Coll North Shore LIJ
$C_{23}H_{31}FO_6$	Mineralocorticoid	Oral tablet	514-36-3	Aug. 1955/ King Pharms

Active ingredient/ trade name	Chemical name	Structure
Flumazenil/ Romazicon	ethyl 8-fluoro-5,6-dihydro-5-methyl-6-oxo-4*H*-imidazo [1,5-*a*](1,4) benzodiazepine-3-carboxylate	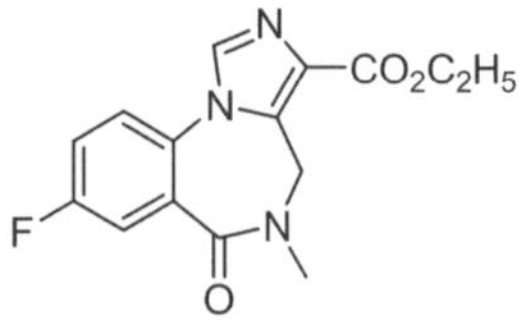
Flumethasone Pivalate[Δ]/ Locorten	[2-[(6*S*,9*R*,11*S*,14*S*,16*R*,17*R*)-6,9-difluoro-11,17-dihydroxy-10,13,16-trimethyl-3-oxo-6,7,8,11,12,14,15,16-octahydrocyclopenta[*a*] phenanthren-17-yl]-2-oxoethyl] 2,2-dimethylpropanoate	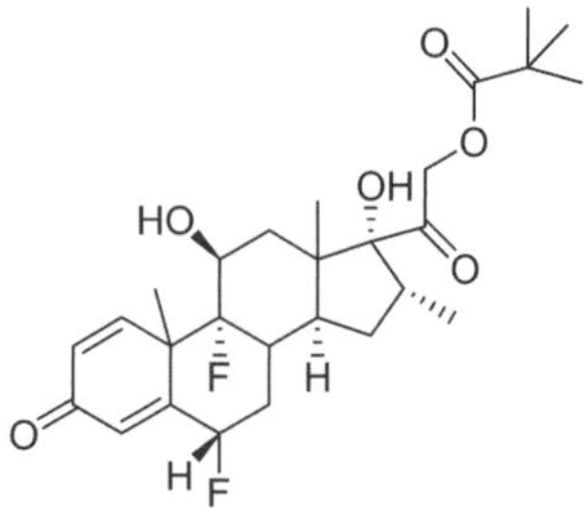
Flunisolide/ Aerospan Aerobid Nararel	(6α,11β,16α)-6-fluoro-11,21-dihydroxy-16,17-[(1-methylethylidene)bis(oxy)]pregna-1,4-diene-3,20-dione	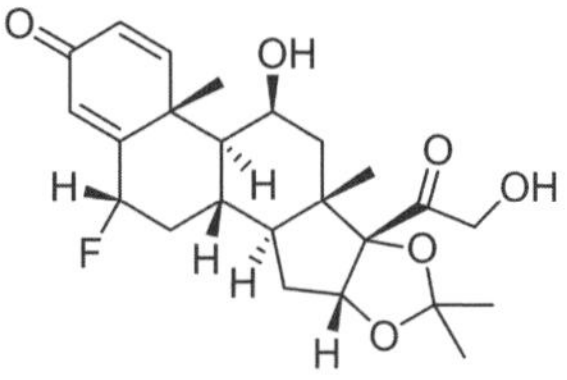
Fluocinolone Acetonide*/ Synalar Retisert Derma-Smoothe FS Shampoo Tri-Luma	(6α,11β,16α)-6,9-difluoro-11,21-dihydroxy-16,17-[(1-methylethylidene)bis (oxy)]-pregna-1,4-diene-3,20-dione	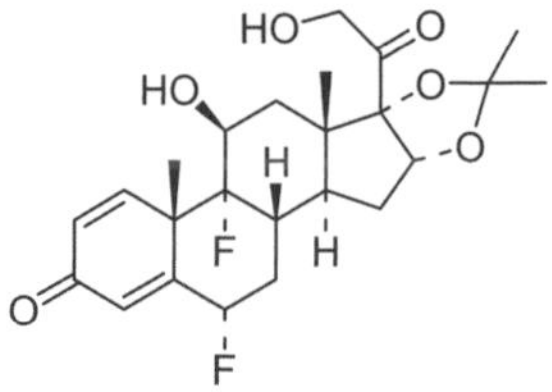
Fluocinonide/ Lidex	6α,9α-difluoro-11β,21-dihydroxy-16α,17α-isopropylidenedioxypregna-1,4-diene-3,20-dione 21-acetate	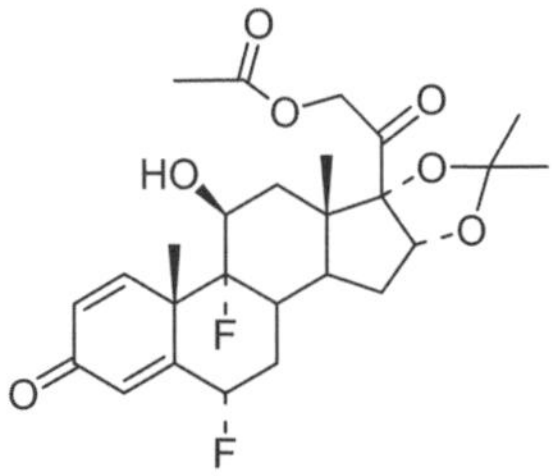
Fluorometholone Acetate/ Flarex Tobrasone	9-fluoro-11β,17-dihydroxy-6α-methylpregna-1, 4-diene-3,20-dione 17-acetate	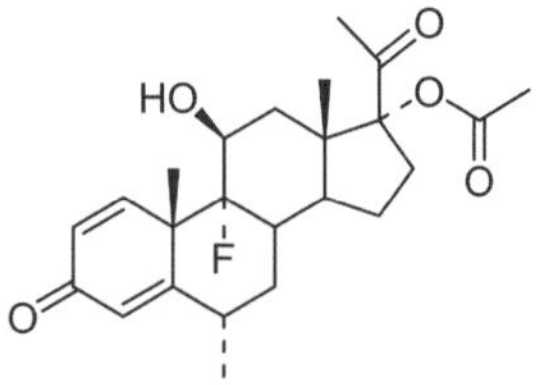

Formula	Pharmaceutical action	Dosage form	CAS No.	Approval date/ drug sponsor
$C_{15}H_{14}FN_3O_3$	Benzodiazepine antagonist	Injectable	78755-81-4	Dec. 1991/ Bedford Labs Abraxis Pharm Apotex Inc. Baxter Hlthcare HLR Sandoz Teva Parental
$C_{27}H_{36}F_2O_6$	Anti-inflammatory Glucocorticoid Antipruritic	Topical cream	2002-29-1	Prior to 1982/ Novartis
$C_{24}H_{31}FO_6$	Anti-inflammatory Anti-asthmatic Glucocorticoid	Nasal spray	3385-03-3	Aug. 1984/ Forest Labs Roche Palo Apotex Inc. IVAX Res LLC Bausch and Lomb
$C_{24}H_{30}F_2O_6$	Anti-inflammatory Anti-asthmatic agent Glucocorticoid	Topical cream and solution Otic drops Intravitreal implant	67-73-2	April 1982/ Fougera G and W Labs Medics Taro Bausch and Lomb Hill Dermac Galderma Labs LP
$C_{26}H_{32}F_2O_7$	Anti-inflammatory Antipruritic Glucocorticoid	Topical cream and solution Otic drops Intravitreal implant	356-12-7	April 1984/ Fougera G and W Labs Medicis Taro Bausch and Lomb Hill Dermac Galderma Labs LP
$C_{24}H_{31}FO_5$	Glucocorticoid	Ophthalmic drops	3801-06-7	Feb. 1986/ Alcon

Active ingredient/ trade name	Chemical name	Structure
Fluorometholone/ FML Fluor-Op	9-fluoro-11β, 7-dihydroxy-6α-methylpregna-1,4-diene-3,20-dione	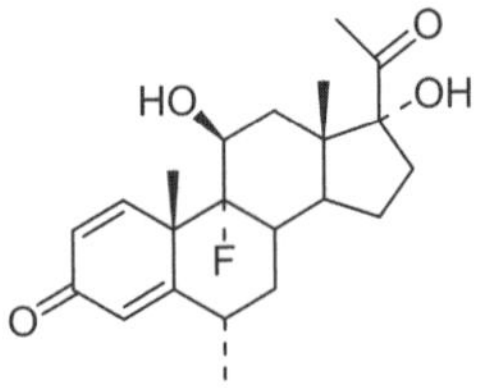 HO, OH, O, F
Fluorouracil/ Fluoroplex Carac Efudex	5-fluoro-2,4 (1*H*,3*H*)-pyrimdinedione	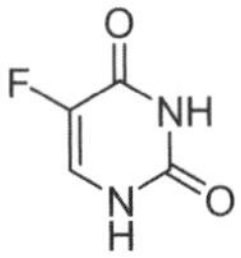 O, F, NH, N, H, O
Fluoxetine Hydrochloride/ Prozac Sarafem Symbax	*N*-methyl-3-phenyl-3-[4-(trifluo romethyl)phenoxy]propan-1-amine hydrochloride	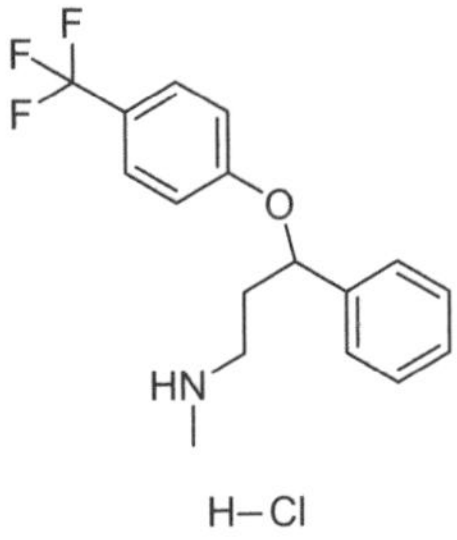 F, F, F, O, HN, H–Cl
Fluoxymesterone/ Halotestin	9-fluoro-11,17-dihydroxy-10,13,17-trimethyl-1,2,6,7,8, 11,12,14,15,16-decahydrocyclopenta[*a*] phenanthren-3-one	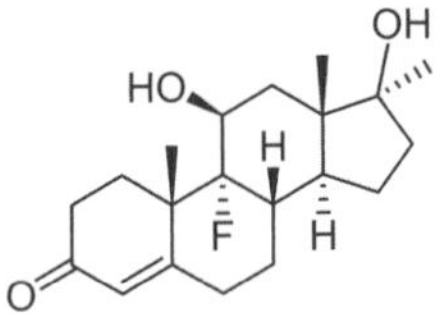 OH, HO, H, F, H, O

Formula	Pharmaceutical action	Dosage form	CAS No.	Approval date/ drug sponsor
$C_{22}H_{29}FO_4$	Anti-inflammatory Glucocorticoid	Ophthalmic drops and ointment	426-13-1	July 1982/ Allergan Novartis
$C_4H_3FN_2O_2$	Antineoplastic Antimetabolite Immunosuppressive agent	Topical solution and cream Injectable	51-21-8	Sept. 1998/ Abraxis Pharm Allergan Herbert Sanofi Aventis US Valeant Pharm Intl Generamedix Teva Parenteral Valeant Taro Valeant Pharm Intl.
$C_{17}H_{18}F_3NO \cdot HCl$	Antidepressant agent Serotonin uptake inhibitor	Oral tablet, capsule, and solution	56296-78-7	Dec. 1987/Barr Lilly Alphapharm Carlsbad Dr Reddys Lab Ltd. IVAX Pharms Mallinckrodt Mylan Par Pharm Pliva Ranbaxy Rxelite Sandoz Teva Actavis Mid Atlanitc Hi Tech Pharma Mallinckrodt Morton Grove Novex Pharm Assoc. Silarx Warner Chilcott
$C_{20}H_{29}FO_3$	Anabolic agent Antineoplastic agents Androgen	Oral tablet	76-43-7	Prior to 1982/ USL Pharma Pharmacia and Upjohn

Active ingredient/ trade name	Chemical name	Structure
Fluphenazine Decanoate	4-[3-[2-(trifluoromethyl)10*H*-phenothiazin-10-yl]propyl]-1-piperazineethanol decanoate	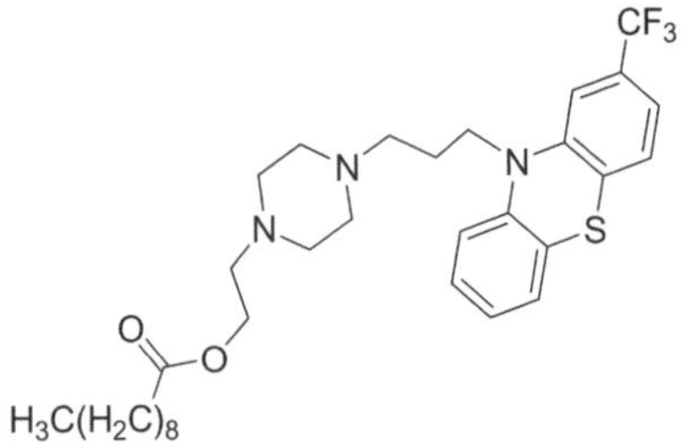
Fluphenazine Enanthate[Δ]/ Prolixin Enanthate	2-[4-[3-[2-(trifluoromethyl) phenothiazin-10-yl]propyl]piperazin-1-yl]ethyl heptanoate	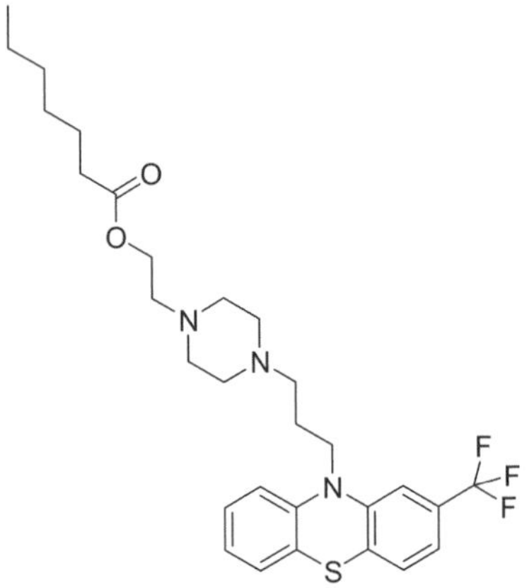
Fluphenazine Hydrochloride	2-[4-[3-[2-(trifluoromethyl) phenothiazin-10-yl]propyl]piperazin-1-yl]ethanol dihydrochloride	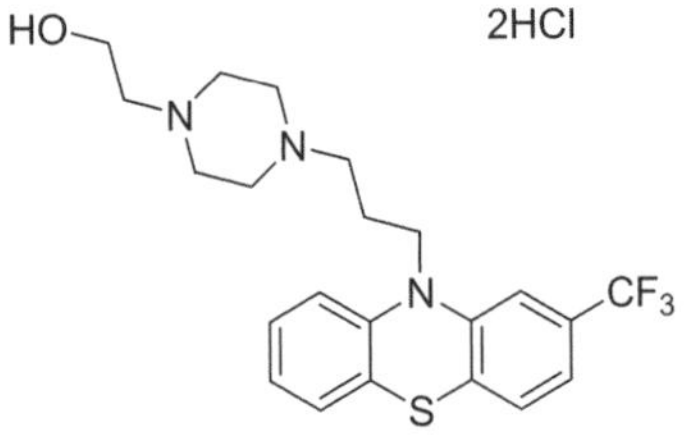
Fluprednisolone[Δ]/ Alphadrol	(6*S*,8*S*,9*S*,10*R*,11*S*,13*S*,14*S*,17*R*)-6-fluoro-11,17-dihydroxy-17-(2-hydroxyacetyl)-10,13-dimethyl-7,8,9,11,12,14,15,16-octahydro-6*H*-cyclopenta[*a*]phenanthren-3-one	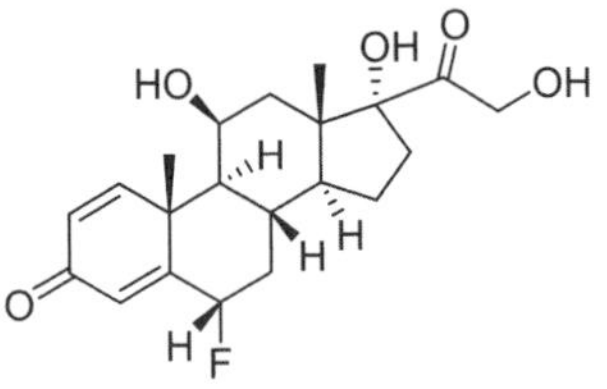
Flurandrenolide/ Cordran	(6α,11β,16α)-6-fluoro-11,21-dihydroxy-16,17-[(1-methylethylidene)bis(oxy)] pregn-4-ene-3,20-dione	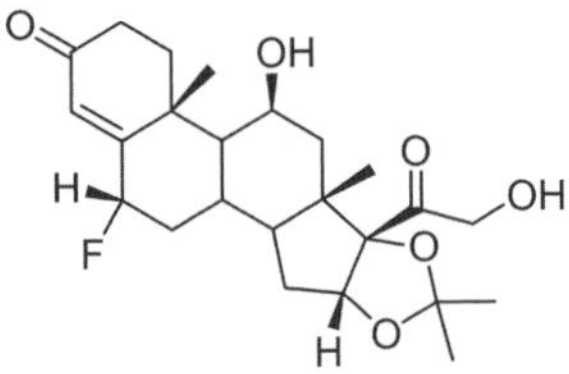

Formula	Pharmaceutical action	Dosage form	CAS No.	Approval date/ drug sponsor
$C_{32}H_{44}F_3N_3O_2S$	Antipsychotic agent	Injectable	5002-47-1	July 1987/ Abraxis Pharm Apotex Inc. Bedford Teva Parenteral
$C_{29}H_{38}F_3N_3O_2S$	Antipsychotic agent	Injectable	2746-81-8	March 1967/ Apothecon
$C_{22}H_{28}Cl_2F_3N_3OS$	Antipsychotic agent	Injectable	146-56-5	June 1960/ Schering$^{\Delta}$ Abraxis Pharm Pharm Assoc. Mylan Par Pharm Sandoz
$C_{21}H_{27}FO_5$	Anti-inflammatory agent Glucocorticoid	Oral tablet	53-34-9	Prior to 1982/ Pharmacia and Upjohn
$C_{24}H_{33}FO_6$	Anti-inflammatory agents Glucocorticoid	Topical cream and ointment	1524-88-5	March 1963/ Oclassen Watson Labs

Active ingredient/ trade name	Chemical name	Structure
Flurazepam Hydrochloride	9-chloro-2-(2-diethylaminoethyl)-6-(2-fluorophenyl)-2,5-diazabicyclo[5.4.0]undeca-5,8,10,12-tetraen-3-one dihydrochloride	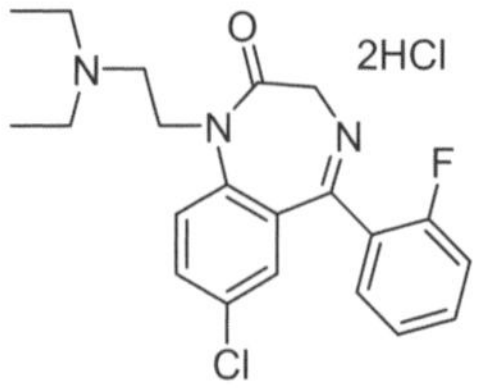
Flurbiprofen Sodium/ Ocufen	sodium (±)-2-(2-fluoro-4-biphenylyl)-propionate dihydrate	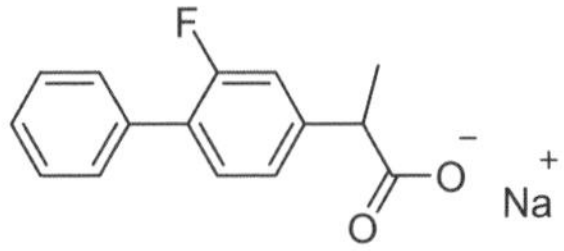
Flurbiprofen/ Ansaid	[1,1′-biphenyl]-4-acetic acid, 2-fluoro-alphamethyl-, (±)-	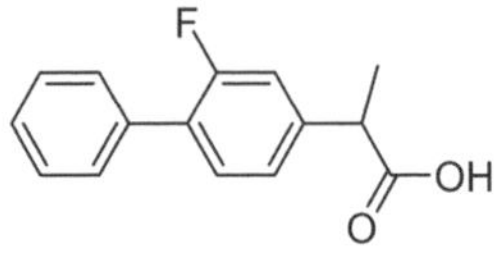
Flutamide	2-methyl-*N*-[4-nitro-3-(trifluoromethyl)phenyl] propanamide	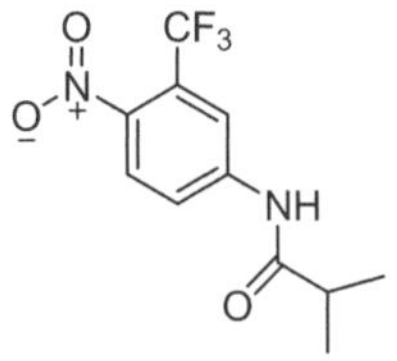
Fluticasone Furoate/ Flonase	(6α,11β,16α,17α)-6,9-difluoro-17-{[(fluoro-methyl) thio]carbonyl}-11-hydroxy-1 6-methyl-3-oxoandrosta-1,4-dien-17-yl 2-furancarboxylate	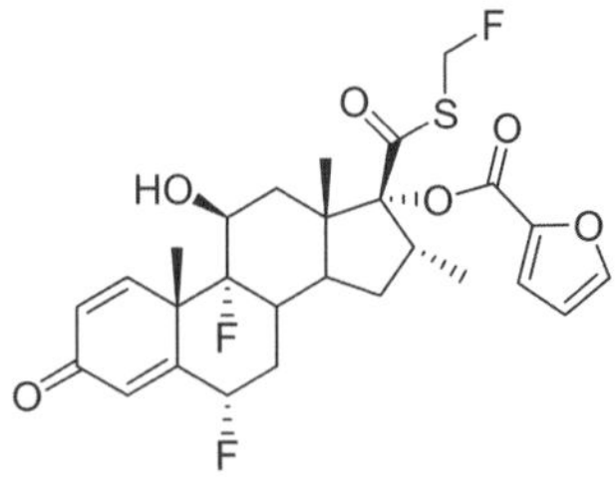
Fluticasone Propionate/ Advair Flovent Cutivate	(6α,11β,16α,17α)-6,9-difluoro-11-hydroxy-16-methyl-3-oxo-17-(1-oxopropoxy) androsta-1,4-diene-17-carbothioic acid, *S*-fluoromethylester	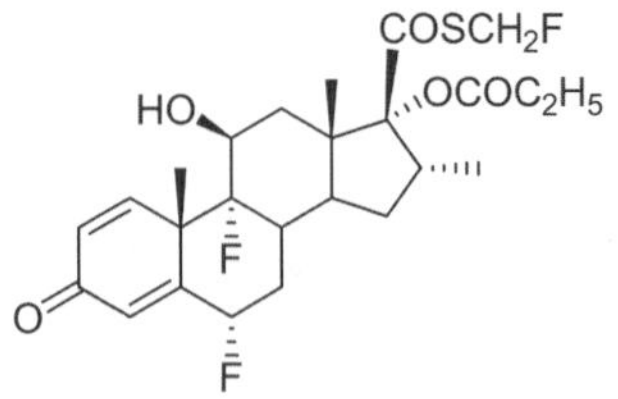

Formula	Pharmaceutical action	Dosage form	CAS No.	Approval date/ drug sponsor
$C_{21}H_{25}Cl_3FN_3O$	Anti-anxiety agent Gaba modulators Hypnotic Sedative	Oral capsule, elixir, and tablet	1172-18-5	Nov. 1985/ Mylan Sandoz Par Pharm Pharm Assoc. Abraxis
$C_{15}H_{12}FNaO_2$	Nonsteroidal anti-inflammatory analgesic	Ophthalmic drops	56767-76-7	Dec. 1986/ Allergan Bausch and Lomb
$C_{15}H_{13}FO_2$	Nonsteroidal anti-inflammatory analgesic	Oral tablet	5104-49-4	Oct. 1988/ Pharmacia and Upjohn Caraco Mylan Pliva Sandoz Teva Theragen
$C_{11}H_{11}F_3N_2O_3$	Androgen antagonist Antineoplastic agent	Oral capsule	13311-84-7	Jan. 1989/ Schering$^{\Delta}$ Barr Genpharm IVAX Pharms Par Pharm Sandoz
$C_{27}H_{29}F_3O_6S$	Glucocorticoid	Inhalation	397864-44-7	April 2007/ GlaxoSmithKline
$C_{25}H_{31}F_3O_5S$	Anti-inflammatory agent Bronchodilator agent Anti-allergic agent Glucocorticoid	Topical ointment and cream Inhalation	80474-14-2	Dec. 1990/ Altana GlaxoSmithKline Roxane Apotex Taro Pharms Inds Perrigo New York G and W Labs Tolmar KV Pharm Glaxo Grp. Ltd.

Active ingredient/ trade name	Chemical name	Structure
Fluvastatin Sodium/ Lescol	sodium (*E*,3*S*,5*R*)-7-[3-(4-fluorophenyl)-1-propan-2-yl-indol-2-yl]-3,5-dihydroxy-hept-6-enoate	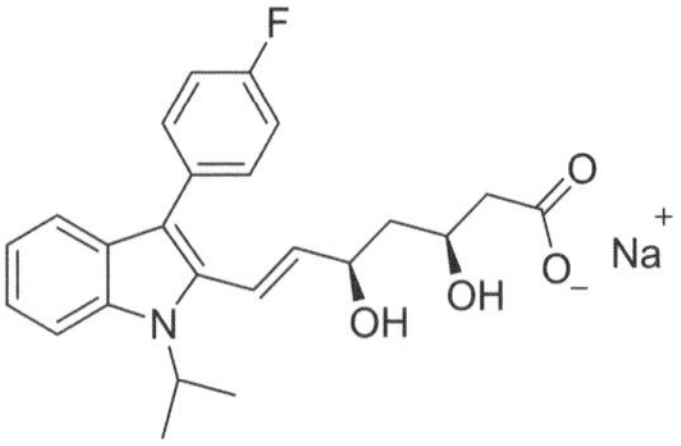
Fluvoxamine Maleate	but-2-enedioic acid; 2-[[5-methoxy-1-[4-(trifluoromethyl)phenyl] pentylidene]amino] oxyethanamine	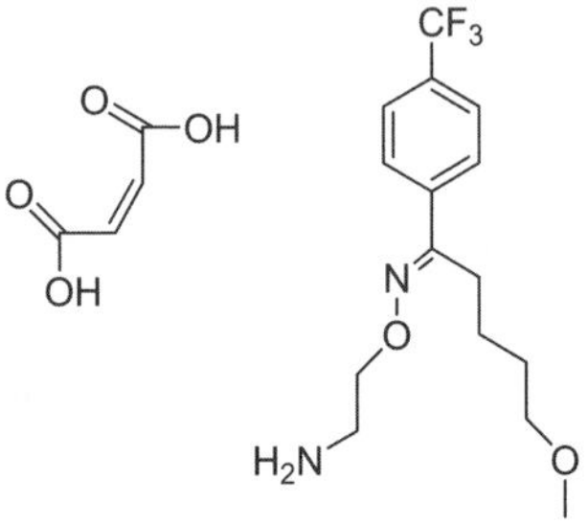
Fulvestrant/ Faslodex	7α-[9-(4,4,5,5,5-pentafluoropentylsulfinyl) nonyl]estra-1,3,5-(10)-triene-3,17β-diol	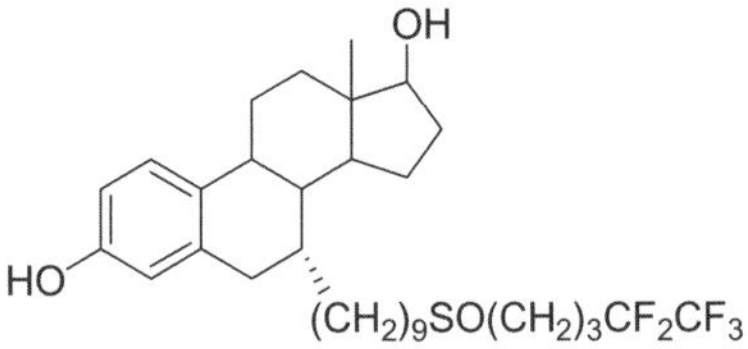 $(CH_2)_9SO(CH_2)_3CF_2CF_3$
Gatifloxacin/ Zymar	1-cyclopropyl-6-fluoro-8-methoxy-7-(3-methylpiperazin-1-yl)-4-oxo-quinoline-3-carboxylic acid	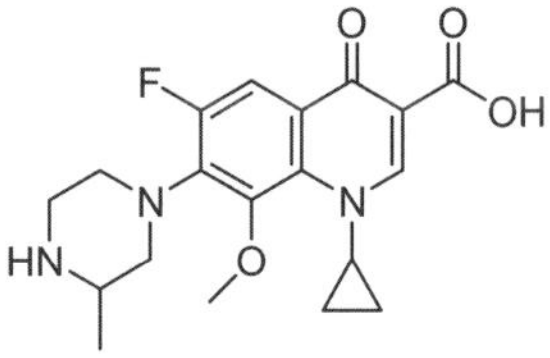
Gefitinib/ Iressa	*N*-(3-chloro-4-fluorophenyl)-7-methoxy-6-[3-4-morpholin) propoxy](4-quinazolinamine)	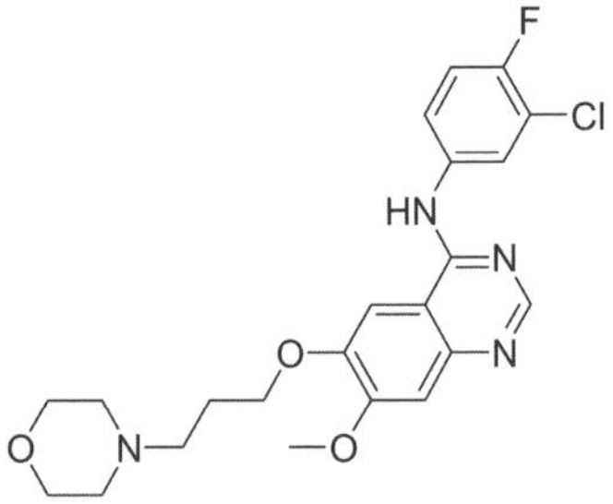
Gemcitabine Hydrochloride/ Gemzar	2′-deoxy-2′,2′-difluorocytidine monohydrochloride (β-isomer)	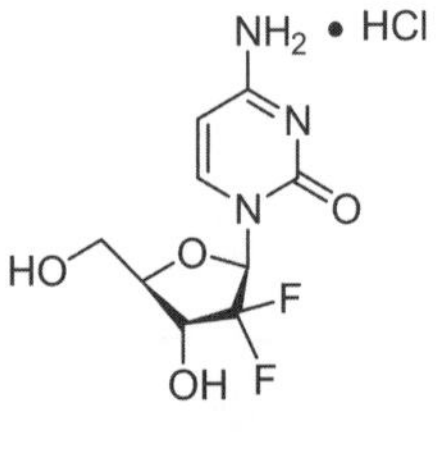

Formula	Pharmaceutical action	Dosage form	CAS No.	Approval date/ drug sponsor
$C_{24}H_{25}FNO_4Na$	Antilipemic Anticholesteremic agent Hydroxymethylglutaryl-CoA reductase inhibitor	Oral capsule and tablet	93957-55-2	Dec. 1993/ Novartis
$C_{19}H_{25}F_3N_2O_6$	Antidepressant Anti-anxiety agent Serotonin uptake inhibitor	Oral tablet	61718-82-9	Nov. 2000/ Mylan Actavis Elizabeth Barr Caraco Genpharm IVAX Pharms Sandoz Synthon Pharms Maleate Teva Maleate Torpharm
$C_{32}H_{47}F_5O_3S$	Estrogen antagonist Antineoplastic agent	Injectable	129453-61-8	April 2002/ Astrazeneca
$C_{19}H_{22}FN_3O_4$	Antibacterial agent	Ophthalmic drops Oral tablet[3]	112811-59-3	March 2003/ Allergan
$C_{22}H_{24}ClFN_4O_3$	Anilinoquinazoline Antineoplastic Agent	Oral tablet	184475-35-2	May 2003/ Astrazeneca
$C_9H_{11}F_2N_3O_4 \cdot HCl$	Antineoplastic agent Enzyme inhibitor Antiviral agent	Injectable	122111-03-9	May 1996/ Lilly

Active ingredient/ trade name	Chemical name	Structure
Gemifloxacin Mesylate/ Factive	(*R*,*S*)-7-[(4Z)-3-(aminomethyl)-4-(methoxyimino)-1-pyrrolidinyl]-1-cyclopropyl-6-fluoro-1,4-dihydro -4-oxo-1,8-naphthyridine-3-carboxylic acid	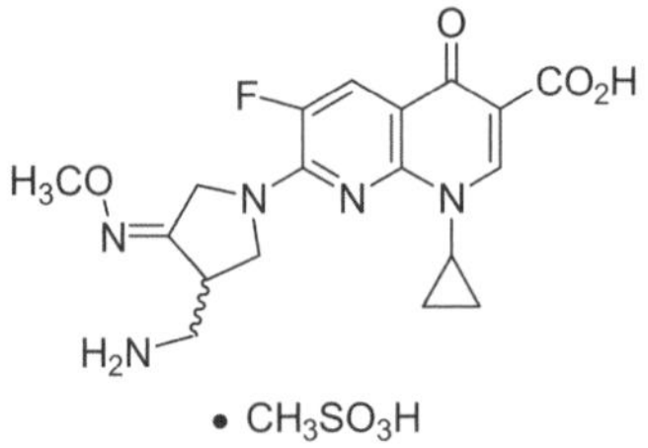 $\bullet$ CH_3SO_3H
Grepafloxacin Hydrochloride$^{\Delta}$/ Raxaar	1-cyclopropyl-6-fluoro-5-methyl-7-(3-methylpiperazin-1-yl)-4-oxo-2,3-dihydroquinoline-3-carboxylic acid hydrochloride	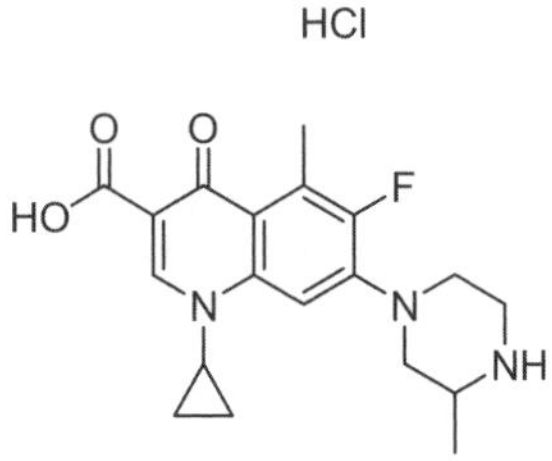 HCl
Halazepam$^{\Delta}$/ Paxipam	7-chloro-5-phenyl-1-(2,2,2-trifluoroethyl)-3*H*-1,4-benzodiazepin-2-one	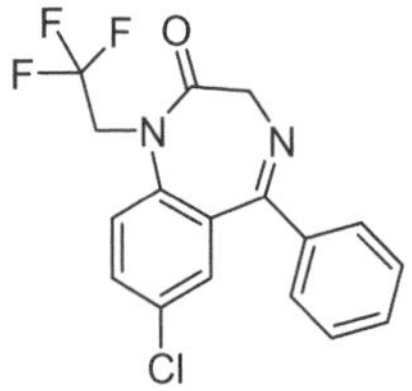
Halcinonide/ Halog	(11β,16α)-21-chloro-9-fluoro-11-hydroxy-16,17-[(1-methylethylidene)bis(oxy)]-pregn-4-ene-3,20-dione	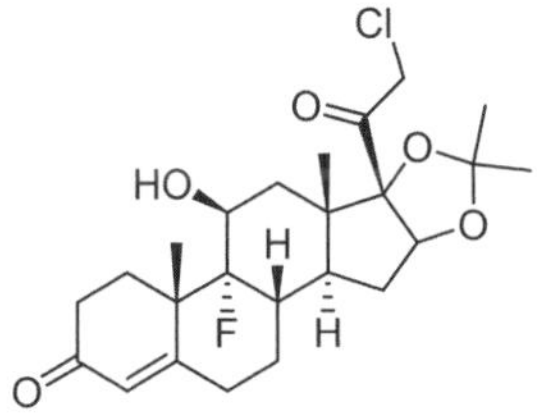
Halobetasol Propionate/ Ultravate	(11β,16α)-21-chloro-9-fluoro-11-hydroxy-16-methyl-17-(1-oxopropoxy)pregn-1,4-diene-3,20-dione	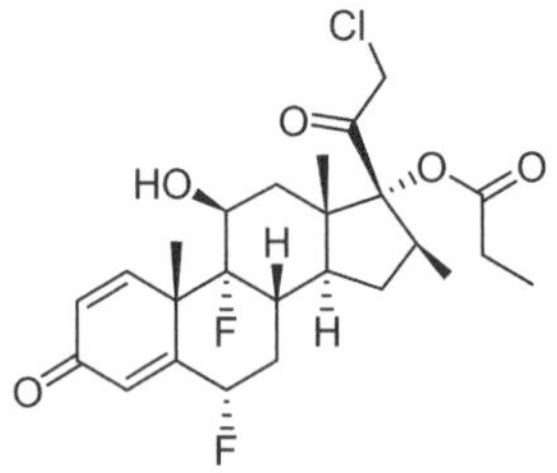

Formula	Pharmaceutical action	Dosage form	CAS No.	Approval date/ drug sponsor
$C_{18}H_{20}FN_5O_4 \cdot CH_4O_3S$	Antibacterial agent	Oral tablet	204519-65-3	April 2003/ Oscient
$C_{19}H_{22}FN_3O_3 \cdot HCl$	Antibacterial agent	Oral tablet	161967-81-3	Nov. 1997/ Otsuka
$C_{17}H_{12}ClF_3N_2O$	Anti-anxiety agent Sedative	Oral tablet	23092-17-3	Sept. 1981/ Schering
$C_{24}H_{32}ClFO_5$	Anti-inflammatory agent Glucocorticoid	Topical cream, ointment, and solution	3093-35-4	Nov. 1974/ Ranbaxy
$C_{25}H_{31}ClF_2O_5$	Glucocorticoid Vasoconstrictor agent	Topical cream and ointment	66852-54-8	Dec. 1990/ Ranbaxy Altana Perrigo Israel Taro G and W Labs

Active ingredient/ trade name	Chemical name	Structure
Halofantrine Hydrochloride/ Halfan$^{\Delta}$	1,3-dichloro-α-[2-(dibutylamino) ethyl]-6-(trifluoromethyl)-9-phenanthrene-methanol hydrochloride	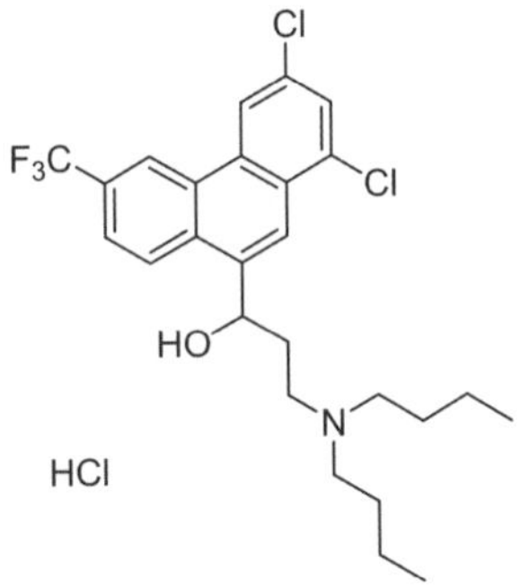 F_3C Cl Cl HO N HCl
Haloperidol	4-[4-(4-chlorophenyl)-4-hydroxy-1-piperidyl]-1-(4-fluorophenyl)butan-1-one	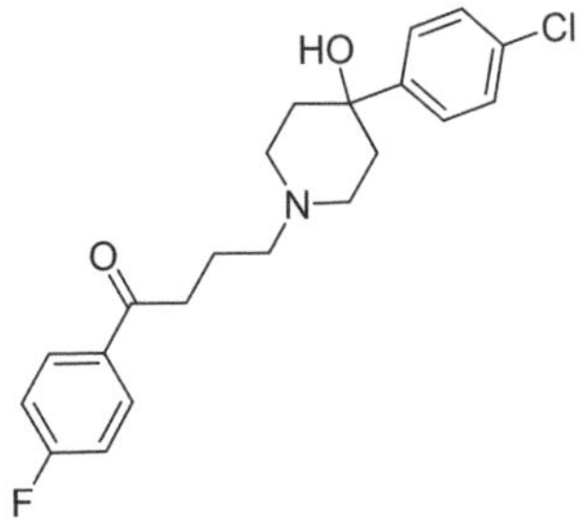 HO Cl N O F
Haloperidol Decanoate/ Haldol	[4-(4-chlorophenyl)-1-[4-(4-fluorophenyl)-4-oxo-butyl]-4-piperidyl] decanoate	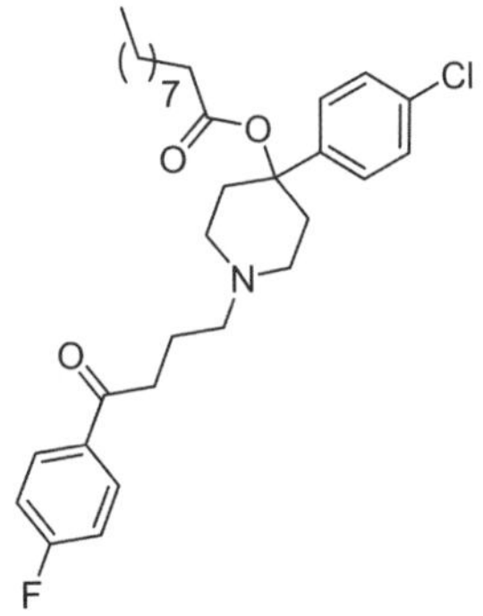 7 O O Cl N O F
Haloperidol Lactate/ Haldol	4-[4-(4-chlorophenyl)-4-hydroxy-1-piperidyl]-1-(4-fluorophenyl)butan-1-one;2-hydroxypropanoic acid	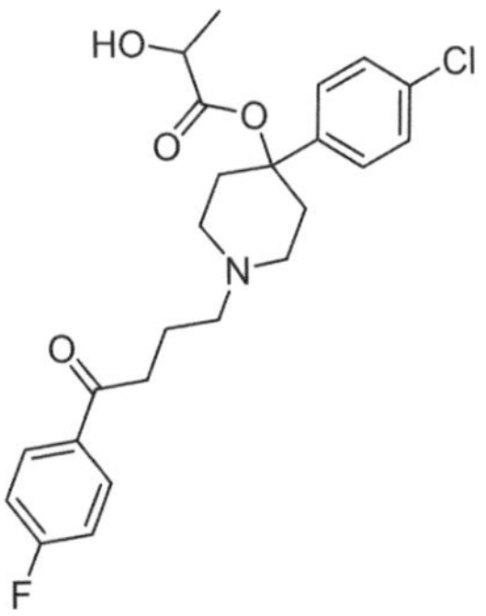 HO O O Cl N O F

Formula	Pharmaceutical action	Dosage form	CAS No.	Approval date/ drug sponsor
$C_{26}H_{30}Cl_2F_3NO \cdot HCl$	Antimalarial	Oral tablet	36167-63-2	July 1992/ GlaxoSmithKline
$C_{21}H_{23}ClFNO_2$	Antiemetics Antipsychotic agent Dopamine antagonist Anti-dyskinesia agent	Oral tablet	52-86-8	April 1967/ Mylan Par Pharm Sandox Watson Labs
$C_{31}H_{41}ClFNO_3$	Antipsychotic agent	Injectable	74050-97-8	Jan. 1986/ Ortho McNeil Abraxis Pharm Apotex Inc. Bedford Sandoz Sicor Pharms
$C_{24}H_{27}ClFNO_4$		Oral concentrate and solution Injectable	75478-79-4	July 1959/ Ortho McNeil Teva Pharms Watsons Labs Pharm Assoc. Silarx

Active ingredient/ trade name	Chemical name	Structure
Halothane[Δ]/ Fluothane	2-bromo-2-chloro-1,1, 1-trifluoroethane	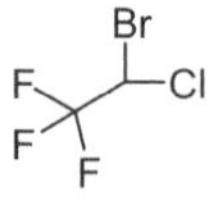
Hydroflumethiazide/ Salutensin [Δ] Saluron Diucardin[Δ]	1,1-dioxo-6-(trifluoromethyl)- 3,4-dihydro-2*H*-benzo[e] [1,2,4]thiadiazine-7- sulfonamide	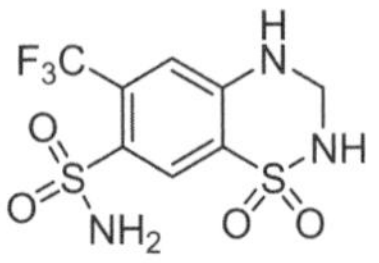
Isoflurane[V]/ Forane	2-chloro-2-(difluoromethoxy)- 1,1,1-trifluoroethane	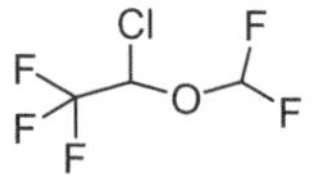
Isoflurophate/ Floropryl[Δ]	2-(fluoro-propan-2- yloxyphosphoryl)oxypropane	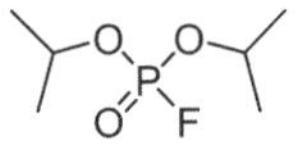
Lansoprazole/ Prevacid	2-[[3-methyl-4-(2,2,2- trifluoroethoxy)pyridin-2-yl] methylsulfinyl]-1*H*- benzoimidazole	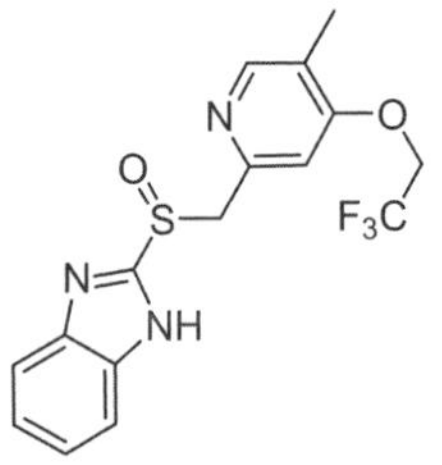
Lapatinib Ditosylate/ Tykerb	*N*-[3-chloro-4-[(3-fluorophenyl) methoxy]phenyl]-6-[5-[(2- methylsulfonylethylamino) methyl]furan-2-yl]quinazolin- 4-amine; 4- methylbenzenesulfonic acid; hydrate	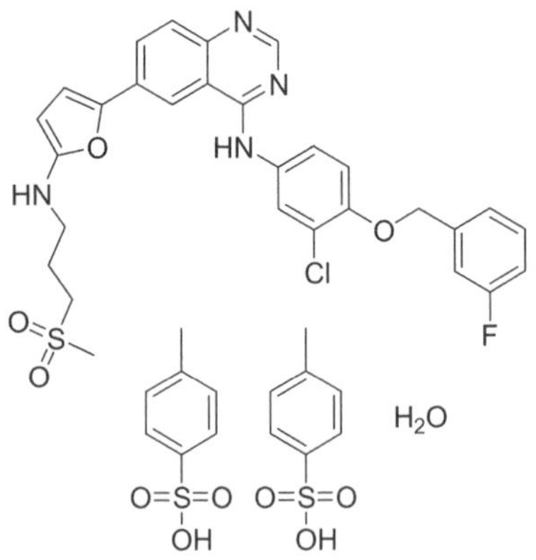

Formula	Pharmaceutical action	Dosage form	CAS No.	Approval date/ drug sponsor
$C_2HBrClF_3$	Anesthetic	Inhalation	151-67-7	March 1958/ Wyeth Ayerst[Δ] Hospira[Δ]
$C_8H_8F_3N_3O_4S_2$	Antihypertensive agent Diuretic Sodium chloride symporter inhibitor	Oral tablet	135-09-1	July 1959/ Shire Wyeth[Δ] Watson Labs[Δ] Par Pharm
$C_3H_2ClF_5O$	Anesthetic	Inhalation	26675-46-7	Dec. 1979/ Baxter Hlthcare Corp. Halocarbon Prods Hospira Marsam Pharms LLCC Minrad Rhodia
$C_6H_{14}FO_3P$	Cholinesterase inhibitor Miotic protease inhibitor	Ophthalmic ointment	55-91-4	April 1957/ Merck
$C_{16}H_{14}F_3N_3O_2S$	Anti-infective agent Anti-ulcer agent Enzyme inhibitor	Oral capsule and tablet Injectable	103577-45-3	May 1995/ Tap Pharm
$C_{43}H_{44}ClFN_4O_{11}S_3$	Antineoplastic agent Protein kinase inhibitor	Oral tablet	388082-78-8	March 2007/ GlaxoSmithKline

Active ingredient/ trade name	Chemical name	Structure
Leflunomide/ Arava	5-methyl-*N*-[4-(trifluoromethyl)phenyl]-1,2-oxazole-4-carboxamide	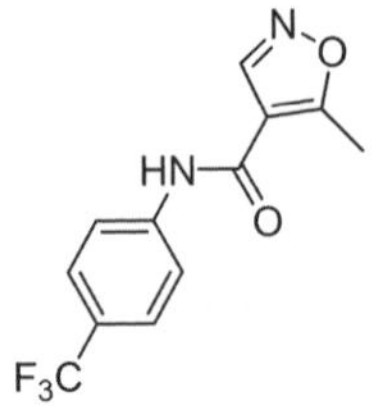
Levocabastine Hydrochloride/ Livostin[Δ]	(3*S*,4*R*)-1-[4-cyano-4-(4-fluorophenyl)cyclohexyl]-3-methyl-4-phenylpiperidine-4-carboxylic acid hydrochloride	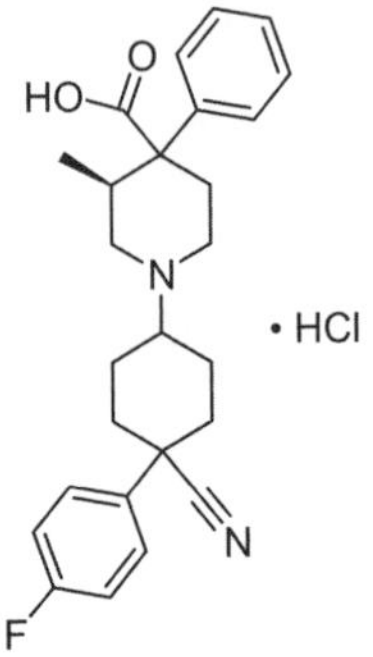 • HCl
Levofloxacin/ Levaquin Quixin Iquix	(-)-(*S*)-9-fluoro-2,3-dihydro-3-methyl-10-(4-methyl-1-piperazinyl)-7-oxo-7*H*-pyrido(1,2,3-*de*)-1,4-benzoxazine-6-carboxylic acid	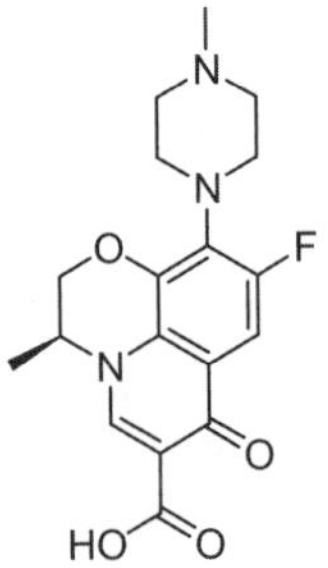
Linezolid/ Zyvox	*N*-[[3-(3-fluoro-4-morpholinylphenyl)-2-oxooxazolidin-5-yl] methyl]acetamide	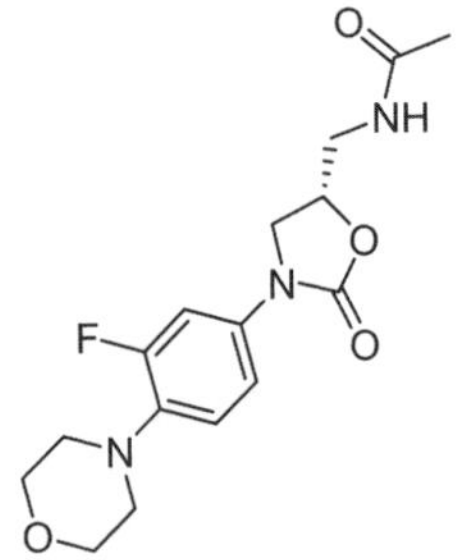
Lomefloxacin Hydrochloride/ Maxaquin	1-ethyl-6,8-difluoro-7-(3-methyl-1-piperazinyl)-4-oxo-3-quinolinecarboxylic acid hydrochloride	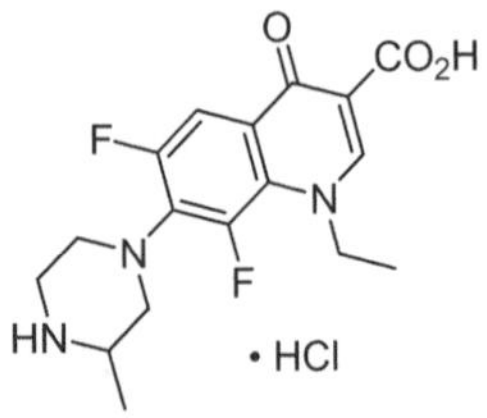 • HCl

Formula	Pharmaceutical action	Dosage form	CAS No.	Approval date/ drug sponsor
$C_{12}H_9F_3N_2O_2$	Enzyme inhibitor Immunosuppressive agent Antirheumatic agent	Oral tablet	75706-12-6	Sept. 1998/ Sanofi Aventis US Aptoex Inc. Barr Kali Labs Sandoz$^{\Delta}$ Teva Products
$C_{26}H_{30}ClFN_2O_2$	Nonsedating antihistaminic	Ophthalmic drops	79547-78-7	Nov. 1993/ Novartis
$C_{18}H_{20}FN_3O_4$	Antibacterial agent	Injectable Oral solution Ophthalmic drops Oral tablet	100986-85-4	Dec. 1996 Aug. 2000/ Ortho McNeil Pharm Santen
$C_{16}H_{20}FN_3O_4$	Anti-infective agent Protein synthesis inhibitor	Oral suspension and tablet Injectable	165800-03-3	April 2000/ Pharmacia and Upjohn
$C_{17}H_{20}ClF_2N_3O_3$	Anti-infective agent	Oral tablet	98079-52-8	Feb. 1992/ Pharmacia

Active ingredient/ trade name	Chemical name	Structure
Lubiprostone/ Amitiza	7-[(1*R*,4*R*,6*R*,9*S*)-4-(1,1-difluoropentyl)-4-hydroxy-8-oxo-5-oxabicyclo[4.3.0] non-9-yl]heptanoic acid	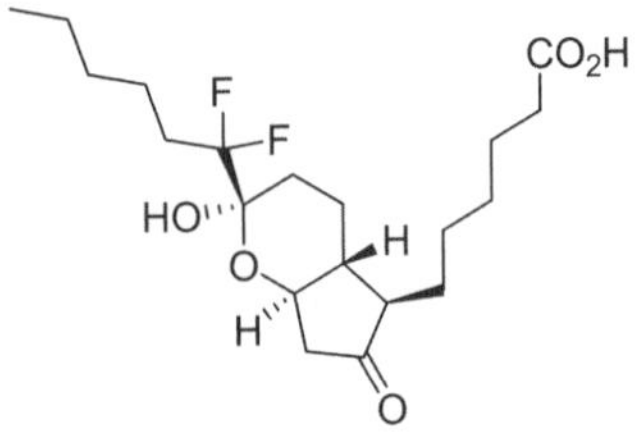
Mefloquine Hydrochloride/ Lariam	2,8-[bis(trifluoromethyl) quinolin-4-yl]-[(2*R*)-2-piperidyl]methanol hydrochloride	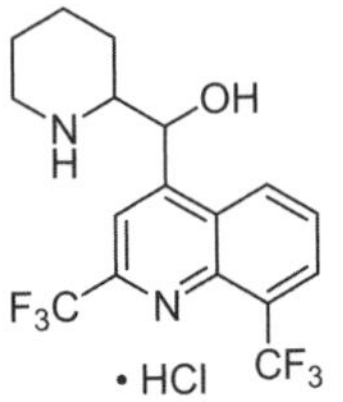
Melperone	1-(4-fluorophenyl)-4-(4-methyl-1-piperidyl)butan-1-one	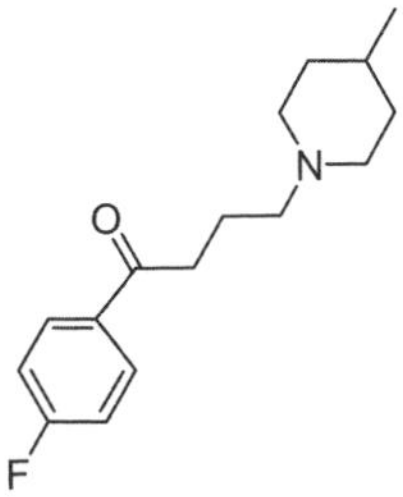
Midazolam Hydrochloride	8-chloro-6-(2-fluorophenyl)-1-methyl-4*H*-imidazo[1,5*a*] [1,4]benzodiazepine hydrochloride	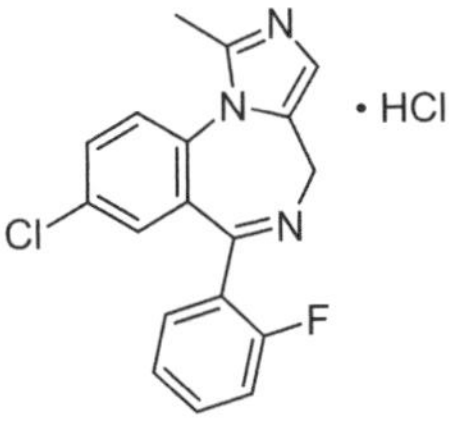
Moxifloxacin Hydrochloride/ Avelox Vigamox	7-[(4a*S*,7a*S*)-1,2,3,4,4a,5,7,7a-octahydropyrrolo[3,4-b] pyridin-6-yl]-1-cyclopropyl-6-fluoro-8-methoxy-4-oxoquinoline-3-carboxylic acid hydrochloride	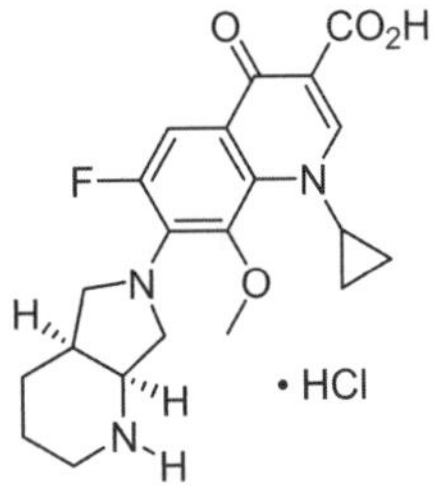

Formula	Pharmaceutical action	Dosage form	CAS No.	Approval date/ drug sponsor
$C_{20}H_{32}F_2O_5$	Laxative	Oral capsule	136790-76-6	Jan. 2006/ Sucampo Pharms
$C_{17}H_{17}ClF_6N_2O$	Antimalarial agent	Oral tablet	51773-92-3	May 1989/ Roche Barr Roxane Sandoz US Army Walter Reed
$C_{16}H_{22}FNO$	Antipsychotic agent	Oral	3575-80-2	Not approved/ Ovation Pharmaceuticals
$C_{18}H_{14}Cl_2FN_3$	Adjuvants, anesthesic Hypnotic Sedative Anti-anxiety agent GABA modulator	Injectable Oral syrup	59467-96-8	June 2000/ Abraxis Pharm Apotex Inc Baxter Hlthcare Bedford Ben Venue Hospira Intl Medicated Intl Medication Mayne Pharma USA Taylor Hi Tech Pharma Paddock Ranbaxy Roxane
$C_{21}H_{25}ClFN_3O_4$	Anti-infective agent	Injectable Opthalmic drops Oral tablet	186286-86-8	Dec. 1999/ Bayer Pharms Alcon

Active ingredient/ trade name	Chemical name	Structure
Nilutamide/ Nilandron	5,5-dimethyl-3-[4-nitro-3-(trifluoromethyl)phenyl] imidazolidine-2,4-dione	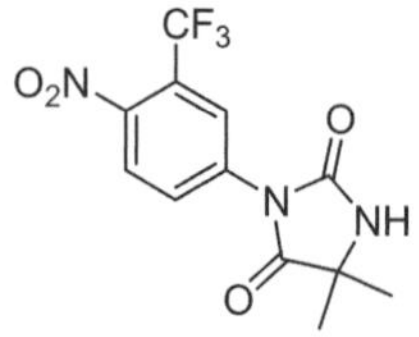
Nitisinone/ Orfadin	2-[2-nitro-4-(trifluoromethyl) benzoyl]cyclohexane-1,3-dione	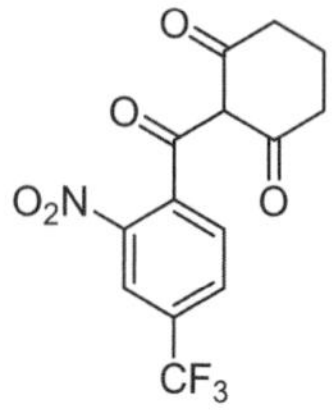
Norfloxacin/ Noroxin	1-ethyl-6-fluoro-4-oxo-7-piperazin-1-yl-quinoline-3-carboxylic acid	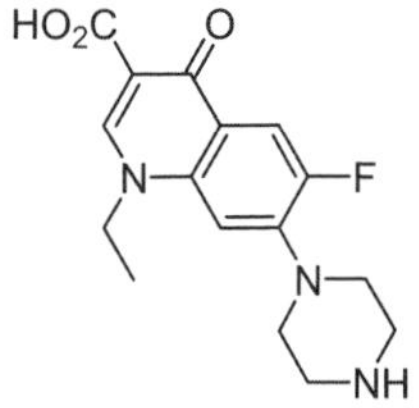
Ofloxacin/ Ocuflox Ofloxacin Floxin	(-)-(*S*)-9-fluoro-2,3-dihydro-3-methyl-10-(4-methyl-1-piperazinyl)-7-oxo-7*H*-pyrido [1,2,3-*de*]-1,4-benzoxazine-6-carboxylic acid	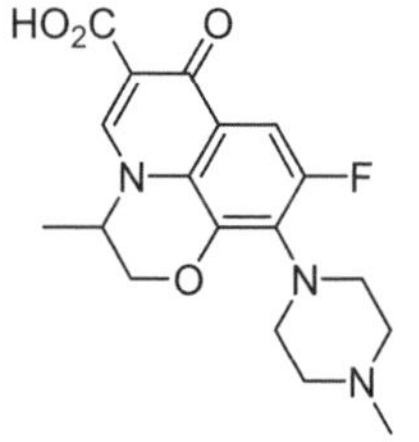
Paliperidone/ Invega	3-[2-[4-(6-fluorobenzo[d]isoxazol-3-yl)-1-piperidyl]ethyl]-7-hydroxy-4-methyl-1,5-diazabicyclo[4.4.0]deca-3,5-dien-2-one	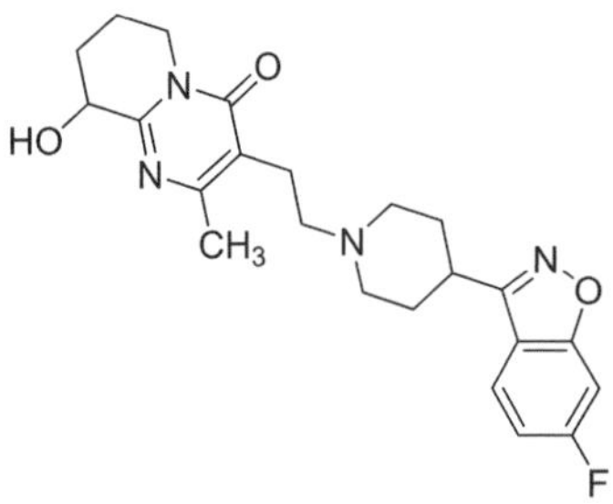

Formula	Pharmaceutical action	Dosage form	CAS No.	Approval date/ drug sponsor
$C_{12}H_{10}F_3N_3O_4$	Androgen antagonist Antineoplastic agent	Oral tablet	63612-50-0	April 1999/ Sanofi Aventis US
$C_{14}H_{10}F_3NO_5$	Enzyme inhibitor	Oral capsule	104206-65-7	Jan. 2002/ Swedish Orphan
$C_{16}H_{18}FN_3O_3$	Anti-infective agent Enzyme inhibitor	Oral tablet	70458-96-7	Oct. 1986/ Merck
$C_{18}H_{20}FN_3O_4$	Anti-infective agent Antibacterial agent	Injectable Opthalmic drops Oral tablet	82419-36-1	July 1993/ Bedford Allergan Bausch and Lomb Hi Tech Pharma Novex Pharmaforce Daiichi Ortho McNeil Pharm Dr Reddys Labs Par Pharm Ranbaxy Teva
$C_{23}H_{27}FN_4O_3$	Antipsychotic agent	Oral tablet	144598-75-4	Dec. 2006/ Janssen

Active ingredient/ trade name	Chemical name	Structure
Pantoprazole Sodium/ Protonix	sodium 5-(difluoromethoxy)-2-[[(3,4-dimethoxy-2-pyridinyl)methyl] sulfinyl]-1*H*-benzimidazole sesquihydrate	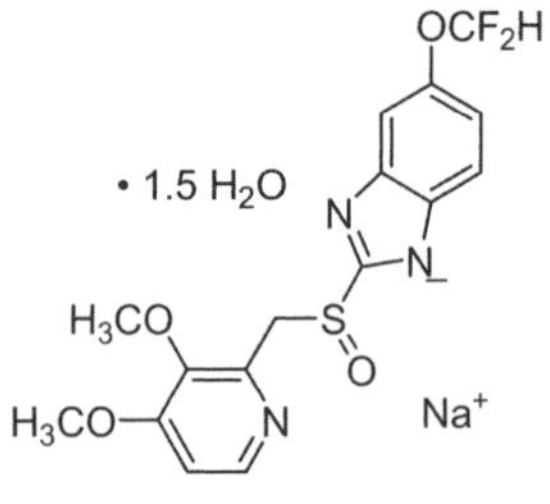 OCF_2H • 1.5 H_2O N N S O H_3CO H_3CO N Na^+
Paramethasone Acetate/ Haldrone[Δ]	[2-[(6*S*,8*S*,9*S*,10*R*,11*S*,13*S*,14*S*,16*R*,17*R*)-6-fluoro-11,17-dihydroxy-10,13,16-trimethyl-3-oxo-7,8,9,11,12,14,15,16-octahydro-6*H*-cyclopenta[a]phenanthren-17-yl]-2-oxoethyl] acetate	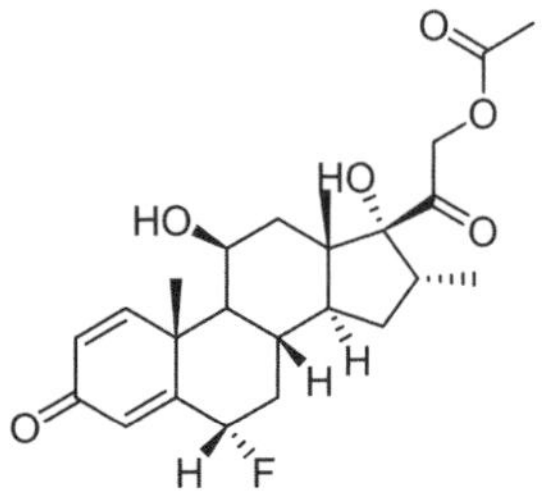 O O HO HO O H H O H F
Paroxetine Hydrochloride/ Paxil	(3*S*,4*R*)-3-(1,3-benzodioxol-5-yloxymethyl)-4-(4-fluorophenyl)piperidine hydrochloride	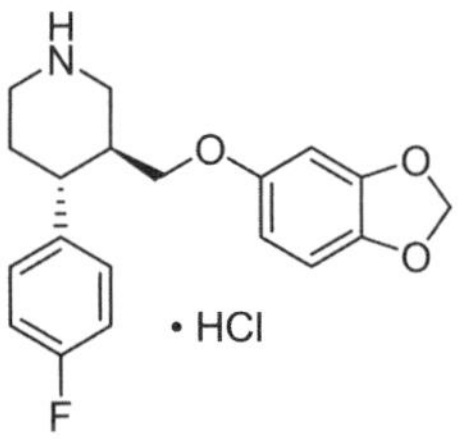 H N O O O F • HCl
Paroxetine Mesylate/ Pexeva	(3*S*,4*R*)-3-(1,3-benzodioxol-5-yloxymethyl)-4-(4-fluorophenyl)piperidine methanesulfonic acid	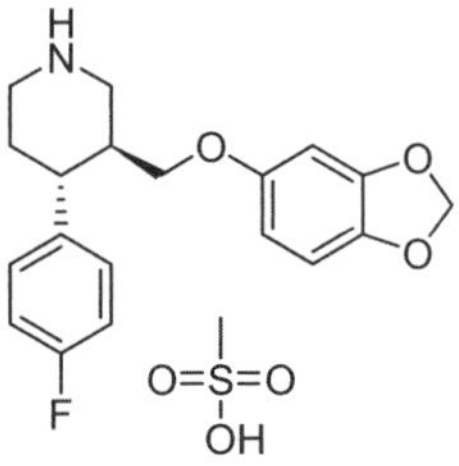 H N O O O F O=S=O OH

Formula	Pharmaceutical action	Dosage form	CAS No.	Approval date/ drug sponsor
$C_{16}H_{14}F_2N_3NaO_4S \cdot 1.5\ H_2O$	Anti-ulcer agent	Injectable Oral tablet	138786-67-1	March 2001/ Wyeth Pharms Inc.
$C_{24}H_{31}FO_6$	Anti-inflammatory agent Glucocorticoid	Oral tablet	1597-82-6	Prior to 1982/ Lilly[Δ]
$C_{19}H_{21}ClFNO_3$	Serotonin uptake inhibitor Antidepressive agent Antiobessional agent	Oral tablet and suspension	78246-49-8	Dec. 1992/ GlaxoSmithKline Torpharm Apotex Inc. Alphapharm Sandoz[Δ] Teva Caraco Mylan Roxane Zydus Pharms USA Aurobindo Pharma
$C_{20}H_{24}FNO_6S$	Antidepressive agent Antiobessional agent	Oral tablet	64006-44-6	July 2003/ JDS Pharms

Active ingredient/ trade name	Chemical name	Structure
Pegaptanib Sodium/ Macugen	((2′-deoxy-2′-fluoro)C-Gm-Gm-A-A-(2′-deoxy-2′-fluoro)U-(2′-deoxy-2′-fluoro)C-Am-Gm-(2′-deoxy-2-fluoro)U-Gm-Am-Am-(2′-deoxy-2′-fluoro)U-Gm-(2′-deoxy-2′-fluoro)C-(2′-deoxy-2′-fluoro)U-(2′-deoxy-2′-fluoro)U-Am-(2′-deoxy-2′-fluoro)U-Am-(2′-deoxy-2′-fluoro)C-Am-(2′-deoxy-2′-fluoro)U-(2′-deoxy-2′-fluoro)C-(2′-deoxy-2′-fluoro)C-Gm-(3′ 3′)-dT), 5′-ester with a, a ′-[4,12-dioxo-6-[[[5-(phosphoonoxy) pentyl]amino]carbonyl]-3,13-dioxa-5,11-diaza-1,15-pentadecanediyl]bis[ω-methoxypoly(oxy-1,2-ethanediyl)], sodium salt.	
Perflexane[Δ]	1,1,1,2,2,3,3,4,4,5,5,6,6,6-tetradecafluorohexane	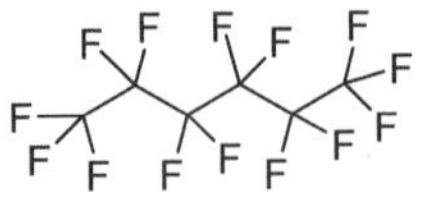
Perflubron/ Imagent[Δ]	1-bromo-1,1,2,2,3,3,4,4,5,5,6,6,7,7,8,8,8-heptadecafluorooctane	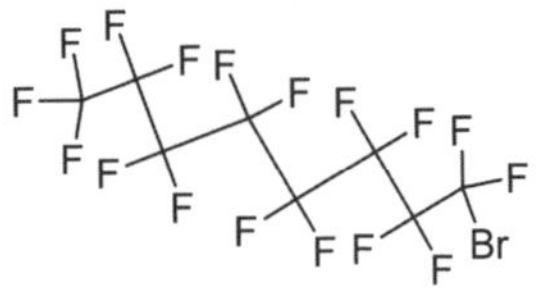
Perfluoropolymethylisopropyl Ether; Polytetrafluoroethylene[Δ]		
Perflutren/ Definity	1,1,1,2,2,3,3,3-octafluoropropane	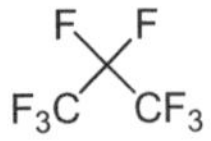 F F F_3C CF_3
Pimozide/ Orap	3-[1-[4,4-bis(4-fluorophenyl)butyl]piperidin-4-yl]-1H-benzimidazol-2-one	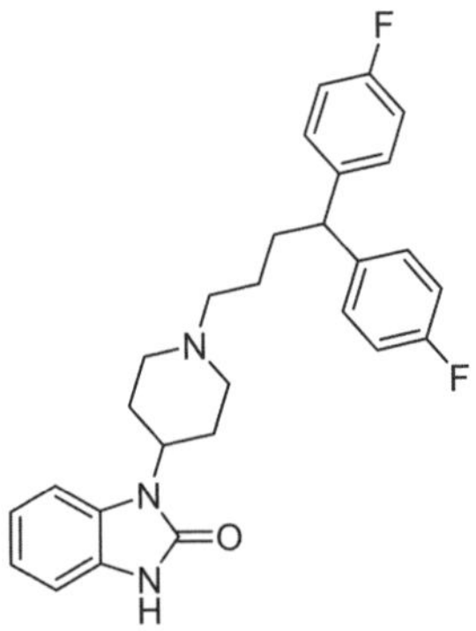

Formula	Pharmaceutical action	Dosage form	CAS No.	Approval date/ drug sponsor
$C_{294}H_{342}F_{13}N_{107}Na_{28}O_{188}P_{28}[C_2H_4O]_n$ (where *n* is approximately 900)	Neovascular age-related macular degeneration (wet) agent	Injectable	222716-86-1	Dec. 2004/ OSI Eye Tech
C_6F_{14}	Preparation of lipid microspheres Imaging agent	Injectable	355-42-0	May 2002/ Imcor PH
C_8BrF_{17}	Contrast media Radiation-sensitizing agent Anti-obesity agent	Oral liquid	423-55-2	Aug. 1993/ Alliance Pharm
	Chemical warfare protectant	Topical		Feb. 2000/ US Army
C_3F_8	Contrast media	Injectable	76-19-7	July 2001/ Bristol Myers Squibb
$C_{28}H_{29}F_2N_3O$	Antipsychotic agent Dopamine antagonist Anti-dyskinesia agent	Oral tablet	2062-78-4	July 1984/ Teva

Active ingredient/ trade name	Chemical name	Structure
Polythiazide/ Renese Minizide	6-chloro-2-methyl-1,1-dioxo-3-(2,2,2-trifluoroethylsulfanylmethyl)-3,4-dihydrobenzo[*e*][1,2,4] thiadiazine-7-sulfonamide	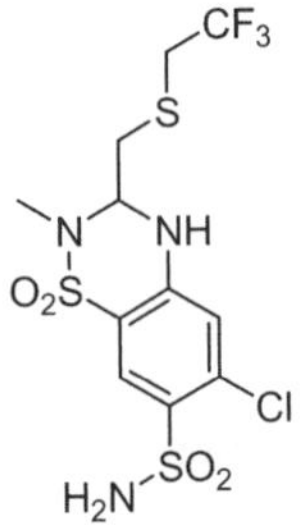
Posaconazole/ Noxafil	4-[4-[4-[4-[[(5*R*)-5-(2,4-difluorophenyl)-5-(1,2,4-triazol-1-ylmethyl)oxolan-3-yl]methoxy]phenyl]piperazin-1-yl]phenyl]-2-[(2*S*,3*S*)-2-hydroxypentan-3-yl]-1,2,4-triazol-3-one	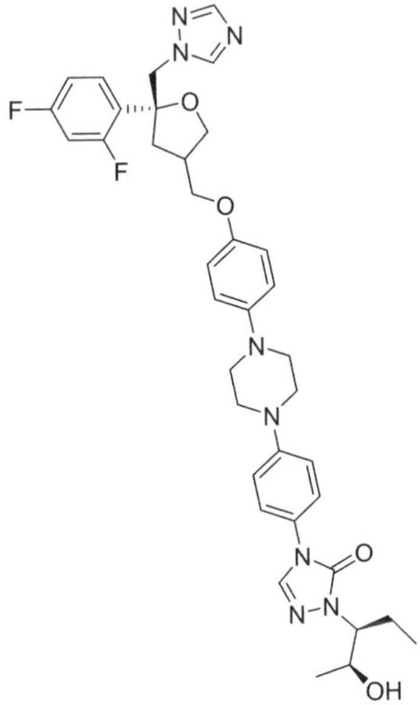
Quazepam/ Doral	7-chloro-5-(2-fluorophenyl)-1-(2,2,2-trifluoroethyl)-3*H*-1,4-benzodiazepine-2-thione	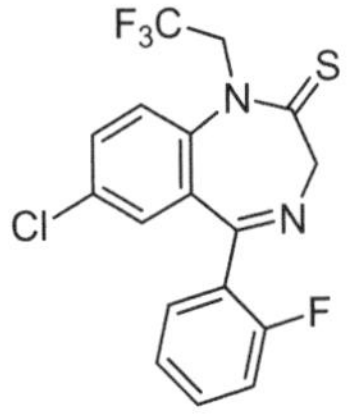
Riluzole/ Rilutek	6-(trifluoromethoxy)-1,3-benzothiazol-2-amine	

Formula	Pharmaceutical action	Dosage form	CAS No.	Approval date/ drug sponsor
$C_{11}H_{13}ClF_3N_3O_4S_3$	Antihypertensive agent Diuretic Sodium chloride symporter inhibitor	Oral tablet	346-18-9	Sept. 1961/ Pfizer
$C_{37}H_{42}F_2N_8O_4$	Antibiotic Antifungal Trypanocidal agent	Oral suspension	171228-49-2	Sept. 2006/ Schering
$C_{17}H_{11}ClF_4N_2S$	Hypnotic Sedative	Oral tablet	36735-22-5	Dec. 1985/ Questcor Pharms
$C_8H_5F_3N_2OS$	Anesthetics Anticonvulsant Excitatory amino acid antagonist Neuroprotective agent	Oral tablet	1744-22-5	Dec. 1995/ Sanofi Aventis Impax Labs

Active ingredient/ trade name	Chemical name	Structure
Risperidone/ Risperdal Risperdal Consta	3-[2-[4-(6-fluoro-1,2-benzoxazol-3-yl)piperidin-1-yl]ethyl]-2-methyl-6,7,8,9-tetrahydropyrido[2,1-*b*] pyrimidin-4-one	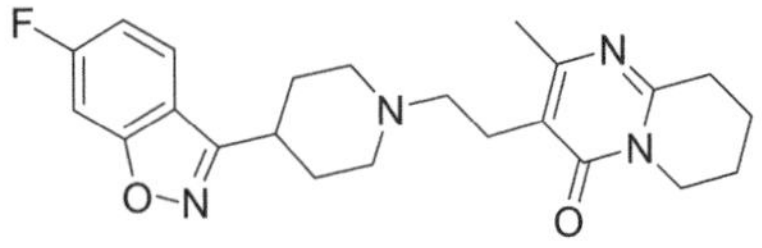
Rosuvastatin Calcium/ Crestor	calcium (*E*,3*R*,5*R*)-7-[4-(4-fluorophenyl)-2-(methyl-methylsulfonyl-amino)-6-propan-2-yl-pyrimidin-5-yl]-3,5-dihydroxy-hept-6-enoate	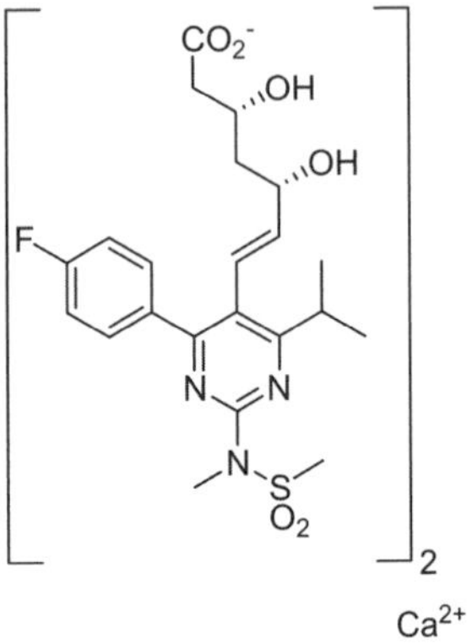
Sevofluran*/ Ultane Sojourn	1,1,1,3,3,3-hexafluoro-2-(fluoromethoxy)propane	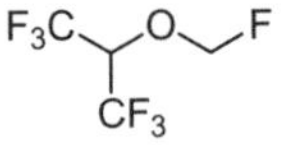
Sitagliptin Phospate/ Januvia Janumet	7-[(3*R*)-3-amino-1-oxo-4-(2,4,5-trifluorophenyl)butyl]-5,6,7,8-tetrahydro-3-(trifluoromethyl)-1,2,4-triazolo[4,3-*a*]pyrazine phosphate (1 : 1) monohydrate	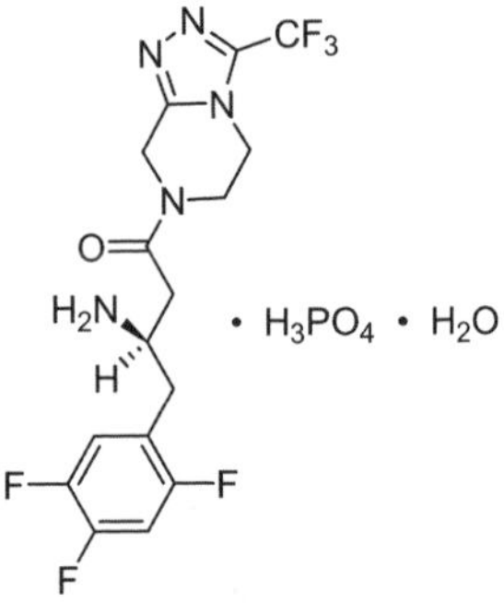
Sodium Fluoride, Triclosan/ Colgate Total		
Sodium Fluroide, F-18		
Sodium Monofluorophosphate/ Extra Strength AIM	disodium fluoro-dioxido-oxophosphorane	

Formula	Pharmaceutical action	Dosage form	CAS No.	Approval date/ drug sponsor
$C_{23}H_{27}FN_4O_2$	Serotonin antagonist Antipsychotic agent Dopamine antagonist	Injectable Oral solution and tablet	106266-06-2	Dec. 1993/ Janssen Pharma
$C_{44}H_{54}CaF_2N_6O_{12}S_2$	Hydroxymethylglutaryl-CoA reductase inhibitor	Oral tablet	147098-20-2	Aug. 2003/ Astrazeneca
$C_4H_3F_7O$	Platelet aggregation inhibitor Anesthetic	Inhalation	28523-86-6	June 1995/ Abbott Baxter Hlthcare Minrad
$C_{16}H_{15}F_6N_5O \cdot H_3PO_4 \cdot H_2O$	DPP-4 inhibitor for type 2 diabetes	Oral tablet	654671-77-9	Oct. 2006/ Merck Co Inc.
NaF	Prevention of dental caries	Dental paste	7681-49-4	July 1997/ Colgate Palmolive Feb. 1972/ GE Healthcare
Na_2FO_3P	Prevention of dental caries	Dental gel and paste	10163-15-2	Aug. 1986/ Chesebrough Brands[Δ]

Active ingredient/ trade name	Chemical name	Structure
Sorafenbin Tosylate/ Nexavar	4-[4-[[4-chloro-3-(trifluorometh yl)phenyl]carbamoylamino] phenoxy]-*N*-methylpyridine-2-carboxamide 4-methylbenzenesulfonate	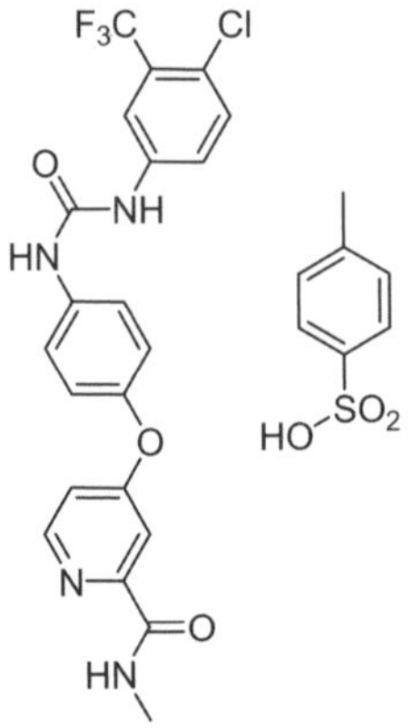
Sparfloxacin/ Zagam[Δ]	5-amino-1-cyclopropyl-7-[(3*S*,5*R*)-3,5-dimethylpiperazin-1-yl]-6,8-difluoro-4-oxoquinoline-3-carboxylic acid	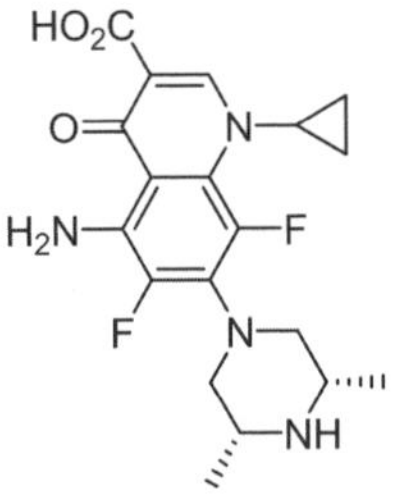
Sulindac/ Clinoril	2-[(3*Z*)-6-fluoro-2-methyl-3-[(4-methylsulfinylphenyl) methylidene]inden-1-yl]acetic acid	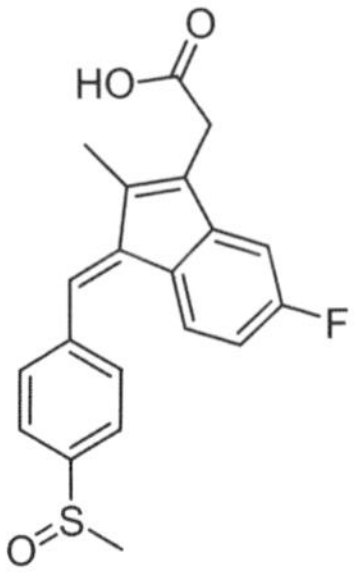
Sunitinib Malate/ Sutent	*N*-(2-diethylaminoethyl)-5-[(*Z*)-(5-fluoro-2-oxo-1*H*-indol-3-ylidene)methyl]-2,4-dimethyl-1*H*-pyrrole-3-carboxamide; (2*S*)-2-hydroxybutanedioic acid	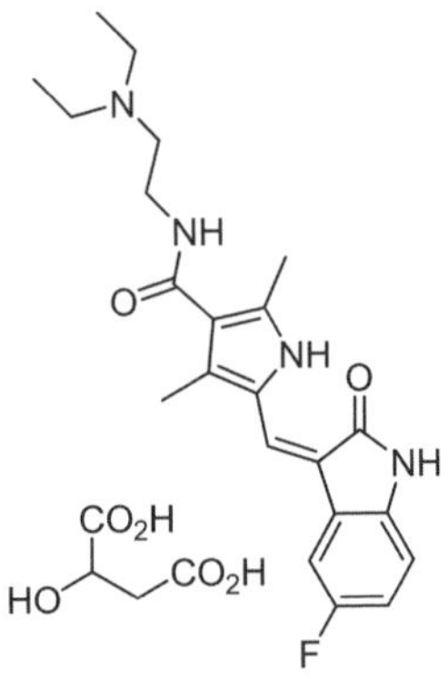

Formula	Pharmaceutical action	Dosage form	CAS No.	Approval date/ drug sponsor
$C_{21}H_{16}ClF_3N_4O_3 \cdot C_7H_8O_3S$	Antineoplastic agent Protein kinase inhibitor	Oral tablet	475207-59-1	Dec. 2005/ Bayer Pharms
$C_{19}H_{22}F_2N_4O_3$	Antibacterial agent	Oral tablet	110871-86-8	Dec. 1996/ Mylan
$C_{20}H_{17}FO_3S$	Nonsteroidal anti-inflammatory agent Antineoplastic agent Cyclooxygenase inhibitor	Oral tablet	38194-50-2	Sept. 1978/ Merck Mutual Pharm Mylan Sandoz Teva[Δ] Watson Labs
$C_{26}H_{33}FN_4O_7$	Antineoplastic agent Angiogenesis inhibitor	Oral capsule	341031-54-7	Jan. 2006/ CPPI CV

Active ingredient/ trade name	Chemical name	Structure
Tipranavir/ Aptivus	*N*-[3-[(1*R*)-1-[(6*R*)-6-(2-cyclohexylethyl)-2-hydroxy-4-oxo-6-propyl-5*H*-pyran-3-yl] propyl]phenyl]-5-(trifluoromethyl)pyridine-2-sulfonamide	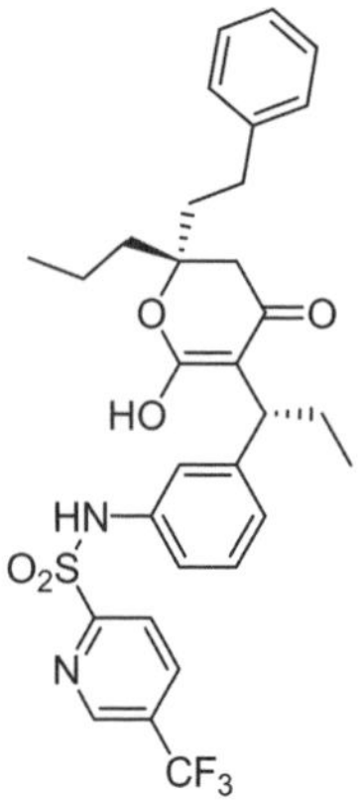
Travoprost/ Travatan Travatan Z	propan-2-yl (*Z*)-7-[(1*R*,2*R*,3*R*,5*S*)-3,5-dihydroxy-2-[(*E*,3*R*)-3-hydroxy-4-[3-(trifluoromethyl)phenoxy] but-1-enyl]cyclopentyl] hept-5-enoate	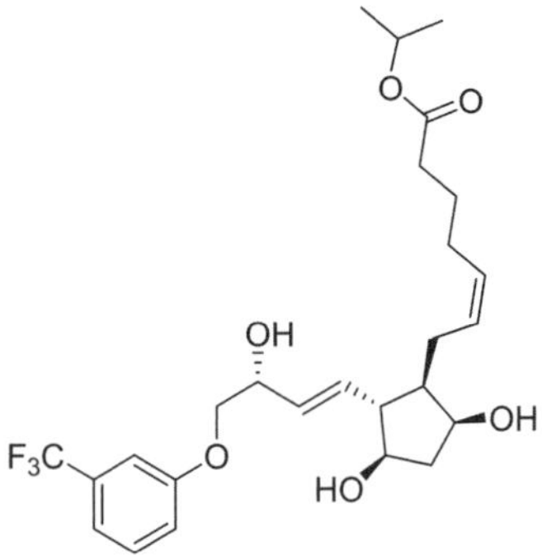
Triamcinolone Acetonide*[1]/ Azmacort Kenalog Triacet Triderm Oracort Oralone Mykacet	(11β,16α)-9-fluoro-11,21-dihydroxy-16,17-[(1-methylethylidene)bis(oxy)]-pregna-1,4-diene-3,20-dione	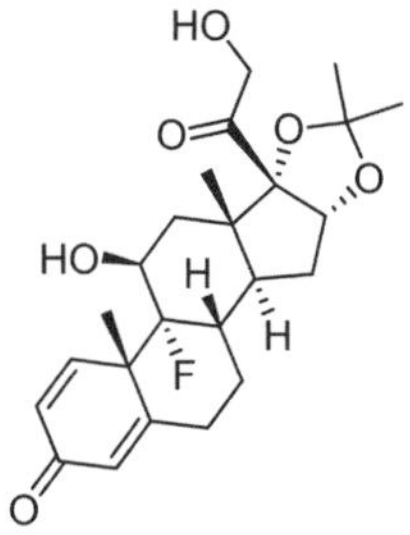

Formula	Pharmaceutical action	Dosage form	CAS No.	Approval date/ drug sponsor
$C_{31}H_{33}F_3N_2O_5S$	Anti-HIV agent Antiviral protease inhibitor	Oral capsule	174484-41-4	June 2005/ Boehringer Ingelheim
$C_{26}H_{35}F_3O_6$	Antihypertensive agent Antiglaucoma	Ophthalmic drops	157283-68-6	March 2001/ Alcon
$C_{24}H_{31}FO_6$	Glucocorticoid Anti-inflammatory agent Immunosuppressive agent	Inhalation (aersol) Topical cream, spray, and ointment Injectable Dental paste Nasal (spray)	76-25-5	March 1959/ Abbott Apothecon Teva Actavis Mid Atlantic Altana G and W Labs Ranbaxy Perrigo New York Taro Vintage Del Ray Labs Morton Grove Carolina Medcl Sanofi Aventis

Active ingredient/ trade name	Chemical name	Structure
Triamcinolone Hexacetonide/ Aristospan	(11β,16α)-21-(3,3-dimethyl-1-oxobutoxy)-9-fluoro-11-hydroxy-16,17-[(1-methylethylidene)bis(oxy)]-pregna-1,4-diene-3,20-dione	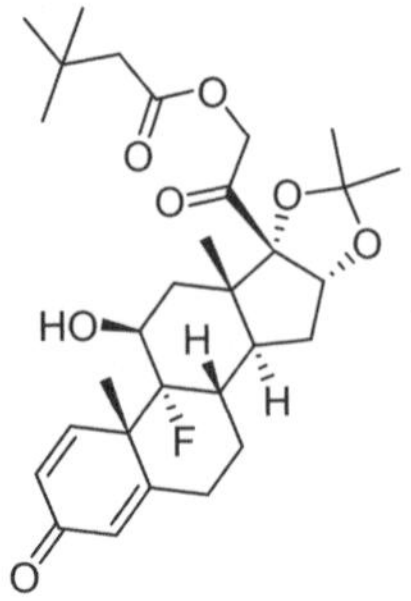
Trifluoperazine Hydrochloride/ Stelazine	10-[3-(4-methylpiperazin-1-yl)propyl]-2-(trifluoromethyl) phenothiazine dihydrochloride	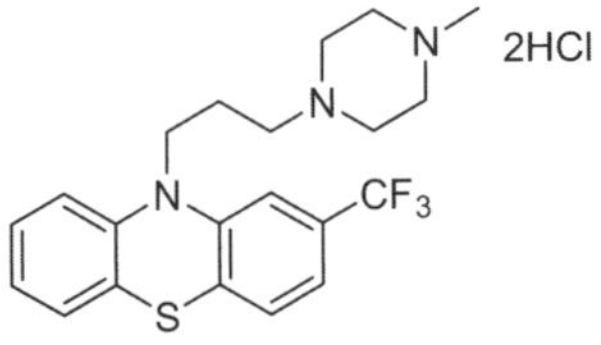
Triflupromazine[2]/ Vesprin[Δ]	*N,N*-dimethyl-3-[2-(trifluoromethyl)phenothiazin-10-yl]propan-1-amine	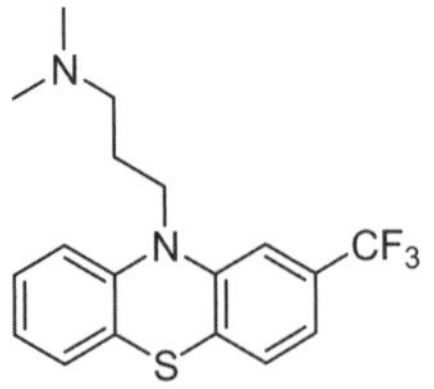
Trifluridine/ Viroptic	1-[4-hydroxy-5-(hydroxymethyl)oxolan-2-yl]-5-(trifluoromethyl)pyrimidine-2,4-dione	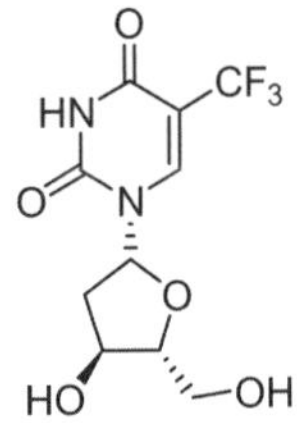
Trovafloxacin Mesylate/ Trovan[Δ]	7-[(1*S*,5*R*)-6-amino-3-azabicyclo[3.1.0]hexan-3-yl]-1-(2,4-difluorophenyl)-6-fluoro-4-oxo-1,8-naphthyridine-3-carboxylic acid; methanesulfonic acid	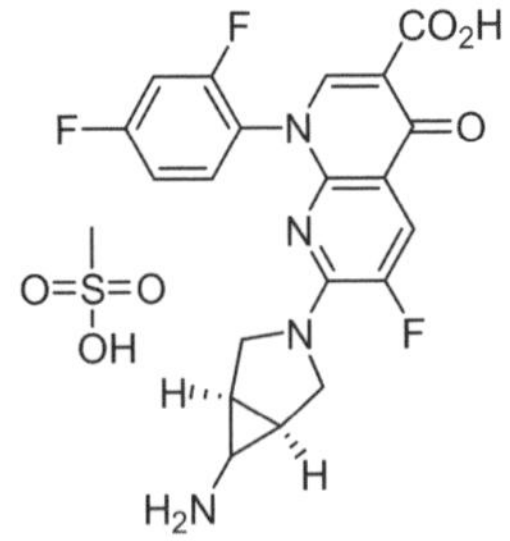

Formula	Pharmaceutical action	Dosage form	CAS No.	Approval date/ drug sponsor
$C_{30}H_{41}FO_7$	Anti-inflammatory agent	Injectable	5611-51-8	July 1969/ Sandoz
$C_{21}H_{26}Cl_2F_3N_3S$	Antiemetic Antipsychotic agent Dopamine antagonist	Oral concentrate and tablet	440-17-5	April 1959/ GlaxoSmithKline Mylan Sandoz
$C_{18}H_{19}F_3N_2S$	Antiemetic Antipsychotic agent Dopamine antagonist	Oral suspension and tablet Injectable	146-54-3	Sept. 1957/ Apothecon
$C_{10}H_{11}F_3N_2O_5$	Antiviral agent Antimetabolite	Ophthalmic drops	70-00-8	April 1980/ Monarch Pharms Alcon
$C_{21}H_{19}F_3N_4O_6S$	Anti-infective agent	Oral tablet	147059-75-4	Dec. 1997/ Pfizer

Active ingredient/ trade name	Chemical name	Structure
Valrubicin/ Valstar	[2-oxo-2-[(2*S*,4*S*)-2,5,12-trihydroxy-4-[5-hydroxy-6-methyl-4-[(2,2,2-trifluoroacetyl)amino]oxan-2-yl]oxy-7-methoxy-6,11-dioxo-3,4-dihydro-1*H*-tetracen-2-yl] ethyl] pentanoate	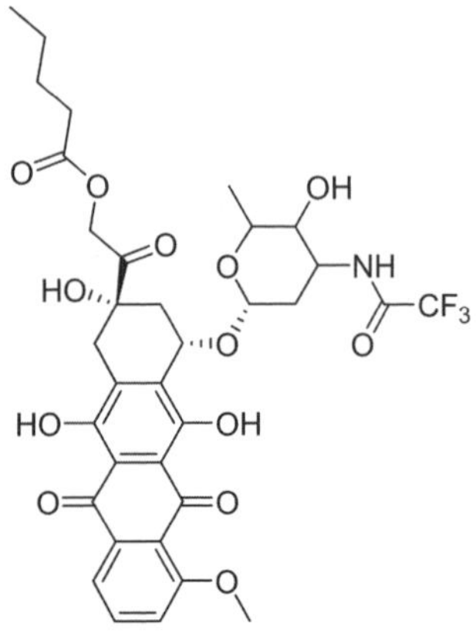
Voriconazole/ Vfend	(2*R*,3*S*)-2-(2,4-difluorophenyl)-3-(5-fluoropyrimidin-4-yl)-1-(1,2,4-triazol-1-yl)butan-2-ol	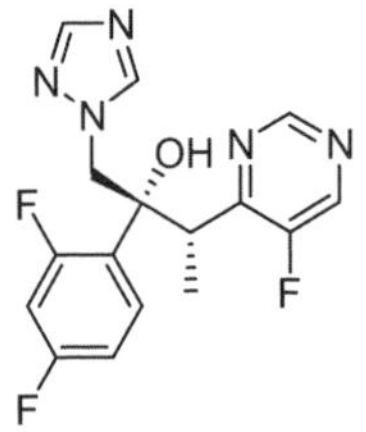

[Δ]Discontinued from the market.
[V]See the veterinary drug list (Section 2).
[1]Triamcinolone and Triamcinolone Diacetate are discontinued from the market.
[2]Triflupromazine and Triflupromazine Hydrochloride are discontinued from the market.
[3]Not approved.

Formula	Pharmaceutical action	Dosage form	CAS No.	Approval date/ drug sponsor
$C_{34}H_{36}F_3NO_{13}$	Antineoplastic agent	Intravescial solution	56124-62-0	Sept. 1998/ Indevus Pharms
$C_{16}H_{14}F_3N_5O$	Antifungal agent	Oral suspension and tablet Injectable	137234-62-9	May 2002/ Pfizer

Section 2. Fluorine-containing Drugs for Veterinary Use Approved by FDA in the United States

Fluorine in Medicinal Chemistry and Chemical Biology Edited by Iwao Ojima

Active ingredient/ trade name	Chemical name	Structure
Azaperone/ Stresnil	1-(4-fluorophenyl)-4-(4-pyridin-2-ylpiperazin-1-yl)butan-1-one	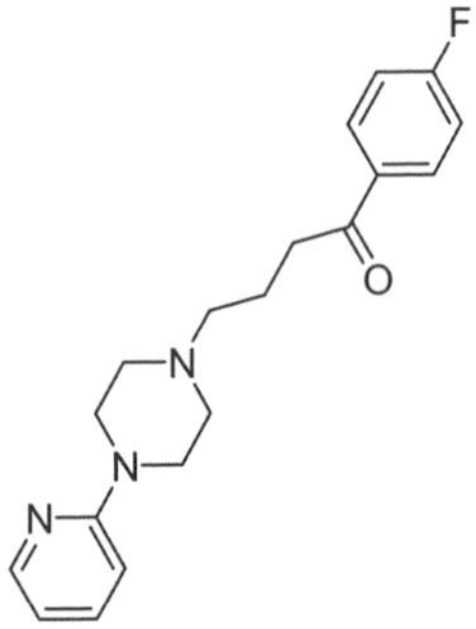
Betamethasone Acetate/ Betavet Soluspan Gentocin Durafilm	[2-[(8*S*,9*R*,10*S*,11*S*,13*S*,14*S*,16*S*,17*R*)-9-fluoro-11,17-dihydroxy-10,13,16-trimethyl-3-oxo-6,7,8,11,12,14,15,16-octahydrocyclopenta[*a*]phenanthren-17-yl]-2-oxoethyl] acetate	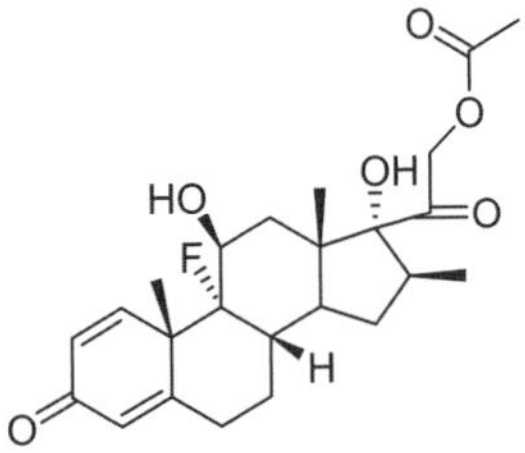
Betamethasone Dipropionate/ Betasone	[(8*S*,9*R*,10*S*,11*S*,13*S*,14*S*,16*S*,17*R*)-9-fluoro-11-hydroxy-10,13,16-trimethyl-3-oxo-17-(2-propanoyloxyacetyl)-6,7,8,11, 12,14,15,16-octahydrocyclopenta[*a*] phenanthren-17-yl] propanoate	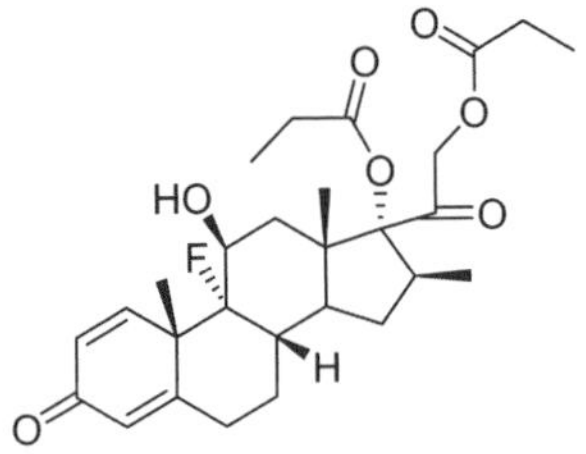
Betamethasone Sodium Phosphate/ Betavet Soluspan Betasone	[2-[(8*S*,9*R*,10*S*,11*S*,13*S*,14*S*,16*S*,17*R*)-9-fluoro-11,17-dihydroxy-10,13,16-trimethyl-3-oxo-6,7,8,11,12,14,15, 16-octahydrocyclopenta[*a*] phenanthren-17-yl]-2-oxoethyl] phosphate	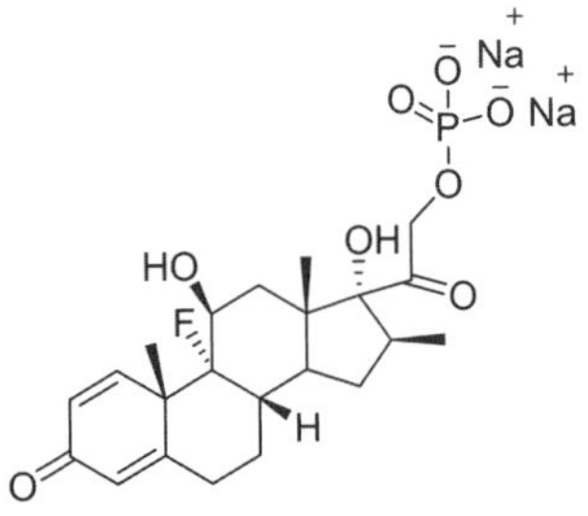
Betamethasone Valerate/ Gentocin Otic Topagen Otomax Gentavet Betagen Tri-Otic Vetro-Max GBC	(1) pregna-1,4-diene-3,20-dione, 9-fluoro-11,21-dihydroxy-16-methyl-17-[(1-oxopentyl)oxy]-, (11β, 16β)-; (2) 9-fluoro-11 β,17,21-trihydroxy-16β-methylpregna-1,4-diene-3,20-dione 17 valerate	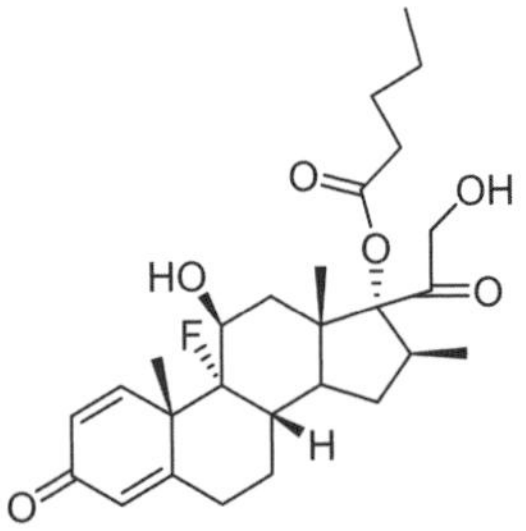

Formula	Pharmaceutical action	Dosage form (species)	CAS No.	Approval date/ drug sponsor
$C_{19}H_{22}FN_3O$	Sedative Tranquilizer	Injectable (Swine)	1649-18-9	Oct. 1983/ Schering-Plough Animal Health Corp.
$C_{24}H_{31}FO_6$	Bronchodilator agent Glucocorticoid	Injectable (Horse) Ophthalmic drops (Dog)	987-24-6	Prior to 1989/ Schering-Plough Animal Health Corp.
$C_{28}H_{37}FO_7$	Anti-inflammatory agent Glucocorticoid	Injectable (Horse, Dog)	5593-20-4	Prior to 1989/ Schering-Plough Animal Health Corp.
$C_{22}H_{28}FNa_2O_8P$	Glucocorticoid	Injectable (Horse, Dog)	151-73-5	Prior to 1989/ Schering-Plough Animal Health Corp.
$C_{27}H_{37}FO_6$	Glucocorticoid	Topical spray, liquid, and ointment (Dog, Cat)	2152-44-5	Prior to 1989 / Schering-Plough Animal Health Corporation Med-Pharmex, Inc. Altana, Inc. IVA Animal Health First Priority

Active ingredient/ trade name	Chemical name	Structure
Danofloxacin Mesylate/ A180	1-cyclopropyl-6-fluoro-7-[(1*S*,4*S*)-3-methyl-3,6-diazabicyclo[2.2.1] heptan-6-yl]-4-oxoquinoline-3-carboxylic acid; methanesulfonic acid	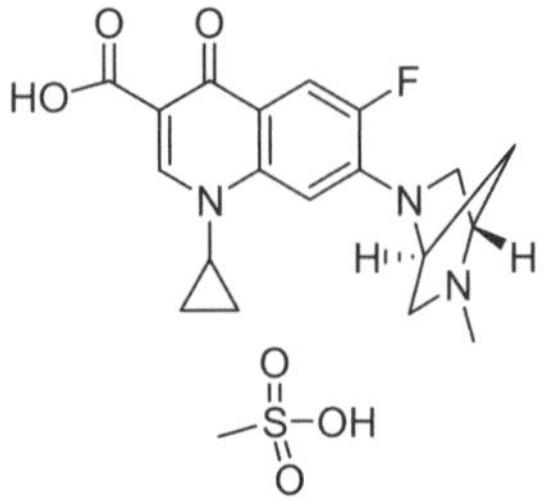
Deracoxib/ Dermaxx	4-[3-(difluoromethyl)-5-(3-fluoro-4-methoxyphenyl)pyrazol-1-yl] benzenesulfonamide	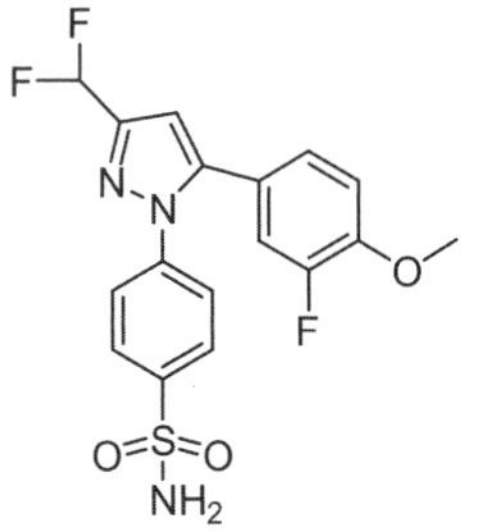
Dexamethasone Sodium Phosphate/ Dex-A-Vet Dexium-SP	[2-[(8*S*,9*R*,10*S*,11*S*,13*S*,14*S*,16*R*,17*R*)-9-fluoro-11,17-dihydroxy-10,13,16-trimethyl-3-oxo-6,7,8,11,12,14,15,16-octahydrocyclopenta[*a*]phenanthren-17-yl]-2-oxoethyl] phosphate	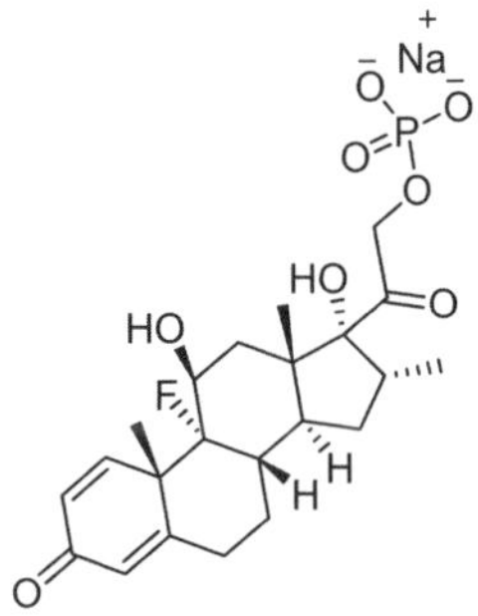
Dexamethasone/ Azium Naquasone Tresaderm Dexium Dexameth-A-Vet Pet Derm Zonometh Dexachel Dexium	9-fluoro-11β,17,21-trihydroxy-16α-methylpregna-1,4-diene-3,20-dione (8*S*,9*R*,10*S*,11*S*, 13*S*,14*S*,16*R*,17*R*)-9-fluoro-11,17-dihydroxy-17-(2-hydroxyacetyl)-10,13,16-trimethyl-6,7,8,11,12,14, 15,16-octahydrocyclopenta[*a*] phenanthren-3-one	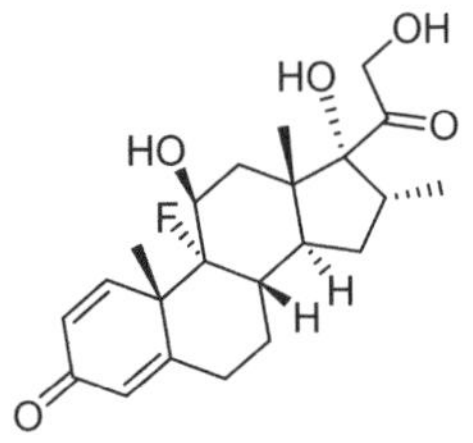

Formula	Pharmaceutical action	Dosage form (species)	CAS No.	Approval date/ drug sponsor
$C_{20}H_{24}FN_3O_6S$	Anti-bacterial agent	Injectable (Cattle)	119478-55-6	Sept. 2002/ Pfizer, Inc.
$C_{17}H_{14}F_3N_3O_3S$	Anti-inflammatory Analgesic	Oral tablet (Dog)	169590-41-4	Aug. 2002/ Novartis Animal Health US, Inc.
$C_{22}H_{28}FNaO_8P$	Anti-inflammatory Glucocorticoid	Injectable (Dog, Horse)	1869-92-7	April 2004/ Cross Vetpharm Group Ltd. Watson Laboratories, Inc. IVX Animal Health
$C_{22}H_{29}FO_5$	Anti-inflammatory Antiemetics Antineoplastic agent Hormonal Glucocorticoid	Oral Tablet and liquid Injectable Bolus Topical liquid (Dog, Cat, Cattle, Horse)	50-02-2	April 1995/ Schering-Plough Animal Health Corp. Merial Ltd. Cross Vetpharm Group Ltd. Pfizer IVX Animal Health Sparhawk Laboratories, Inc.

Active ingredient/ trade name	Chemical name	Structure
Dexamethasone-21-isonicotinate/ Voren	[2-[(8*S*,9*R*,10*S*,11*S*,13*S*,14*S*,16*R*,17*R*)-9-fluoro-11,17-dihydroxy-10,13,16-trimethyl-3-oxo-6,7,8,11,12,14,15,16-octahydrocyclopenta[*a*]phenanthren-17-yl]-2-oxoethyl] pyridine-4-carboxylate	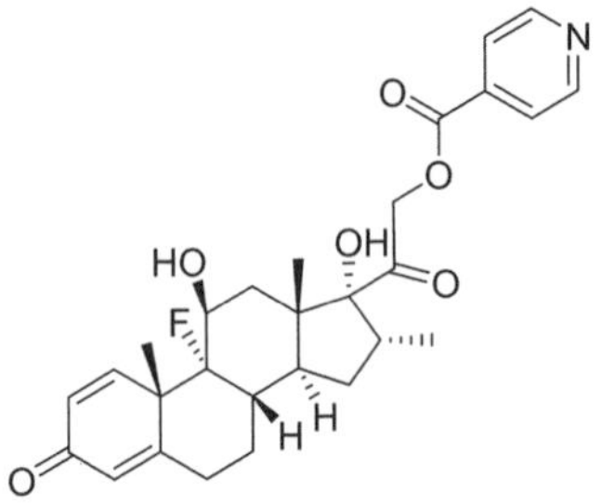
Difloxacin Hydrochloride/ Dicural	6-fluoro-1-(4-fluorophenyl)-7-(4-methylpiperazin-1-yl)-4-oxoquinoline-3-carboxylic acid hydrochloride	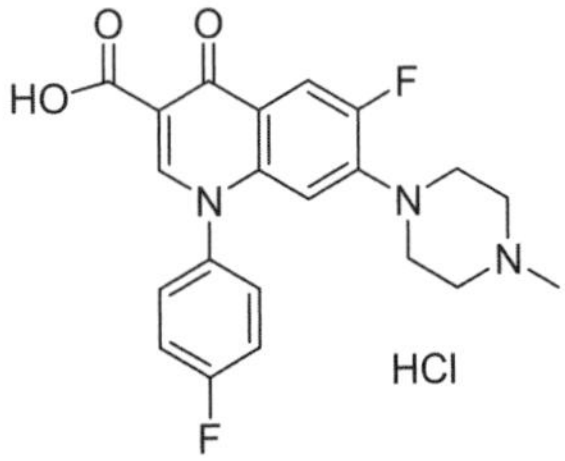 HCl
Dirlotapide/ Slentrol	1-methyl-*N*-[(1*S*)-2-(methyl-(phenylmethyl)amino)-2-oxo-1-phenylethyl]-5-[[2-[4-(trifluoromethyl)phenyl]benzoyl] amino]indole-2-carboxamide	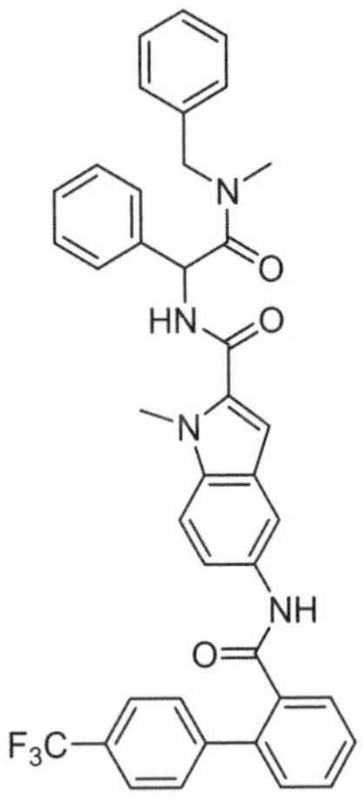
Droperidol/ Innovar-Vet	3-[1-[4-(4-fluorophenyl)-4-oxobutyl]-3,6-dihydro-2*H*-pyridin-4-yl]-1*H*-benzimidazol-2-one	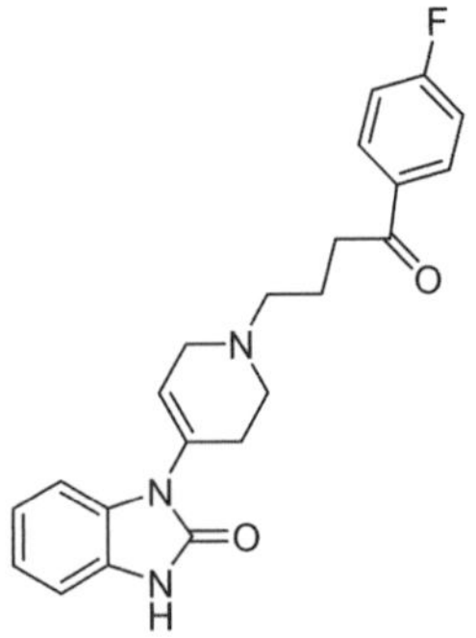

Formula	Pharmaceutical action	Dosage form (species)	CAS No.	Approval date/ drug sponsor
$C_{28}H_{32}FNO_6$	Anti-inflammatory Anti-allergic Glucocorticoid	Injectable (Dog, Cat, Horse)	2265-64-7	Prior to 1989/ Boehringer Ingelheim Vetmedica, Inc.
$C_{21}H_{20}ClF_2N_3O_3$	Antibacterial agent	Oral tablet (Dog)	91296-86-5	Nov. 1997/ Fort Dodge Animal Health, Division of Wyeth
$C_{40}H_{33}F_3N_4O_3$	Anti-obesity agent	Oral liquid (Dog)	481658-94-0	Dec. 2006/ Pfizer, Inc.
$C_{22}H_{22}FN_3O_2$	Anesthetic Tranquilizer	Injectable (Dog)	548-73-2	Prior to 1982/ Schering-Plough Animal Health Corp.

Active ingredient/ trade name	Chemical name	Structure
Enrofloxacin/ Baytril	1-cyclopropyl-7-(4-ethylpiperazin-1-yl)-6-fluoro-4-oxoquinoline-3-carboxylic acid	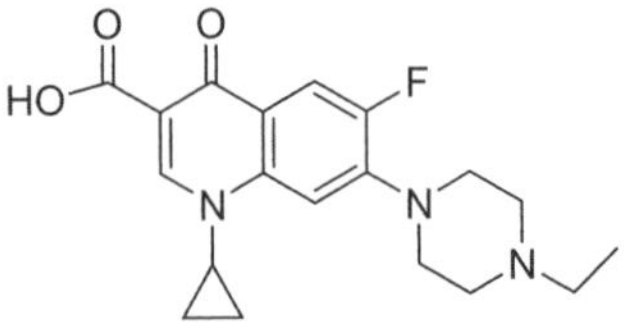
Florfenicol/ Nuflor Aquaflor	2,2-dichloro-*N*-[(1*R*,2*S*)-3-fluoro-1-hydroxy-1-(4-methylsulfonylphenyl) propan-2-yl]acetamide	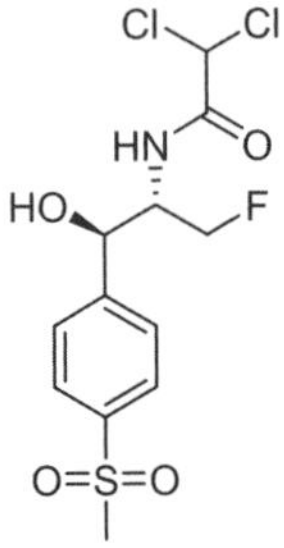
Flumethasone Acetate/ Fluosmin	[(6*S*,9*R*,11*S*,14*S*,16*R*,17*R*)-6,9-difluoro-11-hydroxy-17-(2-hydroxyacetyl)-10,13,16-trimethyl-3-oxo-6,7,8,11,12,14,15,16-octahydrocyclopenta[*a*]phenanthren-17-yl] acetate	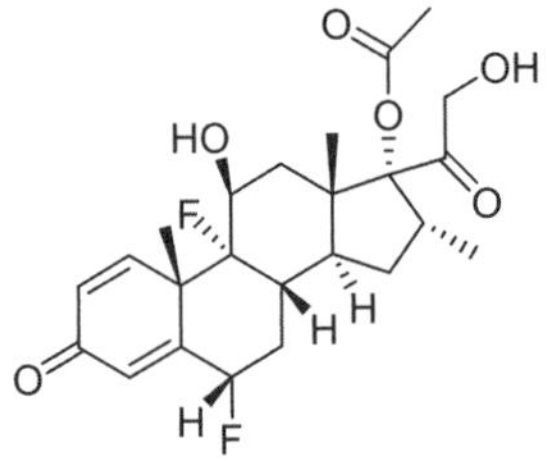
Flumethasone/ Flucort Anaprime Fluosmin Anaprime Toptic$^{\Delta}$	(6*S*,8*S*,9*R*,10*S*,11*S*,13*S*,14*S*,16*R*,17*R*)-6,9-difluoro-11,17-dihydroxy-17-(2-hydroxyacetyl)-10,13,16-trimethyl-6,7,8,11,12,14,15,16-octahydrocyclopenta[*a*] phenanthren-3-one	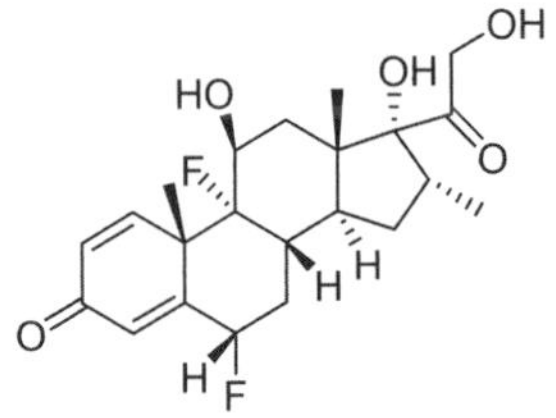
Flunixin Meglumine/ Banamine Flunazine	(2*R*,3*R*,4*R*,5*S*)-6-methylaminohexane-1,2,3,4,5-pentol; 2-[[2-methyl-3-(trifluoromethyl)phenyl]amino] pyridine-3-carboxylic acid	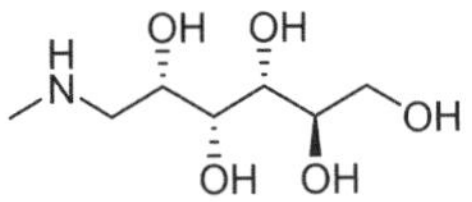 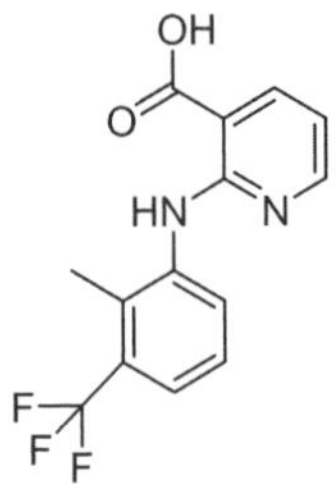

Formula	Pharmaceutical action	Dosage form (species)	CAS No.	Approval date/ drug sponsor
$C_{19}H_{22}FN_3O_3$	Antibacterial agent	Oral tablet Topical liquid Injectable (Dog, Cat, Cattle)	93106-60-6	Dec. 1988/ Mobay Corporation Bayer Healthcare LLC, Animal Health Division
$C_{12}H_{14}Cl_2FNO_4S$	Antibacterial agent	Injectable Oral tablet (Cattle, Swine)	73231-34-2	May 1996/ Schering-Plough Animal Health Corp.
$C_{24}H_{30}F_2O_6$	Glucocorticoid	Injectable (Dog)	2823-42-9	Prior to 1982/ Fort Dodge Animal Health, Division of Wyeth
$C_{22}H_{28}F_2O_5$	Anti-inflammatory Glucocorticoid	Injectable Oral tablet Liquid (Ophthalmic) (Dog, Cat, Horse)	2135-17-3	Prior to 1982/ Fort Dodge Animal Health, Division of Wyeth Elanco Animal Health, A Division of Eli Lilly & Co.$^{\Delta}$
$C_{21}H_{28}F_3N_3O_7$	Anti-inflammatory Analgesic Antipyretic	Injectable Oral granules and paste (Cattle, Horse)	42461-84-7	May 1998/ Schering-Plough Animal Health Corp. Agri Laboratories, Ltd, Fort Dodge Animal Health, Division of Wyeth IVX Animal Health Norbrook Laboratories Ltd.

Active ingredient/ trade name	Chemical name	Structure
Fluocinolone Acetonide/ Neo-synalar Synalar Synotic Synsac	(6α,11β,16α)-6,9-difluoro-11,21-dihydroxy-16,17[(1-methylethylidene)bis(oxy)]-pregna-1,4-diene-3,20-dione	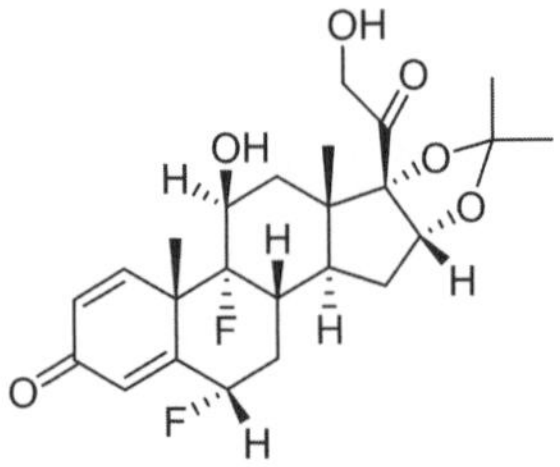
Fluoxetine Hydrochloride/ Reconcile	*N*-methyl-3-phenyl-3-[4-(trifluo romethyl)phenoxy]propan-1-amine hydrochloride	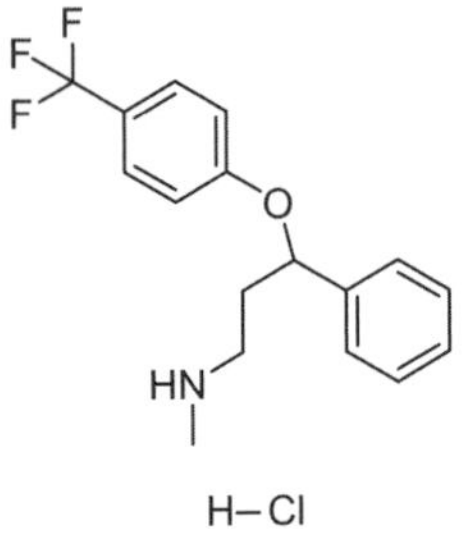
Fluprostenol Sodium/ Equimate	sodium (*E*)-7-[(1*S*,3*S*,5*R*)-3,5-dihydroxy-2-[(*E*,3*S*)-3-hydroxy-4-[3-(trifluoromethyl)phenoxy]but-1-enyl]cyclopentyl]hept-5-enoate,	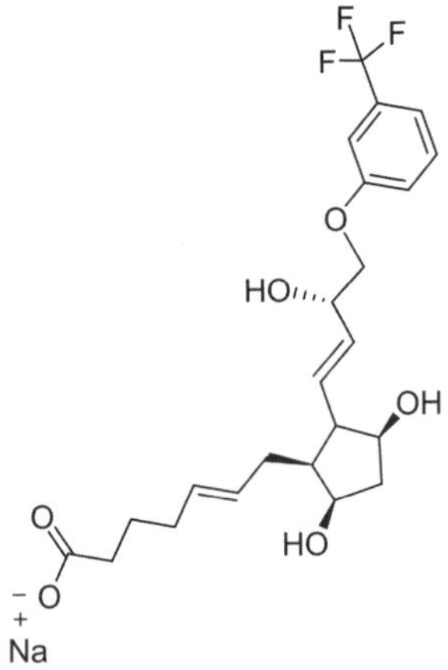
	sodium (*Z*)-7-[(1*R*,2*R*,3*R*,5*S*)-3,5-dihydroxy-2-[(*E*,3*R*)-3-hydroxy-4-[3-(trifluoromethyl)phenoxy]but-1-enyl]cyclopentyl]hept-5-enoate	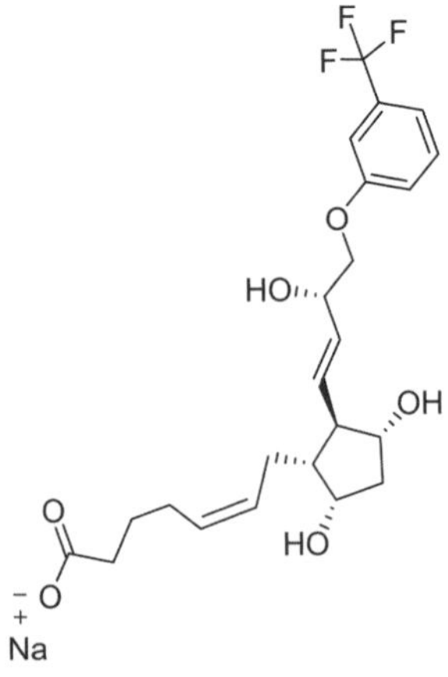

Formula	Pharmaceutical action	Dosage form (species)	CAS No.	Approval date/ drug sponsor
$C_{24}H_{30}F_2O_6$	Anti-inflammatory Glucocorticoid Antipruritic Vasoconstrictive agent	Topical liquid and cream (Dog, Cat)	67-73-2	Prior to 1982/ Medicis Dermatologics, Inc. Fort Dodge Animal Health, Division of Wyeth
$C_{17}H_{19}ClF_3NO$	Serotonin uptake inhibitors Antidepressive agents	Oral tablet (Dog)	59333-67-4	Jan. 2007/ Elanco Animal Health, A Division of Eli Lilly & Co.
$C_{23}H_{28}F_3NaO_6$	Luteolytic agents Estrus synchronization agent	Injectable (Horse)	55028-71-2	Prior to 1982/ Bayer Healthcare LLC, Animal Health Division

Active ingredient/ trade name	Chemical name	Structure
Flurogestone Acetate/ Synchro-Mate	[(8*S*,9*S*,10*S*,11*S*,13*S*,14*S*,17*R*)-17-acetyl-9-fluoro-11-hydroxy-13,14-dimethyl-3-oxo-1,2,6,7,8,10,11,12,15,16-decahydrocyclopenta[*a*]phenanthren-17-yl] acetate	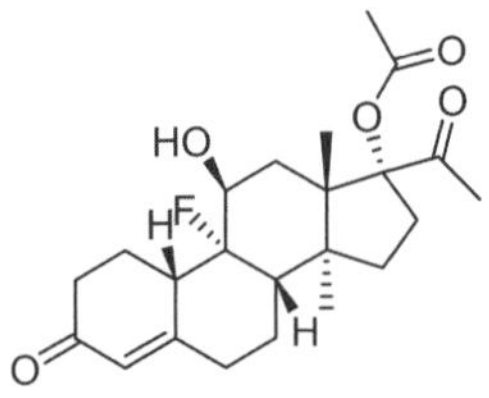
Halothane/ Fluothane	2-bromo-2-chloro-1,1,1-trifluoroethane	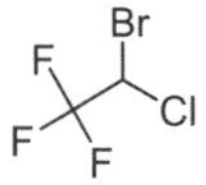
Isoflupredone Acetate/ Predef Neo-Predef	[2-[(8*S*,9*R*,10*S*,11*S*,13*S*,14*S*,17*R*)-9-fluoro-11,17-dihydroxy-10,13-dimethyl-3-oxo-6,7,8,11,12,14,15,16-octahydrocyclopenta[*a*]phenanthren-17-yl]-2-oxoethyl] acetate	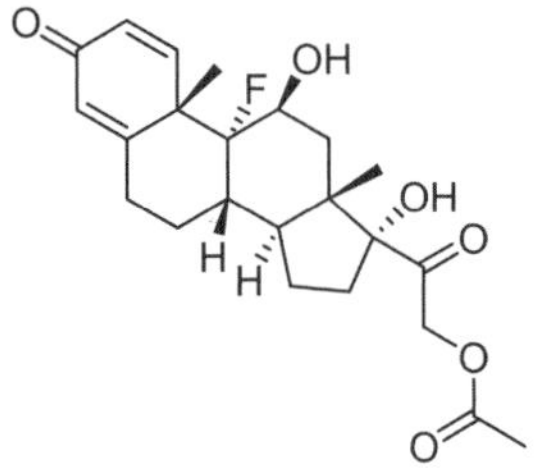
Isoflurane/ Aerrane IsoFlo	2-chloro-2-(difluoromethoxy)-1,1,1-trifluoroethane	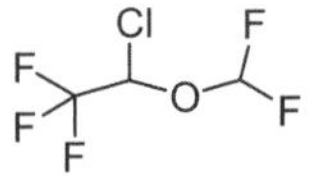
Lufenuron/ Program Sentinel Capstar	*N*-[[2,5-dichloro-4-(1,1,2,3,3,3-hexafluoropropoxy)phenyl] carbamoyl]-2,6-difluorobenzamide	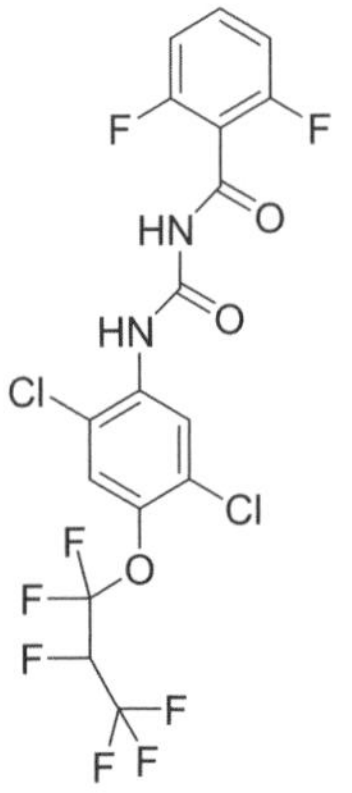

Formula	Pharmaceutical action	Dosage form (species)	CAS No.	Approval date/ drug sponsor
$C_{23}H_{31}FO_5$	Estrus synchronization agent	Vaginal sponge (Sheep)	2529-45-5	Prior to 1982/ G.D. Searle LLC, Pharmacia Corp
$C_2HBrClF_3$	Anesthetic	Inhalation Liquid (Dog, Cat)	151-67-7	Prior to 1982/ Fort Dodge Animal Health, Division of Wyeth Halocarbon Laboratories, Div. of Halocarbon Products Corp.
$C_{23}H_{29}FO_6$	Anti-inflammatory	Injectable Topical ointment (Dog, Cat, Horse, Cattle, Swine)	338-98-7	Prior to 1982/ Pharmacia & Upjohn Co.
$C_3H_2ClF_5O$	Anesthetic	Inhalation (Dog, Horse)	26675-46-7	Jan. 1986/ Abbott Laboratories Halocarbon Laboratories, Div. of Halocarbon Products Corp. Marsam Pharmaceuticals, LLC Minrad, Inc.
$C_{17}H_8Cl_2F_8N_2O_3$	Ectoparasiticide	Oral tablet Injectable (Dog, Cat)	103055-07-8	Nov. 1994/ Novartis Animal Health US, Inc.

Active ingredient/ trade name	Chemical name	Structure
Marbofloxacin/ Zeniquin	9-fluoro-2,3-dihydro-3-methyl-10-(4-methyl-piperazino)-7-oxo-7H-pyrido[1,2,3-ij][1,2,4]benzoxadiazine-6-carboxylic acid	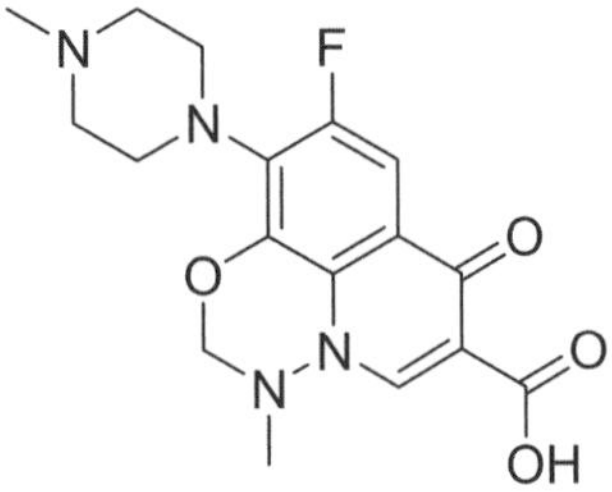
Methoxyflurane/ Metofane	2,2-dichloro-1,1-difluoro-1-methoxyethane	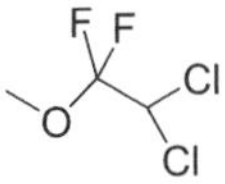
Orbifloxacin/ Orbax	1-cyclopropyl-7-[(3*R*,5*S*)-3,5-dimethylpiperazin-1-yl]-5,6,8-trifluoro-4-oxoquinoline-3-carboxylic acid	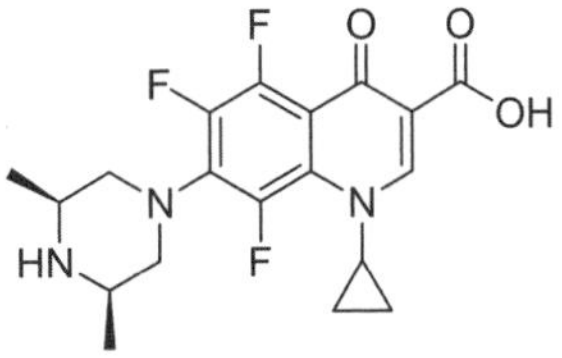
Ponazuril/ Marquis	1-methyl-3-[3-methyl-4-[4-(trifluoromethylsulfonyl)phenoxy] phenyl]-1,3,5-triazinane-2,4,6-trione	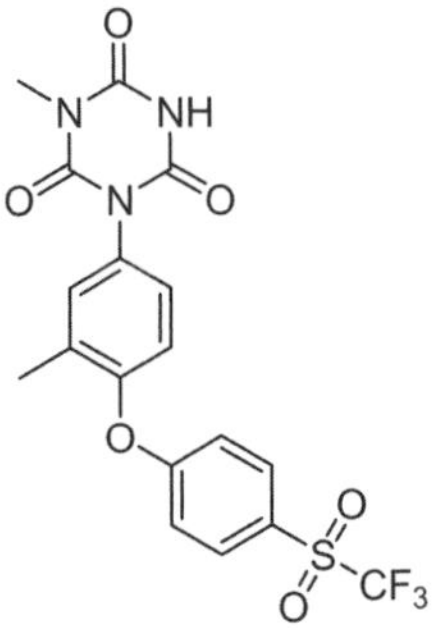
Romifidine Hydrochloride/ Sedivet	*N*-(2-bromo-6-fluorophenyl)-4,5-dihydro-1*H*-imidazol-2-amine hydrochloride	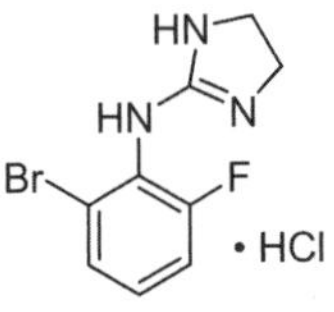
Sarafloxacin Hydrochloride/ SaraFlox$^{\Delta}$	6-fluoro-1-(4-fluorophenyl)-4-oxo-7-piperazin-1-ylquinoline-3-carboxylic acid hydrochloride	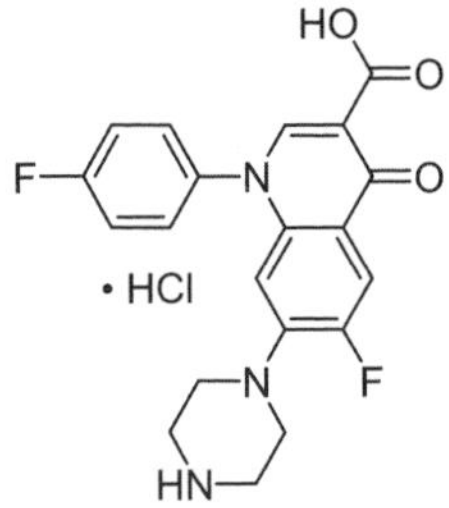

Formula	Pharmaceutical action	Dosage form (species)	CAS No.	Approval date/ drug sponsor
$C_{17}H_{19}FN_4O_4$	Antibacterial agent	Oral tablet (Cat)	115550-35-1	June 1999/ Pfizer, Inc.
$C_3H_4Cl_2F_2O$	Anesthetic	Inhalation (Dog, Cat, Horse, Swine, Sheep, Psittacines)	76-38-0	Prior to 1982/ Schering-Plough Animal Health Corp.
$C_{19}H_{20}F_3N_3O_3$	Antibacterial agent	Oral tablet (Dog, Cat)	113617-63	April 1997/ Schering-Plough Animal Health Corp.
$C_{18}H_{14}F_3N_3O_6S$	Antiprotozoal to *Sarcocystis neurona*.	Paste (Horse)	69004-04-2	July 2001/ Bayer Healthcare LLC, Animal Health Division
$C_9H_9BrFN_3 \cdot HCl$	Adrenergic alpha-agonist Anesthetic	Intravenous (Horse)	65896-14-2	June 2004/ Boehringer Ingelheim Vetmedica, Inc.
$C_{20}H_{17}F_2N_3O_3 \cdot HCl$	Antibacterial agent	Oral Powder Injection (Embryonated broiler eggs, Chicken, Turkey)	91296-87-6	Aug. 1995/ Abbott Laboratories

Active ingredient/ trade name	Chemical name	Structure
Sevoflurane/ SevoFlo	1,1,1,3,3,3-hexafluoro-2-(fluoromethoxy) propane	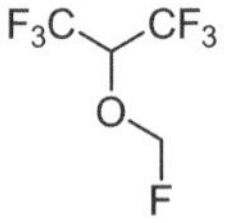
Ticarbodine/ Tribodine $^{\Delta}$	2,6-dimethyl-*N*-[3-(trifluoromethyl) phenyl]piperidine-1-carbothioamide	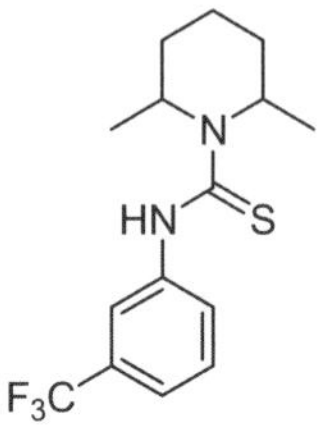
Triamcinolone Acetonide/ Vetalog Panolog Neo-Aristovet$^{\Delta}$ Derma-Vet Panavet Animax Derma 4 Derm-Otic$^{\Delta}$ Genesis Medalon Animax	pregna-1,4-diene-3,20-dione, 9-fluoro-11,21-dihydroxy-16,17-[(1-methylethylidene)bis(oxy)]-, (11.beta.,16.alpha.)-	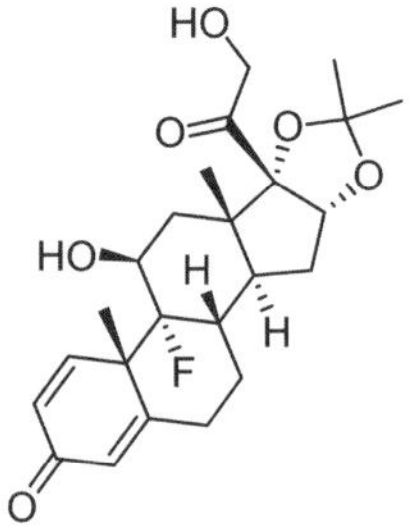
Triamcinolone/ Aristova$^{\Delta}$	9-fluoro-11,16,17-trihydroxy-17- (2-hydroxyacetyl) -10,13-dimethyl-6,7,8,9,10,11,12,13,14,15,16,17-dodecahydrocyclopenta[*a*] phenanthren-3-one	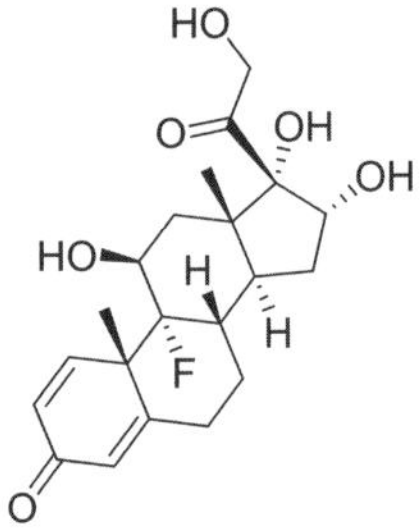
Trifluomeprazine Maleate/ Nortran$^{\Delta}$	*N,N*,β-trimethyl-2-(trifluoromethyl)-10*H*-phenothiazine-10-propiamine maleate	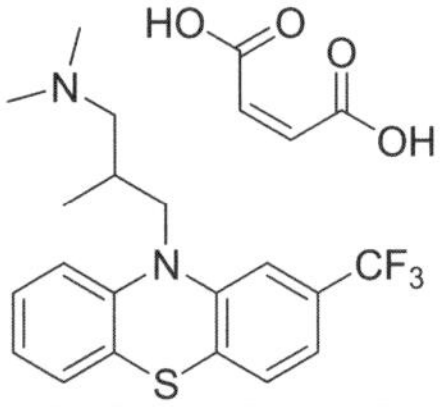

Formula	Pharmaceutical action	Dosage form (species)	CAS No.	Approval date/ drug sponsor
$C_4H_3F_7O$	Anesthetic	Liquid inhalant (Dog)	28523-86-6	Nov. 1999/ Abbott Laboratories
$C_{15}H_{19}F_3N_2S$	Anthelmintic	Oral tablet (Dog)	31932-09-9	Prior to 1982/ Elanco Animal Health, A Division of Eli Lilly & Co.
$C_{24}H_{31}FO_6$	Anti-inflammatory agent Glucocorticoid Immunosuppressive agent	Injectable Ointment Oral tablet and suspension Topical spray (Dog, Cat, Horse)	76-25-5	Prior to 1982/ Fort Dodge Animal Health, Division of Wyeth Holdings Corp. Bayer Healthcare LLC, Animal Health Division Boehringer Ingelheim Vetmedica, Inc. Med-Pharmex, Inc. Altana, Inc. Pfizer, Inc. Biocraft Laboratories, Inc. RMS Laboratories, Inc.
$C_{21}H_{27}FO_6$	Anti-inflammatory agent Glucocorticoid	Oral tablet Injectable	124-94-7	Prior to 1982/ Bayer Healthcare LLC, Animal Health Division
$C_{19}H_{21}F_3N_2S \cdot C_4H_4O_4$	Tranquilizer	Oral tablet	71609-19-3	Prior to 1982/ Norden Laboratories, Inc.

Active ingredient/ trade name	Chemical name	Structure
Triflupromazine Hydrochloride/ Vetame	*N,N*-dimethyl-3-[2-(trifluoromethyl)phenothiazin-10-yl]propan-1-amine hydrochloride	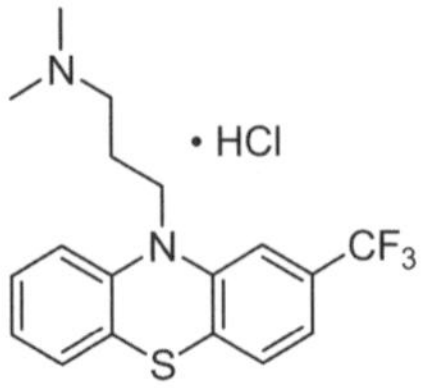 • HCl
Zolazepam Hydrochloride/ Telazol	4-(2-fluorophenyl)-1,3,8-trimethyl-6H-pyrazolo[4,5-*f*][1,4]diazepin-7-one hydrochloride	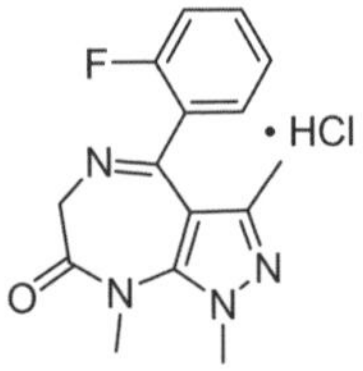 • HCl

[Δ]Withdrawn.

Formula	Pharmaceutical action	Dosage form (species)	CAS No.	Approval date/ drug sponsor
$C_{18}H_{20}ClF_3N_2S$	Antianxiety Tranquilizer Antiemetic	Injectable Oral tablet (Dog, Cat Horse)	1098-60-8	Prior to 1982/ Fort Dodge Animal Health, Division of Wyeth Holdings Corp.
$C_{15}H_{16}ClFN_4O$	Anesthetic	Injectable (Dog, Cat)	33754-49-3	April 1982/ Fort Dodge Animal Health, Division of Wyeth

Index